Amber in the Circum-Adriatic Bronze Age

Mateusz Cwaliński

Amber in the Circum-Adriatic Bronze Age

Acquisition, Circulation and Adaptation

PETER LANG
Lausanne • Berlin • Bruxelles • Chennai • New York • Oxford

Bibliographic Information published by the Deutsche Nationalbibliothek
The Deutsche Nationalbibliothek lists this publication in the Deutsche Nationalbibliografie; detailed bibliographic data is available online at http://dnb.d-nb.de.

Library of Congress Cataloging-in-Publication Data
A CIP catalog record for this book has been applied for at the Library of Congress.

Publication of this work was co-financed from the Publication Programme and the budget of History Department of the University of Gdańsk.

Cover illustration: Courtesy of Mateusz Cwaliński.

ISBN 978-3-631-88857-5 (Print)
E-ISBN 978-3-631-89516-0 (E-PDF)
E-ISBN 978-3-631-89517-7 (E-PUB)
10.3726/b20493

Published by Peter Lang GmbH, Berlin, Germany

info@peterlang.com - www.peterlang.com

This publication has been peer reviewed.

Contents

This publication contains additional information in the form of an appendix. The information is available for download on our website: https://www.peterlang.com/app/uploads/2023/08/Katalog.xlsx

Foreword

The subject of amber is known to have ignited the minds of many archaeologists since the times of H. Schliemann (1822–1890) – the scholar, who discovered the rich collections of beads in the tombs of Mycenae and put the issue of long-distance amber exchange during the Bronze Age on the agenda. To be honest, I was not immediately drawn to this topic, when it was suggested to me as the theme of my PhD thesis by my supervisor Prof. Janusz Czebreszuk. At first I thought that there is not much to add to the discussion beyond what has been said before. It took me a while to find out that there actually are many gaps in the state of knowledge on amber in the prehistory. Finally, at the persistent instigation by prof. Czebreszuk I took up the subject of amber circulation in the Adriatic Basin.

While the raw data I had to gather predominantly came from rather old publications (mid-19th to mid-20th century), in terms of methodology I was inspired by some of the recent works of scholars who already made their names studying amber. In their works they encountered limitations of previously applied methodologies and came up with many useful clues as to how the research on amber should be carried out in the future. Particularly appealing to me was the idea of studying amber's circulation with NA. I gradually became intrigued by what reconsideration of old findings in the light of modern methods and analytical tools could bring. One might say that this work was conceived not out of passion for amber, but a simple curiosity.

Little did I know then that it would be such a laborious task which will take eight years to complete (2012–2021). First four years I spent mostly on searching for source materials and getting acquainted with the Bronze Age archaeology of Italy and the Balkans – a subject not lectured at Adam Mickiewicz University in Poznań, when I was studying there. In 2016 I was granted funding from the National Science Center of Poland which allowed me to conduct the project *Circum-Adriatic branch of the amber route in the Bronze Age* (no. 2015/17/N/HS3/00052). From then on I could more effectively carry out queries in museum and libraries of the countries included in the spatial scope of this thesis. Also in 2016 I had an opportunity to spend five months at the University of Siena as an Erasmus student, deepening my knowledge of Italian archaeology and assembling materials necessary for the project. Another great chance for developing my skills and pushing forward with the study came in 2018, when I was awarded a scholarship from the Vassar College Research Committee (USA). Thanks to it I could spend almost three months learning the chemistry of amber and the craft of its analysis at the Amber Research Laboratory – one of the world's leading scientific institutions with respect to amber studies. Those trips were fundamental in providing me with knowledge and expertise necessary to bring realisation of the project to successful completion.

That being said completing this work would not be possible, if not for the help from numerous people. I owe the deepest gratitude to a huge group of academics, who supported me throughout the process (and who I hope will excuse me for dropping their titles for the sake of brevity). First of all, I need to thank my supervisor Janusz Czebreszuk (Adam Mickiewicz University at Poznań) for acquainting me with the subject and 'opening many doors' for me while I was working on it. Special thanks go to Paolo Bellintani (Superintendence for Library, Archival and Archaeological

Heritage, Autonomous Province of Trento) and Ursula Thun Hohenstein (University of Ferrara), who from the start provided me with invaluable support with respect to Italian amber finds. Many thanks to Ivana Angelini (University of Padova) for taking up the hectic task of analysing multiple samples of amber for the sake of this project. I would like to express endless gratitude to Sarjit Kaur, Edith Stout and all the participating students of Vassar College for making it possible to analyse countless samples of amber at the Amber Research Laboratory and making me feel at home there. My sincere thanks also go to Lucia Sarti, Giovanna Pizziolo, Nicoletta Volante, Chiara de Marco, Gaia Mustone and Alberto Agresti of University of Siena for introducing me to Italian archaeology and making my stay in Siena a real pleasure.

Furthermore, I would like to thank the following scholars for facilitating the access to materials and helping me in various other ways at different stages of the research (in alphabetical order): Lidija Bakarić (Archaeological Museum of Zagreb), Paolo Boccuccia (Pigorini National Museum of Prehistory and Ethnography, Rome), Andrea Cardarelli (Sapienza University of Rome), Michele Cupitò (University of Padova), Katarina Dmitrović (National Museum, Čačak), Vojislav Filipović (Institute of Archaeology, Serbian Academy of Sciences and Arts, Belgrade), Rada Gligorić (Jadara Museum, Loznica), Rovena Kurti (Institute of Archaeology, Academy of Albanological Studies, Tirana), Marija Ljuština (University of Belgrade), Daria Ložnjak Dizdar (Institute of Archaeology, Zagreb), Nicoletta Martinelli (Museum of Natural History, Verona), Janja Mavrović Mokos (University of Zagreb), Kristina Mihovilić (Archaeological Museum of Istria, Pula), Jovan Mitrović (National Museum, Belgrade), Aleksandar Palavestra (University of Belgrade), Aleksandra Papazovska (Archaeological Museum of Macedonia, Skopje), Andrijana Pravidur (National Museum of Bosnia and Herzegovina, Sarajevo), Hubert Steiner (Amt für Bodendenkmäler, Bozen) and Helena Tomas (University of Zagreb).

I should also thank the members of the International Amber Association, in particular Ewa Rachoń (Gdańsk International Fair Co.), Anna Sobecka (University of Gdańsk) and Ewa Wagner-Wysiecka (Gdańsk University of Technology) for giving me many opportunities to introduce the subject of Bronze Age amber to activities of the association.

I would like to thank the reviewers of this work: Giulia Recchia (Sapienza University of Rome) and Martina Blečić Kavur (University of Primorska) not only for the final verdict but also for valuable comments on the earlier papers I wrote on amber.

There is much to thank for with regard to my colleagues from Adam Mickiewicz University in Poznań: Jakub Niebieszczański for helping me prepare countless figures, teaching me GIS and being the second driver during Autumn 2017 trip across the Balkans, Rafał Skrzyniecki for drawing my attention to georeferenced databases and consulting my approach to burial evidence, and Mateusz Jaeger for sharing with me his knowledge of the Bronze Age in the Carpathian Basin. Although I will not mention all of them by their names, I still want to thank all staff members and fellow students of the Institute of Archaeology AMU for all the years spent together on teaching and learning archaeology which became a huge part of my life.

Furthermore, I would like to highlight the role of the University of Gdańsk in turning the manuscript of my thesis into this book. My sincere gratitude goes to my boss Marcin Wąs, head of the Institute of Archaeology, for encouraging me to take up this task as the first endeavour in my new workplace. Publishing this book would not be possible without financial support from Dean's Office of the History Department, and Publication Programme of the Gdańsk University.

Last but not least, I should thank my family and friends in Poznań, Gdynia and many other places for standing by me all these years, Małgorzata Wyka-Cichocka for being always ready to check and correct my English, and my partner Karolina for keeping up with me during the last, hardest months of writing this book.

I would like to dedicate this work to my father Jerzy for his unfailing support in the last years when I have been completely preoccupied with archaeology, but he nonetheless was there for me, and to the memory of my late mother Lucyna who was fully convinced that archaeology is the right path for me.

Gdynia, October 2022

Before we get to the actual part of the book, I would like to briefly discuss some of the principles by which the text was formatted. My research is based largely on the extensive archaeological literature published over a period of over 100 years. I often refer to several sources in one sentence. While I prefer the Harvard Referencing System, being aware of how long in-text citations in parentheses make reading difficult, I decided to transfer some of them to footnotes. Rule of the thumb is that references citing more than two publications were moved to footnotes.

Also, I feel that the reader deserves some explanations as to the guiding principles of using italics and quotation marks. Italics were used to denote phrases, loanwords, proper names, and specialized terms in a language other than English, or to distinguish them from similarly spelled names (e.g., *Apennine* culture and Apennine Peninsula), as well as titles in the original language of publication. Single quotation marks (') were used for proper names in the English translation, for example, names describing archaeological sites, and to distinguish specific terms and words in English. In addition, single quotation marks were used for passages quoted directly from other authors' publications, or to denote colloquial speech. In the case of a quote within a quote, double quotation marks were used.

1. Introduction to the subject

Beginning from the 2nd millennium BC, amber, in the form of jewellery, began to gradually appear among the outfits of various cultural units in Southern Europe, including Italy and the Balkans. Limited availability of fossil resins, largely concentrated in the Baltic zone with few minor sources elsewhere, makes these materials ideal indicators of interregional contacts, mobility and goods exchange. The importance of amber for studying intercultural relations in Europe during the Bronze Age has been a topic discussed among archaeologists for a long time. The fruit of this discourse is a series of concepts discussing the ways of inflow of this raw material from the Baltic coast to civilisation centres located in the Aegean, Asia Minor and Egypt. Traditionally, these roads are called the 'amber routes', among which their 'circum-Adriatic' branch, in the opinion of many researchers, was very important for functioning of the whole process.[1]

Still, some archaeologists have questioned the role of the Adriatic (as a sea route), or Italy and the Balkans (as land routes) in arranging the import of amber in the early stages of prehistory, i.e. before the Iron Age (cf. Harding/Hughes-Brock 1974, 158; Bouzek 1994, 221). What is important, Italy has been usually favoured over the Balkans, because of landscape features presumably facilitating communication and trade: its elongated-peninsular shape, extensive coastline and numerous islands. Not without a significance is the longer history of research on the subject in Italy and a richer publication status in comparison with the Balkans (cf. section 1.4). Furthermore, many scholars tend to interpret the role of Italian and Balkan prehistoric societies as intermediaries in amber transfer from the north to the south of Europe, rather than its consumers. This has to do with the fact that much more attention has been paid to the presence of amber in i.a. Aegean – the *Mycenaean* culture, than to testimonies of its use by i.a. local Apennine or Balkan communities. Finally, there is a conviction in the literature that as an exotic and hardly available material, it was, at least initially, the sole good of the elites, which were the only people able to import it from the distant circum-Baltic areas, where the main sources of amber are located.[2] However, there are also opposing voices that small numbers of finds of this raw material in certain regions (e.g. Western Balkans compared to Greece or the Po Valley) are signs of a low level of development of social structures and the lack of active participation in the far-reaching exchange network (cf. Forenbaher 1995, 276–277).

The unresolved issues mentioned above push the subject of long-distance amber importation into the realm of conjectures, with the term 'amber route' used without a precise definition. Bypassing those problems renders many concepts related to the circulation of amber in the Bronze Age difficult to prove or defend. Still, they are worthy of consideration, as the state of knowledge on amber in Italy and the Balkans has been

1 Cf. De Navarro 1925; Palavestra 1993; Negroni Catacchio et al. 2006; Czebreszuk 2007, 2011; Harding 2013a, 2013b.
2 Cf. Negroni Catacchio 2009; Czebreszuk 2011, 164–171; Maran 2013.

steadily improving over the past years. Thus, I decided to attempt a reassessment of the circum-Adriatic amber route, by subjecting related data to analysis in the light of new findings and theories, while using some of the modern tools available to archaeology.

1.1. Aim of the work

The essential purpose of the presented dissertation is to examine spatial and temporal structure of amber's occurrence in the circum-Adriatic areas during the Bronze Age. The term 'circum-Adriatic' is meant to encapsulate the Adriatic Sea basin and neighbouring lands, from *Caput Adriae* and adjacent catchment systems of Po, Sava and Drava in the north, through the Apennine Peninsula and Western Balkans,[3] to the Ionian Sea in the south. Moreover, Sicily and the Aeolian Islands, due to the documented ties with Bronze Age cultures of the Apennine Peninsula and evidence for the involvement in the Central Mediterranean amber exchange, were included in the study area, even though geographically they lie outside the Adriatic Sea basin. From the geographic point of view, it may seem odd to include Central Balkans,[4] i.e. areas of Vojvodina, Serbia, Kosovo and North Macedonia, in a spatial scope defined in such way. However, this zone has often been considered alongside the Adriatic and Western Balkans in the context of amber circulation (or in other words 'amber routes') in the Bronze Age; hence, there is a need to verify if such an approach is justified. In such delimited spatial scope of the work, waters of the Adriatic play the role of a potential communication artery, enabling access to selected coastal areas. Similar function for the inland transport could have the valleys of large rivers flowing into the Adriatic, or passing in its vicinity, e.g. Po, Krka, Neretva, Drina, Drava, Sava or Morava. Thus, the 'circum-Adriatic' should be understood not as a strict geographic delimiter, but solely as catch-all term coined for the purpose of this work.

The time span of the analysed issues – the Bronze Age – covers the last two centuries of the 3rd millennium BC, 2nd millennium BC in its entirety and the 1st century of the 1st millennium BC.

Key element of this work is the contextual (i.e. regarding chronology and cultural attribution) analysis of amber deposits in the study area, as well as an assessment of its provenance and morphology. These analytical tasks are aimed to determine, whether succinite (Baltic amber) was actually the dominant species of amber brought in and used in the area during the 2nd millennium BC. Furthermore, it is investigated how the influx of amber was subjected to temporary fluctuations and which

3 The term 'Western Balkans' is used here to denote countries located to the west of the Danube – Drina – Lim – Drin rivers line, i.e. Slovenia, Croatia, Bosnia and Herzegovina and Montenegro. However it is worth adding that nowadays the term is often used in a geopolitical sense, to describe Albania and the ex-member countries of Yugoslavia minus Slovenia, already belonging or pending incorporation into the European Union (cf. Ministry of Foreign Affairs Republic of Poland n.d.).

4 The term 'Central Balkans' is used here loosely to denote the area bounded by the Danube Valley and the Dinaric Alps, stretching approximately along the Drina and Morava River Valleys along the meridional axis. While usually this term is used to cover Serbia and adjacent parts of Bosnia and Herzegovina, Bulgaria and North Macedonia, it was decided to include Albania in this zone as its southward extension due to cultural ties connecting above listed regions during the Bronze Age.

communities – cultures were responsible for its importation? Finally, the goal of the study of amber artefacts morphology is to compare the forms present in different regions, recognise patterns in distribution and, on that basis, reconstruct ways – routes of amber circulation within the circum-Adriatic zone. The latter can also help to assess the level of similarity/dissimilarity with amber finds from neighbouring European areas and, consequently, pinpoint most probable exchange partners and detect potential centres of amber processing.

Second, attention is focused on the place of amber in the local societies and their mutual relations. For this purpose the work includes an assessment of the types of ambers deposition, forms of its use and its social attribution. Such considerations are also intended to help determine the level of amber's consumption by the local circum-Adriatic communities vs the idea of them serving mainly as intermediaries. Therefore, presented dissertation tackles the issue of position of these societies in the supra-regional system of goods exchange during the Bronze Age. It approaches this matter through a comparison with European areas surrounding the circum-Adriatic zone, participating in the long-range cultural contacts during the Bronze Age, including amber exchange. Hence, in the following three major issues come to the forefront:

A. Communities participating in amber exchange, sources of amber and ways of its circulation within the circum-Adriatic zone.
B. Material and symbolic status of amber among the Bronze Age societies in the circum-Adriatic zone.
C. The role of the circum-Adriatic territories within the pan-European exchange network of raw materials and goods, including amber.

1.2. Applied methodology

Investigating the three issues, study goals listed in the previous section will be realised in three individual, however, inter-dependant blocks. Each of them is characterised by a different set of methods applied to analyse these various aspects of amber. Technicalities behind specific approaches to studied materials, or different statistical analyses performed at various stages are explained in separate chapters, immediately preceding descriptions of the results of their use. In this place a more general overview of adopted methodology is provided.

With respect to the point A of the above list, of primary importance is determining chronology and cultural affiliation of amber finds in the study area. This will help reconstruct spatial and temporal structure and dynamics of amber occurrence in the circum-Adriatic zone. In consequence, this will allow answering the question which communities and regions took part in the process of amber importation and redistribution at a given moment of the Bronze Age. The task is realised in chapter 2, with revision of amber finds' chronology and cultural affiliation in light of currently proposed periodisations and taxonomies. Second, naming the sources of amber used by the circum-Adriatic communities of the Bronze Age is done by evaluating the results of amber provenience analyses performed up to now. This part focuses on chemical methods used in the studies of amber and how are they employed by archaeology in the assessment of origins of prehistoric amber finds. Conclusions derived from considerations on these matters are summarised at the end of chapter 2.

(Renfrew/Bahn 2007, 337). Moreover, intentions of the main actors of the process, attitudes of its spectators and circumstances in which it takes place can alter perception of the whole act of exchange and render its significance.

The above given description is limited only to the act of giving motivation by positive premises aimed at achieving reciprocal, although not necessarily equal, benefits by the agents. However, one should bear in mind that various forms of reciprocity, that is often considered obligatory in the act of giving (Oka/Kusimba 2008, 343), may be encountered, ranging from balanced benefit to one (delayed reciprocity) or both-sided unrequited gift (conflict of interests) (cf. Sahlins 1972, 191–196).

If exchange can also involve material goods, how then one should understand the meaning of trade? Following to some extent, definition of R. Oka and Ch. M. Kusimba trade is regarded as 'the material-economic component of exchange'; however in opposition to quoted authors, I do not consider it as a 'necessary part of any social exchange' (Oka/Kusimba 2008, 340). It is a certain way of realisation of a basic concept of exchange; however, in this process material value of exchanged goods matters more to the participants than non-economic gains (Doğan 2008, 36; Doğan/Michailidou 2008, 20). Trade does not exclude the presence of factors constitutive to exchange which are related to renegotiation of social relationships by achieving other benefits than simple material enrichment; however, their importance decreases in relation to economic profit (Palavestra/Krstić 2006, 66). Presented approach may be seen as a response to suggestion of K. L. Hutterer, who argued that archaeologists hitherto overstated social side of exchange at the expense of the economic one (Hutterer 1977, 8; after Oka/Kusimba 2008, 340). In reality trade is intrinsic in both social and economic aspect of exchange, which are ever-present in commodity transactions between humans (cf. Oka/Kusimba 2008, 367–368). One can visualise this notion in the example of everyday shopping, when regular buyers are polite or even friendly towards the sellers in exchange for their fast and efficient service, even though their main purpose is simply to buy chosen products. In the world of big markets and economies, merchants aim at conducting their businesses in amicable atmosphere, build by gestures and courtesies, which can help to establish mutual trust (cf. Paterson 1998, 148–149; Morley 2007, 86).

In summary I consider exchange as a more general process, much broader in its possible forms, than trade which is more concrete and conditioned realisation of this basic interaction. We saw the particularity of trade in the way it is narrowed to exchange motivated by certain (rather high) value of some commodity in relation to the price or costs of its purchase. Nevertheless, as we are discussing trade in the broad geographical context concerning the common phenomenon of circulation of a certain material – amber, it is inescapable to regard trade as an organised process. Hence, one has to name at least basic conditions required for the occurrence of large-scale, long-distance trade.

C. Renfrew, who represents minimalistic approach to trade, underlines the factor of distance and importance of reciprocity (Renfrew 1969, 152); however, essence of trade is contained in procurement of materials, by whatever mechanism (Renfrew 1977, 72). It may seem that A. Harding takes the opposite position, as he stresses that trade implies use of markets and money, thus making this term rather inappropriate in relation to prehistory (Harding 2013b, 370). However, it is worth underlining in this place that we are speaking of various levels of economic development, i.e. proto-bureaucratic (institutionalised) mechanisms of trade and exchange in early civilisations of the Near East can hardly be transposed on the Bronze Age realities in most of Europe (cf. Sherratt/Sherratt 1991; Tyborowski 2009).

Big markets, professional traders and other manifestations of economic sophistication have often been exclusively attributed to high commerce which was meant to represent periods of prosperity in the Mediterranean during the antiquity. Lack of large-bulk trade conducted in highly organised infrastructure was regarded as a setback to more primitive economy capable only of small-scale exchange. However, it is argued that 'significant volumes and densities of exchange can exist without such infrastructure' (Horden/Purcell 2000, 144–145). Money, on the other hand, as a fixed token of trade appeared much later than long-distance transfer of goods and therefore should be not viewed as *sine qua non* condition of trade's existence, as barter, for instance, is known from pre-monetary societies, and more traditionalistic communities rely on it even in modern times (Doğan 2008, 36; Doğan/Michailidou 2008, 20).

Contrary opinion to the minimalistic approach to trade is reflected in the definitions invented by K. Polanyi, who perceived trade from institutional perspective. His position is represented by the statements that 'trade is a method of acquiring goods that are not available on the spot' which are moved 'through the market, that is, an institution embodying a supply-demand-price mechanism' (Polanyi 1975, 133–134). Although this definition alone sets an interesting conditioning of trade, the general viewpoint of Polanyi, later adopted by *substantivists*, opted for centrally administered and controlled process of acquisition, redistribution, production and consumption as a domain of political elites (cf. Polanyi et al. 1957; see also Oka/Kusimba 2008, 344). Accordingly, decision-making is entirely socially dependent, operation of the market is fixed and elites have exclusive mandate on exchange of luxury products (cf. Halperin 1984).

It would be a painstaking task to describe in detail the debate between *primitivists/substantivists* and *modernists/formalists*.[10] To close this overview of the meaning of trade, I will limit myself to saying that in the following study a more conciliatory approach is adopted: socio-political complexity is not seen as a precondition for emergence of trade, and the latter is not considered as strictly commercial relationship, nor is it just a non-economic process (cf. Adams 1992).

Apart from the premises for undertaking trade, its infrastructure and relation to the political institutions is the most important to define the actors of exchange. Here they are given a major role as self-motivated agents, capable of independent actions, driven by the knowledge of exchange network and ability to assess rates of gain/loss. Trade could be handled both by individuals (Sherratt/Sherratt 1991, 365; Bryce 2002, 99) or larger entities, such as merchant guilds, e.g. *karum* (Kuhrt 1998, 22; Tyborowski 2009, 138), or political elites, e.g. *Minoan* and *Mycenaean* palaces.[11] In order to engage in trade, it was required of agents to possess sufficient bargaining power and commodities either their own or entrusted to them (Oka/Kusimba 2008, 364). If we accept the 'structuralist' approach to exchange as 'a way to mediate conflict and maintain social structure' (see e.g. Hodder 1980; Earle 1997), or in other words as a subject to structural process between its social, political and ritual components, then the main participants

10 For an overview of the debate see Oka/Kusimba 2008.
11 Cf. Sherratt/Sherratt 1991; Knapp 1998; Berg 1999; Parkinson/Galaty 2007.

may be endowed with Giddensian agency, that would allow them to operate within the structure and occasionally alter it (see Giddens 1979, 1984).

In conclusion to above consideration, it is possible to propose three crucial preconditions for the emergence of trade. First one is similarly perceived system of values, either commercial or symbolic, typical to particular resources or products, which forms a common ground for negotiations.[12] Second precondition is the existence of ideas of demand and supply among the communities, which would induct goods into motion and launch process of trade.[13] Finally, there should be an agent – a middleman, who would be responsible for delivery of goods and will have the power and authority to conduct transactions.[14]

Although exchange and trade most probably were the main ways of obtaining amber, one should not forget that there are other means by which this material could circulate over considerable distances. Amber findspots in the circum-Adriatic zone do not necessarily represent final stages of exchange (e.g. as a gift) or trade (i.e. when objects change hands from supplier to consumer). More distant amber deposits within hoards or single pendants inside warrior graves may have a history of war trophies (Palavestra/Krstić 2006, 65), captured during raids on more prosperous settlements or even trade centres. These spoils of war might be later thrown back into circulation, channelled by lower-scale, more casual trade, as did Suebi according to observations by Julius Caesar (Caesar, IV.2.1, after Hughes-Brock 1993, 223). In case of the Adriatic Sea, piracy cannot be excluded, whereas in its littoral zone, one could expect brigands attacking traders, middlemen or ordinary travellers. Former phenomenon is a well-known element of the Mediterranean seafaring in the antiquity and could range from petty piracy to a more organised criminality, e.g. Cilician corsairs (Horden/Purcell 2000, 134–135, 140–143; see also De Souza 2002). Examples of the latter practice are provided by Amarna letters, describing dangers lurking on the road for the Babylonian merchants (Bryce 2002, 88–89). Taking into account the limitations in access to amber sources, described practices probably played secondary role in comparison to trade. Nevertheless, it should be reminded that social history of exchanged/traded goods 'is of far greater overall importance than the practical significance of the activities that they represent might suggest' (Horden/Purcell 2000, 149).

1.2.2. Biographies, travels, itineraries of things: approach to material culture

Amber is said to have had a number of qualities appealing to people since prehistory. They have been discussed over and over by a constantly growing roster of scholars, however often without explicitly stating theoretical foundations of their reasoning. If for them the issue seems obvious and not worth of expounding, I feel compelled by achievements in the field of anthropological and archaeological theory to clarify my approach to material culture.

From one perspective on the cultural setting of amber, finds made from this material can be simply reduced to the category of imports. Its sources are spatially restricted,

---

12 Sherratt/Sherratt 1991, 354; Doğan 2008, 38; Harding 2013b, 370.
13 Harding 2000, 195; Oka/Kusimba 2008, 344.
14 Doğan 2008, 36; Doğan/Michailidou 2008, 20; Oka/Kusimba 2008, 364.

and in most cases it was necessary to carry it over considerable distance from the place of acquiring to the place of use and subsequent deposition (voluntary or involuntary like in cases of different archaeological contexts, e.g. grave, waste pit, house floor, etc.). In a very basic understanding of the issue, one can say that unique things (materials, items) are likely to have a high status or value. However, things that may seem exotic or valuable to a community in one place could be common and, thus, little appreciated to the other, especially when the latter has it in abundance. One should not overlook significance of the road by which an import travels; the distance from the place of origin or manufacture to the final destination can contribute greatly to an object's value (Cline 2005, 45). Or as J. Chapman puts it, 'by travelling far, a hitherto mundane object was transformed into something special whatever the means of movement' (2008, 335).

However, the definition of import itself is not simple,[15] and assuming its commercial side might unfairly supplant the other aspects that amber could have possessed (as has been done in publications treating amber simply as a commodity European north could offer in exchange for locally unobtainable resources, such as copper, tin, etc. Earle et al. 2015). Still the import, whatever its definitions are, is only 'one side of the coin' in a multidimensional (or 'multivalent') character of amber artefacts, or any other objects that archaeologists are faced with. Too little attention has been paid to inter- and intra-contextual meaning of amber, i.e. the different ways of its adoption and adaptation by people of various places and cultures. While adoption means accepting and using something previously unrelated to its new owner as it is, adapting implies modification to suit better the new conditions and needs. The point of interest here is what value, meaning or functionality – aspects which contribute to the object's status – amber could have for different communities, and how these were created? Arguably what it requires in order to understand the status of an amber object is to pay attention to intra-contextual interplay of humans, things, features and meanings at the place of amber deposition. Approach involving these aspects is advocated by the growing group of archaeologists studying the inflow of goods to the *Mycenaean* Greece (cf. Maran/Stockhammer 2012a). These scholars contest the use of generalising designations such as 'exotic' or 'prestige' imposed on objects, as they 'suggest a semantic stability which cannot be taken for granted' (Maran 2013, 147). One of the issues resurfacing in the debate on amber in the prehistoric communities are the ways of its appropriation – a term nowadays burdened with negative connotations, however not necessarily meaning unlawful or harmful obtention (cf. Maran 2013). As J. Maran writes, 'in dealing with the appearance of foreign traits, the focus of attention must be placed on clarifying the ways of appropriation on a local level and on how, in the course of their integration into existing practices, new cultural forms were created' (Maran 2012, 64).

One of the foundations for the modern archaeological theory is that human individuals are actively using material culture as an expressive symbolic medium for their social strategies and negotiations (Hodder 1982, 1986). Furthermore, it relies on the conviction that primordial societies, as entities constituted by the local transmission of traditions, inherited knowledge and various regional specificities, manifested their culture through materials which by the ways of exchange and trade travelled over

15 For discussion on the definition of import, see e.g. Biehl/Rassamakin 2008.

Above-described approaches to archaeological finds, while certainly full of a thoughtful reflection, may seem difficult to implement – especially when attempting to trace back the origins and follow histories/biographies/travels/itineraries of multiple amber beads, which could have gone a long and a complex way from the source to the final destination. However, within the scope of this work, it seems feasible to try to disentangle and compare some characteristics of studied objects (including meaning, utility, value, status) at the last chapter of their lifetime that is immediately before and in the moment they were deposited in the ground. The idea is to do it by multi-aspect analysis of the amber artefacts themselves, and of their context. Such a procedure is carried out in chapter 2 of this book in relation to chronology and cultural attribution of the amber finds, in chapter 3 regarding the ways of amber circulation within the circum-Adriatic zone, and in chapter 4 aimed at defining social status of amber. The notion of network comes in this place not only as a useful, visual metaphor but also as a theoretical construct and an analytical tool as well.

1.2.3. Amber circulation as a part of a network (analysis)

Idea of analysing social networks is not new to archaeology; already in 1977 C. Irwin-Williams showed the potential of a network model for analysis of prehistoric trade (1977). However in the last years, it received great feedback and gained large group of supporters, applying it in different contexts.[16] The approach was eventually suggested for the study of amber circulation in the circum-Adriatic zone (Harding 2013b).

Social Network Analysis (SNA) derives from sociometry which involves the measurement of interpersonal relations in small groups, often by the means of sociogram – a depiction interpersonal structure of groups as points and lines in two-dimensional space, like graphs used in mathematics (Brughmans 2013, 632; see also Freeman 2004). In fact SNA is partly founded in graph theory: it borrows the concept of a graph as a model of social interactions/relationships, utilises similar terminology and draws from mathematical operations to quantify and measure these properties (Wasserman/Faust 1994, 93, after Brughmans 2013, 628). Nonetheless, it is largely a brainchild of sociologists and still developed as an integral part within more modern sociological approaches.

In a resume, this theory assumes that relationships between entities of various kinds can be pictured as a network, which they form in the real life. So it involves searching for nodes and links among chosen entities on the basis of specific properties connecting them (Knappett 2011, 41–46; 2013b, 3). There are several principles shared by social networks analysis, among them viewing actors and their actions as interdependent rather than independent, autonomous units, and considering relational ties (linkages) between actors as channels for transfer or 'flow' of resources (either material or non-material) (Wasserman and Faust 1994, 4, after Brughmans 2013, 632–633). Thus the networks, at the social level, are created by actors guided by, e.g. the will to reach some material or non-material gains, as in interactions defined in the previous section as exchange and trade. Once the network is established, it facilitates similar

16 See e.g. Knappett 2011, 2013a; Knappett et al. 2008; for the review of Network Analysis application in archaeology see Brughmans 2013.

interactions, however can also be altered by actors endowed with agency, thus contributing to the dynamic character of networks vs static structures.

Hence, the essence of applying this approach in studies of prehistory lays in discovering and defining relational patterns between spatially localised (predominantly although not necessarily) archaeological data, e.g. artefacts, features or sites (Brughmans 2013, 633; Knappett 2013b, 8). Final effect of such analysis is creation of a network model in which selected categories of sources are combined, basing on common quantitative or qualitative features (Harding 2013b, 391). Such models can differ in the level of their randomness, scale and centrality amongst others (cf. Brughmans 2013, 636–645). The model is aimed at reflecting the network which exists in reality between entities entangled in mutual relations. Yet since archaeology deals with past societies, it is more appropriate to speak of reconstructing the networks of human interactions and relationships on the basis of material culture, than real-time observations (Knappett 2013b, 7–8).

Advocates of Network Analysis (NA) argue that by using this approach, a researcher has an ability to derive conclusions directly from an archaeological material instead of imposing *a priori* interpretations dictated by chosen social theory (Harding 2013b, 391). Thus the inference process proceeds from analysis of sources to construction of interpretations (bottom-up), not *vice versa* (top-down) (Knappett 2011, 26–30). Discussed approach assumes to fully reveal specificity of the processes occurring in the micro-scale, e.g. in relation to selected regions of Europe (Knappett 2013a; Knappett et al. 2008). However, it should also be mentioned that some scholars who apply rather large-scale World System Theory took advantage of analytical procedure given above (e.g. Kardulias 1999). In fact the network can be analysed multi-dimensionally, from its constituent parts (nodes and links), through its segments (e.g. dyads, triads, etc.), to a comparison of complete/whole networks built around different entities (cf. Brughmans 2013, 625–626). NA gives a theoretical basis for the use of a wide spectrum of statistical tools borrowed from graph theory that focus i.a. on the detection of temporal and spatial correspondence of objects marked by specific characteristics. It also utilises GIS analytical methods to a great extent, allowing for more refined representation of distribution and interrelations among spatially embedded data.

NA does not enforce a single method or approach to data. There is a plethora of different applications of the concept of network and its analysis to be found in archaeological literature (see e.g. Brughmans 2013; Knappett 2013a). In the following it has been employed in a rather basic format, by visualising data related to amber in the form of a graph (network of linked entities) and analysing the latter by a set of popular but not too-advanced statistical tools. It predominantly found use in chapter 3, where relationships between typologically defined, regional assemblages of amber artefacts are analysed in order to reconstruct circulation of amber in the circum-Adriatic zone. Specific methodological solutions, together with the choice of methods and tools selected for implementation of NA, are described in the section 3.12. Still, the idea of 'network' in its broad theoretical meaning is largely guiding this work through its consecutive stages, up to chapter 5, where it is integrated into the concept of 'system'.

1.2.4. Circum-Adriatic zone from the World System Analysis perspective

Once the local specificity of amber circulation is determined, one can proceed with large-scale analysis that hitherto, i.e. in most of the previous works on amber, has been done in the first place. Such reversed order gives the opportunity to put the previously observed regional phenomenon in a broader geographical context (cf. Cwaliński 2017). Thus, in the final stage of presented dissertation (chapter 5), the network of local connections in which amber circulated within the circum-Adriatic zone is included in a supra-regional axes of interactions, enabling the transmission of knowledge and resources. In order to introduce amber to a broader narrative of European Bronze Age, it was decided to turn to the rich literature on the economic and political aspects of the Bronze Age European and Mediterranean, written in the spirit of the World System Theory (WST).[17]

Not only a multitude of applications but also the continuous advancement of this theory in the field of analytical methods it encompasses caused that nowadays scientists more often use the term 'World System Analysis' (WSA). According to them, WSA assumes that societies did not dwell in complete isolation and cultural or economic contacts were inherent in their development (Hall et al. 2011, 240). Moreover, WSA looks for appropriate models of interaction which would show hierarchy within the network of contacts and cyclic as well as metamorphic nature of their functioning (cf. Chase-Dunn/Anderson 2005). However, this does not mean that the way of inference ignores exceptional cases – outliers from the general trend. WSA aims to show the complexity of these interactions and the multiple roles of individual regions or even sites in the system (Hall et al. 2011, 240). Therefore, it appears as a paradigm rather than a ready-made theory.

The topic discussed here, in the context of the European Bronze Age, is tightly connected with the issues of social stratification, economic surpluses and luxury commodities, display of power, means of governing or cultural mobility. Previous applications of WST on the grounds of European archaeology, predominantly based on the 'core-periphery' idea, interpreted the 2nd millennium BC as a highly innovative period in the social scale during which foundations of internally hierarchical society ruled by elites were formed.[18] For this type of societies, rare raw materials and prestigious goods presumably played a very important role. It is believed that the elites, through skilful manipulation of this kind of resources, gained and stabilised their supreme position inside communities. Amber, as a raw material commonly unavailable in the Mediterranean, was predestined to fulfil this role. Many archaeologists polemicize with the idea of ruling classes controlling long-distance trade, instead stressing local processes and small-scale tribal or segmentary interactions, e.g. the so-called 'down-the-line' chain of exchange.[19] Still, the presence of amber – and other resources such as tin, copper and gold – outside the places of their natural occurrence, sometimes at a very large distance, prompts to think about some kind of a process (network or system), more or less organised, which allowed for translocation of the former.

17 For the definition of the WST see Hall et al. 2011.
18 E.g. Earle 1997; Czebreszuk 2001; Kristiansen/Larsson 2005; Earle et al. 2015.
19 E.g. Harding 2000, 2013b; Kienlin 2012.

Today we may refer to substantial achievements in application of broadly understood WST in the archaeology of the Mediterranean.[20] In the author's opinion, all these works provide a basis for further refinements that will result in a model of interactions, helping us to interpret intercultural and supra-regional relations in the Bronze Age. It is important to note here that the following work does not attempt to build a new European-Mediterranean world-system basing on amber circulation but aims at reviewing the previous concepts in this light of the new results and observations.

That being said, in the presented approach it is not intended to accept these ideas *a priori*, but rather to verify legitimacy of their use in the context of the circum-Adriatic zone. The author wishes to emphasise that he does not dispute the existence of organised elites during the Bronze Age, but at the same time, it is not his goal to show the far-reaching contacts as an exclusive domain of highly hierarchical societies. Presented concept presupposes participation of many cultures of the circum-Adriatic zone at different levels of circulation process that included raw materials, finished products and knowledge which probably accompanied them. Gradation of individual regions involved in exchange/trade should not be based on the classic delimitation 'core-periphery'. Instead, functional division, where different regions are characterised by varying degrees of relevance to the functioning of the system, is proposed. This is a departure from the traditional understanding of the periphery as economically dependent on the core areas. One can benefit from the latest advancements in WSA resigning from the classical model of I. Wallerstein (1974) in favour of more modern applications, especially in the area of the Mediterranean. Particularly important in this place are the works of A. Sherratt and S. Sherratt (Sherratt/Sherratt 1991; Sherratt 1993). According to applications of WSA given above, peripheries can be a source of innovation and their residents more aware of the significant role they play in the process of exchange and transfer of goods. Hence, they exert influence on the network participants from other regions.[21] It is also important to emphasise the relativity of terms such as: centre (destination, point of redistribution), periphery (space of limited acquisition of goods and their temporary inflow) or marginal zone (the area supplying raw materials). In the complex web of cultural and economic ties, along with observer's changing perspective, their definition changes as well ('negotiated peripherality' – cf. Kardulias 2001). The whole can have a very dynamic nature: new centres take the place of those declining (cf. Parkinson/Galaty 2007). Such vision of WSA was sometimes called 'Multicentered World System' in relation to the Bronze Age in South-Western Asia (Kohl 1989) or 'Multi-polar Europe' that is Europe on the threshold of the 2nd millennium BC (Heyd 2007).

1.3. Amber in science and archaeology

1.3.1. What is amber?

The term 'amber' is rather capacious, and there are various opinions on what it designates, or how to define it. This semantic complexity is due to the fact that scholars

20 E.g. Berg 1999; Parkinson/Galaty 2007; Parkinson/Galaty 2010; Papadimitriou/Kriga 2013.

21 Hall et al. 2011, 241, 243; see also Kardulias 1999, 2001.

marker of simetite. The first of the two absorption peaks, the more intense one, appears in the range of 1,230–1,260 cm^{-1} (usually at 1,241 ± 5 cm^{-1}), and the second, less intense peak, is found at an average wave number of 1,181 ± 5 cm^{-1} (Fig. 3). In some samples, the second band was absent and, instead, a wide shoulder was observed – a pattern that visually resembles a 'reversed Baltic shoulder' of succinite. Both fossil resins differ also in the region of 890 cm^{-1}; succinite reveals a weak, but characteristic absorption, while simetite rarely shows such feature. Nonetheless, simetite's spectra tend to be varied; therefore, it is recommended not to rely solely on IRS to distinguish it from other fossil resins.

In 2003 Beck et al. published results on the analysis of a single sample of succinite obtained from National Museum of Natural History in Paris; 0.047 g of amber (about 2.1 % of the sample) was extracted with ether/dichloromethane solution, derivatized by diazomethane, and the methylated filtrate was analysed by GC-MS. As a result the authors were able to identify more than hundred constituents, among which 58 non-terpenoids are absent or appear in different amount in other fossil resins analysed by GC-MS (Beck et al. 2003, 18, Tab. 1). The authors also reported succinic acid level (0.42 %) close to the value reported by Helm (1881). Comparison of compounds/molecules detected using GC-MS in archaeological samples of amber suspected to be simetite, with those found by C. W. Beck and his co-workers may be very useful in proper determination of the discussed fossil resin, thus supplementing FTIR analysis. Finally, the authors came to an important conclusion that Sicilian amber was produced by trees in the family *Leguminosae* (also known as *Fabaceae*). However naming a genus or a species is at this point impossible, because the tree is extinct and/or due to considerable changes occurring during fossilisation leading up to simetite (Beck et al. 2003, 32).

The latest contribution to chemical composition and botanical origin of Sicilian amber is a study by Werf et al. (2016), which has implications for archaeological considerations on simetite's use in prehistory. The study was based on four simetite samples obtained from the Earth Sciences Museum at the University of Bari. The samples were first studied with FTIR-ATR (attenuated total reflectance), to ensure the analysed material is in fact simetite, and subsequently were investigated with Py-GC-MS with thermally assisted hydrolysis and methylation. The level of succinic acid found in all four samples is in agreement with the first investigations performed by Helm (1881) and Beck et al. (2003). Still, in the opinion of the authors, the level of succinic acid is very low and its eventual incorporation within the macromolecular structure seems irrelevant (Werf et al. 2016, 90–91). Moreover, they also found similar occluded diterpenes in the mass spectra that have been previously reported by Beck et al. (2003). Finally they were able to conclude that the polymer fraction of simetite is based on polylabdanoid diterpenes with an 'enantio' configuration, hence the chemical classification of simetite as a Class 1c resinite, even though very low amounts of succinic acid were detected. The research also confirmed botanical origin of Sicilian amber, previously determined by Beck et al. (2003), from the *Fabaceae* family (Werf et al. 2016, 94).

With the application of FTIR in archaeological research, it was possible to determine the origins of some finds as simetite. The earliest cases of such findings are amber beads from mid-3rd millennium BC cemetery of Laterza (Apulia, South-East Italy) (Beck 1971; Beck/Hartnett 1993) and from Vayenas tholos, Pylos (Peloponnese, Greece) dated Middle to Late Helladic I–III (ca. 2200–1600 BC) (Beck 1966b; Beck/Hartnett

1993). Furthermore, simetite was identified among archaeological amber finds from different sites with varying chronology in the Iberian Peninsula, recently summarised in papers by M. Murillo-Barroso and M. Martinón-Torres (2012), and Odriozola et al. (2019). As no local deposits of fossil resins with similar chemical composition and FTIR spectra has been recorded in those regions, all aforementioned finds should be considered as imports from Sicily. Finally a number of Bronze Age amber finds from Sicilian contexts were recently proven to have been made of simetite by I. Angelini and P. Bellintani (2017).

1.3.3.4. Other fossil resins

The description above is limited to European fossil resins that have been recorded in archaeological context dating from the Eneolithic/Copper Age up to the antiquity. There are, however, many more species of amber which were possibly used by prehistoric people, but so far they have not been identified among hitherto collected material. Abundance of fossil resins deposits in various parts of Europe, typically given mineralogical names after the countries or regions of origin, have led some researches in the past to claim that it was not necessary to import amber all the way from the Baltic zone; at least the earliest artefacts were thought to be crafted from locally available resources (cf. Beck 1986, 69). In the light of the abovementioned studies, the use of local fossil resins seem plausible, but the inescapable conclusion is that the vast majority of amber was, in fact, imported from the surroundings of the Baltic Sea.

Archaeologists studying occurrences of amber artefacts in certain regions of Europe, e.g. in Slovakia (Beck/Dušek 1969) or Hungary (Beck/Sprincz 1981), have paid special attention to nearby sources of local fossil resins, as potential alternatives to Baltic amber. In this context one can list i.a. walchowite – a Cretaceous fossil resin appearing in Moravia, and ajkaite occurring in the pit-coal and clays belonging to Upper Cretaceous sediments (dated ca. 100.5–66 million years ago) of the southern Bakona Mountains near Ajka, Hungary, both of which produce very similar IR curves and do not contain succinic acid (cf. Beck 1986, 85; Kosmowska-Ceranowicz 1999, 101–105). Still, in neither case was it proven that these local sources of fossil resin were known and used in the prehistory. It is surprising to note that succinite was registered even in cases of prehistoric sites located in the close vicinity of local amber deposits, such as the Late Palaeolithic hunting station of Mossbühl in Switzerland, located about 25 miles from plaffeite outcrops (Schwab/Beck 1983–1984), or Celtic oppidum of Staré Hradisko in Moravia, located merely 20 miles from walchowite sources (Beck/Dušek 1969).

Aside from already described occurrences of rumanite and simetite among archaeological finds, use of local deposits of amber was ascertained for some Copper Age (Chalcolithic) and Bronze Age sites on the Iberian Peninsula (see Murillo-Barroso/Martinón-Torres 2012; Odriozola et al. 2019). Artefacts analysed from the megalithic tomb of Trikautzi I and the cave of Cova de La Pastora (both dated to the 3rd millennium BC), as well as those from the tholos Los Lagos I, were determined to be made locally from Cretaceous amber (Álvarez-Fernández et al. 2005; García-Puchol et al. 2012). Moreover, there are several cases of archaeological amber artefacts, where the IR spectra do not correspond to succinite, or any other known fossil resins. Such observations were noted for some Bronze Age finds from the Apennine Peninsula, namely Coppa Nevigata and Trinitapoli, Foggia province, Apulia (Angelini et al. 2003;

Angelini/Bellintani 2005). Determination of origin of these samples seems impossible for the time being, despite extensive comparison of their IR spectra with the spectra obtained for various local outcrops of Apennine, Dolomitic and Lessini ambers (Angelini/Bellintani 2005). A similar problem occurred in the case of an amber bead from the Százhalombatta site in Hungary (Vatya settlement), which is not made of Baltic amber, but of some other, still unidentified, fossil resin (Jaeger 2017, 210). To conclude the absence of specific absorption bands in diagnostically useful regions of IR spectra, e.g. the 'Baltic shoulder' of succinite, does not necessarily mean that presumed origin of a sample should immediately be discarded. Archaeological finds, due to long and complicated post-depositional history, are prone to various environmental factors, such as weathering, which can alter the results of laboratory analysis. Hence, in many problematic cases, analysis with multiple methods (e.g. Py-GC-MS, LC-MS, Raman spectroscopy) is recommended, if not requisite.

1.3.4. Significance of amber for archaeology

Archaeologists – amateurs and professionals alike – took interest in amber soon after the birth of archaeology as an academic discipline. During the second *International Congress of Anthropology and Prehistoric Archaeology* in Stockholm in 1874, scholars raised some of the fundamental issues related to amber (term at that time used almost exclusively to denote fossil resin from the Baltic coast), still debated to this day. Another early sign of this fascination were H. Schliemann's reflections on provenience of amber adornments found at Mycenae which he could not determine conclusively, but suspected them to be of Baltic origin (Schliemann 1878). Recurring discoveries of amber in a multitude of spatially and chronologically varied contexts, made archaeologists aware of its importance for tracing long-distance contacts, travels and exchange. Researchers, however, quickly realised that there are resources of fossil resins besides succinite that could have been utilised by prehistoric people, even though Baltic amber has remained the focal point of these considerations.

Restricted access to amber meant that it often had to be moved over considerable distances to its eventual destination and place of deposition (Harding 2017, 155). This, in turn, caused it to be regarded as an indicator of interactions based upon exchange and trade between even very distant societies or places (see e.g. Harding/Hughes-Brock 1974, 145). Aforementioned feature in combination with attractive appearance, owing to colour (varying between shades of yellow, orange and sometimes even red), translucence, light weight and pleasant aroma when burned, contributed to amber being perceived by archaeologists as a luxury good and an emblem of prestige.

There are several important aspects concerning amber that every archaeologist faces when confronted with this inherently fragile material. Traits that make amber so attractive and unique, e.g. lightness facilitating its transport or softness allowing for its processing, are also the reasons for its usually poor state of preservation in archaeological context. Post-depositional factors such as weathering, oxidation, temperature and mechanical damage contribute to degradation and, ultimately, disintegration of amber lumps or artefacts, when left in dry environment. These processes effectively reduce the overall amount of amber buried in the ground that archaeologist can possibly unearth. Also, improper conditions of storage, e.g. in magazines or museums, may affect its state of preservation. Finally, intentional or accidental human actions can cause discussed finds to disappear without a trace; amber is inflammable, hence subjecting it to high temperature, e.g. during rituals such as cremation of the dead including all personal belongings and grave goods, may be very destructive.

In light of the aforementioned issues, some scholars suspect that hitherto assembled collections are, in fact, only a fraction of much larger quantities that were used by prehistoric and early historic societies.[27] Comments warning against underestimating the real number and significance of amber, especially in areas where the finds are scarce, are legit to a certain degree. On the other hand, overreliance on such vague estimations, instead of paying attention to the tangible material and its context, causes a risk of exaggerating. Therefore, when investigating archaeological record of amber, it is not only important to maintain a healthy dose of scepticism but also take into consideration various probable hypotheses.

The abovementioned qualities of amber were already known to Greek and Romans, as can be inferred from ancient writings. The latter provide an excellent source from which to learn a great deal of facts and opinions surrounding amber in the antiquity. For instance, in Greek the word *elektron* meant both amber and an alloy of gold and silver – substances regarded by Greeks as of divine origin, explained in myths such as the legend of Phaëthon or stories about Hyperboreans.[28] The Latin word *succinum* indicates that Romans properly assessed genesis of amber as derivative of tree resin (Czebreszuk 2011, 13). Its actual as well as imaginary therapeutic and apotropaic properties, or even ascribed magical powers were thoroughly reported by Pliny the Elder in *Naturalis Historia* (Pliny, XXXVII.XI; c.f. Negroni Catacchio 2009, 193–194). In 37th volume of his *opus magnum* Pliny, by quoting various authors, attempts to locate sources of amber which he seeks somewhere along southern coasts of Baltic and Northern Seas (Pliny, XXXVII; c.f. Kolendo 1983). Moreover, the same book details an expedition undertook in the reign of Nero (54–68 AD) by a Roman *eques*, who was sent as a merchant officially commissioned to buy amber later to be used during festivities from the barbarians at the mouth of Vistula River (Pliny, XXXVII.XII, c.f. Kolendo 1981). This is the oldest available description of such enterprise which, however, should not be considered as the first journey by an inhabitant of Southern Europe with the intention to acquire amber. It appears that even more common for Romans was to import fossil resin of the north through a chain of barbaric tribes, in some cases militarily and economically dependent on the Empire (see Kolendo 1981). These chains of exchange, or perhaps even trade routes, converged at the head of Adriatic, where the city of Aquileia played the role of the most important centre of amber's acquisition and processing.[29]

1.4. History of archaeological research into amber

1.4.1. 'Amber routes': from birth of the concept to its contemporary perception

Today, with all the scientific methods of proven effectiveness, scholars are usually able to tell apart archaeological finds made of Baltic amber (succinite) from those

27 Harding/Hughes-Brock 1974, 145; Bouzek 1993, 141; Czebreszuk 2007, 363.
28 Bouzek 1993, 141; Negroni Catacchio 2011a, 56; Negroni Catacchio/Gallo 2017, 315–317.
29 Negroni Catacchio 1976; Negroni Catacchio/Gallo 2017, 332; Pasquinucci 2017.

Fig. 5. Maps depicting Bronze and Iron Age amber routes in Europe, as envisioned by N. Negroni Catacchio (A; Negroni Catacchio/Gallo 2017, Fig. 4) and A. Palavestra (B; Palavestra 1993, 283).

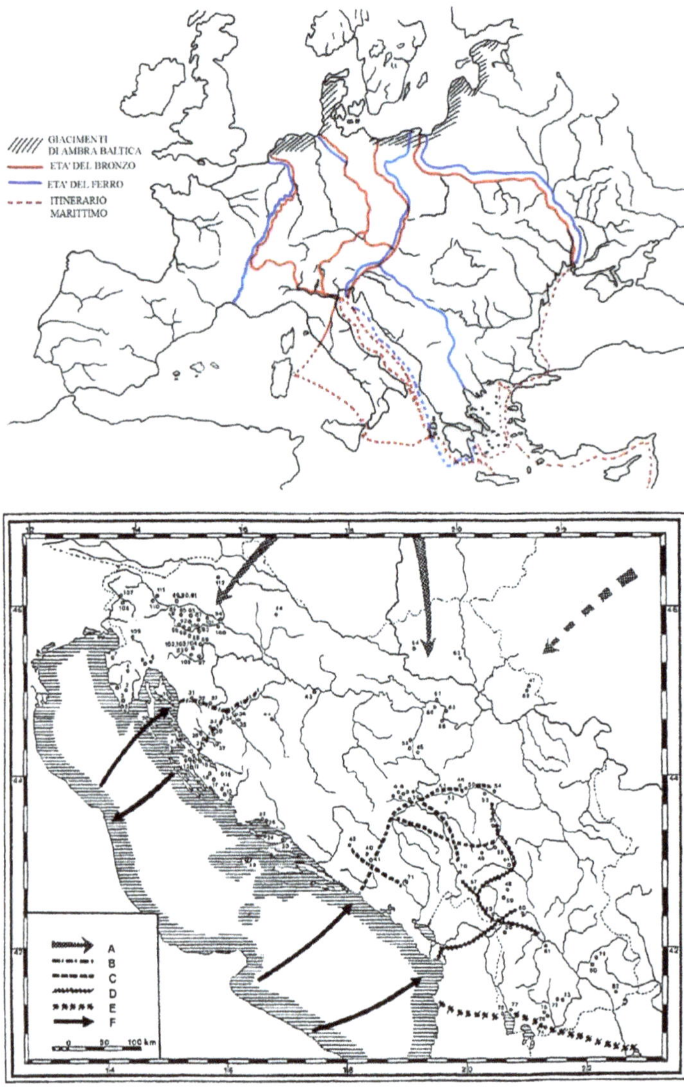

A

B

Presence of prehistoric amber finds in France from the Neolithic to the LBA was extensively presented in the works of C. Du Gardin, who concluded that its import in the 3rd Millennium BC was organised by the *Bell Beaker* culture (Du Gardin 1998). She later included Bronze Age finds of the so-called 'amber spacer plates' in the Western European route, roughly following the Rhône Valley (Du Gardin 2003). This proposition has been received well, and was subsequently adopted in the works of other researchers (e.g. Czebreszuk 2011, 158–159, Fig. 32; Fig. 4A:b). Recently, however, it has been suggested that French examples of spacer plates should be distinguished from British, German and Greek specimens on the basis of different perforation pattern (Harding 2013a, 376; see also Harding 2017, 157–158).

Thanks to the growing number of publications reporting finds of amber, other scholars began to highlight the importance of the Carpathian Basin and the Balkans in the process of amber circulation already since the Early Bronze Age (EBA).[30] The course of this hinterland route has been outlined from the Lower Danube drainage, along the valleys of the Morava/Vardar River, or along the Strymon/Struma River, down to Greece (Czebreszuk 2007, 368; 2009, 107; 2011, 159–160; Fig. 4A:d), even though more sceptic researchers pointed to potential impediments for movement along this way (Bouzek 1994, 221).

Arguably the least investigated branch of prehistoric 'amber routes' is the most easterly one, leading from the south-eastern Baltic shores, along the drainage basins of Vistula and Dniester on one hand and Dvina and Dnieper on the other, all the way to the Black Sea (Czebreszuk 2011, 160, Fig. 32; Fig. 4A:e and ebis). In a similar version, it was proposed already by M. Gimbutas (1965), however, the stimulus for its firm placement came with the discovery of the LBA cemetery of Hordeevka in the upper Boh Valley (Berezanskaja/Kločko 1998). The yield of amber finds of a specific form that have analogies in the *Mycenaean* Greece and even further in Italy indicates long-distance connections between the present-day territory of Ukraine and the Central Mediterranean.[31]

Finally, it is necessary to mention the role of the Adriatic Sea and its drainage basin (or more broadly its surroundings) in the circulation of amber in Europe. The issue has been discussed for a long time and resulted in multiple publications referring to this phenomenon, presenting various, often divergent views (cf. Fig. 4A:c and 4B). Many authors paid special attention to Adriatic, underlining its advantages for presumed seaborne travels and merchant routes that could carry amber. The significance of the Adriatic Sea in this regard seems to have been very prominent during the Iron Age, when both coasts, western and eastern, were occupied by thriving cultures using prolific amounts of amber.[32]

The situation is less clear in the case of the Bronze Age; early authors noted that amber finds from Yugoslavia are few and limited to the areas of Istria and northern Serbia, and predicted that none of them are contemporary with the *Mycenaean* artefacts (Harding/Hughes-Brock 1974, 158). Similar conclusions were drawn with respect to Italy which, in the authors' words, 'apparently played no important part in the dissemination of amber, which was not common there during the Bronze Age' (Harding/Hughes-Brock 1974, 158). Today, it is clear that such assessments were the result of very much preliminary state of research on amber in aforementioned regions.

---

30 E.g. Sprincz/Beck 1981, 1998; Marková 1993, 2003; Palavestra 1993; Beck/Marková 1998; Horváth 1999; Boroffka 2001.

31 Ślusarska-Michalik 2003; see also Czebreszuk 2007, 367–368; 2011, 161.

32 See e.g. Palavestra 1993; Palavestra/Krstić 2006; Palavestra et al. 2009; Negroni Catacchio 2011a, 2014; Bakarić et al. 2006; Bakarić 2017.

Intensive research by numerous scholars interested in prehistoric amber in Italy and the Balkans over the years have produced an impressive body of work. Significant contributions to the state of knowledge about ambers from the Apennine Peninsula came from N. Negroni Catacchio and her colleagues, reporting new discoveries, classifying them within a unified typology, and reconstructing possible ways of its inflow and distribution (Fig. 5A).[33] Although these works by large reflect a typological-chronological approach, focused on tracing the distribution patterns on the map between similarly shaped amber objects, they are of great usefulness for further studies. Among the recent contributions, one should mention M. Cultraro's overview of the Bronze Age amber on Sicily (2007), or works by P. Bellintani, I. Angelini and their collaborators, aimed at proveniencing hitherto registered artefacts and reconstruction of the processes of amber's dispersion in Italy.[34]

On the opposite side of the Adriatic, Yugoslavian scholars from time to time were publishing new discoveries of amber and placing them in a context of 'amber routes'. An important contribution was presented by Todd et al. (1976), who approached this topic from a different angle, paying more attention to geographic localisation of the sites with amber in Western Balkans. In the paper they reported morphological elements that would facilitate or hamper transport of amber in a hilly and wooded landscape of the Balkan littoral and hinterland. It was also the first major publication to discuss provenience of a large assemblage of amber finds, through the results of chemical analyses.

Another important contribution was made by A. Palavestra, who published all prehistoric amber finds in the area of former Yugoslavia known at the time (Palavestra 1993). Moreover, Palavestra argued that previously *en bloc*–treated Western and Central Balkans were characterised by diverse dynamics and routes of amber circulation (Fig. 5B). In subsequent publications, he advocated more territorially oriented approaches in the studies of amber and called for departure from the simplistic and literal use of 'amber routes', which he called 'useful although conditional, archaeological metaphor' (Palavestra/Krstić 2006, 62; cf. Palavestra 2007). Regardless of various interpretations, Adriatic and its surroundings still occupy an important place on maps illustrating 'amber routes', as can be seen in the latest syntheses (Czebreszuk 2011, 159–161, Fig. 32; Harding 2013a, Fig. 20.2).

It is also worthwhile to consider the concept of 'amber routes' from theoretical perspective. Early stages of discussion about the circulation of amber, or in other words 'amber routes' as envisaged by De Navarro, were dominated by the idea of *diffusionism*, strongly promoted in the works of V. G. Childe (1925) and later by M. Gimbutas (1965). Over the years, this simplistic approach in amber studies became the subject of criticism which pointed to the gaps in the current state of knowledge, shortcomings of applied methodology and inadequacy of some assertions in the light of new evidence.[35] Paradigm of *processual* (or *New Archaeology*), calling for functional analyses, testing hypotheses in the light of available data of multiple kind and building models to reconstruct cultural processes, stood in a stark contrast to previously popularized

33 E.g. Negroni Catacchio 1970, 1971, 1972, 1973a, 1973b, 1975a, 1975b, 1978, 1999; Negroni Catacchio et al. 1999, 2006.

34 E.g. Angelini et al. 2003; Angelini/Bellintani 2005; Bellintani 2010, 2014, 2017.

35 See e.g. Stjernquist 1967; Renfrew 1976; Wells 1985; Beck/Shennan 1991.

image of 'amber routes'. Moreover, S. Shennan and C. W. Beck rightly noticed that De Navarro outlined 'amber routes' mostly on the basis of Iron Age finds; therefore, adopting his concepts in the realities of the Bronze Age is incorrect (Beck/Shennan 1991, 128). Those critiques resonated with some scholars, who began to reflect upon theoretical foundations of the 'amber route' concept. Among recent studies one can observe departure from 'joining dots on a map' approach to distribution analysis, or from interpretations of amber circulation solely in economic categories of trade and exchange.[36]

Despite the shortcomings of hitherto applied methodology, amber in the eyes of many archaeologists still represent a commodity whose movement over long distances is an important indicator of intercultural contacts, and continues to raise many interesting questions about the mechanisms – social, economic, technological – by which the movement took place (Harding 2013a, 377). In modern literature one encounters several propositions concerning mechanisms of distribution of prestige products, innovations or knowledge. They vary from long-distance, direct and successive exchanges to gradual transfer known as 'down-the-line' or neighbouring 'prestige chain exchange'.[37] Over the years some authors have envisaged amber transfer to Greece in a few, consecutive, waterborne travels (Harding/Hughes-Brock 1974), while others have favoured sustained, step-by-step process (Bouzek 1993). At present, it is difficult to decide which method was more probable. However, it should be kept in mind that the influences of external cultures (i.e. prestige items) do not have to spread in a particular fashion, each time leaving signs of a single stage (Kadrow 2001, 225). Speculations concerning the nature of amber transfer will surely continue, but one has to remember that amber is a light substance and its lumps usually are not bulky. It has been pointed out that all the amber known in Bronze Age Greece or Croatia could have fitted into one rucksack or pocket (Harding/Hughes-Brock 1974; Forenbaher 1995). Even though varied chronology and multitude of archaeological contexts make it unlikely that all known finds from those countries were brought by a single person at one time, it is quite probable that amber supply had been intermittent rather than regular (Czebreszuk 2007, 368).

Problems inherent in the past approaches to the 'amber routes' are recognised by majority of scholars, but the opinions are far from uniform. Although in the recent years, the research on amber seems to have slowed down and this topic might not be as popular among archaeologists as it used to be, archaeological amber finds still require more investigation. Methodological aspects of the research definitely need to be reconsidered, but certain basic assumptions still hold. Currently, chemical methods allow to credibly distinguish Baltic amber from other fossil resins and, therefore in many cases, it is possible to pinpoint areas from which raw material was acquired. Similarities in the shape among amber beads are treated as a result of unified processing pattern, even though it may be necessary to delve into technological details to refine hitherto existing classifications. Specific forms of some amber artefacts, if widespread, may be very useful in tracking dissemination of products, as well as

---

36 E.g. Palavestra/Krstić 2006; Palavestra 2007; Murillo-Barroso and Martinón-Torres 2012; Jaeger 2017.

37 Harding 2000, 187; Makarowicz 2009, 324; for an overview of possible models of goods transfer see Renfrew/Bahn 2008, 375.

several rock-embedded tombs (nos. 9, 22 and 23; Orsi 1892). The site of Castelluccio is eponymous for the entire cultural unit dominant in Sicily during the EBA. In the following years Orsi was able to recover more amber beads from similar tombs in Syracusan area attributed to the *Castelluccio* culture: tomb no. 10 at Cava Secchiera (com. Melilli) excavated in 1891 (Orsi 1893b), Cava Cana Barbara, tombs no. 3 and 9 (com. Priolo Gargallo) (Orsi 1902a) and Valsavoia, tomb no. 18 (com. Lentini) (Orsi 1902b), the latter sites excavated in the spring 1899 and both characterised by a prolonged use extended to the MBA (cf. Cultraro 2007, 382). Furthermore, Orsi managed to document the presence of amber in slightly younger tombs, nowadays identified with the *Thapsos* culture (MBA), thus giving grounds for the notion of a continuous use of discussed materials in Sicily throughout the longer period. Significant collection of 19 elements was recovered in between 1890 and 1897 from the tombs no. 10, 11, 48 and 49 at the site of Plemmirio (com. Siracusa; Orsi 1891, 1899; Colini 1904). Among them were two flattened oblong beads with tapered ends, perforated at the opposite edges, often described in the literature as 'spacer plates'.[40] Finally a single amber bead was found by Orsi in the tomb no. 61 at the eponymous necropolis of Thapsos located on the Magnisi peninsula (com. Priolo Gargallo; Orsi 1895; Cultraro 2007, 382, Pl. XCVIIIc:5).

Increase in the number of amber artefacts coming from archaeological context has risen questions about provenience of the latter. Scholars debated the origins of amber, simultaneously touching upon the issues of the ability by prehistoric societies to exploit local sources of fossil resins or to enter in a long-range trade. The whole discourse had also political dimension, as it was taking place during the formative years of modern Italian and German states. Science, especially prehistoric archaeology (at that time in Italy called 'palaeoethnology'), played an important part by proving cultural and ethnic continuity with ancient people (considered as ancestors) inhabiting Italy in the distant past. Thus amber was depicted as a cultural emblem of the pre-Roman tribes, while utilisation of local sources of fossil resins manifested independence in civilisational development, specifically from Northern Europe – Nordic cultural zone (see Pizzato 2016, 303–346). The latter notion, if proven, could have considerable repercussions for the debate, as it would imply that proto-Germanic people, which had the access to the largest deposits of amber known in Europe, did not possess a 'monopoly' for export of amber and, therefore, exerted much less influence on Southern Europe and Mediterranean than the German scholars believed.

Differences of opinions on this matter resurfaced for the first time during the seventh *International Congress of Archaeology and Prehistoric Anthropology* in Stockholm, when G. Capellini, renowned Italian geologist and palaeontologist, delivered a lecture questioning the widespread assumption of northern origins of prehistoric amber artefacts from Italy, and advocated the idea of ancient exploitation of local amber resources, known at least since the antiquity (Capellini 1874). Although Capellini's views, presented in the literature already some time before (Capellini 1872), were not groundless, they were met with outright criticism from the vast majority of scholars present at the Stockholm conference, including his fellow Italian L. Pigorini, who

40 Cf. Orsi 1891, 123, Tav. XI:9, 17; Tusa 1999, 490–491, Fig. 29:1; Angelini et al. 2003, 229; Cultraro 2007, 381–382, Pl. XCVIIIc:10.

earlier expressed an opinion that amber finds from the *terramare* have Northern European provenience (Pigorini 1873).

While many took the Northern European provenience of amber beads for granted, a strong argument for their stance offered the results of chemical analyses of amber finds collected from various parts of the Mediterranean by German apothecary from Gdańsk O. Helm, who used the method of calculating succinic acid content in samples by means of dry distillation, with percentage higher than 3 % regarded as reliable succinite indicator (Helm 1877; Beck 1986, 70). Most famously Helm used these criteria to determine the origin of amber beads discovered in the Shaft Graves of the Late Helladic I Grave Circle A in the citadel of Mycenae, which he declared succinite (Helm 1885; cf. Beck 1966b). Prior to this, Helm had an opportunity to analyse four samples of amber native to the province of Emilia and seven amber artefacts from the same region which he had obtained from the Museum of Bologna (Helm 1882). He found the geological samples completely devoid of succinic acid, whereas the archaeological samples, in contrast, contained between 4.8 % and 6.3 % of that compound. This led him to a conclusion that the latter artefacts were not fashioned from local raw material and eventually supported the claim of proponents of the 'allochtonic amber' theory. It should be stressed, however, that Helm's analyses did not consider Bronze Age amber finds from the Apennine Peninsula, and overall amount of samples he used is too small to pronounce that 'all native Apennine amber is devoid of succinic acid' (Beck 1966b, 194).

Majority of the academic world welcomingly accepted Helm's contribution; however, there were some scholars who remained sceptical and continued to advocate the theory of local amber's exploitation in pre-Roman times. P. Strobel, using the method proposed by Helm, found a content of succinic acid in some ambers from Northern Italian Iron Age burials ranging between 2.45 % and 2.50 % – less than the minimal quantity found in Baltic amber, but much higher than that of the native amber (Strobel 1886, 42–44). Strobel, operating from nationalistic stance, in a rather arbitrary manner excluded the possibility of the Baltic origin of studied amber finds, and then argued for Italian provenience, despite the fact that the quantity of succinic acid was more discrepant from the local Apennine fossil resins (demonstrated by Helm to be devoid of this substance), than from Baltic amber.

The presupposition of objectivity stemming from scientific character of the method, as well as its reliability promoted by Helm, attracted many archaeologists, who like H. Schliemann sent him samples of amber artefacts from their collections to verify their origins. Helm also studied Sicilian amber deposits, which he labelled 'simetite', and estimated the main criteria for its identification, i.e. presence of a high oxygen content and a low succinic acid content (0.4 %) (Helm 1881; Helm/Conwentz 1886). In the heat of the debate, P. Orsi also wanted to determine the provenience of some amber beads he discovered during the research on Sicily, including the fragments from the Castelluccio and Plemmirio cemeteries.[41] However, perhaps realising the evidently pro-Germanic posture of Helm, he decided to entrust the materials to A. B. Meyer – another German scientist, however, open towards the idea of local Italian fossil resins' use in the prehistory. He, in turn, handed the samples to chemist F. Oster from Aachen, who by using the methods and criteria applied by Helm proved some, but not all, to be succinite. Total absence of succinic

41 Orsi 1891, 123; 1892, 19, n. 16; cf. Cultraro 2007, 378–379.

on their tops.[42] The works, however, were conducted without proper methodology, and afterwards the materials became dispersed; hence, some finds reported by N. Negroni Catacchio were found in the 'L. Pigorini' National Museum of Prehistory and Ethnography in Rome (Negroni Catacchio 1971, 320–321).

Another pile-dwelling that yielded amber artefacts in that period is Molina di Ledro – site located on the south-east bank of the Ledro lake situated at an altitude of around 655 m a.s.l. in Trentino-Alto Adige region. The settlement was discovered in 1927 after a sudden drop in the water level caused by the construction of the Ponale hydroelectric station (Rageth 1974). First investigations were carried out by Ghislanzoni (1929) on an area of 1,500 m^2, on the southern circumference of the *Isola* – the highest point of the prehistoric deposits. Thanks to a further significant drop in the water level during the winter of 1936–1937, R. Battaglia led excavation campaign on an area of 4,200 m^2. During this campaign, lasting from January to June 1937, considerable assemblage of amber beads was found.

In 1927 U. Calzoni of the Institute of Etruscan and Italic Studies began to investigate one of the most important archaeological sites in Tuscany – the Cetona mountain (1,148 m a.s.l.) with its numerous caves, shelters and other places utilised by man throughout the centuries (Calzoni 1954). The mountain itself situated nearby the border with Umbria, in the commune of Monteriggioni (Siena province), due to strategic position and natural characteristics offered good conditions for habitation. Preliminary field survey was followed by excavations inside several caves that lasted until 1940. Prehistoric site that yielded amber artefacts called Belverde is located on the eastern slope of the mountain, on the altitude of circa 500 m a.s.l. Residues of prehistoric habitation in this place were concentrated mainly in multiple caves, such as Grotta di S. Francesco and Grotta della Noce, both of which yielded amber beads which, however, has never been depicted in the literature.

In the course of the 1930s was revealed another sepulchral site in Apulia, significant i.a. due to the amber finds. It is located inside the cave of Manaccora (it. *Grotta Manaccora*), situated at the western edge of the homonymous bay, on the coast between the towns of Vieste and Peschici (prov. Foggia). The cave was surveyed for the first time in 1929 by U. Rellini (Recchia 1993). Prior to that, the cave had been regularly occupied as a shelter by shepherds and enclosed at the entrance by a drystone wall. Immediately upon the first visit, Rellini established a test trench at the mouth of the cave which yielded abundant prehistoric material, and later during the summer of 1930 began regular excavations at the entrance and in the interior, along the east wall. Afterwards Rellini entrusted the excavations to E. Baumgärtel, who continued the investigations in three successive campaigns from 1931 to 1933. During the last campaign (spring 1933), the excavations moved further inside the cave, where in the part named after abundant skeletal remains *Grotticella funeraria*, three pieces of amber were found.

In the case of Gualdo Tadino (province of Perugia, Umbria), some amber artefacts were discovered as part of a hoard. Found accidentally in 1937 inside a deep recess in the rock, the hoard's content was not inventoried and has remained in storage in the National Museum in Rome until 1962. Afterwards it was moved to the 'G. Bellucci' Prehistoric Museum of Central Italy and published by R. Peroni (1963b). In the same

42 Pigorini 1877, 14; Munro 1890, Fig. 63:9–10; De Marinis 2006, Tav. V:5–6; 2007, 65.

year, amber artefacts occurred for the first time at the cemetery Poggio La Pozza (com. Allumiere, prov. Roma) in Lazio. Even though the site has been known to archaeologist since the late 19th century, large-scale excavations carried out in 1883–1884 which brought to light more than 30 burials did not reveal amber jewellery inside the graves (see Peroni 1960). Research was resumed between 1935 and 1937, and this led to discovery of further burials, among them grave no. 3 excavated in July 1937 which contained three amber beads (Bastianelli 1939).

1.4.2.3. Research in the 2nd half of the 20th century – post-war years

Resumption of archaeological research in the post-war years quickly added new Bronze Age amber finds to the record. Already in spring 1947 archaeologists exploring the cave known as Grotta Misa (com. Ischia di Castro, prov. Viterbo) in Lazio stumbled across anthropogenic layer hiding a single amber bead (Rittatore Vonwiller 1951a). In the subsequent years, F. Rittatore Vonwiller continued surveys in the Fiora River Valley. In 1951 he investigated cemetery discovered accidentally during deep ploughing near the locality Ponte San Pietro (Rittatore Vonwiller 1951b). Among the rubble pulled out by the plough, it was possible to identify remains of seven stone-cist graves and broken ceramic vessels, including biconical urns. Only tomb no. 1 contained additional inventory consisting of two amber beads pertaining to the type Tiryns (cf. Negroni Catacchio 1999, Fig. 3:34). North from those sites, in the south-east Tuscany (com. Sarteano, prov. Siena), further artefacts made of amber came to light, dated roughly to the same period as the finds from Grotta Misa. The site Grotta dell'Orso di Sarteano, located not far away from Monte Cetona, was excavated in 1953 by G. Maetzke, in consequence yielding a single amber bead (Grifoni/Cremonesi 1968). Other in-cave site with amber investigated in that period in Tuscany is the Artofago cave (com. Gavorrano, prov. Grosseto), discovered in 1957 by the Speleological Group of Maremma in collaboration with the Natural History Museum of Grosseto (Aranguren/Perazzi 2004).

Concurrently to the above-described investigations, archaeological excavations were taking place in other parts of Italy. In 1948 C. Drago excavated remains of the prehistoric settlement of Torre Castelluccia (com. Pulsano, prov. Taranto) in Apulia, where along the stone wall enclosing rectangular compartment no. 7 from the west was revealed a place of storage (labelled 'hoard' for the lack of a more suitable term) overbuilt with stones, including ceramic vessel containing numerous elements of bronze jewellery and some amber and glass beads (Harding/Hughes-Brock 1974, Fig. 6:26–30; Gorgoglione et al. 1993). During the works on deepening and widening of the Mincio River channel 5 km south from the town of Valeggio Sul Mincio, a Bronze Age pile-dwelling settlement was discovered, now called Isolone del Mincio (com. Volta Mantovana, prov. Mantova). Excavations conduscted in 1955–1957 by F. Rittatore Vonwiller brought to light considerable assemblage of amber beads, comprising 11 pieces in total.[43] Another site with amber discovered within the province of Mantova in Lombardy at that time is Bande di Cavriana, located between Cavriana and Solferino in a vast depression bounded by a series of hills. The basin, in prehistoric times a lake, was later turned into a bog. Eastern part of the site has been completely

43 Negroni Catacchio 1971, 321; Guerreschi et al. 1985; Negroni Catacchio et al. 1999.

amber beads.[50] In Lazio in 1977 ground works on the road leading from Ardea to the village of Campo del Fico accidentally unearthed a pit, later identified as a grave with cremated burial inside an urn, containing i.a. 12 amber beads resembling the finds known from Allumiere – Poggio La Pozza (Delpino 1978; cf. Negroni Catacchio 1999). Approximately 20 km to the north-east, in the Alban Hills, another fortuitous discovery following the installation of an aqueduct in 1972 allowed to register Bronze Age settlement at Colle della Mola (com. Rocca Priora, prov. Roma). Systematic excavations organised by the 'La Sapienza' University of Rome and the Civic Museum of Albano began already in 1973, but only the second campaign in 1978–1979 revealed sunken structure with materials dated to advanced stages of the Bronze Age, among them a single amber bead (Conati Barbaro 2009). In Calabria, at the locality Bagneria near Santa Domenica di Ricadi (Cosenza province) quarry works performed in January 1977 brought to light a series of tombs embedded in natural cavities in a sandstone bank. Archaeological Superintendence of Calabria intervened promptly, excavating two of them (nos. 4 and 5), the former containing four amber beads (Pacci 1987; Benedetti/ Cardosa 2006).

1.4.2.5. Research in the 1980s and 1990s

The last two decades of the 20th century were the time of a considerable growth in new amber finds. In Northern Italy the most numerous group of finds was recovered from settlements, both newly discovered as well as those subject to reinitiated surveys. Exceptional concentration of the finds, spanning all stages of the Bronze Age, was recorded within the Verona province in Veneto. As early as 1980, two amber beads were found near the southern outskirts of Cavaion, at the site Ca' Nova, unfortunately without defined context (Aspes 1987a). Also from Verona province originates an amber bead collected from the surface near the south-east bank of the Garda Lake, at the locality La Quercia di Lazise – a pile-dwelling settlement reported in 1983 and subsequently excavated by A. Aspes from the Civic Museum of Natural History of Verona (Aspes 1987b). The third finding from the same area are the five beads discovered in the settlement Povegliano Muraiola (com. Povegliano Veronese) in the course of excavations that lasted from 1985 to 1988 (Belemmi et al. 1997). In this case the beads were found in a stratified position (US 107 and 123). Finally towards the end of the decade, in October and November 1989 the Archaeological Superintendence of Veneto carried out excavation campaign at the site Sabbionara (com. Veronella), ending up with a single amber bead recovered from the layers US 117, within the IInd excavation sector (Salzani 1990–1991a). Few hundred metres to the east from the Sabbionara settlement, few groups of cremated burials in pits and urns belonging to the Desmontà necropolis were excavated between 1982 and 2011, with one of the graves (no. 437) containing six amber beads (Salzani 2013).

In the course of the 1990s, occasional collections of surface materials and aerial photographs at the locality Finilone Valle (com. Gazzo Veronese) prompted archaeologists from the Civic Museum of Natural History of Verona led by L. Salzani to conduct a systematic field survey. During the first campaign that took place in the winter of 1995 and in the following year's spring, approximately 600 kg of finds, mostly attributable

50 Nava 1982; Piccoli 1982b; Negroni Catacchio et al. 1999.

to the Bronze Age, were retrieved from the ground after ploughing (Salzani 1996). Since then eight beads found at the site have been considered together with other amber artefacts from the Bronze Age (cf. Negroni Catacchio et al. 1999, 2006).

Not only settlements in the Verona province provided archaeologists with amber finds at that stage of the research. One can list four cemeteries excavated at the time that yielded considerable assemblages of amber artefacts. The most abundant in terms of number of the finds and arguably the most recognisable one out of all four is the necropolis, or complex of necropolises of Olmo di Nogara (com. Nogara; Salzani 2005). In 1987 the Archaeological Superintendence of Veneto executed a preliminary campaign which allowed to circumscribe spatial extent of the cemetery. In the fall of 1990, cadastral subdivision in the area revealed remains of some tombs. A series of excavation campaigns followed, and resulted in systematic exploration of the whole area which ended in 1997. The necropolis extends on a strip of land about 400 m long in the NW–SE direction and about 35 m wide, and has been divided into four sectors labelled A–D. The sectors are separated by roads and modern buildings preventing complete exploration of the burial ground. Consequently, one cannot be absolutely certain whether there is a continuity of burials along the whole extension of the necropolis, or there are empty sections devoid of graves. The necropolis is of bi-ritual type, with 61 cremation and 456 inhumation burials deposited in simple grave-pits, distributed rather sparsely, except for the southern part of the area C, where there is a notable concentration of cremated burials. The Olmo di Nogara necropolis has returned over 60 amber beads, distributed in 29 tombs (cf. Salzani et al. 2006).

Concurrently with the survey at Olmo di Nogara, archaeological investigations were taking place at the chronologically convergent cemetery of Scalvinetto (com. Legnago) some 20 km eastwards, in the vicinity of the Fondo Paviani settlement. Between 1991, when the first excavation campaign was performed, and 2002 when the project was concluded, exactly 535 tombs were uncovered (Salzani 1994, 2004). The burial ritual is largely similar to the one registered at Olmo di Nogara: deceased were either cremated and buried in urns, or inhumed in simple rectangular pits; however, the percentage of burials with equipment is lower, with amber artefacts being recovered from six graves (cf. Salzani et al. 2006). Another cemetery from the same period is Castello del Tartaro (com. Cerea), situated approximately halfway between the sites discussed above. In 1989 brief excavations following construction of a drainage ditch led to recovery of some burials. Preservation state of the graves was poor, caused by levelling of the land that occurred in the 1960s and by deep ploughing which followed shortly after. These factors have almost completely destroyed the most superficial level of burials, consisting of cremation tombs. After the first intervention, other excavations were scheduled in 1994–1995 and then took place continuously from 2004 to 2008, altogether bringing to light 396 graves containing both cremation and inhumation burials. Among the graves, five contained amber artefacts (Salzani 2011). Last but not least, in 1996 the latest campaign at the site Bovolone (com. *loco*) led to discovery of the grave no. 115 – cremation burial in an urn, containing scorched, but not completely burnt piece of amber (Angelini 2010a).

Abovementioned sites are all relevant to the Middle and Recent Bronze Age; however, one can also point to a slightly younger graves with amber in the Verona province. In 1985, in an area located 700 m to north-west from the settlement of Frattesina (com. Fratta Polesine, prov. Rovigo) some burials emerged following preparation of the land for installation of aqueduct pipelines. The site known as Narde is a large necropolis chronologically concurrent with the abovementioned neighbouring settlement. Since 1987 regular excavation campaigns have brought to light around 600 graves, mainly containing cremated burials and dating back to the Bronze Age–EIA transition. Section of the site defined as Narde I yielded two graves containing amber (nos. 169 and 281), while at Narde II amber beads occurred in grave no. 75 and in the part named *Ustrinum*.[51]

Further to the west, in Lombardy, amber finds were reported at Lavagnone (com. Desenzano, prov. Brescia). The latter site, a pile-dwelling settlement, has been a subject of archaeological investigations for a long time (e.g. Perini 1988), but amber beads were revealed during more recent excavations, carried out since 1989 by the Institute of Archaeology of the University of Milan under the direction of R. C. De Marinis (De Marinis 2007; De Marinis et al. 1996, 2005). In 1982 archaeological investigations at the site Lagazzi di Vho (com. Piadena, prov. Cremona) were renewed by the Archaeological Superintendence of Lombardy and the University of Genoa. They lasted until 1986. Thanks to test trenches and core drillings, it was possible to estimate size of the settlement and locate rectangular cabin-type dwellings, placed on wooden decks supported by oak posts (Simone/Tinè 1991–1992, 1996). During the campaign of 1983, in the area attributable to the earliest phase of the settlement, a necklace composed of 17 amber beads was discovered (Baioni et al. 2007b, 27).

In Liguria resumption of excavations at the site Castellaro di Uscio – a hilltop settlement – in 1981 brought to light another amber object (Maggi 1990). Exactly ten years later, following reports of bronze objects collected at the Monte Cavanero near Chiusa di Pesio (Cuneo province) in Piedmont, archaeological excavations set out to verify the context of the finds and eventually led to recovery of the remaining finds attributed to the LBA/EIA hoard including two amber beads (Venturino Gambari 2009).

The third major concentration of amber findings during the 1980s and 1990s occurred in Emilia-Romagna and has to do with exploration of several *terramare* settlements. Group of seven beads comes from the site Santa Rosa di Poviglio (com. Poviglio, prov. Reggio Emilia) – one of the largest settlements of this type, spanning an area of about 7 ha and comprising two main habitation zones.[52] These two zones, in fact, are identified with two phases of the settlement's spatial development. The first one, older and smaller thus called *Villaggio piccolo* (eng. small village), is roughly quadrangular and located to the north. The more spacious *Villaggio Grande* (eng. large village) was created as a result of spatial expansion of inhabited territory. Both are surrounded and separated from each other by a vast moat and delimited by separate embankments. In the course of 13 yearly excavation campaigns, beginning from 1987, a transect covering approximately 5,000 m^2 from north to south was explored. Spatially, it corresponds to about 1/15 of the entire area settled in the Bronze Age, excluding the parts occupied by ditches and embankments. As a result of these works, several amber beads were revealed in various sectors and strata corresponding with consecutive phases of the settlement.

Meanwhile, repeated resurfacing of Roman and prehistoric materials on grounds targeted for investments in the commune Anzola dell'Emilia eventually brought archaeologists' attention in 1992 and ended up enlisting the area as an archaeological

51 Cf. Salzani 1990–1991b; Angelini 2010b; Salzani 2010b; Colonna et al. 2010.
52 Bernabò Brea et al. 1997b, 1997c, 1997d; Bernabò Brea/Cremaschi 2004.

site. Subsequent excavations, which lasted to 2008, revealed a settlement with complex internal structure, including post-holes, pits, ditches and various movable relicts, including five amber beads.[53] Between 1993 and 2001, a century after the last survey, the Civic Archaeological-Ethnological Museum of Modena carried out new research at the *terramara* of Montale (com. Castelnuovo Rangone, prov. Modena) conducted by Andrea Cardarelli and Donato Labate (Cardarelli et al. 1997; Cardarelli/Labate 2004). The research led to the identification of preserved remnants of the settlement, just north of the church, where it ascertained the presence of a perimeter ditch (not identified in the excavations of the 19th century), and verified the state of conservation of the archaeological deposit in the areas saved from the excavations in the last century. Stratigraphic profile uncovered inside the moat allowed to trace the settlement's development throughout the Bronze Age, and assess chronology of further three amber beads. Although not a *terramare* settlement, the site of Riccione (com. loco, prov. Rimini) due to presence of amber, location within the same administrative region and similar chronology, can be added in this place. Detected in a residential district of Riccione, formerly a farmland known as Spina, the site was excavated between 1982 and 1986 and yielded housing structures with concave floors and wooden walls similar to the dwellings at Monte Castellaccio di Imola, investigated by G. Scarabelli (cf. Bermond Montanari et al. 1992).

Growth of data resulting from recent excavations on some of the *terramare* settlements, especially Santa Rosa di Poviglio, provoked an attempt to summarise the state of knowledge on this cultural phenomenon, which resulted in publication of the 1997 Modena exhibition catalogue *Le Terramare: la più antica civiltà padana* (Le Terramare 1997). Among several chapters concerning various aspects of relationships between the *terramare* ecumene and the outside world, it includes also a part devoted to amber (Bergonzi 1997). It offers the reader an introduction to the discourse about sources of amber used in the Bronze Age and routes of its dissemination across Europe, and provides a good overview of the amber finds recovered so far in Italy. It was not the first contribution of such kind by G. Bergonzi, who in collaboration with A. Cardarelli has approached the topic of amber in Northern Italy during the MBA already some time before, considering its status and symbol among other rare materials (Bergonzi/Cardarelli 1991–1992).

The 1980s and 1990s brought to light more artefacts classifiable as the Tiryns and Allumiere types – forms particularly discussed in the literature devoted to the issue of amber. Interestingly, those finds, usually found singularly, have rather wide distribution, spanning most of Northern Italy. In the course of excavation campaign directed between 1981 and 1991 by R. Perini, a single bead representing the Allumiere type was recovered from the bottom layers of a dwelling at the locality Dos Castel near Fai della Paganella (prov. Trento) in Trentino-Alto Adige, where it was found along with FBA pottery (Marzatico 1997, 465; cf. Negroni Catacchio 2014). Single Tiryns-type bead comes from Monte Battaglia in Emilia-Romagna – a hill with remains of Medieval tower at the top (com. Casola Valsenio, prov. Ravenna), which became a subject of archaeological excavations in 1983, and a year later revealed metallurgic workshop dated to the LBA, investigated until 1986 (Bermond Montanari 1987, 1997). More similar finds were discovered at the time in Lombardy: three elements come from

53 Cf. Finotelli et al. 1997a, 1997b; Argentina et al. 2002; Desantis et al. 2011.

limited to the outskirts of the settlement, corresponding spatially with the edge of the hill's plateau. This decision was conditioned by current preservation state of the site; exposure to erosion caused considerable deterioration of the prehistoric relics; however at the western and southern slopes, fragments of fortifications are still present. Five trenches were established and led to the identification of different phases of occupation, spanning major part of the 2nd millennium BC (Cinquepalmi et al. 2004). Among the finds are amber and faience beads, attributed to later phases of the site (Cinquepalmi et al. 2006; Cinquepalmi/Recchia 2010).

During the first two decades of the 21st century, Bronze Age amber finds emerged in regions previously not noted for their occurrence. The site of Rocca di Oratino (Campobasso province, Molise) is located on the slopes of a rocky elevation which dominates the high right bank of the Biferno River. The site has been a subject of systematic excavations since 2005, thanks to the collaboration between the Archaeological Superintendence of Molise and the 'La Sapienza' University of Rome. Throughout the research it has been possible to recognise portions of several Late Medieval structures and a conspicuous late *Subapennine* archaeological deposit, containing three amber beads.[57] The recent-most discovery of amber in Italy comes from Marche, precisely the site Monte Croce Guardia (com. Arcevia, prov. Ancona), located at the top of the namesake mountain (666 m a.s.l.). The presence of archaeological remains is known since the end of 19th century, but only during the subsequent century systematic researches were conducted. Several excavation campaigns have been carried out by the Archaeological Superintendence of Marche from 1961 to 1995. Results of these researches showed that Monte Croce Guardia was intensively settled during the LBA, and inhabited area reached at least 20 ha. The archaeological investigations started again in 2015 under the direction of A. Cardarelli ('La Sapienza' Universty of Rome), in collaboration with M. Bettelli and A. di Renzoni (*CNR-Istituto di Studi sul Mediterraneo Antico*). In trenches 1 and 2 several rectangular huts with gabled roofs and a large circular pit were found; the archaeological materials, among them fragment of an amber bead, suggest chronological attribution to the earliest phases of the FBA, with few elements dated to late stages of the same period (Cardarelli et al. 2017).

Concurrently, amber took a more prominent place in the considerations devoted to Bronze Age Italy by becoming a subject of separate studies. Turn of the centuries saw publication of the first, and so far the only, typological system explicitly devised for the Bronze Age amber finds by N. Negroni Catacchio et al. (1999, 2000, 2006). It incorporated results of the previous studies by the author, especially regarding the Tiryns and Allumiere types (cf. Negroni Catacchio 1999). Even though this typology has never been widely used by the scholars, it set a reference point for any future classification. Furthermore, discussed period brought several monographs devoted to amber in prehistoric Italy, often being an aftermath of exhibitions of the artefacts in various museums. Among them most important are: *Ambre – Transparenze dall'antico* (Nava/Salerno 2007) – a broad and comprehensive overview of amber in Bronze Age Italy – and *Lacrime delle ninfe* (Orsini 2010) – a catalogue of amber exhibits kept in the museums of Emilia-Romagna. Both publications do not delve into typological considerations but touch upon other issues connected with amber, such as routes of its transport into Italy, status and function for the past communities and place in

57 Cazzella et al. 2007; Recchia et al. 2008; Copat/Danesi 2010.

mythology and art since the antiquity. Most of all, they are valuable sources of information concerning history of the research and the present state of knowledge. In many conference proceedings from the last 20 years, one can also find papers approaching amber in a regional scale, e.g. Calabria (Benedetti/Cardosa 2006), Sicily (Cultraro 2007) or Sardinia (Bellintani et al. 2012; cf. Lo Schiavo 1982).

After years of debate on the origins of amber in Italy, sporadically supplemented with laboratory analyses of actual finds, in the last 15 years the IRS was eventually fully incorporated into the research. Thanks to a persistent project involving documenting the finds, sampling them and subjecting to provenience analysis realised by I. Angelini and P. Bellintani in collaboration with various other scholars, the level of recognition of various fossil resins utilised in Italy during the Bronze Age rose tremendously. As an effect, the readers are at disposal of a series of papers discussing origins of the finds from most parts of Italy.[58]

1.4.3. Research on amber in Western and Central Balkans

History of research on the Bronze Age amber in Western and Central Balkans is equally filled with accidental discoveries of spectacular magnitude and painstaking surveys conducted in accordance to archaeological methodology that, however, did not always prove to be as fruitful. Moreover, there are unsolved questions concerning the location of some sites, the appearance of some amber finds or places of their storage. Nevertheless, growing number of publications treating about former investigations (especially those before the 20th century) and addressing abovementioned issues help to clarify the situation to some extent.

1.4.3.1. Research in the 19th century

Many context dated to the Bronze Age and containing amber were revealed already in the second half of the 19th century. The earliest finding comes from the multiphase cemetery near Garica, on the Island of Krk, excavated by the National Museum in Zagreb. Out of the three tumuli excavated by the expedition between 1873 and 1874, two were later published by Š. Ljubić, however, only as a catalogue of finds mostly related to the Bronze Age (Ljubić 1876). Fortunately, in recent years it was possible to shed some more light on this interesting site and artefacts recovered there, including a single amber bead (cf. Blečić Kavur 2014).

Furthermore, abundant discoveries of amber took place during almost decade-long excavations of the burial mounds on the Glasinac plateau, located in south-east Bosnia not far away from Sarajevo. From 1888 to 1897 expedition from the National Museum in Sarajevo, led at different times by Ć. Truhelka, G. Stratimirović and finally F. Fiala, managed to explore over one thousand tumuli, dated to the Bronze and Iron Ages. Many of the burials proved to be equipped with amber beads of various forms which, on the basis of co-occurring bronze objects, were attributed to various periods of development of the cultural phenomenon sometimes described as the *Glasinac* culture (cf. Benac/Čović 1956, 1957). Chronology of contexts revealed throughout the research makes it the longest functioning centre of amber's acquisition in Western

58 E.g. Angelini et al. 2003; Angelini/Bellintani 2005, 2017; Bellintani 2010.

Novi Grad and Žamnjak, both excavated and published by B. Bačić (1957, 1960). Unfortunately, nothing specific is known about the single amber piece recovered at the former site. On the other hand, one of two amber elements detected inside the stone-cist grave of the latter mound retained clearly distinguishable shape, and was later used in provenance analysis (Todd et al. 1976, 320, Tab. 2).

Concurrently to above-described investigations, other Bronze Age amber finds were obtained through field survey conducted in the territories of Serbia and North Macedonia. A single bead was recovered from one of the mounds (no. 3) located at the site Ravnine near the village of Jančići 20 km north-west from the town of Čačak (Nikitović/Vasić 2002). Furthermore, two amber artefacts were collected in January 1959 from previously robbed and destroyed grave at the locality of Popadin Dol near Prilep (Kitanovski 1960). Finally a large assemblage composed of 116 amber elements was revealed during, only partially published, survey of mounds on the site of Jovanin Breg near Banjevac.[60]

Discussed decade also brought discovery of the first amber finds dated to the Bronze Age in Albania. The cemetery of Shtogj located in the forks of the rivers Mat and Fan in central Albania was investigated by an expedition from the Archaeological-Ethnographic Museum of Tirana which from 1952 conducted a series of rescue excavations in the Mat Valley, exactly on an area to be flooded by the waters of artificial lake created for the use of hydroelectric power plant (Islami et al. 1955, 110). Inside the grave 2, tumulus 9 at the site of Shtogj, two complete and multiple fragments of other, damaged amber beads were found (Kurti 2017a, Fig. 7:1; Kurti 2017b, Pl. XCVIIIf:3–4). Despite an early date of this discovery, the complete results of excavations at the site of Shtogj were published only recently (Islami 2013).

In the 1960s archaeological investigations included two in-cave sites discovered by speleologist in Lika. Caves of Bezdanjača located 10 km east of Prozor (Drechsler-Bižić 1979-1980) and Golubnjača situated on the hill-sides of the Velebit mountains between Gospić and Perušić (Drechsler-Bižić 1970), investigated in 1960 and 1962, respectively, yielded only one amber bead each; however, the significance of these findings lies in their rich contexts, comprising many ceramic vessels and bronze objects.

Meanwhile, news of discoveries of further amber artefacts were announced in Albania. Some of them came from a tumulus cemetery near the village of Perlat-Qendër, in the district of Mirdita (middle Mat Valley in geographic terms), where 8 or 9 barrows were excavated by an archaeological team led by S. Islami and H. Ceka (1964). Beads (in total 18) discovered in Perlat, tumulus 2, grave 3 include characteristic forms (i.a. elongated biconical and the Allumiere types; cf. Kurti 2013, Tab. I:8–9) that in the past caused a considerable stir among foreign archaeologist dealing with amber. While some authors correctly describe the form of the beads present in the aforementioned context (Bietti Sestieri/Lo Schiavo 1976, 188–189, Fig. 17), others place among them also the Tiryns-type elements (Harding/Hughes-Brock 1974, 155; Cultraro 2006, 1542) – an error recently revealed by R. Kurti (2013, 100; 2017b, 296).

The turn of the 1960s and 1970s in Albania witnessed even more amber finds being collected during extensive excavations of burials mounds. One should mention the cemetery of Barç in the Korçë Basin, where two tumuli became the subject of

60 Palavestra/Krstić 2006, 51–84; cf. Garašanin 1983c; Vasić 2003, 37, 48.

a systematic research that lasted from 1969 to 1975 (cf. Andrea 1985). Both yielded numerous amber beads, some of which dated to the LBA (tumulus 1, grave 47 and tumulus 2, grave 2; Kurti 2013; Kurti 2017b). Significant collections of amber artefacts come from two sites in the north-east Albania (Kukës district, Upper Drin Valley): Çinamak near the village of Shikaj, and Krumë in the vicinity of the Llakaj reservoir. Regarding the former, the results of excavations, that took place sometime in the late 1960s–early 1970s, remain largely unpublished (cf. Jubani 1969, 1974); however, the inventory of tumulus 16, grave 19 comprising i.a. 14 amber beads was recently given a detailed overview analysing chronology and provenience of the finds (Kurti 2017a, 92–107). Tumulus 1 at Krumë, on the other hand, was published comprehensively already some time ago, including 25 amber beads from the central grave (cf. Jubani 1982; Kurti 2017a, 99).

The late 1960s also mark the second stage of research of the tumuli cemeteries in the Drina Basin, during which more amber beads dated to the Bronze Age were acquired. This time the survey, conducted by M. Kosorić from the Museum of East Bosnia in Tuzla and D. Krstić from the National Museum in Belgrade, was focused on the necropolises placed on the left bank of the river, near Zvornik, and collectively described as the 'Padine-Ročević complex'.[61] Among excavated barrows are those from the localities of Karavlaške kuće and Šundinovača, both situated in the vicinity of Padine village and Jezero located closer to the village of Ročević. Excavations on aforementioned sites lasted from 1968 to 1971, and then again in 1977; however at the second time, they were limited only to the Jezero barrows. In total 55 amber pieces were recovered from six different mounds belonging to the complex (Palavestra 1993, 110–111).

At the time other barrows, with comparable chronology, were excavated further south along the Drina. Single burial mound situated at the site of Vranjani, about 10 km north-west of Požega, surveyed in 1969 by the team lead by M. Zotović from the Užice National Museum, included an amber bead (Zotović 1985). Unfortunately, no detailed information about the bead were published and currently its status is 'missing' (Palavestra 1993, 140).

In the 1970s more reports of further amber finds came from the territory of Kvarner. Laborers working in the northern part of the Pag Island, about 20 m north of the road Novalja – Lun, near the village Vidasi, accidentally stumbled upon remains of a necropolis (Glogović 1991, 17). Single amber bead recovered from the so-called locality *Vidasa napolju* ('Vidas field') represents an elongated biconical form, although D. Glogović defined it as an example of the Tiryns type (Glogović 1991, 19, Tab. 1: 2; cf. Cwaliński forthcoming). It was accompanied by various bronze objects, mostly jewellery, in general datable between LBA and advanced Iron Age. Although nothing certain can be said about the context of the finds (presumably ravaged graves), analysis of the formal traits allowed to divide them into two groups: the LBA one and the Iron Age one, possibly indicating two phases of the site's use, with the amber bead belonging to the former (Blečić Kavur 2014, 172–173).

Another LBA hoard containing amber was discovered in 1977 at the site called Debeli Vrh – a wooded mountain overseeing the village of Predgrad in Slovenian region of Dolenjska (Hirschbäck-Merhar 1984). Found accidently, it was damaged

61 Kosorić/Krstić 1970, 1972; Kosorić 1978.

by mechanical equipment used by the workers which scattered individual objects constituting the hoard up to 80 m distance along the road. Rescue survey under the direction of G. Hirschbäck-Merhar from the Provincial Museum in Kočevje aimed at recovering these finds was carried from 4 May to 25 June 1977. Rich content of the hoard included, among others, 147 bronze items, 612.5 g of crude amorphous bronze and 13 amber beads, including a single Tiryns-type bead (Teržan 1984; Palavestra 1993, 170).

At the end of the 1970s, few more amber finds were recovered from Albanian barrows, this time located in the south-east part of the country. The tumulus of Rehovë situated in the eastern part of the Kolonjë Plateau – one of the many similar tombs spread across the flatland – was excavated in two systematic campaigns in June–August 1979 and 1980 by the Museum of the Kolonjë District (see Aliu 2012). It contained 285 graves, out of which two dated to the Bronze Age included amber beads as grave goods: graves nos. 186 – 1 element and 201 – 18 elements. Further Bronze Age amber finds in the Kolonjë Plateau were found a decade later in the grave 46 of the Luaras mound (Aliu 2004). The tumuli, situated nearby Boshanj stream – a tributary watercourse of the Osum River, was discovered by soldiers during military fieldworks in the area. Subsequent excavations that were conducted within the framework of two campaigns in July–August 1988 and 1989 revealed 203 graves dating between the LBA and the Late Iron Age.

Beginning of the 1980s brought first discoveries of Bronze Age amber in the inland part of Croatia. Archaeological reconnaissance in 1981 registered LBA settlement near the train station of Križevci, 20 km southeast of Koprivnica, of which only peripheral part was subsequently excavated. The area of the site was used for clay extraction by local brickyard – a factor that largely contributed to the site's devastation. The settlement was subsequently excavated and published by Z. Homen from the City Museum of Križevci (Homen 1982). In total 110 m^2 were explored, revealing six furnaces, three waste pits with a lot of soot and ash, as well as several residential compartments. Due to destruction of the area, it is impossible to reconstruct the full extent and internal structure of the settlement. As shown by the deep profile trench almost 80 m wide, excavated residues belonged to a peripheral part of the settlement. The report does not give precise information on the location of individual finds, including two amber beads, but instead characterises them collectively (Homen 1982, 19–22; Palavestra 1993, 170–171).

In the course of underwater surveys, led by N. Logar from the National Museum of Slovenia in Ljubljana, in the upper section of the Ljubljanica River in central Slovenia between 1980 and 1985, a number of LBA pins were collected from the riverbed (Gaspari 2009, 31). Among them is a bronze pin with amber pinhead – a first object of such kind not only in Slovenia but also in the entire Balkans (Turk/Gaspari 2009, 70, Fig. 69).

One of the most abundant collections of amber finds derived from a single spot was discovered inside grave no. 1, at the site of Iglarevo, located within Kosovo territory (Palavestra 1997, 16–18). It is situated in the area of Prekoruplje – a natural border between the plain of Kosovo and that of Metohija – some 20 km east of Peć, on the slopes of Mt. Grebnica descending towards the Klina River (Palavestra 1993, 141). Remains of a Bronze Age burial ground were unearthed there during road construction in 1973. Systematic investigations carried out by K. Luci between 1986 and 1991 revealed two necropolises: Iglarevo I (containing 39 graves) and Iglarevo II (9 graves) (Luci 1998, 126–136). Two graves, one belonging to a female individual (grave no. 1), second to a child (grave no. 4), contained amber. However, majority of it was deposited in the first one (160 beads), while only a single artefact was found in the latter (Palavestra 1997, 16–18).

It is worth stressing that the finds from Iglarevo were not the first discoveries of amber in a Bronze Age context which happened in the Kosovo and Metohija territory. A considerable assemblage of amber beads (30 elements) was recovered from tumulus XIV, grave 19 at the locality Boka near Prčevo village, during excavations conducted by N. Tasić from the Institute for Balkan Studies (Serbian Academy of Sciences and Arts) from 1973 to 1975 (Tasić 1979; cf. Palavestra 1997, 32–33). Out of 24 tombs recorded inside the tumulus, majority have been dated to the Iron Age (ca. 6th – 5th centuries BC). However, reconsiderations of the inventories of graves 19 and 23 allowed to attribute them to an advanced part of the MBA.[62]

Around the same time, an international team under the direction of M. Primas of the Zurich University and J. Martinović of the Department for the Protection of Monuments in Kotor examined large burial mound named Velika Gruda, located in the coastal zone of Montenegro in the centre of the vast alluvial plain Tivatsko Polje (Della Casa 1996a). Excavations lasting from 1988 to 1990 revealed several superimposed clay strata and a thick stone covering. In the course of excavations, two amber beads were recovered from grave no. 9 – a collective, stone-cist grave, with one inhumed and multiple cremated burials. Grave no. 9 belongs to the group of the earliest graves (together with graves 8 and 13) located within the horizon D, although it is suggested that the grave was used for a longer period of time and subsequent burials were placed inside after refurbishment of the cist (Della Casa 1996a, 37–39, 78, Abb. 41–44).

Concurrently, a joint archaeological expedition from the University of Belgrade and the Prijepolje Museum, led under direction of D. Srejović, was working not too far away in north-eastern direction, near the Serbian-Montenegrin border. During August and September 1990, the archaeologists managed to explore one out of three mounds situated at the mountain pass of Savin Lakat, near the old road connecting towns of Prijepolje and Pljevlja (Lazić 2007). Aside from interesting observations regarding the mound's inner structure, the survey documented presence of several objects serving as grave inventory, among them five amber beads.

Another mound that yielded amber was discovered at the time on the territory of Istria, at the site Škicini near Vodnjan, not far from the previously described *gradina* of Vrčin. Site was discovered in May 1989, during earthworks which partly damaged the upper northern part of a tumulus covering a stone-cist grave. The mound was subsequently investigated by an archaeologist, who discovered two amber beads among the rubble and earth dug out from the grave's interior (Buršić-Matijašić 2003, 171–185).

1.4.3.4. Research at the turn of the 20th and 21st century

Last decade of the 20th century marked the beginning of the third stage of research on barrows in the Drina Valley. Archaeological excavation that began in 1989 on a site nearby the village of Brezjak (Paulje), within the administrative borders of Paulje

62 Cf. Luci 1998, 125–127; Bunguri et al. 2006, 96; Ljuština 2019, 95–96.

that followed recovered remains of two residential buildings made of timber, partly sunken in the ground. A single amber bead was found in the smaller building, discovered within test trench E.

Lately, a valuable contribution to the state of knowledge on Bronze Age amber in Serbia appeared (Ljuština 2019). It covers all known-to-date amber artefacts from the country, and includes comprehensive description of their contexts. The publication mentions, among others, a discovery of amber bead that took place in 2008 in eastern Vojvodina near Banatski Dvor, when during rescue excavations on terrains destined for planting a pipeline, archaeological team from the Provincial Institute for the Protection of Cultural Monuments led by D. Anđelić revealed a section of LBA settlement (Hristov 2017).

The last years abounded in scientific meetings and new publications dealing with the topic of amber in the Balkans. One has to mention the *4th International Conference about the Ancient Roads – The Amber Roads* that took place in Novo mesto (Slovenia) on 20–22 April 2017, and the international symposium *Amber: Science and Art* held on the occasion of the 25th *International Fair of Amber, Jewellery and Gemstones 'Amberif'* in Gdańsk (Poland), 22–23 March 2018. Both events gathered many archaeologists from the Balkan countries, who gave papers on the various aspects connected with amber. With respect to the latest publications, apart from the output of the project *Circum-Adriatic branch of the amber route in the Bronze Age*,[65] the contribution by T. Zygouris deserves a mention (Zygouris 2019).

65 I.a. Cwaliński 2017, 2020; Cwaliński/Czebreszuk 2020; Angelini et al. forthcoming; Cwaliński forthcoming; Cwaliński/Pravidur forthcoming; Cwaliński et al. 2018; Cwaliński et al. forthcoming; Kaur et al. 2020.

2. Contextualising amber: distribution, chronology, cultural attribution and provenience

The territorial scope of the work has already been explained in section 1.1. The spatial extent was delimited in a wider manner than the title term 'circum-Adriatic' might suggest. It was done this way in order to analyse, in the largest possible territorial dimension, the concept of amber translocation routes, running in various combinations through the current area of Italy, along the west and east coasts of the Adriatic Sea, and via Western and Central Balkans.

Fig. 6. Distribution of the Bronze Age sites with amber within the study area.

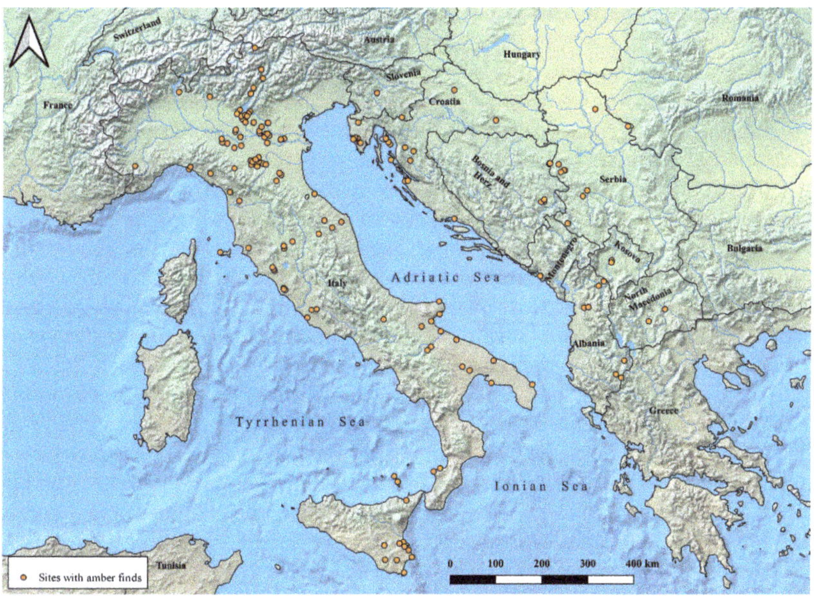

Consequently, the study encompasses areas stretching from the southern sub-Alpine zone, the Sava-Drava interfluve and the Tisa-Danube drainage basin in the north, to the Aeolian Islands, Sicily, southernmost parts of the Apennine Peninsula and the Pindus mountains along Albanian and North Macedonian border with Greece in the south. Northernmost regions can be identified with present provinces of Piedmont, Lombardy, Trentino-Alto Adige, Veneto and Fruli-Venezia Giulia in Italy, Inner and Lower Carniola (Dolenjska) in central and southern Slovenia, Istria and Slavonia in Croatia, and finally Vojvodina in Serbia. Regions delimiting study area to the south are

Sicily, Calabria, Basilicata and Apulia in Italy; Vlorë, Gjirokastër and Korçë in Albania; and Pelagonia and Vardar Valley in North Macedonia (Fig. 6).

Latitudinal extent was drawn along the Ligurian and the Tyrrhenian coastline of the Apennine Peninsula and Sicily in the west, and the Tisa-Danube-Morava-Vardar drainage basins to the east. While western reach of the study area is tantamount to political borders of modern Italy (provinces of Liguria, Tuscany, Lazio, Campania, Calabria and Sicily), the eastern spatial range runs along eastern confines of Banat, Timok, Braničevo and Šopluk districts in Serbia, and eastern borders of North Macedonia.

Between the north–south and the west–east limits of the spatial scope is located a vast area, nowadays belonging to a series of countries (in alphabetical order): Albania, Bosnia and Herzegovina, Croatia, Italy, Kosovo,[66] Montenegro, North Macedonia, Serbia and Slovenia. In this regard the Adriatic Sea forms a central axis which, geographically speaking, separates the area into two parts – western (Italian) and eastern (Balkan).

Above-described delimitation is also reasoned by cultural affinity of communities inhabiting regions in which amber appeared throughout the 2nd and the early 1st millennia BC. In some cases this is visible by durable units occupying vast territories, such as the *Palafitte-Terramare* culture in the Po Valley, or the *Apennine* culture in the central and southern parts of the namesake peninsula. In other instances one can see different sorts of ties, connecting and unifying to a certain level even distanced regions. Degree of similarity and character of relationships between these cultural units vary a lot, but amber is one of the shared features which links them.

In the following, the analysed objects and their contexts are considered regionally, with respect to the latest versions of the Bronze Age periodisations used for respective areas. The provinces of amber's acquisition were distinguished on the basis of visible concentrations of amber finds, coinciding with ecumens of cultural units – cultures, facieses, groups, etc. – as described in the literature. In some cases they are confined, by coincidence, to the spatial limits of modern administrative territorial units, which sometimes have a historical filiation (e.g. Piedmont, Apulia, Istria, Vojvodina/Banat), or span across the borders of several neighbouring districts. In other instances, the provinces were often named after geographic features – zones in which the amber beads are distributed (e.g. the Garda Lake and its inframorenic basin, Emilia/southern Po Plain, Glasinac, Podrinje and Western Serbia, Kolonjë Plateau, Vardar Valley and Pelagonia, etc.). Still, the spatial overlapping between the borders of the provinces of amber's acquisition and the regions to which they are pertaining (when the latter have strictly delimited boundaries) is not always ideal. Nonetheless, the applied nomenclature is predominantly meant to serve as a set of 'landmarks' – hints, suggesting where to look for main concentrations of amber finds.

2.1. Introductory remarks regarding applied chronology

Presented work covers a wide area, characterised by a considerable level of complexity regarding the development of the Bronze Age. When dealing with amber finds in such a broad spatial scope, it is inevitable to face a number of different cultural units and

66 Not recognised by some UN members (predominantly Serbia) as an independent state and, instead, treated as the Autonomous Province of Kosovo and Metohija.

related chronological systems, often in several versions. To be able to discuss issues connected with amber from diachronic perspective, it is required to establish synchronisms between the various periodisations. This matter is especially important, when trying to date specific types of amber artefacts with a wide distribution. Furthermore, switching to calendar dates can help avoiding constant references to phases or periods within regional chronologies.

In the second half of the 20th century, R. Peroni and his collaborators, most importantly G. L. Carancini, laid down the foundations for general chronological sequence of the Bronze Age in Italy.[67] Their aim was to organise the finds mostly from hoards and cemeteries with the help of seriation and on that basis construct a unified chronological sequence for the area of continental Italy. The final effect was division of the period into a series of phases and subphases, beginning from the Early Bronze Age – EBA (it. *Bronzo Antico*; phases BA 1 and 2),[68] through the Middle Bronze Age – MBA (it. *Bronzo Medio*; phases BM 1, 2 and 3, each subdivided into subphases A and B) and the Recent or Late Bronze Age – LBA (it. *Bronzo Recente*; first treated as a whole, but later subdivided into subphases BR 1 and 2) up to the Final Bronze Age – FBA (it. *Bronzo Finale*; subphases BF 1, 2 and 3) (cf. Carancini et al. 1996, Fig. 1–4; Fig. 7). The chronological scheme devised by Peroni and others was largely adopted in Italian Bronze Age studies, but not without words of criticism mostly concerning the method of seriating materials from many unstratified contexts.[69] However, since its invention it has been used more as a general framework: agreed upon with regards to terminology and sequence, but not unanimously accepted in terms of absolute dating. The problem of precise synchronisation and unification of various regional periodisations within Italy partly stems from the character of the country itself: its elongated, peninsular shape with numerous islands, with considerable environmental diversity, inhabited in prehistory by culturally varied communities with specific dynamics of development. Thus, of huge relevance are more locally scoped attempts at systematisation of Bronze Age materials which resulted in a number of new perspectives on chronology in various regions of Italy, published in the last 20 years or so. Among the most significant ones are the works devoted to the Bronze Age in the Alpine Italy,[70] sub-Alpine and northern Po Plain/transpadane Italy,[71] southern Po Plain/cispadane Italy,[72] Central and Southern Italy[73] or Tyrrhenian Italy (Pacciarelli 2001), to name just a few; the list of minor contributions, focused on particular phases, sites or areas is much longer. A separate issue is the Bronze Age chronology of Sicily and the Aeolian Islands – a subject to which L. Bernabò Brea devoted his archaeological career, coming up with syntheses of the problem on several occasions.[74] Since the last publications of Bernabò Brea,

67 Peroni 1959, 1963a, 1967, 1971, 1994, 2004; Carancini et al. 1996; Carancini/Peroni 1999; Peroni/Vanzetti 2005.

68 Alternatively, the EBA – *Bronzo Antico* is divided on the basis of metal/bronze items from hoards into five phases named with capital letters from A to E (Carancini/Peroni 1999, Tav. 35).

69 A good overview of the research history on the Bronze Age chronology in Northern Italy, including problems with Peroni's chronology, is given in Urban 1993, 12–23, 45–48.

70 Perini 1984, 1987, 1990.

71 De Marinis 1997, 1999; De Marinis/Salzani 2005; Cupitò 2006a; Cupitò/Leonardi 2015.

72 Bernabò Brea/Cardarelli 1997; Cardarelli 2009; Bernabò Brea et al. 2018.

73 Cocchi Gennick et al. 1995; Bietti Sestieri 2010; Bietti Sestieri/De Santis 2008.

74 E.g. Bernabò Brea 1958, 1976/1977; Bernabò Brea/Cavalier 1956.

quarter of the 2nd millennium BC corresponds with the following cultural units: classic phases of the *Rhône* culture in the French Alps and Switzerland, *Straubing* culture in Bavaria, *Únětice* culture in northern Lower Austria, Bohemia, central and east Germany and south-west Poland, *Unterwölbling* and *Wieselburg-Gata* cultures in southern Lower Austria, *Encrusted Pottery* in Transdanubia and late MBA I–early MBA II cultures of the Carpathian Basin: *Gyulavarsánd/Otomani-Füzesabony II*, *Vatya II*, late *Hatvan*, early *Wietenberg*, *Corneşti-Crvenka* and *Maros/Mureş/Szöreg-Periam II*.[82]

Final stages of EBA: BzA2b (ca. 1750/1700–1650 BC), BzA2c (ca. 1650–1575/1550 BC), and beginnings of MBA: BzB early (ca. 1575/1550–1550/1525 BC)[83] for many years have been considered as a time of decline of the EBA cultures and subsequent turmoil brought by social changes and ethnic movements in Central Europe. Reconsideration of the problem in recent years puts it in a different light (e.g. Meller/Bertemes 2010; Meller et al. 2013). Nowadays, it is seen more as the heyday of EBA civilisation, which developed largely unaffected by cataclysms happening further to the south, e.g. the Thera eruption. In the French Alps and Switzerland, this period covers the advanced and late stages of the *Rhône* culture (David-Elbiali/David 2009, Fig. 11), while in Southern Germany, it is recognisable through emergence of local cultural groups: the *Sengkofen/Jellenkofen* (post-*Straubing*) and the *Landsberg/Arbon* (Möslein 1998, Abb. 22; David 2006, Abb. 2). In Austria, Bohemia and south-west Poland phase BzA2b is coeval with the post-classical *Únětice* culture, whereas further east along Danube corresponds with the later section of MBA II and related cultural units. In western part of the latter area (west of the Danube) phase BzA2c has brought significant changes with the introduction of a new cycle of cultures named

6 and 7). In the opinion of the study's authors, the results put to the question validity of distinguishing the phase at least on the basis of pins, which show longer periods of use than previously assumed, and provide an argument for considering various types of pins as manifestations of concurrent metallurgic styles – levels of technological prowess in different parts of Central Europe. This conclusion was challenged in the paper by Brunner et al. (2020), which employed almost twice larger pole of metal finds from closed contexts and associated radiocarbon dates from a wider area comprising Switzerland and Southern Germany. Seriating the grave goods using reversed reciprocal averaging and then sequencing the 14C dates with Bayesian method revealed that the overlap between phases BzA1b and BzA2a is brief and amounts to no more than 55 years (1875–1820 cal. BC). Consequently, both phases should be regarded as chronological rather than chorological phenomena in the development of Bronze Age metallurgy and, thus, are in agreement with earlier devised chronologies of bronze pins.

82 Cf. Neugebauer/Neugebauer-Maresch 1989, Abb. 1; David 2002, Abb. 2.8; 2006, Abb. 2; Gogâltan 2005, Abb. 2; 2015, Figs. 1 and 10; Ernée et al. 2009, Abb. 3; David-Elbiali/David 2009, Figs. 1–2, 11; Jaeger/Kulcsár 2013, Fig. 19; Zich 2013, Abb. 13; David-Elbiali/Hafner 2013, Abb. 25; Fischl et al. 2015, Fig. 1:b; Bertemes/Heyd 2015, Fig. 2.

83 Absolute dating of BzA2b–BzA2c–BzB transition is well evidenced by 14C and dendro-dates from various sites in Switzerland and Southern Germany (cf. David-Elbiali/David 2009, Fig. 1; Köninger/Schöbel 2010; Primas 2013), Poland (Kneisel 2012; Müller 2012; Czebreszuk et al. 2015), Bohemia and Moravia (Peška 2013, Abb. 3), Slovakia (Bátora 2000, 2013; Rassmann et al. 2017), or Hungary and Romania (Gogâltan 2015; Jaeger et al. 2018).

Mad'arovce-Věteřov-Böheimkirchen[84] (plus the so-called *Litzenkeramik* or *Drassburg* culture), but in eastern part one can observe further development of the older cultures: e.g. *Vatya III, Wietenberg, Füzesabony,* and introduction of the new ones: e.g. *Suciu de Sus, Gârla Mare.*[85] Phase BzA2c and the subsequent BzB early in the context of many regions is interpreted as a transitional period, visible by changes of material culture, especially bronze metallurgy which gradually breaks with the older tradition and introduces new forms – a process which is also accompanied by changes in settlement patters and funeral rites.[86]

In Central Europe, start of the MBA has traditionally been identified with emergence of the *Tumulus* culture (ger. *Hügelgräberkultur*) in phase BzB. The phenomenon of under-barrow burial spread between the Rhine and the Hungarian plain, leading to creation of numerous territorial groups: i.a. *Bohemian-Palatinate Tumulus* culture (Čujanová-Jílková 1970, 1978, 380–389), *Pre-Lusatian/Silesian-Greater Polish Tumulus* culture (Gedl 1975) or *Middle Danubian Tumulus* culture (Stuchlík 1993, 272–285). Nowadays, the start of the *Tumulus* period (ger. *Hügelgräberzeit*) is shifted to the advanced moment of phase BzB.[87] Development of the *Tumulus* culture can be traced from phase BzB late (ca. 1550/1525–1450 BC), through phase BzC1 (ca. 1450–1400 BC), up to BzC2 (ca. 1400–1325/1300 BC) and in some regions even further into BzD. Absolute chronology of the above-listed phases was stabilised few years ago on the basis of 14C dates from graves of the *Tumulus* culture from Southern Germany (Müller/Lohrke 2009). Yet, it is increasingly more evident that adaptation of the barrow burial rite took place slightly earlier in eastern part of Central Europe, than in the west, as suggested by radiocarbon dates from barrows in Hungary (Fischl et al. 2013, 358, Fig. 6), Czech Republic (Peška 2013, 391, Abb. 3) or south-west Poland (Jaeger et al. forthcoming). The issue definitely requires more research with the aid of absolute chronological determinations. In the Carpathian Basin (Hungary and Romania), Reinecke's phases BzC2 and BzD can be correlated with horizons of hoards often used as chronological denominators: BzC2 – *Forró,* BzD (1) – *Ópályi,* BzD (2) – *Aranyos.*[88]

The LBA in Central Europe, also described as the *Urnfield* period (ger. *Urnenfeldzeit*), is now predominantly understood in terms of relative chronology through the scheme first conceived by P. Reinecke (1911a, 1911b, 1911c) and later subdivided by

---

84 According to new 14C sequences from Moravia and Lower Austria, beginning of the cycle could have taken place already in the 19th century BC (cf. Peška 2013, 391–393, Abb. 3, 6–7).

85 Cf. Neugebauer 1989, Abb. 1; David 2002, Abb. 2.8; Gogâltan 2005, Abb. 2; 2015, Figs. 1, 10; Ernée et al. 2009, Abb. 3; Jaeger/Kulcsár 2013, Fig. 19; Krenn-Leeb 2013, 414–415; Müller 2013, Figs. 4 and 9; Peška 2013, Abb. 3; Zich 2013, Abb. 13.

86 In Central Germany (the Saale-Elbe confluence) it is described as the *Sögel* horizon (Zich 2010, 2013), in Southern Germany and Lower Austria sometimes referred to as the *Bühl-Ackenbach* horizon – phase BzA3 (Innerhofer 2010, 2013; cf. Rittershofer 1984), in south-west Poland is considered as the post-*Únětice*/pre-*Tumulus* horizon (Kneisel 2012; Cwaliński in print), and in the Carpathian basin is termed as the *Koszider* period (Fischl et al. 2013). In the latter case, it is also possible to speak of the *Apa-Hajdusámson* horizon of hoards, following revised classification by W. David (2002).

87 Cf. David 2006, Abb. 2; David-Elbiali/David 2009, 328–334; David et al. 2017, 565–567, 578–581.

88 Mozsolics 1973, 1985; cf. David 2002, Abb. 2.5.

on the Greek mainland (Manning 2012). The approximate timespans of its consecutive phases are largely agreed upon from viewpoints of both the traditional 'historical' dating (inferred from the lists of Egyptian rulers, with their reigns correlated with astronomical observations; cf. Shaw 2000) and scientific-based absolute determinations (mostly radiocarbon and dendro-dates). The concerns may be raised by absolute dating of early Late Helladic phases (i.e. LH I–II) which are connected with the Thera eruption and the *Minoan* phases LM IA, IB and II – itself a matter of the fierce debate between advocates of the so-called 'Low' and 'High' chronologies (in different versions varying between 50 and 120 years).

There is no point in telling all the opinions of various scholars and quoting numerous publications which emerged during the discussion ongoing since the 1980s – its history can be easily found elsewhere (a good overview is given by, e.g. Bietak/Höflmayer 2007; Manning 2007). Suffice to say that in the recent years, strong arguments have been presented for the 'High' chronology by its proponents. Despite some of the absolute dates previously brought into the debate (e.g. Greenland ice cores) having been rejected as insufficiently related with the Thera eruption, the bulk of observations stemming from scientific methods still shows the 2nd half of the 17th century BC to be the most likely time, when the event took place.[92] Bayesian analysis of 14C dates from various contexts associated with the Thera eruption and ensuing destructions on Santorini (volcanic destruction level from Akrotiri) and Crete (Palaikastro Tsunami deposits and other related sites) allows to set the transition LM IA–LM IB–LM II to 1700–1600 –1470/60–1420 BC.[93] Also the critique of the 'High' chronology from the perspective of dendrochronology has been repelled, even though cautionary remarks regarding precision of identifying annual growth increments in olive trees, as in the case of olive branch buried in volcanic tephra on Santorini dated 1627–1600 cal. BC, are substantiated.[94] Furthermore, a number of issues hitherto raised by proponents of the 'Low' chronology as incompatible with the high absolute dates, such as the Thera bowl belonging to White Slip I ware or the Thera pumice deposits in Egypt and Levant, have been shown to fit the higher chronology upon reconsideration, while other problematic questions, i.e. Keftiu-tomb paintings, Egyptian stone vessels or Tell el-Dab'a stratigraphy, were dismissed as not sufficiently researched to be used as testimonies *pro* or *contra* either of the chronologies.[95] Finally, the extensive radiocarbon dating of short-lived plant remains from objects directly associated with Pharaohs reigns or short sections of the historical chronology, and their comparison by means of Bayesian statistics with historically determined accession dates for the individual rulers is an important contribution to settling the discussion (Bronk Ramsey et al. 2010). The results, on one hand, largely confirm the earlier estimations (as put forward by Shaw 2000) and, on the other, prove validity of radiocarbon-based chronological determinations in the context of Thera discussion. However, they also prompt to adjust the dating of the start of New Kingdom and the reign of its first rulers (Ahmose, Amenhotep I, Thutmose III) by raising it by ca. 15–30 years, thus narrowing the gap

---

92 Hammer et al. 2003; Bietak/Höflmayer 2007; Kutschera 2013.
93 Bronk Ramsey et al. 2004; Manning et al. 2006; Bruins et al. 2009; Manning 2009.
94 Friedrich et al. 2006; Cherubini et al. 2014 with debate: Friedrich et al. 2014; Bruins/Plicht 2014; Manning et al. 2014.
95 Manning 2007; Höflmayer 2009, 2012; Manning et al. 2014.

between 14C and the 'Low' dating (i.a. inferred from the Tempest Stela of Ahmose) of the Thera eruption.[96] In light of the evidence, some scholars were prompted to reconsider their stance, and only the most ardent supporters of the 'Low' chronology remain against accepting the need to lengthen the reigns of some Egyptian pharaohs and related chronological phases,[97] even despite the steadily growing number of 14C dates supporting the 'High' chronology (Höflmayer et al. 2016).

In the following it was decided to adopt the 'High' dates for the Aegean mainland ('Helladic') chronology, as presented by Manning (2012, Tab. 2.2), with time intervals for phases MH I–III after Jung 2013, Fig. 5 and 2017, Fig. 6: EH III: 2250–2100/2050 BC; MH I: 2100/2050–1900 BC; MH II: 1900–1800; MH III: 1800–1700 BC; LH I: 1700–1635/1600 BC; LH IIA: 1635/1600–1480/1470 BC; LH IIB: 1480/1470–1420 BC; LH IIIA1: 1420–1390/1370 BC; LH IIIA2: 1390/1370–1330/1315 BC; LH IIIB: 1330/1315–1220/1210 BC (Fig. 10).

Fig. 10. Dating of the relative chronological phases of the Bronze Age in the Greek mainland.

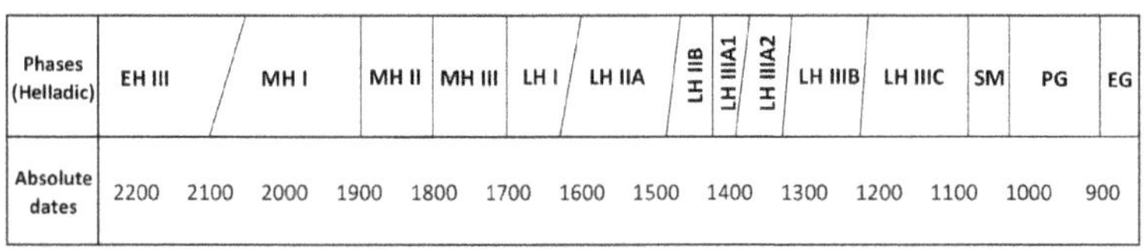

Certain problems arise also in relation to absolute dating of the late- and post-*Mycenaean* periods: LH IIIC, *Submycenaean* (SM) and *Proto-Geometric* (PG) phases. For these phases there are no Egyptian imports in Greece, and vice versa no Aegean imports in Egypt, therefore establishing a concordance between the two regions is not feasible. But Aegean materials pertaining to the abovementioned stages are traceable in Levant and Italy. Still, the synchronisation is difficult as periodistations of the LBA–EIA in both regions are not secure, and numerous uncertainties surround the deposition contexts of imports, especially in the case of Levant (cf. Toffolo et al. 2013). The best chances of estimating duration of the three abovementioned phases, identified with the collapse of the *Mycenaean* culture and the transition to the EIA – *Geometric* period, are offered by absolute dates (radiocarbon and dendrochronological) from Greece itself. The matter still has not been settled, and there are several inherent problems, when attempting to precisely determine duration of the phases on the basis of the scientific methods (e.g. 'old wood' effect and calibration curve plateau between 3000 and 2400 BP). The best indications are provided by absolute dates grounded in stratigraphic sequences of the *toumbes* of Kastanas and Thessalonika, which date the phase LH IIIC (1220/1200–1085/1080 BC) and the following SM (1085/1080–1070/1040 BC) much in accordance with the LBA traditional chronology based on comparisons with Levant (Weninger/Jung

96 Aston 2012–2013; Manning et al. 2014; Ritner/Moeller 2014.

97 For opinions of the 'Low' chronology supporters see e.g. Warren/Hankey 1989; Bietak 2003; Weiner 2003.

2009; Jung et al. 2009).[98] With respect to the end of SM, Bayesian analysis of 15 new radiocarbon determinations from Lefkandi, Kalapodi and Corinth in Greece places it in the second half of the 11th century BC (ca. 1020 BC) (Toffolo et al. 2013).[99] The dating is in line with observation on co-occurrence of finds dated to SM/BF 2 in Italian contexts for which radiocarbon dates are available (cf. Pagliara et al. 2007), and agrees with the synchronisation between SM, BF 2 and HaB1a, and Early/Middle PG, BF 3a and HaB1b postulated by C. F. E. Pare (2008, Fig. 5.4).

The aim of the following two sections (2.2 and 2.3) is to provide the reader with an overview of chronologies and cultural background in the regions of amber's acquisition during the Bronze Age. Furthermore, the sites – contexts of amber deposition – within each region were shortly described, with the focus put on the chronological determination of amber finds. Usually the dating offered by analyses of other scholars and preceding works was valid and sufficient to be quoted as the basis for a context's chronology adopted in this work. However, in cases where the dating was unclear/unspecified, or there are controversies/differences in opinion concerning a context's chronology, a more in-depth discussion with typo-chronological considerations of materials co-occurring with amber is given.

2.2. Italy[100]

2.2.1. Liguria

The north-west part of Italy has so far not produced many amber finds datable to the Bronze Age. A minor concentration of amber artefacts has been noted within the confines of eastern Liguria, in the upper Apennines between Genoa and La Spezia

---

98 Cited sites and related absolute dates are not the only ones pertaining to the discussed period. Radiocarbon and dendro-dates from the Assiros *toumba* in the Greek Macedonia have been used to support a higher chronology of LH IIIC, SM and PG, thus suiting the arguments for an earlier beginning of the Iron Age postulated on the basis of absolute dates from some sites in the Mediterranean (Newton et al. 2005; Plicht et al. 2009). B. Weninger and R. Jung address the dendro-dates from the Assiros toumba, by pointing to uncertain relation between the dated materials (wooden timbers and posts) and conventionally dated strata (phases 3 and 2) (Weninger/Jung 2009, 385–388). However, since then a new publication of absolute dates from the latter site has emerged, maintaining the argument for the 'high' chronology of the phases under consideration (Wardle et al. 2014). Still, the change of chronology postulated by the authors is challenged by other scholars on methodological grounds (Knapp/Manning 2017, 116–117). Hypothesis placing both the transition from LH IIIB to LH IIIC, and from SM to PG about a century earlier is furthermore difficult to reconcile with dates from Greek, Cypriot and Levantine sites as quoted by Manning 2006-2007; Knapp/Manning 2017; Toffolo et al. 2013, 2014.

99 Wardle et al. question the results obtained in the cited publication by naming five problematic issues, but do not develop their arguments *contra* despite a declaration to do so (cf. Wardle et al. 2014, 7).

100 Regarding Italian chronological terminology, the following abbreviations were used: EnR – Eneolitico Recente; BA – Bronzo Antico; BM – Bronzo Medio; BR – Bronzo Recente; BF – Bronzo Finale; PF – Primo Ferro; SF – Secondo Ferro.

(Fig. 11). Amber in that area occurred at three settlements, all defined as the *castelli* – a category of upland/hilltop settlements with natural defensive qualities, often additionally fortified (cf. Maggi 1990, 9). Sites falling into this group are Castellaro di Camogli, Castellaro di Uscio and Castellaro di Zignago.

The first one is located on top of the Mount Castellaro – a 70 m high cliff situated just by the coast line, on the south-east outskirts of the town Camogli. Excavations conducted within three sectors/trenches, named A, B and C lead to recovery of about 7,200 pottery fragments, 700 fragments of daub, 2,600 osteological finds, as well as fragments of polished lithics, some metal objects, two amber beads and an iron slag. Three layers designating consecutive occupation phases of the site, coincident with BM (layer VIII) and BR (layer VII), followed by the second settling of the site at the end of the Iron Age (5th–4th century BC), were distinguished on the basis of the stratigraphy and diagnostic finds (Fossatti/Milanese 1982). Recently, the ceramic material relative to the levels of the Bronze Age and the excavation records were reviewed (Delfino 2005/2006).

The site of Castellaro di Uscio is located on top of the Mount Borgo (728 m a.s.l.), at the intersection of two mountainous ridges: one starting east from the Entella plain, and the second following south from the Portofino Promontory. The settlement occupies summit plateau of the mountain, which has the shape of an irregular quadrilateral, with dimensions of about 50 × 70 m. Over five campaigns, conducted between 1982 and 1985, a total of 152 m^2 were excavated (Maggi 1990). Archaeological deposit was uncovered in a long transect trench, documenting stratigraphy along the slope. The latter was divided into three areas denominated as high, medium and low terraces. The different layers can be grouped, based on archaeological materials, radiocarbon dating and soil analysis, into five horizons, or stratigraphic complexes, named as horizons 1, 2, 3, 4 and 6. They correspond to three phases of the site's occupation: EnR/BA (horizon 4), BF (horizon 3) and the SF (horizons 2 and 1).

Finally, the hilltop settlement of Castellaro di Zignago, located on bare and flat summit of a mountain (945 m a.s.l.) to the east from the namesake village, revealed remains of several huts with drystone walls and *terracotta* floors, built on the terraced southern slope. Excavations were conducted within three zones, among which zone S proved to be the richest in archaeological evidence. It comprised a flat area of quadrangular shape, bounded by stone walls intersecting at right angles. In this area two distinct constructive phases were highlighted, each composed of three phases of structural changes to the hut (numbered I–VI) and related layers: the older one spanning approximately BM and BR (levels IV–VI; layers E–P), and the younger one related to attendance of the site during BF (level I–III; layers A–D) (Mannoni/Tizzoni 1980, 255–256).

Dating of the amber artefacts coming from these sites in general indicates a period from the Middle to the FBA (BM-BF), however, specifying the chronology in terms of absolute dates is largely unfeasible, as the whole region is lacking a well-founded chronology of the Bronze Age (cf. Urban 1993, 19–20, 262–266, Abb. 145; Gambari 1997, 442–444, Tab. 6). In the past, beginning of BM in Liguria used to be linked with the facies *Lagazzi – Farneto – Pollerà*; however, more recently the term came under criticism and was largely abandoned.[101] Materials representative for the phase are

101 Bernabò Brea 1946, 316–318; Tinè 1973, 51; cf. Del Lucchese et al. 1991–1992, 32, 38; Urban 1993, 19; Gambari 1997, 442.

Fig. 11. Distribution of the Bronze Age sites with amber in Northern Italy.

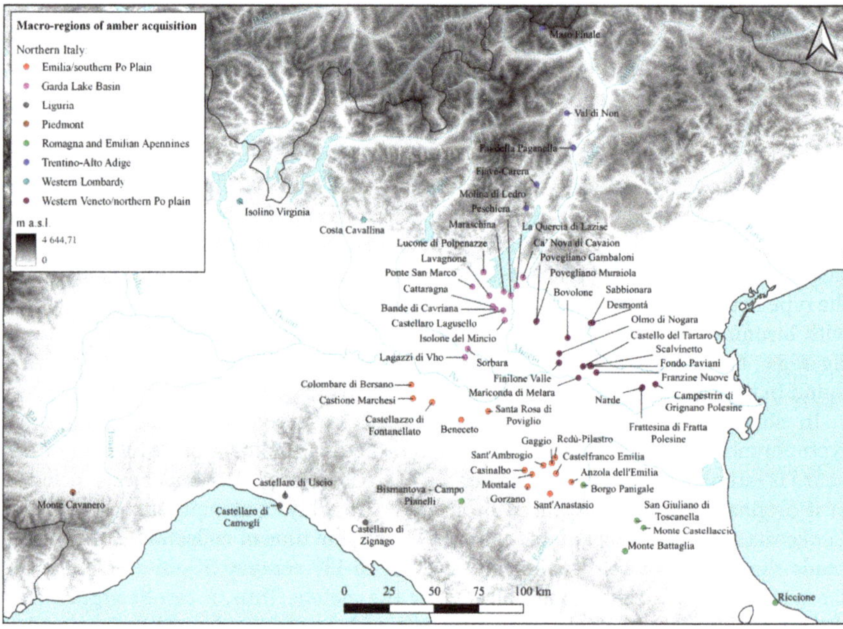

2.2.3. Western Lombardy

Western Lombardy, as opposed to the eastern part, has much fewer amber finds dated to the Bronze Age (Fig. 11). In fact only two sites from the period yielded amber artefacts: Costa Cavallina (Poggiani Keller 1989, 90, Figs. 98 and 99) and Isolino Virginia (Isolino di Varese) (De Marinis 2009c, 143–145, Fig. 5:3). Amber finds from both sites are largely dated to BF – a period corresponding with emergence of the *Protogolasecca* culture in Lombardy, a continuator of the *Canegrate* culture's legacy and at the same time direct ancestor of the Iron Age's *Golasecca* culture.[112] Three chronological subphases of the FBA: BF 1, BF 2 and BF 3 in north-western Italy correspond with individual stages of the *Protogolasecca* culture: *Ascona I* (ca. 1150–1060 BC), *Ascona II* (ca. 1060–1035/1010 BC) and *Ca 'Morta – Malpensa* (ca. 1035/1010–910 BC).[113] All three phases are well defined by the typological evolution of bronze fibulae: BF 1 – twisted violin-bow fibulae raised above the bracket,[114] BF 2 – the oldest simple-arch fibulae, generally with a fully twisted arch[115] and BF 3 – simple-arch fibulae with arch torsion only above the bracket and generally with a slightly raised arch and more robust rod.[116] Other elements belonging to BF 1 are greaves of the type Rinyaszentkirály and axes with elongated median fins of the type Ortucchio, while BF 2 is associated with pins

112 Cf. De Marinis 1972, 2009a, 541–542, Fig. 10; 2009b, 17–20; 2014, 21–22.
113 De Marinis 1972, 94; 2014, Fig. 16; Pare 2008, Fig. 5.2.
114 De Marinis 2000, Fig. 3:1–9; 2014, 21, Fig. 3:2, 12, 19; cf. Eles Masi 1986, 2–3, Tav. 2.
115 De Marinis 2000, Fig. 5–6; 2014, 21; cf. Eles Masi 1986, 14–16, Tav. 3:68, 4:69–74, 76–85.
116 De Marinis 2014, 21, Fig. 4:5, 7; cf. Eles Masi 1986, 16–18, Tav. 4:75, 5:86–97.

of the type Fontanella, and BF 3 with pins of the type Fiavè and bracelets of the type Zerba (De Marinis 2009b, 17–18, Fig. 6).

Burial in the Costa Cavallina cave, located on a mountain overlooking the town of Clanezzo from the north (distance of ca. 500 m) and Botta from the west (distance of ca. 350 m), at the time of discovery was already devastated by illegal excavations. Therefore, no details about the site can be written, apart from the fact that the deceased apparently was inhumed (Poggiani Keller 1989, 88–89). Due to absence of diagnostic elements of inventory deposited with the burial, dating of the context was initially based on the formal characteristics of amber beads. The latter, thanks to presence of decorative motif of a median rib, were classified as the type Tiryns and dated to BF (Poggiani Keller 1989, 89). In N. Negroni Catacchio's opinion shape of the bowl found inside the cave in association with amber allows to place the beads in final BR or early BF (Negroni Catacchio 1999, 245; cf. Poggiani Keller 1989, Fig. 98).

Less problematic is chronological determination of the single amber bead found at Isolino Virginia – an island located near the west bank of the Varese Lake. In the most superficial part of the archaeological deposit in the central and southern parts of the island, mostly between 25 and 65 cm deep, numerous fragments of BF (*Protogolasecca*) and, to a lesser extent, BR (*Canegrate*) pottery were found, testifying to occupation of the island between the 13th and 10th century BC. The sherds are relevant to biconical vases, conical/truncated cone bowls and situla-form vessels. The decorative motifs are predominantly characteristic for the late *Protogolasecca* culture (*Ca' Morta Malpensa* phase): false-cord decorations or wheel-like impressions, incised or fluted bundles of horizontal or zig-zag lines formed into series of triangles ('wolf's teeth' motif), and horizontal lines surrounded by *cuppelle* (De Marinis 2009c, 140–141, Figs. 2 and 4). During I. Regazzoni's excavations in 1885, along with the pottery and the Bismantova-type bronze knife of the variety C, a bead of the Allumiere type was found in the central part of the island, at the depth between 30 and 40 cm below ground level.[117] Among the materials discovered by Regazzoni in the trench A, in the southern part of Isolino Virginia, are two blue glass beads in the shape of a small barrel with a spiral decoration of white filaments, whereas W. K. Foster's trench IV, located about 10 m from the south-eastern shore, yielded two matching fragments of an open bronze bracelet, belonging to the type Zerba.[118] All the aforementioned materials, with the ceramics at the forefront, form a homogeneous assemblage which chronologically can attribute to BF 3 (De Marinis 2009b, 18–19).

117 Negroni Catacchio 1978, 165–169, Fig. 5; De Marinis 2009c, 5:3, 6, fn. 106; cf. Bianco Peroni 1976, 58–61, no. 283.

118 De Marinis 2009c, 5:2, 4–5; cf. De Marinis 2000, 136–137, Fig. 10; Bellintani/Residori 2003, 490, Fig. 4; Bellintani 2012, 21, Fig. 3:13–14, 19, 30, 4.

settlement of Fiavè 6 ended abruptly, as suggested by traces of fire that damaged wooden structures.

There are other archaeological sites in Trentino which yielded amber finds presumably originating from the Bronze Age, but due to lack of determined context or documented stratigraphy, it is difficult to specify their chronology. One of the earliest discoveries of amber in the region comes from the *palafitte* settlement on the south-east bank of the Ledro Lake, near the town of Molina di Ledro.

Seven amber beads were recovered from the top layers of the main part of the site (the so-called *Isola*) during excavations conducted by R. Battaglia in 1937 (Battaglia 1943). The area (about 4,200 m^2) was covered with a net of 4 m^2, divided between individual sectors (trenches named with letters). Locations of the finds were fixed within sectors, squares and layers (Rageth 1974, 80). Battaglia's research brought to light more than 10,000 wooden posts and planks in a 2–3 m deep deposit, not to mention numerous finds of various categories.

Nonetheless, a cautionary note has to be given. Periodical changes of water level in the lake, causing the archaeological deposits on the banks to slip and intermingle, affected preservation of the original stratigraphy (Rageth 1974, 83). Combined with poor documentation of early stages of the research at the site and the turbulent storage history of finds, a full reconstruction of the settlement's structure and chronological sequence is practically impossible (cf. Rageth 1974, 79–88), even despite repeated excavations (cf. Leonardi et al. 1979; Bellintani et al. 2015b). Even though Battaglia's report gives detailed insight into Ledro's stratigraphy, it has to be treated with great caution due to a number of reasons (cf. Rageth 1974, 83–88; De Marinis 1999, 36–37). Furthermore, positioning of many finds, especially pottery, has been neglected in the original publications, and in the years following excavations remaining informations about their findspots, i.e. labels on boxes, were lost or mixed. Hence, one can only rely on stylistic similarity with finds from other sites, to estimate chronology of a bulk of inventory. Moreover, many plans, profiles and a diary from Battaglia's excavations are nowhere to be found, hampering clarification and integration of the available data.

According to the revision of Battaglia's documentation by J. Rageth, seven amber beads were recovered in trenches A and C from strata I–III (Rageth 1974, 186, 240, Taf. 92:6–8, 16–18, 24). However, it should be mentioned that designation of the layers in two major zones of the *Isola* – the western along the lake's shore and the eastern on the elevation further inland – should be accepted with caution, as the labels do not always reflect the same stratum (cf. Rageth 1974, 83–88, Abb. 1). Although the uppermost strata I–II are generally similar in both zones (peat and sand layers), the stratum III in the eastern part contains large amount of charcoal which does not continue in the western part. On the other hand, stratum III in the latter corresponds with layer IV in the eastern zone (pieces of wood and carnelian kernels). Therefore, it is safer to denominate the context of amber beads as layers from I–III/IV.

Layers I–IV yielded chronologically varied materials; however, majority of the finds represents forms characteristics for the advanced/late EBA and the early MBA. That includes pottery (Rageth 1974, Taf. 37:1–2, 38:9, 39:2, 42:3, 5, 45:10, 46:2, 11, 47:5, 58:2–7) and bronzes (Rageth 1974, Taf. 19:4, 27:6, 28:5, 29:6) which can be attributed to layers I–IV in trenches A and C. Finds like these are attributable to the late *Polada* culture and the early stages of the *Palafitte* culture in the Lake Garda area, and find numerous analogies in the horizons 3–5 of the Fiavè settlement.[121] On that basis, 7 out of 15 amber beads can be cautiously attributed to BA 2–BM 1 (Rageth 1974, 186, 207–210). Unfortunately, due to absence of any designations concerning their context, the remaining eight beads cannot be dated (Rageth 1974, Taf. 92:1–5, 19–20, 23).

Chronology of a large assemblage of amber beads from Val di Non is even more difficult to determine. Events that led to discovery of this presumed aquatic deposit/hoard were described by G. Roberti in the article published in 1911 (Roberti 1911). The findspot is nowadays impossible to pinpoint. All that is certain is that the objects were recovered from a peat bog located approximately between the towns of Cles and Tuenno. Amber objects were collected on two occasions. First in 1876, at the depth of 1.5 m, in a layer partly formed by peat and partly by clay, a single spacer and 20 discoid/flattened globular beads came to light. Later (unspecified when exactly), in one of the bogs between Cles and Tuenno, perhaps the same one as previously, another spacer and further four discoid beads were recovered. A total of 26 amber exhibits are now stored in the archaeological collection of the National Museum of Castello del Buonconsiglio in Trento, where they were delivered following the discovery (Angelini et al. 2003, Fig. 2). In the literature one can find different, often contradictory, interpretations of the finding's chronology, ranging from the EIA (Harding/Hughes-Brock 1974, 168), LBA–EIA (Harding 1984, 310) and MBA (Marzatico 1997, 454; Marzatico/Tecchiati 1998, 28–29). Devoid of any information about the context, the amber beads were dated to BM and BR according to results of a comparative analysis of the spacers' formal and technological traits (cf. Angelini et al. 2003, 229).

Amber continued to be used in Trentino-Alto Adige in the final stages of the Bronze Age, as attested by the finds from Fai della Paganella and Maso Finale. End of the 2nd millennium BC in the region is marked first by the *Dos Gustinaci* facies known from the horizon Fiavè 7 (BR) and, subsequently, by phase A of the *Luco* (ger. *Laugen*) culture (BF). There is a genetic link, or rather uninterrupted relationship in evolutionary sense between the two abovementioned units, indicated in particular by aspects of pottery, i.e. predominant limestone admixture and common presence of the truncated cone-shaped vases with everted rims.[122] Basing on stratigraphic observations of the materials connected with the *Luco* culture, it is suspected that its creation took place around BR 2 and early BF, so the turn of 13th and 12th century BC, corresponding with phases BzD and HaA1.[123] The *Luco A* phase continued until the mid-11th century BC (HaA2/B1a), and was followed by the *Luco B* phase designating EIA stage of the culture.

To the phase A of the *Luco* culture is dated the single Allumiere-type amber bead, found at the bottom of the so-called 'Rhaetian house' together with fragments of pottery, at the locality Dos Castel – a promontory situated to the east from the village of Fai della Paganella, located on the right bank of the Adige, just south of Mezzolombardo (Marzatico 1997, 130, n. 528; Negroni Catacchio 2014, Fig. 12). Further north in Alto Adige, an assemblage of 35 amber beads was discovered within the Finail Valley – a tributary valley of the Schnalstal Valley. The archaeological site is situated in a small dell at 2,460 m a.s.l., popularly referred to as the 'Finail pit' (it. *Maso Finale*) or

121 Cf. Perini 1990, Fig. 10; Marzatico/Tecchiati 1998, Fig. 2; De Marinis 1999, Figs. 9, 11–14.
122 Cf. Perini 1990, Fig. 10; Marzatico/Tecchiati 1998, 29, Figs. 4 and 5; Marzatico 2012, 190–191, Figs. 1 and 2.
123 Cf. Stauffer-Isenring 1983; Marzatico/Tecchiati 1998; Marzatico et al. 2010; Marzatico 2012.

the 'lower pit' (Putzer 2012). Amber beads, as well as glass beads, were found scattered within an area of 36 m^2, in a stratum up to 20 cm thick, defined as the 'cultural layer' (ger. *Kulturschicht*). On a background of the layer were distinguished the outlines of four pits, three of them laid out side by side in a row. All of them were filled with charcoals, fragments of charred wood and ash – probably residues of preparation of food offerings. In the immediate vicinity of the pits were numerous post-holes of different dimensions and construction. In the southern part, an oblong-oval 'hearth' of fist-sized burnt stones came to light bordered by larger stone slabs. The site has been interpreted as a ritual place – sanctuary, consisting of an 'altar' built of small stones and offering pits, surrounded by large boulders. After abandonment, the ritual space was sealed with stone slabs. Type of the site matches votive deposits often encountered in the *Luco* culture, either aquatic or connected with sacrificial burning (ger. *Brandopferplätze*),[124] while discovered pottery fits the shapes attributable to BR and BF (BzD and HaA1; cf. Putzer 2012, 159–160). Final argument for dating of the context is provided by three 14C dates; two of them obtained from pits 2 and 4 (LTL8169A and LTL8170A) indicate 1320/1310–1020/980 cal. BC (2σ), while the third one related to pit 5 (LTL8168) slightly lags behind: 1070–800 cal. BC (2σ) (Putzer 2012, 163).[125]

Tab. 4. List of the Bronze Age contexts with amber in Trentino-Alto Adige

Name	Commune	Province	Type	Relative chronology	Absolute chronology	Reference
Fai della Paganella-Dos Castel, 'Rhaetian house'	Fai della Paganella	Trento	Hillfort	BR 2–BF (*Luco A*)	1225–1060/1025 BC	Marzatico 1997, 130, n. 528
Fiavè-Carera, horizon 6, layers D and E	Fiavè	Trento	*Palafitte* settlement	BM 2B–3 early	1475/1450–1350 BC	Perini 1987, Fig. 68
Maso Finale, cultural layer	Senales	Bolzano	Votive deposit (?)	BR–BF (*Luco A*)	1325/1300–1060/1025 BC	Putzer 2012, Abb. 10:1–35
Molina di Ledro, strata I–III/IV	Ledro	Trento	*Palafitte* settlement	BA 2–BM 1	1800/1750–1650 BC	Rageth 1974, Taf. 92:6–8, 16–18, 24
Molina di Ledro, precise context unknown	Ledro	Trento	*Palafitte* settlement	BA–BM (?)	2200–1325/1300 BC (?)	Rageth 1974, Taf. 92:1–5, 19–20, 23
Val di Non, precise context unknown	Cles	Trento	Votive deposit (?)	BM–BR	1650–1175 BC	Angelini et al. 2003, Fig. 2.1–2

124 Cf. Perini 1990, 1159–1160; Marzatico/Tecchiati 1998, 30, 39; Marzatico 2012, 179.

125 Putzer (2012) does not provide the uncalibrated BP measurements; thus, re-calibration of the dates given above could not be carried out by the author.

2.2.5. Garda Lake and its inframorenic basin

Southern surroundings of the Garda Lake (it. *Benaco*) are known for numerous prehistoric lake-dwellings (*palafitte* settlements), many of which yielded amber artefacts (cf. De Marinis 2007). This particular concentration of culturally linked sites, as well as long tradition of amber's use, allows to distinguish an amber's acquisition province, covering eastern banks of the Garda Lake between the shoreline and the Adige River, the inframorenic basin ('amphitheatre') to the west of the Mincio River with several far-flung sites to the south between Mantova and Cremona, and the south-western vicinity of the lake up to Brescia (Fig. 11). Together the area constitutes a tiny portion of westernmost Veneto and eastern part of Lombardy.

The earliest manifestations of amber in the province come from the EBA, when Northern Italy, particularly Benacense area, was a domain of the *Polada* culture.[126] Chronology of this cultural unit, as well as the entire period, is mainly built on the basis of the stratigraphic sequences of materials and dendrochronologically dated wooden construction elements from the settlements Bande di Cavriana, Fiavè, La Quercia, Lavagnone and Lucone.[127] In the classic version of the *Polada* culture chronology, the unit is divided into two phases: *Polada A* (formative stage) and *Polada B* (developed and late stage) (Fasani 1970; Aspes/Fasani 1974). These horizons, in turn, were synchronised with three phases of the EBA: *Polada A* – BA 1, *Polada B* – BA 2–3 (or only BA 2 in an altered version).[128] In a more recent and detailed periodisation according to R. De Marinis, the *Polada* culture's development can be divided into two main phases: BA 1 and BA 2, with the former comprising three subphases: A, B and C (De Marinis 1999, 25–38, Fig. 8). Materials corresponding with the phases distinguished by De Marinis were compared with finds from the circum-Alpine zone, what in turn allowed to correlate them with Reinecke's Central European chronological system.[129] Consequently, BA 1A (ca. 2200–2000 BC) is roughly concurrent with phase BzA1a, BA 1B (ca. 2000–1900/1850 BC) with BzA1b, BA 1C (ca. 1900/1850–1800/1750 BC) with BzA2a, and BA 2 (ca. 1800/1750–1650 BC) with BzA2b.

Estimating the exact moment of amber's first appearance in the region is difficult, due to significant shortfall of data regarding stratigraphic context of the finds. Potentially the oldest artefacts are those found on the south-east bank of the Garda Lake, at the localities Ca' Nova di Cavaion and La Quercia di Lazise. The former site, originally situated at the Ca' Nova Lake – nowadays a pond, provided abundant material, however entirely devoid of stratigraphy (Aspes 1987a, 86–91). Morphological and stylistics traits of a vast majority of materials recovered at the site represent features typical for BA 2, as well as the subsequent phase BM 1 (ca. 1650–1525/1500 BC), e.g. bowls decorated with cross-motifs on the bottom – the *Barche di Solferino* style,

126 For an overview of the Northern Italian EBA and the *Polada* culture see e.g. cf. Laviosa Zambotti 1939; Peroni 1971; Aspes/Fasani 1974; Fasani 1984; Barfield 1994; Bermond Montanari et al. 1996; De Marinis 1997; Rapi 2013; Cupitò/Leonardi 2015; Leonardi et al. 2015; Nicolis 2013.

127 Cf. Rittatore Vonwiller 1954; Soffredi de Camilli 1967; Perini 1984, 1987, 1988; Fasani/Martinelli 1994, 1996; De Marinis et al. 1996, 2005; Baioni et al. 2007a; Fozzati et al. 2015.

128 Cf. Perini 1988, 144; Barfield 1994, Fig. 4.1; Leonardi et al. 2015, Fig. 5.

129 De Marinis 1999; David-Elbiali/David 2009; David et al. 2017.

or with hatched triangles.[130] Furthermore, among the materials are: hemispherical or S-profiled cups with handles slightly raised over the rim and horizontally perforated,[131] bread-loaf idols (it. *oggetti enigmatici*, ger. *Brotlaibidole*)[132] and the Porto di Pancego type pin.[133] Nonetheless, one can notice elements potentially derived from an older phase of the site's attendance: undecorated mug-like or flowerpot-like vessels with simple lugs[134] and bronze daggers with round bases, sometimes decorated with engraved concentric triangles.[135] Earlier occupation of the site is also supported by dendro-dates obtained from a group of samples referable to two felling moments, dated between 2010 and 1962 BC (Fasani/Martinelli 1996, 22). On such basis chronology of the settlement was extended, in relation to the original interpretation, to BA 1B (De Marinis 1997, 406; cf. Aspes 1987a, 86). Still, the two amber beads from the site, as unspecific objects, can be dated only broadly by fitting them into entire time span of the site, from BA 1B to BM 1 (ca. 2000–1525/1500 BC).

The site of La Quercia consists of numerous wooden poles, embedded in the Garda Lake's bottom and shoreline, originally forming the scaffolding for housing structures (Aspes 1987b, 92; 1992, 72–73). Recovery of more than 6,000 poles provided samples necessary for dendrochronology-based sequence of the site which consists of three phases with corresponding felling moments: Lazise – La Quercia I (1939–1844 ± 10 cal. BC), II (1690–1620 ± 10 cal. BC) and III (1588–1558 ± 10 cal. BC) (Fozzati et al. 2015, 247–248, Tab. I). The archaeological deposit, with the depth of about 1 m, is of considerable stratigraphic complexity, with layers VI, IV and II representing three occupation phases, each separated by a sterile alluvial layer corresponding to a period of abandonment. Excavations brought to light finds of various kinds, mainly attributable to the recent phases of the EBA (BA 2) and early stages of the MBA (BM 1–2A) – time span conforming with the time brackets indicated by the dendro-dates (Aspes 1987b, 93–94; 1992, 82–86). Unfortunately, correlation of the dendrochronological curve, and the dated archaeological material, with the stratigraphy is currently beyond possibility; it can only be assumed that the I^st^ phase of the settlement (layer VI) corresponds to BA 1B/1C, the IInd to BA 2 (layer IV), while the IIIrd likely refers to BM 1 (layer II) (De Marinis 1999, 31–32, 42, Fig. 8). As in the previous case, the single amber bead coming from the site was found out-of-context even before the regular archaeological excavations had started, so it can only be ascribed to the entire timespan of the settlement.

Similarly uncertain is the chronology and cultural attribution of the three amber beads found at Bande di Cavriana. In the course of excavations at the site were identified two consecutive settlements (or perhaps stages of horizontal expansion of the inhabited area): one belonging to the EBA (BA 1A–1B) and attributable to the *Polada* culture, and the second placeable in a broader time span, roughly between the MBA and the Recent Bronze Age (Borghesani/Buonopane 1980; Piccoli 1982a). Most of

---

130 Aspes 1987a, Figs. 3:9–12, 4:7, 15 and 5:1–3; cf. Urban 1993, Abb. 142:21, 24; Bermond Montanari et al. 1996, 62, Fig. 3:46, 49; De Marinis 1999, 31, Figs. 9:3 and 14:6–7.

131 Aspes 1987a, Figs. 3:1, 9–12, 14–15 and 4:5; cf. Bermond Montanari et al. 1996, 62, Fig. 3:42, 45, 48.

132 Aspes 1987a, Fig. 3:5; cf. Bándi 1974; Battisti 2015; Martinelli/Marchesini 2015.

133 Aspes 1987a, Fig. 2:1; cf. Rapi 2011.

134 Aspes 1987a, Fig. 3:8, 16–17; cf. Bermond Montanari et al. 1996, Fig. 3; De Marinis 1999, Figs. 2 and 3; Leonardi et al. 2015, Fig. 5.

135 Aspes 1987a, Fig. 2:2–8; cf. Bianco Peroni 1994, Tav. 11–14; Carancini 1997, Fig. 222.

authors admit the impossibility to assign the amber artefacts to either of the two settlements/phases (Negroni Catacchio et al. 1999, 29–30; De Marinis 2007, n. 18 on p. 67). On the other hand, G. Bergonzi dates the discussed finds to the MBA (BM), even though one can hardly find any evidence supporting such hypothesis (Bergonzi 1997, 608).

In the course of the 2nd half of the 19th century, numerous materials belonging to the ancient lake-dwelling settlement dateable to the Bronze Age were reported from the bog of Cattaragna, not far away from Bande di Cavriana. The materials were not recovered in the process of regular excavations but were discovered accidentally on various occasions. The finds later became dispersed in private collections and museums of Rome ('L. Pigorini' Museum), Verona (Civic Museum of Natural History), Brescia (Civic Museums) and Milan (Castle Museum). So far the only published finds belong to the old collections of the Museum of Natural History of Verona. Single amber bead from the site is stored in 'L. Pigorini' Museum in Rome (Negroni Cattacchio 1970, 321, Fig. 12). Interestingly, the pottery collected from the bog comprises vessels typical for the older stage of the EBA (BA 1), whereas bronze artefacts are datable to the later periods (BA 2 – BM – BR).[136] Chronology of the single amber bead in the conical form with V-shaped perforation has been debated, with some authors placing it in an unspecified moment of the EBA (Massari et al. 1996, 620; Negroni Catacchio et al. 2006, 1457–1458), while others accepting a broader timespan covering both BA and BM (Bergonzi/Cardarelli 1991–1992, 218; Bergonzi 1997, 606, 609). R. C. De Marinis states that the bead has to be younger than BA 1 (recalling general absence of a secure context with amber dated this phase), however does not assign it to any particular period (De Marinis 2006, 448). In his opinion, the striking chronological disparity between the pottery, metal items and amber from Cattaragna probably stems from the fact that majority of ceramic vessels have been collected at later times from the waste layers between the wooden poles of the earliest settlement, whereas successive phases of the settlement had to be raised on remediated or dry terrain (as in the case of Lavagnone or Fiavè). There the circumstances does not allow to find intact pottery *in situ*: ceramic fragments from these upper layers, as opposed to bronze objects, did not attract the interest of amateur archaeologists in the 19th century, thus were discarded and not included in the collections (De Marinis 2006, 448). Hence, the amber bead, probably discovered in the superficial layers, at the moment should be dated to ca. BA 2–BM/BR.

In the context of the EBA, one has to consider also the amber finds from Lucone, a site located in the western part of the Garda's inframorenic amphitheatre. Archaeological excavations within the former lake Lucone (now drained, but marshy depression) were at first conducted in its south-west part named 'sector A'.[137] They brought to light two amber beads, found in superficial layers (Negroni Cattacchio 1970, 321, Fig. 14). Later, remaining sections of the site were investigated in the course of surface surveys, leading to distinction of another four areas generally characterised by a copious outcrops of archaeological materials marked with the letters B, C, D and E. Only Lucone A and Lucone D have been excavated so far (cf. Guerreschi 1980–1981; Bocchio 1988). The former area yielded materials attributable to both stages of the EBA (BA 1–2) and not

136 Cf. Aspes 1972; Bebber/Buonopane 1982; De Marinis 2006, 447.
137 Cf. Simoni 1966, 1967, 1969, 1970.

too-advanced phases of the MBA (BM 1–2) (Baioni et al. 2007a, 85–88). Some publications determine chronology of the two amber beads found at Lucone A at the ultimate phase of the settlement's development, that is BM 2 (Bergonzi/Cardarelli 1991–1992, 218; Bergonzi 1997, 608). However, attribution of the finds to the upper, presumably younger layers cannot be used as decisive argument, as the publication quoted in each case does not specify chronology of the artefacts, but instead states that they come from superficial layers and were found out-of-context (cf. Negroni Cattacchio 1970, 321). At the moment it is impossible to precisely assign them to any of the three major horizons of the Bronze Age occupation of the site.

The earliest amber finds from chronologically secure contexts in Northern Italy come from the settlement of Lavagnone. The site is located in a wetland depression (nowadays a drained peat bog), situated about 3.4 km south of the Lake Garda shore, in the past itself occupied by a lake which disappeared after a significant drop of water level (De Marinis et al. 2005, 221, Figs. 1 and 2). The settlement's development can be traced on the basis of stratigraphic order of the layers and associated artefacts, starting from the EBA up to advanced Middle and the LBA. The site, throughout its existence, had the character of a pile-dwelling with individual houses built on wooden poles, in the later stages supported with layers of sand and gravel. From the lake's shore to the settlement led a timber trackway, while the entire complex was surrounded by a wooden fence.[138] The earliest occupation of the site can be divided into several phases which are also equivalent to the individual stages of the Northern Italian EBA (De Marinis 1999, 25–38, Figs. 7 and 8; De Marinis et al. 1999, 2005, 222, Fig. 3). The earliest – Lavagnone 2 – corresponds with BA 1A and is dated dendrochronologically to 2077–1991 cal. BC. It is followed by Lavagnone 3, synonymous with BA 1B (1984–1916 cal. BC), and Lavagnone 4, correlated with BA 2 (ca. 1800/1750–1650 BC). Later, in the sector B, layers US 658–810, it was also possible to distinguish intermediate horizon referring to BA 1C (ca. 1900/1850–1800/1750 BC). Two amber beads were found in the layers of Lavagnone 4 phase: one was found in sector A (layer US441), while the other was recovered from sector B (layer US537); hence, both can be safely attributed to BA 2 (De Marinis 2007, 62, Fig. 2).

Further testimonies of amber presence in the late stage of the EBA come from the plain north and south of the Oglio River, between the towns of Asola and Piadena. The *palafitte* settlement of Lagazzi di Vho, through typo-chronological analysis of the pottery, was divided into two occupation phases: the first falling into BA 2 and the second coinciding with BM 1.[139] To the former phase belong 17 amber beads discovered inside an anthropogenic layer with thickness of 10–15 cm, inside the trench on the west bank of the drainage ditch crossing the site.[140] Due to absence of a compact vertical stratigraphy, dating of particular structures or areas of the settlement could be obtained solely through the analysis of the pottery (Simone/Tinè 1996, 274). Nonetheless, the archaeological layer, from which the amber beads originate, appeared well defined with some concentrations of ashes and homogeneous materials, referable to the end of the EBA.

---

138 Cf. Perini 1988; De Marinis et al. 1996, 2005; De Marinis 1997, 420–421; 2013.
139 Cf. Parazzi 1891; Simone/Tinè 1991–1992, 1996.
140 Simone/Tinè 1991–1992, 293; Baioni et al. 2007b, 27; De Marinis 2007, 62.

Another roughly contemporaneous amber artefacts were found in grave 54 at the site Sorbara, on agricultural fields adjacent to the west bank of the Chiese River. It was a rather large and shallow (average length 3 m; maximum depth 0.25 m) pit with inhumed burial of an adult, furnished with a single ceramic vessel, two flint arrowheads and two amber beads. The grave was located in the southern part of the cemetery, about 160 m away from the most substantial nucleus of prehistoric tombs dated to BA 1. Apart from a distant position, the tomb was quite different from the rest: it was oriented E–W as opposed to prevailing N–S orientation of the graves, and the bones were dispersed in the upper part of the pit. The latter fact, together with a cavity and post-holes observed in the central and E part of the pit, presumably indicate destruction of the grave that might have occurred still in prehistoric times. Nonetheless, homogeneity of the sandy filling did not make it possible to establish with certainty, whether the damages were done at the time or after the burial took place. The elements of the equipment found *in situ* were placed on the bottom of the large pit and not in the deepest part (Baioni 2000, 48–49, 54, 67–69, Figs. 6 and 10:10–13). Comparative analysis of the grave goods from tomb 54 did not provide an unambiguous answer to the feature's chronology. Yet, it was suggested that it cannot be concurrent with the graves from the northernmost cluster and is probably younger (BA 2 proposed as the earliest date). Subsequent considerations of the grave's inventory, mainly focused on the flint arrowheads, apparently support attribution of the context to the last phase of the EBA (De Marinis/Valzolgher 2013, 550).

A significant group of amber finds from the Lake Garda area can be safely assigned to the MBA. The period under consideration saw continuation of many *palafitte* settlements and a steady development of vascular pottery and bronze metallurgy. With the end of the EBA (BA 2) the *Polada* culture in the classical sense ended, its late manifestation being the horizon *Barche di Solferino*. Nonetheless, the habitation mode in dwellings set above or close to the lakes was still practised afterwards, while the material culture characteristic for the region became increasingly widespread and uniform, spanning southern Trentino, east Lombardy and west Veneto north of the Po.[141]

The old concepts invented to describe and systematise material culture of the MBA in the Garda area, such as the *Polada II* (Laviosa Zambotti 1954), *Subpolada* (Rittatore 1960) or the sequential facies *Bor – Isolone – Peschiera* (Aspes/Fasani 1967) were largely rejected. Instead the chronology based on close contexts is favoured among the scholars, with the most accepted periodisations proposed by De Marinis (1999). According to the author, the MBA is divided into four phases: BM 1, 2A, 2B and 2C, the latter eventually renamed to BM 3 (De Marinis/Salzani 2005). The periodisation is synchronised with the Alpine/Central European chronological system in the following way: BM 1 (ca. 1650–1525/1500 BC) is contemporaneous with BzA2c–BzB early, BM 2A (ca. 1525/1500–1475/1450 BC) with BzB late, BM 2B (ca. 1475/1450–1400/1375) with BzC1 and BM 3 (ca. 1400/1375–1325/1300 BC) with BzC2.[142]

There is hardly a consensus regarding cultural affiliation of the MBA lake-dwellings in the Garda Basin – some archaeologists treat them as belonging to an independent unit, while the other accentuate links with the *terramare* settlements of

141 Cf. Urban 1993, 245–250; Barfield 1994, 134; De Marinis 1997, 407–408, 414; 2009a, 537–538; Cupitò/Leonardi 2015, 209; Nicolis 2013, 696.

142 De Marinis 1999, Fig. 44; David-Elbiali/David 2009, Fig. 11; David et al. 2017, Fig. 19.

Emilia-Romagna.[143] Still, many scholars agree that certain similarities in the pottery, bronze and other categories of finds recur both north and south of the Po, testifying to contacts between communities inhabiting cis- and transpadane Italy during the period; however, strength and character of these relationships are debatable (Urban 1993, 258–259; Barfield 1994, 133). De Marinis, a strong proponent of cultural and genetic affinity of *palafitte* and *terramare* inhabitants, in his works highlights aforementioned similarities, and argues that different modes of settlement and economy are simply dictated by differences in the natural environment north and south of the Po. Thus, the scholar coined the term '*Palafitte-Terramare* culture', constituted from the two main regional groups – manifestations, initiated potentially by the same population, but subsequently developing with different dynamics and trajectories.[144]

It is noticeable that together with the continuous habitation of lake-dwellings from BA 2 to BM 1, also amber maintained its popularity among the societies of the Garda Basin. Two beads originate from stratigraphic units US120 and US138 in sector A, referring to phases 5–6 of the Lavagnone settlement (De Marinis 2007, Fig. 3). At that time the houses were built directly on dry land and secured from water with beaten layers of silt, gravel and pebbles overlapping the layers of Lavagnone 4 (BA 2), interleaved with post-holes and remains of hearths.[145] Worth adding is that after BM 1 there are no more traces of habitation in the sector A, as well as sector C. In the subsequent phase, corresponding with BM 2, the settlement shifts towards north-east shore of the lake.

Single amber bead was found during A. Parazzi's excavations at Lagazzi di Vho in 1890, who had recognised three layers within the archaeological deposit, distributed in a series of accumulations (Parazzi 1891). From the upper layer comes a conical button in amber with V-perforation, which is generally dated to BM 1 that is the most recent phase attested at the settlement.[146]

Another lake-dwelling – Castellaro Lagusello – contained amber artefacts in two consecutive horizons of occupation (Piccoli 1982b, 471, 477; Negroni Catacchio et al. 1999, 28–29, Fig. 10:1–7). The first one, known as layer D, corresponds with BM 1, while the second comprising layers C/C1 falls into BM 2A (De Marinis 1997, 407–408; 1999, 42, 52, Fig. 44). Interestingly, observations on typo-chronology of collected materials revealed the continuous occupation of the settlement, only with a sharp break between layers C and B (respectively BM 2A and 2B; cf. Piccoli 1982b).

The largest collection of MBA amber finds from the Garda Basin comes from the settlement of Isolone del Mincio (11 objects in total), yet its precise chronology is difficult to determine. The lake-dwelling comprises three layers containing traces of wooden poles and other timber elements, separated by sterile layers, which indicate existence of three superimposed inhabited settlements, each destroyed by fire (Gaggia/Morelato 1980, 91). The settlement returned a very rich assemblage of finds, including pottery, bronze and amber items; however, due to absence of documentation, they

143 Cf. Säflund 1939; Aspes/Fasani 1967; Fasani 1984; Urban 1993, 12–19, De Marinis 1997; Nicolis 2013, 695–696.

144 Cf. De Marinis 1979, 1981, 1997, 1999, 2009a.

145 Cf. Perini 1988, 135–142, 144; De Marinis et al. 1999, 129–130; 2005, Fig. 3.

146 Parazzi 1891, Tav. II:7; Negroni Catacchio 1971, 321, Fig. 13; Massari et al. 1996, Fig. B:2; De Marinis 2007, 62.

cannot be attached to specific strata or horizon (Guerreschi et al. 1985; cf. Piccoli/Peroni 1991–1992). Although exact place of discovery of amber beads or their position within the stratigraphy is not given, various authors attribute all of them to the phase of maximum development and spatial expansion of the settlement, that is the MBA (BM) or Middle–Recent Bronze Age (BM–BR).[147]

The LBA in the Garda Basin, divided by De Marinis into BR 1 (roughly BzD; ca. 1325/1300–1225 BC) and BR 2 (early HaA1; ca. 1225–1175 BC), still abounds in amber. This was a period of an unprecedented growth in population and density of inhabitants (cf. Barfield 1994, 133–134, 136–137; De Marinis 2009a, 536). Despite the fact that many lake-dwellings in the area experienced permanent collapse, or spatial restructuration in the later stages of MBA (e.g. Castellaro Lagusello, Lagazzi di Vho, Lavagnone and Lucone di Polpenazze), immediately afterwards new and larger settlement were built (e.g. after abandonment of Lagazzi di Vho, the population moved to a newly established settlement of Castellaro del Vhò; cf. De Marinis 1999, 38). These settlements reached the peak of development, both in terms of extension/spatial organisation and capacity/population scale, in the LBA, with Peschiera del Garda *palafitte* being often quoted as the prime example of this process.[148]

In the southern part of the Garda Lake, near the town of Peschiera, remains of at least seven pile-dwelling villages were discovered as a result of construction of Austrian fortifications, during the 19th century. There is a huge numerical disparity between ceramic and metal materials, as the former were largely destroyed during the works or have disappeared since then, while the latter are preserved in much greater numbers. It is clear that the preserved materials were collected selectively, and, due to absence of any documentation, the only possible criterion for the study of the Peschiera *palafitte* complex is the typological one (Aspes 1980, 99). Majority of amber finds originating from the Peschiera area were donated to the Civic Museum of Natural History in Verona in 1921 by Ballardolo family (Negroni Catacchio 1971, 319–320, Figs. 1–8). These artefacts are usually associated with the locality Imboccatura del Mincio, which due to the specific content, comprising mostly bronze items, predominantly pins, daggers and other tools or adornments, in fact might be a votive acquatic deposit (De Marinis 2006, n. 49). The entire assemblage, on the basis of morpho-stylistic features, represents objects typical for BR – the eponymous *Peschiera* horizon. Other amber finds, i.e. the similarly dated pins with inserted amber adornments in the shape of the Tiryns type, possibly come from other localities in the vicinity (Peschiera – *palafitta centrale*?).[149]

Less than 5 km to north-west from the outlet of the Mincio River, at the site Maraschina situated south of the point Gro, was found another pin with the head made of spherical amber element (De Marinis 2006, Tav. V:5; 2007, Fig. 5:a). The lake-dwelling, although being investigated to a limited extent, has a long chronology, spanning from BA to the early stages of BF (cf. Ruggiero et al. 2001). The

147 Cf. Guerreschi et al. 1985, 87; Bergonzi/Cardarelli 1991–1992, 218; Bergonzi 1997, 608, Fig. 349:2, 4, 6, 7; De Marinis 1997, 63; Negroni Catacchio et al. 1999, 31–34, Fig. 11:11–21.

148 Cf. Fasani 1984, 550–557; De Marinis 1997, 411–412; Nicolis 2013, 696.

149 Negroni Catacchio 1971, 320–321, Figs. 9–11; De Marinis 2006, 451, Tav. 5:6; 2007, 65, Figs. 1:10–11 and 5:b.

Tab. 5. Continued

Name	Commune	Province	Type	Relative chronology	Absolute chronology	Reference
Lavagnone, phase 4	Desenzano	Brescia	*Palafitte* settlement	BA 2	1800/1750–1650 BC	De Marinis 2007, Fig. 2
Lavagnone, phase 5–6	Desenzano	Brescia	*Palafitte* settlement	BM 1	1650–1525/1500 BC	De Marinis 2007, Fig. 3
Lucone di Polpenazze, sector A, superficial layers	Polpenazze	Brescia	*Palafitte* settlement	BA 1B–BM 2	2000–1400/1375 BC	Negroni Cattacchio 1970, Fig. 14
Maraschina, precise context unknown	Desenzano	Brescia	*Palafitte* settlement	BR 1–2	1325/1300–1175 BC	De Marinis 2007, Fig. 5:a
Peschiera – Imboccatura del Mincio, precise context unknown	Peschiera del Garda	Verona	*Palafitte* settlement	BR 1–2	1325/1300–1175 BC	Negroni Catacchio 1971, Fig. 9–11; De Marinis 2007, Fig. 5:b
Ponte San Marco, BR–BF horizon	Calcinato	Brescia	Settlement	BR–BF 2	1325/1300–1030/1010 BC	Ruggiero 1994, Fig. 89
Sorbara, tomb 54	Asola	Mantova	Burial – inhumation	BA 2	1800/1750–1650 BC	Baioni 2000, 67, Fig. 10:13

2.2.6. Western Veneto/northern Po Plain

Amber started to occur in western Veneto (here understood mainly as a part of the region between the Adige and the Po) only from the developed MBA but reached considerable quantities immediately upon its introduction (Fig. 11). At that time western Veneto, in particular the territory between the Garda Lake, the Verona plain and the south-western Padua plain, was in the orbit of the *Palafitte-Terramare* culture. Development of cultural identity of the local communities, following the *Camponi di Nogarole Rocca* facies typical for the initial MBA, saw affirmation of the *Tabina di Magreta* and *Monteleoni* aspects of the southern Po Plain, detectable through the changes in pottery.[151] In parallel with the transformation of material culture came selective adoption of the classical *terramare*-type settlements (Cupitò/Leonardi 2015, 210; cf. Balista/De Guio 1997).

151 Cupitò/Leonardi 2015, 209; cf. Cardarelli 1988; Aspes et al. 1991–1992; Carancini et al. 1996; Bernabò Brea/Cardarelli 1997; Bagolan/Vanzetti 1997.

However, later local cultural development proceed differently than in the neighbouring areas; thus, we can speak of a unique regional group of the *Palafitte-Terramare* culture. Some of the growing differences between western Veneto and other territorial units of discussed culture, Emilia/southern Po Plain in particular, are visible on the basis of pottery: territory under examination almost completely lacks handles with frontal discs typical for the late stages of the MBA, decorations with wide grooves and 'grains of wheat' ornamental motif, while, on the other hand, the horned handles, characteristic of the Veneto region, rarely turn up south of the Po.[152] At the same time, in the Po Valley, there are distinct differences evident in the sphere of funerary rites, differentiating Veneto (biritualism) from eastern Lombardy and Emilia (exclusively cremation) (cf. Cardarelli/Tirabassi 1997; De Marinis/Salzani 1997).

From the point of view of the funerary rites, in western Veneto apparent absence of sepulchral testimonies referable to the BM 1 contrasts with truly exceptional density and wealth of evidence attributable to BM 2 and BM 3. In the middle and lower plain of Verona, in significant concomitance with the rise of the *terramare* settlements, one witnesses the birth of large burial grounds characterised by presence of particular nuclei. Among the latter, swords, daggers and defensive elements such as helmets signify adult male burials, whereas ornamental *parures* essentially composed of pairs of large bronze pins, sometimes embellished with amber, form the inventories of women burials. In comparison with Emilia and eastern Lombardy, where cremation was the dominant burial rite throughout the MBA, in western Veneto the habit of burning the dead started to be practised only in the advanced phase of the period (BM 3) (Cupitò/Leonardi 2015, 216).

The next period – Recent Bronze Age (LBA) – was a direct continuation of the *Terramare* culture development in western Veneto. In the sphere of material culture, one can notice a change in the type of handles on ceramic vessels. Subphase BR 1 is characterised by handles with high vertical and oblique appendices in the shape of 'bovine horns' (it. *corna di bovino*) that are clearly linked to BM tradition, whereas in the subsequent BR 2, this form of handles gives way to cylindrical-straight, lobe-shaped or rostrate appendices and strip- and rod-shape handles raised above the rim. Incorporation of the latter, together with some newly introduced bronze items, e.g. early violin-shaped fibulae, manifest influences of the Central–Southern Italian *Subapennine* culture/period.[153] Furthermore, in BR 2 western Veneto saw ubiquitous spread of the brim edge which distinguishes the region from Emilia/southern Po Plain and, together with grooved and *cuppelle* decorations, represents one of the most important elements of cultural continuity between *Palafitte-Terramare* culture and the subsequent *Protovillanovan* culture of the FBA (BF).[154]

Looking at the habitation structure of BR in western Veneto, one can observe widespread adoption of the *terramare*-type settlements, and expansion of some sites in size at the cost of the other, probably reflecting quantitative changes in population and its movements. In the Veronese plain, the exponential increase of the population translates into the formation of territorial system known as the 'polity of *Valli Grandi*

152 Cf. Bagolan/Vanzetti 1997, 356; De Marinis 1997, 411; Cupitò/Leonardi 2015, 209.
153 De Marinis 1997, 411; De Marinis/Rapi 2016, 48–55, Figs. 10–15; Cupitò/Leonardi 2015, 218; cf. Cocchi Genick 2004; Cattani 2009; Damiani 2010.
154 Cupitò/Leonardi 2015, 219; cf. Leonardi 1979; Bagolan/Leonardi 2000.

Veronesi', centred around the site of Fondo Paviani (Cupitò/Leonardi 2015, 219–221; cf. Balista/De Guio 1997, 152–158).

As far as funerary aspects are concerned, the clear dichotomy in burial rite between western Veneto and the Garda Basin, already observed in BM, remained unchanged in BR, even though the percentage of cremation in the former area increased considerably. Nonetheless certain funerary customs previously practised in the Veronese and south-western Paduan plains were abandoned, e.g. deposition of swords in male burials, while other are introduced, e.g. greater diversity in grave clusters including also more modest/poor burials, presumably reflecting profound transformations both in the sphere of ideology and religion, and in the organisation of society. Cremated burials, commonly deposited in urns, are usually devoid of grave goods.[155]

Chronology of the Middle and Recent Bronze Age in western Veneto is closely associated with the necropolises. The latter provide a chance of studying finds in closed contexts, including amber beads which predominantly originate from the graves. In the past, scholars proposed various names and chronologies to explain transformations of the local culture. Already in 1963 R. Peroni saw the possibility of distinguishing three chronological phases on the basis of cemeteries situated between Mincio and Adige: first he defined the facies *Povegliano* with duration lasting through the MBA, but with an emphasis in the earlier stage, then the facies *Monte Lonato* with emphasis in the middle and late parts of the MBA, and finally the LBA facies *Bovolone* (Peroni 1963a). Other authors followed suit and based their chronological observations on the necropolises (e.g. Carancini 1979; Marchesan 1985).

Intensive research on cemeteries and settlements in the region in the last decades enabled formulating a more coherent subdivision of the period into phases. Crucial in this respect is a comprehensive elaboration of materials from the cemetery Olmo di Nogara (De Marinis/Salzani 2005), which was followed shortly after by a complete overview of all contemporaneous cemeteries in the region (Cupitò 2006a). Chronological systems proposed in both publications are to a large extent similar. Still, the differences concerning the dating of certain types of bronzes as well as estimated time frames of the consecutive phases render them not fully compatible.

De Marinis-Salzani's version of the chronology is rooted in the system devised by the former researcher in the vast part basing on the stratigraphic sequences of settlements in the Po Plain and the Garda Basin (cf. De Marinis 1997, 1999). While the subdivision of the MBA into phases BM 1, 2A, 2B and 3 (former 2C) works well with pottery from the settlements (even though critical voices question the legitimacy of distinguishing subphase 2A: cf. Cupitò/Leonardi 2015, n. 64), it is hard to avoid an impression that the separation of the MBA II into two subphases is not fully evidenced by the evolution of bronze items. In fact, presence of horizons corresponding to BM 2A and 2B at Olmo di Nogara is visible only in the case of female burials, and the authors often resort to using the more capacious term 'BM 2' in describing chronology of graves and phases at the cemetery (cf. Salzani 2005; De Marinis/Salzani 2005). Then, there is also the disagreement concerning the length of the Middle and Recent Bronze Age in terms of calendar years.

155 Cupitò/Leonardi 2015, 226; cf. Cupitò/Leonardi 2005; Cupitò 2006a, 2006b; Bietti Sestieri et al. 2013.

The absolute length of the MBA stages proposed by De Marinis (1999, Fig. 44; De Marinis/Salzani 2005, 434–435; see section 2.2.5) is based on the occurrence of interregional bronze objects (mainly swords) in Italian contexts, including Olmo di Nogara graves. Significant level of concordance between Italian BM phases and MBA stages of south-east France, Switzerland and Southern Germany has been reached through comparative studies (cf. David-Elbiali/David 2009; David et al. 2017). Despite this, other scholars dealing with western Veneto refer to the chronological system elaborated within the Rome school of archaeology, adopted in the studies of the *Terramare* in Emilia/southern Po Plain (see section 2.2.7).[156] In the case of western Veneto, adoption of the 'higher' chronology of BR, as applied in the southern Po Plain, seems to be argued by radiocarbon dating of the Fondo Paviani settlement (cf. Whitehouse 1997, 161, Tab. 2; Cupitò et al. 2012). However, absence of 14C dates for BM contexts from the area, especially the cemeteries, is a major handicap in developing independent chronology. As the current state of knowledge prevents from resolving the debate on the MBA chronology in western Veneto, it was decided to adapt in this work a compromise solution by stretching the time limits of consecutive phases to conform chronologies in the neighbouring regions. Such a 'hybrid' chronology corresponds to somewhat intermediary position of western Veneto in the *Palafitte-Terramare* culture, reflected by its specific cultural character incorporating influences both from the Garda area and Emilia. Accordingly, first stage of the MBA – BM 1 occupies the time span from 1650 to 1550/1500 BC, the next – BM 2 covers 1550/1500–1450/1375 and ultimately BM 3 falls into 1450/1375–1350/1300 BC. The LBA (BR) can be dated to 1350/1300–1175/1150 BC, with the boundary between subphases BR 1 and 2 most likely around 1260/1220 BC (cf. Cupitò/Leonardi 2015, n. 142).

Moving on to the sites with amber from western Veneto, the only settlement that can be mentioned here in the context of the MBA is Povegliano Muraiola. Basing on the stratigraphy, it was possible to distinguish four phases of the settlement's occupation (Salzani 1997a). Phase I is known only from pottery attributable to the BM 1. Houses (two huts with a rectangular plan 5.40x3.10 m, separated by 2 m of empty space) and other structures (e.g. walkway, ditch-moat and embankment) appeared for the first time in the phase II, lending the site a *terramare*-character. One can see a certain continuity in the use of the same area, as housing structures of Phases II and III practically overlap, only with small displacements in orientation. These structures are also marked by piles of waste, formed by successive coatings of ash, ceramic fragments and plaster. Phases II, III and IV are all dated to the BM 2, with amber beads coming from the first and the last of them (Salzani 1997b, Tav. 56:14–16, 36, 39). Within that stage one can, nonetheless, distinguish an earlier (phase II) and a later part (phases III-IV); De Marinis considers the phases I–II to be compatible with BM 2A, while phases III-IV fall into BM 2B (Salzani 1997c; De Marinis 1999, Fig. 44).

In the same commune, close to the settlement of Muraiola, is located the Povegliano Gambaloni cemetery which due to its adjacency to the former may be regarded as a part of a single complex (Peroni 1963a, 49–56; Cupitò 2006b). Graves at the cemetery

156 Cupitò/Leonardi 2015; cf. Carancini et al. 1996; Carancini 1997; Carancini/Peroni 1999; Bernabò Brea/Cardarelli 1997; Cardarelli 2009.

stratigraphic levels, including 394 inhumation burials (207 in the first and 187 in the second level) and 188 cremation burials (172 in the first and 16 in the second level). The stratigraphic levels have been interpreted as chronological phases of the cemetery's use: Ist corresponding with the second layer, dated to the *Peschiera* horizon (ca. BR 1), and IInd to transition between the *Bovolone* facies and the initial *Protovillanovan* culture (ca. BR 2–BF 1) (Aspes 1973, 124). However in most cases the graves with amber, despite being situated in the Ist or IInd layer, allow only for a generic classification to BR, or BM 3–BR in the absence of sufficient chronological indicators, but do not reveal evidence of BF. Amber beads from graves 12, 506 and 570 (Aspes 1973, Fig. 1:1–2; 1987, Fig. 2:1–2), the former belonging to the Ist phase, can be dated to BR, thanks to co-occurrence with the Peschiera-type pins. Graves 159, 497 and 523 classified to the IInd phase (Aspes 1973, Fig. 1:3–4; 1987c, Fig. 1:3; 1997, Fig. 420:1–6) contained pins of the types Franzine and Colombare - variant C, therefore can be placed with considerable dose of probability in BR (Aspes 1973, 120–123; cf. De Marinis/Salzani 2005, 421–422). Finally grave 417 included, apart from numerous amber adornments, a single pin of the Boccatura del Mincio type (Aspes 1976, Fig. 1:6–20) whereas tombs 7, 123 and 481 (Aspes/Fasani 1968, Fig. 8:1–3; Aspes 1973, 119, 1987, Fig. 1:1), due to absence of sufficient time markers, can only be presumed to have originated from the same period.

Also the graves with amber discovered at the cemetery of Castello del Tartaro, located less than a hundred metres north-west from the nearby namesake *terramare* settlement, are datable to BR (Salzani 2011). The unearthed tombs are altogether 396, comprising both inhumed and cremated burials; however, their ratio is distorted by the fact that the urns have largely been destroyed by agricultural work. Some skeletal burials were devoid of inventory, hence were interpreted as cenotaphs, while outfits of the furnished tombs almost exclusively consisted of pairs of pins placed on chests of the deceased. At the tips of the pins were sometimes placed large amber beads. Such combination of garment elements was registered in graves 300 and 350, in which amber beads were accompanied by, respectively, the Franzine-type pins and peculiar pins with hemispherical-hollow heads, presumably designed to hold inside an additional (organic?) element (Salzani 2011, 228, Fig. 5). Two pins of the type Peschiera were revealed along amber in tomb 396 (Salzani 2011, Fig. 4:C). Other elements of inventory found in tombs 146 and 166, aside amber, are the rings made of thin bronze wire (*fermatrecce*), openwork pendants with an ogival hole and three biconical buttons at the bottom, and domed plaques decorated with rows of dots and concentric circles (Salzani 2011, Figs. 3:B and 4:B; De Marinis/Salzani 2005, 423–425).

So far all the described graves with amber from western Veneto contained unburnt/inhumed remains of the deceased. In fact the only cremated burial that yielded amber is tomb 115 at the cemetery of Bovolone, where out of the total 200 recovered burials less than 40 were inhumed, whereas all others were cremated (Salzani 2010a). Stratigraphic position of the graves revealed contemporaneity of both funeral rites. The cremated burials were made of an urn, covered with a bowl and laid down on the bottom of a small pit. Inside the urns burnt bones were mostly deposited without any grave goods. Chronology of the necropolis is based especially on the typological analysis of ceramic forms. Cinerary urns and covering bowls allow to assign the necropolis to BM 3–BR, in line with the cemeteries described earlier (Salzani 2010a, 123–129). In the case of tomb 115, attributable to BR, the bowl serving as a cover/lid of the urn was already broken and displaced upon discovery. Apart from the ceramic vessels, the tomb apparently did not contain any element of grave inventory (Salzani 2010a, 101–102, Figs. 205 and 206). However, a fragment of amber was found inside the urn. It was shapeless and heavily damaged; it is possible that the object was exposed to fire during combustion of the deceased body on the funeral pyre (cf. Angelini 2010a, Fig. 1).

Finally, it is necessary to discuss the finds from Finilone Valle – site located 1.7 km south-east from Gazzo Veronese.[177] The site yielded numerous artefacts of various categories, overall weighing ca. 600 kg which, however, are deprived of context as they were mostly retrieved from the topsoil after ploughing (Salzani 1996). Most of the materials chronologically fall between the Early and Recent Bronze Age (BA–BR). At current state of research, it seems that the phase of maximum development of the settlement is to be attributed to an advanced moment of the MBA (BM 2–3). Out of eight amber beads discovered at the site, several were found in close position with two pins of the Monte Lonato type, two tubes of spirally twisted round wire (*saltaleoni*) and a bronze-casted comb (Salzani 1996, Figs. 21:6–7, 9, 16–18, 22–23, 26–28 and 22:5–6). It is speculated that the objects belong to a grave disturbed by agricultural works; however, no features akin to tombs (i.e. grave-pits) have been registered (Salzani 1996, 262). It is possible that the finds were recovered from a secondary position, after being moved from the original location by a plough. However, if they were in fact deposited together at the same time, then the chronology of the assemblage can be estimated between BM 2 and BR, basing on the analogies with the Monte Lonato-type pins from graves 422 and 435 at Olmo di Nogara[178] and the bronze comb from tomb 296 at the same cemetery.[179] Still, the above dating is quite wide and should be approached with a dose of caution.

Considerable amount of amber comes from settlement layers dated to BR. In the first place should be mentioned the collection of beads from Fondo Paviani – site located between the Adige and Tartaro rivers, and generally considered as the focal point in the *Valli Grandi Veronesi* polity.[180] Evidence of amber use at the site comes from all three main phases of its habitation, illustrated by establishment and subsequent structural changes of the *terramare*-like embanked settlement between advanced BR 1/early BR 2 and BF 1–2 (cf. Cupitò et al. 2015a). Aside from an amber button and six beads, including two representing the type Tiryns, the layers yielded four fragments of broken/unfinished beads and a lump of raw amber (Fasani/Salzani 1976; Strafella et al. 2015). Most of the objects come from secure contexts, extensively published in the last years and reinforced with 14C dates; hence, their chronology proposed in the literature does not require verification. It is argued that the settlement played role of an important craft centre, specialised in metal, glass and amber processing (cf. Cupitò et al. 2012, 58–60; 2015a, 366–368).

---

177 Western part of the site is located in the province of Mantova and is known under the name *Cavo Molinella.*

178 Salzani 2005, Tav. XXXV:A–B, XXXVI:D–E; cf. Carancini 1975, 23; 1997, Fig. 223:36–38; De Marinis/Salzani 2005, 416; Cupitò 2006a, 15, 117–118, Fig. 3, Tab. 3.

179 Salzani 2005, Tav. XXIV:A; cf. De Marinis/Salzani 2005, 424–425; Urban 1993, 146–148, Abb. 75.

180 Cupitò/Leonardi 2010, 2015, 221; Cupitò et al. 2012, 58; 2015a.

Amber at the site was discovered in two different contexts. First, a large assemblage of 38 beads comprising various types, i.a. the Tiryns and Allumiere, was found as a part of the so-called *Tesoretto* – hoard 1 (Bellintani/Peretto 1972; Negroni Catacchio 1972). On the basis of typology of bronze and amber objects, the hoard has been dated to BF 2, or BF 2/3 at the latest.[187] Later, three beads were recovered from the settlement layers; a ring-shaped bead is attributed to the 2nd phase (Bietti Sestieri et al. 2015, 429, Fig. 1D:19), while a broken button and unperforated Allumiere-type element have been ascribed to the 3rd phase (Bietti Sestieri et al. 2015, 431, Fig. 2:12–13). Absence of perforation in the latter object gave rise to suspicions that raw amber was processed and turned into adornments also at Frattesina – a function that could have been taken over after cessation of craft activities at Campestrin in BF 1 (cf. Bellintani 2014, 117; Bietti Sestieri et al. 2015, 431).

Upstream the Po, in the *Alto Polesine* area, presence of amber has been documented at the site Mariconda di Melara, situated immediately at the northern bank of the river (cf. Fasani 1966; Salzani 1973). The anthropogenic deposit is about 1.10 m deep and consists of two strata, overlapping with floor levels of the huts. The lower level, roughly contemporaneous with the phase 1 of Frattesina, basing on the types of ceramic and bronzes dates to the initial moment of the *Protovillanovan* culture (BR 2/BF 1) but reveals genetic links with the *Subapennine* culture of Marche. In the upper level was excavated an oak-built hut with terracotta and dirt floor, on which fragments of ceramic, bronze, vitreous pearls and a single amber bead were found (Salzani 1973, Tav. XVIII:22). Discovered ceramic and bronze materials from the upper layer indicate affiliation with BF 1/2 sites located in the *Pianura Padana*, particularly Frattesina and its nearby cemeteries (cf. Bagolan/Leonardi 2000, Fig. 6a; Bellintani 2000, Fig. 5).

With the dawn of the FBA, communities inhabiting western Veneto fully embraced cremation. Yet, the funerary customs, in comparison with those prevailing during BR, changed allowing for more substantial inventories being included in the graves. Among them one can find amber which, together with other grave goods, apparently was added after the cremation, as the beads do not exhibit fire traces (cf. Angelini 2010b, 142). Around 700 m to the north from the settlement of Frattesina are located cemeteries Narde I and Narde II.[188] Excavations brought to light around 600 graves on the first necropolis and 240 at the second. Apart from 20 inhumed burials at Narde II, the main burial rite was cremation. At Narde I the burials formed a large artificial mound, about 30 m long and preserved for a height of just over 1 m, comprising four superimposed levels of burials (marked IV-I from the bottom to the top) and a fifth level in the central area, where a marked concentration of tombs was noted. Simplified chronology of the cemeteries of Narde assumes two main stages of burial deposition: the first one in the initial and full phases of the Bronzo Finale (BF 1–2),

187 Negroni Cattacchio 1972, 81; Bellintani 2000, 65; Pare 2008, 82; Bietti Sestieri et al. 2015, 434.

188 De Min 1982; Salzani 1990–1991b, 2010b; Cardarelli et al. 2015.

and the second in the terminal horizon of the Bronze Age and the beginnings of the Iron Age (BF 3–PF1).

Two graves ascribed to Narde I necropolis contained amber (nos. 169 and 281). Each burial consisted of a cinerary urn, generally of biconical shape, covered by an inverted bowl, placed in a pit, which often has a thin layer of funeral pyre remains on the bottom. The graves were found in layers II and III, respectively. Objects found along amber inside the urns: simple-arch fibulae in variants undecorated and with two knobs on the bow (Salzani 1990–1991b, Figs. 14:11–16, 37:1–6), through analogies (e.g. with Frattesina, hoard 1 *Tesoretto*) allow to date the graves to the first horizon, i.e. to phase BF 2 (cf. Eles Massi 1986, 31–32, n. 243; Pare 1998, 313–314, Abb. 8:5). At Narde II, grave 75, dated to BF 3, yielded a rich collection of amber beads placed inside the cinerary urn, accompanied by glass beads and several other elements of garment (Colonna et al. 2010, 210–212, Tav. 13:B). At Narde II the research has also identified a surface rich in charcoals and ashes (remains of a burining pyre), identified as the *ustrinum* – area destined to incinerate the dead. The stratigraphic layer US1009 within the *ustrinum* contained numerus, mostly broken elements of personal belongings, i.a. single amber bead, which are datable to BF 1 (Colonna et al. 2010, 288–291, Tav. 56:1).

Approximately 500 m to the east from the previously described Sabbionara settlement, on the fields south from Desmontà di Veronella hamlet, has been registered a vast necropolis with numerous graves spanning all three stages of BF (Salzani 1984, 2013). The necropolis is characterised by a wide extension of the burial ground (estimated at ca. 25 ha), and concentration of the graves in three large groups with a more or less dense layout (Salzani 2013, 129–139). The preserved burials amount to 410. The funeral rite practised was exclusively cremation, with a great prevalence of the variant in which symbolic handful of incinerated bones was scattered together with remains of a pyre in a grave pit. The tombs of this type abound to 355 and represent 86.6 % of the entire necropolis. There are also 55 burials in urns, what represents 13.4 % of the total number of tombs on the necropolis. The urns were placed in pits, at the base of which were identified layers of pyre residues. The vases were sealed with lids consisting of bowls or cups placed upside down. Usually, burnt bones were occupying the lower part of the vase and above them was placed the funeral kit. Other times, the kit was found mixed with the bones or at the bottom of the urn. The fact that the bones contained in the urns were devoid of the ashes and charcoals proves that they were collected accurately from the stake. The tombs can be divided chronologically into three groups, corresponding to three phases of the cemetery's development (BF 1, 2 and 3). Grave 437, datable to phase 3 – BF 3, contained burial of a female adult individual (20–40 y.o.) furnished i.a. with six amber beads (Salzani 2013, Tav. XXXII:1–3, 10–11, 35).

Tab. 6. List of the Bronze Age contexts with amber in western Veneto/northern Po Plain

Name	Commune	Province	Type	Relative chronology	Absolute chronology	Reference
Bovolone, grave 115	Bovolone	Verona	Burial – cremation (urn)	BR	1350/1300–1175/1150 BC	Angelini 2010a, Fig. 1
Campestrin di Grignano Polesine, late BR–initial BF layers	Grignano Polesine	Rovigo	Settlement	BR 2–BF 1	1260/1220–1100/1060 BC	Bellintani et al. 2015a, Fig. 2A; 2021
Castello del Tartaro, grave 146	Cerea	Verona	Burial – inhumation	BR	1350/1300–1175/1150 BC	Salzani 2011, Fig. 4:A3
Castello del Tartaro, grave 166	Cerea	Verona	Burial – inhumation	BR	1350/1300–1175/1150 BC	Salzani 2011, Fig. 3:B5
Castello del Tartaro, grave 300	Cerea	Verona	Burial – inhumation	BR	1350/1300–1175/1150 BC	Salzani 2011, Fig. 5:A5–7
Castello del Tartaro, grave 350	Cerea	Verona	Burial – inhumation	BR	1350/1300–1175/1150 BC	Salzani 2011, Fig. 5:B3–4
Castello del Tartaro, grave 396	Cerea	Verona	Burial – inhumation	BR	1350/1300–1175/1150 BC	Salzani 2011, Fig. 4:C3
Desmontà, grave 437	Veronella	Verona	Burial – cremation (urn)	BF 3	1010/1000–910 BC	Salzani 2013, Tav. XXXII:1–3, 10–11, 35
Finilone Valle, subsurface materials	Gazzo Veronese	Verona	Settlement/ grave?	BM 2–BR	1550/1500–1175/1150 BC	Salzani 1996, Fig. 21:16–18, 22–23, 26–28
Fondo Paviani, advanced BR–early BF layers	Legnago	Verona	Embanked settlement	BR 1/2–BF 1–2	1260/1220–1010/1000 BC	Fasani/Salzani 1976, Fig. 2:3–4; Strafella et al. 2015, Fig. 1–2
Franzine Nuove, grave 7	Villa Bartolomea	Verona	Burial – inhumation	BM 3–BR	1450/1375–1175/1150 BC	Aspes/ Fasani 1968, Fig. 8:1–3
Franzine Nuove, grave 12	Villa Bartolomea	Verona	Burial – inhumation	BR	1350/1300–1175/1150 BC	Aspes 1973, Fig. 1:2

Tab. 6. Continued

Name	Commune	Province	Type	Relative chronology	Absolute chronology	Reference
Franzine Nuove, grave 123	Villa Bartolomea	Verona	Burial – inhumation	BM 3–BR	1450/1375–1175/1150 BC	Aspes 1973, 119
Franzine Nuove, grave 159	Villa Bartolomea	Verona	Burial – inhumation	BR	1350/1300–1175/1150 BC	Aspes 1973, 119, Fig. 1:3
Franzine Nuove, grave 417	Villa Bartolomea	Verona	Burial – inhumation	BR	1350/1300–1175/1150 BC	Aspes 1976, Fig. 1:6–20
Franzine Nuove, grave 481	Villa Bartolomea	Verona	Burial – inhumation	BM 3–BR	1450/1375–1175/1150 BC	Aspes 1987c, Fig. 1:1
Franzine Nuove, grave 497	Villa Bartolomea	Verona	Burial – inhumation	BR	1350/1300–1175/1150 BC	Aspes 1987c, Fig. 1:3
Franzine Nuove, grave 506	Villa Bartolomea	Verona	Burial – inhumation	BR	1350/1300–1175/1150 BC	Aspes 1987c, Fig. 2:1
Franzine Nuove, grave 523	Villa Bartolomea	Verona	Burial – inhumation	BR	1350/1300–1175/1150 BC	Aspes 1997, Fig. 420:4–6
Franzine Nuove, grave 570	Villa Bartolomea	Verona	Burial – inhumation	BR	1350/1300–1175/1150 BC	Aspes 1987c, Fig. 2:2
Frattesina di Fratta Polesine, hoard 1 *Tesoretto*	Fratta Polesine	Rovigo	Settlement	BF 2	1100/1060–1010/1000 BC	Negroni Catacchio 1972
Frattesina di Fratta Polesine, phases 2–3	Fratta Polesine	Rovigo	Settlement	BF 2–3	1100/1060–910 BC	Bietti Sestieri et al. 2015, Fig. 1D:19, 2:12–13
Mariconda di Melara, upper layer	Melara	Rovigo	Settlement	BF 1–2	1150–1010/1000 BC	Salzani 1973, Tav. XVIII:22
Narde I, grave 169	Fratta Polesine	Rovigo	Burial – cremation (urn)	BF 2	1100/1060–1010/1000 BC	Salzani 1990–1991b, Fig. 37:6

(*Continued*)

Tab. 6. Continued

Name	Commune	Province	Type	Relative chronology	Absolute chronology	Reference
Narde I, grave 281	Fratta Polesine	Rovigo	Burial – cremation (urn)	BF 2	1100/1060–1010/1000 BC	Salzani 1990–1991b, Fig. 14:14
Narde II, *Ustrinum*, US 1009	Fratta Polesine	Rovigo	Cremation pyre	BF 1	1150–1100/1060 BC	Colonna et al. 2010, Tav. 56:1
Narde II, grave 75	Fratta Polesine	Rovigo	Burial – cremation (urn)	BF 3	1010/1000–910 BC	Colonna et al. 2010, Tav. 13:B3–11
Olmo di Nogara, grave 18	Nogara	Verona	Burial – inhumation	BR	1350/1300–1175/1150 BC	Salzani 2005, Tav. II:C–D
Olmo di Nogara, grave 30	Nogara	Verona	Burial – inhumation	BR	1350/1300–1175/1150 BC	Salzani 2005, Tav. IV:D
Olmo di Nogara, grave 122	Nogara	Verona	Burial – inhumation	BR	1350/1300–1175/1150 BC	Salzani 2005, Tav. XIII:B–D
Olmo di Nogara, grave 190	Nogara	Verona	Burial – inhumation	BR	1350/1300–1175/1150 BC	Salzani 2005, Tav. XVIII:D–E
Olmo di Nogara, grave 208	Nogara	Verona	Burial – inhumation	BR	1350/1300–1175/1150 BC	Salzani 2005, Tav. XX:A
Olmo di Nogara, grave 243	Nogara	Verona	Burial – inhumation	BR	1350/1300–1175/1150 BC	Salzani 2005, Tav. XXI:C, F
Olmo di Nogara, grave 252	Nogara	Verona	Burial – inhumation	BR	1350/1300–1175/1150 BC	Salzani 2005, Tav. XXII:D
Olmo di Nogara, grave 307	Nogara	Verona	Burial – inhumation	BR	1350/1300–1175/1150 BC	Salzani 2005, Tav. XXV:C
Olmo di Nogara, grave 353	Nogara	Verona	Burial – inhumation	BR	1350/1300–1175/1150 BC	Salzani 2005, Tav. XXIX:E
Olmo di Nogara, grave 358	Nogara	Verona	Burial – inhumation	BR	1350/1300–1175/1150 BC	Salzani 2005, Tav. XXIX:C–D
Olmo di Nogara, grave 365	Nogara	Verona	Burial – inhumation	BR	1350/1300–1175/1150 BC	Salzani 2005, Tav. XXIX:B–E

Tab. 6. Continued

Name	Commune	Province	Type	Relative chronology	Absolute chronology	Reference
Olmo di Nogara, grave 441	Nogara	Verona	Burial – inhumation	BR	1350/1300–1175/1150 BC	Salzani 2005, Tav. XXXVI:A, C
Olmo di Nogara, grave 154	Nogara	Verona	Burial – inhumation	BM 3	1450/1375–1350/1300 BC	Salzani 2005, Tav. XVI:D–E, G–H
Olmo di Nogara, grave 411	Nogara	Verona	Burial – inhumation	BM 3	1450/1375–1350/1300 BC	Salzani 2005, Tav. XXXIV:G–H
Olmo di Nogara, grave 493	Nogara	Verona	Burial – inhumation	BM 3	1450/1375–1350/1300 BC	Salzani 2005, Tav. XL:C–D
Olmo di Nogara, grave 32	Nogara	Verona	Burial – inhumation	BM 2–3	1550/1500–1350/1300 BC	Salzani 2005, Tav. V:D–E
Olmo di Nogara, grave 100	Nogara	Verona	Burial – inhumation	BM 2–3	1550/1500–1350/1300 BC	Salzani 2005, Tav. XII:C–D
Olmo di Nogara, grave 155	Nogara	Verona	Burial – inhumation	BM 2–3	1550/1500–1350/1300 BC	Salzani 2005, Tav. XVI:D
Olmo di Nogara, grave 172	Nogara	Verona	Burial – inhumation	BM 2–3	1550/1500–1350/1300 BC	Salzani 2005, Tav. XVII:C–D
Olmo di Nogara, grave 390	Nogara	Verona	Burial – inhumation	BM 2–3	1550/1500–1350/1300 BC	Salzani 2005, Fig. 65, Tav. XXXI:E–F
Olmo di Nogara, grave 85	Nogara	Verona	Burial – inhumation	BM 2	1550/1500–1450/1375 BC	Salzani 2005, Tav. IX:E–F
Olmo di Nogara, grave 89	Nogara	Verona	Burial – inhumation	BM 2	1550/1500–1450/1375 BC	Salzani 2005, Tav. X:D–E
Olmo di Nogara, grave 112	Nogara	Verona	Burial – inhumation	BM 2	1550/1500–1450/1375 BC	Salzani 2005, Tav. XII:D–E
Olmo di Nogara, grave 388	Nogara	Verona	Burial – inhumation	BM 2	1550/1500–1450/1375 BC	Salzani 2005, Tav. XXX:E–F
Olmo di Nogara, grave 422	Nogara	Verona	Burial – inhumation	BM 2	1550/1500–1450/1375 BC	Salzani 2005, Tav. XXXV:C

(Continued)

Tab. 6. Continued

Name	Commune	Province	Type	Relative chronology	Absolute chronology	Reference
Olmo di Nogara, grave 435	Nogara	Verona	Burial – inhumation	BM 2	1550/1500–1450/1375 BC	Salzani 2005, Tav. XXXVI:A–C
Povegliano Gambaloni, sector B, tomb 8	Povegliano Veronese	Verona	Burial – inhumation	BM 2	1550/1500–1450/1375 BC	Negroni Catacchio 1971, Fig. 18–21
Povegliano Muraiola, phase II	Povegliano Veronese	Verona	Settlement	BM 2 early	1550/1500–1500/1450 BC	Salzani 1997b, Tav. 56:39
Povegliano Muraiola, phase IV	Povegliano Veronese	Verona	Settlement	BM 2 late	1500/1450–1450/1375 BC	Salzani 1997b, Tav. 56:14–16, 36
Sabbionara, sector II, US 117	Veronella	Verona	Settlement	BR	1350/1300–1175/1150 BC	Salzani 1990–1991a, Fig. 21:13
Scalvinetto, grave 233	Legnago	Verona	Burial – inhumation	BM 3–BR	1450/1375–1175/1150 BC	Salzani 1994, Tav. 5:5–8
Scalvinetto, grave 282	Legnago	Verona	Burial – inhumation	BR	1350/1300–1175/1150 BC	Salzani 2004, Fig. 3:B
Scalvinetto, grave 315	Legnago	Verona	Burial – inhumation	BM 3–BR	1450/1375–1175/1150 BC	Salzani 2004, Fig. 4:E
Scalvinetto, grave 354	Legnago	Verona	Burial – inhumation	BM 3–BR	1450/1375–1175/1150 BC	Salzani 2004, Fig. 4:B, D, G
Scalvinetto, grave 373	Legnago	Verona	Burial – inhumation	BR	1350/1300–1175/1150 BC	Salzani 2004, Fig. 5:B, D
Scalvinetto, grave 376	Legnago	Verona	Burial – inhumation	BR	1350/1300–1175/1150 BC	Salzani 2004, Fig. 5:B, D

2.2.7. Emilia/southern Po Plain

Presence of amber in Emilia (provinces of Piacenza, Parma, Reggio Emilia, Modena and western part of Bologna) is directly linked to the *Terramare* culture south of the Po during the Middle and Recent Bronze Age (Fig. 11). Nowadays, the unit is commonly understood according to the definition and chronology proposed by M. Bernabò Brea and A. Cardarelli.[189] The basic periodisation proposed by the authors assumes tripartite division of the MBA into BM 1 (1650–1550 BC), BM 2 (1550–1450 BC) and BM 3 (1450–1350/1300 BC), followed by the LBA – BR (1350/1300–1150 BC). Later, thanks to the results of excavations on multi-layered sites i.a. Montale (Cardarelli/Labate 2004) and Santa Rosa di Poviglio (Bernabò Brea/Cremaschi 2004), even finer division into subphases BM 1A, 1B, 2A, 2B, 3A, 3B, and BR 1 and 2 has emerged. Since its inception the chronology has been a subject of polemic with R. C. De Marinis, who came up with an alternative chronological system for Northern Italy (De Marinis 1999; see section 2.2.5). In comparison, Bernabò Brea et al. (2018), partly following R. Peroni (1994), assume a generally higher absolute chronology, whereas De Marinis sets the limits of corresponding phases 50 years later on average (1999; cf. David et al. 2017). In the recent years the offset in dating of the MBA beginning and, to a lower extent, between the corresponding phases has been reduced. It was possible thanks to the growing number of absolute (14C) dates from Italian contexts generally supporting the 'higher' chronology, and reconsideration of the dates in relation to other chronologies of the circum-Alpine zone.[190] Still, there are unsolved problems with correlating individual phases of BM and BR in the two systems, stemming from varying views on dating of guiding contexts. Some of them are tackled below, on the occasion of determining chronology of amber finds.

The area of Emilia, before emergence of the *Terramare* culture, appears to have been sparsely inhabited; the EBA (BA) is known mainly from hoards, loosely found individual bronze objects and scarce ceramic fragments recovered at few sites, although progress of research in recent years brought to light several new funerary and residential contexts.[191] The new discoveries show prolonged presence of the *Bell Beaker* tradition (BA 1 early), which subsequently morphed into an unstable and loosely knit network of communities, culture-wise drawing heavily from the neighbouring areas (BA 1 advanced and BA 2).

The following period – BM 1 – brought more evidence of human activity in the region, but it still presents rather inhomogeneous picture of the local culture. With respect to pottery, the first stage (BM 1A) is marked by a general presence of axe-shaped handles, but in the subsequent phase (BM 1B) eastern Emilia and Romagna came under an influence of the *Protoapennine* culture – *Grotta Nuova* facies from the south, while western Emilia was affected by the *Mercurago* facies (e.g. Chiaravalle della Colomba site).[192] At the same time one can observe appearance of the *palafitte*-type settlements along with the late *Poladian* and *post-Poladian* pottery (e.g. Camponi di

189 Bernabò Brea/Cardarelli 1997; Cardarelli 2009; Bernabò Brea et al. 2018.
190 Cf. David-Elbiali/David 2009; Vanzetti 2013; Capuzzo et al. 2014; David et al. 2017.
191 Cf. Bermond Montanari et al. 1996, 66–68; Bernabò Brea/Cardarelli 1997, 296; Bernabò Brea et al. 2018, 11–13.
192 Aspes et al. 1991–1992, 54; Bernabò Brea/Cardarelli 1997, 297; Cardarelli 2009, 457–458.

figures for this period in Emilia are estimated at 100,000 inhabitants, with the density of 14 people per square kilometre (cf. Cardarelli 2009, 460–463).

To BM 2B the amber bead found in layers relevant to the 2nd phase of Gaggio should most probably be assigned (Orsini 2010, p. 118, no. 8; A. Cardarelli, personal communication). The phase saw rearrangement of the site following a fire which concluded the 1st phase: the embankment was enlarged and moved closer to the ditch/moat, and new houses were built in place of the old ones (cf. Balista et al. 2008, 126–129). Most of the materials found buried in the level of the 2nd phase can be easily fitted into BM 2B – the *Tabina di Magreta* facies, with good comparisons detectable in the phase 2 of Montale (Balista et al. 2008, 126–128, Fig. 20, 21:3, 5–7; cf. Cadarelli/Labate 2004, Fig. 79). In a much smaller number appear features typical for BM 3: bronze pin of the Montale-type (Balista et al. 2008, Fig. 21:4, cf. De Marinis/Salzani 2005, 420–421) and ceramic handles with frontal discs (Balista et al. 2008, Fig. 21:1; cf. De Marinis/Rapi 2016, 46–48, Fig. 9). Therefore, the entire phase 2 of Gaggio can be placed predominantly in BM 2B with continuation in BM 3A.

Passage between the two abovementioned chronological horizons documents changes that took place in the settlement mode in Emilia. A new territorial pattern of settlements emerged, with many of the older *terramare* becoming abandoned, and their inhabitants augmenting the newly established villages, generally larger in terms of occupied space. The cause of this reorganisation is generally seen in a demographic increase and growing pressure for arable land that led to aggregation of population; however, it is possible that this process was accompanied by restructuration of political system as well (cf. Cardarelli 2009, 465–466; Bernabò Brea et al. 2018, 17).

One of the sites which emerged anew during this period is Beneceto – Forno del Gallo, located in the north-east outskirts of Parma (Bernabò Brea et al. 2005, 2008). The area of the site was inhabited for the first time in the EBA (BA 1–2), but was abandoned afterwards. A *terramare*-type settlement was founded here in the advanced MBA (BM 3), and developed, with considerable structural modifications, throughout the LBA. Amber is attested from the beginning of the second habitation episode (BM 3A), as suggested by small fragment of a bead. Two complete beads come from layers US 644S and 5843 dated to BM 3B (Bianchi 2007, 78, Fig. II:34–35). The site's phase IIIa/b, to which the amber beads belong, has been selected as an exemplary context of BM 3B, illustrating evolution of pottery between the late MBA and early LBA (Frontini 2011, 24).

Settlements established during BM 3 frequently continued to function in BR. Many examples of such sites can be mentioned, some of them having yielded amber artefacts datable to the transitional horizon. Particular concentration is notable in the Modena province. At Gaggio, following another destructive event involving fire around BM 3A/3B, some features of the settlement, including moat/ditch and houses, were rebuilt. The 3rd phase, attested by numerous bronzes and pottery characteristic for BM 3B–BR 1 (Balista et al. 2008, Figs. 23 and 24; cf. Frontini 2011, 18–19) ended with two floods, recognisable thanks to different sedimentological characteristics documented in the former moat and, partially, inside the settlement (Balista et al. 2008, 129–133). After the flood that concluded the 3rd phase of the settlement, no housing structures were rebuilt. The abandonment of the site (4th phase – BR 1) coincided with enlargement of the nearby *terramara* Redù (Balista et al. 2008, 133–134; Cardarelli et al. 2013, 246). From the last horizon of Gaggio's attendance (BM 3B–BR 1) come two small fragments of unspecified beads (A. Cardarelli, personal communication).

The other *terramara* in the area, which delivered similarly dated amber finds, is Gorzano, located south-west from Modena. Stratigraphic scheme of the site created by F. Coppi comprises two levels: (A) surface layer, containing remnants of modern times, Medieval and Roman period, with thickness of circa 0.8–1.25 m, and (B) prehistoric deposit, consisting of two main layers (I and II), each with thickness of about 1 m, separated by a thin, corrugated, carbonaceous layer, locally interwoven by a layer of marl (Säflund 1939, 18). Single amber bead was found during F. Coppi's excavations at the depth of ca. 2 m below the ground surface, in the layers IB–IIA which includes materials dated to BM 3–BR 1.[200] In the words of Coppi, the bead 'was discovered at the height of 1 m from the basal soil level' (cf. Säflund 1939, 32). According to the scheme by Coppi, it is a transitional level between the two main stratigraphic units belonging to prehistoric – Bronze Age horizon of the settlement. It is placed at the interface between layer IB (described by Coppi as the *terramara bigia*), consisting of pure marl bed, and layer IIA (described by Coppi as the *terramara verde*). It should be close to, or same as the charred layer separating units I and II. It is not certain, if this layer appeared across the entire area of the settlement. Selection of finds related to this horizon was based on the scheme provided by G. Säflund (1939, 10), who grouped together artefacts discovered at similar depths and later confronted them with typo-chronological considerations.[201]

Similar is the chronology of amber finds from Anzola dell'Emilia, south-east from Modena (Desantis et al. 2011, 35–36). The five beads were recovered from central-eastern section of the site (sector D, layer US 27–26). Inside the trench were also spotted numerous post-holes that, due to regular pattern of their distribution, may be regarded as a residue of former housing structures. These were located within section of the site delimited from the west by the inner moat, and from the east by the Martignone torrent which in the past must have supplied the moat with water. The inner moat circumscribed area of about 2.5 ha (width of 180 m along NW–SE axis and 130 m along SW–NE), whose eastern border was formed by a natural watercourse (Martignone), while the southern edge coincided with the outer perimeter of the settlement. Along the inner bank of the ditch were found traces of a palisade. The site has brought rich evidence of BM 3–BR 1 metallurgy (Desantis et al. 2011, 24–28), as well as similarly dated pottery (cf. Argentina et al. 2002). It is further supplemented with corresponding 14C dates from the layers 26 and 27: respectively LTL4349A: 3092 ± 45 BP, 1415–1295 cal. BC (68.2 %), 1446–1229 cal. BC (95.4 %), and LTL4348A: 2976 ± 45 BP, 1262–1125 cal. BC (68.2 %), 1306–1048 cal. BC (90.8 %) (Desantis et al. 2011, 51).

The transition from BM to BR is a problematic issue of Northern Italian archaeology. Certain Italian scholars disagree on dating of the end of BM and the start of BR, and differently determine the chronology of the key contexts illustrating the process.[202]

---

200 Säflund 1939, 32, n. 154; Miari 2007a, Fig. 3:19; cf. Carancini 1997; Labate 1997a; De Marinis 1999, 56.

201 It is necessary to add that G. Chierci mentions another two amber beads that were included in the second monograph about Gorzano's excavations by F. Coppi, and one more published in an article in a local newspaper and later in the *Annali della Societa dei Naturali in Modena* (Chierci 1877, 29–30, notes 8 and 9). However, the finds were not reported in the more recent literature, nor can be found at the Modena museum.

202 Cf. Bernabò Brea/Cardarelli 1997, 298–299; De Marinis 1999, Fig. 44; Cardarelli 2009, 450; David et al. 2017, Fig. 19; Bernabò Brea et al. 2018, 17–19; see also Frontini 2011, 11–24 for an overview of the discussion.

One of the most controversial aspects of the debate is the dating of the earliest habitation horizon at the site Santa Rosa di Poviglio – *Villaggio Grande*. The 'Large Village' (direct translation of the Italian term) was established adjacent to the 'Small Village' (it. *Villaggio Piccolo*), following the abandonment of the latter at the end of BM 2B and a subsequent period of hiatus at the site. Enclosed by a complex wooden palisade and oriented according to the same alignment as the Small Village, the Large Village is one of the largest *terramare*-type settlements (the entire area of the site is estimated at 7 ha, although only ca. 1/16 of it has been excavated). The stratigraphic sequence is divided into four different phases. Most of the discussion revolves around the phases 1–2, comprising the basal layers and the so-called 'heaps of ash' (Bernabò Brea et al. 1997c). The 1st phase, in the opinion of M. Bernabò Brea and M. Cremaschi, bears evidence of sporadic frequentation of the site without stable structures, datable through scarce pottery to an advanced moment of the MBA – BM 3. Structures determined as representative of the 2nd phase include wooden scaffolding for housing structures marked by rows of post-holes, and a wooden fence which the abovementioned authors place at the initial BR (Bernabò Brea/Cremaschi 1997b, 203–208; Bernabò Brea et al. 1997c, 345). In turn, De Marinis, using the stratified analogies for the bronze objects presented by Bernabò Brea and Cardarelli, treats both phases 1 and 2 as a single horizon and dates it squarely to BR 1 (De Marinis 1999, 65, 71, 88–93, Figs. 45 and 50).

Lately, the basal layer and the so-called 'heaps of ash' layer of Santa Rosa di Poviglio – *Villagio Grande*, in the light of recent re-evaluation of the ceramic materials and relevant contexts representative for BM–BR transition, were assigned to BR 1 by P. Frontini (2011). The basis for such classification is the presence of handles with high vertical appendices ('bovine horns' handles), supplementing the rich bronze evidence pertinent to this period already evaluated by De Marinis (Frontini 2011, 16–18, 23–24; cf. Bernabò Brea et al. 1997c, Fig. 185). Consequently the two amber beads found in the abovementioned layers can be dated to BR 1 (Bernabò Brea et al. 1997c, 347). Nonetheless, one should be aware that in the same layers reportedly have been found, so far unedited, handles with lateral expansions – a feature typical for BM 3 (Frontini 2011, 17, 24, n. 43). Thus slightly earlier beginning of the site cannot be completely excluded, however requires much more thorough research.

Conversely, there are no significant differences in opinions regarding the upper layers of the Large Village, comprising phases 3 and 4 (De Marinis 1999, Fig. 50). The former is related to an event involving burying the previous levels of the settlement under anthropized accumulation of ash and charcoal mixed with pottery (the so-called 'reclamation' layer), on which the new buildings were planted (Bernabò Brea/Cremaschi 1997b, 208). The archaeological material contained in those layers does not seem homogeneous; in addition to many elements similar to those of phase 2 (BR 1), there are other relevant to BR 2 (Bernabò Brea et al. 1997d, 348). Phase 4 sees the last level of housing structures; however, the most superficial part of the stratification connected to this phase is strongly compromised by post-depositional factors affecting the site after its abandonment. The preserved elements are only rare pits and numerous concentrations of materials, including elongated clusters of daub interpretable as collapsed plasters of burnt wooden walls (Bernabò Brea/Cremaschi 1997b, 208–212). The abundant collected material pertinent to this phase, including three amber beads, can be dated to BR 2 (Bernabò Brea et al. 1997d, 349–350).

Also the settlement of Beneceto yielded layers that can be divided into BR 1 and BR 2, both of which contained amber finds. Level of BR 1 seems rather thin, with small features filled with consumption and production waste, as well as layers of ash and burnt clay. The habitation continued largely in the same area as before, with the construction of new housing structures (Bernabò Brea et al. 2008, 96). Two amber objects come from a foundation ditch of one of the rectangular structures discovered in the southern part of the central sector (US 144, UUSS 3; Bianchi 2007, Fig. II:40–41). Another two were found in layers of ash in the north-west part of residential sector (US 674; Bianchi 2007, Fig. II:37, 39). Fourth and fifth bead were registered again in the north-west part, the former in layer superimposed on the phase BM 3B, largely exposed during BR 1 (US 680; Bianchi 2007, Fig. II:38), while the latter near a collapsed domestic structure (US 679; Bianchi 2007, Fig. II:36). During BR 2 the settlement underwent major restructuring, with the abandonment of the western bank and a shift of the housing area to the east and north. One amber bead comes from a collapse layer attributable to huts in the far-east sector of the village (Bianchi 2007, 79; Miari 2007a, Fig. 2:10). Another two were found in levels of abandonment of the settlement in the central sector (US 3; Bianchi 2007, 79, Fig. II:42; Miari 2007a, Fig. 2:9).

Apart from the above described, datable contexts of amber beads, from the area of Emilia/southern Po Plain, at least seven more *terramare* settlements with amber are known. However, in these cases there is no possibility of narrowing the chronology of amber finds beyond the overall lifetime of the sites. The problem mostly stems from the shortcomings of field documentations left by the principal investigators of these sites in the 19th century.

One of the examples is the *terramara* of Casinalbo, located 7 km south-west from the city of Modena, within the Formigine municipality. Most of materials from the site, today kept in the Museum of Archaeology and Ethnology of Modena, including a single amber bead (Säflund 1939, 38; Orsini 2010, 118), are devoid of information about the context. Their dating is therefore based solely on comparisons with similar objects from other sites in the region, pointing to a time span between BM 2 and BR 2 (cf. Carancini 1997; Labate 1997b).

The site of Castellazzo di Fontanellato, located less than 1 km east from the Paroletta village, within the municipality of Fontanellato, returned three amber beads (Mutti 1993, 84, Fig. 57:9–10; Orsini 2010, Fig. 7 on p. 141). Despite extensive research by L. Pigorini, the spatial structure of the site remains poorly documented and only handful of finds recovered there are preserved to this day which only generically can be introduced to a period BM–BR. Moreover, written reports of the excavations together with drawings and sketches made by L. Scotti who assisted Pigorini during fieldworks, allegedly given to the Royal Museum of Prehistory and Ethnography in Rome (now 'L. Pigorini' Museum), are missing. Hence, it is impossible to reconstruct the settlement's structure in detail, or connect the finds with stratigraphic units or excavated sectors (cf. Säflund 1939, 87–95; Mutti 1993, 81–83).

The site of Castelfranco Emilia, belonging to the Predella municipality, revealed some evidence of BM–BR occupation. From unpublished report by E. Brizio, who excavated the site in 1882, stems a general picture of the settlement with remains of huts built on the ground; however, the finds, i.a. single amber bead, cannot be connected to any specific layers or horizons (cf. Scarani 1963). Similarly unknown is the context of the single amber bead from the *terramara* of Colombare di Bersano (com. Besenzone, prov. Piacenza) (Miari 2007a, Fig. 3:1; cf. Säflund 1939, 112–113).

At the settlement of Redù-Pilastro, about 1 km north-west from the village of Redù (com. Nonantola, prov. Modena), most of archaeological materials were recovered

Tab. 7. Continued

Name	Commune	Province	Type	Relative chronology	Absolute chronology	Reference
Gaggio, 1st phase	Castelfranco Emilia	Modena	*Terramare* settlement	BM 1–2A	1650–1500 BC	Miari 2007a, n. 16
Gaggio, 2nd phase	Castelfranco Emilia	Modena	*Terramare* settlement	BM 2B–3A	1500–1400 BC	Orsini 2010, p. 118, no. 8
Gaggio, 3rd and 4th phase	Castelfranco Emilia	Modena	*Terramare* settlement	BM 3B–BR 1	1400–1260/ 1200 BC	Unpublished
Gorzano, IB–IIA layers	Maranello	Modena	*Terramare* settlement	BM 3–BR 1	1450–1260/ 1200 BC	Miari 2007a, Fig. 3:19
Montale, phase II	Castelnuovo Rangone	Modena	*Terramare* settlement	BM 2B	1500–1450 BC	Cardarelli/ Labate 2004, Fig. 79:15; Orsini 2010, 118
Montale, precise context unknown	Castelnuovo Rangone	Modena	*Terramare* settlement	BM 2A–BR 1	1550–1260/ 1200 BC	Cardarelli 1988, Fig. 85; Bergonzi1997, Fig. 352; Orsini 2010, 117
Redù-Pilastro, precise context unknown	Nonantola	Modena	*Terramare* settlement	BM–BR	1650–1150 BC	Orsini 2010, 118, 219
Sant'Ambrogio, precise context unknown	San Cesario sul Panaro	Modena	*Terramare* settlement	BM 3–BR 1	1450–1260/ 1200 BC	Negroni Catacchio 1975b, Fig. 2
Sant'Anastasio, precise context unknown	Savignano sul Panaro	Modena	*Terramare* settlement	BM 2–BR 1	1550–1260/ 1200 BC	Negroni Catacchio 1975b, 380
Santa Rosa di Poviglio, *Villaggio piccolo*, sector S, US 4	Poviglio	Reggio Emilia	*Terramare* settlement	BM 2B	1500–1450 BC	Bellintani et al. 2004, Fig. 274:2
Santa Rosa di Poviglio, *Villaggio Grande*, 'heaps of ash' layer – phase 2	Poviglio	Reggio Emilia	*Terramare* settlement	BR 1	1350/ 1300–1260/ 1200 BC	Bernabò Brea et al. 1997c, 347
Santa Rosa di Poviglio, *Villaggio Grande*, 'summit' layer – phases 3 and 4	Poviglio	Reggio Emilia	*Terramare* settlement	BR 2	1260/ 1200–1150 BC	Bernabò Brea et al. 1997d, 350

2.2.8. Romagna and Emilian Apennines

Similarly to Emilia, amber in Romagna (provinces of Rimini, Forlì-Cesena, Ravenna and eastern part of Bologna; Fig. 11) began to appear in the MBA, despite the fact that the area seems to have been already densely inhabited during the EBA (BA; ca. 2300–1650 BC; cf. Cattani/Miari 2018, 35–37). However, cultural development in Romagna proceeded differently in comparison with Emilia. Starting from BM 1 (ca. 1650–1550 BC), the *Protoapennine* culture began to affect local communities, preceding any influences which may be related to the *Palafitte-Terramare* culture. Typical for the territory in the first subphase of the period – BM 1A – were axe-shaped ceramic handles, linked to the *Belverde* facies of Tuscany, which heralded the emergence of the *Grotta Farneto* facies in Emilia-Romagna (with the centre of gravity in the Bologna area).[205] The next subphase – BM 1B – saw continuing influx of the *Protoapennine* culture to Romagna, but at that point in a new version known as the *Grotta Nuova* facies, with the local autonomous cultural unit described as the *Farneto-Monte Castellaccio* facies.[206] The latter phenomenon remains perceptible until the end of BM 2 (1550–1450 BC); however, it was progressively accompanied by elements of the *Palafitte-Terramare* culture in the version of the *Tabina di Magreta* facies. However, in contrast with Emilia, Romagna never fully absorbed the 'southern Po Plain cultural *koiné*' and during BM 2 elements of the *Palafitte-Terramare* ancestry coexisted with those related to the *Grotta Nuova* facies.[207] Exactly to this period, spanning BM 1B and 2, is dated the settlement of Monte Castellaccio di Imola, where the oldest amber finds in Romagna were revealed.

The site is located at the southern edge of the city of Imola, on flat-top elevation of Monte Castellaccio (76 m a.s.l.). It rises about 30 m above the bottom of the Santerno River Valley. Its isolated and dominant position not only over the Santerno Valley but also the adjacent stretch of the Po Valley gives the site a high strategic value from the viewpoint of territory control (cf. Pacciarelli 1996a). The settlement shows two distinct horizons of the Bronze Age occupation: the first one refers to BM 1B–2 (phase 2), followed by a gap in BM 3, with renewed occupation in BR (phase 3). A serious drawback concerning documentation of the excavations on Monte Castellaccio by G. Scarabelli is the absence of correlation between majority of finds and discovered huts and layers. Fortunately, for the single amber bead known from the site, information about the context of its discovery has been preserved (Pacciarelli 1996c, Tav. 1:F; Miari 2007b, II.43). It was found in basal levels of anthropogenic deposit, just above the layers connected to the Copper Age habitation (phase 1), plentiful with pottery and bronzes pertinent to BM 1–2 (cf. Pacciarelli 1996b, 221–225; 1996c, nn. 3049–3050, Tav. 1). Apart from the finds lending the site its dating in terms of relative chronology, two radiocarbon dates were obtained for the 2nd phase which confirm the results of typo-chronological analysis: MC9 (GX-21870: 3340 ± 50 BP), 1688–1535 cal. BC (68.2 %), 1744–1506 cal. BC (95.4 %), and MC4 (GX-21871: 3265 ± 75 BP), of 1624–1452 cal. BC (68.2 %), 1699–1402 cal. BC (93.1 %) (Pacciarelli 1996a, 145; 1996b, 225).

---

205 Carancini et al. 1996, Fig. 2; Pacciarelli 1997, 423, Tab. 5; Cattani/Miari 2018, 37; cf. Bazzocchi et al. 2009, 194; Cocchi Genick 2018.

206 Pacciarelli 1997, 423–424, Tab. 5; Cattani/Miari 2018, 37–38; cf. Cocchi Genick 1995a, 370–373; Cocchi Genick 2018.

207 Pacciarelli 1997, 424–426, Tab. 5; Cattani/Miari 2018, 39–40; cf. Bazzocchi et al. 2009, 196; Cocchi Genick 2018.

The transition to BM 3 (ca. 1450–1350 BC) is marked by discontinuity in the occupation of the territory. Several settlements active during BM 1–2 were abandoned, i.a. Monte Castellaccio di Imola, while the new ones emerged in previously uninhabited areas and developed continually into BR. In fact, between BM 3 and BR habitation, centres such as S. Giuliano di Toscanella reached dimensions of few hectares and acquired fortifications,[208] assuming pre-eminent role in the organisation of the territory (Pacciarelli 1997, 426). In a sense, this reorganisation of the habitation pattern (concentration of population in large settlements, in parallel to abandoning some of the previously inhabited areas) followed similar trend taking place at the time in Emilia (see section 2.2.7). Nonetheless, the Bolognese-Romagnol area in this phase, unlike in BM 2, differed more substantially in terms of material culture from Emilia, especially its westernmost part. The dissimilarities may be accredited to the absence of pottery shapes characteristic of the *Monte Leoni* facies; decorations such as grooves, bosses or *cuppelle* are absent in Romagna, whereas the horned, crescent-shaped handles or handles with frontal discs are predominantly limited to western Bolognese area, while those found further to the east represent other variants than those encountered in Emilia, e.g. near Parma (Cattani et al. 2009, 218; Cattani/Miari 2018, 40–41). Still, occurrence of the horned appendices seems to testify to a continuous relationship with the *Palafitte-Terramare* culture, commenced in the previous period.

At the same time, (BM 3) decorated vessels with strongly everted handles with expanded extremities, typical of the *Apennine* facies, made a notable appearance in Romagna. The ceramics with the *Apennine*-derived decorations, formed with geometric motifs (triangles, lozenges, spirals, meanders), often in very complex engraved or excised syntaxes, are progressively attested from west to east parts of Emilia-Romagna (Cattani et al. 2009, 218; Cattani/Miari 2018, 40–41). The area of Romagna continued to function as a 'border zone', selectively adopting the traditions of the Po Valley, on the one hand, and the central Apennine Peninsula, on the other. Such selectivity increasingly contributed to a specific character of the local culture which should be attributed to an independent group/unit on the map of the Bronze Age Italy.[209]

Presence of pottery with the *Apennine*-derived decorations, traditionally dated to BM 3, continued in Romagna even in BR (1350–1150 BC). At that time the *Apennine* culture was fluidly replaced by its next evolutionary stadium known as the *Subapennine* culture, exemplified by straight-cylindrical raised handles.[210] Still, the area of Romagna retained a level of cultural individualism, as illustrated by the so-called *a flabello* raised handles (Cattani/Miari 2018, 44–45, Fig. 9).

Concurrently to the uninterrupted evolution of the pottery, one can observe sustained development of the BM 3 settlements, e.g. S. Giuliano di Toscanella, well into BR (cf. Cattani et al. 2009, 220–221; Cattani/Miari 2018, 45–46). The latter site is located in the commune of Toscanella (Imola province), on a flat area, along the alluvial cone of the Sellustra torrent (57 m a.s.l.) (Morabini 1996; Morico/Pacciarelli 1996). The materials excavated by G. Scarabelli indicate a prolonged occupation of the site,

---

208 Presence of fortifications, especially in regard to S. Giuliano di Toscanella, is not unequivocally accepted by all scholars (for discussion see Morico/Pacciarelli 1996; Cattani et al. 2009, 222–223).

209 Cattani et al. 2009, 218; Cattani/Miari 2018, 41–42; cf. Macchiarola 1995, 455.

210 Pacciarelli 1997, 426–427; Cattani/Miari 2018, 43; cf. Damiani 2010.

starting from BM 2 and continuing up to the end of BR. The maximum expansion of the site falls into BM 3, characterised by local appearance of the *Apennine* culture, and into BR 1 (ca. 1350–1250 BC), when the region fully entered the *Subapennine* stylistic *koiné* (Damiani/Morico 1996; Morico 1996). To this phase of the settlement should be most probably attributed the single amber bead which was found in the trench 5 of the 1901 excavation, where the anthropic deposit had the thickness of 30 cm. It included i.a. traces of post-holes and a hearth with fragments of ceramic vessels with cylinder-straight handles, two bronze awls and a cervine horn with traces of workmanship.[211]

At the turn of BM 3 and BR 1 colonisation expanded into the hilly territory along the Apennines ridge, even though the flatlands south from the Po and along the coastline were still primarily selected for residential areas (Cattani et al. 2009, 222). Exactly in the south-eastern, coastal part of Romagna is situated the site of Conti Spina di Riccione, which yielded multiple residues of a BR settlement, similar in structure to that of Monte Castellaccio di Imola (cf. Bermond Montanari et al. 1992). Overall examination of the recovered vascular pottery makes it possible to frame the settlement of Riccione in BR, thanks to numerous references to the *Subapennine* culture pottery. To this chronological horizon refer the numerous faired cups and certain types of handles with zoomorphic appendages, and with rod-like cross-section (Bermond Montanari et al. 1992, 108–110, Figs. 10 and 11). Also the bronze items reflect forms typical for BR, i.a. the Peschiera- and Boccatura del Mincio-type pins pertinent to BR 1, or a violin-bow fibulae signalising BR 2/BF 1 (Bermond Montanari et al. 1992, Fig. 17:1–3).[212] The context of the single amber bead coming from the site is not specified in the literature. Nonetheless, it is listed among other finds datable to BR; therefore, it can be inferred that it was found together in layers with the other objects belonging to this one-phase settlement (Bermond Montanari et al. 1992, 111; cf. Miari 2007a, 68).

The crisis of the 13th/12th century BC, so apparent in Emilia, also affected the territory of Romagna. However, it did not have such grievous consequences there as in the west. It caused the rarefaction of the settlement to a certain degree. Nonetheless, communities in many places occupied earlier managed to survive and maintain their residences until BF 1 (ca. 1150–1100/1060 BC) (Cattani/Miari 2018, 45–47; Cattani/Cavani 2009, 255). The continuing occupation is indicated by presence of metal and ceramic forms signalling late BR and initial BF at the same sites, e.g. Villa Cassarini or Borgo Panigale in the Bologna province.

The latter site revealed amber beads which can be dated to this transitional period (Catarsi Dall'Aglio 1976, 263, Fig. 6:16). The second stratum according to R. Scarani yielded materials which correspond to BR 2 – BF 1 (1962; cf. Catarsi Dall'Aglio 1997). Unfortunately the documentation of the survey is missing; thus, it is impossible to specify many information regarding the internal structure of the settlement. Precise location in which two known amber beads of the Tiryns type were found in June 1957 is unknown. At least one of them, however it is impossible to state which one, probably comes from the south-east part of the site's central sector, where three interconnected

211 Damiani/Morico 1996, 319; Morico/Pacciarelli 1996, 307; Morico 1996, 350, nn. 3498–3495, 3270, Tav. 1:D; Miari 2007c, 80, Fig. II:44.

212 For chronology of the violin-bow fibulae in Italy, see e.g. Eles Masi 1986; Lo Schiavo 2010; Savella 2015.

Tab. 8. List of the Bronze Age contexts with amber in Romagna and Emilian Apennines

Name	Commune	Province	Type	Relative chronology	Absolute chronology	Reference
Bismantova – Campo Pianelli, grave XXXI	Castelnuovo Monti	Reggio Emilia	Burial – cremation (urn)	BF 2	1100/1060–1010/1000 BC	Catarsi/Dall'Aglio 1978, Fig. XXI:1; Miari 2007e, II:16–33
Bismantova – Campo Pianelli, context unknown	Castelnuovo Monti	Reggio Emilia	Burial?	BF	1150–910 BC	Catarsi/Dall'Aglio 1978, Fig. XIII:5–6
Borgo Panigale, 1957 campaign, precise context unknown	Bologna	Bologna	Settlement	BR 2–BF 1	1250–1100/1060 BC	Catarsi Dall'Aglio 1976, Fig. 6:16
Monte Battaglia, hoard	Casola Valsenio	Ravenna	Hoard	BR–BF 1/2	1250–1100/1060 BC	Bermond Montanari 1997, Fig. 128:15; Miari 2007d, II:45
Monte Castellaccio, level/phase 2	Imola	Bologna	Hilltop settlement	BM 1B–2	1600–1450 BC	Pacciarelli 1996c, Tav. 1:F; Miari 2007b, II.43
Riccione, precise context unknown	Riccione	Rimini	Settlement	BR	1350/1300–1150 BC	Bermond Montanari et al. 1992, 111
San Giuliano di Toscanella, excavations of 1901, 5th trench	Imola	Bologna	Hilltop settlement	BM 3–BR 1	1450–1260/1200 BC	Morico 1996, Tav. 1:D; Miari 2007c, Fig. II:44

2.2.9. Tuscany, northern Lazio and western Umbria (northern Etruria)

The group of amber finds described in the following section originates from a wide streak of rugged and forested land between the Tyrrhenian Sea to the west, the Apennines arch to the north and east, and the catchment basin of the Fiora River to the south (Fig. 12). In terms of modern administrative borders, the area comprises the entire region of Tuscany, western part of Umbria (to the west of Perugia) and northern fringes of Lazio (Viterbo province). In terms of historical-geography, the southern part of Tuscany and the northernmost Lazio formed the northern Etruria during the Iron Age and the early antiquity (cf. Negroni Catacchio et al. 2016). The reason for including amber finds from such a vast area under a single province is that throughout the Bronze Age communities inhabiting constituent parts of the abovementioned regions, maintained tight relationships among each other. The result is a series of culturally unified groups (facieses) which they formed in various configurations over the 2nd millennium BC. The whole province itself, partly due to its geography, was pressed between large cultural entities dominating Northern and Southern Italy during the Bronze Age – a factor that lent it the character of a contact zone and contributed to a specific appearance of the local material culture.

Amber turned out in Tuscany possibly already in the early stages of the Bronze Age. The epoch began with the process of transformation and atomisation of the local *Bell Beaker* culture into a series of EBA (BA) regional groups, around the second half of the 3rd millennium BC (cf. Cocchi Gennick 1996). Each of the groups is characterised by a slightly different assortment of pottery which provides the main source of information about them. The evolution of ceramic vessels proceeded from the forms and ornaments reminiscent of the *Bell Beaker* pottery, up to the types precursory of the MBA (BM) *Grotta Nuova* facies. Establishing time limits of BA and its internal subphases is not an easy task, given the scarcity of stratified contexts and almost complete absence of absolute dates. Judging by the 14C dates for the final Eneolithic contexts from Tuscany (used as a *terminus post quem*), the Bronze Age started around 23rd–22nd century BC (cf. Cocchi Gennick 1996, 83). Transition from the late *Bell Beaker* culture to the earliest BA emanations apparently was a fluid one, as attested by continuity in occupation of some of the sites and lack of an evident hiatus between the layers attributable to both periods, e.g. Romita di Asciano, levels 7–8 and 9 (cf. Cocchi Gennick 1996, 83–84). Although the data are not uniform or complete, at the current state of knowledge it is possible to propose a more detailed periodisation (cf. Carancini et al. 1996, Fig. 3; Cocchi Gennick/Sarti 2001). The first subphase of the EBA (BA 1A), still saturated with elements connected to the *Bell Beaker* tradition, is traceable through the levels 7–8 of Romita di Asciano and Lastruccia 1, layer N in northern Tuscany, with 14C date from the latter context (3760 ± 80 BP) indicating 2292–2036 cal. BC (68.2 %) and 2458–1975 cal. BC (95.4 %).[215] The second subphase – BA 1B – is not so well defined; however, it represents the gradual disappearance of the *Bell Beaker* features which has been detected in several contexts (e.g. Lastruccia 1, layer H; Lastruccia 3, layer 6B; Madonna del Piano, layer 7). The subsequent subphase BA 2 signifies complete vanishing of Eneolithic elements. It is recognisable at numerous sites in Tuscany: Lastruccia 3, layer 4, Lastruccia 1, layer E, Riparo delle Felci di Candalla, Poggio Fornello and S. Maria in Belverde, layer 3, for which 14C dates allow to estimate the time frames of the period

215 All the 14C dates cited above were taken from Cocchi Gennick/Sarti 2001. However, the publication does not provide laboratory codes for the determinations, nor it gives any information on the dated materials or the dating method. Hence, only the raw uncalibrated BP dates were used in the following.

at ca. 1950–1650 BC.[216] It is followed by the MBA, with the initial subphase (BM 1A; ca. 1650–1600 BC) being a transitional moment between the EBA groups and those falling into the *Grotta Nuova* facies, particularly evident at S. Maria in Belverde, layer 2-III and Riparo del Capriolo, layer IV (lower part).

Beginning from BA 1A, a series of territorial groups can be distinguished in Tuscany. One of them was situated along the Tyrrhenian coast, approximately within the Grosseto province, hence the name *Grosseto* group (alternatively named the *Scoglietto* group) (cf. Cocchi Gennick 1996, 90; Cocchi Gennick/Sarti 2001, 98). The area is known for residential and sepulchral use of caves both in the Eneolithic and the EBA. One of them – Grotta dell'Artofago, located in the Gavorrano commune – yielded a single amber bead (Aranguren/Perazzi 2004). Other finds recovered from the cave seem to indicate attendance from the Eneolithic to the Middle Ages, with fairly consistent representation of BA and early BM, to which the bead has been ascribed (Aranguren/Perazzi 2004, 474). Some fragments of ceramic vessels in fact illustrate typical forms and ornaments recorded at other sities of the *Grosseto* group, thus can be placed in BA 1 stage, perhaps its later part (BA 1B) for the lack of clear *Bell Beaker* manifestations among the vessels (e.g. Aranguren/Perazzi 2004, Fig. 4:3; cf. Cocchi Genick 1996, Fig. 2C:18–19). However, out of all Bronze Age materials from Grotta dell'Artofago, definitely more pottery sherds illustrate the *Grotta Nuova* facies, both in its early – BM 1B (ca. 1600–1550 BC)[217] and late – BM 2A (ca. 1550–1500 BC) stages of development.[218] These ceramic fragments manifest particular similarities with the eponymous *Grotta Nuova* group from Lazio which in the course of BM absorbed the Grosseto area (Cocchi Genick/Sarti 2001, 103). Unfortunately, as the exact context of the bead's discovery is unknown, it is difficult to specify its chronology beyond the period between BA 1B and BM 2A.

In fact many more amber finds from Tuscany were found in non-stratified contexts, or in layers with mixed materials, as it is often the case of caves frequented in different epochs. The *Grotta Nuova* facies, which lasted approximately from phase BM 1B up to the turn of BM 2 and 3, had a significant impact in the area of the Mount Cetona – centre of the *Belverde* group, characterised by a considerable receptive capacity, connected to a remarkable vivacity of relationships with the other territorial groups belonging to the same facies (cf. Cocchi Genick 1995a, 374–378, Tav. 3:D; Cocchi Genick/Zanini 1997, 445–448). Beginning of the facies is recorded in the lower part of the layer 2 of S. Maria in Belverde, and in the upper portion of the IV level of Riparo del Capriolo sites (Cocchi Genick/Sarti 2001, 94). The final caesura of the facies is indicated by 14C dates from several places,[219] all pointing to ca. 1400 BC at the latest (95.4 % probability) – end of the subphase BM 2B, probably illustrating a transitional stage to a new cultural entity (cf. Cocchi Genick/Sarti 2001, 104).

---

216 Riparo delle Felci di Candalla: (3505 ± 60 BP), 1904–1748 cal. BC (68.2 %), 1979–1685 cal. BC (94.2 %); Lastruccia 1, layer E: (3620 ± 80 BP), 2132–1887 cal. BC (68.2 %), 2201–1756 cal. BC (95.4 %); Poggio Fornello: (3440 ± 70 BP), 1877–1665 cal. BC (68.2 %), 1940–1608 cal. BC (94.1 %) (Cocchi Gennick/Sarti 2001, 100).

217 Aranguren/Perazzi 2004, Figs. 3:1, 3 and 4:1; cf. Cocchi Genick 1995a, 377, 383, 394, Figs. 192:418B and 195F:190; Cocchi Genick/Sarti 2001, Fig. 5:5.

218 Aranguren/Perazzi 2004, Figs. 3:2, 5, 7 and 4:2; cf. Cocchi Gennick 1995, 382, 394, Figs. 192:35B, 38A, 38B, 215, 382 and 193:214; Cocchi Genick/Sarti 2001, Fig. 6:1.

219 Riparo Grande: (3210 ± 35 BP), 1506–1438 cal. BC (68.2 %), 1545–1414 cal. BC (91.9 %); Dicomano: (3280 ± 80 BP), 1643–1455 cal. BC (68.2 %), 1752–1403 cal. BC (95.4 %),

The Mount Cetona remained a focal point of settlement during the subsequent BM 3 corresponding to the *Apennine* facies, when it was incorporated into the *Middle-Tyrrhenian* group (Macchiarola 1995, 448–450, Tav. 12:A). Numerous caves on the mountain and in its vicinity (i.a. Grotta di S. Francesco, Grotta della Noce and Grotta dell'Orso di Sarteano) were used by local communities likely for sepulchral and ritual-cultic practices, according to one interpretation (cf. Cocchi Genick/Sarti 2001, 111). In fact, plentiful accumulation of vascular repertoire led to identification of forms typial both for the early (*Grotta Nuova* facies) and advanced (*Apennine* facies) stages of the MBA, representing various, distant as well as adjacent territorial groups of the both units.[220] All three caves listed above yielded a single amber bead each (Calzoni 1954, 45, 47; Grifoni/Cremonesi 1968, Fig. 31:6). Unfortunately, the finds come from a non-stratified – disturbed deposit, therefore cannot be assigned to any specific phase or cultural unit. Furthermore, results of excavations at Mount Cetona, were published by U. Calzoni in a somewhat incomplete manner: other materials associated with amber were not indicated and figures depicting the finds are missing. Regarding Grotta di S. Francesco, all that can be said about the amber bead is that it was discovered under the entrance causeway leading to the central cavity, in a layer containing finds dating to the Bronze Age (Calzoni 1954, 45). As for Grotta della Noce, the amber bead was found along the left wall of the main cavity, in a clod of earth (Calzoni 1954, 47). To make the matters even more difficult, the caves, apart from rich collections of pottery clearly affiliated with the *Grotta Nuova* and *Apennine* facieses, revealed attestations of attendance during BR and BF in form of bronze items.[221] To conclude, one cannot be certain about the chronology of the amber beads from in-cave sites in Tuscany, and the data at hand only allow to date them broadly to a period from BM 1B to BF.

The *Grotta Nuova* facies took a firm root in northern Lazio, as manifested by establishment of the eponymous group known i.a. from the finds from Grotta Misa in the Fiora Valley. Single amber bead was found in the lateral cavity – side chamber extending from the main hall, where the stratigraphy, sealed with massive travertine blocks detached from the ceiling of the cave, was preserved (Rittatore Vonwiller 1951a, 17). Five layers distinguished in a metre-thick deposit yielded mostly mixed pottery fragments attributable to the advanced *Grotta Nuova* and the *Apennine* facieses.[222] Consequently the chronology of the bead can be determined as BM 2–3.

Following the end of the MBA, the discussed area entered the *Subapennine* stage (BR; ca. 1300–1140/1100 BC). However, recognising the poor state of data related to the period, the region appears sparsely inhabited and remote, from the perspective of main events taking place at the time on the Apennine Peninsula (cf. Cocchi Genick/Zanini 1997, 450–451; Bietti Sestieri et al. 2001, 117–119). Knowledge about

(3270 ± 80 BP), 1630–1451 cal. BC (68.2 %), 1746–1396 cal. BC (95.4 %), (3220 ± 80 BP), 1610–1421 cal. BC (68.2 %), 1687–1370 cal. BC (89.4 %) (Cocchi Genick/Sarti 2001, 104).

220 Calzoni 1962; Grifoni/Cremonesi 1968; cf. Cocchi Genick 1995a, 375–377; Macchiarola 1995, 448–450.

221 Calzoni 1954, Figs. 10–11, 14; Grifoni/Cremonesi 1968, Figs. 31:7 and 32:1; cf. Bianco Peroni 1970, 62–64; 1979, 17–18; Carancini/Peroni 1999, Tav. 27–30; Savella 2015, 20–21.

222 Rittatore Vonwiller 1951a, Figs. 11 and 12; cf. Cocchi Genick 1995a, 365, 382, 388, Fig. 192:35B, 492B, 194:205A–B and 196:37A–B, 402; Macchiarola 1995, 209:181A, 222A.

have been excavated. Unfortunately the site has not been comprehensively published, and only piecemeal information are available in the literature.[229]

Attestations of the *Chiusi-Cetona* facies can be found also in the Fiora Valley, e.g. at the Ponte San Pietro Valle site, where stone-cist graves with burials placed inside urns were found, signalising important change in the burial ritual at the beginning of BF.[230] Only tomb 1 contained additional inventory consisting of two amber beads pertaining to the type Tiryns which, together with the biconical urn, served as the primary indicators of the grave's chronology, estimated at late BR 2 and BF 1 (Negroni Catacchio 1999, 251, Fig. 3:34; Negroni Catacchio et al. 1979, 322–324). The dating is largely in line with the recent evaluation of chronology of the graves in southern Etruria based on morphology of the vessels – urns and associated accessories (De Angelis 2010). According to characteristics of the urn from tomb 1, the entire context can be placed in the phase 1 by S. De Angelis, comprising BF 1 and BF 2a.

Also in the Fiora Valley, but on its northern side, presence of amber has been noted in the BF layers of the Scarceta settlement. Data obtained through the excavations shows chronological sequence of the site spanning from the beginning of BM up to the advanced stages of BF, when the site was finally abandoned (cf. Poggiani Keller 1999). At the turn of BR and BF in the inhabited area (sectors D and E) were established two buildings connected with metallurgy, material-processing activities, and residential functions. The building named 'house-laboratory' (it. *casa-laboratorio*) no. 137, located within sector D, yielded three amber artefacts, all datable to the final horizon of the site (Poggiani Keller 1999, Fig. 104:1–3). Defined for the coexistence of traces of daily life and craftsmanship, the 'house-laboratory' structure was built at the beginning of BF on the ruins, or rather within the previous big elliptical structure no. 61 dated to BR. Chronology of the 'house-laboratory' determined on the basis of materials it contained, is supported by radiocarbon date from the charcoal, taken from the level of abandonment that obliterates the floor level: GX-23268 (2835 ± 175 BP); 1221–821 cal. BC (68.2 %), 1462–741 cal. BC (91.7 %). Bronze objects were collected in large quantities from various levels of the 'house-laboratory': fibulae, rings, arrowheads, knives and numerous working tools such as chisels and awls (Poggiani Keller 1999, Figs. 99–103). Especially the fibulae from level 2 are helpful in defining the context's chronology which can be estimated at BR 2 – BF 3a (Poggiani Keller 1999, Fig. 98:1–6; cf. Savella 2015, 6–7, 16, 18–19, 22, 60, Tav. 72). Pottery from the 'house-laboratory' shows numerous analogies with the *Protovillanovan* forms ascertained in other contexts from the region and representing all three phases of the *Chiusi-Cetona* facies, although there is noticeably fewer element pertaining to the latest horizon.[231] Thus the complete lifespan of the structure, simultaneously indicating time brackets for the amber finds, appears to cover a period from BR 2 to BF, perhaps excluding the subphase BF 3b. It is worth adding that among multiple craft activities, most importantly metallurgy, some evidence potentially testifies to glass and amber processing (cf. Poggiani Keller 1999, 123–124). The processing of amber may be inferred from chips, probably post-production waste, although the small amount of raw material might not be enough to see it as a large-scale activity.

---

229 Cf. Negroni Catacchio 1986, 201; Zanini 2000, 205; Bietti Sestieri et al. 2001, 124.

230 Rittatore Vonwiller 1951b, 1961; cf. Negroni Catacchio et al. 1979; Bietti Sestieri 2013a, 641.

231 Poggiani Keller 1999, 86–101, Figs. 73–83; cf. De Angelis 1979; Zanini 1994.

Tab. 9. List of the Bronze Age contexts with amber in Tuscany, northern Lazio and western Umbria (northern Etruria)

Name	Commune	Province	Type	Relative chronology	Absolute chronology	Reference
Belverde di Cetona, Grotta di S. Francesco	Monteriggioni	Siena	In-cave burial	BM 1B–BF	1600–910 BC	Calzoni 1954, 45
Belverde di Cetona, Grotta della Noce	Monteriggioni	Siena	In-cave burial	BM 1B–BF	1600–910 BC	Calzoni 1954, 47
Fossa Nera di Porcari, precise context unknown	Porcari	Lucca	Settlement	BR 2–BF 2	1220–1035/1010 BC	Andreotti/ Zanini 1995–1996, Fig. 2:12–14
Grotta dell'Artofago, precise context unknown	Gavoranno	Grosseto	In-cave dwelling	BA 1B–BM 2A	2100–1500 BC	Aranguren/ Perazzi 2004, Fig. 4:9
Grotta dell'Orso di Sarteano, precise context unknown	Sarteano	Siena	In-cave burial	BM 1B–BF	1600–910 BC	Grifoni/ Cremonesi 1968, Fig. 31:6
Grotta Misa, side chamber	Ischia di Castro	Viterbo	In-cave dwelling	BM 2–3	1550–1300 BC	Rittatore Vonwiller 1951a, 17
Omo Masso, precise context unknown	Marciana	Livorno	Settlement/ Cemetery?	BR 2–BF 3b/ PF IA	1220–910 BC	Negroni Catacchio 2014, Fig. 8
Panicarola, precise context unknown	Castiglione del Lago	Perugia	Burial?	BF	1140/1100–910 BC	Negroni Catacchio 1986, 201
Ponte San Pietro Valle, tomb 1	Ischia di Castro	Viterbo	Burial – cremation (urn)	BR 2–BF 1/2	1220–1060 BC	Negroni Catacchio 1999, Fig. 3:34
Riparo dell'Ambra, layer 2	Camaiore	Lucca	Rock shelter	BF 2–3a	1060–960/950 BC	Cocchi Genick 1986, Fig. 54:1

(Continued)

Tab. 9. Continued

Name	Commune	Province	Type	Relative chronology	Absolute chronology	Reference
Scarceta, sector D – *casa-laboratorio* no. 137	Manciano	Grosseto	Settlement	BR 2–BF 3a	1220–960/950 BC	Poggiani Keller 1999, Fig. 104:1–3

2.2.10. Marche and eastern Umbria

Area to the east of the Apennine arch, comprising eastern part of Umbria (to the east from Perugia) and the entire province of Marche, have also delivered Bronze Age amber finds, although in terms of numbers they are inferior to those previously described from Tuscany, northern Lazio and western Umbria (Fig. 12). Despite being placed on the opposite sides of the Apennines, both parts of Umbria, as well as Marche, share many similarities in terms of cultural development: throughout the 2nd millennium BC they were encompassed by common phenomena, i.e. *Grotta Nuova, Apennine* and *Subapennine* facieses. Still, they reveal local peculiarities which contribute to subdivision of the abovementioned cultural units into regional groups.

In Marche and eastern Umbria, in comparison with Tuscany or Lazio, there is also apparent delay in arrival of the first amber ornaments. Possibly the oldest find is the single bead discovered at the Ripabianca di Filottrano site, on the right side of the Musone stream, between the villages of Casenuove to the north and Filottrano to the south. Unfortunately, no publication or official report from excavations on the site has ever been released, therefore determining context and chronology of the bead is impossible. Some materials found on the site were published by H. Dumitrescu in a general synthesis of the Bronze Age in Marche (Dumitrescu 1932). Sadly, no precise information about form or state of preservation of the bead are given in the publication (Dumitrescu 1932, 268). Some of the published materials from the site, especially bronze daggers which have analogies among objects from the nearby site of S. Paolina di Filottrano (Dumitrescu 1932, Fig. 23:34–36 cf. Bianco Peroni 1994, 139–140, 142, Taf. 76:1374, 77:1412), allow to date it approximately to phases BM 3-BR (1400–1140/1100 BC). The dating possibly indicates chronology of the bead, however should be approached with caution.

During BM 3 the area of Marche hosted the *Middle-Adriatic* group of the *Apennine* facies (Macchiarola 1995, 442–446, Tav. 12:B), which during BR fully embraced the *Subapennine* culture (Damiani 2010, 388–391). The period overlapping both phases marks a considerable increase in settlements and materials, potentially reflecting a growth of the local population, which enable more refined chronological subdivision in correlation with the neighbouring areas (cf. Baldelli et al. 2005, 550–562). Continuous occupation of many sites from BM 3 to BR, as observed in stratigraphies, as well as fluent transition seen on the basis of pottery, suggest undisturbed development of the local communities.

Among the sites exemplifying these characteristics is the settlement Moscosi di Cingoli located at the locality Piano di Fonte Marcosa, in the upper valley of the Musone River. The site, now submerged under the waters of the Cingoli artificial lake, returned abundant ceramic, lithic, metal and bone finds in a stratigraphic succession divided into several phases from BM 3 to BF (Silvestrini/Pignocchi 1999; Sabbatini/Silvestrini 2005). Phase 4 of the settlement, pertaining to BR 2 (1220–1140/1100 BC), yielded three amber beads (Sabbatini/Silvestrini 2005, 652, Fig. 4:16). Together with the beads were uncovered materials testifying to particularly intense and specialised production, craft and domestic activities, demonstrating a marked prosperity of the settlement. Material evidence proves existence of systematic relations between Marche and the *Terramare* culture. Most of the pottery reflects forms typical for the *Subapennine* period (BR) in Marche, e.g. handles with angled-horns appendices (Sabbatini/Silvestrini 2005, Fig. 1:20–21). However, some particular vessels, e.g. biconical vases decorated with bosses, *cuppelle* and grooves combined in complex syntaxes that have analogies at sites such as Santa Rosa di Poviglio – *Villaggio Grande* or the cemetery of Casinalbo, reflect influences from the north.[232] Also, the bronze items include interregional types common in BR and early BF 1, i.e. numerous bronze daggers of the Pertosa, Bertarina, Torre Castelluccia, Glisente and Miradolo types, violin-bow fibulae with twisted or nodule-decorated arch, and pins of the Boccatura del Mincio and Mezzocorona types.[233]

Roughly at the same time first amber objects appeared in eastern Umbria. They come from the hoard of Gualdo Tadino in the Santo Marzio Valley (Perugia province). The site shares similarities among the finds with the Moscosi di Cingoli settlement. The hoard contained a rich collection of bronze artefacts of various kinds, including i.a. two discs of thin gold foil with embossed decoration, violin-bow fibulae in undecorated as well as decorated variants with two nodules or twisted arch, and a large dagger (it. *daghe*) consisting of a flat tang-grip with three holes for rivets and pronounced median rib on the blade.[234] Among elements possibly forming a necklace were five amber beads (Peroni 1963b, Fig. 5:49–53). Numerous parallels with the most advanced manifestations of the *Peschiera* horizon and the most archaic *Protovillanovan* culture, place the Gualdo Tadino hoard at the turn of BR 2 and BF 1 (1220–1060 BC), in line with the second phase of the *Urnfield* culture – HaA1 (cf. Bietti Sestieri 1973, 389; Pare 2008, 82, 87–88).

During the FBA, Umbria and Marche came under influence of the *Protovillanovan* culture. Chronology of the period was built predominantly on the basis of the Pianello di Genga necropolis, where graves could be divided into three phases Pianello 1, 2 and 3, equivalent with BF 1–3 (Peroni 2005; Bianco Peroni et al. 2010). This tripartite division can be supplemented with observations by C. F. E. Pare, who first dated the beginning of the *Picene* culture (and, thus, the end of the Bronze Age) at ca. 950/920 BC, and later, in an attempt to propose a unified chronological system of BF for continental Italy, distinguished a transitional phase BF 3b/PF IA, coeval with HaB2 (Pare 1998, 322–326, Tab. 2; 2008, Fig. 5.3). According to R. Peroni, the cemetery serves as the focal point of the local eponymous *Pianello* group of BF (Peroni 2001). On the other

232 Sabbatini/Silvestrini 2005, 648, Fig. 3; cf. Bernabò Brea et al. 1997d, Fig. 216; Cardarelli 1997, Figs. 396 and 398.

233 Sabbatini/Silvestrini 2005, Fig. 4:1–4, 6, 9–11, 14–15; cf. Carancini 1997; Bianco Peroni 1994; Savella 2015.

234 Peroni 1963b, Figs. 1–2, 3:3–6 and 4:27; cf. Bianco Peroni 1994, 137, n. 1359; Savella 2015, 4, 9–11, 21, Tav. 72.

had, inventories of the graves at the site show some affinity with the *Chiusi-Cetona* facies, hence it is possible that the area of Marche to some extent also became a part of this interregional cultural unit.[235]

About 10 km from the cemetery of Pianello di Genga is situated the settlement of Monte Croce Guardia which delivered evidence of amber presence in Marche during BF (Cardarelli et al. 2017). The site occupies a 'saddle' adjoining the Monte della Croce (650 m a.s.l.) and the Monte della Guardia (666 m a.s.l.) – two peaks of a single elevation, the namesake Monte Croce Guardia in Arcevia commune (Ancona province). Position of the site is characterised by strong natural defensive qualities and extended visibility range which reaches the Adriatic coasts towards north and east, and the Apennine ridge towards south and west. Results of the excavations conducted up to date indicate continuous occupation of the site which lasted from BR to an advanced moment of BF. The recent-most works have uncovered remains of several huts within sectors 1, 2 and 4, indicating several phases of reconstruction, as well as an evidence of metallurgy and other crafts. An amber bead was found inside US4 – filling of the foundation ditch of the western wall of the hut 1, in the sector 1 (Cardarelli et al. 2017, Fig. 23:4). Majority of the materials recovered during 2015–2016 campaigns from the hut 1 – pottery fragments – are indicative of the initial and developed stages of the FBA (BF 1–2; 1140/1100–1035/1010 BC), although there are some fragments possibly dating back to BR (Cardarelli et al. 2017, 349–354, 367, Figs. 21–22). On the basis of the formal characteristics of the archaeological finds, in particular the pottery, it can be confirmed that the settlement of Monte Croce Guardia shares many similarities with parts of Romagna, Umbria, and inner Tuscany, included in the *Chiusi-Cetona* facies (Cardarelli et al. 2017, 371–372).

Tab. 10. List of the Bronze Age contexts with amber in Marche and eastern Umbria

Name	**Commune**	**Province**	**Type**	**Relative chronology**	**Absolute chronology**	**Reference**
Gualdo Tadino, hoard	Gualdo Tadino	Perugia	Hoard	BR 2–BF 1	1220–1060 BC	Peroni 1963b, Fig. 5:49–53
Monte Croce Guardia, US4, sector 1, hut 1	Arcevia	Ancona	Hilltop settlement	BF 1–2	1140/1100–1035/1010 BC	Cardarelli et al. 2017, Fig. 23:4
Moscosi di Cingoli, phase 4	Cingoli	Macerata	Settlement	BR 2	1220–1140/1100 BC	Sabbatini/Silvestrini 2005, 652, Fig. 4:16
Ripabianca di Filottrano, precise context unknown	Filottrano	Ancona	Settlement	BM 3–BR (?)	1400–1140/1100 BC	Dumitrescu 1932, 268

235 Peroni 2005, 721; Bietti Sestieri et al. 2001, 134–152; Bietti Sestieri 2013a, 641.

Fig. 12. Distribution of the Bronze Age sites with amber in Central Italy.

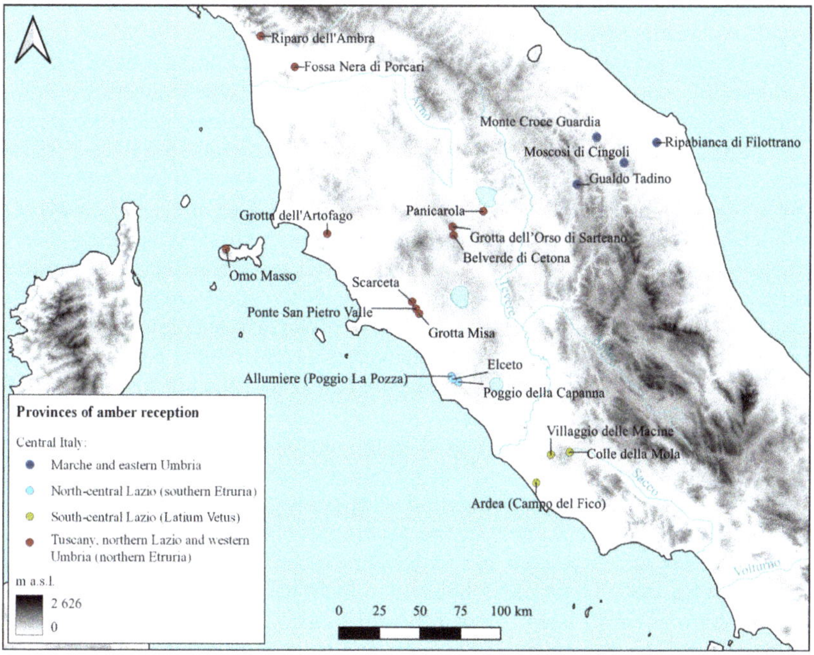

2.2.11. North-central Lazio (southern Etruria)

The area of north-central Lazio, in historical terms often defined as the southern Etruria, during the early and middle stages of the Bronze Age developed in parallel with the neighbouring regions of northern Lazio, Tuscany and Umbria, sharing many traits of material culture. Hence, it is usually treated as a domain of regional group included in larger cultural units, e.g. the *Grotta Nuova* group of the eponymous facies, or the *Middle-Tyrrhenian* group of the *Apennine* facies (cf. Cocchi Genick 1995a; Macchiarola 1995). The trajectories of cultural development followed different paths only in the FBA, when the regions to the north and south of the Fiora River hosted two communities characterised by a different adaptation of the *Protovillanovan* culture (cf. Negroni Catacchio et al. 2016). While the northern parts belonged to the *Chiusi-Cetona* facies discussed earlier, southern Etruria became the centre of the *Tolfa-Allumiere* group of the *Middle-Tyrrhenian* BF facies.[236]

The pottery repertoire of the *Tolfa-Allumiere* group is partly similar to that of the *Chiusi-Cetona* facies but differs in some aspects. In the former the handles with plastic decorations are almost totally absent and the decorations constituted by bundles of parallel grooves were rarely used. Instead, the decorations utilised *cuppelle* and other predominantly engraved, as well as string-impressed motifs. On

236 Di Gennaro/Guidi 2000, 102; Bietti Sestieri 2008, 12–17; 2013, 642–643.

the other hand, there are more elements linking the *Tolfa-Allumiere* group with the *Protovillanovan* facieses of southern Lazio and Campania, visible in bronze metallurgy (Bietti Sestieri 2008, 15).

As the name implies, discussed group developed mainly in the area of Tolfa and Allumiere hills, north-west of Rome; however, it could have reached further north, up to the Fiora River (Fig. 12). Basic chronology of the *Tolfa-Allumiere* group is consisting of two phases: the older named *Tolfa* roughly contemporaneous with BF 1–2, and the younger called *Allumiere* falling into BF 3. Despite generally poor evidence attributable to BF 1, resulting in a difficulty in separating it from BF 2, some scholars propose to distinguish an early horizon from the *Tolfa* phase, on the basis of occupation levels at some settlements.[237] More recently a finer chronology of the sites in the southern Etruria has been elaborated, proposing distinction of four phases within the FBA: I (BF 1–2a), II (BF 2b), III (BF 3a) and IV (BF 3b).[238] This periodisation is essentially aimed at arranging the graves and their inventories within a conventional periodisation, and at the moment cannot be anchored to absolute dates other than those elaborated for the neighbouring regions.

Phases III and IV are best represented on the necropolis of Poggio La Pozza – the largest sepulchral site in the region, located ca. 1.5 km to the north-west of Allumiere town from which the facies and its second stage got its name. The cemetery is sometimes linked with the settlement at the Monte Rovello, at the foot of which it is situated. The site was excavated on numerous occasions over the last two centuries, beginning from 1883, by various scholars and institutions. The place is also known for a collection of amber beads, among them the elongated fusiform ones decorated with transverse grooves which since publishing have been called in archaeological literature of the Allumiere type (cf. Negroni Catacchio 1999).

Four graves at the site yielded amber beads surely pertaining to the Bronze Age. The first one to be excavated was grave 3 (1937), consisting of a pit crudely dug into trachytic rock, containing a hollow, ovoid enclosure sculpted in volcanic grey tuff which, in turn, housed an urn, additional vessels and associated grave goods (Bastianelli 1939). The grave, although not richly furnished, can be broadly assigned to phases III–IV – BF 3, mainly due to not only decorations on the pottery but also the presence of a violin-bow fibula with foliated, rhombus-shaped arch and looped shanks, ascribed to the type A23 by D. Savella.[239]

The remaining three graves were uncovered during the campaign of 1960 (Peroni 1960). Grave 4 in trench D comprised an oval, shaft-like pit with diameter of 1 m and height of 0.75 m. On its south-east margins, it retained a cover, or rather a grave marker, consisting of a small pile of trachytic stones on which rested two large pebbles of *palombina* (type of limestone). At the centre of the grave pit, surrounded by earth mixed with ashes and charcoal from the pyre, was placed a roughly circular tuff enclosure. In its interior was standing a single urn in a mouth-down–bottom-upwards position with a covering bowl below it. Because of such positioning, the bowl was filled with the content of the urn, while the latter was empty. Two other vessels were

237 Di Gennaro 1998, 92; Di Gennaro/Passoni 1998, 130–131; Di Gennaro/Guidi 2000, 100.
238 Barbaro 2013; De Angelis 2010; cf. De Angelis 2006.
239 Bastianelli 1939, Figs. 12:a, c and 13:a; cf. De Angelis 2010, 109, Fig. 37; Savella 2015, 23, Tav. 11:62.

lying crushed behind and below the urn. Amber beads (15 in total) and bronze items were found inside the covering bowl, mixed with ashes and burnt bones (Peroni 1960, 350, Fig. 11:15–17). On the basis of its content, the grave was dated by S. De Angelis to phase III – BF 3a (De Angelis 2010, 109). The other two graves nos. 1 and 2, collectively containing 36 amber beads, were both discovered within trench F. Both were stone-cists tombs housing urns, marked on the surface by piles of stones (*segnacolo*). Bases and walls of the pits were built of trachytic stone plates of *palombina*. The urns were buried under earth mixed with remains of the pyre and broken ceramic vessel, then covered with stones. The urns contained cremated bones halfway up, on top of which rested grave goods, while some other finds and additional vessel were placed around them (Peroni 1960, Fig. 11). Both graves were selected as representative of phase IV – BF 3b by S. De Angelis (De Angelis 2010, 109).

Another testimony of amber in the FBA southern Etruria comes from the site Poggio della Capanna, located on a hill approximately 1 km south from the town of Tolfa. Fragmented amber bead, along with numerous ceramic vessels and several other elements of jewellery mixed with charred bones and ashes, was found accidently during preparations of the ground for a vineyard. According to the landowner, the finds were originally placed inside shaft-like pit enclosed with stones – possibly a grave. Still, it is impossible to determine, whether the finds belonged to a single or more burials (Fugazzola Delpino 1992, 281–291). Morphological and decorative features of the pottery, i.e. tall, biconical vases with everted rims and rounded shoulders, concentric circular *cuppelle* ornament, horizontal and zig-zag bands of parallel grooves, lids with or without an apex made of hemispherical bowls, jugs with S-shaped profiles, by comparison with other contexts allow to date the assemblage to phase IV – BF 3b (Fugazzola Delpino 1992, Figs. 2–9; cf. De Angelis 2010, 109, Fig. 37). Further proof of the late chronology of the finds within the *Tolfa-Allumiere* facies is provided by analogies for the composite bone and ivory *rotella* (pinhead), which north of the Alps appeared in HaB1 contexts of the *Urnfield* culture, while in Italy are known from the last phase (BF 3) of the Pianello di Genga necropolis in Marche (Fugazzola Delpino 1992, 295, Fig. 10:1). Last but not least, the classification seems likely in the light of presence on the site of a serpentine-bow fibula reported by the landowner (crumbled and discarded immediately after the discovery) – a form typical for the ultimate stage of the Bronze Age (Fugazzola Delpino 1992, 286; cf. Savella 2015, Tav. 72).

In southern Etruria amber has also been recorded in a settlement context; unprecise number of beads were discovered at the site of Elceto, on top of the namesake hill immediately south of Allumiere town (Negroni Catacchio et al. 2000, 98). The vast settlement was encircled by imposing system of two stone walls: outer defensive and inner circumscribing the physically distinct summit area interpreted as an 'acropolis'. The majority of finds, predominantly pottery, testify to development of the settlement during phases II and III (BF 2–3a) after which it was abandoned (Belardelli et al. 2007, 18; Barbaro 2013, 245–246). The amber beads were only briefly reported in the literature by the author of excavations O. Toti; hence, their form, number and state of preservation could not be determined (cf. Toti 1972, 1986, Fig. 20).

Tab. 11. List of the Bronze Age contexts with amber in north-central Lazio (southern Etruria)

Name	Commune	Province	Type	Relative chronology	Absolute chronology	Reference
Allumiere – Poggio La Pozza, grave 4, trench D (1960)	Allumiere	Roma	Burial – cremation (urn)	BF 3a	1035/1010–960/950 BC	Peroni 1960, 350, Fig. 11:15–17
Allumiere – Poggio La Pozza, grave 1, trench F (1960)	Allumiere	Roma	Burial – cremation (urn)	BF 3b	960/950–910 BC	Peroni 1960, 351, Fig. 11:6–9
Allumiere – Poggio La Pozza, grave 2, trench F (1960)	Allumiere	Roma	Burial – cremation (urn)	BF 3b	960/950–910 BC	Peroni 1960, 352, Fig. 11:17–24
Allumiere – Poggio La Pozza, grave 3 (1937)	Allumiere	Roma	Burial – cremation (urn)	BF 3	1035/1010–910 BC	Bastianelli 1939, Fig. 13:b, d, e
Elceto, precise context unknown	Allumiere	Roma	Hilltop settlement	BF 2–3a	1060–960/950 BC	Toti 1986, Fig. 20
Poggio della Capanna, grave	Tolfa	Roma	Burial – cremation (urn)	BF 3b	960/950–910 BC	Fugazzola Delpino 1992, Fig. 10:8

2.2.12. South-central Lazio (*Latium Vetus*)

South of the Tiber, in the area of the historical Lazio (*Latium Vetus*), cultural development throughout the Bronze Age was marked by specificity which intensified towards the end of the period (Fig. 12). The EBA is still under-researched here; however, the last 20 years has brought new data which slowly enable to characterise the archaeological facies of BA. Especially useful in this context are the findings from the settlement Lucrezia Romana, where the materials dividable into two horizons reveal analogies with the sites from Tuscany and the southern Etruria (Iaia et al. 2005). In the latter region, first phase of the period (BA 1) continued tradition of the late *Bell Beaker* culture (it. *epicampaniforme*), contributing to the ceramic styles/facieses *Luni-Tre Erici* and *Norchia*, presumably in succession: phase *Tre Erici* – BA 1A; phase *Norchia* – BA

1B and possibly early BA 2.[240] The horizon Lucrezia Romana 1 finds reference to the phase *Tre Erici*, while Lucrezia Romana 2 is roughly parallel with the phase *Norchia* (Angle/Guidi 2007, 148–149). Consecutive stage – BA 2 – again can be characterised through connection with the areas to the north of *Latium Vetus*, where it is possible to distinguish the facies *Belverde-Mezzano* (also known as the *Belverde-Beato Benincasa* facies) continuing up to BM 1A.[241]

During the MBA, the area under discussion functioned as a contact-border zone between the *Grotta Nuova* group/facies and the *Latial* group of the *Protoapennine* facies (Angle/Guidi 2007, 167; cf. Cocchi Genick 2002). The *Protoapennine* facies seemingly had a significant impact on formation of the *Grotta Nuova* facies, as detailed examination of the vessel types shared by sites in south-central Lazio in BM 1 shows clear prevalence of those with a southern origin (Cocchi Genick 1995a, 382). It can be pointed out that the typological repertoire of the *Protoapennine* facies, compared with that of the *Grotta Nuova*, appears in much more substantial numbers already in subphase BM 1A.[242] On the other hand, in the advanced phase of the MBA (BM 2), when consolidation of the *Grotta Nuova* facies and the homonymous group took place, inversion of cultural influences seems to have occurred: pottery for which a northern origin can be assumed began to appear in more substantial quantities in southern Lazio (Cocchi Genick 1995b, 438).

The complex cultural situation of the Bronze Age in Lazio also has chronological implications. Beginnings of the *Protoapennine* facies are generally dated earlier than those of the *Grotta Nuova* facies (see e.g. Carancini et al. 1996, Fig. 3; Pacciarelli 2001, Fig. 38). Basing on stratigraphic observations from Punta Mezzogiorno at the Vivara peninsula, it can be determined that the earliest *Protoapennine* horizon identified with BM 1A is contemporaneous with the first half of LH I, whereas the subsequent BM 1B is coeval with the second half of LH I and first half of LH IIA.[243] Using the synchronisms with the Aegean, the start of BM 1 in southern Lazio can be placed ca. 1700 BC, while its end ca. 1550 BC.[244] More difficult is estimating the boundary between subphases BM 1A and 1B which marks the beginning of the *Grotta Nuova* facies. In light of the 14C dates from Tuscany (see section 2.2.9), the start of BM 1 is unlikely to have occurred before 1650 BC; hence, this date can be used as *terminus post quem* for the *Grotta Nuova* facies. However, if the genesis of the eponymous group and the facies actually began in Lazio, then the formative phase should perhaps be extended

---

240 Cocchi Gennik 1996, 106; cf. Di Gennaro/Pacciarelli 1996; Pacciarelli 2001, 21; Del Fattore 2006.

241 Carancini et al. 1996, 77–78, Fig. 3; Cocchi Gennik 1996, 94–95; Pacciarelli 2001, 21–24; Angle/Guidi 2007, 153.

242 Cocchi Genick 1995b, 438; Pacciarelli 2001, 27–30; cf. Damiani 1995, 402–411.

243 Pacciarelli 2001, 27–30, Fig. 11:A–C; Jung 2013, 236–237, Figs. 1 and 2; 2017, 623–624. It is debatable, whether the emergence of the *Protoapennine* facies can actually be equated with the beginning of the MBA (BM 1), or should it be placed earlier – in final stages of the EBA (BA 2), thus concurrent with MH III (for discussion see Pacciarelli 2001, 27–30, 68, Fig. 38; Angle/Guidi 2007, 154, n. 4; Jung 2017, 624).

244 Cf. Pacciarelli 2001, Fig. 38; Alessandri 2013, Fig. 2; Jung 2017, Fig. 6.

nodules with four ribs (Delpino 1987, Figs. 16 and 18). The fibula is crucial for determining the grave's chronology; through morphologic and stylistic similarity to the fibulae from the hoard of Coste del Marano, it can be assigned to BF 2, or BF 1–2 in the broadest sense.[249]

To the same timespan is attributed the settlement of Colle della Mola, located on the eponymous hill (640 m a.s.l.), within the Alban Hills (Conati Barbaro 2009; cf. Alessandri 2013, 34–44). Excavations at the site led to discovery of two sunken structures – huts defined as houses A and B in the 3rd layer corresponding with an advanced stage of the Recent Bronze Age (BR 2). In the superimposed 1st and 2nd layers were found elements attributable to the final phase of the Bronze Age (BF 1–2), such as the pottery fragment decorated with a false cord and some bowls with a falling edge, thus testifying to the continuity of the settlement in the *Protovillanovan* period (Conati Barbaro 2009, 103). Among the latter layers was found an amber bead (of an undisclosed shape).

Interestingly amber disappeared from Lazio in the following phase – BF 3/Lazio I, even though it is well attested in the neighbouring southern Etruria (see section 2.2.11). However, this apparent break in import of discussed material must have been short, as amber is again documented in the context of the EIA (Lazio II) sites, such as Osteria dell'Osa (Negroni Catacchio 1999, 250; cf. Bietti Sestieri 1992).

Tab. 12. List of the Bronze Age contexts with amber in south-central Lazio (Latium Vetus)

Name	Commune	Province	Type	Relative chronology	Absolute chronology	Reference
Ardea – Campo del Fico, pit-grave	Ardea	Roma	Burial – cremation (urn)	BF 1–2	1175/1100–1050/1010 BC	Delpino 1987, Fig. 18
Colle della Mola, layers 1–2	Rocca Priora	Roma	Settlement	BF 1–2	1175/1100–1050/1010 BC	Conati Barbaro 2009, 103
Villaggio delle Macine, MBA layers	Castel Gandolfo	Roma	*Palafitte* settlement	BM 1–2	1700–1400 BC	Bellintani et al. 2007, Fig. 2

2.2.13. Molise

In comparison with other regions, Molise may appear to have been on the side-track of amber circulation during the Bronze Age (Fig. 13). To date, only a single archaeological site in the area yielded amber artefacts datable to the period – the settlement of Rocca di Oratino on the slopes of the namesake rocky elevation which dominates

249 Cf. Carancini et al. 1996, Fig. 3; Carancini/Peroni 1999, Tav. 30, 32; Pare 2008, 82; Savella 2015, 58–59, Tav. 33:234–235.

over the right bank of the Biferno River. Apart from the Late Medieval remains, it was possible to detect at the site peculiar stone structures accompanied by a conspicuous assemblage of the *Subapennine* pottery (Cazzella et al. 2007). Thanks to comparative analysis of the ceramics, taking into consideration a number of sites from the neighbouring regions, the settlement of Rocca has been dated to the advanced stage of the *Subapennine* culture, coeval with Southern Italian BR 2 (ca. 1220/1210–1130 BC; cf. Cazzella/Ruggini 2010, 553; Jung 2017, 634–637, Fig. 6).[250]

Stratigraphy of the site allows for distinguishing three phases of occupation (Copat/Danesi 2010). During the first two, a wall (perhaps defensive) was built in the eastern part, while in the inner sector, sunken structures lined with stones were constructed. Their function remains unclear; however, these were not residential buildings. Roughly at the same time, a stone cairn – tumulus was heaped up in the vicinity. During the third phase, character of the site's attendance saw a fairly significant change, both for the type of structures found and for the quality of activities identified. The mound was completely obliterated and perhaps partly disassembled, while at least the upper portions of the drystone structures with a quadrangular plan were still visible. In this phase are observable numerous layers characterised by the presence of cooking plates connected with outdoor activities. It was possible to reconstruct domestic activities practised here, such as the preparation and collective consumption of food (Recchia et al. 2008).

At least one out of three amber beads from the site was found within the stone structure 3 dated to the 1st phase of the settlement (Copat/Danesi 2010, Fig. 4:8). The latter was revealed in the central part of the site, consisting of remains of a semi-ellipsoidal stone wall. Another amber bead was found in superficial layers, still pertaining to BR 2, as evidenced by numerous pottery fragments (Copat/Danesi 2010, Fig. 4:9). Finding context of the third, heavily fragmented bead is not specified in the literature (Copat/Danesi 2010, Fig. 4:13).

Tab. 13. List of the Bronze Age contexts with amber in Molise

Name	**Commune**	**Province**	**Type**	**Relative chronology**	**Absolute chronology**	**Reference**
Rocca di Oratino, phases I–III, superficial layers and sunken structures	Oratino	Campobasso	Settlement	BR 2	1220/1210–1130 BC	Copat/Danesi 2010, Fig. 4:8–9, 13

250 Conventional chronology of the site is supported by three recently obtained 14C dates: Poz-117208 (3010 ± 30 BP), 1295–1209 cal. BC (64.6 %), 1310–1157 cal. BC (78.9 %); Poz-117209 (2980 ± 30 BP), 1260–1130 cal. BC (68.2 %), 1297–1112 cal. BC (94.3 %); Poz-125901 (3005 ± 30 BP), 1289–1196 cal. BC (66.3 %), 1306–1127 cal. BC (86.3 %) (analysis performed within the project *Circum-Adriatic branch of the amber route in the Bronze Age* NCN no. 2015/17/N/HS3/00052 conducted by the author).

2.2.14. Northern Apulia and Basilicata (Gargano peninsula and Tavoliere plain)

Abundance of prehistoric amber finds in northern part of Apulia (Fig. 13), particularly on the Gargano peninsula, was noted by archaeologists quite early on (Negroni Catacchio 1975a). Since then, it has been frequently raised in the literature as a proof of the region's participation in the long-distance exchange during the Bronze Age (e.g. Radina/Recchia 2006; Cazzella/Recchia 2009, 2018). Conspicuous concentration of amber finds on the Gargano peninsula and the Tavoliere plain (Foggia province) corresponds with the spatial limits of the culturally unified communities that dwelled in the area for several consecutive stages of the Bronze Age, beginning from BA 2B/ BM 1 – early *Protoapennine* culture/phase (ca. 1900–1550 BC), up to the end of BF (ca. 950 BC), and even afterwards during the EIA (PF). The Ofanto River is often referred to as not only an important communication trail but also a border with culturally slightly different communities of central and southern Apulia, thus is sometimes used as the denominator for the local northern Apulian groups, i.e. the *Ofanto group* of the *Protoapennine* and *Apennine* cultures (cf. Poggiani Keller 1995). Geographical proximity and cultural affinity allow to add to this group several contexts with amber from northern Basilicata, located on the southern edge of the Tavoliere plain, in the catchment basin of the Ofanto River (Potenza province).

Bronze Age chronology of the area north of the Ofanto can be presented in synchrony with the rest of Southern Italy (e.g. using the categorisation of pottery by Cocchi Gennick et al. 1995). However, each phase within the system refers to a specific regional unit, with duration adjusted to the dynamics of local cultural development. A reference point in chronological considerations devoted to northern Apulia is the fortified settlement of Coppa Nevigata, occupied for the majority of the Bronze Age with several episodes of destruction and reconstruction. Habitation at the site began after 1900 BC – a significant caesura in chronology of Southern Italian Bronze Age, roughly coinciding with the Vesuvius's Avellino eruption.[251] The following period, defined as BA 2B, illustrates demise of the *Palma Campania* facies in northern Apulia and a slow birth of the *Protoapennine* culture (Cazzella et al. 2017, 434). However, at this early stage the settlement at Coppa Nevigata must have been unstable, as it lacked defensive structures and larger features built of stone; the latter were eventually constructed during the 17th century BC (Calderoni et al. 2012, 462). Raise of fortifications at Coppa Nevigata corresponds with the development of the *Protoapennine* culture (BM 1) which by ca. 1550 BC entered its matured stage (cf. Recchia 2010, 108–109). There is still some uncertainty as to the synchronisation of the start of this phase; it definitely began no later than LH I (ca. 1700 BC); however, one cannot rule out its inception already during MH III (as previously suggested by Peroni 1994, 205, 207, Fig. 77; Carancini et al. 1996, Fig. 4). According to a different view, the earliest manifestations of the *Protoapennine* culture, e.g. the finds from Vivara (see section 2.2.12), should still be placed in the finale stages of the EBA – BA 2B (Pacciarelli 2001, 27–30, 68, Fig. 38). The question is difficult to solve, because in Italy there are no closed contexts containing exclusively MH pottery, i.e. without accompanying early *Mycenaean* – LH I pottery (Jung 2013, 237; 2017, 624). The end of the *Protoapennine* culture is convergent

251 Calderoni et al. 2012, 459–461; cf. Sevink et al. 2011; Orsi et al. 2013.

with destruction of the first Coppa Nevigata settlement, presumably due to a violent attack that occurred between 1500 and 1450 BC – event also signifying the turn of BM 2A and 2B (Calderoni et al. 2012, 462; Jung 2017, 626). Defensive walls around the settlement were rebuilt and subsequently altered during BM 2B and 3 – phases, respectively, corresponding to the early and advanced stages of the *Apennine* culture.[252] The passage between the *Apennine* and the *Subapennine* cultures/periods is set around 1300 BC; however, the dates from Coppa Nevigata pertaining to the latter period are scarce and not fully compatible due to post-depositional alterations (Calderoni et al. 2012, 463; Cazzella et al. 2017, 435). The time span of the *Subapennine* period in northern Apulia has to be determined by comparison with neighbouring regions, where it has been placed between 1300 and 1130 BC (Cazzella/Ruggini 2010; Jung 2017, Fig. 6). The following FBA with its subphases (BF 1, 2 and 3) can be estimated to have lasted in northern Apulia from ca. 1130 to 950 BC.[253]

The oldest amber finds in the area are almost exclusively known from rock-embedded tombs (i.e. the so-called 'hypogea', or in-cave burial grounds) which constitute one of the most characteristic features in the archaeological panorama of south-east Italy. The group of sites comprises the funerary chamber of the Manaccora cave, 'Hypogeum of Bronzes' and 'Hypogeum of Ivories', both at Madonna di Loreto, and tomb 3 at Toppo Daguzzo. To this group one could add slightly younger hypogeum no. 743 at Lavello. Appart from similar construction and time of use, there are certain aspects of funeral ritual linking these graves, e.g. the custom of furnishing male burials with bronze swords related to the Pertosa type (cf. Recchia 1999).

The Manaccora cave is situated on northern side of the Gargano peninsula by the homonymous bay, with the entrance opening in a steep slope just below the plateau of Punta Manaccora. The latter is known for remains of a settlement related to various stages of the Bronze Age, in part contemporary with the ancient attendance of the cave (cf. Recchia 1993). Internal part of the cave is formed by a large, sub-rectangular cavity, with the entrance facing north. The cave consists of different sectors, named by the excavators in the 1930s. At first, the cave was used for residential purposes; habitation in the hallway (sectors 'Triangle' and 'Annex') started in the late *Protoapennine* period (BM 2A; ca. 1550–1470 BC); however, some pieces of pottery testify to even earlier attendance. To some extent, occupation also included areas of the funerary tunnel-like recess (*Grotticella Funeraria*) and the corridor (*Abris sous roche*), although these parts became more prominent later, in the early *Apennine* period (BM 2B; after ca. 1470 BC). At that moment their function changed from residential to sepulchral, by serving as a large grave chamber, separated with a stone wall from the rest of the cave that simultaneously was had been occupied for the last time. However, it is also probable that this process started already after the abandonment of the 'Triangle' and 'Annex' sectors (see detailed explanation in Recchia 1993, 391–393).[254]

252 Recchia 2010, 109; Cazella et al. 2017, 434; cf. Cazella/Recchia 2012, 2013.

253 Peroni 1967, 115–121; De Juliis 1979; Cazzella/Ruggini 2010; Bietti Sestieri 2013a, 648–649.

254 Previously formulated hypotheses regarding a possible beginning of the funeral depositions in the *Grotticella Funeraria* during the *Protoapennine* period were based largely on the presence among the grave goods of a carinated bowl with a raised handle that should, however, be placed in the 'Annex' layers instead (Recchia 1993, 387).

summit (Cipolloni Sampò 1986). It consists of a rectangular chamber preceded by an uncovered, sloping access corridor (*dromos*) with SE-NW orientation. Its terminal section continues underground, ending with an access to the chamber through an oval opening. Inside the funeral chamber, the excavations revealed a first level of skeletal remains in a poor state of preservation and without grave goods. Once the depositions and the underlying soil layer were removed, another level of burials was identified. The room had been divided by a wooden fence (reconstructed on the basis of four post-holes) into two areas. Fragments of vases and a hearth were found in the part near the access to the corridor, while in the inner part were recorded 11 burials (6 men, 4 women, 1 child) with associated grave goods. The burial no. 3, which contained a single amber bead as a part of inventory (Cipolloni Sampò 1986, Fig. 9:5), belonged to a female, placed on the left side in a curled-up position, facing burial no. 4, which partly overlapped it. Originally, the burials in the inner part of the tomb 3 were dated between the final *Protoapennine* and the early *Apennine* periods (approximately BM 2B–3A, ca. 1470–1350 BC), using as the basis absence of pottery typical for the developed *Apennine* culture on one hand, and the presence of advanced versions of BM armaments, chiefly the Pertosa-type swords and the Capurso-type daggers, on the other.[260] More recently, it has been proposed to narrow down the chronology to BM 2B stage (however without excluding the possibility of continuation into BM 3A), assuming a duration of at least 50 years for the deposition of bodies (Recchia 1999, 27, Tab. 1; cf. Cazzella et al. 2017, 434).

Also in north-west Basilicata is situated the hypogeum no. 743 of Lavello. Since the site has not been fully published, and only fragmentary information are available in the literature,[261] it is difficult to completely characterise the finds from the tomb. This problem is particularly perceptible, when trying to determine chronology of amber finds discovered inside the tomb. On top of that, a part of the grave was removed by a local torrent, whose riverbed passes nearby. Moreover, illegal digging, probably in the early 1980s, disturbed interior of the tomb. From published information, it can be inferred that it was a hypogeum-type tomb comprising a corridor (*dromos*) and an ellipsoidal main chamber with lateral niches, including more than 50 burials accompanied by a kit of ceramic vessels, bronze jewellery and armaments, or amber and glass-paste beads. Out of numerous (it is difficult to say how many) amber beads, only two are depicted in the literature (Cipolloni Sampò 1999, Fig. 10:10–11; Nava/Salerno 2007, Fig. 6:10–11). From the corridor comes a collection of ceramic vessels decorated in the typical *Apennine* style (Cipolloni Sampò 1999, 113, Fig. 9), whereas the chamber yielded mainly bronze artefacts spanning the *Apennine* – BM 2B–3 (Pertosa-type sword), *Subapennine* – BR (violin-bow fibula) and *Protovillanovan* periods – BF 1–2 (simple fibulae with decorated brackets) (Cipolloni Sampò 1999, Fig. 10:1–3; cf. Recchia 1999, Tab. 1). Therefore, it can be hypothesised that the tomb was utilised for a long timespan; however, it is impossible to determine, whether amber was deposited there continuously, or is a one-time addition.

During the advanced stage of the *Apennine* culture (BM 3; ca. 1420–1300 BC), amber appeared for the first time in the settlement of Coppa Nevigata. In the late 15th–early

260 Cipolloni Sampò 1986, 29–31, Figs. 8:7 and 9:1–3; cf. Bianco Peroni 1970, 23–25; 1994, 74–77; De Marinis/Salzani 2005, 410.

261 Cf. Cipolloni Sampò 1989, 1991–1992, 1998, 1999.

14th century BC, the old *Protoapennine* walls were rebuilt and reinforced with square towers.[262] The old gate was shut, and a new entrance was built at another location to the northeast, protected by two foreparts. In front of the walls was dug a moat, about 12 m wide and 4 m deep, covered with stones on the bank towards the settlement. A single amber bead dated to this period was found in the area occupied by western tower of the primary (*Protoapennine*) gate in the north-west section of the walls (Cazzella/Moscoloni 1999, 144, Fig. 11:3).

Further attestations of amber at Coppa Nevigata are known from LBA layers. At the beginning of the *Subapennine* period (early 13th century BC), the walls built about a century before (during the advanced *Apennine* period) were partly dismantled, and defence of the settlement was entrusted solely to the moat (Cazzella et al. 2010, 172–173; Cazzella et al. 2016, 56–57). The only point of access to the inner settlement, so far recognised in the excavation area, is superimposed exactly on the previous *Apennine* gate located in the north-east section of the site. Through the gate led a cobbled road paved on a ramp raised above the surroundings, starting from the edge of the moat, and then tapering towards the built-up area, following the previous *Apennine* path. It seems that at the beginning of the *Subapennine* period, elongated hillocks made of earth mixed with crushed yellow limestone were accumulated on both sides of the access road, partly covering the dismantled walls. In the empty space passed by the access road (square G3A) was found a single amber bead (unpublished – personal communication with G. Recchia).[263]

During the early *Subapennine* period, the internal layout of the settlement underwent significant modifications. Along the southern slopes of the hillocks flanking the entrance were constructed roughly rectangular buildings of stone and wood (cf. Cazzella/Recchia 2012, 296–298, 301). Further south, other remains connected to the period came to light, i.e. oven, hearths and cylindrical pits interpreted as silos for storing cereals (Cazzella/Recchia 2012, 305–307). The most characteristic feature in this area is the widespread presence of daub which suggest existence of elevated structures with walls made with wooden frames covered with clay (Cazzella et al. 2014, 192–193; 2016, 58–59). Another amber bead pertaining to discussed phase was found among remains of a hut-like dwelling (square H3Q) (unpublished – personal communication with G. Recchia).[264]

---

262 Cf. Cazzella et al. 2010, 172; 2018; Cazzella/Recchia 2012, 280–292, Figs. 59–62; 2013; Savino 2014.

263 Attribution of the bead and associated materials to BR is confirmed by two newly obtained 14C dates: Poz-117206 (2970 ± 35 BP), 1257–1127 cal. BC (68.2 %), 1286–1054 cal. BC (95.4 %); Poz-117207 (3025 ± 30 BP), 1373–1221 cal. BC (68.2 %), 1395–1132 cal. BC (95.4 %) (analysis performed within the project *Circum-Adriatic branch of the amber route in the Bronze Age* NCN no. 2015/17/N/HS3/00052 conducted by the author).

264 Attribution of the bead and associated materials to BR is confirmed by two newly obtained 14C dates: Poz-117204 (2975 ± 35 BP), 1259–1128 cal. BC (68.2 %), 1291–1056 cal. BC (94.6 %); Poz-117205 (2945 ± 35 BP), 1218–1110 cal. BC (67.2 %), 1261–1039 cal. BC (95.4 %) (analysis performed within the project *Circum-Adriatic branch of the amber route in the Bronze Age* NCN no. 2015/17/N/HS3/00052 conducted by the author).

With the end of the 12th century BC, situation with regard to the defensive structures and access to the Coppa Nevigata settlement seemingly did not change. In the northern part of the site, however, there are no attested *in situ* features attributable to BF, except for those constituting the filling of older structures, in particular the sub-cylindrical pits dug already at the end of the *Subapennine* period (cf. Onnis 2014, 2017). Ceramic materials pertaining to BF and PF were also found in the filling of the *Subapennine* ditch (the secondary moat) in the north-east part of the site, behind the old *Apennine* walls (Cazzella/Recchia 2012, 308). A single amber bead, potentially belonging to this period, was found during Mosso's excavations in 1909 in superficial/upper layers; however, the exact context is not known. Judging by available figures (the object is considered lost), it was a bead reflecting the Tiryns type, hence could be dated irrespectively of its context (Belardelli 2004, 93, Fig. 31:219). According to the opinion of various authors, the bead can be ascribed to the late *Subapennine* phase of the settlement (BR 2; ca. 1220–1130 BC); however, it cannot be ruled out that it is slightly younger (BF; ca. 1130–950 BC) (cf. Cazzella/Moscoloni 1995, 136–137; Cazzella/Recchia 2016, 362).

There is another Apulian Tiryns-type bead with similarly unsolved problems regarding the discovery context and chronology, with the findspot localised broadly in the province of Capitanata (Negroni Catacchio 1999, Fig. 4:32). The object is stored in the Museum of Reggio Emilia, where it found its way through exchange of materials arranged by G. Chierci. There are two hypotheses forwarded by N. Negroni Catacchio regarding its provenance (cf. Negroni Catacchio 1975a, 316–317). First assumes that the object is in fact the lost Tiryns-type bead discovered during A. Mosso's excavations at Coppa Nevigata in 1909 (the bead is not identical, but nonetheless closely resembles the same variant; cf. Negroni Catacchio 1999, Fig. 4:31). The other hypothesis associates origins of the bead with close contacts between opposite shores of the Adriatic in the FBA, resulting in standardised garments. It posits that the bead comes from the territory of Salapia, or even the necropolis of Salapia itself, given the grave inventories (especially grave no. 14) are similar to those of Vrsi and Privlaka on Dalmatian coast, where the Tiryns-type beads have been found (see section 2.3.2). Still, the history of the Capitanata bead remains unsolved.

Amber finds that can be firmly placed in evolved and late stages of the FBA come from the cemetery of Monte Saraceno, located in the south-western part of the Gargano peninsula. The site is situated on the eponymous promontory overlooking the gulf of Manfredonia, formed by about 2 km long ridge, descending from Monte Sant'Angelo and ending at Capo Grugno (Nava 2001, 276–281, Figs. 1 and 2). Clastic structure of the rock enabled construction of sunken cist tombs with trapezoid (truncated pyramidal) shape and sub-circular, elliptical or rectangular entrances. The vast area used for sepulchral practices seems organised in a series of burial nuclei, consisting of numerous tightly placed tombs, distributed both on the ridge and on the northern slopes of the promontory. The average dimensions of the structures do not exceed 100–120 cm in depth and 90–100 cm in width, while their orientation is mainly E–W. These funeral structures seem to have been closed and covered by heaps of big stones and with intermediate spaces stuffed with smaller stones, although some of the tombs are believed to have been sealed with large slabs, found lying in the secondary position during early stages of the research. The practised funeral rite was exclusively inhumation of the deceased in strongly flexed position on the right or the left side, or in a 'sitting' position. The multiple inhumations of adult individuals of both sexes, as well as adolescents and infants were buried in consecutive stages. The removal or sweeping the bones with intention of 'making space' for successive burials left the skeletons intermingled and without anatomical order. The laboratory reconstruction of bone remains indicates a minimum number of 4/5 and a maximum of 14/15 individuals per funeral structure (Nava 2001, 275–276).

It is difficult to attribute a specific set of materials to each deposition within every tomb. The situation results, as already mentioned, from reutilisation of the structures, as well as disturbances caused by roots and rodents. However, following a re-examination of some of the more significant objects, it was possible to create a relative chronological sequence of the necropolis. Inside each tomb were recorded materials belonging to different chronological phases; thus, distinguished stages of the cemetery partly overlap (Nava 2001, 288–289). Comparative analysis of the inventories, both with the trans-Adriatic as well as Central and Southern Italy, helped to identify three major phases encoded A, B and C, each subdivided into subphases denominated with Arabic numerals 1, 2, 3, etc. Phase A, of relevance to the timeframe of this work, began around BF 2–3 and ended around PF IA (ca. 1100–910 BC), although some graves indicate continuous use during phase B that is until PF IIB (cf. Nava/Preite 1995; Preite 2001).[265]

Graves dated to phase A that contained amber are (in chronological order): 14/IV, 43c/IV, 84/IV, 86/IV and 111/IV (Nava/Preite 1995, 104, 112, Tav. XIII:28, XIV:49, 54; Preite 2001, 311, 315–316, 319). For determination of their chronology, the most telling are simple fibulae: in the undecorated variant – type 35 according to the typology of F. Lo Schiavo, dated to BF 2–3,[266] with twisted central part of the arch classified as type 26, variant A, dated to BF 2–3,[267] and with two nodules and decorated arch, classified as type 20, variety A, dated to BF 2.[268] Furthermore, one should also mention

---

265 Dating of individual phases of the FBA in Apulia is still not strictly determined. Recent studies allowed to narrow the beginning of the period; it appears that BR 2 in Apulia lasted longer than in areas to the north – up to the last decades of the 12th century BC, and partly overlapped the timespan of BF 1 defined for the regions involved in the *Protovillanova* culture, as in Apulia typical pottery of BF 1 (i.e. vessel with oblique grooves – the so-called 'turban' ornament) often appear together with BR 2 – *Subapennine* ceramics in the same layers (cf. Cazzella/Ruggini 2010; Jung 2017; Onnis 2014, 2017). The key to understanding the dynamics and synchronicisms of BR-BF development in Apulia are the stratigraphies of trenches IX and X at Roca Vecchia in Salento (cf. Pagliara et al. 2007, 2008). The first stage of BF can be placed, thanks to the late *Mycenaean* pottery occurring on the sites, within advanced but not final part of LH IIIC (ca. 1130–1100 BC), thus making it shorter in comparison with Italian regions to the north. On the other hand, phase BF 2 is much better represented at Apulian sites both materially and stratigraphically, and its duration can be synchornised with the final LH IIIC and the *Submycenaean* period (ca. 1100–1030 BC). The absolute chronology proposed by M. L. Nava and A. Preite with high dating of BF 2 covering also the 12th century BC is difficult to maintain in the light of recent excavation at Apulian settlements and 14C chronologies obtained for those sites (e.g. Coppa Nevigata and Roca Vecchia); however, one can agree that BF 1 in the south was a brief and 'transitional' period, which was shortly after followed by BF 2.

266 Nava/Preite 1995, Tav. XIV:4; Lo Schiavo 2010, 121–126, nn. 229–229B.

267 Nava/Preite 1995, Tav. XIII:1; Lo Schiavo 2010, 108–110, n. 86.

268 Nava/Preite 1995, 105, Tav. XIII:2; Lo Schiavo 2010, 102–103, n. 69.

found in corresponding levels,[281] the evidence currently at hand allows to place phase 3 of Roca Vecchia in BM 3 (ca. 1420–1300 BC). Analysis of the remains of the *Apennine* fortifications has revealed a crescent-shaped wall with a monumental main entrance and at least five postern gates.[282] For what concerns the inhabited part of BM 3 settlement at Roca Vecchia, there are structures just behind the wall, probably referable to contingency of the siege; however, the only dwelling pertaining to phase 3 is that identified at the bottom of the stratigraphic sequence of SAS X also destroyed around the same time (Scarano 2017, 526–528; cf. Pagliara et al. 2007).

During the LBA (ca. 1300–1130 BC), the fortification wall was rebuilt, using ashlar limestone blocks instead of the unworked slabs used in the MBA. The main gate was reconstructed in the same place, but it was approximately 2 m higher than the original MBA gate (Scarano/Maggiulli 2014, 3). Quite abundant is also the evidence concerning residential structures constructed after the BM 3 destructive event. The succession of layers from BM 3 to BF 2 has been revealed in the trench/sector SAS X, near the inner front of the fortification walls, showing stratigraphical sequence constituted by five building events (Pagliara et al. 2007). Trench SAS IX, also opened just behind the walls but further to the north, brought to light remains of three, superimposed, rectangular huts, occupied between BR 1–2 and BF 2 (Pagliara et al. 2008). Although in the stratigraphic sequence reconstruction of both the wall and the dwellings follows immediately the BM3 destruction horizon, the typo-chronology of related pottery, both the local impasto as well as the *Mycenaean* vessels, suggests a hiatus of several decades before resumption of occupation at the turn of BR 1 and 2 (ca. 1220 BC; cf. Jung 2017, 634). The problem is evident in the case of SAS X, where phase I attributable to BM 3 house destroyed in the late *Apennine* 'siege' is followed by phase 2 that in the broadest sense can be dated to BR 1–2. For the subsequent phase III in the SAS X sequence, consisting of residues of a building with remains of bronze artefacts, local handmade and Aegean-type pottery dated to LH IIIC Early/Middle – BR 2, there are two 14C dates which largely indicate the 12th–11th century BC.[283]

During the FBA (ca. 1130–950 BC), the fortification wall was restructured for the last time using a huge quantity of timber. The main gate remained almost the same as in the Recent Bronze Age, but was completely framed with wood, probably in order to support new covering structures. The fire that destroyed the last Bronze Age settlement of Roca is dated to an advanced phase of BF 2, and it is associated with LH IIIC late and *Submycenaean* Aegean-type pottery (Jung 2006, Pl. 12.1–7; Guglielmino 2007, 96).[284] After the BF 2 destruction (probably sometime in the 11th century BC), the fortification wall has never been rebuilt again and the inhabited area was significantly reduced. Among the best preserved residential structures of the FBA at Roca are the so-called 'Cult Center' dated to BR 2, and slightly younger 'Hut-temple' (BF 1–2), both offering outstanding assemblages of ceramic and metal, and also amber artefacts (Pagliara et al. 2008; Scarano/Maggiulli 2014).

281 Guglielmino 2007, 90, Fig. 2; Jung 2006, 94–101, Pl. 3, 1–8; Scarano 2012, 74.

282 Cf. Scarano 2011, 2012, 2017.

283 LTL1864A (2926 ± 55 BP), 1212–1047 cal. BC (68.2 %), 1281–974 cal. BC (94.3 %); LTL1866A (2947 ± 55 BP), 1224–1056 cal. BC (68.2 %), 1303–1000 cal. BC (93.4 %) (Pagliara et al. 2007, 356–357, Fig. 21cf. Scarano 2011, Tab. 1; Scarano/Maggiulli 2014, 5, n. 6).

284 Available 14 date from the phase V in SAS X (LTL1872A: 2876 ± 60 BP), signifying the ultimate Bronze Age horizon of the settlement, indicates a time span of 1127–974 cal. BC (62.7 %) and 1227–903 cal. BC (95.4 %) (Pagliara et al. 2007, 356–357, Fig. 21).

Some of the earliest Bronze Age amber finds in the region come from the La Chianca dolmen, south of Bisceglie (Cataldo 1995). The tomb had elongated rectangular form, created by placing 10 stone slabs in upright position with the entrance situated in one of the longer sides, and hosted multiple burials with accompanying grave goods. Disturbances noted inside the grave, i.e. lack of anatomical order of deposited skeletons, breakage and scattering of grave goods, and partial filling with topsoil, presumably testify to its looting in the past. It seems that the repeated burials in the tomb caused displacement and damage to the previous depositions. At least two phases of burials could be observed through stratigraphy and typo-chronology of the materials, particularly in the eastern part of the dolmen, where five amber beads were discovered (Cataldo 1995, 269, Fig. XXXVII:8–9, 11–13). There, three skeletons at 80 cm depth were overlying a layer of barren earth, under which was revealed another layer containing five skulls all detached from other bones. The exact position of amber beads and other finds coming from this part of the tomb (clay fusaiola, perforated disc in bone, and the published ceramic vessels) is difficult to determine, but they were all found during the excavations conducted by M. Gervasio in 1910. Thus they probably pertain to the lower level of burials (1st phase?), as the upper one was explored a year earlier without mentioning amber. L. Cataldo, using analogies from stratified contexts in Southern Italy included in the typology by Cocchi Gennick et al. (1995) was able to date few published vessels from the grave (Cataldo 1995). Still the associated pottery does not form chronologically homogeneous group. The oldest vessel pertains to phase 1 of the *Protoapennine* culture – BM 1,[285] while more numerous are the vessels datable to the late *Protoapennine* and the *Apennine* cultures – phases BM 2 and BM 3.[286] On the basis of ceramic material, L. Cataldo proposed to distinguish two phases in the dolmen's use: one referring to the *Protoapennine* culture and the second to the early *Apennine* (Cataldo 1995, 269). Still, it is impossible to decide whether this distinction corresponds with the separation of layers in the eastern part of the tomb, and to which of those should be assigned amber beads with the remaining ornaments. Consequently, using prevalent analogies for accompanying materials in other sepulchral contexts in Apulia, the beads can be broadly dated BM 2–3A, with a slight probability of even earlier chronology – BM 1.

Further south along the coast, similarly dated amber finds were discovered at the settlement of Masseria Chiancudda (municipality of Cisternino, province of Brindisi), occupying top of a hill overlooking the Adriatic, ca. 7 km from the present coastline. The site has suffered considerable erosion which, together with agricultural works, have compromised much of the archaeological deposit, especially on the upper part of the plateau (Cinquepalmi/Recchia 2010, 216–217). Its best preserved archaeological levels are at the western and southern slopes of the hill, probably related to the peripheral areas of the settlement, located in correspondence with remains of the Bronze Age fortifications – two walls made of dry stone, with a width of about 4.5–5 m. Erection of the first wall appears to be related to the *Cavallino-Protoapennine* facies, while its expansion and rearrangement is connected with the *Punta le Terrare* facies of the *Apennine* culture (Cinquepalmi et al. 2004). The walls have common characteristics with those of the most

285 Cataldo 1995, 267, Fig. XXXVII:1; cf. Cipolloni Sampò et al. 1991–1992, Fig. 2:11; Damiani 1995, 405, Fig. 200.

286 Cataldo 1995, 267, Fig. XXXVII:3–7; cf. Tunzi Sisto 1988, Fig. 4:3; 1989, Fig. 4; Recchia 1993, Fig. 15:3; Damiani 1995, 409–410, Figs. 202 and 203:C252, 212; Macchiarola 1995, Fig. 212:I.

materials can be mentioned a skyphos (deep bowl) datable to LH IIIC middle found in US 11322 together with the amber beads (Pagliara et al. 2008, 260, 263, Fig. 14:VI.1).

The youngest, within the timeframes of the Bronze Age, site in the region that yielded amber finds is the settlement of Torre Castelluccia, located on a promontory projecting into the Taranto gulf, approximately 4 km to the south-east from Pulsano. The settlement occupies a plateau measuring around 1.5 ha. Residential area has a terraced structure to fit the slopes, and together with adjacent necropolis has the extension of about 4,000 m^2. It is formed by a tight housing nucleus. Layout of the village is characterised by a series of rooms of different shapes with chronology spanning between the Recent Bronze Age and the EIA, among which the rectangular compartment no. 7 illustrates passage from the FBA to the Iron Age (cf. Gorgoglione et al. 1993).

Between the wall H enclosing the compartment from the west, a layer of beaten-daub floor – feature E to the east, and a small stone wall G to the south, was found a large ceramic vessel kept upright by a group of circular stones (dolium A4), and another vessel partially sunk in the ground (dolium A6). The base of dolium A4 was stuck in the layer US 4, while its mouth was immersed in the younger stratigraphic layers US 2–3 (layers US 2–3-4 document transition from the FBA to the EIA). Therefore, the vessel and accompanying objects must have been deposited during the FBA, perhaps as a hoard called *ripostiglio* (Gorgoglione et al. 1993, 84). Space between the stones revealed reddened soil – most likely a trace of fire. Around the two large vessels were scattered numerous elements of bronze jewellery, as well as glass and amber beads. Due to abovementioned characteristics (composition of the finds and traces of burning), the context is suspected to be a cremated burial, not a hoard. Nonetheless, no bones were discovered, so tentative classification of the finding as a hoard remains. In total six amber beads were recovered (Harding/Hughes-Brock 1974, Fig. 6:26–30; Gorgoglione et al. 1993, 86, Tav. LV:7–10).[298] Chronology of the finds was estimated on final stages of the Bronze Age, with the pottery having prevalent comparisons in the *Ausonian II* culture (Gorgoglione et al. 1993, 87). In claryfing chronology of the context, arguably the most useful are the two simple fibulae: one with two double-nodules on foliated bow which was classified by F. Lo Schiavo to type 18,[299] and the other one with unornamented, but wide leaf-shaped bow, classified by the same author to type 22;[300] both types are dated to BF 2. Such dating seems to be confirmed by other chronologically sensible bronze artefacts from the context: the double-blade razor with semi-circular recess and fixing hole in the handle – type Torre Castelluccia,[301] and the simple, undecorated torc with folded terminals.[302]

---

298 The publication by A. Harding and H. Hughes-Brock, the first one to mention the ambers from Torre Castelluccia, includes illustrations of five beads, although lists six in total with a question mark (cf. Harding/Hughes-Brock 1974, Tab. 1). A full list of amber beads found in the 'hoard', and now kept in the National Archaeological Museum in Taranto (inventory numbers 53922–59924, 53928, 53954–53955), is presented in Gorgoglione et al. 1993; however, the same publication shows figures of only four beads.

299 Gorgoglione et al. 1993, Tav. LVII:8; cf. Lo Schiavo 2010, 99–101, Tav. 6:53.

300 Gorgoglione et al. 1993, Tav. LVI:4; cf. Lo Schiavo 2010, 105–106, Tav. 11:83.

301 Gorgoglione et al. 1993, Tav. LVI:5; cf. Bianco Peroni 1979, 14, Tav. V:67. About chronology of the Torre Castelluccia-type razors, V. Bianco Peroni writes that they should be placed in 'not too advanced moment of the Final Bronze Age' (Bianco Peroni 1979, 14).

302 Gorgoglione et al. 1993, Tav. LVI:1–2; cf. De Juliis 1979, 521, Fig. 4; Vasić 2010, 14–16, Taf. 8:48–54, 44.

Tab. 15. List of the Bronze Age contexts with amber in southern Apulia and south-east Basilicata (Murge plateau and Salento peninsula)

Name	Commune	Province	Type	Relative chronology	Absolute chronology	Reference
Bisceglie, dolmen La Chianca	Bisceglie	Barletta-Andria-Trani	Burial – inhumation	BM 2–3A	1550–1350 BC	Cataldo 1995, Fig. XXXVII:8–9, 11–13
Masseria Chiancudda, layer of *Punta le Terrare* facies	Cisternino	Brindisi	Fortified settlement	BM 2B–3	1470–1300 BC	Cinquepalmi et al. 2006, Fig. 1:C
Murgia Timone, tomb 1	Matera	Matera	Rock-embedded tomb	BM 2B–3	1470–1300 BC	Matarese 2016, Fig. 1:A–C, 2:A1; 2018, Tav. IV:T1C/13–29, XXV: T1CR/1–7
Roca Vecchia, 'Cult Center' (Sector IX, US 11349, Phase II)	Melendugno	Lecce	Fortified settlement	BR 2	1220–1130 BC	Negroni Catacchio 2014, Fig. 6; Guglielmino 2021, Fig. 8.
Roca Vecchia, 'Hut-temple' (Sector IX, US 11322, Phase VI)	Melendugno	Lecce	Fortified settlement	BF 1	1130–1100 BC	Negroni Catacchio 2014, Fig. 13; Iacono/Guglielmino 2021, Fig. 15.3:a
Torre Castelluccia, *ripostiglio*	Pulsano	Taranto	Hoard/burial – cremation(?)	BF 2	1100–1030 BC	Harding/Hughes-Brock 1974, Fig. 6:26–30; Gorgoglione et al. 1993, 86, Tav. LV:7–10

2.2.16. Calabria

The province of Calabria so far has yielded only a handful of amber artefacts which can be connected with the Bronze Age (Fig. 13). They come from two sites of sepulchral character: tomb VI at the site Gallo di Briatico, and tomb IV at Santa Domenica di Ricadi. Both are placed in the southern part of the region, at the Tropea promontory.

The archaeological sources related to the Middle and the LBAs in different areas of Calabria present a somewhat discontinuous and heterogeneous picture. The strong differences between various constituent parts of the province can be presented on the basis of three archaeologically best-recognised areas: the abovementioned Tropea promontory on the southern Tyrrhenian side of the region, the territory of Crotone in the central Ionian coast and the Sibaritide Plain in the northern Ionian coast (Bettelli et al. 2004, 325–326). The Tropea promontory is particularly well researched, thanks to perennial surveys by M. Pacciarelli, M. R. Varricchio and colleagues. Previous research characterised southern Calabria as an 'interface' between several archaeological facieses (cultural groups) which have their focal points in Sicily, the Aeolian Islands and the Apennine Peninsula. The results obtained from these intensive surveys have prime importance in reconstruction of the Bronze Age settlement history on the Tropea promontory.[303]

In short, the EBA in the area can be defined with two peculiar cultural facieses: the *Zungri-Corazzo* pertaining to the first part of the period (BA 1), and the *Cessaniti-Capo Piccolo 1* referable to the late stage (BA 2).[304] Subsequent phases 1 and 2 of the MBA (BM 1–2, considered jointly in the case of Calabria, due to cultural continuity) are marked by appearance of the advanced *Rodì-Tindari* facies, well attested in north-east Sicily and south-east Calabria, where its local version is also described as the *Messina-Ricadi* aspect (Pacciarelli 2001, 30; Pacciarelli/Varricchio 2004, 361, 366–369). This phase is followed by adoption of the *Thapsos* culture of Sicily on the Tropea promontory during BM 3 (Pacciarelli 2001, 32; Pacciarelli/Varricchio 2004, 361, 374).[305] The Recent Bronze Age (BR) on the Tropea promontory is characterised by the *Subapennine* culture, well represented among the materials from Punta di Zambrone (Pacciarelli/Varricchio 2004, 374–377; cf. Jung et al. 2015). The FBA in the area, particularly its phases 1 and 2, is mainly recognised through the graves discovered at Castellace, whereas the EIA (PF IA) began around 950 BC with establishment of the Torre Galli necropolis (phase TG IA).[306]

Absence of radiocarbon dates for the contexts illustrating consecutive phases of the Bronze Age on the Tropea promontory means that one has to refer to the neighbouring regions for absolute timeframe of the period. In this situation the best help is offered by

---

303 Pacciarelli/Varricchio 1991–1992, 2004; Pacciarelli 2001, Fig. 38; cf. Carancini et al. 1996, 80–81, Fig. 4.

304 Marino/Pacciarelli 1996; Pacciarelli 2001, 24–27; Pacciarelli/Varricchio 2004, 360.

305 In a different taxonomical approach, the Tropea promotory, along with the Aeolian Islands and parts of northern Calabria, southern Campania and north-west Basilicata, was included in the *Southern Tyrrhenian* group of the *Apennine* culture by Cocchi Gennick et al. (Poggiani Keller 1995, Tav. 2; Macchiarola 1995, 451–452, Tav. 12, Fig. 210:E). However, by the authors own admission, classification of some of the sites on the promontory, i.a. the tombs of S. Domenica di Ricadi, is nuanced; thus, they cannot be regarded as typical of the *Apennine* culture (Macchiarola 1995, 452).

306 Pacciarelli 1999a, 1999b, 2001, 46–50, 189–196; Fig. 38; Pare 2008, 87, 95, Fig. 5.3.

the already described chronological sequence for Southern Italy, elaborated by R. Jung mainly on the basis of the findings from Campania and Apulia, with the exception of Punta di Zambrone which yielded LH III B–C *Mycenaean* pottery together with typical BR 2 ceramics (Jung 2017, 634–636, Fig. 5). Timing of the FBA can be considered in concordance with other parts of Southern Italy, although there are still too few well-investigated contexts to allow its finer partition (cf. Pacciarelli 1999b; Pare 2008).

Tombs of Gallo di Briatico and S. Domenica di Ricadi are among the most important sepulchral contexts of the Bronze Age on the Tropea promontory (Pacciarelli 2001, 181–188; Pacciarelli/Varricchio 2004, 360). The first site is located within a short distance from the district Gallo – a part of Briatico municipality, south-west from the town and less than 1 km from the coastline. The necropolis most probably served the community of the adjacent settlement of Briatico Vecchio (Betteli et al. 2004, 327; Jung 2013, 244, Fig. 10). Tomb VI contained cremated remains of the deceased, presumably a female individual (very scarce dental fragments), placed inside a large vessel (dolium) covered with a lid. A rich and varied assemblage of grave goods attached to the burial included a bronze awl (Pacciarelli 2001, Fig. 109:3), a silver-wire spiral bracelet (Pacciarelli 2001, Fig. 109:7), a large globular bead or pinhead of crystal (Pacciarelli 2001, Fig. 109:4), a biconical bead of greenstone ophiolitic (Pacciarelli 2001, Fig. 109:5), a carnelian seal of the *Minoan* or *Mycenaean* origin (Pacciarelli 2001, Fig. 109:8) and an amber bead (Pacciarelli 2001, Fig. 109:6). Chronology of the tomb is at best indicated by the lid which represents a form attested in the *Rodì-Tindari* facies, known from some settlements on the Tropea promontory, thus can be dated to BM 1–2 (Pacciarelli 2001, 185–187).

Out of seven tombs located in the locality of Bagneria, near Santa Domenica di Ricadi, two were excavated and one of them (no. 4) yielded four amber beads (Pacci 1987). Unfortunately, at the time of discovery, tomb 4 was already disrupted by earthworks in the nearby quarry, so reconstruction of its interior, as well as original deposition of the burials and related inventory, was seriously hampered. Moreover, it was visibly affected by grave-robbing (Pacci 1987, 35). According to re-evaluation of typo-chronology of ceramic fragments from the grave (Pacci 1987, 44–46), it can be determined that the tomb 4 is most probably concurrent with tomb 5 from the same site. Both can be placed within BM 1–2, so the *Rodì-Tindari* facies, as attested by the bulk of pottery found among the debris (e.g. Pacci 1987, Fig. 2:1–2, 4; cf. Pacciarelli 2001, 30, 182–184). Nonetheless, tomb 4 yielded some sherds pertinent to the *Apennine* (BM 3) and *Subapennine* periods (BR); thus, it seems likely that the tomb was reused after the primary interment (Pacci 1987, Fig. 2:5, 12; cf. Pacciarelli 2001, 189). Prolonged use of the tomb is also indicated by metal items, although few and in a bad shape, i.a. fragment of bronze sheet possibly pertinent to a vessel – vase having LH I–II analogies in Mycenae (Pacci 1987, 47, Fig. 4:1; c.f. Matthäus 1980, pp. 91–92, Taf. 4:25, 12:103), or a drop-shaped, broken pendant, according to L. Vagnetti fragment of a pendulum belonging to the Cypriot-type tripod (Vagnetti 1982, 168–169; Pacci 1987, 47, Fig. 4:2). In light of the abovementioned observations, amber beads from the tomb can be dated only broadly to BM 1–BR, without a possibility of specifying their chronology.[307]

307 G. Bergonzi, as well as M. Cardosa and L. Benedetti, attribute amber beads from the context to BM 1–2 (Bergonzi 1997, 609; Benedetti/Cardosa 2006, 1574), even though following M. Pacci's opinion it is difficult to determine whether the beads were a part of the first burial's inventory or were deposited in the grave at a later date (Pacci 1987, 48–49).

In fact, many settlements appear to have come to an abrupt end indicated by destruction layers apparent at Punta Milazzese (Bernabò Brea/Cavalier 1968, 50), Portella di Salina (Bernabò Brea/Cavalier 1968, 144), or Lipari 'Acropolis' (Bernabò Brea/Cavalier 1980). Still, the destruction and abandonment horizon on individual sites is difficult to correlate with the absolute end of the *Milazzese* culture.

Excavations of Portella di Salina brought to light a single amber bead found in one of the oval structures built on a narrow ridge, called the 'hut F' (Bernabò Brea/Cavalier 1968, 167). The bead appears to have been incorporated into a necklace of carnelian beads (Bernabò Brea/Cavalier 1956, Fig. 42; 1968, Tav. LXXXV:a). There was also a rich collection of faience and glass beads, possibly forming another necklace (Bernabò Brea/Cavalier 1968, 167, Tav. LXXXV:b), and nine fragments of a *Mycenaean* jar, insufficient for a complete reconstruction of the vessel, however enabling classification of the decorative motif to Furumark no. 57.1, and dating them to LH IIB–IIIA (Bernabò Brea/Cavalier 1968, 166; cf. Wijngaarden 2002, 227, 329, cat. I, no. 323).

New research carried out at the Portella settlement on Salina resulted in 20 radiocarbon dates obtained from charred twigs and seeds, pertaining to new huts found at the site (Martinelli et al. 2010; Alberti 2013, Tab. 1). The dates were recently used by two authors to determine the lifespan of the site and see how it relates to a general dating of BM 3, or the Sicilian-Aeolian MBA (Alberti 2015; Jung 2017). G. Alberti, using Bayesian modelling, estimated start of the site's occupation at 1554–1432 cal. BC (95.4 %), while its end at 1386–1187 cal. BC (95.4 %), with the most likely dates being 1490 cal. BC (or a time span between 1490 and 1460 cal. BC according to Alberti 2013, 2510) and 1345 BC, respectively (Alberti 2015, 252, Fig. 5:E–F, Tab. 3). R. Jung, instead, used the dates obtained from legumes ascribed to the last use of the huts V and Z, and estimated their end date on 1320 ± 40 cal. BC (67 %) (Jung 2017, 633). First of all, the results show that beginnings of the *Milazzese* culture should be sought earlier than previously thought, already at the time of LH II, as somewhat suggested by presence of the relevant *Mycenaean* pottery. Second, the analysis indicates that occupation of the Portella settlement ended earlier than presumed transition from the Aeolian Middle to the LBA; thus, the site potentially illustrates a first stage of the *Milazzese* culture. Unfortunately, correlation of the dates with the *Helladic/Mycenaean* chronology is unattainable, because no ceramic imports have been found in the dated huts (Jung 2006, 87).

Although exact timing of the events that affected and eventually led to collapse of the Milazzese settlements is still difficult to specify, their impact must have been severe. While this rough transition has been, as described above, usually dated to ca. 1270/1250 BC, it cannot be ruled out that it took place slightly earlier – ca. 1300 BC, so in parallel with the beginning of the Recent Bronze Age (BR) in Southern Italy. In fact the following culture, known as the *Ausonian I*, shares many similarities with ceramic styles outside the Aeolian Islands, e.g. with the *Subapennine* culture of the peninsular Italy (Bernabò Brea 1979, 582; Malone et al. 1994, 170).

During the *Ausonian I* period, habitation of the archipelago was largely limited to Lipari, where another horizon of buildings has been revealed on the 'Acropolis' (Bernabò Brea 1979, 579–580; cf. Bernabò Brea/Cavalier 1980). The *Ausonian I* structures on the Lipari 'Acropolis' followed much on the same pattern as adopted by the *Milazzese* culture. Moreover, several fragments of the *Mycenaean* ceramics have been collected from these buildings, all attributed to LH IIIB and IIIC, thus allowing to date the *Ausonian I* to the 13th and the first half of the 12th century BC.[318] The most recent Aegean pottery fragments from the *Ausonian I* levels date back to the middle LH IIIC, so around 1170/1160–1150/1140 BC.[319] From the upper *Ausonian I* strata on Lipari (level of destruction that corresponds to the end of the horizon) were recovered ceramic fragments referable to the Sicilian *Pantalica Nord* facies, along with pieces decorated in the *Protovillanovan* style that can be traced back to an initial moment of the peninsular FBA – BF 1 (Bernabò Brea 1979, 584–585; Bettelli et al. 2012, 1095–1096).

Also the *Ausonian I* levels on the Lipari 'Acropolis' revealed evidence of amber. A large oval hut named Beta I was discovered in front of the Lipari cathedral, severely damaged by the walls of the church and Medieval cisterns. Adhering to the wall was a stretch of pavement, made with flat and irregular slabs, resting on small stones. Sunken in this layer was a jar placed upright (Bernabò Brea/Cavalier 1980, Tav. IX:1, 4, CCXVIII:3), with its bottom well below the base of the stone pavement, within the levels of the underlying *Capo Graziano* hut. Inside the vessel were traces of burnt bones and some artefacts, including a single amber and ten glass beads, altogether giving impression of a burial with accompanying grave goods (Bernabò Brea/Cavalier 1980, 114, Tav. CCXVIII:4). Pottery collected above the pavement is characteristic of the *Ausonian I*, thus defining age of the hut and time of the burial's deposition.

Layers of the *Ausonian I* and *Ausonian II* on the Lipari 'Acropolis' are separated by a destruction layer which is interpreted as yet another violent event announcing new settlers to the Aeolian Islands, who not only brought with them many foreign elements but also adopted some aspects of the older local culture (Bernabò Brea 1979, 587–588). A noticeable transformation is illustrated by the housing architectures: huts changed from circular to much larger, elongated oval buildings, still employing drystone walls, but with wooden posts to raise the height of the walls (Malone et al. 1994, 178).

The *Ausonian II* culture appears as a result of synthesis of many components of different origins. Pottery is, in particular, characterised by grooved and *cuppelle* decoration, as well as complex handles (Malone et al. 1994, 170). Evident are connotations with the *Protovillanovan* culture of Central Italy, seen through bronze items as well as pottery (Bernabò Brea 1979, 590). Furthermore, the *Ausonian II* demonstrates a relationship with Sardinia, attested by fragments of the *Nuragic* pottery (Bettelli et al. 2012, 1095). There is also a certain number of pottery fragments pertaining to the final stage of the *Mycenaean* culture, i.e. advanced and late LH IIIC (ca. 1150/1140–1085/1080 BC). The fragments were found in the lower layers at the Lipari 'Acropolis' – an observation which, in combination with dating of the LH IIIC sherds from the *Ausonian I* layers, can help to determine the passage between the two cultural units, which appears to be in line with that of the transition between BR and BF in Southern Italy.[320]

According to Bettelli et al. (2012), two successive phases of the *Ausonian II* culture can be deduced from the typological point of view. The first, earlier phase covers advanced and late LH IIIC associated with some *Submycenaean* and *Protogeometric* ceramics, thus could be placed in correspondence with BF 1–2 (ca. 1130/1100–1030 BC). The second, later phase should be considered coeval with BF 3 and PF IA (ca.

318 Bernabò Brea 1979, 584; cf. Taylour 1958, 40–41; Wijngaarden 2002, 223, Fig. 15.3.
319 Bettelli et al. 2012, 1096; cf. Alberti/Bettelli 2005, 551, Pl. XCXXIV:17; Weninger/Jung 2009, Fig. 14.
320 Alberti/Bettelli 2005; Jung 2006, Abb. 24; Bettelli et al. 2012, 1096.

1030–910 BC), in parallel with the stage *Pantalica II-Cassibile* on Sicily and the oldest phase of the Torre Galli necropolis in Calabria (TG IA).[321]

To the first phase of the *Ausonian II*, culture can be ascribed six burials containing amber (nos. 10, 11, 18, 19, 31 and 34), found inside the layer VII at Piazza Monfalcone in Contrada Diana, Lipari.[322] Remains of deceased were inhumed in large ceramic vessels (pithoi), placed on the side (horizontally) and enclosed with stones. While in four of the graves the amber beads were reasonably well preserved (nos. 18, 19, 31, 34), in the other two (nos. 10 and 11) amber became almost pulverized due to crushing of the pithoi (Bernabò Brea/Cavalier 1960, 149). Also, graves 18, 31 and 34 contained richer inventory, comprising i.a. simple fibulae with flattened-widened (type 22) and twisted central part of the arch (type 26), both finding good analogies among Southern Italian contexts dated to BF 2, and wheel-shape pendants which were found, e.g. in the hoard of Coste del Marano.[323]

As in case of the previous occupation horizons on Lipari, the *Ausonian II* ended due to a violent destruction, with a great fire that incinerated the huts. The town has not been rebuilt and the Aeolian Islands seem to have been depopulated, until the arrival of the Greek settlers in 580 BC (Bernabò Brea 1979, 595).

Tab. 17. List of the Bronze Age contexts with amber on the Aeolian Islands

Name	Commune	Province	Type	Relative chronology	Absolute chronology	Reference
Lipari 'Acropolis', hut Delta XXI	Lipari	Messina	Settlement	*Capo Graziano 2* (LH II)	1600–1490/1460 BC	Bernabò Brea/Cavalier 1980, Pl. 98:i
Lipari 'Acropolis', hut Beta I–tomb III	Lipari	Messina	Burial – cremation (urn)	*Ausonian I*	1300/1250–1130/1100 BC	Bernabò Brea/Cavalier 1980, 114
Portella di Salina, hut F	Lipari	Messina	Settlement	*Milazzese* early (LH IIB–IIIA)	1490/1460–1345/1320 BC	Bernabò Brea/Cavalier 1968, 167
Piazza Monfalcone, tomb 10	Lipari	Messina	Burial – inhumation (*pithos*)	*Ausonian IIa*	1130/1100–1030 BC	Bernabò Brea/Cavalier 1980, 109
Piazza Monfalcone, tomb 11	Lipari	Messina	Burial – inhumation (*pithos*)	*Ausonian IIa*	1130/1100–1030 BC	Bernabò Brea/Cavalier 1980, 109

321 Cf. Bernabò Brea 1979, 591; Carancini et a. 1996, Fig. 4; Pacciarelli 1999a, 42, Fig. 11.

322 Bernabò Brea/Cavalier 1960, 144–150, 159, 161; cf. Negroni Catacchio 1999, 250; Nava/Salerno 2007, 91–92.

323 Bernabò Brea/Cavalier 1960, 160, Tav. XLI:2b, XLII:5a-b, 8; cf. Bietti Sestieri 1973, Fig. 7:11; Lo Schiavo 2010, 105–106, 108–110, n. 80, 89, Tav. 11:80–81, 12:86–90.

Tab. 17. Continued

Name	Commune	Province	Type	Relative chronology	Absolute chronology	Reference
Piazza Monfalcone, tomb 18	Lipari	Messina	Burial – inhumation (*pithos*)	*Ausonian IIa*	1130/1100–1030 BC	Bernabò Brea/Cavalier 1960, 111, Tav. XLII:8
Piazza Monfalcone, tomb 19	Lipari	Messina	Burial – inhumation (*pithos*)	*Ausonian IIa*	1130/1100–1030 BC	Bernabò Brea/Cavalier 1960, 113
Piazza Monfalcone, tomb 31	Lipari	Messina	Burial – inhumation (*pithos*)	*Ausonian IIa*	1130/1100–1030 BC	Bernabò Brea/Cavalier 1960, 119, Tav. XLIII:1
Piazza Monfalcone, tomb 34	Lipari	Messina	Burial – inhumation (*pithos*)	*Ausonian IIa*	1130/1100–1030 BC	Bernabò Brea/Cavalier 1960, 121

2.2.18. Sicily

There is at least one Copper Age site on Sicily containing amber finds: Grotta di Calafarina (Pachino commune, Siracusa province), where an ovoid ('tear drop-shaped') pendant was found (Orsi 1907, Fig. B; Massari et al. 1996, Fig. A:8). The context of the find was dated basing on co-occurring pottery which can be considered as a local variety of the *San Cono-Piano Notaro* ware and shows a particular decorative syntax with incised wavy lines filled with red ochre, known as the *Calafarina* style (Cultraro 2007, 379; Crispino/Cultraro 2014, 189). Nonetheless, it has to be mentioned that the cave also returned evidence of later occupation (cf. Orsi 1907). Interestingly, the find, along with some small amber fragments present in the Syracuse Archaeological Museum and reportedly coming from the cave, was recently classified as simetite (Angelini/Bellintani 2017, 676–677, Fig. 1:a–b, 4:5–6, Tab. 1:CF-A, CF-B), confirming earlier speculations (Cultraro 2007, 380).

Many amber finds have been recovered from the sites related to Sicily's emblematic EBA cultural facies – the *Castelluccio* culture (Fig. 13). Sadly, this one of the longest known prehistoric units of Italy is still ill-recognised in many aspects, most of all its chronology, which inevitably impairs precise dating of the amber finds. First described by P. Orsi, who coined the name, at the end of the 19th century, some of the most important sites of the *Castelluccio* culture have remained fragmentarily published. Fortunately, in the recent years, there has been a noticeable comeback to the *Castelluccio* culture's studies, aimed at re-analysis of the archival data. Arranging materials related to the *Castelluccio* culture in chronological order is complicated by the fact that many of its sites, chiefly cemeteries with rock-embedded tombs recognised as the culture's trademark, were robbed. Second, there is a shortage of investigated stratified settlements, documenting evolution of pottery over a longer period.

P. Orsi placed the *Castelluccio* culture on the timeline, by assigning it to the *Ist Siculan period* in his periodisation of the Sicilian pre- and protohistory (Orsi 1892, 1893a). The first attempt at internal periodisation of the *Castelluccian* materials was carried out by L. Bernabò Brea, who used east Sicilian contexts, some of them still partly unpublished; the scholar developed a three-partite scheme: (1) 'proto-*Castelluccio*', (2) 'typical *Castelluccio*' and (3) 'lingering phase' (Bernabò Brea 1968–1969, 44–46; 1991–1992, 108–110). The periodisation was revisited by M. Cultraro, who added a fourth phase, drawing from his observations on the evolution of ceramic vessels (Cultraro 1996). R. Peroni, in an attempt to homogenize the Sicilian prehistoric chronology with the rest of Italy, established a different two-partite periodisation by attributing the *Castelluccio 1* phase – the classic stage of culture – to the EBA, while correlating its later manifestations called the *Castelluccio 2*, or the *Castelluccio Bruno* (also including the *Rodì-Tindari-Vallelunga* pottery style), with not advanced stage of the MBA (Peroni 2004, 92–96, 192). In the recent years, an overall typo-chronology of the *Castelluccian* pottery created by the means of seriation was elaborated by R. Gennusa, who enclosed the culture's development in four phases marked 0, 1, 2 and 3, with the latter two further divided into subphases A and B (Gennusa 2015). Roughly at the same time, at the symposium *Prima di Thapsos. The central-southern Sicily between the Final Eneolithic and the Early Bronze Age* held in Syracuse (2011), F. Iannì presented a periodisation of the *Castelluccio* culture in central Sicily based on stratigraphic data accompanied by radiometric dates from the sites of Tornambè (Pietraperzia commune, Agrigento province) and Case Bastione (Villarosa commune, Agrigento province). Thanks to the stratigraphies of the two abovementioned sites, three groups of painted ceramics could be isolated (Iannì forthcoming). Typological characteristics allow to parallelise these pottery groups with the first three phases of the Gennusa sequence; there is a good match with the Gennusa's phases 0 and 1, while a little less well fitting with Gennusa's phase 2 (Iannì 2017, 278).

Transition of the *Castelluccio* culture through the phases is observable, first and foremost, on the basis of pottery with characteristic painted decorations, which throughout the Bronze Age remained largely unaffected by the trends in ceramic production in the peninsular Italy (Malone et al. 1994, 170, 186). In general, the *Castelluccio's* pottery can be said to have progressed from handmade impasto pottery with usually black-on-red, matt-painted decorations in geometric patterns, derivative of the Late Copper Age (LCA) tradition, to introduction of new decorative motifs in more and more complex syntaxes during the developed stage of the culture, e.g. high-handled carinated cups, ovoid amphorae and distinctive pedestalled vessels, lavishly decorated in the so-called style of *horror vacui*, until the appearance of new, scantly ornamented, predominantly carinated forms in the final stages, often associated with the prevailing acrome and grey ceramics of the *Rodì-Tindari-Vallelunga* facies that herald the MBA *Thapsos* culture.[324] Along with chronology, scholars have also used geographic variability as a key for interpreting stylistic variation, arguing in particular that it is possible to identify different

324 Cf. Bernabò Brea 1991–1992, 109; Cultraro 1996; Bietti Sestieri 2013b, 654–655; Gennusa 2015.

stylistic 'sub-provinces' within the area of the *Castelluccio* culture (Copat et al. 2017, 111–113; cf. Cultraro 1996).

Using the latest publications regarding the *Castelluccio* culture, most of all the works of Gennusa and Iannì, it is possible to outline development of the unit both in terms of relative and absolute chronologies. On the basis of 14C dates obtained from the stratified sites, beginning of the *Castelluccio* culture can be dated to ca. 2200 BC. The most meaningful contexts illustrating the *Castelluccian* early and middle stage are La Muculufa, Tornambè and Case Bastione, all of which delivered sequences of 14C dates useful in establishing absolute chronology for the culture's development;[325] however, there are a number of other sites which delivered fewer but nonetheless helpful radiocarbon determinations (cf. Pacciarelli et al. 2015, 267–269, Appendix 1). On the abovementioned sites, the earliest manifestations of the *Castelluccio* culture are preceded by the strata assigned to the local LCA *Malpasso* and *Sant' Ippolito* cultures, particularly the so-called *Naro-Partanna* style (ca. 2700/2600–2300/2200 BC). The subsequent early stages of the EBA, corresponding with the 'Sanctuary area' at La Muculufa and the huts of Case Bastione, can be dated between ca. 2200 and 1800/1700 BC (Giannitrapani/Iannì 2011, 278; Pacciarelli et al. 2015, 269–270).

The first phase of the *Castelluccio* culture (Gennusa 0, Iannì 1) in fact includes pottery pertinent to what in literature is called the *Sant' Ippolito* culture/style, stratigraphically associated with elements of the *Bell Beaker* culture and the late *Malpasso* culture, thus falls into the final stage of LCA (Gennusa 2015, 217–218; Iannì 2017, 279). Thanks to 14C dates from the La Muculufa settlement, hut 1 of Tornambè and area α of Case Bastione, the period can be estimated to have lasted approximately between 2500/2400 and 2200 BC.[326] The next phase (Gennusa 1, Iannì 2) illustrates emergence of the *Castelluccio* culture in its fully shaped EBA variety, with the La Muculufa sanctuary and hut 1 of area β at Case Bastione as its prime, radiocarbon-dated contexts.[327] With regard to the third phase (Gennusa 2, Iannì 3), correspondences among pottery from the south-western and south-eastern parts of the *Castelluccio's* ecumene are less punctual. Potentially, it is a consequence of growing regionalisation of ceramic production, reflecting atomisation of the *Castelluccian* society (Iannì 2017, 279). The

---

325 Cf. Holloway et al. 1990; McConnell 1995; Giannitrapani 2013; Giannitrapani et al. 2014.

326 Cf. McConnell 1995, 13, 16–17, 99–100; Giannitrapani/Iannì 2011, 275–278; Giannitrapani 2013, 190–191; Giannitrapani et al. 2014, 193–194, Tab. 1. The full list showing 14C dates and contexts used in assessing the timing of LCA–EBA transition is given in Pacciarelli et al. 2015, Appendix 1.

327 Case Bastione, hut 1, phase 1: LTL3657A (3699 ± 45 BP), 2190–2030 cal. BC (68.2 %), 2205–1951 cal. BC (95.4 %); Rome-2055 (3645 ± 35 BP), 2120–1950 cal. BC (68.2 %), 2135–1921 cal. BC (95.4 %); LTL3656A (3612 ± 35 BP), 2025–1930 cal. BC (68.2 %), 2121–1886 cal. BC (95.4 %). Case Bastione, hut 1, phase 2: Rome-2059 (3650 ± 40 BP), 2120–1950 cal. BC (68.2 %), 2140–1914 cal. BC (95.4 %); Rome-2061 (3600 ± 40 BP), 2020–1905 cal. BC (68.2 %), 2127–1784 cal. BC (95.4 %); Rome-2051 (3585 ± 40 BP), 2010–1890 cal. BC (68.2 %), 2113–1777 cal. BC (95.4 %); LTL-3655A (3552 ± 35 BP), 1950–1780 cal. BC (68.2 %), 2014–1771 cal. BC (95.4 %) (Giannitrapani et al. 2014, 196, Tab. 1). La Muculufa, sanctuary: A-3967 (3730 ± 90 BP), 2286–1981 cal. BC (68.2 %), 2457–1914 cal. BC (95.4 %); A-3959 (3640 ± 80 BP), 2135–1911 cal. BC (68.2 %), 2276–1771 cal. BC (95.4 %) (Pacciarelli et al. 2015, Appendix 1).

horizon is reflected by materials from hut 2 placed in the area α of Case Bastione which was radiocarbon dated.[328] By way of comparing the start and end dates for the two phases at the 68.2 % and 95.4 % probability intervals, and rejecting the outlying observations, one can figure the range of ca. 2200–2000/1900 BC for phase Gennusa 1=Iannì 2, and 2000/1900–1800/1700 BC for phase Gennusa 2=Iannì 3, which does not contradict the datings suggested by Giannitrapani et al. (cf. Giannitrapani et al. 2014, 196–198).[329] The last phase of the *Castelluccio* culture is the least defined, because it lacks solid stratigraphic references and, consequently, few contexts and ceramic forms connected to it (Gennusa 2015, 219; Iannì 2017, 279). Also, suggesting a final caesura for the culture's development is burdened with a risk of error. Nowadays, most authors suggest an overall time span of 2300/2200–1600/1500 BC for the *Castelluccio* culture, with its end falling sometime around LH I and II transition.[330] Still, this chronological boundary is not clear when looking at currently available data and requires more research to be defined with bigger precision.

The *Castelluccio* culture was first recognised and is still predominantly known from its cemeteries. Among them are sites which yielded amber finds: tombs 9, 22 and 23 at the eponymous site of Castelluccio (Orsi 1892, 30, Tab. III:15, 18, V:17; Massari et al. 1996, Fig. B:5–7), tombs 3 and 9 at Cava Cana Barbara (Orsi 1902a, 186, Tab. VI:11), tomb 10 at Cava Secchiera (Angelini/Bellintani 2017, Fig. 1:c), tomb 5 at Monte Sallia (Orsi 1923, 17) and tomb 18 at Valsavoia (Orsi 1902b, 114), all situated in the south-east corner of Sicily, thus placeable in the group 'B. Calatina-Syracusa' by M. Cultraro (1996, 170, Fig. 3:24–26, 31). Exact chronology of the tombs is very difficult to determine, as practically all of them were looted in the past (or disturbed in one way or another), and many of them did not deliver the key chronological indicator which is the *Castelluccio* culture's pottery. In fact, their affiliation with the unit largely relies on similarities in construction of the tombs and among the non-clay grave goods. The cemeteries, like the ones listed above, comprise multiple rock-embedded or in-cave tombs which often represent complex architectural systems of corridors, antechambers, main chambers and niches housing collective burials (up to 50 buried individuals), always inhumed.[331] Grave goods, apart from amber, included flint blades, stone and shell beads and pendants (sometimes of semi-precious rocks such as alabaster or obsidian) and rarely bronze items, which were only occasionally accompanied by fine painted locally made ceramic vessels.[332]

---

328 Rome-2060: 3550 ± 40 BP, 1950–1780 cal BC (1σ), 2016–1756 cal BC (2σ); Rome-2050: 3540 ± 40 BP, 1940–1780 cal BC (1σ), 2009–1751 cal BC (2σ); Rome-2056: 3530 ± 40 BP, 1920–1780 cal BC (1σ), 1963–1745 cal BC (2σ); LTL-3654A: 3511 ± 45 BP, 1890–1770 cal BC (1σ), 1952–1695 cal BC (2σ) (Giannitrapani et al. 2014, 196, Tab. 1).

329 Newly obtained 14C dates from the eponymous Castelluccio settlement, classified by R. Gennusa to her phases 1 and 2, fit well into the period indicated above; however, precise correlation of the dates and typochronology of the materials from dated contexts (huts 2 and 8) yet needs to be assessed (cf. Gennusa 2015, 25; Pacciarelli et al. 2015, 270, Appendix 1).

330 E.g. Carancini et al. 1996, 80, Fig. 4; Cultraro 1996, 170; Bietti Sestieri 2013b, 654; Gennusa 2015, 11.

331 Cf. Orsi 1892, Tav. I; 1893b, Pl. II; 1902a, Figs. B–E; 1902b, Figs. B–E; 1923, Fig. 9.

332 Orsi 1892, Tav: III:8–22, V:2–24; cf. Bietti Sestieri 2013b, 657; Crispino/Cultraro 2015; Matarese et al. 2015.

As a result of the classification by R. Gennusa, some of the *Castelluccian* graves with amber were dated within her four-partite periodistation. Accordingly, the tombs from the Castelluccio cemetery belong to phases 1–2, thus possibly indicating the oldest horizon of amber use in the culture, while the finds from tomb 5 of Monte Sallia[333] and tomb 10 of Cava Secchiera fall into slightly later time – phases 2A and 2B–3A, respectively (Gennusa 2015, 21, 25, 27–29). More difficult to estimate is the chronology of amber finds from Cava Cana Barbara and Valsavoia. Tombs 3 and 9 at the first site did not contain any reconstructable pottery, classifiable within the typology devised by R. Gennusa, and can be dated only through a reference to tomb 4 which the author rather broadly attributes to phases 1–3 (Gennusa 2015, 21). Moreover, main chamber of tomb 9, where two amber beads were found, contained objects possibly of later origin, i.e. fragment of a bronze dagger and faience beads; thus, chronology of the whole assemblage from this part of the tomb was prolonged beyond the temporal limits of the *Castellucio* culture, to the Sicilian MBA – *Thapsos* culture (Orsi 1902a, 188, Tab. VI:9, 12; cf. Matarese et al. 2015, 112, Tab. 1, n. 343; I. Matarese, personal communication).[334] Similar is the case of tomb 18 of Valsavoia, where the main chamber, containing most of grave goods, did not return any pottery, while ceramic fragments found in the antechamber are comparable with the *Thapsos* culture's earthenware (Orsi 1902b, 114; cf. Matarese et al. 2015, 112, Tab. 1, n. 343). Still, the pottery from tombs 3, 6, 7 and 22 at the site clearly resembles forms and ornaments of the *Castelluccio* culture, and allows to place the graves in phases 3A and 3B, giving a clue about chronology of the remaining graves (Gennusa 2015, 19).

Even though majority of the *Castelluccio* culture's sites are confined to the south, central and south-east Sicily, its domain expanded to the north around Etna and, particularly, north-west towards Marsala and Trapani (Malone et al. 1994, 173). The north-eastern border of the *Castelluccio* culture can be drawn along the Alcantara River Valley which constitutes a natural watershed between the Etnean region and the Tyrrhenian coast. As for the borders with the western regions, sporadic attestations of painted ceramics in the Molara cave (Palermo province) and in funerary complexes of the Egadi document the farthest reach of the *Castelluccio* culture in the north-west direction (Cultraro 1996, 163). In the latter area, the *Castelluccio* culture took part in the complex formation process of the *Rodì-Tindari-Vallelunga* (RTV) facies/culture which manifests the advanced EBA of northern Sicily. In this part of the island, LCA is defined by interactions between the *Bell Beaker* culture of the Western Mediterranean with the local cultures of *Malpasso, Sant'Ippolito* and *Naro-Partanna* between the end of the 3rd millennium and the first centuries of the 2nd millennium BC, which in turn led to creation of the local pottery style named *Moarda* (Nicoletti/Tusa 2012, 105–106; Bietti Sestieri 2013b, 656). A general view is that the *Naro-Partanna* culture was followed by RTV; however, the complexes which possibly document the formation of the latter are devoid of sufficient stratigraphic data (Nicoletti/Tusa 2012,

333 In the case of Monte Sallia, the dating is indirectly confirmed by the matt-painted carinated cup with high-swung handle – MH import of the Aegean origin, found at the site (cf. Taylour 1958, 54–56, Pl. 16:1a-b).

334 Furthermore, lateral niches inside tomb 9 contained even younger elements, such as a serpentine fibula dated to the FBA (Orsi 1902a, 188).

Alberti 2007). The new periodisation has been largely accepted and applied in numerous newer publications.[339]

During MBA there was a notable trend in moving the settlements closer to the coast, especially along the eastern shore of Sicily, progressing in parallel to the rarefication of habitation in the inland parts, e.g. around Etna (Bernabò Brea 1991–1992, 113). Distribution of the *Thapsian* sites, settlements and cemeteries alike, close to the sea is interpreted as a sign of dependence of local communities on seafaring and trade, in particular with the Aegean, Cyprus and with the Maltese *Borg-in-Nadur* culture.[340] Particular concentration of LH imported pottery, along with items made of different exotic materials, e.g. glass and amber, gave a way to interpretations of the Thapsos site as a sort of trade emporium, visited by Aegean sailors – merchants (Malone et al. 1994, 190–191; Bietti Sestieri 2013b, 659).

The site is located north of Syracuse, on the Magnisi peninsula, connected with the mainland by a narrow isthmus. It comprises two districts: a residential one and a sepulchral one. The settlement is located in the area where the narrow isthmus joins the peninsula, whereas the main cemetery is situated near the lighthouse in the northern part of the peninsula, with smaller groups of graves placed in central and southern parts. The complex was investigated in the last two centuries on several occasions: first by P. Orsi at the end of the 19th century (Orsi 1895), and then from the late 1960s up to mid-1980s first under the direction of L. Bernabò Brea and later G. Voza.[341] The results of the later works are still largely unpublished. The cemetery is made up of chamber tombs dug into the soft limestone, hosting multiple burials simply laid inside the cells or inside ceramic pithoi. Some of the graves have more articulated structures with antechambers and lateral niches, or tholos-like roofs, potentially inspired by the Aegean LH tombs. Orsi estimated that some 300 of such tombs were present at Thapsos, of which more than 100 have been excavated.[342] Many of the tombs were richly furnished with grave goods, comprising a mix of indigenous and imported products, with fine pottery, bronzes that were almost certainly imported or copied from the Aegean and mainland Italy, amber (tombs 61 and D), faience, *Mycenaean* cups, etc.

Tomb 61 had a small, dome-shaped burial chamber, preceded by a long canal-corridor opening towards the sea, inside of which skeletal remains of 19 individuals were almost all piled up in the right half of the cell (Orsi 1895, 134). Fragments of numerous ceramic vessels, mostly of local origin excluding a small piriform jar of the *Mycenaean* provenience – LH IIIA2–B1, were collected in the right half of the cell, (Orsi 1895, Fig. 49; cf. Wijngaarden 2002, Catalogue IX:16). Notable is the assemblage of glass beads and a single amber bead (Orsi 1895, Fig. 49; Cultraro 2007, Pl. XCVIIIc:5). The tomb was assigned to the IInd phase of the Thapsian cemeteries (Alberti 2004, 135; Alberti 2006, 389–390).

---

339 E.g. Tanasi 2008, 2009, 2010; Tanasi et al. 2014.

340 Bernabò Brea 1991–1992, 115; Tanasi 2008, 104; 2010, 105.

341 For the full evidence of publications reporting various excavation campaigns at Thapsos, see e.g. Wijngaarden 2002, 229; Alberti 2006, 2007; Voza 2008.

342 Orsi 1895, 94; Bernabò Brea 1991–1992, 114; Malone et al. 1994, 190–191; Leighton 1999, 162–165; Bietti Sestieri 2013b, 659.

Inside tomb D were discovered skeletal remains of about 49 individuals, accompanied by a rich and varied inventory, including i.a. nine *Mycenaean*, three Cypriot and one Maltese imported vessels, as well as some imitations of the *Mycenaean* clay containers, not to mention variety of local ceramics.[343] There was also a dagger of the same shape as the Thapsos-type swords.[344] Impressive is the assemblage of beads of various kinds: 6 in gold, 11 in carnelian, 300 in bone, 48 in glass, 1 in jasper and 4 in amber.[345] The tomb was assigned to the group of graves with prolonged use, spanning phases I, II and III (Albert 2004, 135; 2006, 393–395).

Research on the Plemmyrion peninsula, south of Syracuse, revealed a similar *Thapsian* site: a wall is marking the settlement's perimeter, inside which was recorded a single hut along with other traces of habitation, as well as ditches linking the site with the sea, while the cemetery extends to the north and north-east from the settlement, on a gentle slope ascending towards the central part of the promontory.[346] The rock-embedded tombs were provided with vertical, shaft-like accesses, and consist of antechambers, main chambers and lateral niches, all used for deposition of burials. Human remains were mostly inhumed, with the skulls laid on small platforms and the legs visibly bent. The burials were often accompanied by rich mortuary outfits, consisting i.a. of imported objects (bronze swords, faience, amber). Four graves yielded amber: 10, 11, 48 and 49. Unfortunately, due to lack of major publication summarising archaeological excavations at Plemmyrion, only limited amount of data, scratched from various publications, can be presented.

Tomb 10 is notable for discovery of two amber adornments identified as 'amber spacer-beads', probably of Aegean origin.[347] Tomb 11 reportedly contained a single 'globular disc of fossil resin' (Cultraro 2007, 381; cf. Orsi 1891, 122–124). Among all the four graves, the richest, and most covered in the literature, is tomb 48. Inside the tomb were identified skeletal remains of approximately 50 individuals. At one place were found ivory comb and 140 beads of different materials, including 15 in amber,[348] 1 in jasper and remaining made of glass paste (faience), presumably originally forming a necklace.[349] Residues of a necklace were discovered near one of the skeletons also in the case of tomb 49; they comprised 14 beads in glass paste, 3 beads of sedimentary material (including one in limestone) and a single biconical amber bead (Colini 1904, 253, Fig. 55; Matarese et al. 2015, SE18.3–5). Tomb 48 was dated to Alberti's I[st] and IInd phases; however, remaining three graves were not included in the analysis, presumably due to absence of ceramic materials suitable for dating (Alberti 2004, 136). Thus tombs 10, 11 and 49 can be only generally attributed to the whole time span of the *Thapsos* culture development.

---

343 Alberti 2004, 128–129, 131–134, Tav. IX:6, 11; 2006, 385, Tab. 3–5; cf. Voza 1973.

344 Voza 1973, 40, n. 116, Tav. 6:116; Militello 2004, Fig. 3:A; cf. Jung 2009, 131–132, n. 11.

345 Cf. Voza 1973, p. 40, nn. 105–115; Militello 2004, Fig. 4: A, C; Alberti 2006, Tab. 2; Cultraro 2007, 381–382, Pl. XCVIII:c9; Matarese et al. 2015, SE19.2–13.

346 Orsi 1891, 1899; Colini 1904, 232; Malone et al. 1994, 175.

347 Orsi 1891, 121–122, Pl. XI:9; Tusa 1999, Fig. 29:1; Cultraro 2007, Pl. XCVIIIc:10; cf. Cornaggia Castiglioni/Calegari 1978.

348 Of the original 15 amber beads (Colini 1904, 253), new publications mention only seven conserved in the museum (Angelini/Bellintani 2017, Tab. 1).

349 Orsi 1899, 30–33, Fig. 7; Colini 1904, 253, Fig. 54; Genovese 2015, 121–123, Figs. 6–8; Matarese et al. 2015, SE18.2.

During the Sicilian LBA (ca. 1270/1250–1035/1010 BC), corresponding with development of the local facies called *Pantalica I – North*, amber became absent on Sicily for a while. It is currently difficult to explain this in any way other than simply by the state of research, taking into account the abundant testimonies of continuing relationships with oversea communities, of the Aegean or the Italian mainland (cf. Cultraro 2007, 383; Tanasi 2010, 110). In the next phase – FBA (ca. 1035/1010–910 BC) – amber reappeared, even though in limited number and spatial extent.

In the periodisation devised by L. Bernabò Brea, Sicilian FBA is synonymous with the term *Pantalica II – Cassibile*, which does not denominate any coherent cultural unit, but rather a loosely connected group of communities inhabiting different parts of Sicily, predominantly known from cemeteries (Nicoletti 2006, 142–143). Dating of the phase to the 10th and the first half of the 9th century BC is generally still valid, but has to be shortened in line with chronology of BF 3 in other Italian provinces. C. F. E. Pare adopted division of the *Pantalica II* phase into two subphases proposed by M. Turco, and used bronze elements of grave inventories, namely the stilted, arched and the Cassibile-type elbow/knee fibulae, as well as certain armaments, to establish a concordance between BF 3a and *Pantalica 2a*, and BF 3b and *Pantalica 2b* (cf. Pare 2008, 85–87). Terminological reference to the *Pantalica* culture of the previous period serves to underline a certain continuity regarding use of the namesake necropolis. However, the phase is better recognised and understood on the basis of other cemeteries which emerged at the time, i.e. Cassibile, Dessueri and Madonna del Piano/Molino della Badia. The abovementioned cemeteries are in fact remarkably different with regard to the *Pantalica* cemetery; the unifying element is substantially reduced to a few metal types, including in particular the Cassabile-type fibula – a new model of fibula which hallmarks this period (Bietti Sestieri 1979, 622, n. 35). Each of the abovementioned cemeteries can be considered as belonging to a separate community, receptive of influences arriving from different sources.

The Cassibile necropolis is characterised by introduction of many novelties, in case of earthenware seen above all through the spread of the 'plumed' decoration. In bronze metallurgy significant is the appearance of arched fibulae with an elbow/knee and a straight pin – the Cassibile type; stilted fibulae; simple fibulae with thick, decorated arch; curved knives with looped handles; or the quadrangular razors of the Italic type and Iberian leaf-shaped razors. Analogies for some of the bronzes, particularly the fibulae, in a vast Mediterranean sphere extending from the Levant up to the Iberian Peninsula, prompted L. Bernabò Brea to speak of Sicily's involvement in the Phoenician trade – an idea which, however, is disputed.[350]

Presence of numerous common elements of inventory links the cemeteries of Cassibile with that of Madonna del Piano/Molino della Badia. The latter, nonetheless, has its unique character, attributable to strong ties, perhaps even genetic ones, with the *Ausonian II* culture of the Aeolian archipelago.[351] Other authors highlight affinities with the *Castellace* group in Calabria and speak in favour of an incentive for setting the cemetery as arriving from Italian mainland (Albanese Procelli/

---

350 Bernabò Brea et al. 1969, 211–212; cf. Bietti Sestieri 1979, 622; Malone et al. 1994, 191; Leighton 2000, 34–38.

351 Bietti Sestieri 1979, 619, 622–623; 2014, 665; Nicoletti 2006, 145.

Lo Schiavo 2004, 412). The cemetery is located in Catania province, north of the Grammichelle town, and borrows its name from localities, where the first tombs were discovered between 1898 and 1905. Professional archaeological excavations conducted at various times in the last century overall brought to light 337 tombs, dating by large to BF 3a and BF 3b/PF 1A.[352] The funeral rite includes inhumation of the deceased in supine position in rectangular pits cut in bedrock, or with the bodies deposited inside pithos vessels (*enchytrismòs* burials), but there are also several cases of cremation. According to seriation of the graves discovered up to 1969, cremation was used in the early stages of the cemetery's use, followed by the burials in pithos vessels and finally the burials in the pit (Nicoletti 2006, 163, 176, Tab. 2). The grave kits mostly include a single ceramic vessel, usually single-handle container (jug or cup, sometimes *askos*) with recessed rim, distinct neck and plumed decoration that resemble FBA forms of Calabria and the Aeolian Islands, but not those of the *Pantalica* group.[353] The variety of bronze items added to the burials include the earlier mentioned fibulae, knives, razors, armaments (e.g. swords) and jewellery (rings, pendants, etc.) (cf. Bernabò Brea et al. 1969, 227–243, Figs. 15–22; Giumlia-Mair et al. 2010). At least three graves contained amber: 11, 46 and 211 (exact number of artefacts and contexts is difficult to quantify due to incomplete state of publication regarding the necropolis). Reported amber(s) from tomb 11 are still to be verified in terms of the quality and number, as founder of the grave P. Orsi noted only 'two dozens of discoid beads made of marine fossil material', however without giving more specific descriptions or providing any figures (Orsi 1905, 106). Judging by other elements of the inventory, primarily a serpentine-bow fibula with flattened, leaf-shaped bracket and grooved ornament (false twisting), classified by F. Lo Schiavo to the type 286A, the grave can be placed in the early stage of the necropolis – *Pantalica 2a*/BF 3a (cf. Giumlia-Mair et al. 2010, 476, Fig. 11; Pare 2008, 85). Tomb 46 yielded a single amber bead (Bernabò Brea et al. 1969, Fig. 22:b), accompanied by a rich and varied set of objects. Among them particular attention deserves the ivory comb with partly preserved handle in trapezoidal shape, decorated with concentric circles, which finds contemporaneous analogies in the mainland Italy (cf. Bernabò Brea et al. 1969, 243–244, Fig. 22:g; Albanese Procelli/Lo Schiavo 2004, 411–412). Dating to the BF 3, perhaps its later part – *Pantalica 2b* – is warranted by two fibulae: one with simple arch decorated with the 'herringbone' motif (Bernabò Brea et al. 1969, Fig. 22:c) and the other with threaded elbow arch – early variant of the Cassibile type, assigned by Lo Schiavo to types 31, var. E and 288 respectively (Giumlia-Mair et al. 2010, 476, Fig. 11; Lo Schiavo 2010, 118–119, n. 174). Grave 211, also containing a single amber bead, had a less rich inventory, but still can be placed within the phase BF 3 (Giumlia-Mair et al. 2010, 470).

Amber continued to be used in the following period known as the *Pantalica 3 – South*, corresponding with PF IB of the Italian mainland sequence, as indicated by, e.g. graves 6 and 29 from the namesake cemetery (cf. Orsi 1912, 309–311, P. VII:43).

352 Orsi 1905; Bernabò Brea et al. 1969; Albanese Procelli 1992, 1994; Albanese Procelli/ Lo Schiavo 2004, 404.

353 Bernabò Brea et al. 1969, 244–250, Figs. 24–33; Bietti Sestieri 1979, 617, Fig. 6:1–9; Albanese Procelli/Lo Schiavo 2004, 411.

Tab. 18. List of the Bronze Age contexts with amber on Sicily

Name	Commune	Province	Type	Relative chronology	Absolute chronology	Reference
Castelluccio, tomb 9	Noto	Siracusa	Rock-embedded tomb	*Castelluccio 1–2*	2200–1800/1700 BC	Orsi 1892, Tab. III:15, 18; Massari et al. 1996, Fig. B:5–7
Castelluccio, tomb 22	Noto	Siracusa	Rock-embedded tomb	*Castelluccio 1–2*	2200–1800/1700 BC	Orsi 1892, 30
Castelluccio, tomb 23	Noto	Siracusa	Rock-embedded tomb	*Castelluccio 1–2*	2200–1800/1700 BC	Orsi 1892, Tab. V:17
Cava Cana Barbara, tomb 3	Priolo Gargallo	Siracusa	Rock-embedded tomb	*Castelluccio 1–3?*	2200–1600/1500 BC	Orsi 1902a, 186
Cava Cana Barbara, tomb 9	Priolo Gargallo	Siracusa	Rock-embedded tomb	*Castelluccio 1–3* (*Thapsos?*)	2200–1300/1250 BC	Orsi 1902a, Tab. VI:11
Monte Sallia, tomb 5	Comiso	Siracusa	Rock-embedded tomb	*Castelluccio 2A*	2000/1900–1900/1800 BC	Orsi 1923, 17
Cava Secchiera, tomb 10	Melilli	Siracusa	Rock-embedded tomb	*Castelluccio 2B–3A*	1900/1800–1700/1600 BC	Orsi 1893b, Pl. II; Angelini/Bellintani 2017, Fig. 1:c
Valsavoia, tomb 18	Lentini	Siracusa	Rock-embedded tomb	*Castelluccio 3* (*Thapsos?*)	1800/1700–1300/1250 BC	Orsi 1902b, 114
Longane, tomb 21	Rodì Milici	Messina	Rock-embedded tomb	*Rodì-Tindari-Vallelunga*; BA2	1900–1420 BC	Bernabò Brea 1967, Fig. 34:1
Plemmirio, tomb 10	Siracusa	Siracusa	Rock-embedded tomb	*Thapsos I–III*	1420–1270/1250 BC	Orsi 1891, Pl. XI:9; Cultraro 2007, Pl. XCVIIIc:10
Plemmirio, tomb 11	Siracusa	Siracusa	Rock-embedded tomb	*Thapsos I–III*	1420–1270/1250 BC	Cultraro 2007, 381
Plemmirio, tomb 48	Siracusa	Siracusa	Rock-embedded tomb	*Thapsos I–II*	1420–1330/1315 BC	Colini 1904, Fig. 54; Genovese2015, Fig. 8

Tab. 18. Continued

Name	Commune	Province	Type	Relative chronology	Absolute chronology	Reference
Plemmirio, tomb 49	Siracusa	Siracusa	Rock-embedded tomb	*Thapsos I–III*	1420–1270/1250 BC	Colini 1904, Fig. 55
Thapsos, tomb D	Priolo Gargallo	Siracusa	Rock-embedded tomb	*Thapsos I–III*	1420–1270/1250 BC	Voza 1973, 40, nn. 105–106; Cultraro 2007, Pl. XCVIII:c9
Thapsos, tomb 61	Priolo Gargallo	Siracusa	Rock-embedded tomb	*Thapsos II*	1390/1370–1330/1315 BC	Orsi 1895, Fig. 49; Cultraro 2007, Pl. XCVIIIc:5
Madonna del Piano, tomb 11	Grammichele	Catania	Burial – inhumation	*Pantalica II – Cassibile*; FBA 2a	1035/1010–960/950 BC	Orsi 1905, 106; Cultraro 2007, 384
Madonna del Piano, tomb 46	Grammichele	Catania	Burial – inhumation (*pithos*)	*Pantalica II – Cassibile*; FBA 2b	960/950–910 BC	Bernabò Brea et al. 1969, Fig. 22:b
Madonna del Piano, tomb 211	Grammichele	Catania	Burial – inhumation	*Pantalica II – Cassibile*; FBA 2	1035/1010–910 BC	Giumlia-Mair et al. 2010, 471

2.3. Western and Central Balkans

2.3.1. Istria

The Istrian peninsula in Croatia presents one of the most difficult cases of amber finds in terms of chronological determination. Some of the amber-bearing contexts from Istria (Fig. 14) have been interpreted as the oldest sites with amber in the entire Western Balkans (Čović 1983a, 122, 124, Palavestra 1993, 46, 272). New data from the region, especially 14C dates, allow to verify this notion.

Chronological framework of the Bronze Age in Istria devised by B. Čović assumes a tripartite division of the period into phases *Istra I* (BzA1), *Istra II* (BzA2–B) and *Istra III* (BzC1–C2), followed by transition from the Bronze to the EIA (also called the Istrian LBA) with its dating being a subject of debate (cf. Čović 1983a, 1983b; Batović 1983). In a recent elaboration of the *Istrian/Histrian* group of the Iron Age by K. Mihovilić, beginning of the period was set at the mid-12th century BC and labelled as the phase Ia (roughly corresponding with late HaA1, HaA2 and early HaB1) (Mihovilić 2013a, 114–144). Estimating absolute time frames of the Bronze Age phases of Istria is a much more problematic task, mainly due to scarcity of radiometric determinations. The phases and relevant contexts are predominantly dated mainly by reference to Central European modified Reinecke's system, in the parallelisation given above.

Nonetheless, dating the Istrian sites in relation to Italian Bronze Age chronology is also possible. Present state of research allows to include Bronze Age Istria within the spatial scope of the so-called *Castellieri-Gradine* culture – cultural facies prevailing in *Caput Adriae* during early and middle stages of the 2nd millennium BC. In terms of chronology, this cultural phenomenon can be enclosed within the period starting roughly from phase BA 2 in Northern Italian, and BzA2 in Central European periodisations, up to phase BR and BzD, respectively.[354] Within the above-outlined time span can be placed all Bronze Age amber finds from Istria, discovered on the following sites: Krmedski Novi Grad, Laganiši, Mušego, Škicini, Vrčin and Žamnjak.

The most prominent element of the *Castellieri-Gradine* culture are the eponymous hillforts (it. *Castellieri*, hr. *Gradine*), usually situated on elevations and promontories dominating over the coastland and valleys of upper Adriatic karst. They are surrounded by one or two drystone ramparts, enhancing defensive qualities of the terrain. In cases where there were no usable elevations, other natural features of the landscape, such as sinkholes, were used for ensuring security of the inhabitants (Mihovilić 2013b, 867). The walls often have several gates, sometimes with very complex structure, built to hinder the access to settlements by potential invaders. Inner space of the hillforts was arranged in various ways, with housing structures, subsistence facilities and production areas separated into zones, sometimes delimited by walls or other features. In the entire coastal area of Primorska, Notrjanska, Istria and Kvarner have been registered over 400 *gradine*, most of them concentrated on the Istrian peninsula (Mihovilić 2013b, Fig. 46.1). Despite a long history of research, dating back to the second half of the 19th century, only few *gradine* were thoroughly excavated.

In the funeral sphere, the *Castellieri-Gradine* culture is represented by rare graves, mostly found in the settlements' vicinity. Three basic forms of burial have been recognised (from the most to the least common): inhumations inside stone tumuli, located some distance away from the settlements; second, inside stone-cist graves arranged around entrances to the settlements; third, in simple, stone-enclosed grave-pits situated within the settlements' walls (Buršić-Matijašić/Žerić 2013, 68–69). Common for the first two burial practices is construction of graves in the form of a chamber, made of carefully hewn stone slabs (Hänsel et al. 2007–2008, 95, Sl. 12–15, 18–21). The graves usually held remains of multiple interments, deposited in lying or sitting position, with their leg and hands bent. They were rather poorly furnished, by adding only the simplest elements of bronze jewellery, rarely supplemented with other, more exotic materials, such as amber, glass (faience) and sea snail shells. Burial customs described above refer to almost all contexts with amber in Istria, with the sole exclusion of Laganiši, where a mass grave was discovered inside a cave.

Estimating the exact chronology of amber artefacts from Istria is a difficult task. The problem in this regard results from insufficient number of accurately datable archaeological finds, accompanying amber in the contexts. Modern-time robberies of the graves, in which Bronze Age Istrian amber appears exclusively, are only partly to take the blame (cf. Mihovilić et al. 2011; Buršić-Matijašić/Žerić 2013). The

354 Cf. Cardarelli 1983; Čović 1983a; Bianchin Citton/Vitri 1991–1992; Mihovilić 2013b; Borgna et al. 2018.

general modesty of grave inventories seems to have been an effect of conscious and deliberate cultural behaviour. This leaves a group of chronological markers confined to merely several weapons and tools, supplemented with numerous elements of jewellery commonly fashioned from a bronze wire. In a few cases fragments of ceramic vessels may additionally serve as a referential material in determining the chronology.

The contexts with amber that can be satisfactorily dated on the basis of their content are: grave 1 at the Novi Grad hillfort near Krmed, Laganiši-sepulchral cave and, with some reservations, also the tumulus of Žamnjak. Grave 1 of Krmedski Novi Grad is placed on a saddle connecting two hills, raising above the village of Krmed in the south of Istria, which locals call the 'old' and the 'new town' (Bačić 1957). The 'old town' is a prehistoric *gradine*-type settlement, while the 'new town', lower by 21 m, has a cemetery on its plateau. Grave 1, situated on the latter, comprised a stone mound covering a rectangular grave chamber built of stone slabs. Inside were found skeletons of two adult individuals originally placed inside the tomb in a 'sitting' position, leaning against the walls. In between the chamber, skeleton of a child was found. Bronze and amber finds were discovered nearby the child's skull. The context owes its dating to the bracelet made of a single turn of thick bronze wire with semi-circular cross-section, with its ends coiled up in a pair of opposed spiral disc (Bačić 1957, Fig. 4; Batović 1983, Tab. XLII: 6). Similar bracelets, or armlets in a different terminology, probably evolved from earlier EBA varieties, made of thin and long, repeatedly coiled strip of bronze tape. They subsequently became widespread across the regions influenced by the *Tumulus* culture,[355] where they are usually dated to BzC1–BzD – time that fits the earlier proposed dating of the Krmed specimen.[356] In the Balkan area similarly dated analogies are known from Western Serbia, e.g. Grotnica in Guča (Dmitrović 2016, Fig. 113:6), whereas most probably the earliest finds of such kind come from the Lovas hoard dated to the *Koszider* horizon, roughly contemporary with phase BzB.[357]

The site of Laganiši, located in the municipality of Oprtalj, on the edge of a rocky plateau, is a large cave consisting of the main corridor-like chamber with numerous side-rooms (Mihovilić 2008). Among them is one filled with remains of the deceased, separated from the rest of the cave by a drywall blocking the narrow passage. The bones of 12 individuals (4 adults – 3 females and 1 male, and 8 children – all less than 10 y.o.) were scattered around the room, so it is considered that the dead were not buried, but laid on the floor. This so-called 'sepulchral cave' yielded two well-datable bronze items: an axe with short median wings/fins and recess at the heel, akin to the type Feudenberg generally dated to BzD–HaA1, and a dagger pertinent to the *Peschiera* family of daggers from Italy, in particular the type Teno dated to BR.[358]

355 Hochstetter 1980, Beilage 1:40; Gedl 1975, Tab. XIX:1–2, 7–9, 15, XX:8–10, XXI:5–6; Čujanová-Jílková 1978, Fig. 129:10–11.

356 Čović 1983a, 117; 1983b, 236; Batović 1983, 290.

357 David 2002, Taf. 113:2, 3; Taf. 187:8c; Ložnjak Dizdar, Potrebica 2017, 57, Fig. 16.

358 Mihovilić 2008, 45; cf. Meyer 1977, 93–95; Bianco Peroni 1994, 156, nos. 1552–1563.

The Žamnjak tumulus, located approximately 12 km east from Rovinj, on the hill immediately south of the village Sošići, was built of large, heaped up stones, with the base circumscribed by a ring of vertically mounted stones (Bačić 1960). In the mound's centre was located a chamber-tomb, constructed of six stone slabs. Fragmented skeletal remains indicate that it was a family tomb, where bodies of three individuals (two adults and a child) were successively buried in a cramped position. The grave, apart from amber beads and tubes of spirally twisted bronze wire (*saltaleoni*) did not contain chronologically unambiguous finds. However, on the outskirts of the mound was found a small bowl with hemispherical body and funnel-shaped neck, reportedly having analogies in the *Maros/Moriš* culture.[359] On that basis the mound was dated to the late BzA2 and BzB (Palavestra 1993, 46), although, due to questioned validity of associating the vessel with to the grave, the dating remains controversial among the scholars (J. Mavrović Mokos, personal communication).

Fortunately, absolute chronology chiefly based on 14C measurements is increasingly becoming a standard, and thanks to an ample series of radiocarbon determinations from the Monkodonja hillfort and the nearby Mušego cemetery, archaeologists were given fix-points for chronological considerations of other sites in the area. While habitation layers of Monkodonja have not revealed any evidence of amber, three amber-like beads were recovered from the so-called 'tomb A' composed into the western gate of the hillfort (Hänsel et al. 2015, Fig. 153:5–7). However, the beads were lately analysed with IRS as well as gas chromatography with mass spectrometry, and the results revealed they are not made of fossil resin, nor of any organic material (Cwaliński et al. 2018; Cwaliński et al. forthcoming).

Still, three mounds at the abovementioned cemetery of Mušego yielded positively verified amber beads (nos. 6, 7 and 9).[360] The site is located about 1 km to southeast from the hillfort of Monkodonja. The tumuli, in total 13, are arranged over the hillside in a number of clusters counting two to four mounds on average, with some of the larger cairns raised in isolated positions. The mounds were erected on the plateau and on the upper parts of the slopes. Stone-cist graves under the tumuli contain remains of several individuals on average, in some cases deposited over longer stretches of time (e.g. tumulus 7). Radiocarbon measurements obtained for all the three barrows in total indicate a time span roughly between 1500 and 1100 cal. BC; however, the earliest and the latest dates should be omitted according to the authors.[361] Remaining dates give the results falling in the range between 1400 and 1200 cal. BC on average, so corresponding with phases BzC2–D and BM 3–BR.[362]

---

359 Čović 1983a, 131, Tab. XIII:9; cf. Garašanin 1983a, Tab. LXVII:5, LXVIII:18.

360 Mihovilić et al. 2009, 2011; Cwaliński et al. forthcoming.

361 KIA 33494 (3245 ± 30 BP), 1535–1456 cal. BC (55.7 %), 1611–1447 cal. BC (95.4 %); KIA 37976 (2910 ± 35 BP), 1190–1031 cal. BC (68.2 %), 1214–1006 cal. BC (95.4 %) (Hänsel et al. 2015, 434, 437, 449–450).

362 Tumulus 6: KIA 33493 (3065 ± 25 BP), 1389–1285 cal. BC (68.2 %), 1410–1261 cal. BC (95.4 %); KIA 33495 (3020 ± 25 BP), 1292–1220 cal. BC (65.5 %), 1388–1194 cal. BC (94.4 %). Tumulus 7: KIA 36728 (3080 ± 25 BP), 1399–1301 cal. BC (68.2 %), 1415–1274 cal. BC (95.4 %); KIA 36729 (3050 ± 25 BP), 1382–1265 cal. BC (68.2 %), 1397–1230 cal.BC (95.4 %); KIA 36727 (3025 ± 25 BP), 1296–1226 cal. BC (63.0 %), 1391–1196 cal. BC (95.4 %); KIA 36730 (3015 ± 20 BP), 1281–1221 cal. BC (68.2 %), 1379–1196 cal. BC

Although the final *caesura* seems safe, the beginnings of the Mušego cemetery might be slightly earlier, as the primary graves within dated tumuli were already robbed at the time of the excavations.[363]

Very little can be said about the chronology of amber finds from the Škicini tumulus. The grave itself did not lend many clues regarding the dating, and its inventory lacks precise chronological markers (cf. Buršić-Matijašić 2003). In the face of general scarcity of accurately datable materials and absolute dates, one has to resort to other kinds of data to estimate chronology of the sites. Similarities in the burial rite practised in Istria during the Bronze Age, as seen in the case of analysed graves, may serve as indicators of contemporaneity. Stone cairn covering cist tomb, the latter overbuilt with large, roughly rectangular slabs, is a recurring grave form (Krmedski Novi Grad, Mušego, Škicini, Žamnjak), regarded as a local manifestation of the wider European phenomenon of tumulus cemeteries, especially prominent during the MBA (BzB–C2).[364] The tumuli generally housed singular graves, although, as the case of Mušego shows, the habit of adding secondary burials was not uncommon. The tombs usually comprised skeletal remains of several individuals, buried in a few, consecutive interments, with the bodies arranged in a flexed position to fit constrained compartment. All these features help to estimate the chronology of the Škicini tumulus, and other alike mounds, on phases *Istra II–III*, coeval with stages BzB–HaA1 (1575/1550–1150 BC), what generally confirms the dating proposed by K. Buršić-Matijašić (Buršić-Matijašić 2003, 175–176).

To date the largest collection of amber finds has been encountered on the site of Vrčin near Vodnjan. The site consists of a hillfort including within its walls a group of stone-cist graves, and an adjacent tumuli cemetery to the east. The settlement is surrounded by double stone-made wall with several entrances. Emergence of adherent necropolis, located along the walls and explored from 1925 to 1929 by R. Battaglia and B. Tamaro-Forlati (Buršić-Matijašić 1989, 477), is connected with the third phase of the fort's construction, when one of the gates was shut. Number of excavated graves varies in the literature from 17 to 20. Those were mostly flat graves with stone enclosures, inside which deceased were buried in a sitting position. Stone walls of the fort separated the graves into groups. Almost every grave contained three to four burials. Analysis of collected bones showed that they belonged to 34 skeletons of which 11 were males, another 11 females and 12 children.

Approximate chronology of the necropolis at the Vrčin hillfort, where a number of graves yielded amber (Buršić-Matijašić 1989; Palavestra 1993, 47–48), may be inferred *per analogiam* from the chronology of the Monkodonja's cist graves A and B. Similarities in the location of the graves within both settlements, occupying adjacent position to the defensive walls, their akin construction in the form of rectangular chamber built of large stone slabs and, last but not least, strikingly complementary

(95.4 %). Tumulus 9: KIA 39759 (3020 ± 20 BP), 1287–1223 cal. BC (68.2 %), 1383–1208 cal. BC (95.4 %); KIA 39758 (2965 ± 30 BP), 1223–1126 cal. BC (68.2 %), 1268–1056 cal. BC (95.4 %) (Hänsel et al. 2015, 434–437).

363 Hänsel et al. 2007–2008, 92; Hänsel et al. 2015, 451.

364 See e.g. Čović 1983a, 124–125; 1983b 234–236; Mihovilić et al. 2011, 371–372; Buršić-Matijašić/Žerić 2013, 70–71.

2.3.2. Kvarner and northern Dalmatia

Numerous sites with amber known from the eastern Adriatic coast, including regions of Kvarner, Dalmatia and Montenegro (Fig. 14), have already been a subject of several studies in the past.[367] Almost all of them, in contrast to Istria, are dated to the LBA. Unfortunately, due to distant and obscure research history, there are still many unsolved issues about the finds localisation, chronology or composition. Moreover, their cultural context, despite rich accumulation of sites and finds, remains poorly understood.

Studies of the LBA–EIA relicts coming from the Adriatic coastline (formerly belonging to Yugoslavia) by Š. Batović resulted in a proposal to separate the latter into several cultural groups, with names referring to Illyrian tribes inhabiting these areas in the antiquity. Thus, the area stretching from the Raša river in Istria to the Krka river in northern Dalmatia was included in the *Liburnian* group, central Dalmatian coastal belt spanning between the Krka and Neretva rivers became the *Dalmatian* group (or the *Middle-Dalmatian* group – cf. Čović 1987), and finally the southernmost part of the coast from the Neretva to Montenegro was labelled as the *Južnoprimorska grupa*, loosely translated as the *Southern-coastal* group (Batović 1983, 272–273). Batović also developed chronological systematisation of the LBA and Iron Age for the entire coast, comprising all the three groups. According to its latest version, the ending of the 12th and 11th century BC is denominated as the LBA, whereas the 10th century BC is considered as the transitional period preceding full inception of the EIA in the 9th century BC (Batović 1983, 273–274, 329–330; Batović 1987, 346–351).

In the recent study devoted to the Bronze-Iron Age transition in the Kvarner Bay, hitherto functioning periodisation for the *Liburnian* group has been revised, and adjusted to the re-evaluated chronology of local sites and particular types of finds (Blečić Kavur 2014). Consequently, the dating of the Ist phase (including subphases IA, IB and IC) was set roughly between 1200 and 950 BC, whereas the IInd phase was synchronised with the EIA (Blečić Kavur 2014, 162–165, Fig. 90), therefore in contemporaneity with the final stage of the Bronze Age and beginning of the Iron Age in Dalmatia according to C. F. E. Pare (1998, 327–332, Tab. 3). Certain similarities in the composition of the inventories from Northern Dalmatian and Kvarnerian contexts, along with the presence of typologically compatible amber beads, allow to treat them as two subgroups of culturally related communities participating in the same exchange network (cf. Blečić Kavur 2014, 45, 57–58, 152, n. 815), hence the decision to include them in a single province of amber's acquisition.

All contexts from the area can be placed in a similar time span roughly between the 13th and 9th century BC. Two graves containing amber were discovered in the vicinity of Nin: Privlaka, grave 87 and Vrsi, grave 89 – both well covered in the literature, although due to the circumstances of the discovery, largely undetermined in terms of localisation (Batović 1960). Chronological consideration of their inventories, comprising such characteristic items for discussed zone as: arched fibulae with two nodules on the bow, ribbed-cuff bracelets and closed bracelets with V-shaped cross-section[368] resulted in their assignment to the Ist phase of the *Liburnian* group, corresponding with

367 E.g. Lo Schiavo 1970; Palavestra 1993; Forenbaher 1995; Blečić Kavur 2009, 2014.
368 Batović 1983, T. XLIV:7–8, XLIV:13–15, XLIV:1–6, 16–18.

phases HaA2–HaB1.[369] The latest revisions of the typo-chronology of aforementioned fibulae tend to shift their initial emergence to the 12th century BC (LH IIIC; HaA1)[370] – a caesura more fitting also in the case of Tiryns-type beads with which they sometimes co-occur, e.g. Vrsi, grave 89.

Furthermore, numerous amber artefacts, most likely coming from two contexts on the Island of Krk, are currently stored in the Museum of Natural History in Vienna. Again, the exact location of the sites cannot be specified, as the circumstances of discovery are not clear (they are probably graves). First one is traditionally attributed to the town of Baška (Lo Schiavo 1970, 424–426; Palavestra 1993, 62–63). Recent investigations allowed to verify to some extent location of the site, which should be looked for in the vicinity of Draga Bašćanska, and to reconstruct composition of the finds (Mader 2006, 253–264; Blečić Kavur 2014, 46). As a result, part of amber collection and correlated finds are currently tentatively defined as the 'reconstructed grave from Baška', as they originally might have belonged to richly furnished grave with cremated female burial that was destroyed due to agricultural works (Blečić Kavur 2014, 169–170, Fig. 20).

Remaining part of the collection kept in the Viennese museum under the label 'Island of Veglia' (Italian name for the Krk Island) is left without determination of the exact place of their discovery. Despite well-grounded suspicion that they were also found by E. Nowotny in 1895 near Bescavalle (modern village of Draga Bašćanska) (cf. Mader 2006, 260), it seems the most reasonable to leave it as a separate context, as already proposed by F. Lo Schiavo.[371] The assemblage, probably coming from a single grave, contained i.a. Tiryns- and Allumiere-type beads (Blečić Kavur 2014, 171–172, Fig. 21). Thanks to comprehensive analysis of the co-occurring finds, among them closed bracelets with V-shaped cross-section, anthropomorphic and ring pendants, both contexts from Krk can be dated approximately to subphases IA–IC (HaA1–HaB1) (Blečić Kavur 2014, Fig. 90).

Another site from the Krk Island is known under the name of Garica – a town near Vrbnik on the north-east part of the island, where a large collection of materials including a single amber bead representing type 9 was discovered during excavations between 1873 and 1874. These finds, originally not included in the Palavestra's catalogue, were recovered from two tumuli containing traces of cremation; however, it is impossible to assign individual artefacts to a specific mound (Ljubić 1876; Blečić Kavur 2014, 189–191, Tab. 1). Overview of the rich inventory, comprising numerous bronze elements of jewellery (i.a. ribbed-cuff bracelets, Velika Gorica-type pin and pin with an onion-shaped head and profiled neck), allows to place the site around phase IB (Blečić Kavur 2014, Tab. 1:155–156, 164); however, presence of some typologically younger items, e.g. two-part serpentine fibulae[372] or Oblekovice/Fontanella-type razor[373] prolongs its chronology to the beginning of phase IIA.

---

369 Batović 1959, 74–75, n. 75; Pare 1998, 327–328, Abb. 15:2, 6–7; Glogović 2003, 11, Taf. 71.

370 Teržan 2007, 160–162; Blečić Kavur 2014, 48–51, Figs. 87 and 90.

371 Lo Schiavo 1970, 424–427; see also Blečić Kavur 2014, n. 188; cf. Palavestra 1993, 62–63.

372 Glogović 2003, 76, Taf. 47:355–363; Blečić Kavur 2014, 133–134, Tab. 1:162.

373 Bianco Peroni 1979, 58–60, Tab. 112b, 113a; Weber 1996, 234–240, Tab. 50–51; Blečić Kavur 2014, 137–138, Tab. 1:168.

The last site – Vidasi – is located in the northern part of the island of Pag, near the Novalja church. Section of the site, quite possibly a necropolis, was discovered accidentally in the 1970s on the so-called locality *Vidasa napolju* (en. Vidas field) (Glogović 1991). On the basis of chronology and stylistic traits, the material from the site can be roughly divided into two groups: LBA (HaA1–HaB), and Iron Age (HaC1–HaD/La Tène A), possibly indicating two phases of the necropolis. Finds attributed to the first phase, apart from a single amber bead, include arched fibulae with two knobs on the bow, ribbed-cuff and closed bracelets with V-shaped cross-section, and the Šula-type pin which put it quite firmly into HaA1–HaB1 horizon; however, certain types of fibulae indicate its continuation up to HaB2/3 (Blečić Kavur 2014, 172–173, Fig. 23).

Tab. 20. List of the Bronze Age contexts with amber in Kvarner and northern Dalmatia

Name	Commune	Province	Type	Relative chronology	Absolute chronology	Reference
Baška, 'reconstructed grave'	Baška	Primorsko-goranska županija	Burial – cremation	HaA1–HaB1	1200–950 BC	Blečić Kavur 2014, Fig. 20:10–11
Garica, precise context unknown	Vrbnik	Primorsko-goranska županija	Burial?	HaA2/B1–HaB2/3	1100–910/850 BC	Drechsler-Bižić 1970, Tab. V:1
Island of Krk, precise context unknown	?	Primorsko-goranska županija	Burial?	HaA1–HaB1	1200–950 BC	Blečić Kavur 2014, Fig. 21:22–24
Privlaka, grave 87	Privlaka	Zadarska županija	Burial – cremation	HaA1–HaB1	1200–950 BC	Batović 1983, T. XLIV:19–21
Vidasi, precise context unknown	Novalja	Ličko-senjska županija	Burial?	HaA1–HaB2/3	1200–910/850 BC	Glogović 1991, T. 1:2
Vrsi, grave 89	Vrsi	Zadarska županija	Burial – cremation	HaA1–HaB1	1200–950 BC	Batović 1983, T. XLIV:9–12

2.3.3. Middle Dalmatia

From the coastal belt stretching between Krka and Neretva rivers attributed to the *Dalmatian* group, but here tentatively named as middle Dalmatia, comes a single site which yielded amber artefacts – Trcela near Vranjic (Marović 1960; Palavestra 1993, 65; Fig. 14). The site is located on the areas surrounding ancient Salona (nowadays Solin), approximately few hundred metres south-west from the estuary of the Jadar River and the Vranjic Island located at its mouth. Site's exact location is difficult to specify; however, it must be somewhere on the land presently included within the split city limits. It was discovered accidently in 1925; however, circumstances of this discovery are unknown. First short report was published in 1927 by F. Bulić, who described the finds by relying on the testimonies of witnesses present at the site at the time of discovery. Context of the finds reportedly was akin to a stone-cist grave. It contained skeletons of two adults and a child. It is difficult to determine whether the tomb was flat or hid under a stone cairn, because it was located in an area disturbed by marl extraction.

During the LBA in middle Dalmatia, dominating burial rite was inhumation of the deceased in stone cists enclosed with slabs, designed as flat graves or covered with mounds, but burials in caves are also known (Batović 1983, 334; Ložnjak Dizdar/Potrebica 2017, 121). The layout of bodies inside the grave was usually in contracted/flexed position on stone boards. Most common are singular burials, but multiple ones are also known, as in the case of Trcela-Vranjic. With respect to tumuli, it was observed that the cairns were heaped up of broken limestone, located in elevated positions, on the slopes or summits of hills, at different altitudes, more or less distant from settlements. They occur individually or in groups. They usually cover single, primary grave in the centre with additional graves on the sides, placed without any regularity, what potentially indicates prolonged use of these mounds.

Crucial for determination of the site's chronology is presence of the Golinjevo-type fibulae (Marović 1960, Fig. 3) – variant of arched fibulae with two knots on the bracket, characteristic for Dalmatian coast and southern Bosnia with singular examples known from Albania, Lika and lower Sava Valley (Batović 1983, 338–339). In comparison with the other Golinjevo-type fibulae, the two specimens from Trcela can be considered as prototypes (Glogović 2003, 12–13, Taf. 71). Primarily, the difference lies in cross-section of the bracket: almost all other specimens were made from longitudinally grooved bronze wire, while the Trcela examples have polygonal cross-section. On that basis, the latter can be dated to an earlier part of the Golinjevo-type timeline which occupies HaB1 and early HaB2 (Glogović 2003, 13). Other finds from Trcela that fit the chronology outlined above are two ribbed-cuff bracelets[374] and an imitated twisted torc.[375]

---

374 Marović 1960, Fig. 1:3, 5. Regardless of the attempts to distinguish specific variants of ribbed-cuff bracelets basing on the shape of their endings – straight or rolled up – the entire category can be dated from HaA onwards up to early stages of the Iron Age (Blečić-Kavur 2014, 86). C. F. E. Pare treats these bracelets as a typical indicator of the final horizon of the Bronze Age in Dalmatia, dating the contexts containing them before 950 BC (Pare 1998, Abb. 15:6).

375 Marović 1960, Fig. 2:1. The torc from Trcela is made of massive bronze rod with rhombus cross-section, seemingly corrugated in its middle section (form of decoration performed by engraving), with the endings rolled up, therefore more akin to younger versions dated to HaA2/B1. C. F. E. Pare places them in an even later period, i.e. in the first phase of the Early Iron Age in Dalmatia (HaB2 – after 950 BC; Pare 1998, 330, Abb. 17:10); however it should be kept in mind that these objects have rather lengthy chronology, with first appearance on the territories of the *Urnfield* culture, already around BzD/HaA1 and much younger specimens from Macedonia dated to the 7th century BC (Vasić 2010, 42–43). The example from Trcela can be correlated with other examples of twisted torc present in a large number in the Glasinac mainly in younger graves of phase IIIC (HaB1; Vasić 2010, 42) or, in Pare's chronology, dated to *Glasinac* IA phase (Pare 1998, 334, Abb. 19:3). On the other hand, very similar specimen from Konjuša in Western Serbia has been dated to a time circa Ha2–HaB1 (Vasić 2010, Taf. 28:185, Taf. 44), so quite similar to the discussed object from Trcela.

Tab. 22. List of the Bronze Age contexts with amber in Montenegro and southern Dalmatia

Name	Commune	Province	Type	Relative chronology	Absolute chronology	Reference
Velika Gruda, grave 9	Kotor	Kotor	Burial – inhumation and cremation	BzC2–BzD	1400–1200 BC	Della Casa 1996a, 44:18

Fig. 14. Distribution of the Bronze Age sites with amber in Western and Central Balkans.

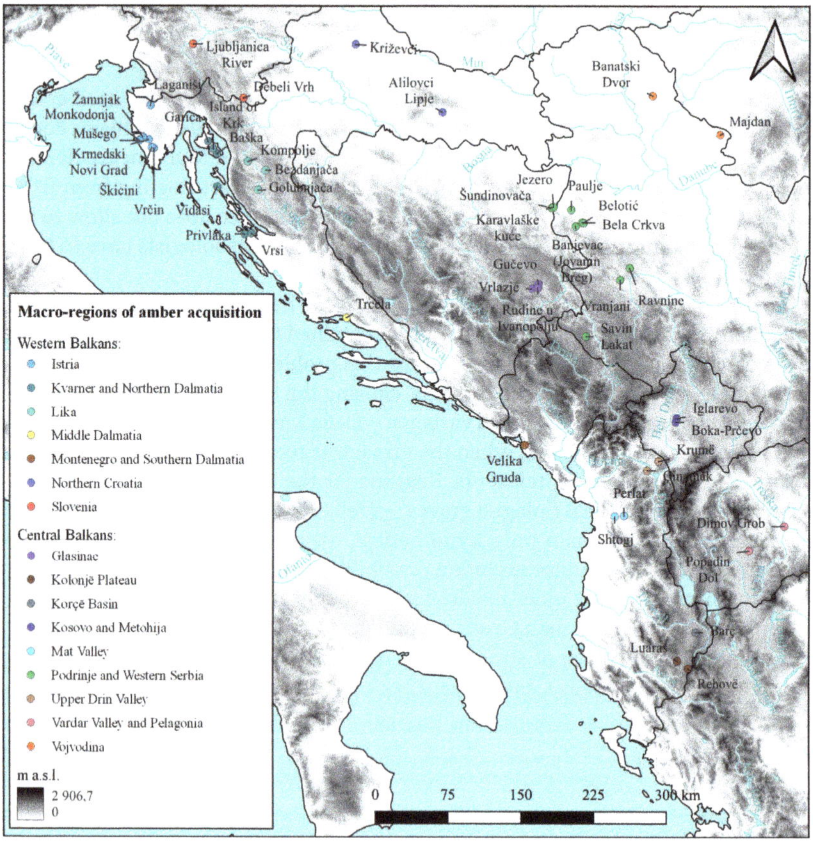

2.3.5. Lika

Three Bronze Age sites with amber are known from the territory of Lika – traditionally distinguished region occupying large part of the so-called Croatia proper, stretching between the Velebit mountain range in the south-west and Plješevica Mountain in the north-east (Fig. 14). Among them are two caves – a category of archaeological sites that has yielded the biggest amount of data on the Bronze Age in the region. Chronologically, the earliest of them is the Bezdanjača cave located 10 km east of

Prozor, between the Zalužnica and Vrhovine villages, at the Vatinovac hill (Drechsler-Bižić 1979–1980). Entrance to the cave is located at the height of 740 m a.s.l. The upper part of the cave's entrance is funnel shaped, changing into a vertical shaft, ending in a scree of fallen material. The entire entrance section is 31 m in depth. At the base of the vertical entrance, the cave branches out in two directions. A short and broad channel leads to the west, while a considerably longer and morphologically more complex main channel extends to the south-east, with numerous extensions on the sides. The total length of all explored channels measures 1,176 m, but only the first 190 m are interesting in archaeological terms (Malinar 1998, 147–148).

South-east channel has been interpreted as a cemetery, where the dead were buried mostly in natural depressions along the walls. Single, double and group graves were identified, while the total number of buried individuals is estimated at 200. Majority of graves was disturbed in the past, either by removing the bones to obtain space for new burials, or by natural factors, such as erosion. Apart from burials, the cave included also traces of fireplaces (hearths), animal bones (probably offerings) and remains of wooden structures of unknown function (excavators suggest they could have been used for hanging vessels) that create a picture of a special place, perhaps used to perform rituals. Single amber bead was found in the mass grave 4 – skeletal burial consisting of dislocated bones, belonging to nine individuals, located in the south-east, final part of the south-east channel (Drechsler-Bižić 1979–1980, Tab. XXV: 3–11).

Significance of Bezdanjača cave lies in the fact that it yielded materials of varied chronology, spanning several centuries, thus has been used as a focal point in the chronology of the Middle and LBA in Lika. Sequence of the burials was separated into two horizons. The first one belongs to advanced and late stages of the MBA (BzC and early BzD), whereas the second one is contemporary with early stages of the LBA (advanced and late parts of BzD and early HaA) (Drechsler-Bižić 1983a, 243–246). The mass grave no. 4 was assigned to the I[st] horizon (Drechsler-Bižić 1979–1980, 33; Palavestra 1993, 67–68). Entire time span of the cave's use as a burial ground (BzC–HaA1) is supported by three 14C dates (out of seven in total – the remaining ones relate to features not linked directly to the burials), out of which two indicate a period between 1400 and 1100 BC, while the third seems to document start of the Bezdanjača necropolis beyond phase BzC that is before 1450 BC.[378] However, the dates were measured in the pre-AMS era, so their precision leaves much to be desired.

Although chronological division proposed by R. Drechsler-Bižić has been accepted ever since, one has to be aware that it was done on the basis of 'horizontal' stratigraphy. Even though there are visible patterns in distribution of selected categories of finds (cf. Della Casa 1996a, Abb. 158–159), this separation is not stark, and the zones covered by materials of different chronology are not mutually exclusive. The problem also stems from the fact that most of datable ceramic vessels and bronze objects comes from collective ('mass') graves at the back of the cave (among them grave 4), which cannot be regarded as closed features.

Overall, ceramics and bronze finds from the I[st] horizon of Bezdanjača show affinity with some indigenous cultures of the central and south-eastern Bosnia and

378 Z-186/I (2986 ± 75 BP), 1377–1111 cal. BC (68.2 %), 1411–1011 cal. BC (95.4 %); Z-186/II (3299 ± 60 BP), 1638–1505 cal. BC (68.2 %), 1695–1446 cal. BC (93.1 %); Z-219 (3060 ± 60 BP), 1406–1236 cal. BC (68.2 %), 1443–1155 cal. BC (93.2 %) (Sliepcević/Srdoč 1979–1980, 81–82).

Herzegovina (hillforts of Varvara and Pod, Glasinac), as well as those of the middle part of Pannonia and partially western Carpathian Basin. Among the latter regions, the most manifested seems to be the influences of the *Tumulus* culture and the *Virovitica* group.[379] Pottery can be further supplemented with some of the bronzes belonging to the I[st] horizon of Bezdanjača cave, e.g. an open bracelet with expanded ends, decorated by parallel ribs which has analogies in the *Tumulus* culture of the Central Europe (BzC1) and the Carpathian Basin (BzC1–C2/D).[380] These observations contribute to the general picture of Northern Croatia and Lika as a transitional zone between the Adriatic coast, the Pannonian plain and the Danube Basin, not only in terms of geography but also due to a mixture of various cultural traditions.[381]

Another in-cave site with amber from Lika is Golubnjača, located on the hillside of the Velebit Mountains, approximately 20 km north-west of Gospić and 10 km west of Perušić. Due to the absence of any remains of graves or buried skeletal remains, the site has been interpreted as a temporary shelter. Archaeological material, belonging to several periods, spanning from the Neolithic to Roman to Slavic times, was scattered in layers, without firm stratigraphy (Drechsler-Bižić 1970). Single amber bead was found within an exploratory surface in a narrow part of the cave facing north, named as the block 4 (Drechsler-Bižić 1970, 115, Tab. V: 1). Pottery and other related relics found there virtually represent all periods, and while the stratigraphy could not be determined, the place was considered as a waste pit. The amber bead exhibiting two parallel, horizontal ledges was classified as type 3 by Palavestra (Palavestra 1993, 68). Thanks to its characteristic shape, the bead can be dated *per analogiam* with the only other example of the same type (Baška, 'reconstructed grave' on Krk Island) ca. HaA1–B1, corresponding with the LBA in Lika, especially its IInd phase which, according to R. Drechsler-Bižić, reflects the influences arriving from the *Liburnian* group especially prominent in jewellery (Drechsler-Bižić 1983b, 376). Such chronology is further confirmed by certain fragments of pottery from the cave, characterised by the presence of oblique fluting – form of decoration typical for the *Urnfield* culture and related units (Drechsler-Bižić 1970, Tab. III: 8, IV: 3–4, 6). Particularly one of the vessels with biconical shape and everted rim, decorated on the belly with oblique flutings, bears resemblance to ceramic containers of the *Dalj* group (Drechsler-Bižić 1970, Tab. IV: 1; cf. Ložnjak Dizdar, Potrebica 2017, Sl. 48).

Similar analogy can be used to determine chronology of amber items from grave 293 (1903–1905) at the Kompolje necropolis. Absence of detailed description regarding the situation and stratigraphy of the grave, as well as its modest inventory, apart from amber beads comprising a single spectacle pendant with raised middle part, prevents from placing it in either of the two horizons observed at the cemetery (Bakarić 2017, 26; cf. Drechsler-Bižić 1961). In the face of absence of auxiliary datable materials in the context, the dating can be estimated through morphological comparison of the amber beads which represents variants classifiable as the types Tiryns and Allumiere (Bakarić 2017, 26, cat. no. 9). The closest counterparts come from northern Dalmatian

379 Drechsler-Bižić 1979–1980, 34–37; 1983, 251–258; Karavanić 2007, 43–44.

380 Drechsler-Bižić 1979–1980, Table XXVI:2; cf. Hänsel 1968, 102, Taf. 42:7; Gedl 1975, XIX:12, XXVI:8; Čujanová-Jílková 1978, Fig. 106:13; Hochstetter 1980, Beilage 1:46; Stuchlík 1993, Fig. 172:3.

381 Cf. Karavanić 2007; Ložnjak Dizdar 2007, 2011, 2014.

and Kvarnerian sites of Baška, Island of Krk, Privlaka and Vrsi.[382] The only reasonable option seems to be to adopt chronology of above-listed contexts. Therefore, grave 293 should be dated to circa HaA1–HaB1 that would correspond with phases I–III of the LBA and perhaps the beginning of phase Ia of the *Iapodean* group.

Tab. 23. List of the Bronze Age contexts with amber in Lika

Name	Commune	Province	Type	Relative chronology	Absolute chronology	Reference
Bezdanjača cave, mass grave nr 4	Vrhovine	Ličko-senjska županija	In-cave burial	BzC2–BzD1	1400–1250 BC	Drechsler-Bižić 1979–1980, Tab. XXV:6
Golubnjača, block 4	Perušić	Ličko-senjska županija	In-cave dwelling/shelter	HaA1–HaB1	1200–950 BC	Drechsler-Bižić 1970, Tab. V:1
Kompolje, grave 293 (1903–1905)	Otočac	Ličko-senjska županija	Burial	HaA1–HaB1	1200–950 BC	Bakarić 2017, 26, cat. No. 9

2.3.6. Slovenia

So far, the territory of Slovenia has not delivered many attestations of amber in the Bronze Age contexts (Fig. 14), what stands in a striking contrast to the abundance of this material in the area during the Iron Age (cf. Palavestra 1993). One of the two Bronze Age sites with amber in Slovenia – the hoard of Debeli Vrh – is located within the traditional borders of Lower Carniola (sl. *Dolenjska*), in the southernmost part of the country (Hirschbäck-Merhar 1984). The site, named after the mountain towering above the village of Predgrad on which it is located, yielded abundant collection of bronze items of various categories, as well as 13 amber beads, among which one pertains to the Tiryns type.[383] Due to its composition, the hoard was included in the IInd horizon of the LBA hoards in Slovenia by P. Turk (1996, 110). It is a regional manifestation of the wider phenomenon of hoarding in the Balkans, known under similar denomination in Hungary (Mozsolics 1985), Northern Croatia (Vinski Gasparini 1973), Serbia and Vojvodina (Vasić 1982; Garšanin 1983b), or Bosnia and Herzegovina (König 2004). It can be generally placed within HaA1 phase with some objects of inventory connecting it to following stage of HaA2. Nevertheless, complete typo-chronological differentiation of Slovenian hoards into subphases of HaA is impossible due to the absence of sufficient evidence (Turk 1996, 114–115).

382 Cf. Palavestra 1993, 50–51, 62–63, Tab. I:1–4; Blečić-Kavur 2014, Sl. 20:10–11, 21:22–24; Cwaliński forthcoming.
383 Hirschbäck-Merhar 1984, T. IX:17; cf. Teržan 1984; Palavestra 1993, 170.

also appear resembling forms of the *Belegiš I* culture (Mavrović Mokos/Pavličić 2015, 13–17). While most of the vessels discovered at the site reflect late development of forms related to the *Litzen*-type pottery, such as bowls and pitchers with characteristic ornament of cord impressions on the neck, morphology of some vessel, e.g. conical neck can be traced back to EBA cultures, such as the *Kisapostag* or *Vatin* (Mavrović Mokos/Pavličić 2015, 16–17).

Most of ceramic material, testifying to the development of the Alilovci Lipje settlement at the beginning of the MBA, comes from the so-far published 'house A'. The context yielded also a clay bread-loaf idol (ger. *Brotlaibidol*; it. *oggetto enigmatico*) which, due to limited spatial occurrence of such items and their presumed mercantile function, documents participation of the local community in a long-distance exchange.[387] Within the scope of Croatia and the neighbouring area, it is an isolated finding, far from the main concentrations of similar idols, not counting the collections from the Istrian hillforts of Monkodonja, Vrčin and Nezakcij (Mihovilić et al. 2017). The idol from Alilovci Lipje is rather simply decorated with round, impressed markers with a central point and radial notches (marker/sign C6 *con punto centrale e campitura raggiata* according to Piccoli/Zanini 1999), arranged without particular pattern, thus can be inserted into group 3 in by S. Carafa (2013, 136–139). This group of clay idols is richly represented in Italy, also on sites which yielded amber artefacts,[388] however they are also not missing in other parts of Europe, e.g. Southern Germany (cf. David 2016) or the Carpathian Basin (cf. Bándi 1974).

The context at Alilovci Lipje from which fragments of a single amber bead were recovered (Mavrović Mokos/Pavličić 2015, 21, Fig. 16), named as the 'test trench E/small house', is still analysed in terms of chronology and cultural affinity of the finds. Nonetheless, four 14C dates of charcoals and bones, acquired from the layers below and above the feature with amber, indicate time span roughly between 1600 and 1400 cal. BC.[389] Thanks to the radiocarbon dates, the amber bead and its immediate context can be placed in the intermediate horizon of the settlement's habitation around BzB–C1, therefore roughly contemporaneous with emergence of the *Belegiš I* culture.

The LBA in Northern Croatia is marked by appearance of several cultural units genetically linked to the complex of *Urnfield* cultures: *Virovitica* group, *Barice-Gređani* group and *Belegiš II* culture. The former unit emerged at the beginning of BzD phase (ca. 1300 BC), on the substratum of the *Tumulus* culture in the upper parts of the Sava-Drava interfluve.[390] Around that period emerged the settlement of Križevci-Ciglana which yielded two amber beads (Homen 1982, Tab. II:7–8). Due to destruction of the

387 Mavrović Mokos/Pavličić 2015, 24, Fig. 18; cf. Barfield 1971, 77; Bándi 1974, 247–248; Primas 1997, 37.

388 E.g. Gaggio, 1st phase of the settlement (Balista et al. 2008, Fig. 19:8).

389 Materials for radiocarbon dating were provided by the courtesy of Janja Mavrović Mokos, University of Zagreb. The dates are: Poz-92324 (3275 ± 30 BP), 1608–1511 cal. BC (68.2 %), 1626–1497 cal. BC (93.8 %); Poz-92323 (3270 ± 35 BP), 1608–1506 cal. BC (68.2 %), 1627–1454 cal. BC (95.4 %); Poz-92398 (3170 ± 35 BP), 1496–1416 cal. BC (68.2 %), 1511–1388 cal. BC (92.9 %); Poz-92397 (3170 ± 30 BP), 1495–1416 cal. BC (68.2 %), 1505–1396 cal. BC (95.4 %) (analysis performed within the project 'Circum-Adriatic branch of the amber route in the Bronze Age' NCN no. 2015/17/N/HS3/00052 conducted by the author).

390 Cf. Vinski Gasparini 1983c, 551–566, karta 15; Ložnjak Dizdar 2011, 37–40; 2014.

site in consequence of brickworks, its character and organisation cannot be fully reconstructed, apart from several waste pits and sunken features, which may have been dwellings or working places (Dizdar et al. 2011, 77). Singular bronze objects attest to some activities in the vicinity of the site already in the MBA (Vinski Gasparini 1983b, 500, 503; Ložnjak Dizdar 2011, 39, Map 1), whereas first evidence of occupation of the settlement can be linked with the *Virovitica I* group (BzD; ca. 1300–1200 BC) (Dizdar et al. 2011, 76). Nonetheless, the most dynamic period of the settlement's development falls into the second and third phase of the *Urnfield* culture in Northern Croatia and, thus, can be attributed to the *Virovitica II* and *Zagreb* groups (HaA1–A2; 1200–1060 BC) with some evidence of continuous use in the following, younger stages (up to HaB1).[391] Even though the two amber beads come from undefined context, in the original, as well as subsequent publication, they have been associated with the materials representative for the main phase of the settlement, hence their dating (cf. Palavestra 1993, 170–171).

Tab. 25. List of the Bronze Age contexts with amber in Northern Croatia

Name	Commune	Province	Type	Relative chronology	Absolute chronology	Reference
Alilovci Lipje, test trench E/ small house	Kaptol	Požeško-slavonska županija	Settlement	BzB–BzC1	1600–1400 BC	Mavrović Mokos/ Pavličić 2015, 21, Fig. 16
Križevci-Ciglana, precise context unknown	Križevci	Koprivničko-križevačka županija	Settlement	HaA1–HaA2	1200–1060 BC	Homen 1982, Tab. II:7–8

2.3.8. Vojvodina

Two sites with amber are known from the autonomous province of Vojvodina, more precisely its part situated on the east bank of the Tisa, called Banat (Fig. 14). Both are dated to the same period, spanning the turn of the 2nd and 1st millennia BC.

A rich collection of amber jewellery, numbering 45 beads of various forms, was recovered from accidentally discovered hoard at the locality Majdan, on north-west slopes of the Vršacka Planina, about 5 km away in north-eastern direction from Vršac (Rašajski 1988, Figs. 58–66). Majority of morphologic and stylistic features of the abundant inventory belonging to the hoard allow to date it to HaA1 phase (ca. 1200–1100 BC), making it roughly contemporary with the *Veliko Nabrđe* horizon of hoards in Northern Croatia and the *Cincu-Sušeni* horizon in Romania.[392] Among the finds

391 Homen 1982, Tab. I:5–5a; Dizdar et al. 2011, 78–79; see also Vinski Gasparini 1983c, 566–581; Dular et al. 2002, 197–205, Abb. 48.

392 Rašajski 1988, 27–28; cf. Vasić 1982, 282; Garašanin 1983b, 691.

diagnostic of aforementioned horizons are arched fibulae with spiral discs, wheel-shape pendants of the Bingula Divoš type, round decorative plaques with spikes (ger. *Stachelscheibe*), decorated bracelets with tapered ends and bronze bowls representing the Handlova type (Rašajski 1988, Figs. 2, 14, 30–31, 33–42, 57).

Other finds, however, are characterised by prolonged use continuing from phase HaA1 up to HaB1, hence testify to a possibility of later chronology for the deposition of the Majdan hoard. Among them the most significant are spectacle fibula without eight-shaped loop belonging to the type Suchdol,[393] elements (leaf-shaped pendants) of *Posamenteriefibeln* of the group C – types Sviloš or Suseni[394] and small, undecorated, socketed axes with an eyelet on the side.[395] Last but not least, the final argument for a slightly younger chronology of the hoard may be the presence of scrapped fragments akin to the Hajdúböszörmény-type situla – a hallmark of the eponymous HaB1 horizon of hoards in Hungary (Rašajski 1988, 25–26, Figs. 51–54; cf. Mozsolics 2000).

Regardless of the exact moment when the hard was buried in the ground, it is possible to estimate timeframe for the objects that it contained, including amber beads, which would be HaA1 at the earliest and HaB1 at the latest (ca. 1200–950 BC). Both the dating, as well as the actual content of the hoard fit into the chronological division of the hoards in northern Serbia, namely the 2nd, 3rd and 4th horizons (Vasić 1982; Garašanin 1983b). The temporal position and spatial distribution of the hoards in Vojvodina correspond with the *Belegiš II-Gava* culture – a unit that evolved under the influences from the *Urnfield* complex/*Gava* culture from the preceding stadium known as the *Belegiš I*.[396]

This *Belegiš II-Gava* culture, generally included in the milieu of the *Urnfield* cultures, is regarded either as equivalent of the 1st phase of the Iron Age in Vojvodina (later also called 'transitional period'; ca. 1300–800 BC; BzD–HaB2) according to the chronology of M. Garšanin (cf. Garšanin 1983d), or the LBA (ca. 1200–900 BC; HaA1–HaB1/2) if the chronological scheme of N. Tasić is adopted (Tasić 1984, 2003–2004). Internal division of the *Belegiš* culture is based on the stratigraphy of the Gomolava settlement which shows progressive changes in the materials during continuous occupation of the site (Tasić 2006). According to the most widespread proposition regarding its periodisation, each of the two major stages of the *Belegiš* culture can be subdivided into three phases: *Belegiš Ia–Ic* (BzB2/C1–BrD) and *Belegiš IIa–IIc* (HaA1–HaA2) (Tasić 2001, 2002).[397] The main difference separating the two main stages of the *Belegiš* culture is chiefly visible through the evolution of pottery from sepulchral contexts: disappearance of comb-decoration towards the end of BzD and parallel introduction of black-burnished pottery, decorated with circular or garland grooves on the neck or obliquely on the rim (the so-called 'turban motif').

---

393 Rašajski 1988, Fig. 3; cf. Vasić 1982, 281; 1999, 29, Tab. 8:102.

394 Rašajski 1988, 22, Figs. 12 and 13; cf. Vasić 1982, 276; Bader 1983, 55, Taf. 9–10, 43B.

395 Rašajski 1988, 25, Figs. 44 and 45; cf. Vasić 1982, 278–280, Žeravica 1993, 75–81, Taf. 20:253–264, 21:265–280.

396 See Garašanin 1983c; Tasić 1984, 2003–2004.

397 Following the proposition of S. Morintz (1978, 40–45), Romanian archaeologists largely denominate materials from Romania related to discussed unit as the *Cruceni-Belegiš* culture, and adopt a three-partite periodisation consisting of the phases *Cruceni-Belegiš I* (BzB1–BzB2/C1), *Cruceni-Belegiš II* (BzC2–D) and *Cruceni-Belegiš III* (HaA1–HaA2/HaB1) (Szentmiklosi 2006, 2009).

Settlements of the *Belegiš* culture throughout the entire development were similar, and can be divided into open and fortified. The former type of settlement is typical to Vojvodina, where it was conditioned by the mostly flat terrain (Garašanin 1983c, 677). The other type, comprising settlements built on dominant positions granting defensive qualities and often additionally protected by moats and/or ramparts, is represented by fewer sites in the region, e.g. Feudvar in Mošorin (Falkenstein et al. 2016).

To this small group of defensive settlement perhaps should be classified the newly discovered sites no. 21 and 22 near Banatski Dvor in Serbian Banat (Hristov 2017). The sites are located on a slightly elevated position, on the eastern edge of the Novi Bečej – Zrenjanin loess terrace, approximately 5 km to the west from the Banatski Dvor village. What can be inferred from the data recovered from a thin transect excavated up to now is that the settlement was fortified with a ditch; however, remains of earthen ramparts or palisades have not been recognised. Inside the settlement zone, there were two smaller ditches with unknown function. A number of waste pits were excavated, as well as four shallow objects (4×4 m) with sunken floors, rectangular in shape, presumably hut dwellings, some of them with wooden roof supports. The waste pits and ditches contained numerous fragments of daub from housing structures (Hristov 2017, 7–8). Inside object no. 38 (a large waste pit) on the site no. 22, along with fragments of the *Belegiš II-Gava* pottery, was found a small amber bead (Hristov 2017, 8, 12, T. 16:23). Pottery from the object, as well as from almost all remaining features on the site excluding pit 26, bears characteristics of the advanced/late *Belegiš II-Gava* culture, presumably its phases *IIb–IIc* (HaA1–HaA2) (Hristov 2017, 12–13). However, a full classification of the materials is unfeasible due to the fact that N. Tasić's chrono-typology is based predominantly on ceramic vessels from cemeteries, and substantial assemblages of pottery from *Belegiš II-Gava* settlements are still lacking.

Tab. 26. List of the Bronze Age contexts with amber in Vojvodina

Name	Commune	Province	Type	Relative chronology	Absolute chronology	Reference
Majdan, hoard	Vršac	Južno-banatski okrug	Hoard	HaA1–HaB1	1200–950 BC	Rašajski 1988, Figs. 58–66
Banatski Dvor, site 22, object 38	Žitište	Srednjo-banatski okrug	Settlement	HaA1–HaA2	1200–1060 BC	Hristov 2017, T. 16:23

2.3.9. Podrinje and Western Serbia

Geographically, the most extensive province of amber's acquisition in the Balkans can be delimited along the course of the Drina river and adjacent part of Western Serbia (Fig. 14). Sites from this territory that yielded amber are exclusively of funeral character. Due to shared burial customs (use of barrows as the main grave form), as well as similarities in bronze metallurgy (visible on the basis of jewellery found among grave inventories), and to lesser extent pottery, these sites can be interpreted as belonging to a single cultural unit, spanning over a vast area.[398] The term 'Podrinje' refers to the catchment basin along both sides of the Drina River, including the easternmost parts of Bosnia and Herzegovina and the westernmost fringes of Serbia. However, on the western (Bosnian) side of the Drina, the sites that yielded amber are concentrated in a single cluster near Zvornik (the Padine-Ročević tumulus complex[399]), in the case of Serbia the contexts have a wider distribution that stretches in the south-east direction along the Jadar River, up to the upper reaches of the West Morava.

Tumuli necropolises on the western bank of the Drina form the sepulchral complex of Padine-Ročević. All sites belonging to this complex are stretched along the road from Zvornik to Bijeljina, on a gentle slope descending in the eastern direction towards the river. Amber objects were found in the following contexts: Jezero (mound IV, graves 1–3; mound III, grave 3 and 5; mound VI, grave 1; mound IX, graves 4 and 6), Karavlaške kuće (mound 1, grave 2) and Šundinovača (mound III, grave 2). Cemetery of Jezero is located close to the village Ročević, on the right (east) side of the village road, which forms the border with the village Padine. Barrows of Karavlaške kuće and Šundinovača, on the other hand, are located closer to the village Padine, at the distance of about 300 m from the Jezero necropolis (Kosorić/Krstić 1970, 23–24; Kosorić 1978, Karta I).

On the Drina's east bank are located following sites with amber: Paulje (tumuli A, K and N) near the village of Brezjak, some 14 km northeast of Loznica,[400] necropolises of Belotić-Bela Crkva complex[401] and tumulus 2 at Jovanin Breg near Banjevac.[402] Regarding the Belotić-Bela Crkva complex, amber beads were discovered on two sites attributed to the complex. First, inside two barrows (mound XI, graves 3 and 4; mound VII, grave 1), located within the administrative borders of Belotić village, at the locality named Šumar, just beside the border with the Bela Crkva district. Mounds were situated on a densely forested ridge stretching along NE-SW axis, characterised by mild depressions on its both longitudinal sides. Within the administrative district of Bela Crkva can be found several concentrations of prehistoric mounds. Amber beads were discovered in a single tumulus (mound III, grave 1), situated at the locality Cerik, close to yet another group of mounds defined in the literature as the Bandera site. Both sites are placed approximately 1 km north-west from the locality Šumar (Belotić), on a hillside overlooking the valley and the village of Bela Crkva.

Another concentration of barrows, among which some included amber as grave goods, is situated in the vicinity of Čačak and Užice, more specifically along the upper

398 See Garašanin 1983d; Filipović 2008, 2013; Dmitrović 2016; Ljuština/Dmitrović 2016; Bulatović et al. 2017.

399 Kosorić/Krstić 1970, 1972; Kosorić 1978.

400 Madas 1996; Canić-Tešanović/Gligorić 2001; Palavestra/Krstić 2006, 310; Gligorić/Canić-Tešanović 2010; Bulatović et al. 2017, 116–137.

401 Garašanin/Garašanin 1958, 1962; Bulatović et al. 2017, 109–114.

402 Palavestra/Krstić 2006, 291; Exact location of the site is unknown, due to absence of publication. From available information one can infer that the site has to be looked for between the villages of Tolisavac and Banjevac (see also Bulatović et al. 2017, 108–109).

course of the West Morava River. Singular amber beads come from tumulus 1 of Vranjani (Zotović 1985) and tumulus 3 of Ravnine (Nikitović/Vasić 2002). The former site is located about 10 km north-west of Požega and about the same distance from Užice, near the village of Vranjani (Zotović 1985). Barrows of Ravnine are situated on the Kablar range, on the levelled slope extending from the village of Rošci to the area of Gornja Dobrinja and Gojna Gora, ca. 20 km north-west from the town of Čačak.

The most distant site, however, strikingly similar in character to those above-mentioned, that delivered amber is located in the fork-like watershed zone between the Lim and the Drina Rivers, coinciding with the Zlatibor district. The mound of Savin Lakat is named after a winding mountain pass in the south-east fringes of the Gradina – a massif separating the valleys of the Lim and the Cehotina rivers, and can be reached via an old gravel road connecting the towns of Prijepolje and Pljevlja (Lazić 2007).

In terms of chronology, all sites with amber in Podrinje and Western Serbia can be placed within a time span of ca. 300 years between 1500 and 1200 BC, corresponding with BzB–BzD.[403] This period in Serbia is defined according to different authors as the LBA,[404] 2nd phase of the MBA,[405] or the developed Bronze Age.[406] Question concerning how to define discussed period is strictly connected with the problem of cultural and chronological attribution of the barrow cemeteries in Podrinje and Western Serbia. In 1958 M. Garašanin, based on his excavations at the necropolises of Dobrača, Belotić and Bela Crkva, singled-out cultural phenomenon in this part of Serbia called the 'West Serbian variant of the *Vatin* culture' (Garašanin 1973, 359). The reasons for adopting such name were drawn from apparent similarities between local pottery and the ceramics of the *Vatin* culture.

Currently, with the revision of chronology, spatial extent and subdivision of the *Vatin* culture, it is no longer acceptable to maintain terminology proposed by M. Garašanin.[407] Instead, it appears that the relevant materials recovered from barrows in Podrinje and Western Serbia have more in common with the *Belegiš I* culture in Syrmia (sr. Srem), as well as are reminiscent of the *Tumulus* culture in metallurgy which in southern Pannonia (Vojvodina) marks a horizon succeeding disintegration of the *Vatin* culture.[408] Despite accepting the need to reject the term

---

403 Bulatović et al. date the discussed phenomenon roughly between 1400 and 1200 BC, using several radiocarbon dates, as well as analogies of metal and ceramic finds from the barrows of Podrinje and Western Serbia with the *Belegiš I* and the *Tumulus* culture (Bulatović et al. 2018, 125–126). These considerations can be supplemented with a fragment of wood, possibly remaining after a wooden coffin from tumulus K of Paulje, that has recently been 14C-dated, giving following results: Poz-110879 (3195 ± 30 BP), 1497–1440 cal. BC (68.2 %), 1517–1414 cal. BC (95.4 %); Poz-110880 (3165 ± 30 BP), 1494–1414 cal. (68.2 %), 1503–1394 cal. BC (94.8 %); Poz-110881 (3095 ± 35 BP), 1414–1302 cal. BC (68.2 %), 1432–1267 cal. BC (95.4 %) (analysis performed within the project *Circum-Adriatic branch of the amber route in the Bronze Age* NCN no. 2015/17/N/HS3/00052 conducted by the author).

404 Garašanin 1967; cf. Filipović 2008, 2013.

405 Dmitrović 2016; Dmitrović/Ljuština 2016.

406 Garašanin 1973, 378; 1983d, 737–738; Bulatović et al. 2017.

407 See e.g. Ložnjak Dizdar 2007; Ljuština 2011; Ljuština/Dmitrović 2015.

408 Tasić 1983, 60, 97; Filipović 2008, 99–100; Dmitrović/Ljuština 2016, 47.

'*West Serbian* variant of the *Vatin* culture', scholars came up with various propositions how to replace it with a new denomination. Major suggestions are those of V. Filipović who suggests the name '*Brezjak* culture', derived from the eponymous cemetery at Brezjak (Paulje) near Loznica (Filipović 2013, 76), while M. Ljuština and K. Dmitrović opt for the term 'West-Serbian group of the Middle Bronze Age' (Ljuština/Dmitrović 2016, 45).

In case of many graves containing amber adornments, the presence of chronologically 'sensitive' bronze artefacts, such as pins, bracelets and torcs, allows for more precise dating with respect to Reinecke's system.[409] Among bronze objects occurring together with amber in the Eastern Bosnian and Western Serbian contexts, in the first place, should be listed pins with spindle- (cf. Vasić 2003, 47–48),[410] stamp- (cf. Vasić 2003, 37–39)[411] and nail-head (cf. Vasić 2003, 31–33).[412] The former were assigned to the MBA phases BzC2–D and in terms of general form are regarded as an indicator of the *Tumulus* culture influences (Vasić 2003, 48, T. 70). However, Central European variants differ in terms of ornamentation and lack the swelling in the upper part of the neck; thus, Bosnian and Serbian specimens should be considered as products of local metallurgic tradition (Filipović 2013, 66–68). Slightly earlier dating (BzC) but similar cultural attribution can be applied to stamp-head pins (Vasić 2003, 38–39, T. 70), whereas nail-head pins with perforated necks, also dated to BzC–D (Vasić 2003, 33, T. 70), can be stylistically linked to the *Dubovac-Žuto Brdo* culture (Filipović 2013, 59, Tab. 1), although similar pins were found also in Hungary (Říhovský 1983, Taf. 3:35–36).

Second, one should mention open bracelets made from bronze bar of semi-circular or flattened triangular cross-section, with first tapered and then widened (stamp-shaped) terminals, lately divided according to ornamentation into types I-A (decorated with garland and fish-bladder patterns),[413] I-B (spiral motifs)[414] and I-C (hatched triangles)[415] (Dmitrović 2016, 166–168, Fig. 113:1–4). They were all used at various phases of advanced and late MBA (BzC1–D), unifying Western Serbia and Eastern Bosnia along the Drina, however sometimes appear even further south, as in the case of grave 1 at the Iglarevo cemetery in Kosovo. Bracelets of the type I-A bear

---

409 In assessing chronology of many contexts considered in this section, very helpful is seriation of the tumuli and related finds from the Drina drainage basin by P. Della Casa (cf. 1996a, Tab. 23).

410 Banjevac, tumulus 2 (Vasić 2003, T. 17:254); Jezero, mound 3, grave 3 (Kosorić/Krstić 1972, Tab. V, Fig. 6) and mound 9, grave 6 (Kosorić 1978, Tab. VI, Fig. 1); Karavlaške kuće, tumulus 1, grave 2 (Kosorić/Krstić 1970, Tab. IV, Fig. 2); Paulje, tumulus K (Bulatović et al. 2017, T. XVIII:29).

411 Banjevac, tumulus 2 (Vasić 2003, T. 12:182); Jezero, mound III, grave 3 and mound IV, graves 1 and 3 (Kosorić/Krstić 1972, Tab. V:6, VI:3, VII:1).

412 E.g. Belotić, mound VII, grave 1 (Garašanin/Garašanin 1962, Fig. 6a–b); Paulje, tumulus K (Bulatović et al. 2017, T. XVIII:49).

413 Jezero, mound IV, grave 1 (Kosorić/Krstić 1972, Tab. VI, Fig. 3), grave 2 (Kosorić/Krstić 1972, Tab. VI, Fig. 4); Ravnine, mound 3 (Nikitović/Vasić 2002, Tab. II:5); Vranjani, tumulus 1 (Zotović 1985, T. IX:1, 3).

414 Jezero, mound III, grave 3 (Kosorić/Krstić 1972, Tab. VI, Fig. 6), grave 5 (Kosorić/Krstić 1972, Tab. VI, Fig. 2); Jezero, mound IV, grave 3 (Kosorić/Krstić 1972, Tab. VI, Fig. 4); Paulje, tumulus A (Madas 1996, Fig. 1:5).

415 Paulje, tumulus K (Bulatović et al. 2017, T. XVII:26–27).

striking similarity to the contemporaneous bracelets from the Carpathian Basin and Central Europe influenced by the *Tumulus* culture.[416] Bracelets decorated with the spiral motif (type I-B) are generally considered as younger forms in the development of open armbands; while their particular way of decoration, according to some authors, is interpreted as an influence of the *Dubovač-Žuto Brdo* culture (Kosorić/Krstić 1970, 31), others argue that it was derived from the *Hajdúsámson-Apa* style (Nikitović/Vasić 2002, 29).

Similar ornament appears on torcs with circular/oval cross-section and coiled endings – another element typical for some of discussed contexts.[417] They can be included in the group of smooth, decorated torcs by R. Vasić (ger. *Glatte verzierte Halsringe*; Vasić 2010, 22–31) and type II by K. Dmitrović (2016, 162, Fig. 112:3–4). Basing on associated elements of inventory, examples of such torcs found in the contexts with amber should be dated to BzC1–D (Vasić 2010, 28). It is considered that they were probably created under the influence from Central Europe and the Carpathian Basin, as evidenced by ornamentation resembling the *Hajdúsámson-Apa* style (Vasić 2010, 27–28).

This description should be supplemented by bracelets with narrowed or hammered endings, decorated on the outer surface with incised lines or the 'herringbone' pattern, rarely hatched triangles and spirals (type II; Dmitrović 2016, 170, Fig. 113:5).[418] Their dating roughly corresponds with chronology of aforementioned open bracelets as does their cultural affiliation, to support which one can list analogies from the Hungarian territory where they appeared particularly early, because already in the *Koszider* horizon – phase MD I,[419] and from Central European materials of the classic to late *Tumulus* culture (BzC1–D).[420]

Finally, to the group of bronzes typical for West Serbian and East Bosnian area can be added heart-shaped pendants of variants 6 and 7 by B. Hänsel (1968, 117–118)[421] and round plaques with the middle-thorn (ger. *Stachelscheibe*).[422] Heart-shaped pendants of the abovementioned variants can be safely included in the phase MDII–III (Hänsel 1968, Beilage 4:55, 5:45), roughly corresponding with phases BzB–C1. While

---

416 See e.g. Willvonseder 1937, Figs. 7:7 and 8:18; Mozsolics 1967, Taf. 53:5a–b; 1973, Taf. 4:3a, 134:6–8; Gedl 1975, Tab. XI:11, XII:1–3; XIII:15; Čujanová-Jílková 1978, Fig. 106:37; Stuchlík 1993, Figs. 172:18–21 and 177:11, 13.

417 Jezero, mound IX, grave 6 (Kosorić 1978, Tab. VI, Fig. 2); Paulje, tumulus K (Bulatović et al. 2017, T. XVII:25); Ravnine, mound 3 (Nikitović/Vasić 2002, Tab. II:4).

418 Jezero, mound VI, grave 1 (Kosorić/Krstić 1972, Tab. VII, Fig. 2); Jezero, mound IX, grave 6 (Kosorić 1978, Tab. VI, Fig. 4); Paulje, tumulus K (Bulatović et al. 2017, T. XVII:28).

419 Mozsolics 1967, 77–78, Abb. 21:1–2, Taf. 52:2–5, 60:7, 63:3–6, 69:5; Hänsel 1968, 92–94, Beilage 4:18.

420 Gedl 1975, Tab. XLII-XLIII; Hochstetter 1980, Beilage 1:42–45, 48–52; Stuchlík 1993, Fig. 179:4–7.

421 Bela Crkva, mound III, grave 1 (Garašanin/Garašanin 1958, 40, Fig. 15g); Paulje, tumulus K (Bulatović et al. 2017, T. XIX:33–41); Ravnine, mound 3 (Nikitović/Vasić 2002, Tab. II:3).

422 Bela Crkva, mound III, grave 1 (Garašanin/Garašanin 1958, 40, Fig. 15b); Savin Lakat, cremated burial (Lazić 2007, T. IV:2–5).

widespread in Central Europe during the *Tumulus* period,[423] they are quite rare in the territory of Western and Central Balkans, and their biggest concentration is situated in North-Western Serbia, in the lower Drina Basin, while they are less frequent around Čačak, Požega and Užice (Dmitrović 2016, 184). While pendants of variants 6 and 7 may be regarded as imports from the northern regions, those found in tumulus K in Paulje (Canić-Tešanović/Gligorić 2001, Cat. No. 51–63), due to atypical form, belong to yet another variant, perhaps of local origin. Plaques with the middle-thorn in the central part and 3–5 concentric ribs are very rare in Western Serbia and Bosnia, while more frequent in Syrmia and Slavonia (Dmitrović 2016, 178). Due to their singular appearance in Western and Central Balkans, they are generally regarded as imports (Vasić 1997, 39).

During discussed period, the dominating burial rite in Podrinje and Western Serbia was burying the dead under earthen tumuli, often by using mounds from earlier periods. Analysis of 86 graves from the Bronze Age on the territory of north-west Serbia revealed that relative proportions of inhumed and cremated individuals amount to circa 17.5 % versus 82.5 %, respectively (Filipović 2013, 55). On the other hand, this relationship is inversely proportional on the other bank of the Drina in eastern Bosnia probably inhabited by the same population, as can be inferred from archaeological material. It seems that the areas of Eastern Bosnia and the Glasinac plateau retained and maintained the tradition of inhumed burials, even at the time when the custom of incineration prevailed in the surrounding areas. Similarly, in the upper West Morava Valley cremation dominated over inhumation. In some mounds were registered both types of burial, such as barrow XI at Belotić and III at Bela Crkva. Deceased were incinerated on the spot, as well as away from barrows; however, in the latter case no traces of pyres have been detected so far (Filipović 2008, 96). Remains of cremated bodies were usually deposited in urns; however, in some cases earthen embankments of barrows were raised directly over cremation places. In the tumuli necropolises to the west from Čačak, amphorae and beakers of greater dimensions were dominantly used as containers for the burnt bones (Ljuština/Dmitrović 2016, 45). Some urns were placed inside small stone cists, whereas others lacked any outer structure surrounding them. Grave goods, mainly bronze jewellery, were placed either on the pyre prior to cremation, or afterwards beside the urns. Regarding inhumed burials, it can be said that they were arranged in supine position on the back or flexed, often on the left side. In many cases the skeletons were placed on stone pavements, constructed of pebbles, rarely stone slabs. Also, towards the end of the 2nd millennium BC, when in the north appeared spacious flat necropolises, the population in Western Serbia and Eastern Bosnia continued the earlier custom of under-tumulus burials.

423 See e.g. Čujanová-Jílková 1978, Figs. 106:38 and 129:18; Wels-Weyrauch 1978, 63–65, Tab. 16; Hochstetter 1980, Tab. 80:5, 83:16; Stuchlík 1993, Fig. 177:2–3.

Tab. 27. List of the Bronze Age contexts with amber in Podrinje and Western Serbia

Name	Commune	Province	Type	Relative chronology	Absolute chronology	Reference
Banjevac (Jovanin Breg), tumulus 2	Krupanj	Mačvanski okrug	Burial – inhumation	BzC1–BzD	1450–1200 BC	Palavestra/ Krstić 2006, 291
Bela Crkva, mound III, grave 1	Mionica	Kolubarski okrug	Burial – inhumation	BzB/ C1–BzC2	1500–1300 BC	Garašanin/ Garašanin 1958, 40, Fig. 15:f
Belotić, mound XI, grave 3	Mionica	Kolubarski okrug	Burial – inhumation	BzB/ C1–BzC2	1500–1300 BC	Garašanin/ Garašanin 1958, 23
Belotić, mound XI, grave 4	Mionica	Kolubarski okrug	Burial – inhumation	BzB/ C1–BzC2	1500–1300 BC	Garašanin/ Garašanin 1958, 24, Fig. 5:c–d
Belotić, mound VII, grave 1	Mionica	Kolubarski okrug	Burial – cremation (urn)	BzB/ C1–BzC2	1500–1300 BC	Garašanin/ Garašanin 1962, 51, Fig. 7:a–b
Jezero, mound III, grave 3	Zvornik	Republika Srpska	Burial – inhumation	BzC	1450–1300 BC	Kosorić/ Krstić 1972, 14
Jezero, mound III, grave 5	Zvornik	Republika Srpska	Burial – inhumation	BzC	1450–1300 BC	Kosorić/ Krstić 1972, Tab. VI:2
Jezero, mound IV, grave 1	Zvornik	Republika Srpska	Burial – inhumation	BzC	1450–1300 BC	Kosorić/ Krstić 1972, 14, 21–22
Jezero, mound IV, grave 2	Zvornik	Republika Srpska	Burial – inhumation	BzB/ C1–BzC2	1500–1300 BC	Kosorić/ Krstić 1972, Tab. VI:4
Jezero, mound IV, grave 3	Zvornik	Republika Srpska	Burial – inhumation	BzC	1450–1300 BC	Kosorić/ Krstić 1972, 15
Jezero, mound VI, grave 1	Zvornik	Republika Srpska	Burial – inhumation	BzC2–BzD	1400–1200 BC	Kosorić/ Krstić 1972, 16
Jezero, mound IX, grave 4	Zvornik	Republika Srpska	Burial – cremation (pit)	BzC2–BzD	1400–1200 BC	Kosorić 1978, 24–25, Tab. V:2

(Continued)

Tab. 27. Continued

Name	Commune	Province	Type	Relative chronology	Absolute chronology	Reference
Jezero, mound IX, grave 6	Zvornik	Republika Srpska	Burial – inhumation	BzC2–BzD	1400–1200 BC	Kosorić 1978, 25, Tab. V:7
Karavlaške kuće, mound 1, grave 2	Zvornik	Republika Srpska	Burial – inhumation	BzC2–BzD	1400–1200 BC	Kosorić/ Krstić 1970, Tab. IV:3
Paulje, tumulus A	Loznica	Moravički okrug	Burial – cremation (pit)	BzD	1300–1200 BC	Madas 1996, Fig. 1:7; Bulatović et al. 2017, T. XIV:16
Paulje, tumulus K	Loznica	Moravički okrug	Burial – cremation (pit)	BzC2	1400–1300 BC	Canić-Tešanović/ Gligorić 2001, Cat. No. 66–71; Filipović 2013, T. VI:6
Paulje, tumulus N	Loznica	Moravički okrug	Burial – cremation (pit)	BzC1–BzD	1450–1200 BC	Palavestra/ Krstić 2006, 310; Bulatović et al. 2017, T. XIX:46
Ravnine, mound 3	Čačak	Moravički okrug	Burial – cremation (pit)	BzB/ C1–BzC2	1500–1300 BC	Nikitović/ Vasić 2002, Tab. III:1
Savin Lakat, cremated burial	Prijepolje	Zlatiborski okrug	Burial – cremation (pit)	BzB–BzC1	1600–1400 BC	Lazić 2007, T. IV:16
Šundinovača, mound III, grave 2	Zvornik	Republika Srpska	Burial – inhumation	BzC2–BzD	1400–1200 BC	Palavestra 1993, 111
Vranjani, tumulus 1	Požega	Zlatiborski okrug	Burial – cremation (pit)	BzC2	1400–1300 BC	Zotović 1985, 40

2.3.10. Glasinac

An abundant collection of prehistoric amber artefacts comes from the tumuli located at the Glasinac plateau in the eastern part of Bosnia and Herzegovina, approximately between Sarajevo and the Drina River (Fig. 14). The tumuli along with their inventories are considered as evidence of a very long habitation of this area, spanning almost the entirety of the Bronze and Iron Ages. The phenomenon of continuous occupation of the same land as a burial ground for more than one thousand years led B. Čović to distinguish the *Glasinac* group and then the *Glasinac* culture (Čović 1983c, 1987). In the 1950s the barrow cemeteries from the Glasinac area were divided by A. Benac and B. Čović into five phases: I (roughly concurrent with the EBA), II (MBA), III (LBA), IV (EIA) and V (Late Iron Age), which were further subdivided into shorter stages (Benac/Čović 1956, 1957). This chronological system was later revised by Čović several times, with the final version of the periodisation being elaborated in the fourth volume of *Praistorija Jugoslavenskih Zemalja – bronzano doba* (Čović 1983c). An exhaustive overview of the development of the Glasinac's Bronze Age chronology has recently been published in a paper by the author and A. Pravidur (Cwaliński/Pravidur forthcoming), so below a condensed description is presented.

In establishing concordance between the periodisation of the Glasinac necropolises and a general chronology of the Bronze Age in Central Europe, Benac and Čović adhered to the Reinecke's system. Consequently, the *Glasinac* II phase was synchronised with the MBA phases BzB(late)–BzC(1–2), whereas the subsequent subphase IIIa was linked to the end of phase BzC(2) and the majority of phase BzD (Čović 1981, 111; Drechsler-Bižić 1983a, 261–262). Next, the phase IIIb was synchronised with late BzD and HaA, while IIIc was correlated to HaB1–2 (Čović 1983c, 415–418). Subsequent phases IV and V signify the Iron Age period. Absolute time frames of the individual phases were estimated as follows: phase II was approximated with a period spanning between 1500 and 1300 BC; phase IIIa was equated with 1300–1200 BC, IIIb with 1200–1000 BC and IIIc with 1000–800 BC.

This basic version of the Glasinac periodisation underwent several minor changes. First of all Čović split phase IIIb into two subphases: IIIb-1 and IIIb-2. The earlier subphase became identified with the introduction of the violin-bow fibulae – the so-called *Peschiera* fibulae horizon in Italy, corresponding with the final part of BzD and HaA1 (Čović 1981, 111–112; 119–124; 1983c, 418, 424–425). As for the subphase *Glasinac* IIIb-2, Čović, with considerable reservations, was able to select only a small number of graves which potentially originate from this period (roughly coeval with HaA2). It has to be emphasised that the finds chosen by Čović as representative for this stage are mainly simple elements of bronze jewellery in the form of bracelets, rings and pendants (Čović 1983c, 425, Fig. 29:10–12, T. LXIII:13–15). The author expressed doubts about the usefulness of those objects as indicators of a separate chronological phase; however, at the same time, he underlined that such finds are absent from graves attributed to *Glasinac* IIIb-1 or IIIc (Čović 1981, 124–126).

Furthermore, Čović also split the final phase of the *Glasinac* LBA into two subphases: IIIc-1 and IIIc-2.[424] The basis for this solution was separation of the

424 Čović 1981, 126–129; 1983c, 418, 426–427; Gabrovec/Čović 1987, 909.

Golinjevo-type arched fibulae into older and younger variants. Specific feature of the specimens from the Glasinac is the absence of the two nodules on the arched bracket in comparison with other examples of the Golinjevo type from the Balkans. When it comes to observing the development of the Golinjevo fibulae design, an important chronological indicator is the cross-section of the bracket: for example, the fibula with an octagonal bracket from Brezje, tumulus V, grave 1 (1896) is closer to the older variant, whereas the fibulae from Taline, tumulus XX, grave 3 (1895)[425] and Vrlazje, tumulus IV, grave 2 (1894)[426] with round or square bracket are representative of a more recent version. In a comparative approach *Glasinac* phase IIIc was equated with Central European phases HaB1 (IIIc-1) and HaB2 (IIIc-2) (Čović 1983c, 426, Fig. 29; Gabrovec/Čović 1987, 909).

In the years following the publication of the chronological scheme of Benac and Čović, several authors dealing with materials from the Glasinac made corrections to the original periodisation. C. F. E. Pare proposed to incorporate the finds hitherto linked with phase IIIc (i.a. older variants of the Golinjevo-type fibulae) into phase IA, signifying the beginning of the EIA, whereas the subsequent phase IVa was renamed IB, covering materials previously classified by N. Lucentini (Pare 1998, 333–336; cf. Lucentini 1981, 67–162).

The latest review of the *Glasinac* chronology was undertaken by M. Gavranović, who adjusted the timeframes of phases III and IV (Gavranović 2011, 32–38, Abb. 1). Gavranović criticised Pare's periodisation of the Glasinac for the lack of differentiation between the older and younger Golinjevo-type fibulae (cf. Gavranović 2011, 34). In contrast to Pare, Gavranović argued that the Golinjevo-type fibulae, apart from a few examples of the youngest variant, should remain attributed chiefly to phase IIIc. According to his version of the periodisation, phase IIIa largely corresponds to the end of phase BzC and the majority of BzD (ca. 1350–1250 BC), while IIIb is contemporaneous with late BzD and HaA (ca. 1250–1060 BC). The author shifted phase IIIc so that it is concurrent with stage HaB1, with the first subphase (IIIc-1) approximating 1100–1000 BC, and the second (IIIc-2) falling in the period 1000–900 BC (Gavranović 2011, Abb. 1). In this approach, the temporal boundary between stages HaB1 and HaB2/B3 at around 950 BC is recognised as the transition from phase IIIc to IVa, so between the Bronze and Iron Ages in this part of the Balkans.

While the vast majority of amber finds from the Glasinac is dated to the Iron Age, there are some finds that, on the basis of co-occurring finds in the graves, apparently originate from the Bronze Age and, as such, were determined typologically and catalogued by A. Palavestra (Palavestra 1993, 112–113). According to the latter author, six tumuli document the presence of amber in the Bronze Age stage of the *Glasinac* development (designated as phases II and III).

In the light of a more recent investigations (Cwaliński/Pravidur forthcoming), only two out of six abovementioned graves with amber can be reliably dated to phase *Glasinac* II, whereas three other possibly document transitory horizon between final stage of the Bronze Age (phase IIIc-2) and the EIA (phase IVa-b). Three other contexts, hitherto considered relevant to discussed period, were dropped from this group, because they revealed insufficient evidence for such classification or were proved to

425 Benac/Čović 1956, 24, Pl. XXXXIV:5.
426 Benac/Čović 1956, 10, Pl. IV:5.

be younger: Borovsko, tumulus II (1892), Crvena Lokva, tumulus II (1890) and Osovo, tumulus I (1895).

Among the positively classified contexts, the oldest is grave 1, tumulus IV at Gučevo (1895) which, as the reassessment of chronology of its content has shown, should be most likely connected with phase *Glasinac* IIa (cf. Benac/Čović 1956, 46, 69–71, Tab. V: 27–33). Tumulus IV is one of six mounds excavated at the site by an archaeological expedition led by F. Fiala and V. Ćurčić in 1895. According to Fiala's report, only two graves were found inside this tumulus: grave 1 with an interred burial, and grave 2 containing cremated bones and residues of a pyre (Fiala 1895, 539–540, Fig. 7). Fiala did not specify which objects belonged to which grave. Nevertheless, on the basis of combustion traces – a premise lately found unconvincing, Benac and Čović divided the finds into two assemblages, with some artefacts – including two amber beads – being ascribed to the cremation burial (Benac/Čović 1956, 10, Pl. V:28–33). The recently discovered Fiala's diary from the excavations of 1895 lists amber artefacts by the central, inhumed burial (grave 1), thus proving the attribution of the amber beads to grave 2, previously suggested by Benac and Čović, wrong (Cwaliński/Pravidur forthcoming). The bronze bracelet with buffer terminals, decorated with a 'fish-bladder' motif and bands of parallel lines, serves as a good determinant of the assemblage's chronology (Benac/Čović 1956, Pl. V:1). The specimen and other similar bracelets from Glasinac are unanimously dated to phase IIa, corresponding with Central European stages BzB–C1.[427] Such dating is confirmed in light of analogies known from, e.g. the Padine-Ročević funeral complex in Eastern Bosnia (Kosorić/Krstić 1988, 44–45, T. XI: 2–3), as well as from numerous sites throughout Western Serbia (E.g. Zotović 1985, T. IX:1, 3; Dmitrović 2016, 166, Fig. 113:1). Consequently it is necessary to shift the dating of the two amber beads from tumulus IV at Gučevo to phase IIa, thus making them the earliest find of their kind within the area of Glasinac.

Slightly younger are the finds from tumulus I (1894) from the site Rudine u Ivanopolju, which represents phase *Glasinac* IIb; however, there are hints of a more prolonged chronology of the context potentially extending to phase IIIa. The mound is one of the two barrows excavated at the site by an archaeological expedition led by F. Fiala in 1894 (Fiala 1894, 750). Inside were discovered several unburnt human bones, lying together with some grave goods. However, Fiala's report does not clearly state whether or not these belonged to the grave of a single person. Despite this uncertainty, materials found inside the barrow were later assigned to a single skeletal burial, tentatively named as grave 1 (Benac/Čović 1956, 14, 63–64). Tumulus I yielded several finds, among which the most significant is an open, bronze bracelet with buffer terminals, decorated with carved ornamentation of hatched triangles (a so-called 'wolf's teeth' motif), bands of lines and rows of dots (Fig. 2:2).[428] The remaining items consist of three *saltaleoni* (Benac/Čović 1956, 14, Pl. XIV:8) and a single amber bead in flattened, biconical form, later classified by Palavestra as an example of type 12a (Benac/Čović 1956, 14, Pl. XIV: 7; Palavestra 1993, 112, 116, 123, Pl. 1:12a). Open bracelets with a triangular cross-section and, occasionally, expanded buffer terminals are considered

---

427 Benac/Čović 1956, 63–64; Drechsler Bižić 1983a, 261–262; cf. Čović 1983c, T. XXXVIII:11.

428 Benac/Čović 1956, 14, Pl. XIV:6, 6a, 6b; Čović 1981, 106, 108, Pl. II:9; Drechsler-Bižić 1983a, 262, Pl. XL:7.

to be characteristics of bronze objects belonging to phase IIb. They are reminiscent of the bracelets known from the previous stage (IIa); however, during phase IIb they became increasingly narrow, with the ornamentation composed almost exclusively of incised lines, formed in bands, rhombuses, triangles and combined in metopes.[429] Such a style of decoration replaced ornamentation made of dashed lines, garlands and 'fish-bladder' motifs known from the previous phase (Benac/Čović 1956, 26–27). This stylistic change is also evident in bracelets known from neighbouring area of Western Serbia dated to and around phase BzC2, although it must be mentioned that elsewhere similar decorative patterns sometimes appear on slightly later open bracelets with triangular cross-sections, dated from BzD up to HaA1.[430] On the basis of the aforementioned bracelet, the grave was chronologically defined as belonging to *Glasinac* phase IIb in the first overview by Benac and Čović (Benac/Čović 1956, 28) – a classification that has been upheld in subsequent publications (cf. Čović 1981, 106, 108; Drechsler-Bižić 1983a, 262).

More problematic is the identification of amber materials that could be linked with *Glasinac* phase III, signifying the LBA in this area. Chronological revision of the contexts with amber hitherto attributed to this phase revealed insufficient evidence for such classification in almost all cases (Cwaliński/Pravidur forthcoming). Due to absence of any other indicators, the Tiryns-type bead from grave 3, tumulus III (1894) at Vrlazje,[431] thanks to numerous analogies in well-dated contexts,[432] is considered as a fix-point in dating the grave to the developed LBA (*Glasinac* IIIb–IIIc). The recentmost tombs with amber that can be linked with the Bronze Age at the Glasinac are grave 2, tumulus IV (1894), and possibly also grave 1, tumulus XI (1894) from the site of Vrlazje – the former on the basis of a late variant of Golinjevo-type fibula,[433] whereas the latter due to similarity of the fragmented amber bead to the Tiryns type.[434] Nonetheless, both graves contained also indications of slightly later chronology in the form of iron-made tools and weapons, extending to the initial stages of the Iron Age. In both the cases, these are heavily corroded blades of iron knifes[435] which, although not lending themselves to a precise typological determination, may prolong the chronology of the context to phase IVb, when such objects became common in the Glasinac (cf. Čović 1987, 591–592, Fig. 34). In the case of grave 2, tumulus IV (1894), there are also pieces of a spectacle fibula classifiable to the types Suseni or Rudovci, so chronologically between phases IIIc–IVb.[436] Still, the temporal position

429 Benac/Čović 1956, Pl. XIV:1, 1a, 1b; Drechsler Bižić 1983a, 262, Pl. XXXIX:1–2, Pl. XL:7, 11, 12.

430 Zotović 1985, T. X:14–15; Dmitrović 2016, Fig. 113:4; cf. Čović 1981, 102, 106–108; Dmitrović 2016, 168.

431 Fiala 1894, 724, Fig. 9a; cf. Palavestra 1993, 116; Cwaliński/Pravidur forthcoming.

432 See e.g. Harding/Hughes-Brock 1974; Negroni Catacchio 1999, 2014; Czebreszuk 2011; Bellintani 2014.

433 Fiala 1894, 725, Figs. 10 and 11; Benac/Čović 1956, 46, Tab. IV:2–8; see also Pare 1998, 333–334; Gavranović 2011, 34.

434 Fiala 1894, 728; cf. Palavestra 1993, 116; Cwaliński/Pravidur forthcoming.

435 Fiala 1894, 725, 728; Benac/Čović 1956, 10, Pl. IV:7; Cwaliński/Pravidur forthcoming, Figs. 5:11 and 6:2–3.

436 Benac/Čović 1956, 10, Pl. IV:3; Cwaliński/Pravidur forthcoming, Fig. 5:2; cf. Bader 1983, 56–60, Taf. 11–13; Vasić 1999, 29–36, Taf. 8–11; Pabst 2008, 624, Abb. 11:2, 12:2.

of the Golinjevo-type fibulae and the Tiryns-type amber beads stands quite firmly in the literature; accordingly, they are definitely types invented and widespread during the final stage of the Bronze Age. However, at least in several cases, their chronology extends to the earliest moments of the Iron Age (Negroni Catacchio 1999, 251–252; Glogović 2003, 13). In the light of this evidence, it would be reasonable to extend the chronology of the grave 3, tumulus III (1894) at Vrlazje as well at least to phase IVa (as the context lacks any iron objects), in order to accept the possibility of protracted occurrence of the Tiryns type at the Glasinac. As a consequence, the best solution is to consider graves 2 from tumulus IV and 1 from tumulus XI at Vrlazje as features documenting the transition from the LBA (subphase IIIc-2) to the EIA (phases IVa and IVb), and at the same time as the final tombs containing amber that can be linked with the Bronze Age in the area of Glasinac.

Tab. 28. List of the Bronze Age contexts with amber at the Glasinac

Name	**Commune**	**Province**	**Type**	**Relative chronology**	**Absolute chronology**	**Reference**
Gučevo, tumulus IV, grave 1 (1895)	Rogatica	Republika Srpska	Burial – inhumation	*Glasinac IIa*; BzB	1550–1450 BC	Benac/Čović 1956, Tab. V:31, 33
Rudine u Ivanopolju, tumulus I (1894)	Rogatica	Republika Srpska	Burial – inhumation	*Glasinac IIb–IIIa*; BzC	1450–1300 BC	Benac/Čović 1956, Tab. XIV:7
Vrlazje, tumulus III, grave 3 (1894)	Rogatica	Republika Srpska	Burial – inhumation	*Glasinac IIIb–IVa*; BzD–HaB3	1250–800 BC	Fiala 1894, 724, Fig. 9a–b
Vrlazje, tumulus IV, grave 2 (1894)	Rogatica	Republika Srpska	Burial – inhumation	*Glasinac IIIc-2–IVb*; HaB1b–HaC1	1030/1010–700 BC	Benac/Čović 1956, Tab. IV:6
Vrlazje, grave 1, tumulus XI (1894)	Rogatica	Republika Srpska	Burial – inhumation	*Glasinac IIIc-2–IVb*; HaB1b–HaC1	1030/1010–700 BC	Cwaliński/Pravidur forthcoming, Pl. 1:6

2.3.11. Kosovo and Metohija

Up to date the largest assemblage of amber beads dated to the Bronze Age in the Balkans has been found in the grave 1 at the cemetery of Iglarevo on the territory of Kosovo and Metohija. Specifically, the site is situated at the border between Kosovo Polje and the plain of Metohija, some 20 km east of Peć, by the road Priština – Peć, on the slopes of Mt. Grebnica descending towards the Klina River. Discovered accidently

2.3.12. Vardar Valley and Pelagonia

A small number of amber artefacts dated to the Bronze Age was detected within the borders of the Republic of North Macedonia. The older one of the two sites which yielded amber – Dimov Grob – is located near the Ulanci village, in the Middle Vardar Valley (gr. *Axios*), less than 500 m from the river itself, on its east bank. This vast cemetery, spatially divided into two parts in respect to the relief of terrain, up to this day yielded 125 graves, with unburnt bodies of the deceased placed on the side in rectangular stone cists, arranged in rows organised in N–S direction.[440]

Amber was recovered from three graves: nos. 54, 73 and 120, each containing burial of a female individual with rich offerings (Papazovska 2017). In the first case it was found by the burial of 40-year-old female, richly furnished with grave goods, including gold earrings, glass beads of yellow paste, two large bronze pins, labrys belt of Aegean origin and another belt made of sea snail shells (Videski 2004, 26, 78, Figs. 33–34, T. XXX-XXXII; 2007b, 315, Figs. 2–4, 7). By the head of the buried woman was laid one matt-painted alabastron (Mitrevski 2007, Fig. 4). Grave 73 contained a female individual, deceased in the age of 40–45 years old, buried with a necklace made of amber, glass and bone beads, two bronze spiral earrings, two large bronze pins found on the chest and, again, a matt-painted alabastron (Videski 2004, 31–32, 78, T. XLI-XLII). Finally, a single, almost completely disintegrated amber bead was recovered from the grave 120. Most probably it was part of a necklace, composed together with glass, faience and bone beads, and combined with small, circular bronze pendants (Videski 2004, 78–79, Fig. 34). Skeletal remains of the deceased person buried in this grave were assigned to a 10-year-old female which additionally was given a single spiral bronze earring (Videski 2004, 46, T. LXXI).

Among the grave goods, particular attention should be paid to amber, bone, faience and glass beads,[441] large bronze pins with analogies in Kosovo and Banat,[442] labrys belt of Aegean origin[443] and matt-painted alabastra stylistically deriving from the *Mycenaean* pottery.[444] Basing on the chronology of materials co-occurring with amber in the contexts, the latter can be dated to the LBA in the North Macedonia. Typology and stylistics of the pottery correspond with the materials of the *Ulanci* group, named after eponymous settlement near the Ulanci village, located less than 1.5 km north-west from the necropolis of Dimov Grob (Mitrevski 2003, 46–52, T.V, T.VI; 2007, 444–445, Figs. 2 and 3). This unit developed approximately in the time span between 1400/1300 and 1100 BC (roughly BzC2/BzD–HaA1), thus immediately precedes transitional period to the EIA. Archaeological sites containing materials relevant to the *Ulanci* group have been found all along the Vardar Valley, with minor extension towards the west in Pelagonia and east in the Struma (gr. *Strimonas*) Valley (Mitrevski 2003, 50; 2007, Fig. 1).

---

440 Mitrevski 2003, 185, Fig. 6; Videski 2007a, 211, Pl. LIII:b–e; Papazovska Sanev 2014, 8–10, Fig. 1.

441 Videski 2007a, Pl. LV:a.

442 Mitrevski 2003, T. V:8–9; Videski 2007a, Pl. LV:b–c; see also Vasić 2003, 35, Taf. 12:170–175.

443 Mitrevski 2007, Fig. 4; Videski 2007a, Pl. LV:e.

444 Mitrevski 2007, Fig. 4; Videski 2007a, 212, Pl. LIV:b–c, e; Mitrevski 2003, 49, T. V:7.

At the turn of the 12th and 11th centuries BC, development of the *Ulanci* group was abruptly terminated, as seen on the basis of burnt layers on many sites in the North Macedonia, testifying to destruction after which the settlement have not been rebuilt (Mitrevski 2003, 51; 2007, 447–448). Around the same time a shift in burial practices is noted, with the inhumation in stone cists being terminated and replaced with deposition of cremated remains in urns (Mitrevski 2007, 447). Under the influence of the late *Brnjica* group (*Brnjica* II/III), already affected by channelled pottery of the *Belegiš II-Gava* culture, a completely new culture emerged in the Upper Vardar Valley.[445] Interestingly these cultural changes did not affect western parts of the North Macedonia (to the west from the Vardar Valley – in Pelagonia), where earlier Bronze Age traditions were continued (Mitrevski 2007, 449).

This process is evidenced i.a. by graves from the site of Popadin Dol, located approximately 2 km north-west from Prilep. Stone-cist grave 1, containing skeletal burial of a single individual, yielded two amber beads as well as a single glass bead made of dark-blue paste (Kitanovski 1960, Fig. 14; Palavestra 1993, 161–162). Remaining grave offerings include two arched fibulae with two discs on the bow, ornamented with short lines in the 'herringbone' pattern, and open bracelets with semi-circular cross-section and thickened ends (Kitanovski 1960, Figs. 2–4, 7–12). Both fibulae from Popadin Dol have close analogies at several sites in the North Macedonia, for instance, in the graves from Demir Kapija near Negotino, Prilepec, Rapeš and Trojaci (Vasić 2003, 45, Tab. 24: 274, 278–281). Although formally they can be correlated with the Dalmatian/Liburnian variants of fibulae of the same group, they still differ in details, such as shaping of the nodules or decoration (Bietti Sestieri/Lo Schiavo 1976, 185). R. Vasić suggest to date given examples to the second half of the 10th century BC, as the argument recalling similar chronology of geographically close examples from Vergina and Pateli (Vasić 2003, 46). These fibulae, together with other pieces of jewellery such as open bracelets also found in Popadin Dol, were classified to revised phase IIIA-B of the EIA in Macedonia by C. F. E. Pare which corresponds in his perspective to a period between 950/920 and 800 BC (Pare 1998, 336–337, Abb. 21, Tab. 3). By the decoration of bow, the fibulae from Popadin Dol definitely resemble some specimens from northern Greece; however, there is a difference in appearance of the feet holding the needles – in the latter fibulae those are larger and decorated on the edge with embossed dots.

B. Teržan, while reconsidering the chronology of Aegean examples of arched fibulae with two nodules on the bow, tackles the problem of chronology and origins of the aforementioned Pelagonian examples (Teržan 2007, 161–162). She noticed that strikingly similar sets of jewellery including amber, fibulae and bracelets, as documented in the case of Popadin Dol and Demir Kapija, are recurring in the grave inventories from Dalmatia already in the 12th century BC. Moreover, the latter date stands in line with the first appearance of the arched fibulae with two nodules on the bow in the Aegean during LH IIIC. Whether the fibulae from Pelagonia should be treated as imports either from the Aegean or Adriatic coast, or local products still needs to be solved, but at the moment it seems necessary to dilate their chronology from 10th to 11th century BC at the very least.

Time span between 1100 and 900 BC in Pelagonia can be regarded as transitional period from the LBA to the EIA (see Mitrevski 2003, 2007). In contrast to neighbouring

445 Mitrevski 2007, 448; Papazovska Sanev 2014, 11–12; see also Bulatović 2007.

regions, here inhumation was still practised as a leading burial rite, while rectangular stone cists, similar to the ones from necropolises of the *Ulanci* group, were used as a main grave form. Nevertheless, they never appeared in large concentrations like the cemeteries of Dimov Grob; therefore, one can only speak of small, isolated sites. Similarities in grave construction and inventory prompted M. Garašanin to place these tombs in a single horizon termed as the 'Popadin Dol-type cist graves' (Garašanin 1983f, 792). This group, apart from the eponymous grave, includes following sites: Rakida, Prilepac, Rapeš-Karanka and Demir Kapija – all of them containing similar inventory, comprising fibulae, bracelets, rings and sometimes also glass beads. Exceptional presence of amber, resemblance of the graves and potentially earlier chronology than previously assumed allow to suspect partial contemporaneity, or at least some relationship between the *Ulanci* group and the 'Popadin Dol-type cist graves'.

Tab. 30. List of the Bronze Age contexts with amber in the Vardar Valley and Pelagonia

Name	Commune	Province	Type	Relative chronology	Absolute chronology	Reference
Dimov Grob, grave 54	Gradsko	-	Burial – inhumation	BzD–HaA1	1300–1100 BC	Videski 2004, Figs. 33–34, T. XXX-XXXII
Dimov Grob, grave 73	Gradsko	-	Burial – inhumation	BzD–HaA1	1300–1100 BC	Videski 2004, Figs. 33–34, T. XLI-XLII
Dimov Grob, grave 120	Gradsko	-	Burial – inhumation	BzD–HaA1	1300–1100 BC	Videski 2004, 78–79, Fig. 34
Popadin Dol, grave 1	Prilep	-	Burial – inhumation	HaA2–HaB2	1100–910 BC	Kitanovski 1960, Fig. 14a–b

2.3.13. Albania

Thanks to the recent contributions, the topic of amber in Albanian prehistory, essentially unfathomable not a long time ago, slowly goes out of obscurity and becomes more intelligible to foreign scholars. In total seven sites with amber in the territory of Albania, all of the cemeteries comprising tumuli can be reliably dated to the Bronze Age (Kurti 2013, Fig. 2; 2017b, Pl. CI). Most of them (6) are generally dated to the LBA – a period in Albania dated between ca. 1500 and 1100 BC, while a single context is placed in the transitional period between the Bronze and EIA, covering a time span between ca. 1100 and 900 BC (cf. Prendi 1982, 214; Bodinaku 1995, 259). The sites are distributed in four concentrations which coincide with the largest clusters of prehistoric tumuli known in Albania: (1) at the confluence of the White Drin and Black Drin rivers in north-east part of the country, (2) in the middle of Mat Valley and the synonymous district, (3) in the mountainous areas of the Kolonjë Plateau and (4) the

Korçë Basin, the two latter placed in south-eastern part of the country, not far away from the border with Greece. First two regions can be grouped, according to their location and prevailing character of the material culture, into the northern variety of the Albanian LBA, whereas the latter two represent the southern variant.

The cultural differences between northern and southern Albania that started to emerge in the MBA, but became pronounced in the subsequent period, are largely due to a varying ratio of foreign influences radiating from opposite directions. The two main external milieus that have left their marks on the indigenous culture of Albania during the LBA, as evidenced by numerous imports or rather imitations in pottery and metallurgy, are the *Mycenaean* culture (or generally speaking Aegean zone) on one hand, and the inner Balkans, transmitting the influences arriving from Central Europe, on the other.[446] Basically, materials connected or originating from these two centres often reached the same locations, in effect becoming mixed in contexts creating very eclectic compositions, as in the case of tombs with amber. Still, on the basis of prevalence of the Aegean/*Mycenaean* or Central European/Balkan factor, Albania can be divided roughly in half, with the valley of Mat River forming a sort of cultural boundary, not impermeable but visibly demarcating spatial extents of these two traditions.

The LBA, that developed from the previous MBA background, is at best observed on the basis of archaeological material from southern Albania. Apart from continuing development of the local cultural substratum, this new stage in Albanian prehistory is conditioned by influences of the foreign cultures, among which the most evident and powerful seems to be the Aegean zone and the *Mycenaean* culture in particular, in contrast to northern Albania, where Balkan and Central European influences seem to have played a more important role. The Kolonjë Plateau, Korçë Basin and the areas of lakes Prespa and Ohrid represent three regions geographically clearly distinguished in the south-east Albania, but throughout the Bronze Age culturally uniform. The development of the local culture in this part of the country can be observed through stratigraphy of the Maliq settlement (Prendi 1966). The site located in an upland basin near the city of Korçë was formerly a lake that in the course of irrigation and drainage works dried up and eventually disappeared. There is a well-grounded suspicion that the settlement at least in part had a *palafitte* character – the buildings were set up on long wooden posts driven into the marshy ground (Harding 1992, 17). It provided the reference stratigraphy for the definition of southern Albanian prehistory, with a long succession of layers that produced pottery of types that enable comparisons with areas to north and south. In short, the Bronze Age part of the Maliq sequence can be described as follows: phases Maliq IIIa-b – EBA; phase Maliq IIIc – MBA; phase Maliq IIId – LBA (cf. Prendi 1982).

Of equal significance is the stratigraphy from the neighbouring site of Sovjan. It is located north-east from the Korçë depression, on the western shore of the ancient lake Maliq, not far away from the eponymous settlement. It is defined in the literature as a 'submerged tell' with a culmination at 818 m a.s.l., but in relative height, just 2 m above the ambient plain (Korkuti 2003, 225). Stratigraphy of Sovjan comprises eight main layers, out of which four are related to the Bronze Age occupation of the site. These are layer 8 signifying the EBA horizon (corresponding with phase Maliq

446 See e.g. Prendi 1982, 1995; Bodinaku 1995; Jubani 1995; Kurti 2017a, 2017b.

strengthening ties with the Apennine Peninsula (Prendi 1982, Fig. 13:7–8).[452] Still evident are contacts with Central Europe, at that time being under influence of the *Urnfield* cultures, documented by bronze objects, whereas in the pottery it is important to mention vessels decorated with the so-called 'turban' motif (wide, oblique grooves placed on the belly) (Prendi 1982, Fig.13:13, 17). Later (10th–9th century BC) the strongest evidence for sustained link between Albania and areas to the north come in a form of spectacle fibulae with eight-shaped loop mostly representing the type Vergina (Prendi 1982, Fig. 13:9; cf. Pabst 2008). Other artefacts testify to maintained connection with eastern Adriatic coastal zone, at the turn of the 2nd and 1st millennia BC flourishing with local metallurgical industry.[453]

2.3.13.1. Upper Drin Valley

The north-east part of Albania, near the White Drim (alb. *Drini i Bardhë*) and the Black Drim (alb. *Drini i Zi*) confluence, is known for numerous prehistoric tumuli distributed on the various river terraces of the Kukësi Basin at an altitude of ca. 300–400 m a.s.l., out of which around 70 were excavated (Kurti 2017a, 88). The most extensive use of tumuli falls into developed phase of the Iron Age (7th–6th century BC); however, there are many burials dated to the Bronze Age, among them grave 19 from tumulus 16 of Çinamak and the central grave from tumulus 1 of Krumë, both of which yielded numerous amber beads. Both sites are situated in the Upper Drin Valley (initial section of the river, right where it is formed by merging waters of the White and the Black Drim), within administrative borders of the Kukës district.

Grave 19, tumulus 16 of Çinamak, situated near the village of Shikaj, included a pair of pins (one with sleeve-like head and the other with spindle-head) and a sewing needle; a belt hook and around six disc-plate ornaments; three bracelets (including two open with stamp-shaped terminals); a pair of bronze hair spirals; and a necklace of 14 amber beads (Kurti 2017a, Fig. 4). Originally the assemblage was included in the final phase of the Bronze Age or the 12th–11th century BC (Jubani 1990); however, reconsideration of the chronology of its components led to shifting the dating to the 14th century BC (BzC2 – a period of occurrence of both the open bracelets with geometric and fish-bladder decorative motifs, and the spindle-head pins),[454] therefore at the beginning of the Albanian LBA (Kurti 2017a, 92–102).

A necklace of 25 amber beads, belonging most probably to the same period, was found in the central grave (no. 1) of tumulus 1 of Krumë, in the vicinity of the Llakaj reservoir, where upon deforestation of a heavily forested area 24 tumuli were identified

452 Among them it is worth mentioning a bronze fibula belonging to the variant of the Cassibile type (Jubani 1995, Fig. 2:4), or the arched fibulae found in Dukati near Vlora (Jubani 1995, Fig. 2:5) dated to the 11th–10th century BC (cf. Lo Schiavo 2010, Tav. 23–26, 374). Furthermore, some spearheads might be originating from the area of modern-day Italy, although there are lances that reflect inner Balkan types (Prendi 1982, 230, Fig. 12:10–15; Jubani 1995, 363–364).

453 This particularly documented with the distribution of the so-called Albano-Dalmatian axes (Prendi 1982, 220, Fig. 10:1; Harding 1992, 20), stretching along the littoral zone from Albania to southern and middle Dalmatia (cf. Batović 1983, 360).

454 Cf. Kosorić/Krstić 1988, 44–45; T. XI/2, 3; Vasić 2003, 48, T. 70; Dmitrović 2016, 166–168.

(Jubani 1982). The tumulus 1 was 0.90 m in height and 14 m in diameter, and contained 5 graves in total, located mainly in its central and eastern parts. Grave 1 was found in the centre of the mound, at a depth of 0.80 m. The earth at the south-western edge of the tomb was visibly scorched and some charcoals were found in the space within a stone enclosure, along with elements of the inventory (jewellery), all in bad preservation state. The latter comprised two open, massif bracelets with D-shaped cross-section and slightly thickened (stamp-shaped) terminals; a spiral bracelet made of bronze tape decorated in the middle with longitudinal rib; two bronze buttons; piece of a bronze pin shaft; two glasses and 25 amber beads (Jubani 1982, Tab. III; cf. Kurti 2017a, 100–101). Composition of the grave goods, similar to those from grave 19, tumulus 16 of Çinamak, warrants attribution of the central grave of tumulus 1 of Krumë to the Albanian LBA (Kurti 2017b, Pl. CI). However, due to poor preservation of the finds, particularly the bracelets, it is not possible to narrow the time span as in the previous case.

Tab. 31. List of the Bronze Age contexts with amber in the Upper Drin Valley

Name	Commune	Province	Type	Relative chronology	Absolute chronology	Reference
Çinamak, tumulus 16, grave 19	Kukës	Kukës	Burial – inhumation	BzC2; LH IIIA	1400–1300 BC	Kurti 2017a, Fig. 4:5
Krumë, tumulus 1, grave 1	Has	Kukës	Burial – cremation (pit)	BzC2–BzD; LH IIIA–B	1400–1200 BC	Jubani 1982, Tab. III:3

2.3.13.2. Mat Valley

The Mat Valley is a hilly area densely covered with forests and agricultural land, mainly situated on the terraced slopes on the right side of the river. From a large number of excavated tumuli in the middle Mat Valley (around 53 in total), only a few can be safely assigned to the LBA (Kurti 2017a, 107; see also Islami 2013). One of them is tumulus 2 at the site of Shtogj, inside of which grave 2 yielded several amber beads, along with bronze adornments and weapons of foreign provenience, thus is considered as an elite burial.[455] Chronology of the materials forming the grave's inventory allows to place it in the 14th–13th century BC. It contained i.a. a leaf-shape spearhead of northern origin[456] and a sword of Aegean provenience (or inspired by Aegean sword-making) dated circa LH IIIA/early IIIB, being a local variant of the type C swords, or a hybrid

455 Islami 2013, 116–117, 217, Tab. LVI:9; Kurti 2017a, 107–108, Fig. 7; 2017b, Pl. XCVIIIf:4–5; CI.

456 Kurti 2017a, Fig. 7:7; 2017b, 289; cf. Harding 1972, 218–220.

of the types A and C.[457] There is also a disc-head pin, initially interpreted as handle of a knife, which has a slight biconical and symmetrical swelling, perforated in the centre (Kurti 2017a, Fig. 3:10). This type has several parallels in the tumuli of the Kolonjë Plateau, in the southeast Albania, dated roughly the same period.[458]

The tumulus of Perlat – a single context with amber belonging to the Bronze and Iron Age transitional period – is located on the south bank of the Mat River, near the village of Perlat-Qendër, in the district of Mirdita. The grave contained a simple fibula with two nodules on the bow, associated with an amber bead related to the Allumiere type and several other in the form of a barrel.[459] The fibula in many ways resembles the type Golinjevo: on the bow above the nodules, there are two zones decorated with two rows of triangles, fitted in-between two bands of thin grooves. Below the nodules there are other bundles of grooves. A part of the bracket and the needle are missing, while the spring is broken. The slight elbow of the bow above the undecorated, triangular foot, the bow with octagonal cross-section at the centre, shaping of the nodules, and the rhomboid cross-section of the spring make the discussed example most akin to the fibulae found at Trcela near Vranjic (cf. Marović 1960, Fig. 3:1–2; Glogović 2003, Taf. 6:33–34). There are slight differences in decoration; the pattern of hatched triangles and incised bands near the nodules at the fibula from Perlat is reflected on other examples of the Golinjevo type, but in different combinations. Hence, the morphological features seem to be of key importance and, in the absence of any other analogical specimens, the fibulae from Vranjic should be currently considered as the closest counterparts. Summing up, dating of the Perlat finds to HaB1 phase – late 11th and early 10th century BC, as in the case of remaining examples of the Golinjevo type (cf. Glogović 2003, 13, Taf. 71), seems to be justified.

Tab. 32. List of the Bronze Age contexts with amber in the Mat Valley

Name	Commune	Province	Type	Relative chronology	Absolute chronology	Reference
Perlat, tumulus 2, grave 3	Mirditë	Lezhë	Burial – inhumation	*Submycenaean–Protogeometric*; HaB1	1060–950 BC	Kurti 2013, Tab. I:8–9
Shtogj, tumulus 2, grave 9	Mirditë	Lezhë	Burial – inhumation	LH IIIA–B; BzC2–BzD	1400–1200 BC	Kurti 2017a, Fig. 7:1; Kurti 2017b, Pl. XCVIIIf:4–5

457 I. Kilian-Dirlmeier considers it as an imported blade provided with a handle made in local workshop, similar to the A-type sword from Iglarevo – an opinion shared by A. Harding (Kilian-Dirlmeier 1985, 253; Harding 1995, 22). In the opinion of N. Bodinaku, the sword should be regarded as a special link in the evolution of Aegean swords towards the type C: while the form and the technology of the blade resembles the type A, the handle is more akin to the type C. The author excludes the possibility of a local production, and supposes it is an import from highly specialised Aegean workshops (Bodinaku 1995, 261). R. Kurti describes it as a distinct and probably a local variant of an Aegean type C sword (Kurti 2017a, 108).

458 Aliu 2012, 193–194, 199; cf. Kurti 2017a, 107–107. Identical pins were found in the coeval tumuli of the Kolonjë Plateau, e.g. Rehovë, graves 81 and 192, and were classified by S. Aliu as the type IIa (Aliu 2012b, 300, Tab. II:42, 44, XVI:205; Kurti 2017b, Pl. XCIX).

459 Bietti-Sestieri, Lo Schiavo 1976, Fig. 17:1–2; Kurti 2013, Tab. I:8–9; 2017b, Pl. CI.

2.3.13.3. Kolonjë Plateau

Three contexts with amber (Luaras tumulus, grave 46 and Rehovë, graves 186 and 201) are located in the Kolonjë Plateau – a highland area in the form of plateau (ca. 900m a.s.l.) in south-east corner of Albania, surrounded by mountains on all sides. The area is known for numerous burial mounds (about 70 in total) of variable size, usually measuring between 2 m in height and 20 m in diameter (Prodan) to 5 m in height and 50 m in diameter (Rehovë). All the tumuli discovered at the Kolonjë Plateau contained large numbers of graves: 72 graves (Prodan), 203 graves (Luaras) and 285 graves (Rehovë). In general, these mounds are built of soil mixed with small stones, usually heaped up over the oldest, central burial in the cist grave, dating to the 13th century BC (as in the case of Luaras tumulus) or the 15th century BC (as in the case of Rehovë tumulus) (Aliu 2007, 240). Most of the finds from the Kolonjë tumuli include different types of jewellery made of bronze, and ceramic vessels, often indicating foreign provenience or stylistic influence. In general, jewellery from these graves can be divided into several categories, such as pins, plaques, rings, necklaces, brooches, bracelets, etc. Chronologically and culturally most indicative seem to be pins, which were classified into various types and variants, ordered to reflect their chronology (see Aliu 2007).

Participation of southeast Albania in the north–south exchanges along the route linking the Aegean with Central Balkans is well evidenced in the most distinct LBA grave contexts. Together with the northern types of pins, in the same grave assemblages, items of *Mycenaean* provenience also appear, such as necklaces purely of carnelian, or incorporating carnelian and amber beads, and pairs of gold appliqués probably of the same origin (Kurti 2017a, 87).

The tumulus of Rehovë is located in the eastern part of the Kolonjë Plateau, nearby Erseke town. It was a lone tumulus erected on a flatland, however not far away from other barrow cemeteries on the plateau. It measured 5 m in height and 50 m in diameter. It included 285 graves, belonging to the different periods from the Bronze Age, antiquity up to the Middle Ages (cf. Aliu 2012). Inside the tumulus of Rehovë, two graves contained amber: nos. 186 and 201 (Kurti 2013, Fig. 2; 2017b, Pl. CI). On the basis of their inventory, both of them can be dated approximately to the 14th–12th century BC.

Grave 186 was discovered in the south-western section of the tumulus. It was a simple grave pit with the depth of 3.5 m and dimensions of 1.20 × 0.50 m, and with north–south orientation. Inhumed skeleton partly preserved inside the grave pit was laid on the back in supine position. It contained, apart from amber, multiple carnelian beads of varying shape (mostly spheroidal, discoid and globular), ceramic jug with S-shaped profile and a handle protruding above the rim, finished with horn-like appendix, and two pins with conical heads and swollen shafts pierced in the upper parts (Kurti 2017b, Pl. XCIX:f; CI). The two pins were defined as type VA by S. Aliu, and described as of 'northern' that is Central European origin (Aliu 2007, 240, Pl. LXIIa; 2012, 193–199). Similar specimens in Central Europe are usually regarded as the type Paarstadl (Pabst 2013, 155–157, Abb. 1). Development of these northern types in south Albania and western Macedonia seems to correspond to their last period of use in Central Europe, where most of them appeared between the 16th and 14th centuries BC (Kurti 2017a, 86). The vessel with S-shaped profile was attributed to the variant C of jugs, comprising hemispherical bodies and short cylindrical necks, marked by slight or sharp angular profile (Aliu 2012, 280–281). Similarly shaped vessels, especially in

a pair of double wire gold spirals (Kurti 2017b, 289, Pl. XCVIb/4).[460] The inventory is completed by a set of glass beads (Andrea 1985, Tab. VI:1–4). Basing on the overview of the finds chronology, particularly the stirrup-jar, it was suggested to date the entire context to LH IIIC – 12th century BC (Kurti 2017b, 295).

The tumulus 2 was located on the left side of the Korçë–Bilisht road, opposite the tumulus 1, about 100 m away from it. The mound had irregular shape; the north-eastern and south-eastern sectors were heavily damaged due to works on a nearby motorway, while the surface was constantly eroded by ploughing. The most regular, semi-spherical shape was retained by north-western and south-western sectors. The average height of the tumulus was 1.20 m (varying between 1 and 1.66 m depending on a measurement point), while its diameter varied between 20 and 22 m. The tumulus was built of earth mixed with small stones. Larger stones were used to build enclosures from some of the graves. Unlike tumulus I, during the excavation and study of the longitudinal profiles of tumulus II were discerned three different soil layers; thus, it is possible to define initial and subsequent phases of the tumulus enlargement. According to preserved stratigraphy, the development of the mound can be reconstructed in a following way: originally a large earthy mound was built surrounded by a white limestone ring. The original – primary tumulus contained 11 graves. Later the tumulus underwent an increase in height, its circumference was supplemented with a secondary ring, and 10 more graves were added. The secondary stone ring was built over the primary one and had a diameter of 11.50 m. Later one more grave was dug in, so in total the mound included 22 graves (Andrea 1985, 12–14).

Grave 2 of tumulus 2 at Barç, due to a more modest inventory, does not allow an easy determination of its chronology. Apart from 15 amber beads and a single kantharos-type vessel, it contained two 'headless' bronze pins with rounded upper terminals similar to awls, reminiscent of the *Urnfield* culture pins widespread since BzD, but due to a lack of characteristic features difficult to use as a safe indicator of chronology and cultural affiliation (Andrea 1985, 85, Tab. XVII:1–2; Kurti 2017b, Pl. CI). The kantharos – cup with hemispherical body, thickened rim and two raised handles (Andrea 1985, Tab. XVII:3) – can be compared with numerous similar vessel from Central Balkans, usually dated to the LBA (cf. Bulatović 2011, 124–127, Fig. 1; Bulatović/Filipović 2017, Fig. 5). Still, the specimen from grave 2 lacks any particular feature which could aid determination of its chronology. Thus it is suggested to use the beginning of the *Urnfield* period (BzD; 13th century BC) – the moment of introduction of the 'headless' pins – as the *terminus post quem*, and the end of the Albanian LBA (HaA1/A2; ca. 1100 BC) as the *terminus ante quem* in dating of the grave, in line with the proposition of R. Kurti (Kurti 2017b, Pl. CI).

460 This type of ornament of Central European ancestry and probably origin is widely distributed during the second part of the 2nd millennium as far south as Southern Italy and the Aegean (Kurti 2017a, 108). Gold spirals are the earliest objects in gold to appear in Albania, in grave 7 from tumulus 1 at Pazhok in central Albania, associated with a sword of Sandars A type and an imported Vapheio cup dating to the LH IIA (Kurti 2017b, Pl. XCVIa, b/1). They consist of a double wire of gold with round cross-section, forming a loop at both extremities. Even though golden spirals become more numerous in Albania during LH III and stayed in use even during the Early Iron Age, the question of their provenance remains uncertain (see Kurti 2017b, 289).

Tab. 34. List of the Bronze Age contexts with amber in the Korçë Basin

Name	Commune	Province	Type	Relative chronology	Absolute chronology	Reference
Barç, tumulus 1, grave 47	Korçë	Korçë	Burial - inhumation	LH IIIC; HaA1	1200–1100 BC	Andrea 1985, 28; Kurti 2013, Tab. 1:11; 2017b, Pl. XCVIIIf:8
Barç, tumulus 2, grave 2	Korçë	Korçë	Burial - inhumation	LH IIIB–C; BzD–HaA1	1300–1100 BC	Andrea 1985, 48; Kurti 2013, Tab. 1:1–2, 5, 7, 4:7–9; 2017b, Pl. XCVIIIe; XCVIIIf:6–7, 9

2.4. Provenience of amber finds

Already in the late 19th–early 20th century scholars, both chemists and archaeologists took interest in the origins of amber finds from Southern Europe. They understood the eventual implications of discovering a way to determine provenience of amber for reconstructing movement of people and materials across prehistoric Europe. However, it took some time until a reliable scientific method of determining species of amber was invented. Therefore, results of the earliest analyses of archaeological amber artefacts, such as those performed by O. Helm at the turn of the last two centuries, have to be taken with a grain of salt and need to be verified with modern methods and tools. In fact one can speak of credibility in assessing provenience of amber only since the 1960s, when IRS started to be utilised for this purpose. For those reasons, the following sections will be chiefly concerned with the latest studies on prehistoric amber artefacts from the circum-Adriatic zone.

2.4.1. Applied methods for determining species of fossil resins

Chemical properties of amber are known, thanks to a long tradition of research on this substance. Even though pieces of amber taken from various deposits may seem similar at the first glance, there are characteristic traits, both in their physical appearance as well as chemical composition, that can help distinguishing between certain species of fossil resins. Nevertheless, for a long time, there was no reliable and objective method of differentiating them. Finally in 1962 C. W. Beck and his students at the Amber Research Laboratory, Vassar College (NY) proved the usefulness of IRS in this regard (Beck et al. 1964, 1965), and began an extensive compilation of the IR spectra of fossil resins as means of determining the origin of prehistoric amber finds (see e.g. Beck 1966a, 1966b; Beck et al. 1968, 1970, 1972). An early result was the observation that the spectra of succinite can be empirically identified by a characteristic horizontal shelf, or 'shoulder', in the IR spectrum between 1,250 and 1,175 cm^{-1}, followed by a sharp peak between 1,160 and 1,150 cm^{-1} (Beck et al. 1965, 103; Beck 1986, 73–74).

This absorption pattern, named as the 'Baltic shoulder', is assigned to succinic acid ester (Langenheim/Beck 1965, 52–53; Beck 1986, 75). Although this feature is not shared by any other European resins, outside, most notably in the north-eastern part of the North American continent, there are Cretaceous fossil resins deposits whose IR spectra are identical to those of Baltic amber (Beck 1986, 81). Still, they are irrelevant for considerations on the use of amber in prehistoric Europe.

Beginning from the 1960s, discussed method has been applied to numerous types of ambers by a rapidly increasing group of authors (e.g. Savkevich 1970). In the course of following years, many other amber types have been recognised on the basis of characteristic IR spectra: e.g. simetite (Beck 1966a; Beck/Hartnett 1993); French pseudosuccinite (Beck/Liu 1976); gedanite and gedano-succinite (Stout et al. 1995); and rumanite (Stout et al. 2000). Such studies could be performed in empirical manner, utilising certain regions of IR spectra as 'fingerprints' of fossil resins (cf. Kosmowska-Ceranowicz/Vávra 2015). Besides this more common way of applying IRS, there has been a growing application of pattern-recognition programmes that were shown to be a valuable extension of the possibilities offered by spectroscopical studies (Beck et al. 1971).

However, because of the complexity of the formation process (maturation) and the heterogeneity of the original resin material, the IR spectra are not always unambiguously interpreted, and as a matter of fact many amber samples do not show readily distinctive spectra (Angelini/Bellintani 2005, 442). Moreover, extreme weathering, often encountered in the case of archaeological ambers, can render the identification impossible as it affects diagnostic regions of the spectra. Possible way to avoid this problem is to take a sample from the core of an object, instead from the opaque crust on its surface. The more serious problem presents contamination by conservation treatment which can occur with many older archaeological finds stored in museums (Beck 1986, 80).

Despite these drawbacks, IRS remains the most widely used and practical method for the structural characterisation and discrimination of amber types. Its effectiveness was enhanced with the introduction of Fourier transform infrared spectroscopy (FTIR) which provides high spectral resolution and short collection times, while requiring only a tiny portion of analysed object as a sample (Angelini/Bellintani 2005, 442; Caldararo et al. 2013, 47). The latter factor seems to be very important from the standpoint of archaeologists and museum curators, for whom integrity of stored artefacts is often a key issue. Thanks to the cost-effectiveness of FTIR, the sample required for obtaining the spectrum is very small, usually ranging between 1 and 2 milligrams (mg). Therefore, the method is invasive only to a very limited degree, and often it is possible to avoid leaving any trace of the sampling procedure, e.g. by utilising already broken-off fragments or cutting out small pieces from the holes and cracks in the surface.

All it takes to produce a spectrum is to obtain ~1 mg of amber sample, and then ground it and mix with potassium bromide (KBr) to prepare fine powder. The substance is then pressed into a piston and compressed under vacuum conditions for several minutes until it turns into translucent pellet, ready to be subjected to IR. In the course of the latter, an IR spectrum of the sample in the region of 4,000–400 cm^{-1} is recorded. Then the analyst, aided with the database-containing software, proceeds to look for characteristic absorption patterns ('fingerprints') that would allow to determine origin of the sample. A single analysis can be performed in circa 20 min, whereas the pellet can be preserved for future verification (Beck 1986, 77).

Another method frequently applied in amber analysis is attenuated total reflectance spectroscopy (ATR) which utilises amber samples in the solid state. More recent studies have showed that diffuse-reflectance infrared Fourier transform (DRIFT) spectroscopy can be efficiently used as an alternative to FTIR in discriminating various types of amber (Angelini/Bellintani 2005). Its main advantage is that it can be applied directly to the surface of solid amber (non-invasive mode), or by taking very low amounts of sample (0.2–0.1 mg). In spite of the advantages, DRIFT has been used in the studies of amber only for a short while and the database of results relevant for the deposits of various fossil resins is still under construction. Furthermore peaks' shape and intensity are often found to be distorted in respect to the same peaks recorded with FTIR. Therefore, direct comparison of DRIFT and FTIR spectra is rather difficult, and caution is required in interpretation of the data (Angelini/Bellintani 2005, 442).

Apart from analytical techniques based on IRS, there are other methods successfully applied in amber studies, showing a potential for archaeologists in determinations of amber artefacts origins. One of these is gas liquid chromatography (GLC), especially its modern variety in combination with mass spectroscopy (GLC/MS, usually performed by means of a 'mass sensitive detector' – MSD). Its application reportedly yielded an incredible amount of data especially in respect to the compounds contained in various soluble fractions of amber (Vávra 2009b, 449). Gas chromatography, possibly coupled with pyrolysis and/or mass spectroscopy (Py-GC-MS), is another technique that has frequently been applied in amber studies. It is extremely appropriate to investigate the chemical composition both of the amber soluble fraction (Mills et al. 1984/1985; Clifford et al. 1997) and of the bulk sample (Boon et al. 1993). Py-GC-MS is a good technique to correlate the amber type to the botanical source of the fossil resin so that the analytical data may be effectively used as a 'fingerprint' of a specific amber type (cf. Mills et al. 1984/1985; Boon et al. 1993; Stout et al. 2000). GC-MS analysis requires generally 1–3 g of amber sample to be used for the extraction of the soluble fraction (cf. Clifford et al. 1997); therefore, its application to archaeometry is rather limited if conservation of an artefact is an issue.

2.4.2. Provenience of amber finds from Italy

Until recently a total of 47 Bronze Age contexts with amber finds from Italy, accounting for over 150 samples,[461] have been included in the provenience analyses. Lately, a batch of 38 samples from 11 Italian contexts included in this dissertation has been submitted to the provenience analysis with FTIR method by Prof. I. Angelini from Dipartimento dei Beni Culturali, Università di Padova, as a part of the project *Circum-Adriatic branch of the amber route in the Bronze Age* (NCN no. 2015/17/N/HS3/00052) (Angelini 2020). Thus, in total one can report provenience of circa 200 samples of amber from 58 Italian Bronze Age contexts.

461 It is difficult to determine the exact number of samples subjected to provenience studies in the past, as many publications do not give this detail and only inform about general results of the analysis.

Apart from simetite, there is a chance that some other fossil resins were used to a limited extent in Italy during the Bronze Age; in as many as four contexts, the provenience studies detected presence of at the moment undefined fossil resins (Tab. 35). Some samples taken from Coppa Nevigata, Recent Apennine levels and Madonna di Loreto, Hypogea of Bronzes and Ivories revealed differences from the Baltic amber spectrum.[464] Moreover, they do not resemble spectra of local Italian fossil resins analysed so far with IRS (cf. Angelini/Bellintani 2005, 452). It has to be stressed that the term 'undefined' (as used in here, as well as in Tab. 35) does not imply that all the samples which it was given to belong to a single, yet undetermined resin. Their spectra also differ among themselves, and there is no single pattern common to all of them and any of the fossil resins of known origin. Nonetheless, a conspicuous concentration of finds made of fossil resin(s) other than succinite is evident in northern Apulia. Interestingly, in Madonna di Loreto 'Hypogeum of Bronzes' beads made of currently undefined fossil resin(s) were found besides those of succinite, what confirms simultaneous import of amber from different sources at least in the case of northern Apulia and Basilicata (Gargano peninsula and Tavoliere plain). The 'undefined' amber objects might originate from the circum-Adriatic zone or other areas of Europe. It is possible that they pertain to some, by now, completely exploited sources of hitherto undocumented fossil resins. However, one should not rule out that they also come from the Baltic zone, but due to specific diagenesis and/or post-depositional alterations, their spectra are irreconcilable with that of succinite. It is worth mentioning here other species of fossil resins co-occurring with succinite, whose spectra are far less standardised and often evade easy recognition, but on the outside they all look strikingly similar: e.g. gedanite or gedano-succinite (cf. Kosmowska-Ceranowicz 1999; Stout et al. 1995). The subject definitely requires more studies, especially in the recognition of spectral patterns of various European fossil resins.

In the case of nine contexts (Allumiere – Poggio La Pozza, grave 2, trench F; Bismantova – Campo Pianelli, grave XXXI; Bovolone, tomb 115; Coppa Nevigata, *Subapennine* stage; Franzine Nuove, graves 7, 12 and 123; Plemmirio, tomb 48; Torre Castelluccia, *ripostiglio*) provenience analysis of some of the finds was inconclusive due to their state of conservation. Some of the factors noted in the samples are exposition to high temperature, extreme degree of weathering, contamination with preservatives (e.g. cellulose nitrate) or glue, etc.[465] Consequently, a total of 16 samples were classified as indeterminable of the fossil resins they belong to. One cannot completely rule out that they are in fact composed of succinite, but it is recommended to subject those samples to re-analysis with IRS as well as other methods.

464 Angelini et al. 2003, Tab. 1; Angelini/Bellintani 2005, Tab. 2; Bellintani 2010, 144.

465 Cf. Angelini et al. 2003, 233, Fig. 7:1; Angelini 2010a, 209–210; Angelini/Bellintani 2017, 680; Angelini 2020, 59–65; Angelini et al. forthcoming.

Tab. 35. Italian Bronze Age amber finds subjected to provenience analyses

Region	Context name	Relative chronology	Absolute chronology	Variety of resin (number of samples)	Reference
Aeolian Islands	Portella di Salina, hut F	*Milazzese* early (LH IIB–IIIA)	1490/1460–1345/1320 BC	Succinite (1)	Beck 1970
Emilia/ southern Po Plain	Anzola dell'Emilia, central-eastern section of the site (sector D, trench?, layer US 27–26?)	BM 3–BR 1	1450–1260/1200 BC	Succinite (?)	Desantis et al. 2011
Emilia/ southern Po Plain	Beneceto (all contexts)	BM 3B–BR 2	1400–1150 BC	Succinite (3)	Angelini/ Bellintani 2005
Emilia/ southern Po Plain	Sant'Ambrogio, precise context unknown	BM 3–BR 1	1450–1260/1200 BC	Succinite (1)	Guerreschi 1975
Emilia/ southern Po Plain	Sant'Anastasio, precise context unknown	BM 2–BR 1	1550–1260/1200 BC	Succinite (1)	Guerreschi 1975
Garda Lake and its inframorenic basin	Lagazzi di Vho, Bronzo Antico 2 phase	BA 2	1800/1750–1650 BC	Succinite (?)	Instituto Donegani, Novara (unpublished results)
Garda Lake and its inframorenic basin	Peschiera – Imboccatura del Mincio, precise contextunknown	BR 1–2	1325/1300–1175 BC	Succinite (7)	Guerreschi 1970a; Negroni Catacchio 1971
Liguria	Castellaro di Uscio, horizon 3	BF 3	1035/1010–910 BC	Succinite (2)	Mello 1990
Marche and eastern Umbria	Monte Croce Guardia, US4, sector 1, hut 1	BF 1–2	1140/1100–1035/1010 BC	Succinite (1)	Angelini 2020; Angelini et al. forthcoming
Molise	Rocca di Oratino, phases I–III, superficial layers and sunken structures	BR 2	1220–1130 BC	Succinite (2)	Angelini 2020; Angelini et al. forthcoming

(*Continued*)

Tab. 35. Continued

Region	Context name	Relative chronology	Absolute chronology	Variety of resin (number of samples)	Reference
North-central Lazio (southern Etruria)	Allumiere – Poggio La Pozza, grave 1, trench F (1960)	BF 3b	960/950–910 BC	Succinite (1)	Angelini 2020; Angelini et al. forthcoming
North-central Lazio (southern Etruria)	Poggio della Capanna, grave	BF 3b	960/950–910 BC	Succinite (1)	Fugazzola Delpino 1992
Northern Apulia and Basilicata (Gargano peninsula and Tavoliere plain)	Grotta Manaccora, *Grotticella Funeraria* – SS layers	BM 2B–3A	1470–1350 BC	Succinite (2)	Angelini et al. 2003; Angelini/ Bellintani 2005
Piedmont	Monte Cavanero, hoard	BR–PF I	1300–800 BC	Succinite (2)	Angelini 2009
Sicily	Cava Secchiera, tomb 10	*Castelluccio 2B–3A*	1900/1800–1700/1600 BC	Succinite (1)	Angelini/ Bellintani 2017
Sicily	Plemmirio, tomb 10	*Thapsos I–III*	1420–1270/1250 BC	Succinite (1)	Angelini/ Bellintani 2017
Sicily	Thapsos, tomb D	*Thapsos I–III*	1420–1270/1250 BC	Succinite (4)	Angelini/ Bellintani 2017
Sicily	Thapsos, tomb 61	*Thapsos II*	1390/1370–1330/1315 BC	Succinite (1)	Angelini/ Bellintani 2017
South-central Lazio (Latium Vetus)	Villaggio delle Macine, MBA layers	BM 1–2	1700–1400 BC	Succinite (1)	Bellintani et al. 2007
Southern Apulia and south-east Basilicata (Murge plateau and Salento peninsula)	Murgia Timone, tomb 1	BM 2B–3	1470–1300 BC	Succinite (?)	Kaur et al. 2013

Tab. 35. Continued

Region	Context name	Relative chronology	Absolute chronology	Variety of resin (number of samples)	Reference
Southern Apulia and south-east Basilicata (Murge plateau and Salento peninsula)	Masseria Chiancudda, layer of *Punta le Terrare* facies	BM 2B–3	1470–1300 BC	Succinite (1)	Cinquepalmi et al. 2006
Trentino-Alto Adige	Maso Finale, cultural layer	BR–BF (*Luco A*)	1325/1300–1060/1025 BC	Succinite (11)	Angelini 2020; Angelini et al. forthcoming
Trentino-Alto Adige	Molina di Ledro, precise context unknown	BA–BM (?)	2200–1325/1300 BC (?)	Succinite (3)	Angelini/Bellintani 2005
Trentino-Alto Adige	Molina di Ledro, strata I–III/IV	BA 2–BM 1	1800/1750–1650 BC	Succinite (2)	Angelini/Bellintani 2005
Trentino-Alto Adige	Fiavè-Carera, horizon 6	BM 2B–3 early	1475/1450–1350 BC	Succinite (3)	Angelini/Bellintani 2005
Trentino-Alto Adige	Val di Non, precise context unknown	BM–BR	1650–1175 BC	Succinite (4)	Angelini et al. 2003; Angelini/Bellintani 2005
Western Veneto / northern Po Plain	Campestrin di Grignano Polesine, late BR–initial BF layers	BR 2–BF 1	1260/1220–1100/1060 BC	Succinite (9)	Bellintani et al. 2015a
Western Veneto / northern Po Plain	Desmontà, grave 437	BF 3	1010/1000–910 BC	Succinite (6)	Angelini 2013
Western Veneto / northern Po Plain	Franzine Nuove, grave 523	BR	1350/1300–1175/1150 BC	Succinite (3)	Angelini 2020; Angelini et al. forthcoming
Western Veneto / northern Po Plain	Franzine Nuove, grave 570	BR	1350/1300–1175/1150 BC	Succinite (2)	Angelini 2020; Angelini et al. forthcoming

(*Continued*)

Tab. 35. Continued

Region	Context name	Relative chronology	Absolute chronology	Variety of resin (number of samples)	Reference
Western Veneto / northern Po Plain	Olmo di Nogara, grave 30	BR	1350/1300–1175/1150 BC	Succinite (13)	Salzani et al. 2006
Western Veneto / northern Po Plain	Olmo di Nogara, grave 85	BM 2	1550/1500–1450/1375 BC	Succinite (5)	Salzani et al. 2006
Western Veneto / northern Po Plain	Olmo di Nogara, grave 89	BM 2	1550/1500–1450/1375 BC	Succinite (2)	Salzani et al. 2006
Western Veneto / northern Po Plain	Olmo di Nogara, grave 122	BR	1350/1300–1175/1150 BC	Succinite (?)	Salzani et al. 2006
Western Veneto / northern Po Plain	Olmo di Nogara, grave 154	BM 3	1450/1375–1350/1300 BC	Succinite (7)	Salzani et al. 2006
Western Veneto / northern Po Plain	Olmo di Nogara, grave 190	BR	1350/1300–1175/1150 BC	Succinite (?)	Salzani et al. 2006
Western Veneto / northern Po Plain	Olmo di Nogara, grave 252	BR	1350/1300–1175/1150 BC	Succinite (?)	Salzani et al. 2006
Western Veneto / northern Po Plain	Olmo di Nogara, grave 493	BM 3	1450/1375–1350/1300 BC	Succinite (4)	Salzani et al. 2006
Western Veneto / northern Po Plain	Scalvinetto, tomb 282	BR	1350/1300–1175/1150 BC	Succinite (?)	Salzani et al. 2006
Western Veneto / northern Po Plain	Scalvinetto, tomb 354	BM 3–BR	1450/1375–1175/1150 BC	Succinite (?)	Salzani et al. 2006

Tab. 35. Continued

Region	Context name	Relative chronology	Absolute chronology	Variety of resin (number of samples)	Reference
Western Veneto / northern Po Plain	Narde II, grave 75	BF 3	1010/1000–910 BC	Succinite (9)	Angelini 2010b
Western Veneto / northern Po Plain	Fondo Paviani, advanced BR–early BF layers	BR 1/2–BF 1–2	1260/1220–1010/1000 BC	Succinite (3)	Strafella et al. 2015
Western Veneto / northern Po Plain	Frattesina di Fratta Polesine, hoard 1 *Tesoretto*	BF 2	1100/1060–1010/1000 BC	Succinite (16)	Bietti Sestieri et al. 2015
Sicily	Castelluccio, tomb 9	*Castelluccio 1–2*	2200–1800/1700 BC	Simetite (3)	Angelini/ Bellintani 2017
Sicily	Madonna del Piano, 'from grave'	*Pantalica II – Cassibile*; FBA 2	1035/1010–910 BC	Simetite (1); Succinite (2)	Angelini/ Bellintani 2017
Northern Apulia and Basilicata (Gargano peninsula and Tavoliere plain)	Coppa Nevigata, *Recent Apennine* stage	BM 3	1420–1300 BC	Undefined (1)	Angelini et al. 2003; Angelini/ Bellintani 2005; Bellintani 2010
Northern Apulia and Basilicata (Gargano peninsula and Tavoliere plain)	Madonna di Loreto, 'Hypogeum of Bronzes'	BM 2B–BR 1	1470–1220 BC	Succinite (2); undefined (7)	Malagodi 1999; Mazzeo et al. 1999; Angelini/ Bellintani 2005; Bellintani 2010
Northern Apulia and Basilicata (Gargano peninsula and Tavoliere plain)	Madonna di Loreto, 'Hypogeum of Ivories'	BM 2B–BR 1	1470–1220 BC	Undefined (1)	Angelini/ Bellintani 2005; Bellintani 2010

(*Continued*)

Tab. 35. Continued

Region	**Context name**	**Relative chronology**	**Absolute chronology**	**Variety of resin (number of samples)**	**Reference**
North-central Lazio (southern Etruria)	Allumiere – Poggio La Pozza, grave 2, trench F (1960)	BF 3b	960/950–910 BC	Succinite (7); indeterminable (2)	Angelini 2020; Angelini et al. forthcoming
Northern Apulia and Basilicata (Gargano peninsula and Tavoliere plain)	Coppa Nevigata, *Subapennine* stage	BR	1300–1130 BC	Indeterminable (2)	Angelini 2020; Angelini et al. forthcoming
Romagna and Emilian Apennines	Bismantova – Campo Pianelli, grave XXXI	BF 2	1100/1060–1010/1000 BC	Succinite (2); indeterminable (1)	Angelini et al. 2003; Angelini/ Bellintani 2005
Sicily	Plemmirio, tomb 48	*Thapsos I–II*; LHIIIA	1420–1330/1315 BC	Succinite (4); indeterminable (2)	Angelini/ Bellintani 2017
Southern Apulia and south-east Basilicata (Murge plateau and Salento peninsula)	Torre Castelluccia, *ripostiglio*	BF 2	1100–1030 BC	Succinite (2); indeterminable (4)	Amber Research Laboratory, Vassar College (unpublished results)
Western Veneto / northern Po Plain	Bovolone, tomb 115	BR	1350/1300–1175/1150 BC	Indeterminable (1)	Angelini 2010a
Western Veneto / northern Po Plain	Franzine Nuove, grave 7	BM 3–BR	1450/1375–1175/1150 BC	Succinite (2); indeterminable (2)	Guerreschi 1970a, 1970b; Angelini 2020; Angelini et al. forthcoming
Western Veneto / northern Po Plain	Franzine Nuove, grave 12	BR	1350/1300–1175/1150 BC	Succinite (2); indeterminable (1)	Angelini 2020; Angelini et al. forthcoming
Western Veneto / northern Po Plain	Franzine Nuove, grave 123	BM 3–BR	1450/1375–1175/1150 BC	Indeterminable (1)	Angelini 2020; Angelini et al. forthcoming

2.4.3. Provenience of amber finds from Western and Central Balkans

In the case of former Yugoslavia up to recently provenience analysis by means of IRS was performed on selected Bronze Age amber finds merely from 13 archaeological contexts, located on 10 sites.[466] Interestingly all of them, save for two objects: one from the tumulus of Žamnjak in Istria (Todd et al. 1976, 320) and another from grave 9 of Velika Gruda in Montenegro (Beck 1996, 136), proved succinite to be a primary fossil resin used by Bronze Age communities inhabiting Western and Central Balkans (Tab. 36). Therefore, the notion of importing majority of raw amber or ready-made products from the circum-Baltic zone to the Balkans during the Bronze Age seems the most plausible hypothesis. On the other hand, until lately Albanian amber finds have not been considered in any provenience studies.

Fig. 17. FTIR spectra of selected archaeological finds from Western and Central Balkans identified as succinite and geological samples of succinite of known origin: 1 – reference spectrum of succinite; 2 – Ravnine, mound 3; 3 – Vrlazje, tumulus IV, grave 2 (1894); 4 – Garica, precise context unknown; 5 – Mušego, tumulus 9; 6 – Kompolje, grave 293; 7 – Belotić, mound XI, grave 4 (courtesy of the Amber Research Laboratory, Vassar College).

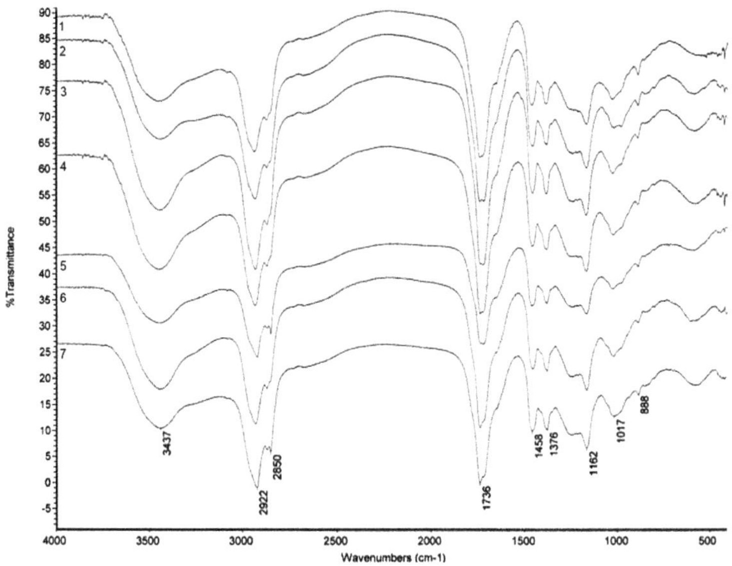

466 Jaworski et al. 1972; Todd et al. 1976; Hadži/Orel 1978; Drechsler-Bižić 1979–1980, 42, nn. 63a; Rašajski 1988, 27, n. 57; Beck 1996.

Lately, a set of 51 samples of selected finds from the museums of Sarajevo (National Museum of Bosnia and Herzegovina), Pula (Archaeological Museum of Istria), Zagreb (Archaeological Museum), Belgrade (National Museum), Čačak (National Museum), Loznica (Jadar Museum) and Tirana (Institute of Archaeology) was analysed at the Amber Research Laboratory, Vassar College (NY) and the laboratory of International Amber Association (Gdańsk) in order to verify assumed 'Baltic' origin of amber imported to the Balkans during the Bronze Age (Cwaliński et al. 2018; Kaur et al. 2020). The study was performed as a part of the project *Circum-Adriatic branch of the amber route in the Bronze Age* (NCN no. 2015/17/N/HS3/00052), and some of its results have been described in several papers.[467] Thanks to the project, a total number of analysed Bronze Age contexts with amber from the Balkans was raised to 41. In the Amber Research Laboratory the samples were predominantly analysed with FTIR, whereas laboratory of the International Amber Association utilised ATR for this purpose. Some of the strongly weathered samples for which IRS proved inconclusive (i.e. from Banjevac, tumulus 2, Belotić, mound VII, grave 1 and mound XI, grave 3, Shtogj, tumulus 2, grave 9 and Škicini, tumulus) were additionally analysed with GC-MS at the Amber Research Laboratory.

In the light of the study's results, previous conclusions regarding the provenience of Bronze Age amber finds from the Balkans can be sustained – the analysis confirmed that Bronze Age amber artefacts from Western and Central Balkans were made of Baltic amber (Tab. 36). A vast majority of the samples could be attributed to the succinite basing on the presence of the Baltic shoulder in their IR spectra (Fig. 17). Thus, it can be said that Western and Central Balkan communities in the Bronze Age were dependant on the inflow of amber from the north. In case of the samples that lack clearly defined infrared spectra, GC-MS pinpointed the class of fossil resins they belong to as Ia and, furthermore, allowed to validate supposed Baltic origin of most of them. Additionally, one of the two discoidal beads from the Žamnjak tumulus previously found indeterminable in terms of fossil resin species, although deemed possibly 'Baltic' (Todd et al. 1976, 320, Tab. 2) was sampled again and this time IRS allowed to determine it surely as succinite, thanks to the clearly readable Baltic shoulder (Cwaliński et al. 2018; Cwaliński et al forthcoming) (Fig. 18).

467 Kurti et al. 2020; Cwaliński/Pravidur forthcoming; Cwaliński et al. forthcoming.

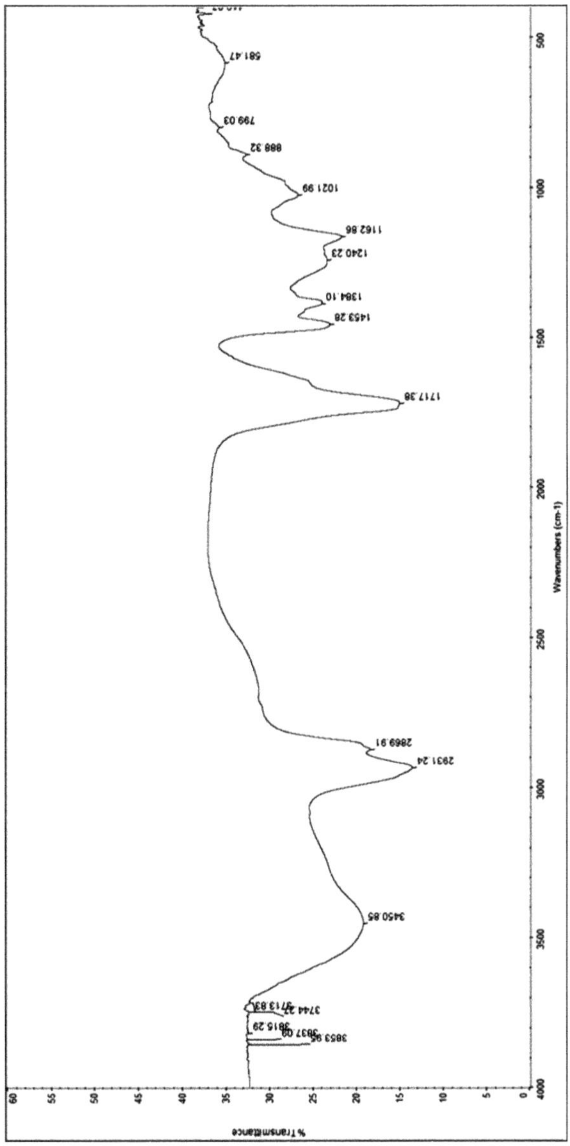

Fig. 18. New FTIR spectrum of the sample from the tumulus of Žamnjak obtained in the course of the 2018' study (Cwaliński et al. 2018).

Some of the samples could not be classified to any of the known fossil resins, because they revealed atypical infrared spectra upon the analysis. Origin of the irregularly oval bead from grave 9 of Velika Gruda still remains indeterminable; however, by C. W. Beck's own admission, its spectrum is 'compatible with that of weathered and contaminated Baltic amber', possibly also affected by contamination (Beck 1996, 136). Similarly indeterminable in terms of origin is the sample obtained from fragments of discoidal and spheroidal beads from Perlat, tumulus 2, grave 3. Both the initial analysis with ATR as well as the FTIR test conducted subsequently showed spectra in general similar to succinite, however lacking the clear Baltic shoulder (instead they present a flattened band between 1,300 and 110 cm^{-1} with a weakly discernible peak at 1,175 cm^{-1}) (Kurti et al. 2020, Figs. 6 and 11). It is worth underlining that the sampled material was already in a poor condition, strongly fragmented and weathered. Consequently, it cannot be decided to which fossil resin the sample should be classified; however, the Baltic origin should not be excluded, especially as the second sample from Perlat, taken from one of the barrel-shaped beads produced in the typical succinite spectrum. Finally, one of the newly analysed amber finds – discoid bead from Škicini – exhibited a spectrum unlike other European fossil resins, such as succinite, simetite, gedanite, gedano-succinite or rumanite (Cwaliński et al. forthcoming, Fig. 4) (Fig. 19). The sample was studied further by GC-MS analysis. Although there was only a small amount of the sample left for the analysis, several key components detected lend support that Škicini artefact possibly is a Baltic amber; these compounds were methyl succinite, succinic derivatives, several long chain fatty esters and, importantly, methyl dehydroabietate and its oxidized products (Cwaliński et al. forthcoming, Tab. 2). Still on this basis alone the question cannot be solved definitely, and the only thing that can be surely admitted is that the sampled bead belongs to the Class Ia Resinites. The above-discussed examples of amber finds of doubtful origin, however, do not alter significantly the overall picture that suggest the succinite was the major fossil resin used in Western and Central Balkans during the Bronze Age.

Fig. 19. Comparison of FTIR spectrum of the undefined sample from Škicini tumulus and reference spectra of selected fossil resins belonging to Class Ia Resinites. From top to bottom: (a) Škicini, (b) gedanite, (c) Baltic, (d) gedano-succinite and (e) rumanite (Cwaliński et al. forthcoming).

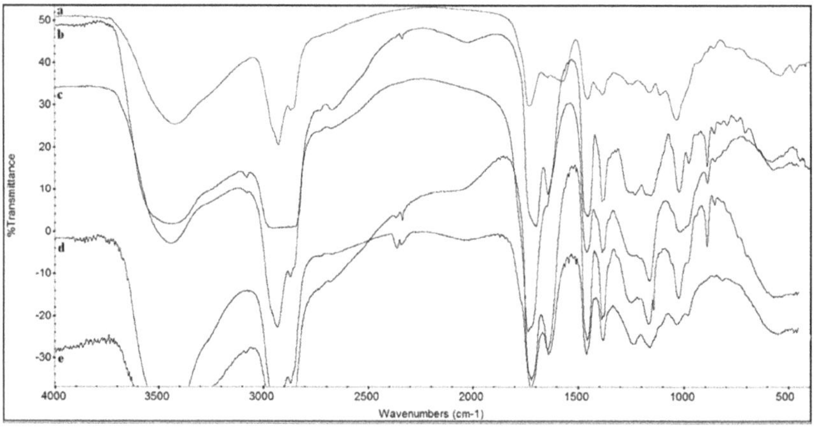

Tab. 36. Western and Central Balkan Bronze Age amber finds subjected to provenience analyses

Region	Context name	Relative chronology	Absolute chronology	Variety of resin (number of samples)	References
Glasinac	Vrlazje, tumulus III, grave 3 (1894)	*Glasinac IIIb–IVa*; BzD–HaB3	1250–800 BC	Succinite (1)	Cwaliński et al. 2018; Cwaliński/ Pravidur forthcoming
Glasinac	Vrlazje, tumulus IV, grave 2 (1894)	*Glasinac IIIc-2–IVb*; HaB1b–HaC1	1030/1010–700 BC	Succinite (1)	Cwaliński et al. 2018; Cwaliński/ Pravidur forthcoming
Istria	Vrčin, grave 3	*Istra II–III*; BzA2–BzD	1700–1200 BC	Succinite (3)	Jaworski et al. 1972; Cwaliński et al. 2018; forthcoming
Istria	Vrčin, grave 7	*Istra II–III*; BzA2–BzD	1700–1200 BC	Succinite (1)	Cwaliński et al. 2018; forthcoming
Istria	Vrčin, grave 8	*Istra II–III*; BzA2–BzD	1700–1200 BC	Succinite (1)	Cwaliński et al. 2018; forthcoming
Istria	Vrčin, grave 12	*Istra II–III*; BzA2–BzD	1700–1200 BC	Succinite (1)	Cwaliński et al. 2018; forthcoming
Istria	Vrčin, grave 13	*Istra II–III*; BzA2–BzD	1700–1200 BC	Succinite (2)	Cwaliński et al. 2018; forthcoming
Istria	Vrčin, tumulus B	*Istra II–III*; BzA2–BzD	1700–1200 BC	Succinite (5)	Todd et al. 1976
Istria	Žamnjak, tumulus	*Istra II–III*; BzA2/B-BzC2	1600–1300 BC	Succinite (2)	Todd et al. 1976; Cwaliński et al. 2018; forthcoming
Istria	Mušego, tumulus 6	*Istra III*; BzC2–BzD	1400–1200 BC	Succinite (3)	Cwaliński et al. 2018; forthcoming

(*Continued*)

Tab. 36. Continued

Region	**Context name**	**Relative chronology**	**Absolute chronology**	**Variety of resin (number of samples)**	**References**
Istria	Mušego, tumulus 9	*Istra III*; BzC2–HaA1	1350–1150 BC	Succinite (3)	Cwaliński et al. 2018; forthcoming
Kolonjë Plateau	Rehovë tumulus, grave 186	LH IIIA–B; BzC2–BzD	1400–1200 BC	Succinite (1)	Kaur et al. 2020; Kurti et al. 2020
Kolonjë Plateau	Rehovë tumulus, grave 201	LH IIIA–B; BzC2–BzD	1400–1200 BC	Succinite (1)	Kaur et al. 2020; Kurti et al. 2020
Kolonjë Plateau	Luaras tumulus, grave 46	LH IIIB–C; BzD–HaA1	1300–1100 BC	Succinite (1)	Kaur et al. 2020; Kurti et al. 2020
Korçë Basin	Barç, tumulus 1, grave 47	LH IIIC; HaA1	1200–1100 BC	Succinite (3)	Kaur et al. 2020; Kurti et al. 2020
Korçë Basin	Barç, tumulus 2, grave 2	LH IIIB–C; BzD–HaA1	1300–1100 BC	Succinite (3)	Kaur et al. 2020; Kurti et al. 2020
Kvarner and northern Dalmatia	Garica, precise context unknown	HaA2/ B1–HaB2/3	1100–910/850 BC	Succinite (1)	Cwaliński et al. 2018
Lika	Bezdanjača cave, mass grave nr 4	BzC2–BzD1	1400–1250 BC	Succinite (?)	Drechsler-Bižić 1979–1980
Lika	Golubnjača, block 4	HaA1–HaB1	1200–950 BC	Succinite (1)	Cwaliński et al. 2018
Lika	Kompolje, grave 293 (1903–1905)	HaA1–HaB1	1200–950 BC	Succinite (5)	Cwaliński et al. 2018
Mat Valley	Shtogj, tumulus 2, grave 9	LH IIIA–B; BzC2–BzD	1400–1200 BC	Succinite (1)	Kaur et al. 2020; Kurti et al. 2020
Middle Dalmatia	Trcela kod Vranjica, grave	HaB1	1060–950 BC	Succinite (1)	Todd et al. 1976

Tab. 36. Continued

Region	Context name	Relative chronology	Absolute chronology	Variety of resin (number of samples)	References
Northern Croatia	Alilovci Lipje, test trench E/ small house	BzB–BzC1	1600–1400 BC	Succinite (1)	Unpublished analysis performed by Centar za istraživanje materijala Istarske županije METRIS
Podrinje and Western Serbia	Banjevac, tumulus 2	BzC1–BzD	1450–1200 BC	Succinite (1)	Kaur et al. 2020
Podrinje and Western Serbia	Belotić, mound VII, grave 1	BzB/C1–BzC2	1500–1300 BC	Succinite (1)	Kaur et al. 2020
Podrinje and Western Serbia	Belotić, mound XI, grave 3	BzB/C1–BzC2	1500–1300 BC	Succinite (3)	Kaur et al. 2020
Podrinje and Western Serbia	Belotić, mound XI, grave 4	BzB/C1–BzC2	1500–1300 BC	Succinite (1)	Kaur et al. 2020
Podrinje and Western Serbia	Karavlaške kuće, mound 1, grave 2	BzC2–BzD	1400–1200 BC	Succinite (1)	Todd et al. 1976
Podrinje and Western Serbia	Jezero, mound III, grave 3	BzC	1450–1300 BC	Succinite (1)	Todd et al. 1976
Podrinje and Western Serbia	Jezero, mound III, grave 5	BzC	1450–1300 BC	Succinite (3)	Todd et al. 1976
Podrinje and Western Serbia	Jezero, mound VI, grave 1	BzC2–BzD	1400–1200 BC	Succinite (1)	Todd et al. 1976
Podrinje and Western Serbia	Paulje, tumulus A	BzD	1300–1200 BC	Succinite (1)	Cwaliński et al. 2018

(*Continued*)

Tab. 36. Continued

Region	Context name	Relative chronology	Absolute chronology	Variety of resin (number of samples)	References
Podrinje and Western Serbia	Paulje, tumulus K	BzC	1450–1300 BC	Succinite (4)	Cwaliński et al. 2018
Podrinje and Western Serbia	Paulje, tumulus N	BzC1–BzD	1450–1200 BC	Succinite (1)	Cwaliński et al. 2018
Podrinje and Western Serbia	Ravnine, mound 3	BzB/BzC1–BzC2	1500–1300 BC	Succinite (1)	Cwaliński et al. 2018
Slovenia	Debeli Vrh, hoard	HaA1–A2	1200–1060 BC	Succinite (?)	Hadži/Orel 1978
Upper Drin Valley	Çinamak, tumulus 16, grave 19	BzC2; LH IIIA	1400–1300 BC	Succinite (2)	Kaur et al. 2020; Kurti et al. 2020
Vojvodina	Majdan hoard	HaA1–HaB1	1200–950 BC	Succinite (2)	Rašajski 1988
Mat Valley	Perlat, tumulus 2, grave 3	*Submycenaean–Protogeometric*; HaB1	1060–950 BC	Succinite (1); indeterminable (1)	Kaur et al. 2020; Kurti et al. 2020
Montenegro and southern Dalmatia	Velika Gruda, grave 9	BzC2–BzD	1400–1200 BC	Succinite (1); indeterminable (1)	Beck 1996
Istria	Škicini, tumulus	*Istra II–III*; BzB–HaA1	1575/1550–1150 BC	Undefined (1)	Cwaliński et al. 2018; forthcoming

2.5. Spatial and temporal dynamics of amber influx to the Circum-Adriatic zone – assessment

Summarising the earlier considerations regarding amber on the micro-regional level allows drawing a number of general conclusions regarding the dynamics of the discussed material's inflow to the circum-Adriatic zone during the Bronze Age. In the following summary, the study area was divided into five sub-zones, based on the geographical and cultural criteria: Northern Italy, Central Italy, Southern Italy, Western Balkans and Central Balkans. Such approach allows to notice some general trends in the process of amber inflow to the circum-Adriatic zone in a macroscale.

With respect to Italy, the tripartite division is often used to characterise cultural situation during the Bronze Age.[468] Accordingly, the following provinces of amber's acquisition were ascribed to the three Italian sub-zones: Northern Italy – Emilia/southern Po Plain, Garda Lake basin, Liguria, Piedmont, Romagna and Emilian Apennines, Trentino-Alto Adige, western Lombardy and western Veneto/northern Po Plain; Central Italy – Marche and eastern Umbria, north-central Lazio, south-central Lazio and Tuscany, northern Lazio and western Umbria (northern Etruria); and Southern Italy – Aeolian Islands, Calabria, Molise, northern Apulia and Basilicata, Sicily and southern Apulia and south-east Basilicata.

Denotations of the terms 'Western Balkans' and 'Central Balkans' have already been tackled at the beginning of this work (cf. footnotes 3 and 4). Similar to the tripartite division of Italy, the distinction between Western and Central Balkans stems from the modern efforts to order geographic space in political, administrative and demographic-statistical terms. However, it also bears significance for the Bronze Age archaeology of the Balkans, although the allocation of provinces of amber's acquisition may not always correspond to a general perception of geo-political or cultural divisions in this part of Europe. The border between the two sub-zones runs roughly meridionally, in parallel to the eastern fringes of the Dinaric Alps, along the Drina River, then the modern Croatian-Serbian border to the west of Fruška gora, and the Danube. Thus, the sub-zones can be colloquially equated with Western Balkans (comprising the provinces of Istria, Kvarner and northern Dalmatia, middle Dalmatia, Montenegro and southern Dalmatia, Lika, Slovenia and Northern Croatia) and Central Balkans (comprising the provinces of Vojvodina, Podrinje and Western Serbia, Glasinac, Kosovo and Metohija, Vardar Valley and Pelagonia and all four provinces distinguished in the territory of Albania).

Each of the sub-zones is discussed separately, with chronological synopsis of amber's influx presented in the form of tables. Marking the horizons of amber occurrence in individual provinces of amber's acquisition (that is columns in the aforementioned tables) visually aids their dating in both relative and absolute terms. Furthermore, the regional chronologies were juxtaposed against the two outer Bronze Age periodisations: *Mycenaean* – Helladic on one hand and Central European – adjusted Reinecke's on the other.

General observations on the dynamics of amber influx into the circum-Adriatic zone allow to distinguish several phases in this process which largely correspond to

468 E.g. Barfield 1994; Barker/Stoddart 1994; Malone et al. 1994; Carancini et al. 1996; Bietti Sestieri 2010.

the traditional four-partite division of the Italian Bronze Age that, nonetheless, can by transposed onto the Balkans as well. Consequently, the analysed contexts with amber are divided into EBA (ca. 2150 ± 50 – 1650 ± 50 BC), MBA (1650 ± 50 – 1300 ± 50 BC), LBA (1300 ± 50 – 1150 ± 50 BC) and FBA (1150 ± 50 – 950 ± 50 BC). As can be inferred from the regional descriptions of amber finds above, some of the contexts, however, cannot be assigned to a specific phase, either due to insufficient data or due to prolonged use. Some of the sites have yielded evidence of continuous occupation or frequentation over several consecutive periods; thus, they do not lend themselves well to a detailed analysis. Contexts from which amber possibly belongs to two adjacent phases, by way of a compromise, were placed in the intermediary stages: EBA/MBA, MBA/LBA, LBA/FBA and FBA/EIA.

Furthermore, it was decided to supplement this summary with considerations on the changes in availability and commonness of amber in the provinces over time. In order to quantify them, two measures were applied. First, to determine commonness of amber in a province at a given stage, the relevant contexts were summed up. Nonetheless, it needs to be stressed that intra-regional comparison of the coefficient may not be fully reliable, because it does not take into account a total number of coeval archaeological contexts within a region. Thus, the larger provinces (identified with cultural units) have an advantage over the smaller ones. To have a comparable measure of amber's commonness, the numbers of contexts with amber should be divided by a total number of contexts in a region. However, such solution is hardly feasible, as the recognition of archaeological sites attributed to a cultural or territorial unit is strongly dependant on the state of research in a region (which varies a lot), and such numbers are difficult to find in the literature.

Second, the availability of amber in a region was measured by calculating total amount of amber per chronological stage. Still, the task is no less difficult than the one described above. A scholar needs to decide which measure of amber quantity she or he is going to use. Arguably, the most reliable one is the total weight of amber finds. However, calculating the volume of amber artefacts on the basis of the most frequently noted dimensions (height and diameter) is not practical, due to irregularity of the finds. Furthermore, these measures are not always given in the literature, and their precision can vary from artefact to artefact. The only feasible solution is to sum up the number of artefacts from a region dated to consecutive phases of the Bronze Age. This coefficient is not ideal, as the finds differ in terms of size, i.e. a bead from a region may have a volume of two other from a different region, but in absolute terms it would still be counted as 'one'. Hence, the availability coefficient given here should be approached with caution as well.

2.5.1. Northern Italy

In the entire Northern Italy (Fig. 20), the earliest amber finds come from the lake-dwellings (*palafitte*) of the *Polada* culture in the vicinity of the Garda Lake and in Trentino. Even if indication of a context indisputably belonging to the early stage of EBA (BA 1) is impossible in the light of currently available data, the finds from Lavagnone, phase 4 safely set the *terminus ad quem* for the initial appearance of amber in Northern Italy at BA 2 (1800/1750–1650 BC).[469] Similar chronology, with a certain dose of probability, can be ascribed to the finds from the earlier phase of the Lagazzi di Vho settlement and grave 54 from the Sorbara d'Asola cemetery (see section 2.2.5). According to unpublished results of IR spectroscopy performed by Instituto Donegani in Novara (P. Bellintani, personal communication), the beads from Lagazzi di Vho were made of succinite, thus proving that Baltic amber was imported to Northern Italy already in the EBA. This hypothesis is reinforced by similar conclusions derived from IRS analysis of five samples of amber beads from Molina di Ledro, dated between BA 2 and BM 1 (cf. Angelini/Bellintani 2005). In comparison with the later stages, commonness and availability of amber in Northern Italy during EBA were quite limited, even if the sites from the EBA/MBA intermediary stages are added (Figs. 21 and 22).

Fig. 20. Chronological synopsis of the occurrence of amber in Northern Italy between the 22nd and 9th c. BC.

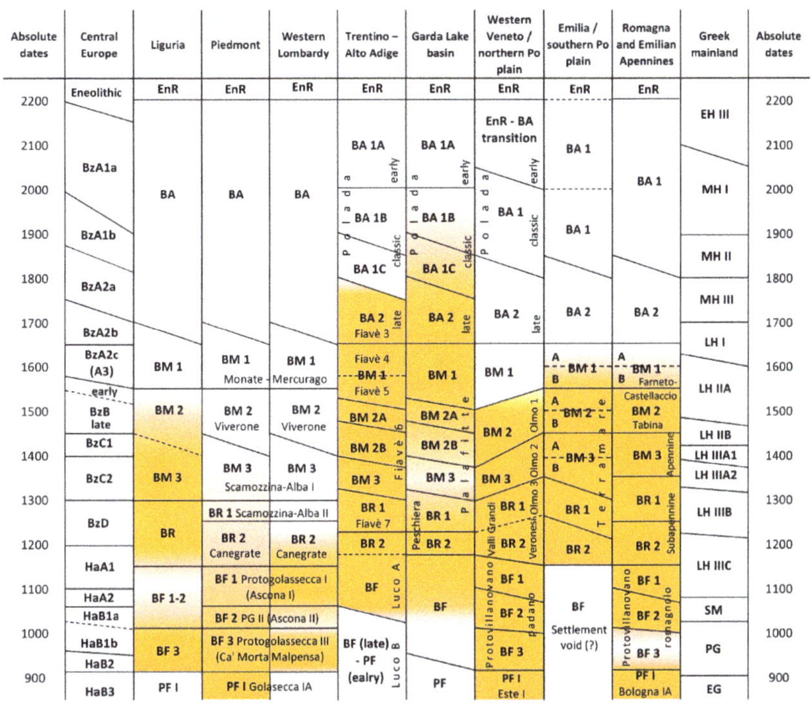

469 There still is a question if dating of the deposition context of an amber artefact should be considered tantamount to the time of its arrival and the period of its use in a region. It is possible that amber jewellery was kept for a long time, being passed as a part of inheritance or dowry (i.e. trousseau) from generation to generation. On the other hand, amber, upon retrieving from its natural source – deposit, looses its original qualities quite rapidly, in comparison to other gemstones: it is prone to weathering, changes of colour or mechanic damages due to its inherent fragility, when not conserved. Thus, it seems unlikely that discussed amber artefacts remained in use for several centuries in a row.

Fig. 21. Fluctuations in number of contexts – commonness of amber in Northern Italy.

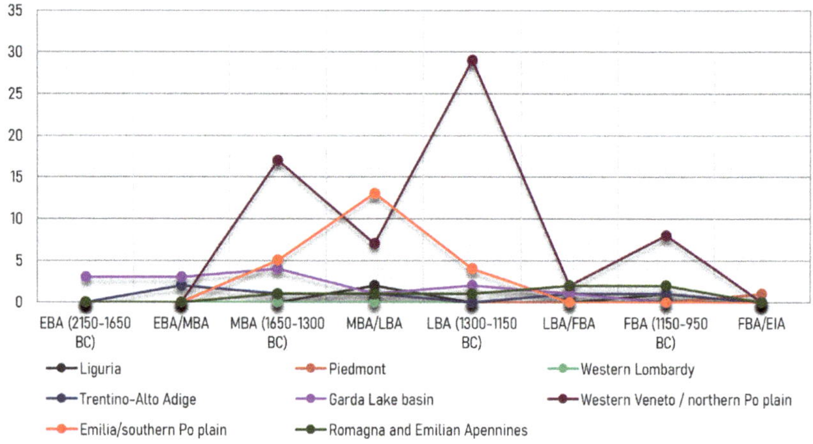

Fig. 22. Fluctuations in the quantity of artefacts – availability of amber in Northern Italy.

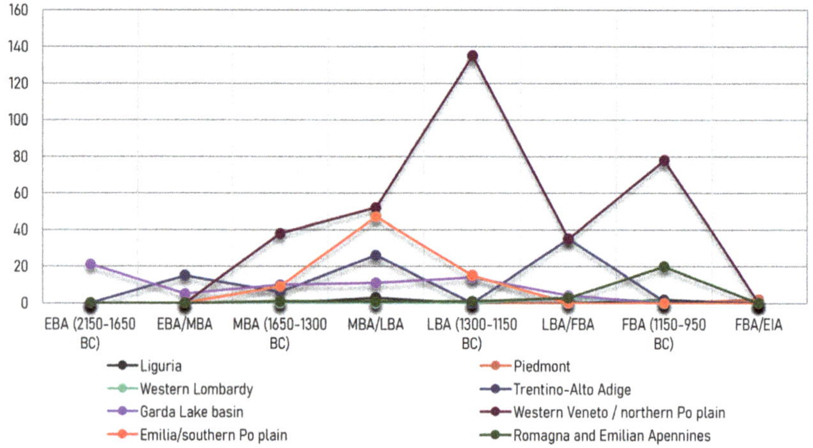

While during the initial MBA (BM 1) presence of amber was still largely confined to the Garda Lake basin and Trentino, beginning from BM 2 amber was becoming increasingly widespread across Northern Italy, arriving in Liguria, western Veneto, Emilia and Romagna (Fig. 20). Amber appears there in a multitude of contexts dated to successive stages of MBA and LBA, with only the Garda Lake basin lacking amber testimonies datable to BM 3 – indication of a short break in amber import to the province which as well can be just a gap in the state of research. This first amber 'boom' in Northern Italy is best expressed by the measures of commonness and availability (Figs. 21 and 22): while in the case of Trentino and the Garda Lake basin these coefficients stay at roughly the same level as in EBA, in Emilia and western

Veneto the rates are significantly higher immediately upon introduction of amber. In the case of the latter region, they seemingly reached the peak around LBA. It is difficult to say if the same is a fact with respect to the neighbouring regions, due to large numbers of contexts classified to MBA/LBA intermediary stage. Nonetheless, superiority of western Veneto/northern Po Plain in terms of amber's commonness and availability over remaining Northern Italian regions is striking. Whereas in provinces such as Liguria or Romagna presence of amber during MBA and LBA may be described as incidental, in the northern Po Plain it became a prominent feature of material culture, predominantly observable among sepulchral contexts (e.g. Olmo di Nogara necropolis). In Trentino, the Garda Lake basin and Emilia/southern Po Plain amber in that period chiefly appears at settlements; however, it has to be considered that cremation, practised in those regions as the main burial rite at the time, potentially erased evidence of amber from cemeteries. A considerable number of analysed amber samples dated to this period supports the notion that Baltic amber remained the predominant source of fossil resin for Northern Italian Bronze Age communities (cf. section 2.4.2).

Presence of amber in Northern Italy continued up to the end of LBA (BR), when various factors caused the previous social organisms, most notably the *Palafitte-Terramare* culture, to collapse (cf. sections 2.2.5–7), bringing amber inflow to a halt in several regions. In Liguria amber seems to have disappeared from material culture until BF 3 (Fig. 20); however, its resurgence at the end of FBA must have been of small scale, considering just two beads coming from one site dated to the period (Figs. 21 and 22). In Trentino-Alto Adige and the Garda Lake basin, amber is still observable in the first half of FBA, but then it disappears around the 11th century BC. With respect to the area around Garda, the number of sites with amber between LBA and FBA fell down to one. In Trentino-Alto Adige only two contexts with amber dated to FBA are known, but in total they yielded 36 amber beads; therefore, one can argue that this region retained its significance for the inflow of amber even at this late stage. Amber import to Emilia/southern Po Plain definitely stopped ca. 1150 BC, with the region probably becoming a settlement void for a while, but it continued in the adjacent provinces of western Veneto/northern Po Plain and Romagna.

Surprisingly, in the case of the northern Po plain available data suggest that amber import went on (Fig. 20), although a slight translocation of its reception centre has definitely occurred. Some settlements which were established in the area during LBA (BR), i.e. Fondo Paviani, show uninterrupted presence of amber around the transition to FBA. Under the *Protovillanovan* culture, new settlements developed and craftsmanship flourished, including amber processing for which evidence has been recorded at the sites of Campestrin di Grignano Polesine and Frattesina di Fratta Polesine. The former site yielded an astonishing amount of amber production waste, unquantifiable in number of pieces but weighing in total ca. 6 kg, which shows the grand scale of this enterprise. Number of contexts dated to this period implies that the commonness of amber dropped, but its availability still was high, and even increased in certain regions, i.e. Romagna and Emilian Apennines – parts of Italy, where amber was yet to truly mark its presence during the Iron Age (Figs. 21 and 22). Furthermore, at the turn of LBA and FBA, amber emerged for the first time in western Lombardy, although its manifestations there are rather modest, coming down to two sites and four artefacts. Finally, a single context proves that amber reached also Piedmont; however, determining the time of its arrival there is very difficult. Some of the finds from Monte Cavanero potentially originate already from LBA; however, the hoard was definitely deposited at the beginning of PF I, and thus it is attributed to the group of FBA/EIA contexts. As with the previous periods, IR analyses indicate that Baltic amber was the main, if not the only, fossil resin utilised by the Northern Italian communities during FBA.

Fig. 23. Chronological synopsis of the occurrence of amber in Central Italy between the 22nd and 9th century BC.

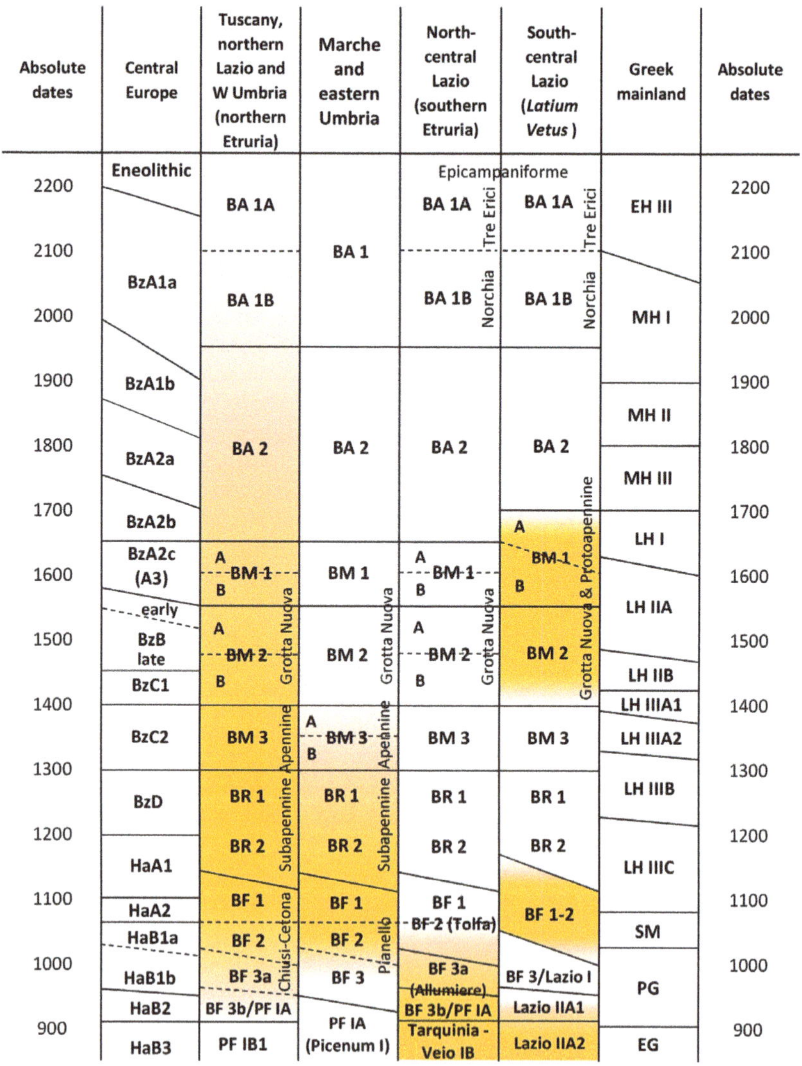

2.5.2. Central Italy

The area of Central Italy so far has delivered least evidence of amber in the Bronze Age, in comparison with the remaining sub-zones. A single bead from Grotta dell'Artofago gives a certain credibility to the idea of amber reaching Tuscany already in EBA (Fig. 23). Still, the context remains poorly defined in terms of chronology, and one has to accept possibility that the finds were re-deposited due to prolonged occupation, as with many other in-cave sites.

More secure in terms of dating is the situation of the bead with confirmed Baltic origin from Villaggio delle Macine in south-central Lazio, falling into BM 1–2 (Bellintani et al. 2007). This could mark the starting point for a continuous inflow of amber to Central Italy; however, it is difficult to point to other contexts with unquestionably MBA amber artefacts. Finds from Grotta di S. Francesco and Grotta della Noce at Belverde di Cetona, or Grotta dell'Orso di Sarteano are all missing safe indications of stratigraphic positions, while co-occurring materials chronologically range from MBA to FBA. Thus, they are useless in detailed considerations on chronology of amber import to Central Italy. Certain attestation of amber in BM 2–3 come from Grotta Misa in northernmost Lazio. Also, it is possible that the bead from Ripabianca di Filottrano in Marche belongs to MBA; however, later dating to LBA is also possible. Still, even if the abovementioned contexts of uncertain chronology are assigned to MBA, the overall number of artefacts indicates low amber availability in the period (Figs. 24 and 25).

Amber truly came to prominence in Central Italy during FBA. Around LBA–FBA transition there is a noticeable increase both in number of contexts and artefacts in Tuscany, Umbria and Marche (Fig. 24). This growth in amber's acquisition started no earlier than BR 2 – the initial caesura for the sites relevant to this period, but most probably reached its peak before the late FBA (prior to BF 3). Even more conspicuous is the increase in amber quantities in Lazio at that time (Fig. 25). After a several-centuries-long break, amber made its comeback in south-central Lazio, where it has been recorded on two sites dated to BF 1–2, together containing 13 beads. Still, the leading recipient of amber during FBA in Central Italy must have been the *Tolfa-Allumiere* group in north-central Lazio. Various sites belonging to the group, predominantly cemeteries dated to BF 3, yielded a considerable number of 55 amber artefacts which places it as the richest province of amber's acquisition in Central Italy (Fig. 25). It is to be underlined that this period of increased amber import took place after remaining regions of Central Italy apparently ceased to participate in amber exchange. Nonetheless, for Lazio it marks the beginning of the heyday of amber inflow which continued into the Iron Age and the antiquity.

Fig. 24. Fluctuations in number of contexts – commonness of amber in Central Italy.

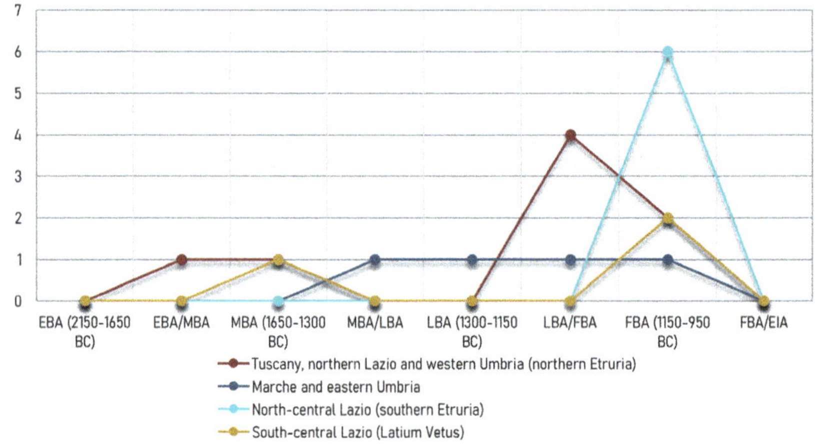

Fig. 25. Fluctuations in the quantity of artefacts – availability of amber in Central Italy.

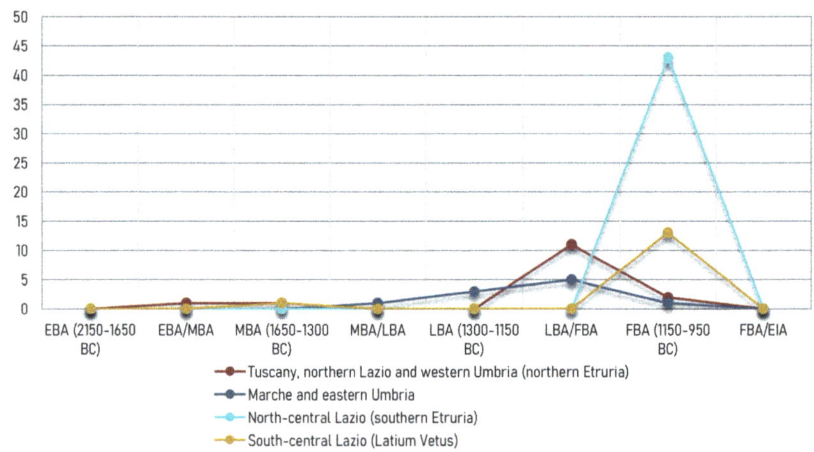

2.5.3. Southern Italy

Amber began to be used in Southern Italy already in the Late Eneolithic/Copper Age (i.e. before 2200 BC); there are two contexts with amber falling into this period: Grotta Calafarina in Sicily (Orsi 1907) and Laterza in southern Apulia (Biancofiore 1967). Interestingly, at this early stage the local fossil resin – simetite could have been the only known source of amber, what is indicated by the results of IRS performed on samples from the both sites (Beck 1971; Angelini Bellintani 2017).

Fig. 26. Chronological synopsis of the occurrence of amber in Southern Italy between the 22nd and 9th century BC.

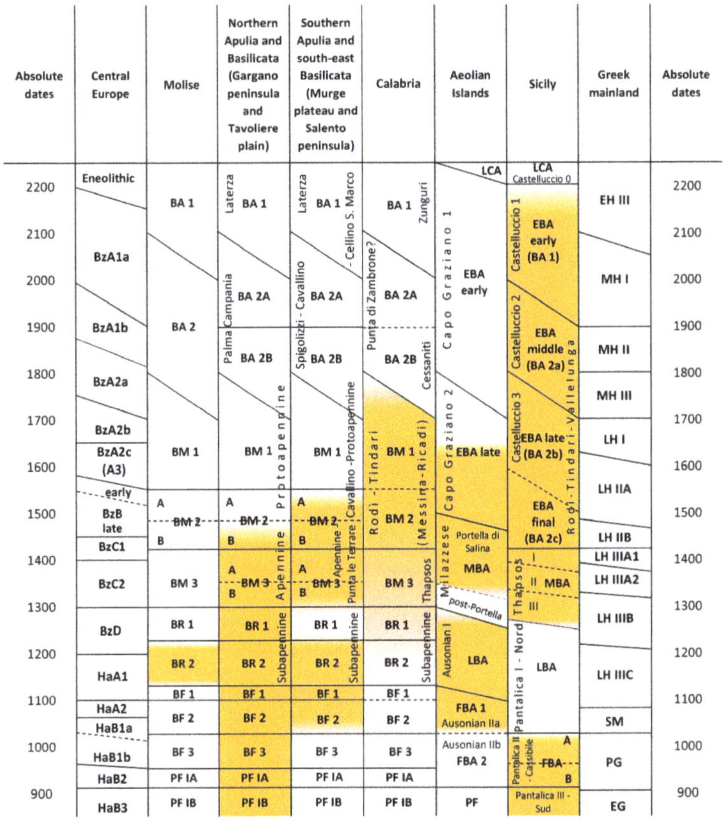

With respect to the Bronze Age, Sicily has returned the earliest evidence of amber presence of all the regions in Southern Italy (Fig. 26). The first amber beads can be dated as far back as the 1st phase of the *Castelluccio* culture (2200–2000/1900 BC), as documented by some of the graves from the eponymous cemetery. To the 2nd phase of the culture can be ascribed the bead from Monte Sallia (2000/1900–1800/1700 BC). In the earliest phase of the Bronze Age, simetite most likely was still the only fossil resin known in Sicily – conclusion based upon the IRS results of the samples from Castelluccio, tomb 9 (cf. section 2.4.2). However, it apparently was used by the *Castelluccian* communities for production of amber jewellery only for a brief time, as IRS results show that the bead from Cava Secchiera, not much younger than those from the Castelluccio cemetery because attributable to late 2nd–early 3rd phase of the culture, was made of succinite. In fact, almost all analysed amber beads from subsequent stages of the Bronze Age indicate 'Baltic' provenience, whereas simetite resurfaced only once – in the amber fragments from Madonna del Piano cemetery assigned to FBA. Four Sicilian contexts were classified to EBA/MBA category, while another six firmly belong to MBA, thus proving that amber was quite widespread for the greater part of the Bronze Age (Fig. 27). In terms of availability, it reached a peak during MBA with the total number of 24 artefacts belonging to the *Thapsos* culture (ca. 1420–1270/1250 BC) (Fig. 28).

Fig. 27. Fluctuations in number of contexts – commonness of amber in Southern Italy.

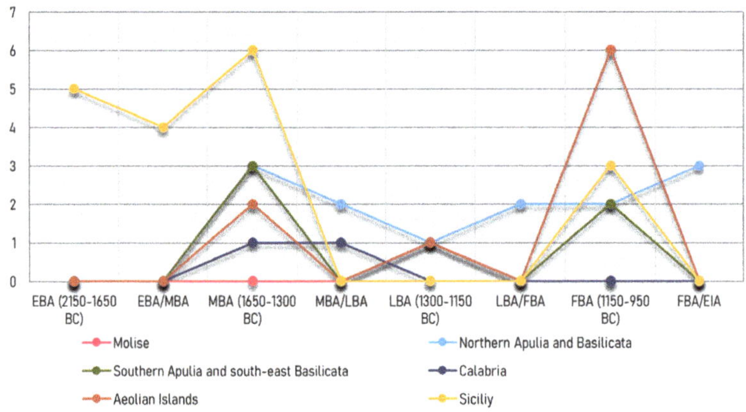

Fig. 28. Fluctuations in the quantity of artefacts – availability of amber in Southern Italy.

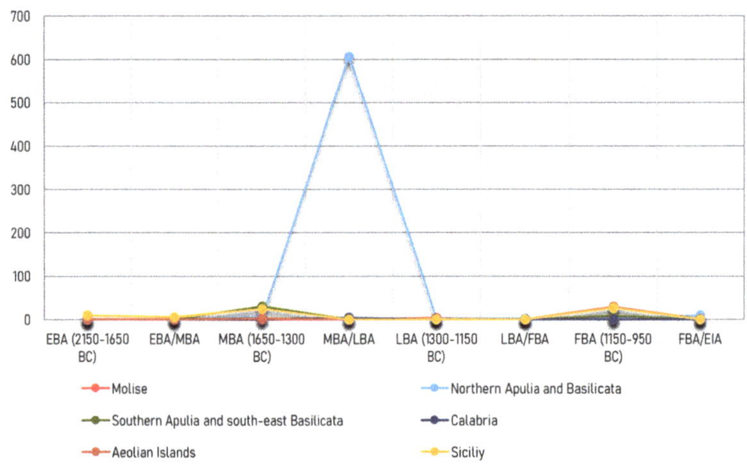

By that time amber must have already been known in the southern parts of the Apennine Peninsula and on the Aeolian Islands (Fig. 26). In Calabria it appeared around BM 1–2, when the province, specifically its southern part, has been culturally re-oriented towards north-east Sicily, at the time dominated by the *Rodì-Tindari-Vallelunga* culture (*Messina-Ricadi* facies). There is some probability that amber continued to be imported to Calabria during LBA (BR), but in any case the spatial scope of its occurrence is limited to the Tropea Promotory, and quantitatively does not exceed five beads. Singular beads are also known from the levels of the *Capo Graziano 2* and the *Milazzese* cultures on the Aeolian Islands (Figs. 27 and 28).

A definitely more marked is MBA occurrence of amber in Apulia and Basilicata. Northern parts of these provinces are well known for rock-embedded tombs, utilised from the advanced phases of MBA (BM 2) up to the early LBA (BR 1), many of which yielded amber beads. The most striking examples are the hypogea of Madonna di Loreto near Trinitapoli which altogether returned a stunning number of 606 amber pieces, assembled there while the tombs were used for burial practices between BM 2B and BR 1. None of other tombs from Apulia and Basilicata revealed such abundance of amber (Fig. 28), although the commonness of resin-made jewellery must have been quite limited anyway, what is suggested by relatively low number of contexts (Fig. 27). Even more interesting is the fact that while succinite has certainly been attested at several sites, some of the analysed amber samples from Madonna di Loreto, 'Hypogeum of Bronzes', as well as from *Recent Apennine* layers (BM 3) at the nearby settlement of Coppa Nevigata, turned out to have not been made of succinite, or any other fossil resin with recognised spectrum.[470] The issue of origins of those amber finds has not been solved yet. In this situation it is justified to presume that communities which attended the hypogea of Madonna di Loreto – the same which perhaps inhabited Coppa Nevigata, apart from importing Baltic amber, used other sources of fossil resin. Some of the indigenous Apennine ambers are typified as the most likely candidates; however, it is possible that the deposits of this mysterious fossil resin(s) have been exhausted already in the past.

Southern Apulia and south-east Basilicata also delivered considerable testimonies of amber use during BM 2 and 3, recovered from cemeteries as well as settlements (Fig. 26). However, the influx of amber to this province, unlike the northern parts of this regions, experienced a short pause which approximately lasted for almost a century (BR 1). Nonetheless, amber returned in the subsequent phase BR 2. The settlements of Coppa Nevigata and Roca Vecchia, through presence of amber beads in well-stratified contexts, document a constant inflow of amber to Apulia at the turn of LBA and FBA. Commonness of amber during LBA seems to have been low (Fig. 27), but at the same time one can observe a spread of discussed material to Molise. Moreover, after a seemingly short break corresponding to post-Portella horizon of the *Milazzese* culture, there was a revival of amber on the Aeolian Islands, evidenced by the bead from the *Ausonian I* levels at Lipari.

Commonness of amber grew significantly during FBA, when it gained a particular popularity on the Aeolian Islands (Fig. 27). The *Ausonian IIa* cemetery of Piazza Monfalcone on Lipari yielded 28 amber beads. Although these numbers are no match to the quantities of amber known in the hypogea of Madonna di Loreto, they nonetheless express considerable availability of amber for the Aeolian communities during FBA, especially when comparing them with other provinces in that time span (Fig. 28). Elsewhere in Southern Italy, the occurrence of amber was much more limited. Only towards the end of the Bronze Age, amber in Apulia once again became more abundant. Here, one can mention assemblage of six beads from the hoard of Torre Castelluccia near Taranto, or the collection of 12 ambers adornments from the graves at the Monte Saraceno necropolis on the Gargano Promontory. In these final stages of FBA, amber found its way back to Sicily after a period of absence coeval with the *Pantalica I – Nord* period, and remained in use also during EIA.

---

470 Angelini et al. 2003; Angelini/Bellintani 2005; Bellintani 2010.

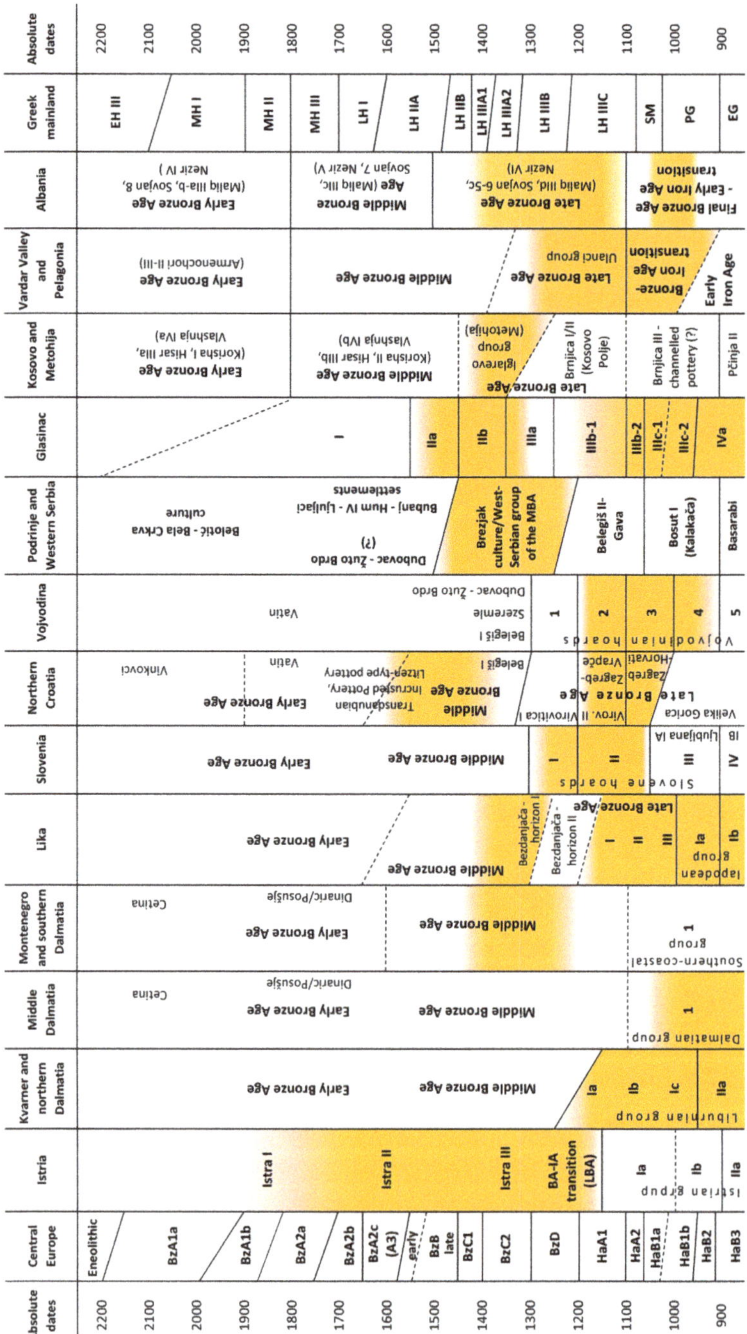

Fig. 29. Chronological synopsis of the occurrence of amber in Western and Central Balkans.

2.5.4. Western Balkans

In short, one can discern two major phases or horizons of amber occurrence in Western and Central Balkans (Fig. 29). The first one lasted circa from the earliest influx of amber in the first half of the 2nd millennium BC, and ended between 1200 and 1150 BC, so around LBA–FBA transition. As revealed by IRS analyses conducted to this day, Baltic amber (succinite) was most probably the only fossil resin known to and used by the Balkan communities during the Bronze Age.

Fig. 30. Fluctuations in number of contexts – commonness of amber in Western Balkans.

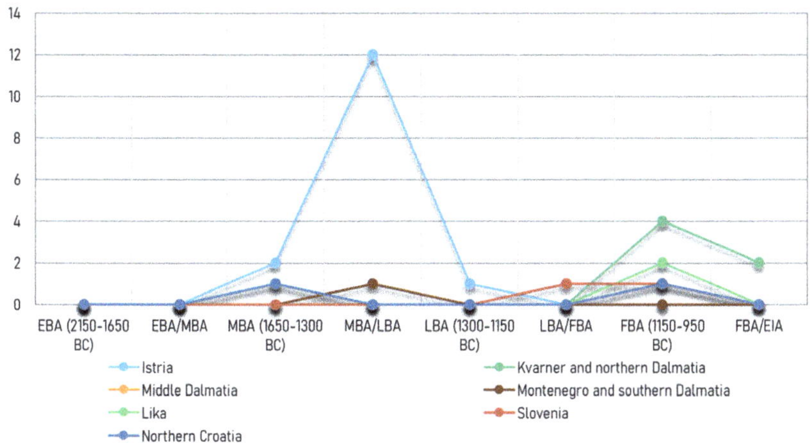

Potentially, the earliest evidence of amber presence comes from Istria. There are only two 14C dates available for the cemetery of Vrčin; however, they deliver plausible grounds for dating amber beads found in the graves already to the 17th century BC (cf. Cupitò et al. 2018). Still, the lower boundary of the dates is placed around 1200 BC, so they cannot be used as a definite indicator of the earliest occurrence of amber in the Balkans. Other contexts from Istria, dated with conventional methods as well as radiocarbon, firmly support the hypothesis that amber was known and used in this region between advanced MBA and final LBA. Moreover, statistics show that Istrian communities had relatively easy access to amber and possessed it in considerable amounts (Figs. 30 and 31). None other Western Balkan province enjoyed such a high commonness and availability of amber during MBA and LBA. Among them, the earliest attestation of amber is known from the Drava-Sava interfluve, where four 14C dates allow to assign the single bead from Alilovci Lipje to the 16th–14th century BC. Sometime afterwards amber, although in a very limited amount, arrived in Montenegro and Lika. Thus, it occurs that MBA introduction of amber in most Western Balkan provinces (except Istria) was incidental and did not lead to its fuller incorporation into material culture.

Fig. 31. Fluctuations in the quantity of artefacts – availability of amber in Western Balkans.

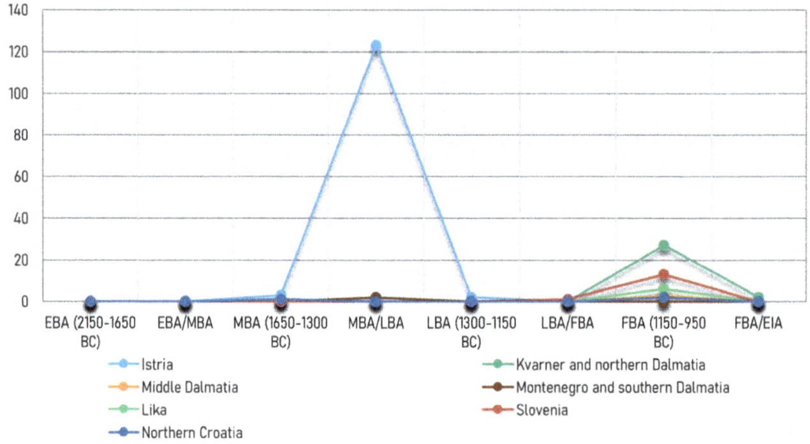

Turn of the 13th and 12th century BC marks a watershed in the spatial structure of amber's acquisition in Western Balkans (Fig. 29). After this date no more amber occurs in Istria. The reason for this may lie in the changes that accompanied introduction of the *Urnfield* culture on the peninsula, probably involving social and religious transformations as shown by a new burial rite, although it could have economic implications as well. Apparently no amber came to Montenegro and southern Dalmatia during FBA. On the other hand, LBA and early FBA saw a short-lived horizon of amber's appearance in Slovenia. Conversely, around the same time there occurred a hiatus in amber's inflow to inland provinces of Lika and Northern Croatian, lasting ca. 150–100 years. The material returned to the aforementioned regions during the 12th century BC, in quite different cultural circumstances connected to the spread of *Urnfield* influences. Still, the levels of amber's commonness and availability in both Lika and Northern Croatia are low in relation to FBA (Figs. 30 and 31). Nevertheless, in Lika the second wave of amber import was a long-lasting phenomenon; the FBA finds are merely a prelude to the far more numerous and technologically superior amber artefacts (e.g. fibulae with amber adornments) linked to the *Iapodean* communities of EIA.[471] FBA was also a time of amber's introduction to the communities of the Kvarner Bay and northern as well as middle Dalmatia. Although the indexes of commonness and availability may not suggest it, Kvarner and northern Dalmatia became the second-most important centre of amber's acquisition, besides Lika, which lasted well into EIA, when the former area was inhabited by the *Liburnians* (cf. Palavestra 1993; Blečić Kavur 2009).

471 Cf. Palavestra 1993; Bakarić et al. 2006; Bakarić 2017.

2.5.5. Central Balkans

The 16th century BC can be regarded as a *terminus post quem* for the vast majority of amber finds in Central Balkans (Fig. 29). Amber appeared almost simultaneously in Podrinje and Western Serbia, Glasinac, Metohija and Albania. However, out of all these regions, it definitely has the greatest representation, both in terms of contexts and finds, in the Drina Valley and adjacent parts of Serbia (Figs. 32 and 33). Almost equally large assemblage of amber finds was discovered in Metohija, but it comes only from two contexts, while in the former province these reached over 20. Presence of amber in MBA at the Glasinac was even more sporadic and confined to singular graves. Similarly to Metohija, it ceased with the end of MBA, while in Podrinje and Western Serbia amber continued to be used during LBA, although to a somewhat lesser extent than before. Albania also produced evidence of prolonged use of amber extending over MBA and LBA; however, when its provinces of amber's acquisitions are considered individually, it appears that resin-made jewellery was not so common or available as in the more northerly parts of Central Balkans. Slightly later, around the 13th century BC, amber reached the Vardar Valley; however, it found only limited use there, as seen on the basis of merely three graves from the Dimov Grob cemetery, in total containing four beads.

Fig. 32. Fluctuations in number of contexts – commonness of amber in Central Balkans.

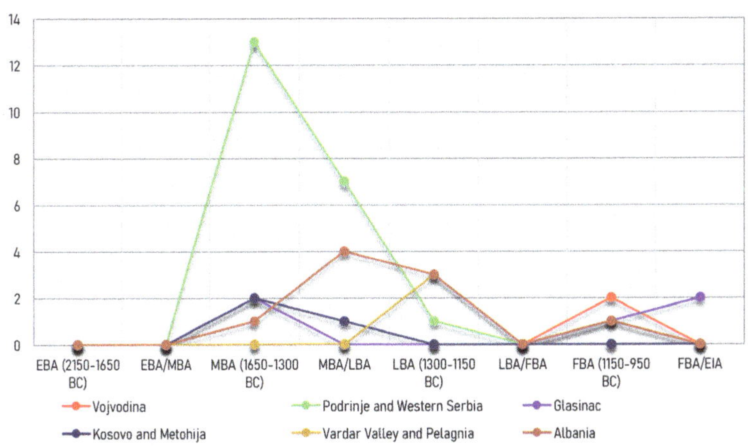

Interestingly, the influx of amber to Western and Central Balkans experienced an intermission about the same time, i.e. at the turn of the 13th and 12th century BC (Fig. 29). With several exceptions, all the previously active provinces of amber's acquisition stopped importing this material – a break that lasted at least until EIA. Hitherto main recipients of amber, such as Podrinje – Western Serbia, have fallen out of circulation, but similarly to Lika and Northern Croatia amber somehow managed to come back to Glasinac after a horizon of no imports, coeval with *Bezdanjača II*, *Virovitica I* and *Glasinac IIIa/b-1*.

The boundary, however, cannot be observed in all the regions taken into consideration; in North Macedonia there seems to have been a continuity in amber influx between LBA and FBA; however, the destination at which amber arrived certainly shifted from the Vardar Valley (*Ulanci* group) to Pelagonia. In Albania this break in amber inflow happened slightly later, in the course of the 11th century BC; however, in this case the old centres of amber's acquisition became defunct as well, with only the Mat Valley yielding evidence of amber during FBA. Hence, it is apparent that amber inflow in Central Balkans was quickly resumed after the LBA/early FBA intermission, although in much fewer quantities and distribution scope (Figs. 32 and 33).

Fig. 33. Fluctuations in the quantity of artefacts – availability of amber in Central Balkans.

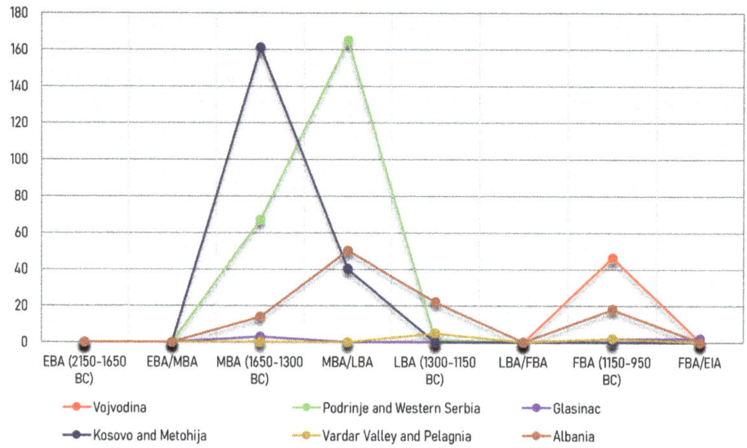

It is difficult to determine whether this interruption in amber inflow to the Balkans was one phenomenon, common in its roots and with similar consequences for the entire zone, or was it more likely a series of roughly contemporary but independent events. Regardless of the cause(s), amber influx around the 12th century BC apparently changed its destinations; while the previous main centres of its acquisition ceased to exist, the new ones emerged. This is manifested by numerous finds from Dalmatia and Kvarner, Slovenia and Vojvodina – all these regions hitherto lacking any evidence of discussed material, but since its first occurrence in LBA obtaining it almost incessantly until the antiquity. Even though amber in these provinces apparently never reached the same levels of commonness or availability as in Istria or Podrinje and Western Serbia in their heyday, it still retained considerable representation during FBA, and continued to grow in numbers during EIA.

3. From typology to networks: amber finds analysis

The following section is aimed at discussing the Bronze Age amber finds collected from the circum-Adriatic zone in relation to their morphology. Detailed consideration of the artefacts' appearance led to their division into a series of groups and types. The latter were subsequently analysed in terms of chronology and spatial distribution. A separate sub-section is devoted to each group of finds. In the end the created typology was used as the basis of NA which seeks to understand relationships between the circum-Adriatic provinces of amber's acquisition.

Up to this point prehistoric amber finds in the Central Mediterranean and the Balkans were classified within several regional typologies, which are mutually incompatible. Therefore, in order to proceed with analysis of amber in a broader scale, one is confronted with the need for a uniform system of description and classification of morpho-stylistic features characterising the amber artefacts. Such system was devised for the purposes of this work. Before moving on to explain the applied typology, it is worthwhile to discuss previous major contributions in this regard which provided an inspiration and, in some cases, concrete methodological solutions for building the new classification.

3.1. Previous typologies of Bronze Age amber artefacts

Overview of the available literature on amber in the Bronze Age Europe allows to distinguish at least several typologies. Although all of them concern artefacts of similar function and form, they differ in approaches, mainly in terms of selecting the criteria of classification and the level of detail applied in the morphological analysis.

The most common approach in the hitherto applied classifications of amber artefacts is the use of functional criteria. As the vast majority of objects collected so far is interpreted as adornments – elements of jewellery, their examination is focused on supposed ways of wearing them. Thus, these finds are usually divided into beads – simplest forms with a single bore used to suspend them – and more sophisticated elements with multiple perforations, often termed as the 'spacers'. Such demarcation is encountered in numerous publications treating about amber.[472]

In early stages of research, so much attention was paid to the significance of spacers as potential indicators of long-distance contacts between the *Mycenaean* Greece and north-west Europe that the remaining finds were essentially left aside from the discourse. Morphology of the latter was usually discussed collectively, and even though they were not separated into types *per se*, the customary names reflecting their forms became established in the process, e.g. discoid, globular, biconical, etc. Nonetheless, it has to be said that due to their similar, sphere-derived form, the beads did not lend

472 E.g. Hachmann 1957; Sandars 1959; Harding/Hughes-Brock 1974.

themselves to a neat classification, and early on the use of abovementioned terminology was deemed potentially misleading (Harding/Hughes-Brock 1974, 154; Sprincz/Beck 1981, 470).

A notable exception was roughly cylindrical or concave-biconical beads with a median swelling in the form of a rib/ledge, sometimes described as *astragalus*-like (e.g. Teržan 1984, 110; Glogović 1991, 19), but otherwise known as the Tiryns-type beads – name derived from eponymous hoard on the Peloponnese, where several such objects were threaded on the spokes of golden wheels (Karo 1930, 127, Beilage XXXA, XXXII; Maran 2013, 159–160). This specific decorative feature is relatively easily recognisable on amber beads, and early on, scholars realised the time span and spatial range within which this form appeared.[473] The Tiryns type is often considered in relation to the Allumiere type – oblong fusiform or conical beads decorated on the surface with carved, parallel grooves, named after the cemetery of Allumiere – Poggio La Pozza in Lazio (Italy) where these artefacts were discovered for the first time (Peroni 1960; Negroni Catacchio 1972, 15–18). Particularity of these types of beads, as chronologically and geographically definable objects, made them stand apart in the studies of amber artefacts in the Central Mediterranean and the Balkans.[474] But as these two, undeniably important types were intensively discussed in the literature, the scholars rarely took a closer look at other amber finds. The complete typological overviews of amber beads started to appear later, approximately since the 1980s, but were often limited to certain areas.[475]

Among several noteworthy attempts to build typologies for local assemblages of amber beads is an application of modified classification by H. C. Beck (1928) in the context of Hungarian Bronze Age finds by E. Sprincz and C. W. Beck (Sprincz/Beck 1981). The authors, being aware of the terminological difficulties in describing forms of amber objects, used the cross- and longitudinal sections of the beads, as well as the ratio between the dimensions along the perforation and the diameter. Thus, they were able to distinguish ten types (denominated with Roman numbers I–X), with subtypes labelled with capital letters (A, B, C, etc.) (Sprincz/Beck 1981, Fig. 6, Tab. 1).

In the recent typology of amber finds in the *Únětice* culture by M. Ernée (2012), number of perforations is the main criterion of classification, followed by examination of the holes' shape and their arrangement on an object's surface. Thus, the author separated analysed artefacts into two major classes: beads with a single perforation and those with multiple, more complex orifices. In the case of simple, singularly perforated beads, the evaluation of their overall form led to division into four groups (labelled A–D), subsequently separated into types based on the cross-sections (e.g. B1, C2). The category of multi-perforated objects includes more variants; however, as these are characteristic almost exclusively for Bohemia during the EBA, their classification has distinctly regional scope and can hardly be transposed over other areas.

From the perspective of this work of huge relevance are typological analyses of amber beads from Greece associated with the *Mycenaean* culture. The latest and most

473 Cf. Harding/Hughes-Brock 1974, 155; Negroni Catacchio 1972, 18–20; 1973b, 47–49.
474 See e.g. Negroni Catacchio 1999, 2014; Cultraro 2006; Bellintani 2014.
475 E.g. Sprincz/Beck 1981; Mazurowski 1983; Beck/Shennan 1991; Marková 1993; Palavestra 1993; Hughes-Brock 1998; Negroni Catacchio et al. 1999, 2006; Bukowski 2002; Stahl 2006; Czebreszuk 2011; Ernée 2012; Woltermann 2016a.

up-to-date typology was elaborated by J. Czebreszuk (2011) who, in turn, based his classification on earlier works written by many distinguished specialists in this field.[476] The author divided analysed objects into five categories according to the functional criterion: (1) Beads, (2) Spacers, (3) Pendants, (4) Discs, (5) Other. Each of these groups comprises types characterised by different morphology and additional features, such as arrangement of perforations or presence of decorations. For instance, beads were separated into a series of types, first numbered and then denoted with customary names, e.g. type 101. Spherical, 102. Flattened spherical, 103. Lentoid, etc., furthermore divided into subtypes with respect to their size: 1 – small, 2 – standard, 3 – large (Czebreszuk 2011, 77–92). Apart from beads and spacers, thoroughly elaborated using hitherto written literature, the work includes an interesting category of pendants – rare objects usually with elongated, oblong shape and the suspension hole placed close to the edge at one end (Czebreszuk 2011, 94).

There are two typologies that directly preceded the current effort in respect to the area under study. These are A. Palavestra's typology summarised in the catalogue of prehistoric amber finds from Western and Central Balkans (Palavestra 1993), and the classification system of Italian Bronze Age amber artefacts elaborated by N. Negroni Catacchio and her co-workers in a series of publications.[477] These are described below in a greater detail.

3.1.1. A. Palavestra's typology of Western and Central Balkan amber finds

Thanks to the already-mentioned A. Palavestra's catalogue (Palavestra 1993), the majority of prehistoric amber artefacts hitherto discovered in Western and Central Balkans were analysed and classified in terms of morpho-stylistics. Beads from additional 30 contexts dated to the Bronze Age, mostly discovered after 1993 or coming from the territory of Albania and originally not included in the latter work, have recently been compared against the Palavestra's typology (Cwaliński forthcoming).

As Palavestra was creating the framework for classification of nearly all then-known prehistoric amber finds from Western and Central Balkans, numbering circa several thousands of objects spanning nearly two millennia, he was faced with a difficult task. Formal and technological diversity of the dataset, that comprised both the simplest, undecorated beads with a single perforation, as well as skilfully crafted anthropo- and zoomorphic adornments, required creation of comprehensive and detailed typological system. Palavestra's achievement is a meticulous analysis of the finds and their attribution to types and their variants on the basis of specific traits, such as shape, size or manners of perforation and decoration. These criteria allowed to differentiate even between seemingly identical forms, e.g. discoid beads belonging to the type 8 which has been divided into five variants labelled with letters a–e (Palavestra 1993, Tab. I).

Out of 146 main types of amber artefacts in total distinguished by A. Palavestra (not including their variants/subtypes), only a small number refer to the Bronze Age finds. The fact stems from numerical inferiority and morphological simplicity of these

476 I.a. Karo 1930; Blegen 1937; Blegen et al. 1973; Harding/Hughes-Brock 1974; Hughes-Brock 1998, 1999.

477 Negroni Catacchio 1999; Negroni Catacchio et al. 1999, 2006.

artefacts in comparison with the finds from later periods. First four types and their respective variants (1a-b, 2, 3, 4a-b) all relate to cylindrical or barrel-shaped beads, decorated on the surface with carved horizontal ledge(s) or a series of transverse grooves (Palavestra 1993, Tab. I). Palavestra treated them collectively as representatives of the Tiryns type, yet according to presently valid classification, they include forms akin to both the Tiryns and the Allumiere types (Palavestra 1993, 251; c.f. Negroni Catacchio 1999; Cultraro 2006). Moreover some of the types, such as 2 and 3, are rather unique, as no direct analogies have ever been found outside Western Balkans (cf. Cwaliński forthcoming). Regardless of denomination, separation of these seemingly similar beads into types is an important contribution by the author.

Among the Balkan beads attributable to the Bronze Age, one can find spherical and annular forms, classified by Palavestra as types 5 and 7, respectively. Yet the most represented among the Balkan Bronze Age finds are types 8 and 9, comprising beads with a round or oval cross-section, and an ellipsoid/oval profile with various degree of flattening, thus giving them forms ranging from discoid to flattened spheroid (Palavestra 1993, Tab. I). In fact, attribution of the finds to adequate varieties may be problematic, as the distinctions between them are not sharp but rather intuitive. For example, Palavestra explains that 'the type 8a is a rather small, flattened, oval bead, whereas the type 8b is somewhat larger and more irregular' (Palavestra 1993, 31).

Another quite common form are biconical beads, which Palavestra included under types 12–13. Type 12 refers to rather small and low varieties, with the flattened (12a) or average (12b) variants only slightly differing on the figures provided by the author (Palavestra 1993, Tab. I). Type 13, on the other hand, defines biconical beads in an elongated variant. Apart from abovementioned forms, Palavestra distinguished lenticular/lentoid beads (type 16), cylindrical beads (type 18) in three variants: elongated (a), average (b) and flattened (c), and finally beads with quadrangular/rectangular cross-section and profile, resembling cubic forms (types 35b and 37) (Palavestra 1993, Tab. I–II).

It has to be stressed that in many cases the differences between the variants, as depicted on the reference figures, are barely noticeable, and the absence of full description explaining the principles governing applied classification makes the apprehension of the typology quite difficult. These factors hinder utilisation of the latter in future studies, in some cases making it necessary to come up with propositions of new types. Despite the impediments, what the typology offers is a good starting point for formal considerations of newly discovered or previously unclassified finds.

3.1.2. Italian Bronze Age amber finds in the typology of N. Negroni Cattacchio et al.

Arguably, the biggest contributions to the research on prehistoric amber finds in Italy are the merit of N. Negroni Catacchio and her co-workers at the Institute of Archaeology of the University of Milan. The result of these ongoing studies is a series of publications reporting new finds of amber, describing their morphology and putting them on distribution maps which altogether are an important source of information in the discussions on amber exchange in a wider, European scale.[478] Having gathered

478 E.g. Negroni Catacchio 1970, 1971, 1972, 1973a, 1973b, 1975a, 1975b, 1976, 1978, 1999, 2009, 2014; Negroni Catacchio et al. 1999, 2000, 2006.

plenty of data and experience throughout the years of research allowed N. Negroni Catacchio, A. Massari and B. Raposso to create a framework for a typology of Italian amber finds. Implementation of this classification in regard to Bronze Age finds was conducted in stages, depending on a region of Italy and chronology (Massari et al. 1996; Negroni Catacchio et al. 1999, 2000), or a character of the finds (Negroni Catacchio 1999, 2014). The most recent and updated presentation of this typology (Negroni Catacchio et al. 2006) explains classification criteria applied by the authors in all their complexity.

The authors divided all artefacts from Italy relevant to the Bronze Age, with respect to the location, number and shape of the holes for suspending/threading the objects, into five functional groups: beads, pendants, spacers, buttons and elements with multiple functions (Negroni Catacchio et al. 2006, 1444). Beads – the largest and the most documented group – do not only contain elements of the necklaces or bracelets, as the name suggests. It also includes pin and fibulae adornments. Critical feature is the presence of a single hole, perforated more or less through the centre of an object's cross-section (along the axis of rotation). If this condition is not fulfilled, i.e. the single hole passes off the centre, close to the edge, or through an additional loophole, then the object was classified as a pendant. Category of spacers was distinguished in accordance to commonly understood definition explained earlier, therefore consists of items with multiple, parallel or perpendicular holes. Buttons, in turn, comprise objects with the hole, often V-shaped, drilled perpendicularly to the cross-section/axis of rotation. The last group – elements with multiple functions – is explained as covering all objects characterised by the presence of at least two perforations, usually longitudinal and transverse or eccentric.

Apart from differences stemming from arrangements of the perforations, the groups vary also in terms of shape. Pendants usually have oblong form with longitudinal sections in the shape of a rectangle (type 1), trapeze (2), axe-shape (3) or bicone (4) (Negroni Catacchio et al. 2006, 1449–1450, Fig. 3). Buttons, on the other hand, mostly have low profile, V-shaped perforation and conical, semi-spherical, spherical, or biconical form (types 1, 2, 3 and 5, respectively); rarely the perforation is cylindrical (types 4 and 6) (Negroni Catacchio et al. 2006, 1450, Fig. 4). Spacers and other elements, e.g. not entirely perforated, oblong-cylindrical/conical elements used as pinheads, due to their rarity, were not divided into types, but they were treated separately as unique cases (Negroni Catacchio et al. 2006, 1450–1452, Figs. 4 and 5).

The greatest variety of forms was observed with regard to the most common group – the beads. For this category of finds, Negroni Catacchio et al. used elaborate set of criteria (Negroni Catacchio et al. 2006, Fig. 1). First of all, each bead was considered in a position with the hole oriented vertically. The description begins with the definition of the cross-section (surface perpendicular to the hole), using geometric figures as a reference: circular, elliptical, oval, triangular, etc. (Negroni Catacchio et al. 2006, Fig. 1:1). The authors came up with 13 reference shapes in total, however left the list open in case other varieties are found in the future. Although such approach allows for quite precise determination of a bead's cross-section, some of the reference forms, e.g. oval, elliptical and sub-elliptical, or circular and sub-circular, may seem difficult to tell apart, when it comes to examination of the real objects.

Afterwards the analysis proceeded with assessment of the left- and right-side profiles, and the upper and lower bases (Negroni Catacchio et al. 2006, Fig. 1:2–3, 5–6). Following five variants of lateral surfaces were distinguished: rectilinear, convex, concave, rectilinear-convex and irregular. Additionally, if the profile was strongly convex – carinated (e.g. in biconical forms), the edge was determined either as sharp or rounded (Negroni Catacchio et al. 2006, Fig. 1:4). As for the bases, the authors listed five variants: flat, flat and inclined, convex, concave and irregular. This detailed classification procedure enables to grasp the form of a bead with a high level of accuracy.

Furthermore, the analysis considered also shape of perforation (cylindrical, truncated-conical, V-shaped) and presence of any recesses around its outlets (here there are three variants: rectilinear, concave and convex; Negroni Catacchio et al. 2006, Fig. 1:7–8). Interestingly, the authors also proposed to assess the profile of the perforation, by classifying it as straight or rounded, and the edges of the recesses with the same denomination (Negroni Catacchio et al. 2006, Fig. 1:9). Nonetheless, it must be realised that determination of these two features is, in most cases, impossible without the help of microscope or X-ray photography.

Using above-described criteria, with the emphasis put on lateral profiles and bases, the authors defined a series of types. The cross-section was considered less relevant, except in the cases in which the association between the lateral surfaces and the cross-section seemed to be particularly characteristic. In the end Negroni Catacchio et al. were able to distinguish 13 types, described as combinations of certain features (Negroni Catacchio et al. 2006, 1446–1449, Figs. 2 and 3). For instance type 1 – discoidal was defined as having circular, sub-circular or elliptical cross-section; convex side profiles; and flat, oblique, convex or concave bases, whereas type 3 – cylindrical has variable, but not polygonal cross-section, straight lateral profiles and flat or oblique bases. Within each type, with the exception of the discoidal and spheroidal, some variants characterised by a greater or smaller height have also been identified, called: (a) medium (similar height and width), (b) elongated (height greater than or equal to doubled width), (c) flattened (height lower than the width), (d) very flattened (height equal to or less than half of the width).

Unlike most of the classification systems utilised so far in the studies of amber, the typology of Negroni Catacchio et al. is very complex, but at the same time carefully explained. Yet, there are two issues with its implementation. The first problem is the generality of the types – final products of the classification, in comparison with very elaborate analytical process based on numerous criteria. The types, as a result, may seem too vague, as they are still intended to reflect geometric solids as referential models (e.g. cylindrical, conical, etc.), thus imposing a certain way of visualising them, despite actually having more varied form. In such approach, the use of terms derived from solids may also be misleading, or awkward at least: e.g. type 3 – cylindrical, covers also the beads with triangular cross-section. The second problem with the discussed typology is that it largely remained in a state of proposition, as only a part of Italian Bronze Age amber finds was introduced to it. Undoubtedly, its biggest advantage is the comprehensive set of criteria ready to be used in further studies.

3.2. New typology of amber artefacts – material basis, applied criteria and structure

During preliminary research it was possible to gather data on and put into the catalogue a total number of 2,239 amber artefacts from the study area, out of which 1,181 could be classified typologically (see Catalog). Remaining 1,058 objects were not eligible to the typological analysis, due to absence of figures in the literature, their inaccessibility (often being considered as missing) or, otherwise, inability to study them in person. Evaluated objects are finished (sometimes semi-finished or, in rare cases, barely processed pieces) elements of jewellery or, otherwise, attire. Most of them can be described as beads. Regardless of their shape or ornamentation, a standard feature of their morphology is perforation – a hole presumably serving to suspend an object on a string, e.g. as a part of necklace. Usually these perforations are singular; however, multiple holes also occur, although number of such objects is definitely smaller.

Creation of the typology began with distinguishing all possible morphologic and stylistic features characterising the finds. In the following manner, analysed objects were so to speak 'decomposed into individual features', with each aspect of their appearance treated separately.

Every object was evaluated in two projections, named here as cross-section and profile. Such approach was applied in order to reduce the objects under consideration from three-dimensional to two-dimensional representation, instead of referring to geometric solids which they resemble as in the older typologies. While not all artefacts were analysed directly (thus enabling positioning of an object according to the hereby adopted rules), these two projections are usually corresponding to figures depicting analysed objects in the literature. Cross-section is defined here as two-dimensional representation of an object from the top perspective (vertical projection), with the figure usually showing the upper or lower base (i.e. surface on which an object is resting) and transversely the suspension hole. Profile, on the other hand, illustrates the lateral section of an object (horizontal projection) with the perforation presented longitudinally (Fig. 34:A1).

Two dimensions were used to describe the objects as visible in the profile: height (H) and diameter (D)/length (L), the latter depending on the cross-section of an analysed object. As these dimensions can be defined differently in case of each object, depending on what perspective is assumed (or how an object is positioned towards the viewer), it was necessary to adopt several basic rules. Therefore, H is a dimension calculated from an object's base (the surface on which it can be stably set free-standing) to the opposite surface (Fig. 34:A1). As a rule of thumb, H is almost always running along the perforation, which is usually shown vertically in the profile. Because the vast majority of the finds have round or oval cross-section, the height usually coincides with the axis of rotation. In the case of objects based on cuboids, in order to avoid confusion in naming the dimensions depending on positioning of an artefact, it was decided to measure H strictly along perforation (Fig. 34:A2).

Diameter (D) is used in case of objects with round or oval cross-sections and means dimension measured between surfaces on which an object cannot be positioned free-standing; on figures it can be viewed as a plane perpendicular to the perforation's axis (Fig. 34:A1). Length (L) is regarded as a measurement along the same plane, but it is used in reference to cuboid objects (i.e. those with non-round or oval cross-sections; in their cases surfaces lateral to the perforation can be flat) (Fig. 34:A2).

If the perforation passes through an object non-perpendicularly, but in parallel to the surfaces on which the latter can be placed in a free-standing position (here called the 'bases'), then it is assumed that the perforation reflects, or in other words runs along D. In such instances H is measured between the bases, or from the base to the highest opposite point of an object, as represented in the profile (Fig. 34:A3).

Described rules are well suited for the majority of analysed objects which have a rounded form, varying from circle to oval in cross-section, thus having two (upper and lower) bases perpendicularly connected with perforation. It is more of a problem to apply them in case of objects with a non-circular, but triangular, rectangular or polygonal cross-sections, which have more than two parallel bases. In these instances proper determination of H and L in respect to location of the hole could sometimes be difficult, but eventually is feasible. Nonetheless, it should be remembered that these rules have conventional character and their application might be subjective to some extent.

Fig. 34. Graphic presentation of the criteria used in the typological analysis of amber artefacts.

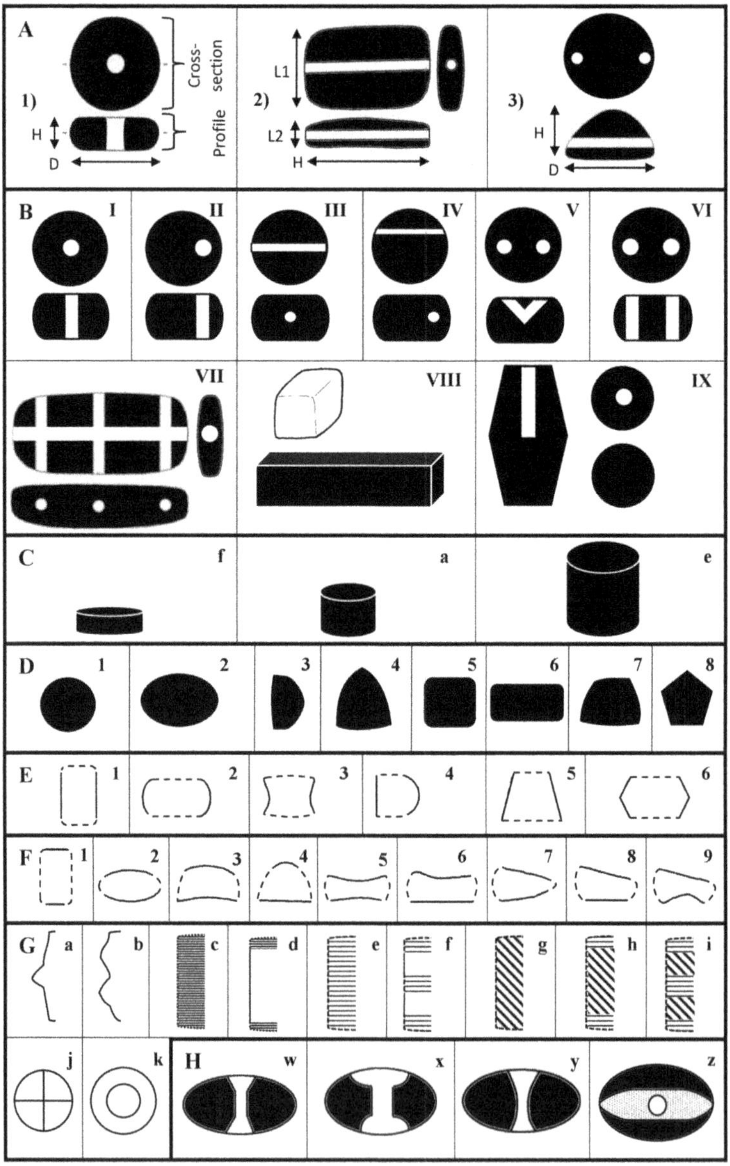

Typological analysis of assembled data was based on a series of categories according to which each object was analysed.

In the first place location of the perforation – the hole serving to suspend the objects – was considered as the criterion deciding the manner in which the objects were worn. Initially, four groups designated with Roman numbers were distinguished: perforated concentrically along H (I), perforated eccentrically along H (II), perforated concentrically along D/L (III) and perforated eccentrically along D/L (IV) (Fig. 34:BI-IV). In all four above-listed groups of beads, the perforation resembles more or less a cylindrical hole. There is, however, a small group of artefacts with perforation in the shape of 'V' letter, as visible in the lateral section. Its specificity and significance was frequently discussed in the literature;[479] therefore, it is necessary to include artefacts bearing such perforation in an independent group (V) (Fig. 34:BV).

Another distinct category is formed by objects with secondary or multiple holes, testifying to more complex ways of suspending them, or otherwise showing evidence of repairs or modifications to which they were subjected. Secondary perforation may run along or perpendicularly, rarely diagonally to the primary hole. Beads with double perforation were described as hybrids and included in a separate group (VI) (Fig. 34:BVI). Yet another group was devised for objects with multiple (or more than three) perforations (VII) (Fig. 34:BVII). Regardless of exact positioning of the holes, we are dealing here with large objects perforated multiple times, usually along one axis, in the literature customarily called the 'spacers' (see section 3.1).

Finally, it was necessary to discriminate objects without or with semi-finished (incomplete) perforations (groups VIII and IX, respectively). Finds assigned to these groups may be tentatively called 'semi-processed'; however, the overall level of processing of an object is of secondary importance. In certain cases inside an uncompleted hole were found pieces of other materials (e.g. bronze or bone) – probably fragments of stems of pins or needles; therefore, such artefacts can be regarded as finished, only their function differing from the standard beads. On the other hand, it cannot be ruled out that the aforementioned fragments inside the holes are broken off pieces of tools (i.e. drills) used for perforating amber. The main criterion of assigning an object to group VIII is absence of any traces of perforation (Fig. 34:BVIII), while in respect to group IX the perforation is not passing through entire section of an object, but only its part (Fig. 34:BIX).

Above-described division of the beads on the basis of number, location and shape of the perforations often indicates the functional nature, as it determines how an object was worn and what was its position or function within a larger, more complex piece of attire or jewellery. Even though one can suspect with certain level of probability the function of an object (as disclosed by customarily given name, such as bead, pendant, spacer, etc.), it has to be realised that the analysed artefacts were predominantly found dispersed without forming any clearly discernible composition, e.g. necklace. In very few cases, an object was attached to a piece of material on which it was suspended or held: sometimes there are pieces of bronze wire still stuck inside the perforation, while just four objects were found slid onto bronze pins. Nonetheless, in the following considerations, the shape and location of the perforation is given the primary role as a criterion dividing the dataset into groups.

479 E.g. Du Gardin 1998; De Marinis 2007, 60; Czebreszuk 2011, 41–44; Negroni Catacchio et al. 2006, 1457–1458.

After the first stage of analysis focused on perforations, examination progressed to comparison of H and D/L of an object (criterion used i.a. by Sprincz/Beck 1981). Ratios of the dimensions taken into account were not reflected in absolute (scalar) values. The point of such comparison was merely to assess proportions of an object, i.e. is its height larger or smaller than diameter/length? In case of objects with irregular cross-section, i.e. with varying diameter or height, the largest (maximum) dimensions (measured along the longest axis) were compared (such approach was also used by Sprincz/Beck 1981, 471). It should be added that real (absolute) measurements of amber finds are rarely given in publications, or represented in the figures by including referential scales. Therefore, comparison of scalar values would require complete measurement of every object taken into consideration – process hampered by the fact that many of the finds are no longer accessible, or preserved in the same state as at the time of discovery or in the original publication. For those reasons, the author decided to resort to a relative comparison of the studied objects and their dimensions. As a result, analysed objects could be assigned to one of the three categories, somewhat resembling the division applied by A. Palavestra (1993, Tab. 1–2) and J. Czebreszuk (2011, 77): elongated (H larger than D/L; name abbreviated to 'E'), average (H roughly equal to D/L with accepted deviation of 1–3 mm; abbreviated to 'A') or flattened (H smaller than D/L; abbreviated to 'F') (Fig. 34:C).

Third, classification process considered the shape of an object's cross-section and profile. In order to reflect great variety of forms represented in the dataset, adopted method was to examine each object in two-planar way, by comparing cross-sections and profiles to geometric shapes. Categories used here were chiefly borrowed from the works of Negroni Cattachio et al. (1999, 2006), however with certain adjustments made by the author.

Cross-section was divided into several variants based on discernible geometric shapes (ordered by Arabic numerals): (1) circular, (2) oval/elliptical, (3) semi-circular/semi-oval, (4) triangular, (5) quadrangular, (6) rectangular, (7) trapezoidal and (8) polygonal (Fig. 34:D).

To aptly characterise an object's profile required a more developed set of determinants and separate consideration of lateral surfaces (sides) as well as bottom and top surfaces (bases). Following variants were devised for the purpose of describing the sides (ordered by Arabic numerals): (1) rectilinear, (2) convex (rounded outwards), (3) concave (rounded inwards), (4) straight-convex, (5) pitched, (6) double-pitched (Fig. 34:E). In case of the bases, following variants were used (ordered by Arabic numerals): (1) rectilinear, (2) convex-convex, (3) convex-concave, (4) convex-flat, (5) concave-concave, (6) concave-flat, (7) skewed-skewed, (8) skewed-flat, (9) concave-skewed (Fig. 34:F).

Multiplanar evaluation of amber artefacts is driven by the desire to characterise their form in a greatest possible detail. This, in turn, can shed light on variations in the level of technical proficiency of amber processing over time and space. It is important to realise that appearance of an amber object is not only a result of conscious and intentional crafting of raw amber but also the knowledge and skills of an artisan. Despite its softness, amber is not a material easy to work with – an observation stemming from several published experiments on manufacturing amber adornments modelled after prehistoric and early historic sources.[480] There are a number

480 E.g. Mazurowski 1984; Popkiewicz 2012, 2016; Popkiewicz/Czebreszuk 2016.

of techniques and tools which could be applied by Bronze Age amber craftsmen in shaping the surfaces of amber lumps to acquire desired effect (cf. Popkiewicz 2016, 57–58; Popkiewicz/Czebreszuk 2016, 82–83). Usually, larger amber blocks were divided into smaller pieces, either by chopping them or cutting, e.g. with saws made of linen thread attached to a wooden stick. In the most rudimentary approach, small pieces of raw material were merely perforated to turn them into beads. More advanced processing began with mechanical removal of the cortex with a sharp tool, e.g. a bronze knife or flint blade. The next step would be grinding the surfaces with grinding stones of varying grain thickness, from coarse ones to very fine ones, not to mention various assorted materials such as chalk, river mud or amber dust. Final product was polished with all sorts of textiles, leathers, hides or even ash. More sophisticated processing techniques involved primitive lathes, thermal treatment to soften a lump, change its colour or increase its translucence, or lubrication with fat or oil to give it a gloss. Thus, all the visible features in amber artefacts potentially inform about the techniques known to and used by a craftsman; the level of treatment indicates amount of experience in amber processing, while flaws and repairs suggest errors committed in the process.

Additional criteria of evaluation were taken into consideration regarding alterations of the surfaces. In the first place were considered rarely present but very distinct patterns of ornamentation. On the basis of examined dataset, it was possible to distinguish following modes of decoration (ordered by lowercase letters of the alphabet): a. single transverse rib/ledge, b. multiple transverse ribs/ledges, c. multiple and evenly spaced transverse grooves, d. multiple transverse grooves at the extremities, e. sparse and evenly spaced transverse grooves, f. sparse and unevenly spaced transverse grooves, g. oblique, regular grooves, h. oblique, regular grooves bounded by transverse grooves, i. oblique, regular grooves, alternating with transverse ones, j. cross cuts/criss-crossing incisions and k. carved concentric circles (Fig. 34:G).

The single transverse rib, or ledge for the lack of better word, is usually identified as the Tiryns-type beads, with the finds from the Tiryns citadel generally accepted as the standard to which all others are compared (cf. Karo 1930). The name, even if implying the sole presence of the decorative motif, however, is often used in respect to any kind of amber artefact, regardless of form. Thus, it is often confusing when strikingly different artefacts are given the same label, just for the sake of the single decorative feature. It is therefore suggested to abandon using the term 'Tiryns type' for an object as a whole, but reserve it only for the decorative element consisting of transverse rib/ledge (variant 'a' of decorative motifs).[481] Moreover, in the course of typological analysis of the dataset were discerned few cases where the transverse rib/ledge seems to be multiplied up to two or three parallel ribs, thus forming a specific group of artefacts (variant 'b' of decorative motifs). In the past typological examinations, these were included in a separate class (type 3 in the Palavestra's typology); however, instead of treating them as a specific case, in many publications they were being referred to as the Tiryns-type beads (cf. Palavestra 1993).

481 A process of manufacturing beads with the Tiryns-type decoration – horizontal ledge in the middle – was reconstructed step-by-step by Bellintani et al. on the basis of finds from the site Campestrin di Grignano Polesine in Northern Italy (Bellintani et al. 2015a).

Abovementioned remark leads to the following types of ornaments. Seven variants of decoration ('c'–'i'), comprising grooves carved in the surface with different orientation, width and spacing, were distinguished to accurately characterise varieties of decorated objects encountered in the dataset. Presence of such grooves, despite obvious differences in the way they were performed, has widely been considered as the indicator of the Allumiere type – the name borrowed from the eponymous site of Allumiere – Poggio La Pozza, where beads decorated with such carvings, albeit of varying form, were discovered in a large quantity (cf. Peroni 1960). Such approach may lead to over-simplification by neglecting the varieties that potentially reflect regional styles. Hence, it was decided to stop using the term 'Allumiere type' *en bloc* to all the beads decorated with transverse grooves, but divide it into seven variants.

Finally, in reference to a certain group of objects could be noted modifications of the perforations' outlets. They were performed in order to widen the outlets, presumably to facilitate threading an object onto a string or wire. Following modes of outlets expansion, or in other words forms of recesses around the holes, were recorded (ordered by lowercase letters of the alphabet): w. slanting edges of the hole's outlet, x. concave edges of the hole's outlet, y. convex edges of the hole's outlet, z. elongated groove across the hole's outlet (Fig. 34:H).[482]

Perforating amber lumps was itself a challenging activity. Technically, this could have been done at any stage of processing; however, the most 'ergonomic' way was to do it at the beginning, when the pre-selected amber piece was still large enough to facilitate its manipulation in the hands of the craftsman (Popkiewicz/Czebreszuk 2016, 83–84). It would also be a more 'economical' solution, as in case of breaking the piece, there is a smaller loss in material and time invested in its processing (Popkiewicz 2012, 92). Nonetheless, observations on the *chaîne opératoire* of beads with the Tiryns-type decoration from the LBA amber workshop at Campestrin di Grignano Polesine show that perforation, in this particular case, was made in the final stages of processing (Bellintani et al. 2015a, 422). Several manners of perforating amber have been identified among prehistoric finds, leaving distinct traces as classified above. An artisan could simply use metal knife or flint blade to make rough holes.[483] However, most of Bronze Age amber finds, as well as ancient iconography, suggest the use of metal (bronze) drills with different tips mounted on wooden handles (Popkiewicz/Czebreszuk 2016, 84, Figs. 1 and 2). Metal drills, in contrast to flint ones, allow for more precision and can be applied to even small pieces of amber (Popkiewicz 2016, 59). They

---

482 Variants of expansions of perforations' outlets were included in a single series of designators together with decorative motifs, however, starting from the end of the alphabet; both are not common to all analysed artefacts, thus can be described as secondary features (in contrast to traits such as profiles or bases). Their designators (consecutive lowercase letters of the alphabet) were ascribed in the course of typological analysis of the finds, once a specific variant of perforation's outlet or decorative motif was recorded.

483 Experiments with bone drill showed that it actually melts amber due to increase of temperature caused by turning of the drill. However, the drastic change of material's temperature (first warming and then cooling) apparently impacted its durability, causing the lump to fall apart moments after finishing the drilling (Popkiewicz 2016, 56–57).

were either held in hand, or driven by a strand (Popkiewicz/Czebreszuk 2016, Fig. 3). Precise perforation needed the surface to be traced that is marked in the place where the hole shall be made (Popkiewicz 2012, 92). Drilling could be carried out in one direction by passing through the entire width of an object, or from two opposite directions. Perforations can be divided into cylindrical, semi-hourglass or hourglass-shaped – the latter two corresponding to the abovelisted variants 'w', 'x', 'y' and 'z' of the recesses. Simple, cylindrical perforations were usually performed with a drill with narrow and straight tip, moving in one direction. However, they provoked a danger of tilting the drill sideways from the desired perpendicular course (Popkiewicz/Czebreszuk 2016, 85). Also, pressure exerted on a piece of amber while drilling could cause it to break apart. The semi-hourglass and hourglass perforations take their shapes from the use of drills with wide base and narrow tip, or/and from preliminary preparation of the place for perforation.[484] They possibly indicate technologically more advanced amber processing, because when the hole is expanded towards the outlets physical forces inflicted by the drill are distributed differently, lowering the danger of breaking an object. Furthermore, hourglass perforations help to avoid damaging the opposite surface of an object, when a drill eventually cuts through. Hourglass perforations are characteristic for beads made with a lathe (Popkiewicz/Czebreszuk 2016, 85).

Above-described criteria of the objects examination are aimed to give reader the most detailed and comprehensive picture of the analysed dataset. Nonetheless, the number and complexity of applied variants of every feature (category of description) make it difficult to grasp all the information in one term that would describe a given artefact. For the sake of brevity and clarity, it was decided to reduce the definitions of every variant of applied criteria to designators (either numerical or alphabetical), and combine them in alphanumerical codes, simultaneously denoting a type to which given artefact belongs. The codes are formed by strings of Roman and Arabic numerals, as well as capital and lowercase alphabet letters, symbolising the variants as depicted on Fig. 34. Such codes facilitate description of the artefacts and their inclusion in statistical analyses of the dataset, conducted in the subsequent stages of the work. The description below offers a condensed overview of the types, without naming the contexts which yielded their individual representatives. For the latter information, readers are advised to check the catalogue.

3.3. Group I

Group I covers all amber artefacts perforated concentrically along the height (H). It is the largest group in the entire dataset, in total comprising 1,099 objects. Being the most numerous, it is also greatly varied internally, with the objects being characterised by different sizes and vast array of shapes spanning all variants of cross-sections and profiles. This variability calls for a detailed overview of the objects and ordering them into a set of types, according to the criteria described in section 3.2. In the first place, finds belonging to group I were divided with respect to the dimensions ratio into three subgroups: flattened (D>H), average (D=H) and elongated (D<H).

484 A wider outlet of perforation with rounded edges has an advantage of not causing the string to wear down quickly and, if beads are separated by knots, allows to conceal the spacings (Popkiewicz/Czebreszuk 2016, 85).

3.3.1. Flattened

The largest subset of artefacts within group I represents flattened forms (808 objects). The lion's share of this subgroup are those with circular or oval cross-section (numbering 395 and 247 finds, respectively). Among them, very characteristic are the beads with rectilinear sides and bases, as seen in the profile, which together with round cross-section lend them appearance of a flattened cylinder (type IF111) (Fig. 35:1).[485] Type IF111 in total comprises 35 elements. Within this group were recognised two beads decorated with single transverse rib/ledge (IF111a),[486] one with slanting edges of the hole's outlet (IF111w) and one incorporating both features (IF111aw) (Fig. 35:2–3). Furthermore, there are some cylindrical beads with irregular bases, either convex-flat (IF114; 2 beads) or skewed-flat (IF118; 1 bead) (Fig. 35:4–5). As flattened cylindrical beads, 15 beads can be considered with rectilinear surfaces in the lateral section, but oval cross-sections (IF211) (Fig. 35:6). Among them are three specimens decorated with single median ledges/ribs (IF211a) (Fig. 35:7) and one with slanting edges of the hole's outlet (IF211w). Yet, there are others that exhibit slight deviations from the cylinder model, namely those with skewed (IF218; 7 beads), convex (IF214; 1) or concave bases (IF216; 1), some of them ornamented with decorative motif 'a' (Fig. 35:8–10). Artefacts decorated with this motif, consisting of a single horizontal rib, are not numerous in IF subgroup, amounting only to 12 beads.

All flattened beads with convex sides, and circular or oval cross-sections can be considered collectively, as they largely are variations on a disc-like form. In literature one can find several terms used to describe similar beads, such as 'discoidal', 'annular', 'ovoidal', 'ellipsoidal', etc.[487] They are usually the largest category of amber beads found in various regions, and the circum-Adriatic zone is not an exception

485 The term 'cylindrical' is used to describe similar beads by N. Negroni Catacchio et al., who place flattened cylindrical beads under type 3, variant C (Negroni Catacchio et al. 2006, 1446, Fig. 2), while A. Palavestra attributes them to the type 18c (Palavestra 1993, Tab. II). They are similarly defined by authors of amber typologies designed for the areas neighbouring the circum-Adriatic zone: e.g. E. Sprincz and C. W. Beck classify such beads from Hungary as type IA (Sprincz/Beck 1981, Fig. 6, Tab. 1), C. Du Gardin designates French specimens as type 1a (Du Gardin 1986, 547, Fig. 1:1–3) and G. Woltermann places them within type 7a of prehistoric amber beads from Germany (Woltermann 2016a, 42, Abb. 18).

486 In these particular cases, we are dealing with beads commonly attributed to the Tiryns type in a flattened variant: type 1b by Palavestra (1993, Tab. I) and type 4c-d by Negroni Catacchio (1999, Fig. 2).

487 A. Palavestra classifies such beads as types 8–9, the former with two variants: 'a' and 'b' (Palavestra 1993, Tab. I). N. Negroni Catacchio et al. place them collectively under type 1, without distinguishing variants (Negroni Catacchio et al. 2006, 1446, Fig. 2). E. Sprincz and C. W. Beck worked a more elaborative division into several types: IB, II, III and IV (Sprincz/Beck 1981, Fig. 6, Tab. 1). Also J. Czebreszuk distinguished several varieties: 102. Flattened spheroidal, 104. Ellipsoidal and 108. Annular (Czebreszuk 2011, 78–88). G. Woltermann, following Beck/Sprincz 1981, included such beads within types 1–4 (Woltermann 2016a, 40–41, Abb. 18). C. Du Gardin limits them to type 1b, although highlights the differences between the varieties (Du Gardin 1986, 547).

in this aspect. Some of them differ from the model-like flattened cylindrical beads (IF111), only by having convex sides (IF121; Fig. 35:11). The latter type was identified in 96 cases among collected data.[488] It was quite frequently (17 cases in total) outfitted with a recess widening the perforation's outlets, either in slanting (IF121w; 3 beads) or convex shape (IF121y; 13) (Fig. 35:12–13).[489] On one bead it was possible to observe incised ornament on the surface (IF121j) (Fig. 35:14). The variant with oval, instead of circular, cross-section appeared 49 times (IF221) (Fig. 35:15).[490] Again, some of the beads revealed alterations of the perforations' outlets (IF221w, IF221y; 2 and 5 beads, respectively) (Fig. 35:16–17). Then, there are beads with convex walls and bases which give them appearance of a flattened spheroid (IF122), or ellipsoid, if the cross-section is oval not circular (IF222) (Fig. 35:18–19).[491] These types form a considerable part of the dataset, numbering 59 and 52 examples, respectively. As in the previous types, there are also cases of the beads with widened perforation's outlets (IF122w, IF122y, IF122z, IF222w) (Fig. 35:20–23). Eventually, there is a variety of beads with convex sides differing in shape of the bases, ranging from convex-flat to skewed-flat.[492] Among those with circular cross-sections, it was possible to distinguish types IF123 (4 beads), IF123y (1), IF124 (6), IF124y (2), IF125 (1), IF127 (3), IF127y (1) and IF128 (32, including one with recess 'w' and one with recess 'y') (Figs. 35:24–30, 36:1). Analogically, non-parallel bases were encountered among amber finds with convex sides and oval cross-sections: IF223 (7), IF224 (8), IF224z (1), IF226 (1), IF227 (1) and IF228 (30) (Fig. 36:2–6). Some of the latter beads have slanting (IF228w) or convex edges of the perforation's outlets (IF228y), and/or bear incised cross-cuts (IF228jy, IF228j) (Fig. 36:7–9).

488 It corresponds to: type IB by Sprincz/Beck (1981, Fig. 6, Tab. 1), type 8a by Palavestra (1993, Tab. I), type 1 by Negroni Catacchio et al. (2006, 1446, Fig. 2), type 102 by Czebreszuk (Czebreszuk 2011, 78) and type 2 by Woltermann (2016a, 40, Abb. 18).

489 Discoidal amber beads with wide perforation are customarily called in the literature as 'annular'. They can be linked to type II by Sprincz/Beck (1981, Fig. 6, Tab. 1), type 7b by Palavestra (1993, Tab. I), type 108 by Czebreszuk (Czebreszuk 2011, 88) and type 2 by Woltermann (2016a, 40, Abb. 18).

490 Authors of various typologies of amber artefacts usually do not differentiate the discoidal beads between those with round and those with oval cross-section. Thus, type IF221 can be referenced to the same types by other authors, as type IF121.

491 They can be linked to type III by Sprincz/Beck (1981, Fig. 6, Tab. 1), types 8b-9 by Palavestra (1993, Tab. I), type 104 by Czebreszuk (Czebreszuk 2011, 79) and type 3a by Woltermann (2016a, 41, Abb. 18).

492 Beads with skewed-skewed or skewed-flat profiles fit type IV by Sprincz/Beck (1981, Fig. 6, Tab. 1), type 7a by Palavestra (1993, Tab. I) and type 4 by Woltermann (2016a, 41, Abb. 18).

Fig. 35. Examples of the types of amber finds belonging to flattened category of group I: 1 – Scalvinetto, tomb 354 (Salzani 2004, Fig. 4:G); 2 – Vrlazje, tumulus III, grave 3 (1894) (Cwaliński/Pravidur forthcoming, Pl. 1:1); 3 – Peschiera – Imboccatura del Mincio, precise context unknown (Negroni Catacchio 1971, Fig. 3); 4 – Castione Marchesi, layer 3/phases 1–2 (Mutti et al. 1988, Fig. 81:5); 5 – Molina di Ledro, strata I–III/IV (Rageth 1974, Taf. 92:6); 6 – Maso Finale, cultural layer (Putzer 2012, Abb. 10:9); 7 – Vrsi, grave 89 (Batović 1983, T. XLIV:9); 8–9 – Castello del Tartaro, tomb 146 (Salzani 2011, Fig. 4:A3); 10 – Torre Castelluccia, ripostiglio *(Harding/Hughes-Brock 1974, Fig. 6:27); 11 – Isolone del Mincio, precise context unknown (Negroni Catacchio et al. 1999, Fig. 11:14); 12 – Povegliano Muraiola, phase IV (Salzani 1997b, Tav. 56:16); 13 – Fiavè-Carera, horizon 6 (Perini 1987, Fig. 68:a1); 14 – Olmo di Nogara, grave 388 (Salzani 2005, Tav. XXX:E); 15 – Ravnine, mound 3 (Nikitović/Vasić 2002, Tab. III:1); 16 – Vrsi, grave 89 (Batović 1983, T. XLIV:11); 17–18 – Fiavè-Carera, horizon 6 (Perini 1987, Fig. 68:a2, a6); 19 – Olmo di Nogara, grave 307 (Salzani 2005, Tav. XXV:C); 20 – Isolone del Mincio, precise context unknown (Negroni Catacchio et al. 1999, Fig. 12:17); 21 – Olmo di Nogara, grave 172 (Salzani 2005, Tav. XVII:C); 22 – Grotta dell'Artofago, precise context unknown (Aranguren/Perazzi 2004, Fig. 4:9); 23 – Castione Marchesi, layer 3/phases 1–2 (Mutti et al. 1988, Fig. 81:8); 24 – Isolone del Mincio, precise context unknown (Negroni Catacchio et al. 1999, Fig. 11:16); 25 – Villaggio delle Macine, US 6/185 (Bellintani et al. 2007, Fig. 2); 26 – Castello del Tartaro, tomb 350 (Salzani 2011, Fig. 5:B3–4); 27 – Castellaro Lagusello, layer D (Negroni Catacchio et al. 1999, 28, Fig. 10:1); 28 – Olmo di Nogara, grave 435 (Salzani 2005, Tav. XXXVI:B); 29 – Campestrin di Grignano Polesine, late BR–initial FBA layers (Bellintani et al. 2015a, Fig. 2A:9); 30 – Franzine Nuove, grave 7 (Aspes/Fasani 1968, Fig. 8:3).*

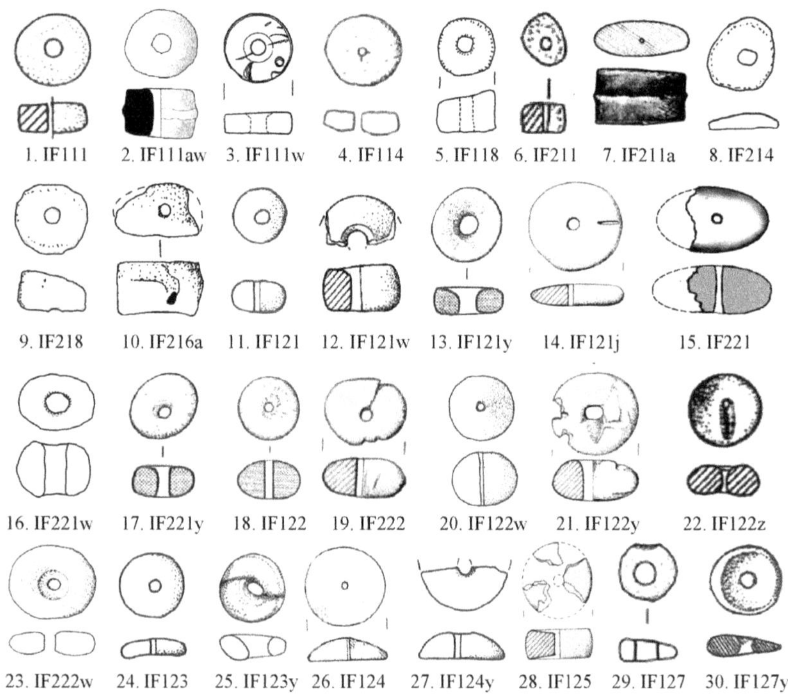

Fig. 36. Examples of the types of amber finds belonging to flattened category of group I: Fig. 36. 1 – Povegliano Muraiola, phase II (Salzani 1997b, Tav. 56:39); 2 – Shtogj, tumulus 2, grave 9 (Kurti 2017b, Pl. XCVIIIf:4); 3 – Maso Finale, cultural layer (Putzer 2012, Abb. 10:5); 4 – Lavello, ipogeo no. 743 (Nava/Salerno 2007, Fig. 6:10); 5 – Scalvinetto, tomb 233 (Salzani 1994, Tav. 5:5); 6 – Olmo di Nogara, grave 190 (Salzani 2005, Tav. XVIII:D); 7 – Finilone Valle, subsurface materials (Salzani 1996, Fig. 21:27–28); 8 – Olmo di Nogara, grave 365 (Salzani 2005, Tav. XXIX:B); 9 – Madonna di Loreto, 'Hypogeum of Bronzes' (Tunzi Sisto 1999a, Tav. X:1213); 10 – Bismantova – Campo Pianelli, grave XXXI (Catarsi/ Dall'Aglio 1978, Fig. XXI:1); 11 – Iglarevo, grave 1 (Palavestra 1997, Pl. 1:17); 12 – Olmo di Nogara, grave 493 (Salzani 2005, Tav. XL:C); 13 – Maso Finale, cultural layer (Putzer 2012, Abb. 10:17); 14 – Madonna di Loreto, 'Hypogeum of Bronzes' (Tunzi Sisto 1999a, Tav. X:649); 15 – Peschiera – Imboccatura del Mincio, precise context unknown (Negroni Catacchio 1971, Fig. 8); 16 – Iglarevo, grave 1 (Palavestra 1997, Pl. 1:15); 17 – Vrčin, grave 3 (Buršić-Matijašić 1989, Fig. 1:2); 18 – Olmo di Nogara, grave 353 (Salzani 2005, Tav. XXIX:E); 19 – Peschiera – Imboccatura del Mincio, precise context unknown (Negroni Catacchio 1971, Fig. 6); 20 – Castello del Tartaro, tomb 146 (Salzani 2011, Fig. 4:A3); 21 – Rudine u Ivanopolju, tumulus I (1894) (Cwaliński/Pravidur forthcoming, Pl. 1:3); 22 – Finilone Valle, subsurface materials (Salzani 1996, Fig. 21:18); 23 – Olmo di Nogara, grave 422 (Salzani 2005, Tav. XXXV:C); 24 – Isolone del Mincio, precise context unknown (Negroni Catacchio et al. 1999, Fig. 12:19); 25 – Peschiera – Imboccatura del Mincio, precise context unknown (Negroni Catacchio 1971, Fig. 2).

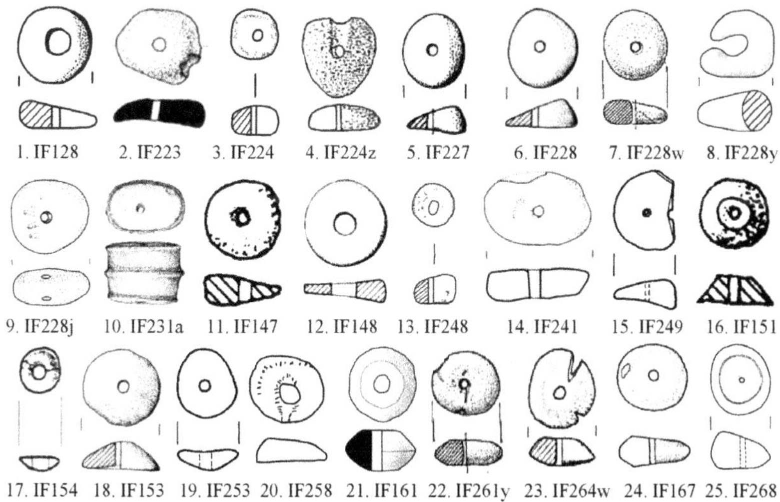

Fig. 37. Examples of the types of amber finds belonging to flattened category of group I: 1 – Olmo di Nogara, grave 172 (Salzani 2005, Tav. XVII:D); 2 – Mariconda di Melara, upper layer (Salzani 1973, Tav. XVIII:22); 3 – Isolone del Mincio, precise context unknown (Negroni Catacchio et al. 1999, Fig. 11:11); 4–5 – Castione Marchesi, layer 3/phases 1–2 (Mutti et al. 1988, Fig. 81:1–2); 6 – Olmo di Nogara, grave 243 (Salzani 2005, Tav. XXI:F); 7 – Çinamak, tumulus 16, grave 19 (Kurti 2013, Tab. I:15); 8 – Gučevo, tumulus IV, grave 1 (1895) (Cwaliński/Pravidur forthcoming, Pl. 1:8); 9 – Madonna di Loreto, 'Hypogeum of Bronzes' (Tunzi Sisto 1999a, Tav. X:648); 10 – Campestrin di Grignano Polesine, late BR–initial FBA layers (Bellintani et al. 2021, CGP 19); 11–12 – Iglarevo, grave 1 (Palavestra 1997, Pl. 1:2, 5); 13 – Madonna di Loreto, 'Hypogeum of Bronzes' (Tunzi Sisto 1999a, Tav. X:646); 14 – Iglarevo, grave 4 (Palavestra 1997, Pl. 1:21); 15 – Paulje, tumulus K (Canić-Tešanović/Gligorić 2001, Cat. No. 66); 16 – Iglarevo, grave 1 (Palavestra 1997, Pl. 1:1); 17–18 – Campestrin di Grignano Polesine, late BR–initial BF layers (Bellintani et al. 2015a, Fig. 2A:7–8); 19 – Iglarevo, grave 1 (Palavestra 1997, Pl. 1:3); 20 – Jezero, mound IX, grave 6 (Kosorić 1978, Tab. V, Fig. 7); 21 – Madonna di Loreto, 'Hypogeum of Bronzes' (Tunzi Sisto 1999a, Tav. X:927); 22 – Murgia Timone, tomb 1 (Matarese 2016, Fig. 1:A); 23–24 – Madonna di Loreto, 'Hypogeum of Bronzes' (Tunzi Sisto 1999a, Tav. X:944, 1767); 25 – Jezero, mound IX, grave 6 (Kosorić 1978, Tab. V, Fig. 7); 26 – Madonna di Loreto, 'Hypogeum of Bronzes' (Tunzi Sisto 1999a, Tav. XI:1283); 27 – Iglarevo, grave 1 (Palavestra 1997, Pl. 1:6); 28 – Gučevo, tumulus IV, grave 1 (1895) (Cwaliński/Pravidur forthcoming, Pl. 1:7).

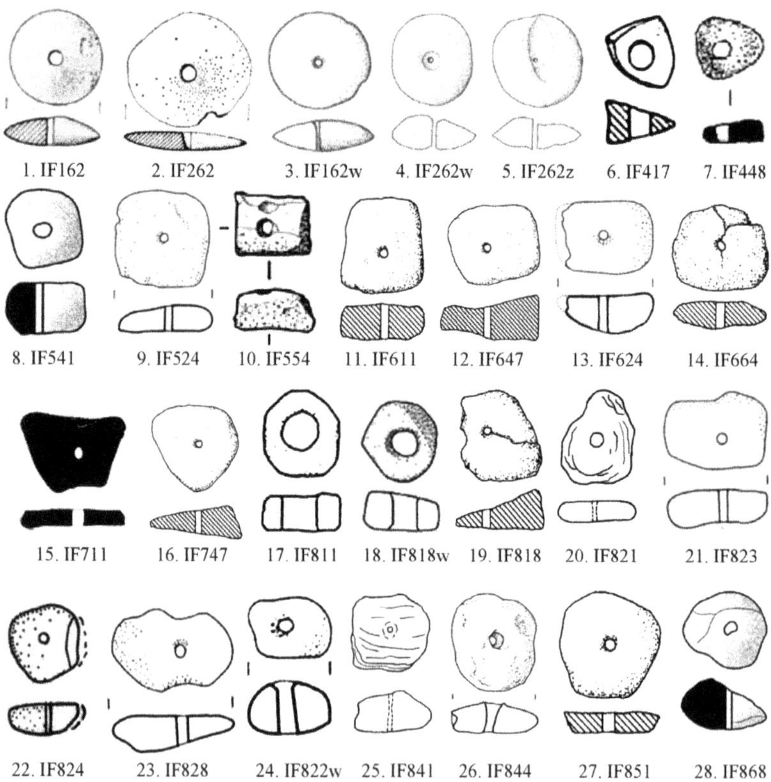

There certainly are two examples of amber beads with oval cross-sections, convex sides and decorative motifs 'a' (IF231a; Fig. 36:10), but apart from that concave profiles are rare among flattened beads. The only other similar case within the discussed group has polygonal cross-section and convex-flat bases (IF834).

A small group is formed by beads with straight-convex sides. When it comes to the varieties with circular or oval cross-sections, it can be noticed that they largely consist of beads with skewed-skewed (IF147; 4 beads) or skewed-flat bases (IF148, IF248; 5 and 11 beads, respectively) (Fig. 36:11–13). At the first glance they can be mistaken for the beads with skewed bases, but convex sides, discussed in the previous paragraphs. The difference is more apparent in case of few objects with straight-convex sides and rectilinear bases (IF241; 2 beads) (Fig. 36:14). Only few of the finds with convex-straight sides exhibit traces of expanding the perforation outlets (IF244y and IF248w; 1 bead each). There is also one example in this group with concave-skewed bases (IF249) (Fig. 36:15).

Another small, but characteristic group of finds is formed by beads with pitched sides which in three-dimensional projection reflect the cone-shaped form. The ones which visually correspond to the classic conical amber beads, as illustrated in archaeological literature, are those with circular and oval cross-section, the former numbering 19 objects and the latter 7.[493] Two leading variants can be distinguished among the collected finds: one with the truncated cone profile (IF151, IF251; 10 and 1 bead, respectively) and one with pyramidal-like profile (IF154, IF254; 7 and 4, respectively), also known in version with recesses 'w' or 'y' (Fig. 36:16–17). There are also cases of beads with conical profiles having slightly convex-concave (IF153; IF253; 1 bead each) or skewed-flat bases (IF158, IF258; 1 and 2 beads, respectively) (Fig. 36:18–20).

Next category of flattened beads discussed here comprises artefacts with, first and foremost, biconical profile and, secondarily, circular or oval cross-sections. The double-pitched shape of the lateral surfaces gives those beads characteristic look; however, it is sometimes difficult to judge, whether the cusp is sharp enough for the profile to be considered 'biconical' or is more rounded, thus akin to 'convex'. In comparison with classic flattened biconical amber beads depicted in the literature, the closest type among the dataset would be the one with circular cross-section and rectilinear-flat bases – IF161 (76 beads in total, twice with recess 'w') (Fig. 36:21).[494] There is a much smaller group of similar beads, but with oval cross-section – IF261 – counting six beads, including one with convex recess 'y' (Fig. 36:22). As with other categories of beads discussed earlier, there are some biconical forms with non-parallel bases, i.e. convex-flat (IF164, IF264; 3 and 2 beads, respectively), skewed-skewed (IF167, IF267; 5 and 2) or skewed-flat (IF168, IF268; 13 and 6) (Fig. 36:23–25), including those with alterations of the hole's outlets. Specific cases of biconical beads are those with convex bases (IF162, IF262; 29 and 34 beads, respectively), due to which objects look

---

493 Similar beads with conical profile were included in type X by Sprincz/Beck (1981, Fig. 6, Tab. 1), type 146 by Palavestra (1993, Tab. XIII), type 6 by Negroni Catacchio et al. (2006, 1446, Fig. 2) and type 10 by Woltermann (2016a, 43, Abb. 18).

494 Flattened biconical beads were included in types IXA and IXB by Sprincz/Beck (1981, Fig. 6, Tab. 1), type 1d by Du Gardin (1986, 547, Fig. 1:19), type 12a by Palavestra (1993, Tab. I), type 8, var. 'c' and 'd' by Negroni Catacchio et al. (2006, 1446, Fig. 2), type 105 by Czebreszuk (2011, 79) and types 9a and 9b by Woltermann (2016a, 43, Abb. 18).

lenticular in the profile (Fig. 37:1–2).[495] Many among them exhibit traces of widening the perforation's outlets in various manners: IF162w (6 beads), IF162x (2), IF162y (2), IF262w (24), IF262y (3), IF262z (1) (Fig. 37:3–5).

Remaining objects placed in the group of flattened, centrally perforated beads have non-round cross-sections, spanning shapes such as triangular, quadrangular, rectangular, trapezoidal and polygonal. Triangular cross-section is represented in five types of beads: those with rectilinear sides and bases (IF411; 3 beads), rectilinear sides and both bases skewed (IF417; 1), convex sides and convex-flat bases (IF424; 2), convex sides and skewed-flat bases (IF428; 2) and, finally, those with straight-convex sides and skewed-flat bases (IF448; 3). Aforementioned types, despite the cross-section, in the profile often resemble typical discoidal beads discussed earlier (Fig. 37:6–7).[496]

There is only a handful of beads with quadrangular cross-section (6 in total).[497] One of them has straight-convex sides and rectilinear bases, thus is the closest to the cubic form (IF541) (Fig. 37:8). The other ones have less regular form; they usually combine quadrangular cross-section with flat-convex or skewed bases, but shape of the sides is far less uniform among the discussed specimens (IF514, IF524, IF528, IF548 and IF554; one bead each) (Fig. 37:9–10).

Only little more common are the beads with rectangular cross-sections (7 examples). Four beads have a fairly regular form (IF611; Fig. 37:11), while the remaining five artefacts substantially differ between each other. One of them has straight-convex sides but skewed bases (IF647), three exhibit convex sides and convex-flat bases (IF624), while the fifth has similar bases to the latter but in connection with biconical lateral profile (IF664) (Fig. 37:12–14).

Among the objects with trapezoidal cross-section, also not numerous (4 beads in total), one can notice more regular finds with rectilinear sides and bases (IF711; 2 beads in total) (Fig. 37:15), as well as those with less uniform profiles. Namely, these are beads with straight-convex sides, and skewed bases (IF747; 2 beads in total) (Fig. 37:16).

Beads with polygonal cross-sections constitute a large and varied group of finds.[498] Some of them show certain degree of regularity in execution, especially those with rectilinear sides and bases (IF811; 6 beads) (Fig. 37:17). Then, there are more numerous beads which do not have such neatly executed profiles. One of them has already been mentioned earlier – the bead with concave sides and convex-flat bases (IF834). However, there are many other types which incorporate uneven lateral and bottom/

---

495 Discoidal beads with lenticular profile were included in type V by Sprincz/Beck (1981, Fig. 6, Tab. 1), type 1d by Du Gardin (1986, 547, Fig. 1:17), type 16 by Palavestra (1993, Tab. I), type 103 by Czebreszuk (2011, 78) and type 5 by Woltermann (2016a, 41, Abb. 18).

496 Beads with triangular cross-section were included in type 30 by Palavestra (1993, Tab. II), group D1 in the typology by Ernée (2012, Obr. 7) and type 12c by Woltermann (2016a, 47, Abb. 18).

497 Beads with quadrangular, rectangular and trapezoidal cross-sections correspond with type 35B by Palavestra (1993, Tab. II), type 112 by Czebreszuk (2011, 88), group D4 by Ernée (2012, Obr. 7) and types 13 and 14 by Woltermann (2016a, 48, Abb. 18).

498 Beads with polygonal cross-section, especially when having non-parallel bases and sides, are usually disregarded in formal classifications of amber beads, or are treated as irregular beads (as a separate category of finds).

top surfaces. One can wonder, if it is justified to distinguish those amber finds as separate types. The beads with polygonal cross-section and irregular profiles may appear as a result of inferior craftsmanship – roughly chopped pieces of amber, perforated so they can be worn for the sole fact of being amber, regardless of form. Even so, they are still carrying this important information which, in comparison with other types of amber beads, can shed light on varying levels of technological skill in amber processing.

To the variant of beads with polygonal cross-section and rectilinear sides can be added those with flat-concave (IF816; 1 bead) and skewed-flat bases (IF818, IF818w; 2 and 1 bead, respectively) (Fig. 37:18–19). Quite frequent are the beads with polygonal cross-section and convex sides, but differing according to bases: rectilinear (IF821; 5 beads), convex (IF822; 2), convex-concave (IF823; 4), convex-flat (IF824; 8) and skewed-flat (IF828; 4) (Fig. 37:20–23). Perforations in those beads rarely reveal any traces of expanding; in fact only four cases were registered: IF821y (3) and IF822w (1) (Fig. 37:24).

Among the beads with polygonal cross-section, numerically preponderant are those with straight-convex sides (98 specimens in total). Between them, the greatest share have the beads with rectilinear bases (IF841), with 60 out of 62 such elements being recovered from a single context – Banjevac-Jovanin Breg, tumulus 2 (Fig. 37:25). Types with convex-flat (IF844) or concave-flat (IF846) are considerably less represented (2 cases each) (Fig. 37:26). Again, quite numerous are the beads with skewed-flat bases (IF848), counting altogether 32 specimens. The other types included in the group of beads with polygonal cross-sections, i.e. those with conical sides and rectilinear bases (IF851), with biconical sides and rectilinear bases (IF861), with biconical sides and convex bases (IF862) and with biconical sides and skewed-flat bases (IF868) occurred marginally, being represented by a single bead each (Fig. 37:27–28).

3.3.1.1. Chronology

From the chronological perspective, the amber finds classified to subgroup IF can be divided on the basis of their contexts' dating. The phases distinguished in section 2.5: EBA (2150–1650 BC), MBA (1650–1300 BC), LBA (1650–1300 BC) and FBA (1150–950 BC), along with the intermediate stages (EBA/MBA, MBA/LBA, LBA/FBA and FBA/EIA), were applied for that purpose in the following sections. Thus, in total there are eight chronological groups of contexts to which types of amber finds can be attributed.

In terms of numbers, there is a clear preponderance of MBA and LBA both in contexts as well as finds (Fig. 38). A significant contribution to this observation are the 41 contexts with 362 finds assigned to the intermediate stage MBA/LBA, meaning they have a long chronology spanning both periods, or cannot be unambiguously dated to one or the other stage. Either way, these numbers only augment already large assemblages of contexts and finds from MBA and LBA. The number of contexts datable to FBA are definitely lower (21), than in the previous two stages; however, it is still accompanied by a considerable quantity of amber artefacts with flattened profile (89). Decisively less numerous are the contexts and associated finds from intermediate stages EBA/MBA and LBA/FBA which would not significantly alter the general picture, when added to any of the preceding or succeeding periods. Finally, chronologically the earliest (EBA) and the latest contexts (FBA/EIA) delivered only a small portion of the finds classified to subgroup IF. Therefore, the long-term fluctuation of amber beads belonging to subgroup IF can be described as low in numbers at the beginning of the Bronze Age, then ascending from EBA to MBA, holding a considerable level during LBA and falling during FBA.

Fig. 38. Temporal fluctuations in numbers of contexts and finds classified to subgroup IF during the Bronze Age.

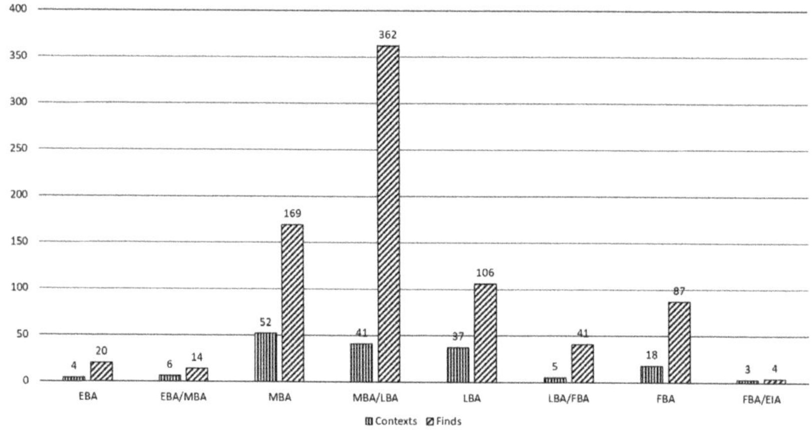

One of the most pressing questions posed in this work is if the amber artefacts' morphology was subject to change over time. In order to find an answer to this question, one seeks a way that would show whether there was an evolution pattern, beginning with one, simple form of amber bead and ending with other, more complex designs. The method mostly favoured by archaeologists to meet this task is seriation, also adopted in this section. The basis for its application is a contingency table, containing contexts in the columns and types in the rows with '1' or '0' at their intersections indicating, respectively, presence or absence of a type in a context. The analysis was performed in PAST 3.0 software, using the algorithm described by Brower/Kile (1988), in the 'constrained' mode. It means that the array of columns is ordered according to contexts' chronology, beginning with the oldest ones on the left and ending with the youngest ones on the right, and does not change during the analysis, as opposed to rows which exchange their positions to achieve the most regular range plot. Thus, the already known chronology of the contexts is used to date specific types of amber beads. This 'reversed' way of applying seriation is different from most of its applications in archaeology, where it is usually employed to disclose relative chronology of contexts. The solution proposed here is more characteristic for environmental scientists, who sometimes use seriation to observe changes in occurrence of taxa along a known gradient.

Fig. 39. Original range plot resulting from seriation of types and contexts classified to subgroup IF. (Figure in full resolution can be downloaded from https://www.peterlang.com/app/uploads/2023/09/Fig.39.jpg).

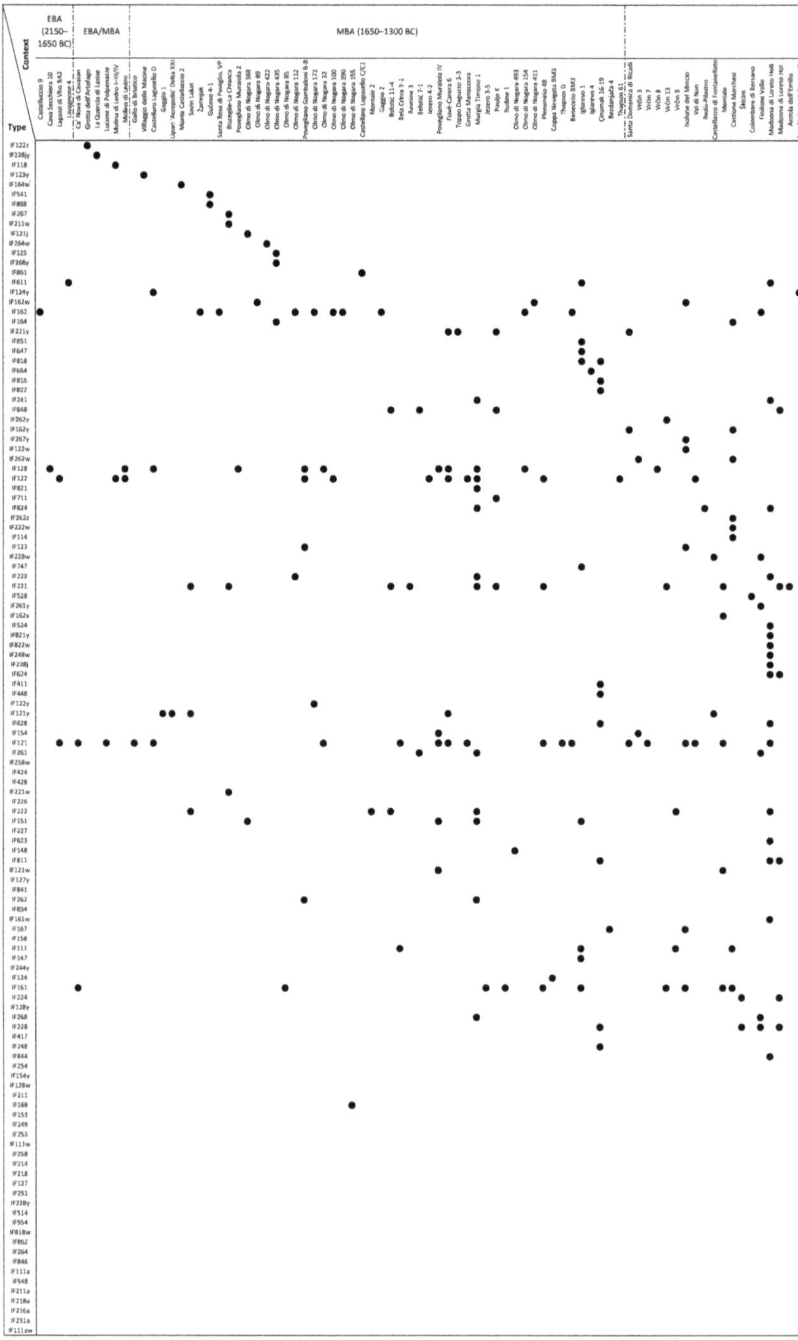

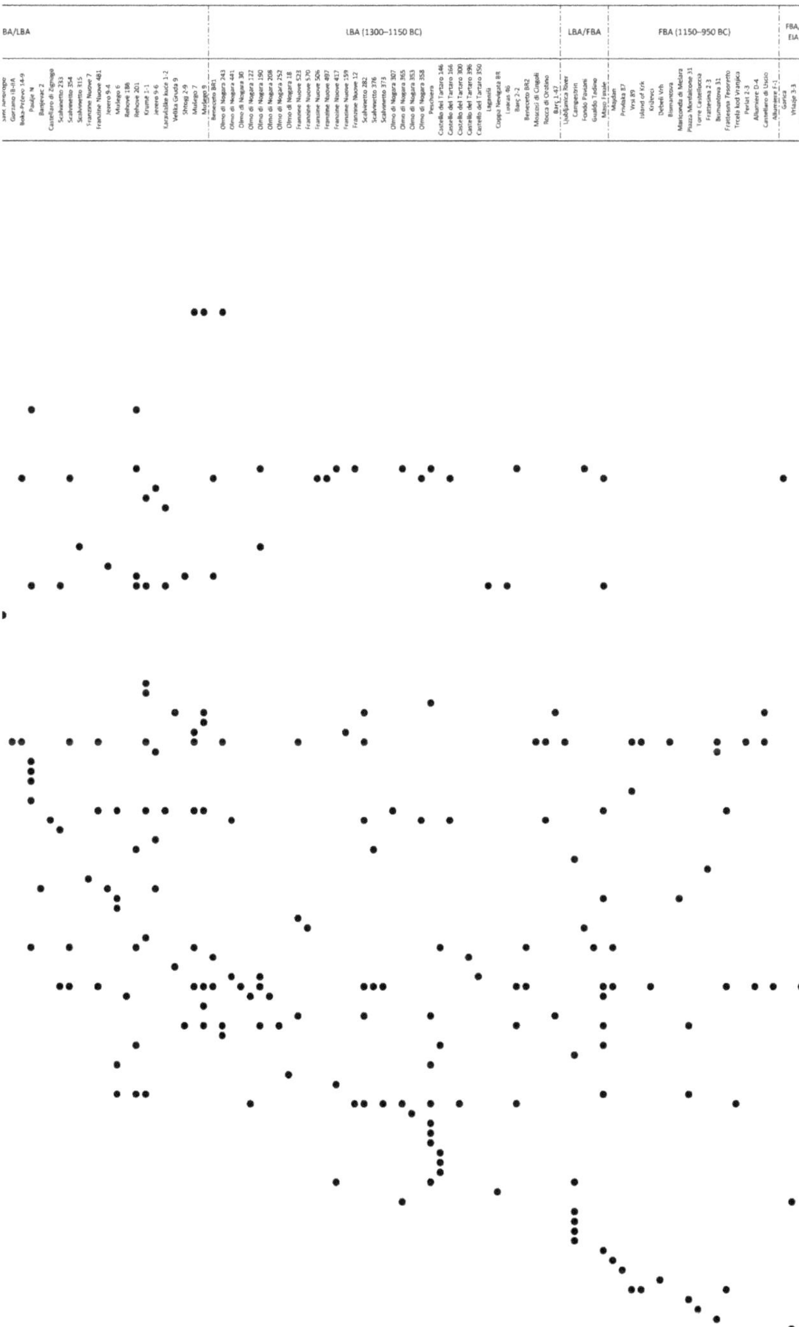
BA/LBA
LBA (1300–1150 BC)
LBA/FBA
FBA (1150–950 BC)
FBA/EIA

Fig. 40. Modified range plot resulting from seriation of types and contexts classified to subgroup IF. (Figure in full resolution can be downloaded from https://www.peterlang.com/app/uploads/2023/09/Fig.40.jpg).

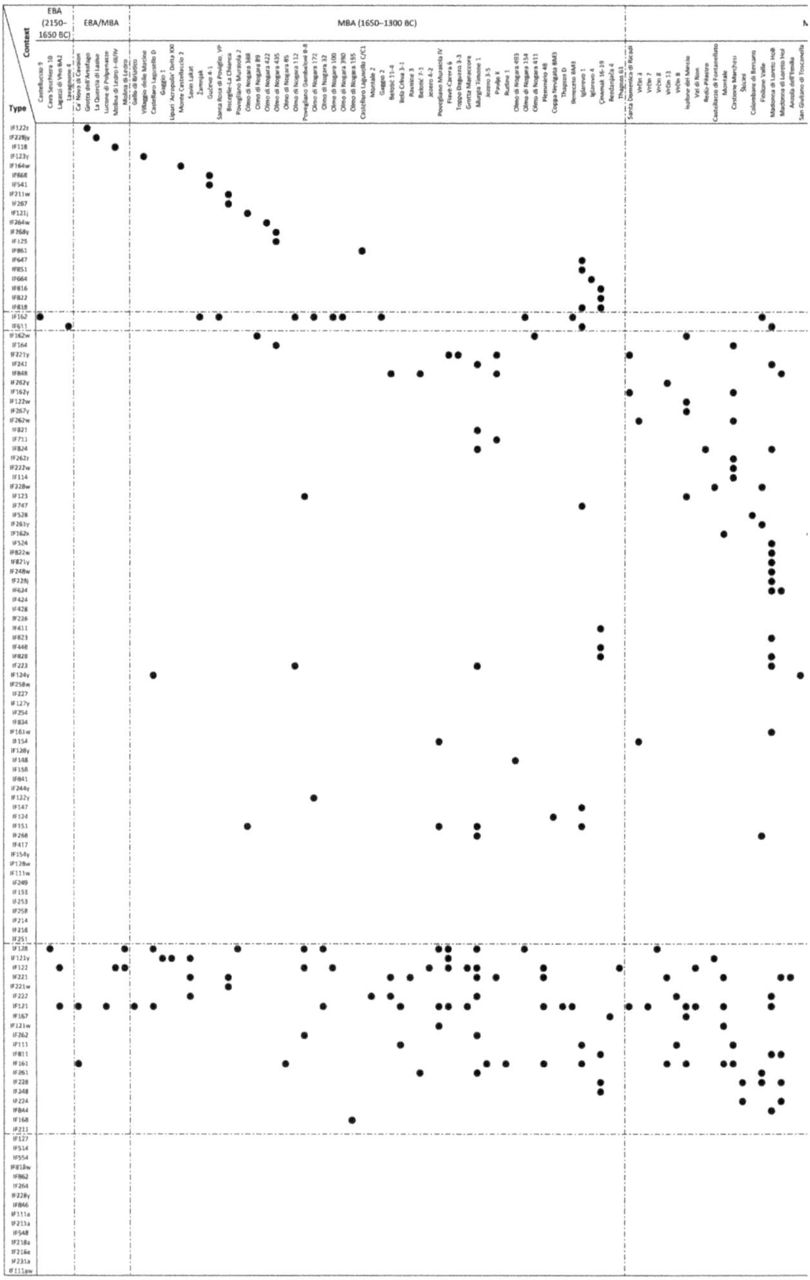

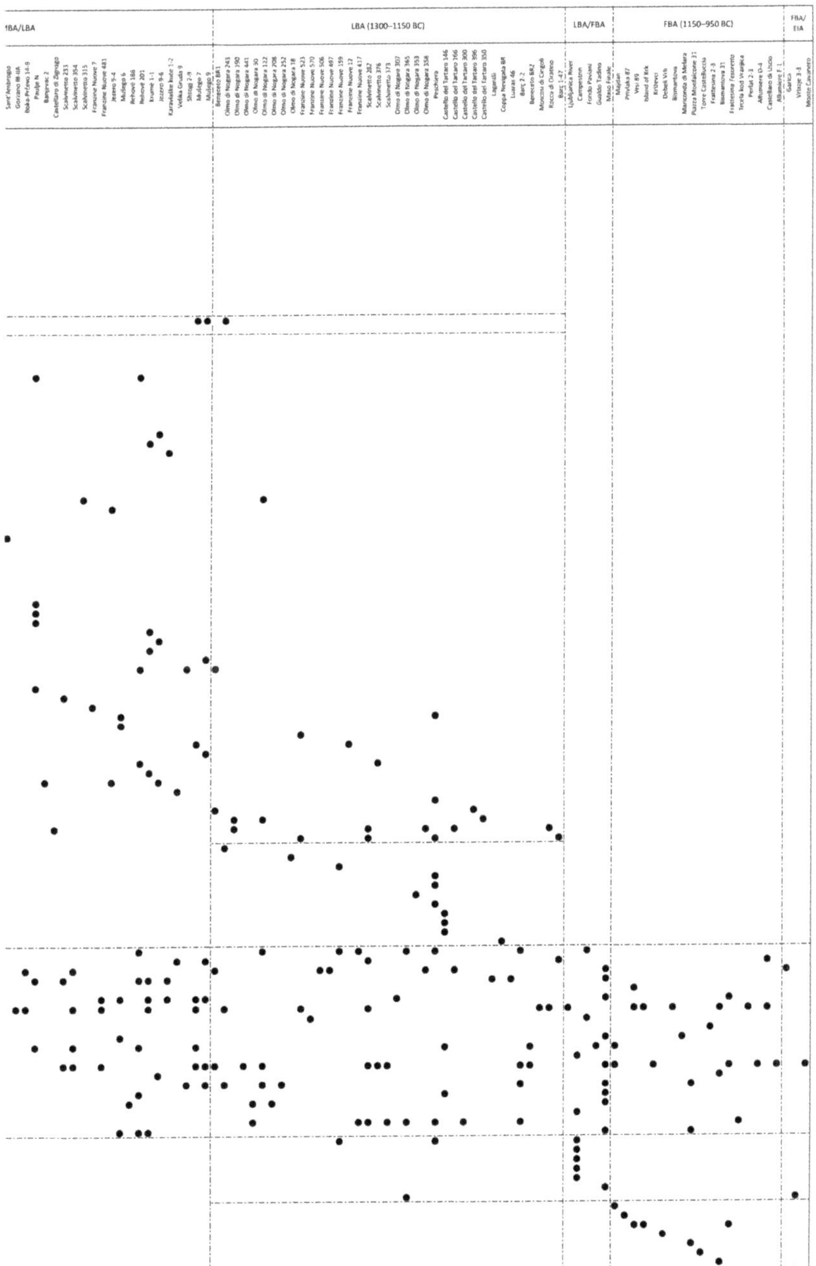
MBA/LBA
LBA (1300–1150 BC)
LBA/FBA
FBA (1150–950 BC)
FBA/EIA

The initial results of seriation of contexts and types belonging to subgroup IF form a scattered range plot: there is a weak but discernible diagonal trend in distribution of the incidences in the matrix, but majority of the types are spread horizontally across the table (Fig. 39). So, despite the general observation that there apparently was a steady process of change in morphology of IF amber beads during the Bronze Age, one cannot avoid an impression that a large portion of the types were used for longer periods, extending beyond the time frames of a single phase. There is a lack of linearity or unidirectionality in Bronze Age evolution of IF beads, and a clear-cut division of the types in line with consecutive stages of the epoch is rather difficult. Nonetheless, a slight manipulation of the results by moving the rows with types by few positions up or down allows to attain a picture which, although with a lower goodness of fit (lower similarity 'criterion') than the original solution produced by the programme, gives a clearer distinction of the types into chronological groups (Fig. 40). By comparing distribution of the points signifying occurrences in the range plot and time boundaries of successive phases, it was possible to distinguish seven groups of types (Fig. 41).

Fig. 41. Division of IF types of amber artefacts into chronological groups of the Bronze Age.

EBA–MBA	EBA–LBA	MBA–LBA		LBA	EBA–FBA	LBA–FBA	FBA
IF118	IF162	IF114	IF248w	IF111w	IF111	IF127	IF111a
IF121j	IF611	IF122w	IF254	IF128w	IF121	IF228y	IF111aw
IF122z		IF122y	IF258w	IF153	IF121w	IF264	IF211a
IF123y		IF123	IF261y	IF154y	IF121y	IF514	IF216a
IF125		IF124	IF262w	IF214	IF122	IF554	IF218a
IF164w		IF124y	IF262y	IF218	IF128	IF818w	IF231a
IF211w		IF127y	IF262z	IF249	IF161	IF862	IF548
IF228jy		IF128y	IF267y	IF251	IF167		IF846
IF264w		IF147	IF268	IF253	IF168		
IF267		IF148	IF411	IF258	IF211		
IF268y		IF151	IF424	IF417	IF221		
IF541		IF154	IF428		IF221w		
IF647		IF158	IF448		IF222		
IF664		IF161w	IF524		IF224		
IF816		IF162w	IF528		IF228		
IF818		IF162x	IF624		IF248		
IF822		IF162y	IF711		IF261		
IF851		IF164	IF747		IF262		
IF861		IF221y	IF821		IF811		
IF868		IF222w	IF821y		IF844		
		IF223	IF822w				
		IF226	IF823				
		IF227	IF824				
		IF228j	IF828				
		IF228w	IF834				
		IF241	IF841				
		IF244y	IF848				

Proceeding from the top-left corner of the chart (Fig. 40), the types in the first 20 rows are chronologically confined to EBA–MBA (Fig. 41). Three of them appeared already in the contexts dated broadly to EBA/MBA (IF118, IF122z, IF228jy), while the rest are typical solely to MBA. Within the group one can notice peculiar variants of flattened cylindrical, spheroidal or discoidal beads of circular or oval cross-section, with various expansions of the perforation's outlets (IF122z, IF211w, IF222y), sometimes enriched by simple decorations of the surface (IF121j, IF228jy). Traces of recesses around suspension holes are also recognisable in biconical beads with non-parallel bases (IF164w, IF264w, IF268y). In fact many of the types seem to exhibit a certain degree of morphological irregularity, e.g. there are flattened cylindrical beads with skewed-flat bases (IF118) or flattened biconical beads with skewed-skewed bases (IF267). Nine types manifest angular cross-sections, ranging from quadrangular to polygonal, and different shapes of sides and bases (IF541–IF868). It is difficult to say whether this formal heterogeneity was intentional or resulted from shortcomings of amber-processing skills at this early stage. A cautionary remark is that each of those types has been registered only in single contexts (and in a small number of artefacts); therefore, sureness of distinguishing above-discussed group is limited from the statistical standpoint.

Two types can be ascribed to a lengthy period spanning EBA, MBA and LBA: IF162 and IF611. First of them can be compared to a form typically described in archaeological literature as lenticular. The chronological attribution mentioned above confirms earlier observation by the author that lenticular beads in Western and Central Balkans (type 16 by Palavestra) have not been registered after LBA (Cwaliński 2020, 141, Fig. 4). Type IF611, so cuboid beads with rectangular cross-section, also seem to fall within the same time frames.

The largest category of IF types is the one covering MBA to LBA. Among them are further variants of flattened spheroidal and discoidal beads with expanded perforation's outlets (IF122w, IF122y, IF221y, IF222w). Numerous are discoidal beads, both with circular and oval cross-sections, with less regularly formed sides or bases, in some cases also fitted with recesses (IF123, IF124, IF124y, IF127y, IF128y, IF147, IF148, IF223, IF226, IF227, IF228j, IF228w, IF241, IF248w). Notable is the appearance of conical beads with truncated or convex top-bases (IF151, IF154), although less regular varieties were also present (IF158, IF254, IF258w). While the occurrence of lenticular beads (IF162) has already been noted in EBA, during MBA and LBA formal complexity of this type grew with addition of recesses (IF162w, IF162x, IF162y). Furthermore, lenticular beads with oval cross-sections are also to be placed around the same time (IF262w, IF262y, IF262z). Other types with biconical sides include IF161w, IF164, IF261y, IF267y and IF268. Introduction of so many forms of amber beads with biconical sides during MBA–LBA corresponds with their widespread use in Central Europe beginning from the *Tumulus* period (after ca. 1550 BC).[499] Finally, an abundant subgroup of artefacts with non-round cross-sections is to be mentioned here. These are finds i.a. with triangular or trapezoidal cross-sections with rectilinear sides and bases (IF411, IF711), as well as more rounded profiles (IF424, IF428, IF448, IF747). Some amber beads exhibit quadrangular or rectangular cross-sections with rounded sides and flat-convex or flat-skewed bases (IF524, IF528, IF624). Still, the largest among them is the subgroup of finds with polygonal cross-sections. Some of those beads, by their rounded edges and/or rectilinear bases, reflect a more careful processing (IF821, IF821y, IF822w), while other represent visually more roughly made examples (IF823, IF824, IF828, IF834, IF841, IF848). So, again, they seemingly suggest a different level of technical prowess among amber craftsmen: next to relatively regular artefacts (those that can be described as possessing a concrete form) are finds which do not easily lend themselves to such formal classification.

---

499 Cf. Sprincz/Beck 1981, 482–483, Fig. 5:4–15, 22–42, Tab. 2; Czebreszuk 2011, 51, Fig. 13:1–3; Chvojka 2017, Figs. 2:6, 3:13–14 and 4:3–4; Woltermann 2016a, 152–183.

Some of the types can be attributed exclusively to LBA. As in the case of EBA–MBA group, these types are known only from a single context each. Quite prominent in LBA group are the beads with conical profile, with either circular or oval cross-sections, and varying bases (IF153, IF154y, IF251, 253, IF258).

Then, there is a small group pertaining to LBA–FBA, comprising seven types in total: IF127, IF228y, IF264, IF514, IF554, IF818w and IF862. Its size results from the scarcity of contexts and finds which can be placed at the turn of LBA and FBA.

The most striking change in the IF beads' repertoire is visible in the case of FBA. Even though this period covers a considerably large assemblage of finds (118 artefacts), the number of types they represent drops in comparison with the preceding MBA and LBA stages. The lower variety of types potentially means a growing standardisation in production of amber adornments that took place over the Bronze Age. Formally, the flattened beads attributable to FBA replicate shapes already known from the previous phases. However, there is one significant addition – the variant 'a' of ornament that is a single horizontal rib/ledge, otherwise known as the Tiryns-type decoration (IF111a, IF111aw, IF211a, IF216a, IF218a, IF231a). At the same time there is a noticeable decrease in the number of types that have recesses around the perforations. The latter observation may reflect a change in the perforation-drilling techniques of amber beads which no longer left distinct traces. Among the newly introduced forms are IF231, IF548 and IF846. From this list can be inferred that the beads with angular cross-sections largely ceased to be have been made. This might be interpreted as a sign of general improvement in the amber-processing skills of the craftsmen.

Fig. 42. IF types ascribed to stages EBA–FBA supplied with frequencies of occurrence in consecutive phases.

	EBA (2150–1650 BC)	EBA/MBA	MBA (1650–1300 BC)	MBA/LBA	LBA (1300–1150 BC)	LBA/FBA	FBA (1150–950 BC)	FBA/EIA
IF111	0	0	7	10	2	1	12	0
IF211	0	0	0	8	0	2	1	0
IF811	0	0	1	4	0	1	0	0
IF121	12	3	12	26	5	1	20	0
IF121w	0	0	1	1	0	0	1	0
IF121y	0	0	5	4	2	0	1	0
IF221	0	0	9	23	2	8	0	0
IF221w	0	0	1	0	0	0	1	0
IF122	5	6	15	16	7	2	0	1
IF222	0	0	7	37	1	4	1	0
IF224	0	0	0	4	2	2	0	0
IF128	1	1	7	7	13	1	0	0
IF228	0	0	1	11	6	3	2	0
IF248	0	0	4	1	1	4	0	0
IF161	0	1	8	19	13	1	31	1
IF167	0	0	1	1	1	2	0	0
IF168	0	0	1	0	11	0	1	0
IF162	1	0	11	6	1	0	0	0
IF261	0	0	2	2	0	0	1	0
IF262	0	0	2	1	0	2	1	0
IF844	0	0	0	1	0	1	0	0

Still, there is one more important group of types to be discussed. It includes forms with the longest duration, spanning almost the entire Bronze Age (EBA–FBA). In the first place, one should mention types IF111 and IF211 – commonly known as flattened cylindrical – introduced sometime during MBA. Same group also includes a variety of flattened cylindrical beads with polygonal cross-section (IF811). Then there is type IF121, together with varieties with slanting and convex recesses (IF121w and IF121y, respectively), which can be defined as a standard discoidal bead. It continuously held a considerable representation over the Bronze Age (Fig. 42). Third, there are forms akin to flattened spheroidal beads (IF122) or discoidal beads with skewed-flat bases (IF128). Above-listed spheroidal/discoidal beads were registered already in EBA contexts and stayed in use until the end of the Bronze Age, forming quantitatively the largest categories of finds classified to subgroup IF (Fig. 42). Slightly younger, because appearing for the first time in MBA, are the variants of the discoidal/spheroidal beads with oval cross-sections (IF221 and IF222) (Fig. 41 and 42). Next to them are types IF224, IF228 and IF248, signifying similar flattened discoidal beads, but with convex-flat or skewed-flat bases. Finally, the flattened biconical beads (IF161, IF167, IF168) should be listed here, with their most typical example (IF161) appearing around the turn of EBA and MBA and growing in numbers towards the end of the Bronze Age (Fig. 41 and 42). Also, the lenticular beads with oval cross-sections (IF262), in contrast to IF162, seem to have a longer chronology, lasting up to FBA. Abovementioned observations prompt a conclusion that the general models of IF beads were already invented during EBA or MBA at the latest, and throughout the Bronze Age were constantly modified by adding or changing some features (e.g. decorations) which resulted in types – in practice subtypes – with varying lifespans. The latter can be treated as indicators of the levels of their popularity among the circum-Adriatic Bronze Age communities.

Fig. 43. Frequency seriation of the IF types belonging to EBA–FBA segment.

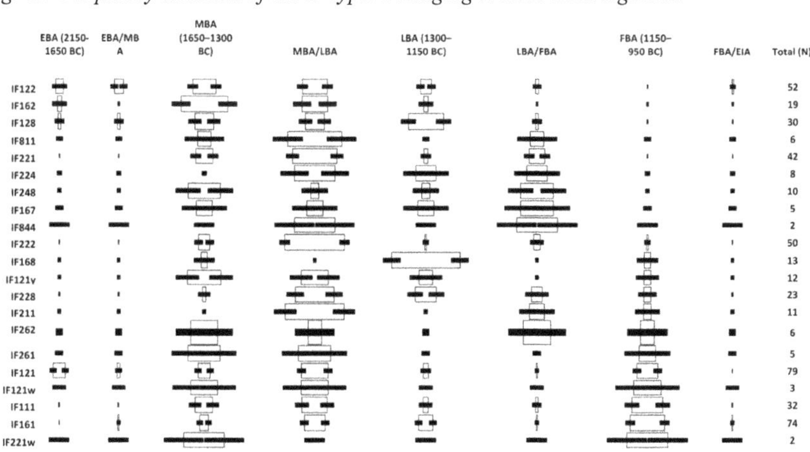

Quantities per phase of the above-discussed IF types were used in frequency seriation to assess if their numerical fluctuation reveals any particular ordering across time, so in other words if some types were more typical for one phase of the Bronze Age than the others. The analysis was carried out in Frequency Seriation Tool 3.0 macro for MS Excel developed by T. Hunt and C. P. Lipo with the idea behind it explained first in Lipo et al. (1997) and at the latest in Lipo et al. (2015). While in the original application by Lipo et al. frequency seriation was used to arrange assemblages of archaeological finds in the most probable sequence from the cultural, temporal and geographical viewpoints, here the assemblages have already been divided into groups of known chronology. Furthermore, rather than being carefully sampled according to a single method (as advised by the authors), objects of seriation – types of amber beads – were simply collected under the assumption that they form an exhaustive and representative record of prehistoric reality. The results of frequency seriation for the IF finds belonging to EBA–FBA segment are depicted in the chart with length of horizontal white bars equivalent to the relative proportion of the type in a chronological stage, and the smaller black bars reflecting statistical uncertainty of the proportions calculated using 0.95 confidence interval (Fig. 43). Upon sorting the types according to EBA (descending order) and FBA (ascending order), the bars illustrate the low level of chronological ordering of the used data. Only few types seem to be tightly linked to particular phases, judging by their quantities. These are, e.g. IF122 and IF128 having steady representation from EBA to LBA, but afterwards declining in numbers, IF162 typical for MBA and LBA after which it disappeared, IF211 and IF222 predominantly present in MBA/LBA and IF168 most characteristic for LBA. Also IF161, and to a lesser extent IF111, shows a gradual increase in numbers from MBA to FBA, reaching their respective maxima in the latter stage. For the remaining types, statistical uncertainty, resulting from their low total quantities, stops from making more far-reaching conclusions.

3.3.1.2. Spatial distribution

The next step of the analysis was to determine spatial distribution of the types belonging to subgroup IF within the circum-Adriatic zone. A contingency table was again used to perform the task, but this time the contexts were grouped into regions, as described in sections 2.2 and 2.3. Aggregated incidences of artefacts within individual regions were normalised to balance the differences between the least and the most frequent types, by changing them to percentages of the total sums of each type. Relationships between the types and regions were depicted in a graph, using Gephi 0.9.5 graph visualisation software and Yifan Hu's layout (2005) (Fig. 44). The graph is built as a two-mode network with both the regions as well as the IF types serving as nodes (visualised as circles). The former were divided into five macro-regions, in line with description in section 2.5: Northern Italy (dark green), Central Italy (turquoise), Southern Italy (red), Western Balkans (light blue) and Central Balkans (olive). The types were categorised in correspondence with chronological groups distinguished in the previous section: EBA–MBA (orange), EBA–LBA (light green), MBA–LBA (yellow), LBA (pink), EBA–FBA (brown), LBA–FBA (purple) and FBA (dark blue). Percentages of types' occurrences in the regions are expressed through varying thickness of the links connecting both categories of nodes.

On the graph the types having distribution confined to single regions tend to be placed at the outer rim, while those pertaining to more than one region are located in the inner part. Types represented in two or three regions adjacent on the graph are placed in-between the nodes. It is easily visible that types exclusive for single regions are bounded with the thickest links. Interestingly, all of the latter have chronology limited to one or two phases at most; thus, it can be said that not only have they shorter duration but also confined distribution. On the other hand, types in the centre of the graph, i.e. those with the most widespread occurrence, are mainly belonging to the groups with longer time spans: EBA–LBA and EBA–FBA, although there are also some types from MBA–LBA and few from LBA–FBA. Accordingly, they are connected with thinner links, depending on the percentages in individual regions.

By comparing the regions and the macro-regions, one can spot morphological differences which characterise assemblages of IF amber beads in various parts of the circum-Adriatic zone. The richest and formally most varied are collections from Northern Italian regions, in particular western Veneto, the Garda Lake basin and Emilia. Apart from types with a wide spatial occurrence, e.g. IF121, IF161 or IF222, they include a whole series of more limited forms. Among the latter many MBA–LBA types seem to be shared either by western Veneto and the Garda Lake basin, or western Veneto and Emilia. Vast majority of them have circular or oval cross-sections with rectilinear, convex, pitched or double-pitched sides and various bases. They are often characterised by recesses around perforations' outlets. There is particular preponderance of types with hourglass or semi-hourglass perforations in Northern Italy, matched only by northern Apulia and Basilicata, which potentially suggests use of specific techniques of drilling. Shared similarities in the forms of amber artefacts allow to perceive the abovementioned three areas as a sub-network – well-connected triad of nodes. Western Veneto, the Garda Lake basin and Emilia share many similarities with Trentino-Alto Adige, yet they are based on the co-occurrence of rather common and long-lasting forms (coloured brown). On the other hand, there are noticeably fewer commonalities between Liguria, Piedmont or Romagna and Emilian Apennines and remaining parts of Northern Italy; however, this results from the generally small assemblages of amber finds from those regions.

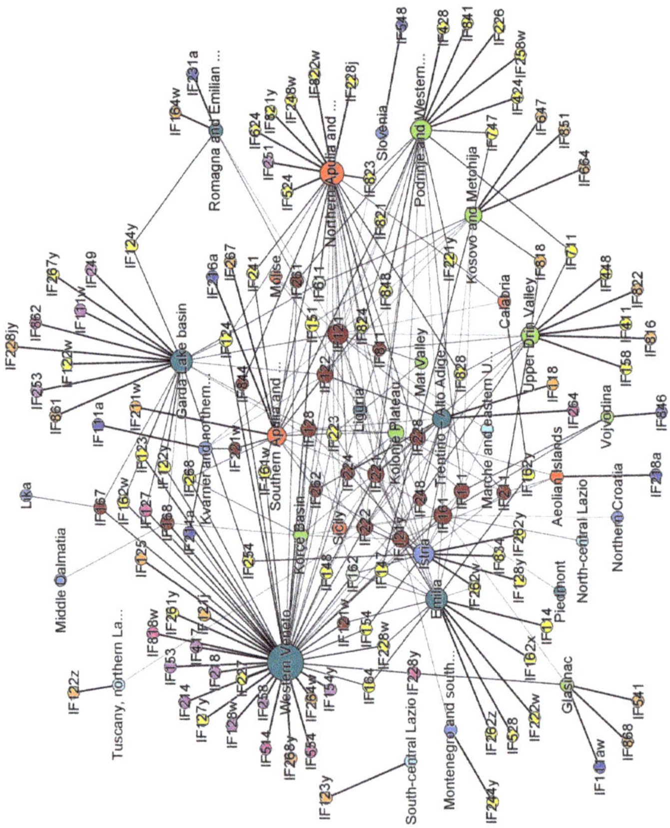

Fig. 44. Graph presenting two-mode network of spatial relationships between regions and IF types of amber artefacts. Thickness of ties expresses percentage of types within individual regions. Size of the nodes expresses degree centrality. Regions are grouped into larger geographical units (micro-regions): Northern Italy (dark green), Central Italy (turquoise), Southern Italy (red), Western Balkans (light blue) and Central Balkans (olive). Types are chronologically classified: EBA–MBA (orange), EBA–LBA (light green), MBA–LBA (yellow), LBA (pink), EBA–FBA (brown), LBA–FBA (purple) and FBA (dark blue).

There is a distinct similarity in the composition of IF forms in Northern Italy and Western Balkan provinces what holds them close to each other on the graph. Istria and then Kvarner and northern Dalmatia are the most interconnected, whereas remaining Western Balkan regions are linked to the rest of the network by only one type each. Istrian assemblage of IF beads is particularly close to collections from Northern Italy. Kvarner and northern Dalmatia itself is represented by only five types: apart from already mentioned IF211a, there are IF111a, IF221w and more common IF121 and IF222, the former two characteristic only for FBA. Other regions included in Western Balkan macro-region have even fewer IF types: Slovenia and Lika two each, and Northern Croatia, middle Dalmatia and Montenegro one each. Again, these are mostly popular forms, except for Lika which shares types IF167 only with the Garda Lake basin and western Veneto. In contrast, Istria has typologically much more varied set of types. Four types are exclusively present there, among them IF262y. Istrian assemblage, furthermore, includes IF154 shared only with western Veneto, IF254 elsewhere appearing only in the Garda Lake area and IF262w present also in Emilia. Furthermore, IF162 and IF262 have occurrence limited to Istria, Northern and Southern Italy, so there clearly is a pattern in distribution of beads with flattened lenticular profile.

Southern Italy is also internally varied when it comes to typology of IF amber beads. Apulia and Basilicata with its northern and southern parts yielded a large group of finds which do not appear elsewhere. There are beads with polygonal, quadrangular or rectangular cross-sections, and convex or straight-convex sides and bases (IF524, IF624, IF821y, IF822w). Present are also beads with oval cross-section and conical (IF251) or rectilinear, convex or straight-convex side, among them one directly linking those two neighbouring provinces (IF241). There are also types quite regular in appearance which reveal similarities between Apulia and Basilicata, and Northern Italy (IF124, IF224, IF161w), or those more angular in form (polygonal) which were found predominantly in Central Balkans (IF811, IF821, IF823, IF848). In contrast, remaining parts of Southern Italy do not indicate such geographically direct connections. Molise, due to small total number of amber artefacts, delivered only two, rather common types: IF121 and IF151. Sicily and the Aeolian Islands by large reveal rather unremarkable set of IF types, with exception of IF162 in the former and IF218a in the latter case. Calabria manifests a certain adherence to Northern Italy and Central Balkans through types IF162y and IF221y. In the case of Southern Italy, notable is low representation of classic flattened biconical beads (IF161 and derivatives). On the other hand, presence of decorated FBA types IF216a and IF218a deserves a mention.

Central Balkans, particularly its inner-most parts, Glasinac, Podrinje and Western Serbia, Kosovo and Metohija and Upper Drin Valley, yielded a rich variety of flattened beads with non-round cross-sections. Without going into details, one can clearly see a number of types with triangular, quadrangular, rectangular, trapezoidal and polygonal cross-sections which rarely have regular parallel or rounded surfaces. These types are almost entirely dated to EBA–MBA and MBA–LBA. This 'angularity' of amber artefacts can be considered as a characteristic trait of Central

Balkans during the Bronze Age, which could have equally resulted from intentional processing of the beads dictated by local tastes, as well as the level of craftsmanship that did not allow for production of more regular forms. This tradition seems to have continued also in FBA, as indicated by type IF846 from Vojvodina. Further to the south, in the Albanian provinces of Mat Valley, Korçë Basin and Kolonjë Plateau, these 'cuboid' forms are replaced by more common flattened discoidal and spheroidal beads as revealed by positions within the graph of their respective nodes. Nonetheless, there is no shortage of other characteristic forms of amber adornments in Central Balkans. Among them one can mention cylindrical (IF111) or biconical (IF161) beads.

Finally, it is necessary to characterise Central Italy. It is striking how few types from IF subgroup are represented in the constituent parts of this macro-region. Most of them are resembling classic flattened cylindrical (IF111), discoidal (IF121), spheroidal (IF222) and biconical (IF161) forms. Still, there are some peculiarities; type IF122z has been registered only in Tuscany. South-central Lazio is detached from the rest of the network. The region yielded a type of flattened amber beads which has not been found in other provinces (IF222y), even though they are not all that different from the artefacts from Northern as well as Southern Italy.

One of the more general observations is that the regions differ in terms of the number of types ascribed to them. Those differences can be illustrated better with a centrality measure called *degree* – one of the analytical measures used commonly in NA, calculating number of ties connected to a node, which in this case is expressed by sizes of the nodes (Fig. 44). It is possible to see that in every macro-region, with the exception of Central Italy which itself is poorly connected with the rest of the network, there is one region with greatest *degree* centrality. In case of Northern Italy it is western Veneto, in Southern Italy it is northern Apulia and Basilicata, in Western Balkans it is Istria, while in Central Balkans the position is taken by Podrinje and Western Serbia. Other regions are marked by gradually smaller nodes. Therefore, one can divide the nodes in relation to occurrences of IF beads, into those with the primary, secondary, tertiary rank, etc. Similar approach can be used in reference to nodes symbolising the types. Among the latter, the largest *degree* centrality characterises IF121, while slightly lower level is manifested by IF161. Other types with considerable number of ties are IF111, IF122, IF121y, IF128, IF221, IF223, IF222, IF228 and IF151. All of them represent classic flattened discoidal, spheroidal, cylindrical, conical and biconical beads with circular or oval cross-sections and rectilinear or convex, rarely convex-concave or skewed-flat bases. Thus, it can be posited that these IF forms were most frequently channelled into interregional circulation, perhaps because they were the most desired amber adornments in that subgroup. If that is true, then the types with distribution confined to single regions (and usually low in numbers) were presumably perceived as 'imperfect', burdened with a defect stemming from faulty craftsmanship, thus were destined only for the local use. Nevertheless, this should be treated as one of the possible explanations, as less regular beads, albeit in small number and spatial scope, were also streamed into circuit, albeit to a lesser extent.

3.3.2. Average

In comparison with the flattened (IF) subgroup, the artefacts perforated concentrically along H in average variant are significantly fewer, numbering 93 objects. A lion's share is composed by beads with round – circular or oval – cross-sections (84). Majority of finds included in this group replicate popular forms described in the previous section, albeit in a more shapely – symmetrical manner. The first example is IA111 – colloquially described as 'average cylindrical', represented by four examples (Fig. 45:1).[500] Furthermore, there are two such beads decorated with single horizontal ledge – IA111a, while one of them has also concave edges of the hole's outlet – IA111ax (Fig. 45:2).[501] Average cylindrical beads occurred more frequently in the variant with oval cross-section, either non-decorated – IA211 (2 beads), as well as with single – IA211a (2), or double horizontal ribs – IA211b (3) (Fig. 45:3–4).[502] Moreover, there are four cases of beads with oval cross-sections, rectilinear sides, skewed-flat bases and single decorative horizontal ribs – IA218a (Fig. 45:5). Same ornament appears on four out of five beads with oval cross-sections, concave sides and rectilinear bases – IA231 and IA231a (Fig. 45:6–7). Another cylindrical form among the analysed set on which the Tiryns-type ornament (motif 'a') occurs in one of the three cases has triangular cross-section combined with rectilinear sides and bases – IA411 and IA411a (Fig. 45:8–9).

Artefacts with convex sides and rectilinear bases are fewer, than in subgroup IF. Altogether seven beads pertaining to IA121 (including one with convex edges of the hole's outlet – IA121y) and four to IA221 were registered (Fig. 45:10–12).[503] More numerous are spheroidal beads, i.e. those with convex sides and bases; type IA122 numbers 16 specimens, while IA222 numbers 18 in total (Fig. 45:13–14).[504] There are also few beads with less regular form, visible mainly through concave or skewed bases, e.g. IA125 (2 beads), IA128 (2) and IA226w (2) (Fig. 45:15–16).

Among subgroup IA, a number of finds exhibit biconical profile. The most typical form with circular cross-section and rectilinear bases – IA161 – is represented by six beads (Fig. 45:17). Variety with oval cross-section – IA261 – appeared only once (Fig. 45:18). Some of the finds, while having biconical sides and circular cross-section, manifest skewed-skewed or skewed-flat bases – IA168, IA267 and IA268 (Fig. 45:19–21). There are also two cases of beads with biconical profiles, decorated with single horizontal rib: IA261a and IA268a (Fig. 45:22). The ultimate bead from IA subgroup to possess the Tiryns-type decoration is IA451a (Fig. 45:23). The number of beads

---

500 Average or short cylindrical beads were classified to type VIIA by Sprincz/Beck (1981, Fig. 6, Tab. 1), type 1a by Du Gardin (1986, 547, Fig. 1:2), type 18b by Palavestra (1993, Tab. II), type 3, var. A by Negroni Catacchio et al. (2006, 1446, Fig. 2) and type 7a by Woltermann (2016a, 42, Abb. 18).

501 N. Negroni Catacchio classifies similar 'Tiryns type beads' into variant A (1999, 243, Fig. 2).

502 Equivalents of IA211b were classified by A. Palavestra as type 3 (1993, Tab. I).

503 Those beads can be correlated with type VIIB by Sprincz/Beck (1981, Fig. 6, Tab. 1), type 5a by Palavestra (1993, Tab. I) and 7b by Woltermann (2016a, 42, Abb. 18).

504 Direct counterparts to average spheroidal beads are type VI by Sprincz/Beck (1981, Fig. 6, Tab. 1), type 1c by Du Gardin (1986, 547, Fig. 1:14), type 5b by Palavestra (1993, Tab. I), type 2 by Negroni Catacchio et al. (2006, 1446, Fig. 2) and type 6 by Woltermann (2016a, 41, Abb. 18).

Fig. 45. Examples of the types of amber finds belonging to average category of group I: 1 – Molina di Ledro, precise context unknown (Rageth 1974, Taf. 92:2); 2 – Vrlazje, tumulus XI, grave 1 (Cwaliński/Pravidur forthcoming, Pl. 1:6); 3 – Monte Battaglia, hoard (Miari 2007a, Fig. 3:24); 4 – Baška, 'reconstructed grave' (Blečić Kavur 2014, Fig. 20:10); 5 – Torre Castelluccia, ripostiglio *(Harding/Hughes-Brock 1974, Fig. 6:26); 6 – Ardea – Campo del Fico, pit-grave (Delpino 1987, Fig. 18:14); 7 – Costa Cavallina, cave burial (Poggiani Keller 1989, Fig. 98:1); 8 – Bismantova – Campo Pianelli, grave XXXI (Catarsi/Dall'Aglio 1978, Fig. XXI:1); 9 – Kompolje, grave 293 (1903–1905) (Bakarić 2017, 26, cat. no. 9); 10 – Sorbara, tomb 54 (Baioni 2000, Fig. 10:13); 11 – Molina di Ledro, strata I–III/IV (Rageth 1974, Taf. 92:8); 12 – Franzine Nuove, grave 417 (Aspes 1976, Fig. 1:6); 13 – Molina di Ledro, strata I–III/IV (Rageth 1974, Taf. 92:18); 14 – Narde I, grave 169 (Salzani 1990–1991b, Fig. 37:6); 15 – Franzine Nuove, grave 417 (Aspes 1976, Fig. 1:7); 16 – Madonna di Loreto, 'Hypogeum of Bronzes' (Tunzi Sisto 1999a, Tav. X:1391); 17 – Iglarevo, grave 1 (Palavestra 1997, Pl. 1:12); 18 – Scalvinetto, tomb 282 (Salzani 2004, Fig. 3:B); 19 – Peschiera – Imboccatura del Mincio, precise context unknown (Negroni Catacchio 1971, Fig. 5); 20 – Frattesina di Fratta Polesine, hoard 1* Tesoretto *(Negroni Catacchio 1972, Fig. 3); 21 – Scalvinetto, tomb 282 (Salzani 2004, Fig. 3:B); 22–23 – Frattesina di Fratta Polesine, hoard 1* Tesoretto *(Negroni Catacchio 1972, Fig. 20–21); 24 – Scalvinetto, tomb 354 (Salzani 2004, Fig. 4:G).*

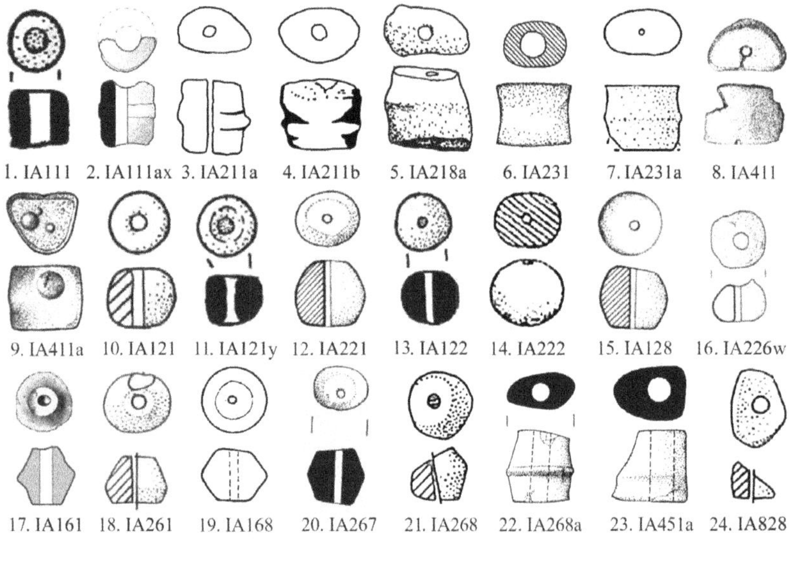

decorated with motif 'a', amounting to 16, is slightly larger than in IF subgroup (12). Finally, a handful of beads can be ascribed to types with angular cross-sections. Types IA418 (1), IA448 (1), IA511 (1), IA811 (1) and IA828 (1) are represented by a single bead each (Fig. 45:24).

3.3.2.1. Chronology

From diachronic perspective, there is a noticeable gradual increase in quantities of contexts and finds belonging to IA subgroup from EBA to FBA (Fig. 46). Between EBA and MBA, the numbers are on the same level, amounting to four contexts and four finds, plus two and six, respectively, from EBA/MBA. With respect to MBA/LBA and LBA, there is a significant raise in both the contexts and the finds, while there are again few in the intermediate stage LBA/FBA. Finally, in FBA both the contexts and the finds reach the highest levels of 12 and 41, respectively. Therefore, it is a different trend to the one observed in the case of IF amber artefacts.

Procedure applied to the study of IA artefacts from chronological perspective is essentially the same as in the case of previously discussed IF subgroup: the types and contexts were first compiled in the form of contingency table which then served as the basis for seriation. Again, the constrained mode was chosen, imposing ordering of the columns according to the contexts' chronology. As the types and contexts are fewer in this case, the resulting diagram is easier to read (Fig. 47). The range plot is more regularly diagonal, suggesting a linear evolution of amber beads in average size. Most of the types seem to have their periods of occurrence limited to single phases; however, there is also a small group of them spanning more than one stage.

Fig. 46. Temporal fluctuations in numbers of contexts and finds classified to subgroup IA during the Bronze Age.

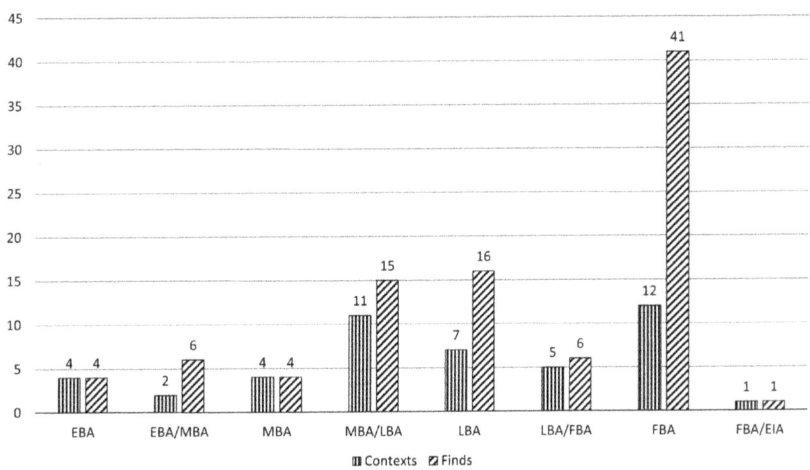

The oldest types within IA subgroup are IA511 and IA125, both known exclusively from the Castelluccio cemetery on Sicily dated to the initial part of EBA (ca. 2200–1800/1700 BC). In a more advanced part of the period (ca. 1800/1750–1650 BC), two sites from the Garda Lake basin: Lavagnone, phase 4 and Sorbara di Asola, tomb 54 produced first examples of the most long-lasting types in the entire IA subgroup: IA122 and IA121. Of these two types, IA122 – classic spheroidal bead – is the most common form in the subgroup, appearing at various sites from all periods. Similar is the case of IA111 introduced sometime in EBA/MBA. There are other types of average beads which were worn during MBA–LBA: IA128, IA221, IA226w, IA418, IA448 and IA828. However, apart from IA122 and IA121, they all stopped being used by the end of LBA.

During MBA–LBA average biconical beads (IA161, IA168, IA261, IA261a) found a large popularity in the circum-Adriatic zone – an observation that confirms the trend in use of the flattened biconical beads in the study area. Among them, the most interesting is IA261a, represented by a single bead from Peschiera – Imboccatura del Mincio, inserted on a bronze pin. This potentially could be the earliest case of application of 'a' decorative motif; however, the find comes from an unprecise context and can be dated only broadly to the general lifespan of the Peschiera *palafitte* settlements attributed to LBA (cf. De Marinis 2006, 451, Tav. 5:6). It is worth mentioning that an object of type IA122 from Marashina, also in the Garda Lake shoreline, was too found as an adornment on a pin (cf. De Marinis 2006, Tav. 5:5). Even if IA261a from Peschiera is not the 'earliest' amber bead in the average variant decorated with the Tiryns-type motif, there are several other similarly ornamented artefacts with LBA–FBA dating: IA111a from Omo Masso, IA211a from Monte Battaglia hoard or IA231a from Costa Cavallina cave burial and Ponte San Pietro Valle, tomb 1. Still, in case of all above-listed amber artefacts, the contexts were found disturbed or lacking precise time indicators among the finds (see sections 2.2.3, 2.2.8 and 2.2.9). Thus, their attribution to LBA cannot be sufficiently proven, nor it can be rejected outright.

Fig. 47. Range plot resulting from seriation of types and contexts classified to subgroup IA. (Figure in full resolution can be downloaded from https://www.peterlang.com/app/uploads/2023/09/Fig.47-scaled.jpg).

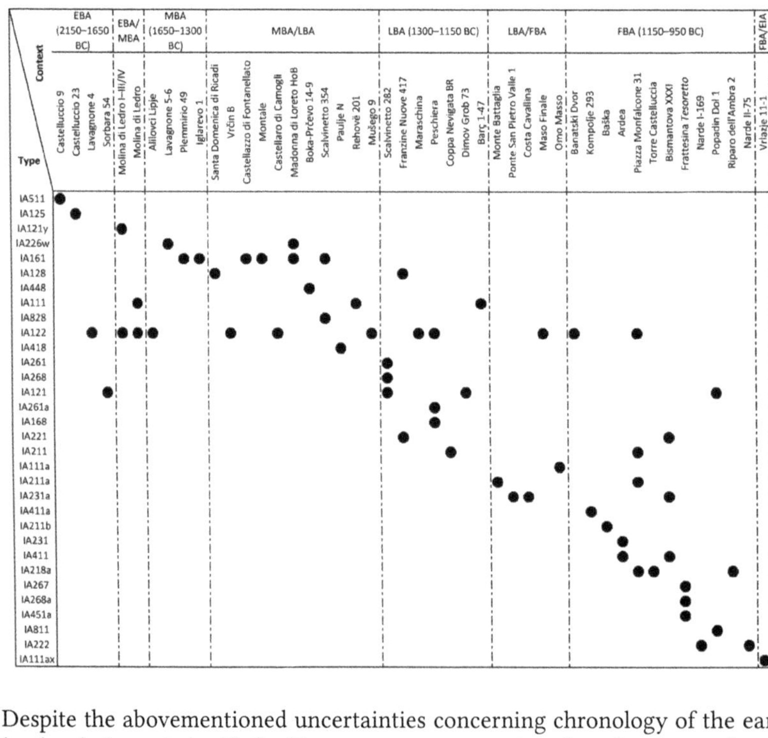

Despite the abovementioned uncertainties concerning chronology of the earliest amber beads decorated with the Tiryns-type ornament, it is clear that decorative motif 'a' among IA subgroup became common during FBA. To other forms exhibiting variant 'a' of decorations in this period – IA211a, IA218a, IA231a, IA268a, IA411a and IA451a can be added three beads from the 'reconstructed grave' of Baška documenting variant 'b' – IA211b. This peculiar and rare decorative motif reveals a level of stylistic affiliation with the Tiryns-type ornament, and can be dated to roughly the same period. There is also a single example of IA111ax dated to FBA–EIA from Glasinac cemetery of Vrlazje. As mentioned above, ornamented beads in subgroup IA number in total 17 specimens, in comparison to 12 in subgroup IF. However, in the former they compose much larger percentage of LBA–FBA and FBA beads, than in the latter (33 % and 9 %, respectively). In IA subgroup ornamented beads were used during FBA alongside few undecorated types: IA211, IA222, IA231, IA267, IA411 and IA811.

3.3.2.2. Spatial distribution

Spatial distribution of IA objects was, again, depicted in the form of a graph to facilitate the analysis (Fig. 48). It is the same two-mode network type of graph, constructed with the Yifan Hu's layout as in the section 3.3.1.2 (Fig. 44). Coloured nodes stand for regions (grouped into macro-regions), while white nodes with black outline stand for types (this time they were not divided into phases, as their chronology is clearly explained in Fig. 47). Thickness of the ties signifies percentage of artefacts representing a type in a region, whereas size of the nodes signifies *degree* centrality.

It is easily discernible that the graph, next to the main network in the centre, contains several isolated 'branches'. The latter are formed mostly by types limited in their occurrence to single regions and, logically, regions with representation of IA artefacts narrowed to one type, together forming dyads (associations of two nodes).

Fig. 48. Graph presenting two-mode network of spatial relationships between regions and IA types of amber artefacts. Thickness of ties expresses percentage of types within individual regions. Size of the nodes expresses degree *centrality. Regions are grouped into larger geographical units (micro-regions): Northern Italy (dark green), Central Italy (turquoise), Southern Italy (red), Western Balkans (blue) and Central Balkans (olive).*

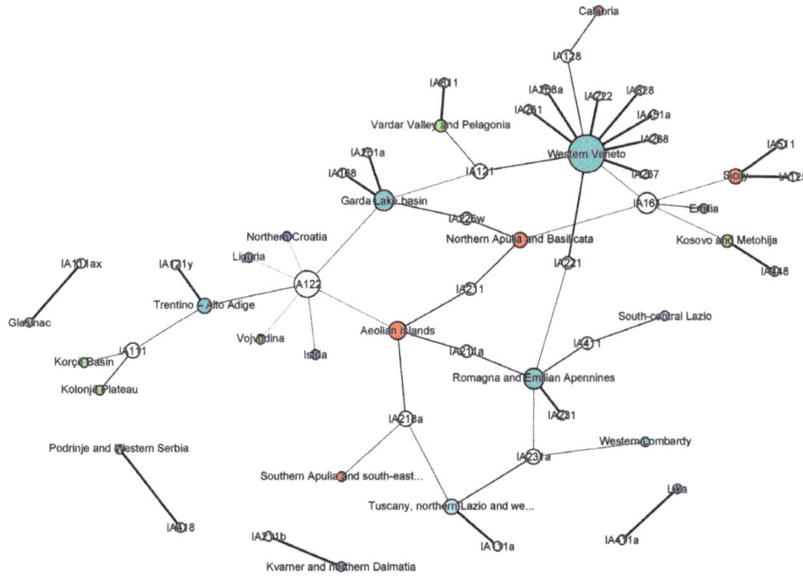

There are four dyads in the graph: IA111ax – Glasinac, IA211b – Kvarner and northern Dalmatia, IA411a – Lika, and IA418 – Podrinje and Western Serbia. All of the abovementioned types are rather specific forms with narrow lifespans, but most of them are characterised by presence of decoration with carved horizontal ribs, known from LBA/FBA–FBA/EIA. There are many other decorated types in the graph, but most of them can be described as 'pendants' – nodes with a single tie. Only three types exhibiting decorative motif 'a' among IA subgroup have interregional distribution (IA211a, IA218a and IA231a). Its structure is quite interesting: IA218a appears to be present only in selected parts of Southern and Central Italy (Aeolian Islands, Southern Apulia and Tuscany), whereas IA231a is represented exclusively in Tuscany and two provinces from Northern Italy (Romagna and Emilian Apennines and western Lombardy). Only IA211a link Southern and Northern Italy as the Tiryns-decorated type of bead in the average variant.

Foregoing observations show how difficult it is to trace a single, universal 'Tiryns-type bead'. Morphology of those beads differs, and the only common element between them is the horizontal rib, either single (var. 'a') or multiplied (var. 'b'). Thus, it seems justified to speak of regional variants of applying discussed decorations, perhaps using other forms which were known and manufactured in different parts of the circum-Adriatic zone. As a potential example can be mentioned presence of single horizontal ribs on beads with biconical profiles (IA268a) in western Veneto – region known for several biconical types (IA161, IA168, IA261) and LBA–FBA amber workshops. Similar beads appeared in the nearby Garda Lake basin (IA261a), either as consequence of local processing based on flattened biconical beads or through exchange with the neighbours; however in the latter case, disparities in chronology should be noted. With respect to other regions, it can be reminded that similarly decorated beads, albeit in the flattened variant, also occurred there during FBA: Southern Apulia – IF216a, the Aeolian Islands – IF218a, Kvarner and Northern Dalmatia – IF111a and IF211a, Romagna and Emilian Apennines – IF211a, Glasinac – IF111aw.

Apart from miscellaneous types which have distributions clearly confined to single regions, often converging with similar forms but in flattened variants, e.g. the Aeolian Islands – IA211 and IF211 or Trentino-Alto Adige – IA121y and IF121y, there are few undecorated types with interregional range of distribution. They are characterised by the largest *degree* centrality: IA122 and IA161 in the first place, followed by types such as IA111, IA121 and IA221, just as in the case of their counterparts in flattened form.

3.3.3. Elongated

The number of concentrically perforated beads in elongated version (subgroup IE) are as many as 198 artefacts. As in the previous two subgroups, most of them have circular or oval cross-sections. Examples which can be described as elongated cylindrical with circular cross-sections (IE111) amount to 5.[505] Four of them are decorated: 2 with single horizontal rib (IE111a), 1 with multiple and evenly spaced transverse grooves (IE111c) and 1 with sparse and evenly spaced transverse grooves (IE111e) (Fig. 49:1–2).

505 Elongated cylindrical beads were classified to type VIIC by Sprincz/Beck (1981, Fig. 6, Tab. 1), type 18a by Palavestra (1993, Tab. II), type 3, var. B by Negroni Catacchio et al. (2006, 1446, Fig. 2), type 107 by Czebreszuk (2011, 79), type C2 by Ernée (2012, Obr. 7), and type 7c by Woltermann (2016a, 42, Abb. 18).

More numerous are elongated cylindrical beads with oval cross-sections (30 objects); all of them exhibit various modes of decoration. A total of seven beads have single transverse horizontal ribs carved in the middle of their profiles (IE211a) (Fig. 49:3).[506] Another eight are ornamented with multiple and evenly spaced transverse grooves (IE211c) (Fig. 49:4).[507] Further 10 beads bear sparse and evenly spaced transverse grooves (IE211e) (Fig. 49:5),[508] while a single specimen is covered with sparse and unevenly spaced transverse grooves (IE211f) (Fig. 49:6).[509] One object shows oblique, regular grooves (IE211g) (Fig. 49:7). Three more beads can be added in the context of above-listed ornamented objects, relatively similar but with flat-skewed bases – IE218a (Fig. 49:8). It is worth stressing that decorations pertaining to Allumiere type (var. 'c', 'd', 'e', 'f', 'g', 'h' and 'i') appear exclusively on elongated concentrically perforated beads (IE), while they are completely absent in subgroups IF and IA.

Same types of decorations appear on amber artefacts with rectilinear sides, but angular cross-sections. Aside from a single non-decorated bead with triangular cross-section (IE411), there is also one adorned with sparse and unevenly spaced transverse grooves (IE411f) (Fig. 49:9).[510] Among beads with trapezoidal cross-sections, there is one which exhibits rectilinear sides and convex-flat bases (IE714) (Fig. 49:10). Beads with polygonal cross-sections include one, however, non-decorated bead with rectangular/rectilinear profile (IE811) (Fig. 49:11). As in the previous cases, also here there are two cases with irregular/non-parallel bases (IE418) (Fig. 49:12).

The next characteristic family of finds within subgroup IE is formed by objects with barrel-shaped or, in other words, fusiform profiles.[511] They are characterised by convex sides, rectilinear bases and circular or oval cross-sections: IE121 and IE221 (Fig. 49:13–14). The variant with circular cross-section in the circum-Adriatic zone has been registered in smaller quantity (7 cases in total), than the latter one (30 objects). Furthermore, IE221 has a number of decorated varieties, including specimens with horizontal rib and slanting edges of the hole's outlet (IE221aw; 1); multiple and evenly spaced transverse grooves (IE221c; 10); oblique, regular grooves (IE221g; 1); oblique, regular grooves bounded by transverse grooves (IE221h; 1); and oblique, regular grooves, alternating with transverse grooves (IE221i; 1) (Fig. 49:15–19). Interestingly,

506 All elongated beads with single horizontal rib (Tiryns-type decoration) were included in type 1A by Palavestra (1993, Tab. I), type 4, var. A–B by Negroni Cataccio et al. (2006, 1446, Fig. 2) and type 113 by Czebreszuk (2011, 88).

507 Elongated beads with transverse grooves (Allumiere-type decoration) were included in type 4B by Palavestra (1993, Tab. I), type 5, var. A–B by Negroni Catacchio et al. (2006, 1446, Fig. 2) and type 109 by Czebreszuk (2011, 88).

508 This variant was classified as type 4A by Palavestra (1993, Tab. I).

509 Elongated beads decorated with sparse and unevenly spaced transverse grooves were classified by Palavestra as type 2 (1993, Tab. I).

510 Elongated beads with triangular cross-sections were classified to type 34 by Palavestra (1993, Tab. II).

511 Beads with barrel-shaped/fusiform profiles were classified as type VIIC by Sprincz/Beck (1981, Fig. 6, Tab. 1), type 1c by Du Gardin (1986, 547, Fig. 1:15), type 6c by Palavestra (1993, Tab. I), type 9 by Negroni Catacchio et al. (2006, 1449, Fig. 3), type 111 by Czebreszuk (2011, 88), type C1 by Ernée (2012, Obr. 7) and type 8 by Woltermann (2016a, 42, Abb. 18).

the IE221aw bead is the only artefact in the entire IE subgroup with widened perforation's outlet. In the context of elongated beads with convex sides, one can also list the single specimen with oval cross-section and skewed-flat bases (IE228) (Fig. 49:20). At this point it can be concluded that decorations more frequently appear on artefacts with oval cross-sections (58 cases), than the remaining variants of elongated beads (21 cases in total). Perhaps, the fact has to do with the technique of applying decorations through carving or incisions on lateral surfaces of amber lumps which often left them round but uneven in cross-section.

Fig. 49. Examples of the types of amber finds belonging to elongated category of group I: 1 – Ponte San Marco, BR–BF horizon (Ruggiero 1994, Fig. 89); 2 – Narde II, grave 75 (Colonna et al. 2010, Tav. 13:B10), 3 – Fondo Paviani, advanced BR–early BF layers (Fasani/Salzani 1976, Fig. 2:3); 4 – Ardea – Campo del Fico, pit-grave (Delpino 1987, Fig. 18:8); 5 – Allumiere – Poggio La Pozza, grave 2, trench F (Peroni 1960, Fig. 11:18); 6 – Baška, 'reconstructed grave'(Lo Schiavo 1970, Fig. XXIII:13a); 7 – Isolino Virginia, central part, superficial layers (De Marinis 2009c, Fig. 5:3); 8 – Frattesina di Fratta Polesine, hoard 1 Tesoretto *(Negroni Catacchio 1972, Fig. 17); 9 – Kompolje, grave 293 (1903–1905) (Bakarić 2017, 26, cat. No. 9); 10 – Allumiere – Poggio La Pozza, grave 4, trench D (Peroni 1960, Fig. 11:15); 11 – Torre Castelluccia,* ripostiglio *(Harding/Hughes-Brock 1974, Fig. 6:30); 12 – Allumiere – Poggio La Pozza, grave 1, trench F (Peroni 1960, Fig. 11:6); 13 – Narde II, grave 75 (Colonna et al. 2010, Tav. 13:B11); 14 – Castello del Tartaro, tomb 166 (Salzani 2011, Fig. 3:B5); 15 – Capitanata, precise context unknown (Negroni Catacchio 1999, Fig. 4:32); 16 – Narde II, grave 75 (Colonna et al. 2010, Tav. 13:B4); 17–19 – Desmontà, grave 437 (Salzani 2013, Tav. XXXII:3, 10, 35); 20 – Franzine Nuove, grave 7 (Aspes/Fasani 1968, Fig. 8:1); 21 – Castello del Tartaro, tomb 166 (Salzani 2011, Fig. 3:B5); 22 – Allumiere – Poggio La Pozza, grave 2, trench F (Peroni 1960, Fig. 11:17); 23 – Allumiere – Poggio La Pozza, grave 1, trench F (Peroni 1960, Fig. 11:7); 24 – Desmontà, grave 437 (Salzani 2013, Tav. XXXII:1).*

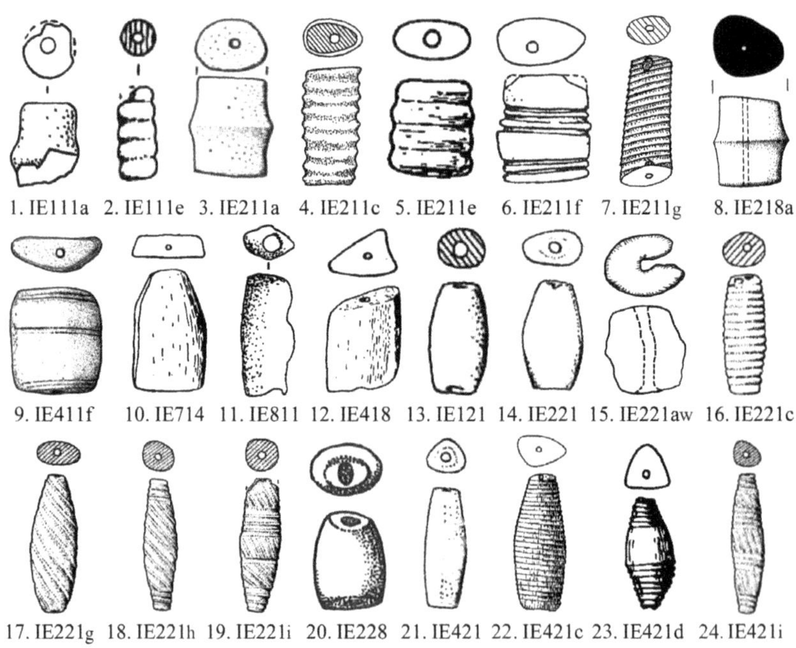

Barrel-shaped profiles and many of the abovementioned decorative motifs, in contrast to flattened and average beads, are recurring in objects with non-round cross-sections in IE subgroup. It likely shows that cross-section probably had a secondary significance in processing of amber, and (a) was determined by the form or raw amber lumps and/or (b) was a side-product of mechanical shaping of the surfaces. In fact some of the finds included in this category manifest certain degree of coarseness, as if they were unfinished. To this group of beads can be assigned those with triangular cross-sections, convex sides and rectilinear bases (9 in total). Among them five are undecorated (IE421), two are covered with multiple and evenly spaced transverse grooves (IE421c), one is covered with multiple and unevenly spaced transverse grooves (IE421d) and another one with oblique, regular grooves, alternating with the transverse ones (IE421i) (Fig. 49:21–24). Motif of regular grooves appears also on one artefact with triangular cross-section, straight-convex sides and roughly parallel bases (IE441c) (Fig. 50:1). Other two elongated beads with triangular cross-sections have flat-skewed bases, but no decoration (IE428) (Fig. 50:2). There is also a bead with more slender fusiform profile and approximately square cross-section (IE521) (Fig. 50:3).[512] Finally, there are three artefacts which overall depart from a classic barrel-shaped form due to polygonal cross-sections, but still retain convex sides, although differ in terms of bases: the first one has convex upper and lower surfaces (IE822), the second one convex-flat (IE824) (Fig. 50:4)[513] and the third one skewed-flat (IE828).

A small, but distinct, set is formed by elongated beads with concave sides, adorned with decorative motif 'a' – single transverse rib.[514] All five beads reflecting this form have oval cross-sections and rectilinear bases (IE231a) (Fig. 50:5).

A total of 29 artefacts belonging to subgroup IE exhibits straight-convex sides in profile. They are quite diverse with respect to cross-sections. Only one is circular from this perspective (IE141) (Fig. 50:6). Five beads reflect oval cross-sections; among them four have rectilinear bases (IE241), including one decorated with multiple and evenly spaced transverse grooves (IE241c), while a single object has skewed-flat bases (IE248) (Fig. 50:7). Out of three elongated beads with semi-circular cross-sections, two possess straight-convex sides: one of them has rectilinear bases (IE341), while the other one flat-convex (IE344) (Fig. 50:8). Furthermore, there are two objects merging triangular cross-sections with straight-convex sides and rectilinear bases (IE441), one of which is decorated (Fig. 50:1). In two artefacts straight-convex sides appear in connection with trapezoidal cross-sections and skewed-flat bases (IE748), in one of the cases additionally decorated with single transverse rib (IE748a) (Fig. 50:9–10). Finally, straight-convex sides were noted in as many as 17 objects with polygonal cross-section. Most of them (16 objects) have rectilinear bases (IE841) (Fig. 50:11), while only a single bead differs from the others by having skewed-flat bases (IE848).

512 The bead was classified by Negroni Catacchio et al. as type 12, var. B (2006, 1456–1457, Fig. 3).

513 The latter bead was classified by Negroni Catacchio et al. as type 13 (2006, 1457, Fig. 3).

514 Elongated Tiryns-type beads with convex sides were classified as type 1A by Palavestra (1993, Tab. I) and type 4, var. B by Negroni Catacchio et al. (2006, Fig. 2).

Fig. 50. Examples of the types of amber finds belonging to elongated category of group I: 1 – Allumiere – Poggio La Pozza, grave 2, trench F (1960) (Peroni 1960: 352; Fig. 11:17); 2 – Franzine Nuove, grave 7 (Aspes/Fasani 1968, Fig. 8:2); 3 – Allumiere – Poggio La Pozza, grave 2, trench F (Peroni 1960, Fig. 11:19); 4 – Allumiere – Poggio La Pozza, grave 4, trench D (Peroni 1960, Fig. 11:15); 5 – Privlaka, grave 87 (Batović 1983, T. XLIV:20); 6 – Castello del Tartaro, tomb 146 (Salzani 2011, Fig. 4:A3); 7 – Torre Castelluccia, ripostiglio *(Gorgoglione et al. 1993, Tav. LV:8); 8 – Castellaro di Camogli, precise context unknown (Fossati/Milanese 1982, Tav. V:90); 9 – Castello del Tartaro, tomb 146 (Salzani 2011, Fig. 4:A3); 10 – Kompolje, grave 293 (1903–1905) (Bakarić 2017, 26, cat. No. 9); 11 – Majdan, hoard (Rašajski 1988, Fig. 60); 12 – Scarceta, sector D –* casa-laboratorio *no. 137 (Poggiani Keller 1999, Fig. 104:3); 13 – Frattesina di Fratta Polesine, hoard 1* Tesoretto *(Negroni Catacchio 1972, Fig. 14); 14 – Narde II, grave 75 (Colonna et al. 2010, Tav. 13:B6); 15 – Campestrin di Grignano Polesine, late BR–initial BF layers (Bellintani et al. 2021, CGP 41); 16 – Campestrin di Grignano Polesine, late BR–initial BF layers (Bellintani et al. 2015a, Fig. 2A:1); 17 – Bismantova – Campo Pianelli, grave XXXI (Catarsi/Dall'Aglio 1978, Fig. XXI:1); 18 – Perlat, tumulus 2, grave 3 (Kurti 2013, Tab. I:9); 19 – Campestrin di Grignano Polesine, late BR–initial BF layers (Bellintani et al. 2021, CGP 18); 20 – Castello del Tartaro, tomb 300 (Salzani 2011, Fig. 5:A7); 21 – Gualdo Tadino, hoard (Peroni 1963b, 5:49); 22 – Vidasi, precise context unknown (Glogović 1991, T. 1:2); 23 – Narde I, grave 281 (Salzani 1990–1991b, Fig. 14:14); 24 – Frattesina di Fratta Polesine, hoard 1* Tesoretto *(Negroni Catacchio 1972, Fig. 4); 25 – Borgo Panigale, 1957 campaign, precise context unknown (Catarsi Dall'Aglio 1976, Fig. 6:16).*

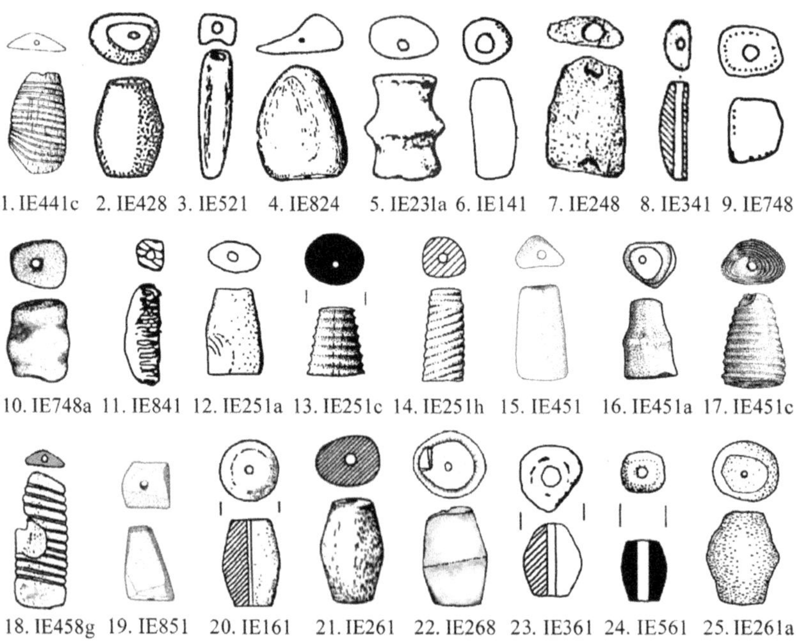

Some of the beads may appear as crudely made, with some surfaces lacking in regularity. A general impression is that the beads with straight-convex sides, in contrast to other elongated forms, are considerably varied in terms of morphology what makes it difficult to compare them to a single model, as in the case of cylindrical or barrel-shaped forms. Attention is drawn especially by varying cross-sections and occasional presence of decorations. Perhaps, the cases listed in this paragraph were originally intended to resemble other more common forms, such as the abovementioned ones, but eventually ended as rather untypical adornments.

A total of 20 objects can be described as having elongated conical profile. Seven of them actually resemble truncated cone in form, due to oval cross-sections, rectilinear or slightly skewed bases, and sides converging towards the upper base. Moreover, all of them are decorated on the surface with various motifs. Consequently, one bead represents IE251a, three IE251c and one IE251h (Fig. 50:12–14).[515] Remaining objects ascribed to 'elongated conical' category have more 'pyramidal' appearance. Eight beads have triangular cross-sections, conical sides and rectilinear bases, and are additionally decorated: one is not decorated (IE451), two have single transverse ribs (IE451a) and five are covered with multiple and evenly spaced transverse grooves (IE451c) (Fig. 50:15–17). Other two beads have more convex-rounded bases (IE452), while three objects, aside from skewed-flat bases, are decorated with straight (IE451c; 2) or oblique, regular grooves (IE458g; 1) (Fig. 50:18). Finally, conical profile was recorded in one bead with rectangular cross-section and convex-flat bases (IE654), and another with polygonal cross-section and flat bases (IE851); however, in both cases the beads seem roughly hewn (Fig. 50:19).

The last category of artefacts within subgroup IE is distinguished on the basis of biconical profile.[516] This feature is reflected in 37 objects. Among them, 14 have circular cross-sections and rectilinear bases (IE161) (Fig. 50:20). Other 16 beads have oval cross-sections and rectilinear bases (IE261) (Fig. 50:21). Two beads with oval cross-sections, in connection with biconical sides, exhibit skewed-flat bases (IE268) (Fig. 50:22). Finally, there are two more artefacts with biconical profiles which do not possess round cross-sections. One of them is semi-circular (IE361), while another one is quadrangular (IE561) (Fig. 50:23–34).[517] Additionally, three similar beads are decorated with single transverse rib (IE261a) (Fig. 50:25).

3.3.3.1. Chronology

Out of all size variants of concentrically perforated amber beads, those with elongated form show the most distinct chronological trend. In terms of numbers, after a

---

515 Elongated beads with conical profiles and oval cross-sections were classified as type 32 by Palavestra (1993, Tab. I). The same author, while not applying it to any of the Balkan Allumiere type beads, distinguished type 27 for elongated conical beads decorated with transverse grooves (Palavestra 1993, Tab. II). Negroni Catacchio et al. classify undecorated conical beads to type 6, whereas those ornamented with transverse grooves to type 7 (Negroni Catacchio et al. 2006, Fig. 2).

516 Elongated biconical beads were classified to type IXE by Sprincz/Beck (1981, Fig. 6, Tab. 1), type 1d by Du Gardin (1986, 547, Fig. 1:18), type 13 by Palavestra (1993, Tab. I), type 8, var. B by Negroni Catacchio et al. (2006, 1446, Fig. 2) and type 9e by Woltermann (2016a, 43, Abb. 18).

517 The latter two beads were classified by Negroni Catacchio et al. to type 11, var. A (2006, 1456, Fig. 3).

period spanning EBA–MBA when elongated beads were almost non-existent, there is clearly observable increase of both the representative contexts and the finds from MBA/LBA to FBA (Fig. 51). For the first three phases (EBA, EBA/MBA and MBA), one can mention only singular IE artefacts and contexts which yielded them. However, beginning from the turn of MBA and LBA, these numbers show a steady growth (remembering that context and finds attributed to MBA/LBA and LBA/FBA stages are in fact chronologically overlapping both periods each of them contains in their respective names). While already LBA shows a significant representation of elongated beads (11 contexts and 50 finds), a majority of IE objects fall into FBA (28 contexts and 112 finds). Thus, in terms of chronology IE beads show similar trend to IA subgroup, however with larger quantitative disproportions between the phases. Consequently, it can be concluded that concentrically perforated beads in elongated variant started to be 'mass-produced' not before LBA, and reached the greatest popularity during FBA, when they surpassed the quantity of flattened beads (respectively 112 to 89 finds).

This important observation potentially reveals technological advancement in amber processing in the study area: as mentioned earlier, perforating is the most difficult part of making a bead; therefore, mastering of drilling techniques enabled larger pieces of amber to be turned into adornments. Acquisition of proficiency in drilling surely required time and practice, thus could explain why elongated beads appeared in the circum-Adriatic zone gradually towards the end of the Bronze Age.

Fig. 51. Temporal fluctuations in numbers of contexts and finds classified to IE subgroup during the Bronze Age.

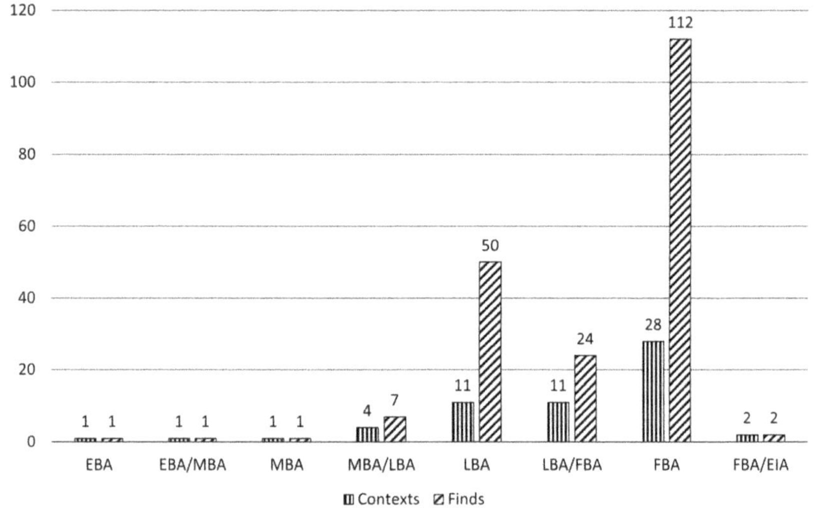

Seriation of the types belonging to IE subgroup, performed in the same way as before, revealed well-serried range plot, indicating gradual evolution of elongated beads in terms of morphology and ornamentation (Fig. 52). Incidences of types in contexts align along the diagonal, from the oldest (Castelluccio, grave 9) to the youngest site (Monte Saraceno, grave 86/IV). Most of the types have occurrences limited to one, thus are chronologically easy to determine. However, some types comprising more numerous examples are spanning several phases, suggesting prolonged use. The latter are placed in the centre of the graph, where the range plot widens.

Proceeding with the description from the top-left corner of the graph (Fig. 49), it is clear to see that EBA, EBA/MBA and MBA are all characterised by a single type each. The oldest of them, found in tomb 9 of the Sicilian cemetery of Castelluccio, is IE441 – a rather crude-looking artefact (Orsi 1892, Tab. III:15). It is followed by more regular, barrel-shaped types IE221 and IE121, dated to EBA/MBA and MBA, respectively. Interestingly, they were also found on Sicily, what possibly shows that this region developed a skill of manufacturing or importing elongated beads quite early, although it is hard to say whether these adornments were common at the time, taking into account their low number. Nonetheless, these types became the two longest lasting forms of elongated beads, continuously manufactured up to FBA. It is worth mentioning that flattened and average variants of these forms (IF121, IA121, IF221, IA221) were also among the first to be introduced to the repertoire of amber adornments. This proves a certain concurrence in applying features or general designs to raw amber lumps of varying sizes.

Fig. 52. Range plot resulting from seriation of types and contexts classified to subgroup IE. (Figure in full resolution can be downloaded from https://www.peterlang.com/app/uploads/2023/09/Fig.52-scaled.jpg).

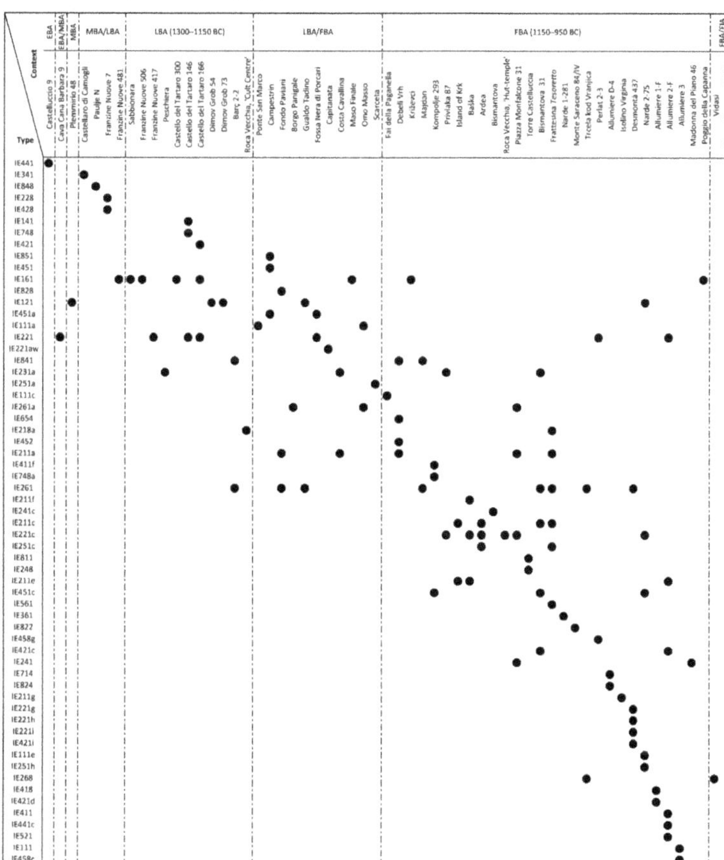

During the period from MBA to LBA emerged a variety of new forms of elongated beads. Types such as IE161, IE228, IE341, IE428 or IE848 appeared roughly at the same time as their flattened or average counterparts; however, only the first of those proved to be a more durable design. Variety of elongated beads grew further in LBA, with introduction of types such as IE141, IE261, IE421, IE748 and IE841. Most of them, with the exception of IE261 and IE841, have limited occurrence and chronology, showing that they were not commonly used. A notable novelty during LBA are decorated beads of types IE218a and IE231a – the first elongated forms decorated with a single horizontal rib. Possibly the earliest representative of this type comes from Peschiera – Imboccatura del Mincio (De Marinis 2007, Fig. 5); however, one has to remember about the problems with chronology of this context (already described in section 3.3.2.1). Apart from the carved rib, a distinct feature of the bead are concave sides; perhaps technique of applying decorative motif 'a' at the beginning involved slimming the lateral surfaces around the middle and leaving just a ledge, instead of removing excess body from the solid. Examples of the type are known also from LBA/FBA and FBA, proving that this design was quite durable. Definitely safer is the dating of IE218 bead from the so-called 'Cult Center' of Roca Vecchia (Sector IX, US 11349, Phase II) to BR 2 (1220–1130 BC) supported with radiocarbon dates as well as co-occurring Aegean pottery (see section 2.2.15).

The IE231a bead from Peschiera and IE218 from Roca Vecchia cannot be much older than other artefacts decorated with horizontal rib, predominantly dated to LBA/FBA. The turn of these two phases saw application of decorative motif 'a' on forms of beads mostly invented earlier: IE211a, IE221aw, IE261a and IE451a. One can add to the list singular cases of IE111a or IE251a. It is yet another argument which shows that the beads known in the older literature as the Tiryns type were not a sudden innovation, but the characteristic decoration of horizontal rib was added gradually to forms of amber beads already known for some time. It seems that with the growing ability to produce longer beads, the desire for further diversification of amber adornments also grew. Wider lateral surfaces allowed for applying more and more sophisticated ornaments which lead to a variety of ornamented beads dated to LBA/FBA.

The last phase of the Bronze Age brought first examples of ornaments based on bundles of parallel grooves covering the sides of beads, more commonly recognised as the Allumiere type. Its basic form can be described as variant 'c' (multiple and evenly spaced transverse grooves) which appears on numerous, yet morphologically varied finds: IE211c, IE221c, IE241c, IE251c, IE421c, IE441c, IE451c and IE458c. Most of them were created before the late FBA (ca. pre-1000 BC). Again, one can see adaptation of the ornament to already existing forms of elongated amber beads. It is difficult to chronologically determine the first occurrence of decorative variants 'e' (sparse and evenly spaced transverse grooves) and 'f' (sparse and unevenly spaced transverse grooves). Most types bearing those ornaments (IE211e, IE211f and IE411f) come from sites with imprecise chronology, spanning the entire FBA. However, singular cases of IE111e and IE211, recovered from contexts with more finely adjusted chronology (Allumiere – Poggio La Pozza, grave 2, trench F and Narde II, grave 75), point to the later stages of FBA (BF 3 – late 11th and 10th century BC). With much more certainty such dating can be ascribed to all specimens adorned with oblique grooves – variants 'g', 'h' and 'i', along with variant 'd', showing that these variations on the Allumiere type entered into the amber craft towards the end of the Bronze Age. Thus, it is possible to discern a certain evolution of decorative patterns throughout LBA and FBA. Apart from these decorated artefacts, there is a series of non-ornamented types of elongated amber beads created in the same period: IE241, IE248, IE268, IE344, IE361, IE411, IE418, IE452, IE521, IE561, IE654, IE714, IE811, IE822 and IE824; however, they are usually represented by fewer examples, than the decorated ones. All these beads were used concurrently to some of the forms invented earlier, such as the common types IE121, IE221, IE161 or IE261 mentioned earlier.

Fig. 53. Division of IE types of amber artefacts into chronological groups of the Bronze Age.

EBA–LBA	EBA/MBA–FBA	LBA/FBA–FBA	FBA–FBA/EIA	
IE141	IE121	IE111a	IE111	IE411
IE228	IE161	IE211a	IE111c	IE411f
IE341	IE218a	IE221aw	IE111e	IE418
IE421	IE221	IE251a	IE211c	IE421c
IE428	IE231a	IE261a	IE211e	IE421d
IE441	IE261	IE451	IE211f	IE421i
IE748	IE841	IE451a	IE211g	IE441c
IE848		IE828	IE221c	IE451c
		IE851	IE221g	IE452
			IE221h	IE458c
			IE221i	IE458g
			IE241	IE521
			IE241c	IE561
			IE248	IE654
			IE251c	IE714
			IE251h	IE748a
			IE268	IE811
			IE344	IE822
			IE361	IE824

Judging by the time span during which the types belonging to IE subgroup were used, it is possible to divide them into four chronological groups (Fig. 53). The first one comprises types invented and used between EBA and LBA. The second consists of more durable designs, first created around the turn of EBA and MBA, and used up to FBA. The third group comprises types which turned up for the first time around LBA/FBA, but remained in use also during FBA (these are types decorated with motif 'a'). The last – fourth group is the largest one, consisting of the types ascribed to FBA–FBA/EIA.

3.3.3.2. Spatial distribution

Determination of distribution patterns among concentrically perforated elongated beads, performed with the same network analysis tools as in sections 3.3.1.2 and 3.3.2.2, produced graph with a complex structure (Fig. 54). Nodes in the graph signify provinces of amber's acquisition and types of IE beads, marked with colours according to their location within the five macro-regions or attribution to one of the four chronological groups (Fig. 53). Size of the nodes expresses *degree* centrality (the number of ties with other nodes), whereas thickness of the ties grows proportionally to the percentage of artefacts in a region representing a type. As in the previous sections, the Yifan Hu's algorithm was used to produce layout of the graph. Placement of the nodes in the graph resembles the one applied in the case of IF subgroup (Fig. 44): types that have occurrence limited to single regions are placed on the outer rim of the graphs, whereas those more widespread were fitted into its inner part.

Fig. 54. Graph presenting two-mode network of spatial relationships between regions and IE types of amber artefacts. Thickness of ties expresses percentage of types within individual regions. Size of the nodes expresses degree *centrality. Regions are grouped into larger geographical units (micro-regions): Northern Italy (dark green), Central Italy (turquoise), Southern Italy (red), Western Balkans (blue) and Central Balkans (olive). Types are chronologically classified: EBA–LBA (violet), EBA/MBA–FBA (orange), LBA/FBA–FBA (yellow), FBA–FBA/EIA (white).*

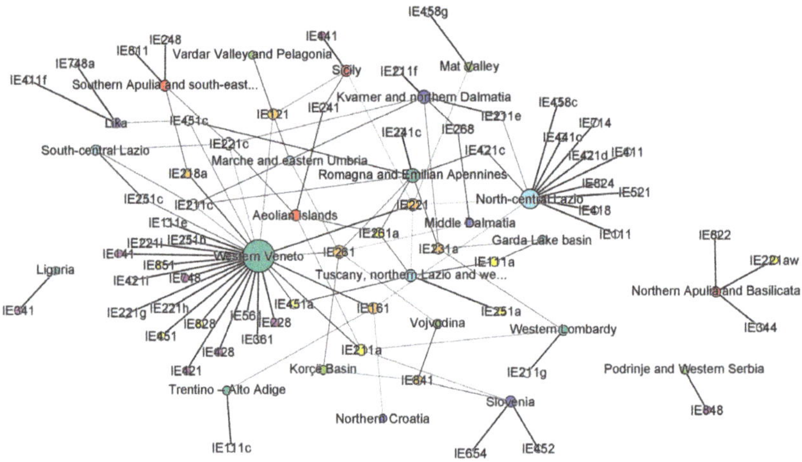

Basing on the level of *degree* centrality, the regions can be ordered into a hierarchy, from the most to the least representative in terms of different types of IE beads. Western Veneto appears to be having the most varied assemblage of discussed artefacts (27 types in total). North-central Lazio takes the second place, however, with only half as much ties linked to the different types (13 in total). Further in that order are Romagna and Emilian Apennines (7), and Kvarner and northern Dalmatia (6). Behind them is Tuscany, northern Lazio and western Umbria, representing five types. Remaining regions have no more than four types of IE beads detected among their amber assemblages. Macro-regionally speaking, Central Balkans reveal the least diverse set of IE beads, with each of its constituent regions linked to no more than two types. Also, among Northern and Central Italian or Western Balkan regions, there are some areas rather poorly representing IE beads as well, e.g. Northern Croatia, middle Dalmatia, Trentino-Alto Adige, the Garda Lake basin, Marche and eastern Umbria or south-central Lazio. Nevertheless, the network of ties seems denser among the regions from the western parts of the circum-Adriatic zone, specifically from the Apennine Peninsula. Furthermore, the latter tend to have more 'exclusive' types, i.e. forms of IE beads that occurred only in a single region. There are also three regions detached from the rest of the network: Liguria, northern Apulia and Basilicata, and Podrinje and Western Serbia. It is interesting to note that, unlike in the case of IF group (Fig. 44), the constituent parts of macro-regions are not interconnected by IE types with the 'macro-regional' scope of distribution. The graph reveals a more interregional specificity of IE beads, linking areas from different macro-regions of the circum-Adriatic zone.

The older types of IE amber beads (group EBA–LBA) are limited in their distribution to western Veneto, Liguria, Sicily, and Podrinje and Western Serbia. Yet, it is easy to notice that none of them is shared by any of the four abovementioned regions; they are ascribed exclusively to a single region and, moreover, represented usually by one or two artefacts. Thus, they can be considered either as early attempts at creating elongated amber beads, undertaken locally, or incidental imports from outside the circum-Adriatic zone.

Amber beads from the following chronological groups definitely have more widespread distribution. Types classified to EBA/MBA–LBA horizon, especially IE261, IE221, IE121 and IE161 (in a decreasing order), manifest the highest *degree* centrality, in this way showing that biconical and fusiform (barrel shaped) elongated beads had the widest appeal. From the other two groups (LBA/FBA–FBA and FBA–FBA/EIA), IE211a and IE221c deserve a mention. Both reveal the highest *degree* centrality in their respective groups. It proves that cylindrical beads with oval cross-sections were the most common ones to be decorated, either with the Tiryns-type (var. 'a') or Allumiere-type (var. 'c') decorations. Slightly more restricted scope of occurrence characterises IE211c, IE261a, IE231a and IE451c, yet they still can be placed in the group of the most widespread decorated amber beads in the study area.

Other variants of decorations within subgroup IE have more regional character: e.g. var. 'f' is spatially limited to Western Balkans, specifically Kvarner and Lika, while var. 'd' occurred exclusively in north-central Lazio. Variant 'e' appears, apart from IE111e found only in western Veneto, on type IE211e shared by north-central Lazio, on one hand, and Kvarner and northern Dalmatia, on the other. There is a number of other decorated types known from only two or three regions: e.g. IE111a, IE451a, IE421c or IE451c. Quite peculiar is the distribution of amber artefacts adorned with decorative motif 'g', known from western Veneto, western Lombardy and the Mat Valley. In some cases it could be suspected that the grooves were meant to be perpendicular (like in var. 'c'), but their oblique course results from a faulty craftsmanship, or adjustment of the ornament to the form of an amber lump (Figs. 49:7, 50:18). However, in other examples they appear well performed, as if they were intentional (Fig. 49:17). In any case, distribution of decorative motif 'g' is largely limited to the northern Adriatic Basin, minus the specimen from Perlat in the Mat Valley.

Important to note is the morphological similarity between variously decorated beads found in many regions, i.e. same forms of beads were possibly produced in series and adorned with different decorative motifs, e.g. IE211a and IE211c. This characteristic is particularly evident in the case of western Veneto, but becomes apparent in other regions with high *degree* centrality as well upon a closer look.

Other types of IE beads, especially the non-decorated ones, often have distribution confined to single regions. Many of them exhibit irregularities in forms, thus could be regarded as products considered not worthy of channelling into wider circulation, and kept for personal use. The exception is IE841 widespread in some of the Balkan regions, i.e. Slovenia, Vojvodina and the Korçë Basin.

3.4. Group II

Group II covers amber artefacts perforated along H, but with the perforation located eccentrically, at an object's edge. Although many finds classified to group I exhibit deviations from the central axis of rotation in location of the hole, when looking at their cross-section, this could be blamed upon faulty craftsmanship (i.e. errors committed during the process of drilling). In those cases, the hole usually is not further away from the centre than one-fourth of the diameter. In some cases, however, the hole is visibly distanced from the centre, and one can hardly think of an error. Instead, it looks like perforation was drilled at the objects' edge on purpose. Such amber finds are usually referred to as 'pendants'.[518] Types of amber adornments discussed here usually replicate forms already discussed in group I, however only in the flattened variant. Group II in total includes 11 amber beads from eight contexts (Tab. 37).

Tab. 37. List of the finds classified to group II

Context	Type	Quantity	Phase	Region
Banjevac, tumulus 2	IIF221	1	MBA/LBA	Podrinje and Western Serbia
Belotić, mound XI, grave 4	IIF222	1	MBA (1650–1300 BC)	Podrinje and Western Serbia
Beneceto, early Recent Bronze Age (BR1) layers (sector central/ part NW – US 674, 679, 680; part SE – US 144, UUSS 3)	IIF261	1	LBA (1300–1150 BC)	Emilia/southern Po plain
Grotta Manaccora, *Grotticella Funeraria* – SS layers	IIF222	1	MBA (1650–1300 BC)	Northern Apulia and Basilicata (Gargano peninsula and Tavoliere plain)
Iglarevo, grave 1	IIF261	4	MBA (1650–1300 BC)	Kosovo and Metohija
Madonna di Loreto, 'Hypogeum of Ivories'	IIF221	1	MBA/LBA	Northern Apulia and Basilicata (Gargano peninsula and Tavoliere plain)
Monte Cavanero, hoard	IIF225	1	FBA/EIA	Piedmont
Paulje, tumulus N	IIF222	1	MBA/LBA	Podrinje and Western Serbia

518 Similar Bronze Age amber finds from various parts of Europe were classified as pendants, e.g. by C. Du Gardin (1986, 549–553, Figs. 3:1–11 and 4:4–5), N. Negroni Catacchio et al. (2006, 1449–1450, Fig. 3), J. Czebreszuk (2011, 94) or G. Woltermann (2016a, 49–51, Abb. 18). Amber pendants are also known in considerable numbers from Neolithic sites in the Baltic coast (see e.g. Mazurowski 1984, 7; Butrimas et al. 2018).

Although sometimes broken or damaged, thus hampering classification, most artefacts in group II exhibit oval cross-sections and convex sides which altogether give them oblong form. The difference between them predominantly lies in the shape of bases. Hence, they could be distinguished oblong pendants with rectilinear (IIF221; 2 beads), convex (IIF222; 3) and concave bases (IIF225; 1) (Fig. 55:1–3). The remaining five pendants exhibit double-pitched sides and rectilinear bases, resulting in biconical profiles: IIF261 (Fig. 55:4). None of the artefacts in group II revealed any traces of decorating the surfaces. Also the perforations lack expansions of the outlets.

Fig. 55. Examples of the types of amber finds belonging to group II: 1 – Banjevac, tumulus 2 (drawing by the author); 2 – Belotić, mound XI, grave 4 (Garašanin/Garašanin 1958, Fig. 5d); 3 – Monte Cavanero, hoard (Rubat Borel 2009b, Fig. 74:1); 4 – Iglarevo, grave 1 (Palavestra 1997, Pl. 1:8).

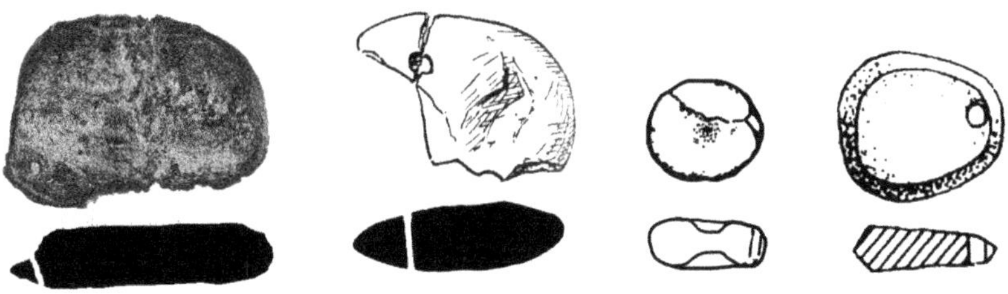

1. IIF221   2. IIF222   3. IIF225   4. IIF261

It is worth pointing out that the artefacts from group II considered commonly as pendants are usually larger than the regular amber beads from flattened category. The difference in size is particularly visible, when an object like that was found in the company of other concentrically perforated amber adornments in a single context, e.g. grave. In those cases it can be suspected that the pendants served as special were components of a necklace or bracelet, and by an altered perforation an object could be exposed in a row of beads. A good example to illustrate such situation is grave 1 from Iglarevo, in which 160 amber adornments of various sizes and forms were discovered (Palavestra 1997). Among them, some of the largest objects are eccentrically perforated (Palavestra 1997, Pl. 1:8–11).

3.4.1. Chronology

From the chronological perspective, types included in group II appear to be almost exclusively dated to MBA and LBA, although in some cases proper assignation of eccentrically perforated objects to either of the phases is impossible at the moment due to imprecise dating of their contexts (Tab. 37). The oldest of them are dated to MBA: finds from Belotić and Grotta Manaccora represent type IIF222, while those from Iglarevo were classified as IIF261. The latter form stayed in use until LBA, as documented by the single find from the BR layers at Beneceto. Two artefacts illustrating type IIF221 can be dated only broadly to MBA/LBA. The youngest in the entire group is the single IIF225 object from Monte Cavanero, which is an isolated case of a pendant in the post-LBA times in the entire circum-Adriatic zone.

3.4.2. Spatial distribution

Spatial distribution of the finds belonging to group II shows highly localised character. In terms of macro-regions, they are predominantly represented in Central Balkans (7 finds), while Northern and Southern Italy constitute smaller parts of the assemblage, both having delivered only two artefacts each. When considering the regions, the most prominent in this regard are Podrinje and Western Serbia and northern Apulia which are linked by the co-occurrence of two types IIF221 and IIF222. On the other hand, Kosovo and Metohija is linked to Emilia by presence in both regions of artefacts reflecting type IIF261. Only IIF225 from Piedmont does not have any analogy in the circum-Adriatic zone. Thus, it seems that the use of pendants was not very common and did not spread across larger areas.

3.5. Group III

Group of amber finds perforated concentrically along D/L is similar in numbers to the one discussed above, encompassing 11 artefacts in total (Tab. 38). They include mostly objects in the flattened size variant and are quite varied morphologically. The most numerous are the artefacts characterised by conical profile. Five of them have flattened forms, and circular (IIIF154; 3 finds), or oval cross-sections (IIIF254; 2) (Fig. 56:1–2). A similar object, however with average dimensions ratio (IIIA254), was registered as well (Fig. 56:3). Conical profile has also been noted among some of the less regular finds. One of the objects is characterised by rectangular cross-section (IIIF654) (Fig. 56:4). There is also one flattened artefact that combines conical sides with trapezoidal cross-section and skewed-flat bases (IIIF758) (Fig. 56:5). Numerical preponderance of conical forms in group III potentially suggest that this kind of adornments were worn in a different manner to majority of amber beads. Perhaps they were hanged in a specific position within necklaces, or were strapped to the clothing similarly to buttons – word often used to describe such amber objects.[519] The last two objects from group III have oblong form, combining oval cross-section with convex side. They differ in terms of bases: one has rectilinear upper and lower surfaces (IIIF221), while the other flat-convex and, additionally, exhibits criss-crossing incisions on the surface (IIIF224j) (Fig. 56:6). Although the latter object's base is broken off, the perforation is still visible upon a closer look as a straight groove.

519 Cf. Negroni Catacchio et al. 2006, 1457–1458; Bellintani 2014, 120; Strafella et al. 2015, 850.

Tab. 38. List of the finds classified to group III

Context	Type	Quantity	Phase	Region
Allumiere – Poggio La Pozza, grave 1, trench F (1960)	IIIF224j	1	FBA (1150–950 BC)	North-central Lazio (southern Etruria)
Allumiere – Poggio La Pozza, grave 4, trench D (1960)	IIIF154	1	FBA (1150–950 BC)	North-central Lazio (southern Etruria)
Debeli Vrh, hoard	IIIF221	1	FBA (1150–950 BC)	Slovenia
Debeli Vrh, hoard	IIIF654	2	FBA (1150–950 BC)	Slovenia
Fondo Paviani, advanced BR–early BF layers	IIIF154	1	LBA/FBA	Western Veneto / northern Po Plain
Maso Finale, cultural layer	IIIF154	1	LBA/FBA	Trentino-Alto Adige
Maso Finale, cultural layer	IIIF254	2	LBA/FBA	Trentino-Alto Adige
Narde II, grave 75	IIIA254	1	FBA (1150–950 BC)	Western Veneto / northern Po Plain
Paulje, tumulus N	IIIF758	1	MBA/LBA	Podrinje and Western Serbia

Fig. 56. Examples of the types of amber finds belonging to group III: 1 – Fondo Paviani, advanced BR–early BF layers (Strafella et al. 2015, Fig. 1:F); 2 – Maso Finale, cultural layer (Putzer 2012, Abb. 10:26); 3 – Narde II, grave 75 (Colonna et al. 2010, Tav. 13:B8); 4 – Debeli Vrh, hoard (drawing by the author); 5 – Paulje, tumulus N (drawing by the author); 6 – Allumiere – Poggio La Pozza, grave 1, trench F (Peroni 1960, Fig. 11:9).

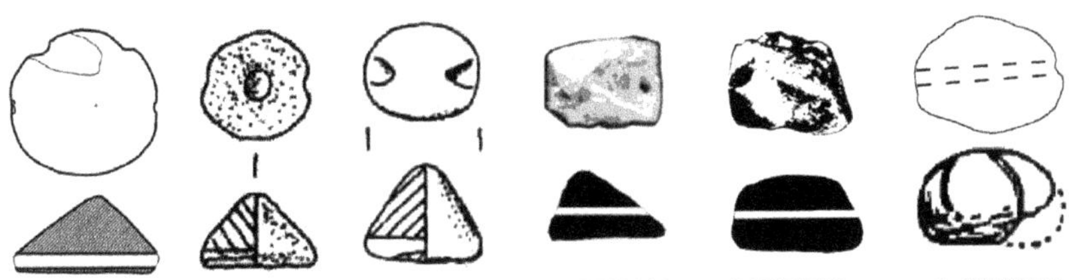

3.5.1. Chronology

In terms of chronology, all artefacts assigned to group III belong to advanced stages of the Bronze Age, with the majority falling into FBA (1150–950 BC). The earliest example of concentrical perforation along D/L is IIIF758 from Paulje, tumulus N, dated to BzC1–BzD (ca. 1450–1200 BC). This first specimen characterised by conical profile, however somewhat irregular, only slightly precedes other similar adornments. Those with more regular conical profiles can be firmly dated to the final stage of the Bronze

Age (12th–9th century BC). Same chronology is shared by the two oblong elements (IIIF221 and IIIF224j). This shows that the manner of perforating amber adornments perpendicularly began to be practised quite late in the Bronze Age.

3.5.2. Spatial distribution

Spatial distribution of the finds belonging to group III also does not seem random. Apart from single specimen pertaining to IIIF154 found in north-central Lazio, majority of the objects with conical profiles (IIIF154, IIIF254, IIIA254) come from Northern Italian provinces of western Veneto and Trentino-Alto Adige. Thus, the three above-listed types can be considered as typical for this macro-region. Types IIIF221 and IIIF654 were discovered not too far away from that area, because in southern Slovenia, but their form slighlty differs from the rest of the group. The easternmost find classified to group – IIIF758 – comes from Western Serbia. By its remote findspot and somewhat early chronology, it seem detached from the rest of the similarly perforated finds.

3.6. Group IV

Group IV – amber artefacts perforated eccentrically along D/L – forms a small set of finds (Tab. 39). Location of the hole for suspending these objects away from their middle sections validates labelling them as pendants, as in the case of group II. Among the total of seven objects classified to group IV, one can find six different types. Two specimens are characterised by circular cross-sections, and convex sides and bases with average size ratio that lend the objects a globular form (IVA122) (Fig. 57:1). The third find in the group features oval cross-section and flattened profile, hence has more of a discoidal form (IVF222) (Fig. 57:2). Finally, there are four elongated artefacts with the form akin to a 'drop-shape'. One of them is characterised by rounded rectangular cross-section, convex sides and flat-convex bases, but is additionally decorated with sparse grooves carved in the upper part, around the U-shaped perforation (IVE624f) (Fig. 57:3). The second elongated object, in turn, has oval cross-section, biconical sides and rectilinear bases (IVE261) (Fig. 57:4).[520] The final two artefacts in the group are only partially preserved, and in fact are missing the perforated section; thus, their classification to group IV is not entirely certain.[521] From available figures, one of them can be described as having oval cross-section, pitched sides and flat-convex bases (IVE254); however, its poor state, with the lower and upper parts damaged, does not allow for classification beyond any doubts (Fig. 57:5). The other artefact combines triangular cross-section with rectilinear sides and bases, while its surface is covered with a dense pattern of criss-crossing grooves (IVE411j) (Fig. 57:6).

---

520 The find representing type IVE261 was classified by N. Negroni Catacchio et al. as a biconical pendant (2006, 1457).

521 The objects were classified as pendants by Negroni Catacchio et al. based on the premise that they most likely were perforated from the side, close to one of the terminals, or had appendices allowing for suspension in the form of lug-like extension with an eyelet (2006, 1449–1450, 1457, Fig. 3).

Tab. 39. List of the finds classified to group IV

Context	Type	Quantity	Phase	Region
Allumiere – Poggio La Pozza, grave 2, trench F (1960)	IVA122	2	FBA (1150–950 BC)	North-central Lazio (southern Etruria)
Desmontà, grave 437	IVF222	1	FBA (1150–950 BC)	Western Veneto / northern Po Plain
Frattesina di Fratta Polesine, hoard 1 *Tesoretto*	IVE254	1	FBA (1150–950 BC)	Western Veneto / northern Po Plain
Frattesina di Fratta Polesine, hoard 1 *Tesoretto*	IVE261	1	FBA (1150–950 BC)	Western Veneto / northern Po Plain
Monte Saraceno, tomb 86/IV	IVE411j	1	FBA/EIA	Northern Apulia and Basilicata (Gargano peninsula and Tavoliere plain)
Narde II, grave 75	IVE624f	1	FBA (1150–950 BC)	Western Veneto / northern Po Plain

Fig. 57. Examples of the types of amber finds belonging to group IV: 1 – Allumiere – Poggio La Pozza, grave 2, trench F (1960) (Peroni 1960: 352; Fig. 11:22); 2 – Desmontà, grave 437 (Salzani 2013, Tav. XXXII:11); 3 – Narde II, grave 75 (Colonna et al. 2010, Tav. 13:B7); 4–5 – Frattesina di Fratta Polesine, hoard 1 Tesoretto *(Negroni Catacchio 1972, Fig. 7–8); 6 – Monte Saraceno, tomb 86/IV (Nava/Preite 1995, Tav. XIV:49).*

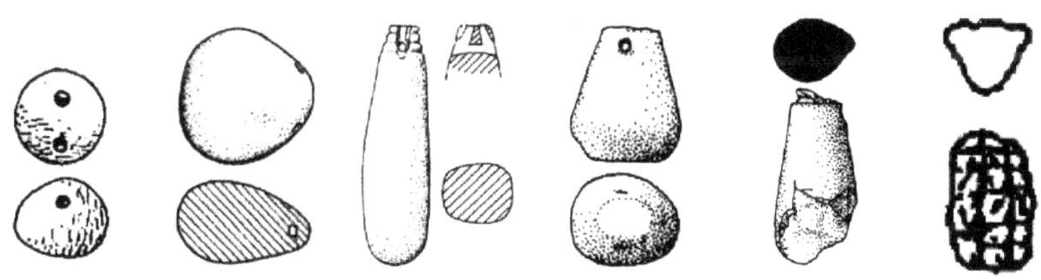

1. IIA122 2. IVF222 3. IVE624f 4. IVE261 5. IVE254 6. IV411j

3.6.1. Chronology

All finds classified to group IV are dated to the final phase of the Bronze Age – FBA (1150–950 BC), with a slightly longer chronology reaching EIA also possible in one case (Tab. 39). Thus, they largely are similar in terms of chronology to the finds belonging to group III (cf. section 3.5.1). From this observation it can be inferred that the idea of perforating amber adornments concentrically or eccentrically along their diameter/length instead the height, although possibly invented slightly earlier, was developed and practised in the circum-Adriatic zone in the final centuries of the 2nd millennium BC.

3.6.2. Spatial distribution

The finds classified to group IV also show similar distribution to those belonging to group III (Tab. 38). Most of them (IVF222, IVE624f, IVE254 and IVE261) were found in contexts located in western Veneto, while two (IVA122) in north-central Lazio. Only single specimen comes from northern Apulia (IVE411j). It seems that the practice of perforating amber adornments concentrically or eccentrically along their diameter/length was used in western Veneto more than in any other province of amber's acquisition. Presumably, this resulted from the creative work of local craftsmen, whose activity in the field of amber processing is confirmed for this region during the FBA (cf. Bellintani et al. 2015a). Nonetheless, the examples from outside Veneto seem quite unique in their own right, especially IVE411j due to the quite specific decorative motif. Thus, one should not overlook the possibility of indigenous manufacturing of adornments classifiable to group IV to the south of Northern Italy, albeit in smaller numbers.

3.7. Group V

Group V includes objects perforated in a very specific manner. In the lateral section, the bore appears as having a 'V'-shape with two inlets located on the same plane.[522] It was achieved by drilling two oblique holes, starting from two apertures placed close to each other which finally converge inside an object (cf. Popkiewicz/Czebreszuk 2016, 90–91). In total nine of the typologically analysed finds possess this kind of perforation (Tab. 40). All of them have either circular or oval cross-sections, and flattened or average size ratio. In terms of general appearance, they can be divided into those with a conical or biconical profile and those with discoidal-spheroidal profile. The first group comprises three types: VF154 (2 artefacts), VF154z (1) and VF264 (1) (Fig. 58:1–3). The second one is more varied and includes two objects with spheroidal form – VA122, and three flattened discoidal finds: VF122 (1), VF221f (1) and VF228z (1) (Fig. 58:4–8). The artefacts sometimes additionally reveal ornamentation of the surfaces: sparse and unevenly spaced transverse grooves (decorative motif 'f'), or widened distal outlets in the form of elongated grooves (recess 'z').

522 Amber adornments with V-shaped perforation from Italy were classified within four variants of buttons by N. Negroni Catacchio et al. (2006, 1457–1458). Similar objects from France were included in type 4 by C. Du Gardin (1986, 553, Fig. 2:4–9), while those from Germany in type 10b by G. Woltermann (2016a, 43–44, Abb. 18). Amber buttons with V-shaped perforations are known from some of the Neolithic culture of southern and eastern Baltic zone, particularly the *Rzucewo* culture (Mazurowski 1984), the *Globular Amphora* culture (Szmyt 1996), the *Bell Beaker* culture (Czebreszuk/Makarowicz 1993), as well as various cultural units from the territory of Lithuania, Latvia and Estonia (Butrimas 2017; Loze 2017).

Tab. 40. List of the finds classified to group V

Context	Type	Quantity	Phase	Region
Cattaragna, precise context unknown	VF264	1	EBA–LBA (2150–1150 BC)	Garda Lake and its inframorenic basin
Fossa Nera di Porcari, precise context unknown	VF154	1	LBA/FBA	Tuscany, northern Lazio and western Umbria (northern Etruria)
Frattesina di Fratta Polesine, hoard 1 *Tesoretto*	VA122	1	FBA (1150–950 BC)	Western Veneto / northern Po Plain
Frattesina di Fratta Polesine, hoard 1 *Tesoretto*	VF221f	1	FBA (1150–950 BC)	Western Veneto / northern Po Plain
Lagazzi di Vho, *Bronzo Medio 1* phase	VF154	1	MBA (1650–1300 BC)	Garda Lake and its inframorenic basin
Narde II, *Ustrinum*, US 1009	VA122	1	FBA (1150–950 BC)	Western Veneto / northern Po Plain
Olmo di Nogara, grave 154	VF228z	1	MBA (1650–1300 BC)	Western Veneto / northern Po Plain
Paulje, tumulus A	VF154z	1	LBA (1300–1150 BC)	Podrinje and Western Serbia
Vrsi, grave 89	VF122	1	FBA (1150–950 BC)	Kvarner and Northern Dalmatia

Fig. 58. Examples of the types of amber finds belonging to group V: 1 – Fossa Nera di Porcari, precise context unknown (Andreotti/Zanini 1995–1996, Fig. 2:12); 2 – Paulje, tumulus A (Bulatović et al. 2017, T. XIV:16); 3 – Cattaragna, precise context unknown (Massari et al. 1996, Fig. B:1); 4 – Frattesina di Fratta Polesine, hoard 1 Tesoretto *(Negroni Catacchio 1972, Fig. 9); 5 – Narde II,* Ustrinum, *US 1009 (Colonna et al. 2010, Tav. 56:1); 6 – Vrsi, grave 89 (Batović 1983, T. XLIV:10); 7 – Frattesina di Fratta Polesine, hoard 1* Tesoretto *(Negroni Catacchio 1972, Fig. 10); 8 – Olmo di Nogara, grave 154 (Salzani 2005, Tav. XVI:E).*

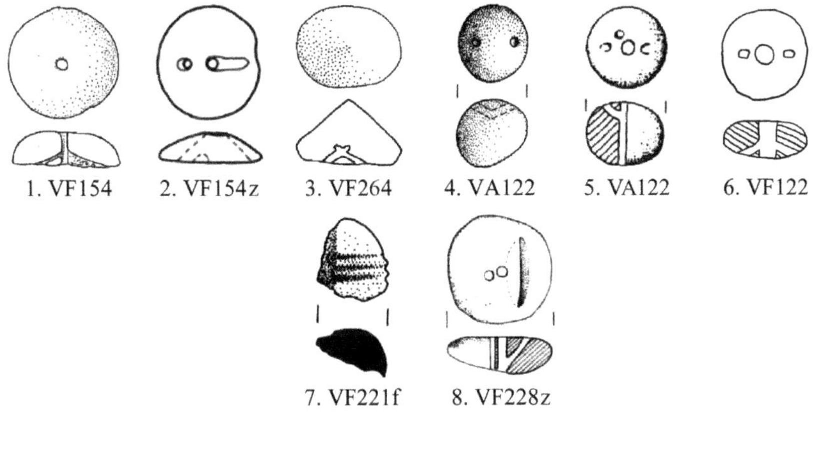

It is worth noting that in some of the abovementioned finds the perforations are more complex than in the others. The typical V-shaped perforations are represented by types VF264 and VA122 (Fig. 58:3–4). According to N. Negroni Catacchio, traces of V-shaped perforation are recognisable in the broken fragment of VF221f (Negroni Catacchio 1972, 15; Fig. 58:7). By its form and presence of decoration, it presents a unique artefact in the entire dataset. However, the poor state of preservation allows only for a piecemeal reconstruction of its appearance. In the remaining five objects classified to group V, the V-shaped perforation is combined with concentric and straight perforations along H, in the case of VF228z actually doubled (Fig. 58:1–2, 5–6, 8). The oblique bores diverge laterally from the straight ones, visually resembling 'Y' letter (or the 'peace symbol'). This feature prompts to include those four objects in a subgroup within group V.

3.7.1. Chronology

It is difficult to determine when the first objects with V-shaped perforation occurred in the circum-Adriatic zone. Some Italian authors were eager to attribute the buttons with V-shaped perforation from Cattaragna (VF264) and Lagazzi di Vho (VF154) to the EBA (BA), on the basis of analogies from the areas further to the north – southern Baltic zone, where such amber adornments have been known since the Neolithic (Bergonzi 1997, 606; Negroni Catacchio et al. 2006, 1457). However, the former artefact comes from unprecise contexts and, since the site yielded materials datable to a long time span (BA–BR), it can be only provisionally placed between phases EBA and LBA (2150–1150 BC) (cf. De Marinis 2006, 448; 2007, 62). Furthermore, the second find should be attributed to the BM 1 phase of the Lagazzi di Vho settlement, not the earlier BA 2 horizon from which originate other amber artefacts from the site (cf. De Marinis 2007, 62; Baioni et al. 2007b, 27). Thus, there is not a single amber find with V-shaped perforation in Italy that could be safely dated to the EBA. On the other hand, most of the finds from group V originate from verified LBA–FBA (1300–950 BC) contexts, therefore contrary to what some authors posit (Bergonzi 1997, 606). This dating seems particularly relevant for the finds with Y-shaped perforation, discussed earlier. Only the object from Olmo di Nogara, grave 154 (VF228z) is earlier, falling into MBA (1650–1300 BC), more precisely BM 3 (1450/1375–1350/1300 BC); however, it differs from the remaining three artefacts by the fact that the second oblique bore is missing and there are two parallel perpendicular-straight holes.

3.7.2. Spatial distribution

With one exception, all of the finds included in group V were discovered in regions of the northern part of the circum-Adriatic zone. Even the southernmost of the Italian finds – the bead from Fossa Nera di Porcari – was recovered near the northern fringes of Tuscany. Most of the V-perforated objects originate from the adjacent provinces of western Veneto and the Garda Lake basin. Two artefacts were found on the opposite coast of the Adriatic – the buttons from Vrsi, grave 89 and Paulje, tumulus A. As opposed to the formal similarity with the Northern Italian counterparts, the latter object is, geographically speaking, quite distanced from the remaining finds. It is worth noting that the decorative motif 'f', within subgroup IE recognised only on the finds from the eastern Adriatic provinces (cf. section 3.3.3.2), appears on two specifically perforated artefacts from western Veneto: IVE624f from Narde II, grave 75 and VF221f recovered from hoard 1 *Tesoretto* of Frattesina di Fratta Polesine. Still, it needs to be stressed that the abovementioned finds by their form do not resemble types IE211f or IE411f (cf. Figs. 49:6, 9; 57:3; 58:7).

3.8. Group VI

Group VI is among the least homogeneous sets of amber adornments. It includes 'hybrid forms', called that way due to presence of secondary or even tertiary perforations (excluding the V-shaped ones discussed in the previous section), in addition to the main one, which means that they must have been worn in a specific manner. However, one cannot rule out that sometimes those secondary perforations result from failed attempts at drilling a bead, subsequent repairs, or modifications, e.g. done to correct positioning of the bead within a necklace. Due to scanty evidence offered by the finds themselves, it is difficult to tell in each case what were the exact reasons for drilling the second or third hole. Still, these objects form a peculiar group which deserves to be considered separately.

In total there are 21 objects assigned to group VI (Tab. 41). Morphological variety and different placement of the perforations make almost every artefact a specific case. Out of the two elongated beads in the group, one is characterised by rounded triangular cross-section, rectilinear sides and bases, and multiple and evenly spaced transverse grooves decorating the sides (VIE411c) (Fig. 59:1). The other one has roughly rectangular cross-section, rectilinear sides and bases, and is decorated with sparse and evenly spaced transverse grooves (VIE611e) (Fig. 59:2). Both beads are characterised by the presence of the main perforation along the axes of rotation, and the secondary holes (or traces of attempts of drilling thereof) on the lateral surfaces.

Among the three artefacts with average size, two have oval cross-sections, convex sides and bases (VIA222), while one combines oval cross-section with rectilinear sides and bases, decorated with double horizontal ribs (VIA211b). All three objects possess three perforations: one along H and two parallel along D (Fig. 59:3–4). Interestingly, the latter type includes two parallel ribs circumscribing the sides and is one of the only four artefacts ornamented with decorative motif 'b' (the other three classified as IA211b; cf. Fig. 45:4).

Flattened objects are definitely the most numerous category in group VI, numbering 17 finds in total. Three finds with circular cross-sections, convex sides and rectilinear bases (VIF121) have secondary or even tertiary perforations, running perpendicularly to the main hole (Fig. 59:5). There are several flattened discoidal beads with oval cross-sections that also bear traces of secondary perforations (VIF221, VIF222, VIF222z, VIF224). In some of them, the two perforations are placed alongside on the same plane (e.g. VIF222; Fig. 59:6), whereas in others the perforations are perpendicular (e.g. VIF222z; Fig. 59:7). Some of the finds are in a poor, fragmentary state of preservation; therefore, reconstruction of their original appearance is seriously hampered. The perpendicular layout of perforations is also present in one of the finds decorated with single horizontal rib (VIF218a) (Fig. 59:8).

Tab. 41. List of the finds classified to group VI (broken fragments were marked grey)

Context	Type	Quantity	Phase	Region
Allumiere – Poggio La Pozza, grave 2, trench F (1960)	VIF121	1	FBA (1150–950 BC)	North-central Lazio (southern Etruria)
Allumiere – Poggio La Pozza, grave 2, trench F (1960)	VIF161	1	FBA (1150–950 BC)	North-central Lazio (southern Etruria)
Allumiere – Poggio La Pozza, grave 2, trench F (1960)	VIA222	2	FBA (1150–950 BC)	North-central Lazio (southern Etruria)
Allumiere – Poggio La Pozza, grave 2, trench F (1960)	VIE611e	1	FBA (1150–950 BC)	North-central Lazio (southern Etruria)
Beneceto, early Recent Bronze Age (BR1) layers (sector central/part NW – US 674, 679, 680; part SE – US 144, UUSS 3)	VIF222	1	LBA (1300–1150 BC)	Emilia/southern Po Plain
Beneceto, advanced Recent Bronze Age (BR2) layers (sector central/part NE – US 3; sector E)	VIF224	1	LBA (1300–1150 BC)	Emilia/southern Po Plain
Castellaro Lagusello, layer C/C1	VIF161	1	MBA (1650–1300 BC)	Garda Lake and its inframorenic basin
Castellaro Lagusello, layer C/C1	VIF267x	1	MBA (1650–1300 BC)	Garda Lake and its inframorenic basin
Castione Marchesi, layers 2–3/phases 1–4	VIF121	1	MBA/LBA	Emilia/southern Po Plain
Frattesina di Fratta Polesine, phases 2–3	VIF121	1	FBA (1150–950 BC)	Western Veneto / northern Po Plain
Golubnjača, block 4	VIA211b	1	FBA (1150–950 BC)	Lika
Isolone del Mincio, precise context unknown	VIF162w	1	MBA/LBA	Garda Lake and its inframorenic basin
Kompolje, grave 293 (1903–1905)	VIF218a	1	FBA (1150–950 BC)	Lika
Madonna di Loreto, 'Hypogeum of Bronzes'	VIF222	1	MBA/LBA	Northern Apulia and Basilicata (Gargano peninsula and Tavoliere plain)
Madonna di Loreto, 'Hypogeum of Ivories'	VIF221	1	MBA/LBA	Northern Apulia and Basilicata (Gargano peninsula and Tavoliere plain)
Montale, phase II	VIF162y	1	MBA (1650–1300 BC)	Emilia/southern Po Plain

Tab. 41. Continued

Context	Type	Quantity	Phase	Region
Narde I, grave 169	VIF222z	1	FBA (1150–950 BC)	Western Veneto / northern Po Plain
Narde II, grave 75	VIE411c	1	FBA (1150–950 BC)	Western Veneto / northern Po Plain
Scalvinetto, tomb 354	VIF161	1	MBA/LBA	Western Veneto / northern Po Plain
Vrlazje, tumulus IV, grave 2 (1894)	VIF714	1	FBA/EIA	Glasinac

Fig. 59. Examples of the types of amber finds belonging to group VI: 1 – Narde II, grave 75 (Colonna et al. 2010, Tav. 13:B5); 2–3 – Allumiere – Poggio La Pozza, grave 2, trench F (Peroni 1960, Fig. 11:17, 21); 4 – Golubnjača, block 4 (Drechsler-Bižić 1970, Tab. V:1); 5 – Allumiere – Poggio La Pozza, grave 2, trench F (Peroni 1960, Fig. 11:23); 6 – Madonna di Loreto, 'Hypogeum of Bronzes' (Tunzi Sisto 1999a, Tav. X:1132); 7 – Narde I, grave 169 (Salzani 1990–1991b, Fig. 37:6); 8 – Kompolje, grave 293 (1903–1905) (Bakarić 2017, 26, cat. no. 9); 9 – Allumiere – Poggio La Pozza, grave 2, trench F (Peroni 1960, Fig. 11:24); 10 – Castellaro Lagusello, layer C/C1 (Negroni Catacchio et al. 1999, Fig. 10:5); 11 – Scalvinetto, tomb 354 (Salzani 2004, Fig. 4:B); 12 – Isolone del Mincio, precise context unknown (Negroni Catacchio et al. 1999, Fig. 11:13); 13 – Montale, phase II (Cardarelli/Labate 2004, Fig. 79:15); 14 – Castellaro Lagusello, layer C/C1 (Negroni Catacchio et al. 1999, Fig. 10:4); 15 – Vrlazje, tumulus IV, grave 2 (1894) (Cwaliński/Pravidur forthcoming, Pl. 1:5).

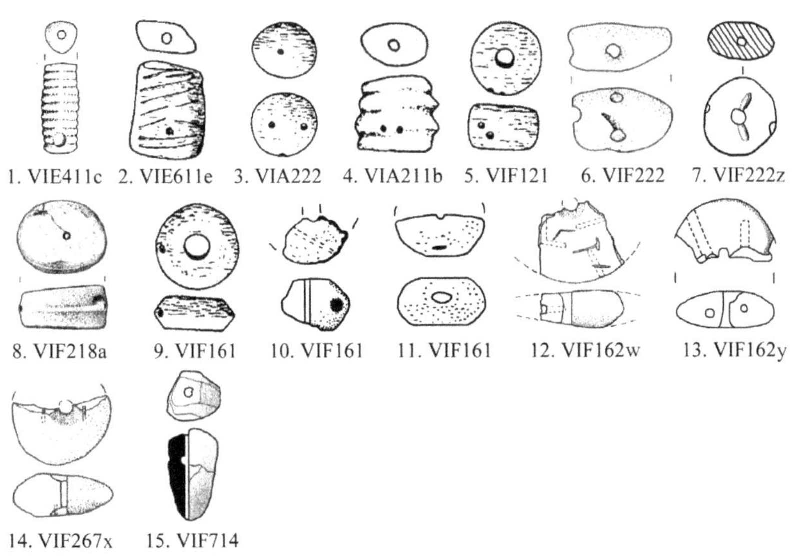

After the flattened discoidal beads, the second largest subgroup within group VI is formed by artefacts with flattened biconical profile. In three beads representing type VIF161, the two perforations are situated perpendicularly (Fig. 59:9–11). Interestingly, two of the finds are preserved only in fragments, so it is possible that in those cases the secondary perforations were made after the beads were broken. Similarly interesting are the three beads with lenticular profiles: types VIF162w, VIF162y and VIF267x which are all broken approximately in half with only one part preserved in each case (Fig. 59:12–14). Each of the beads has two holes: one on each side of the main perforation which became unusable upon cracking of the adornment in half. Worth highlighting is the fact that each one of the beads representing the above-listed types features recess around the perforation's outlet in one of the following variants: slanting, convex or concave. The lateral holes are perpendicular in respect to the primary perforation, and are apparently unfinished – the drillings do not pass through the entire objects. In the case of VIF162w, there is additionally a fourth hole, starting from the opposite side of the primary perforation. All three objects were found in settlement layers; perhaps they were discarded after failed attempts to bring them back to use by drilling new perforations. Yet, the layout of perforations is quite peculiar and uncommon, meaning that the beads were potentially intended to be strung in a manner different to the beads from subgroup IF. The last object classified to group VI has trapezoidal cross-section, rectilinear sides and flat-convex bases (VIF714), with the two perforations perpendicularly crossing each other (Fig. 59:15).

3.8.1. Chronology

It seems correct to suspect that broken amber adornments were 'recycled' by drilling secondary perforations, since the beginning of amber influx into the circum-Adriatic zone. This presumption is apparently confirmed by charted temporal fluctuations of the finds ascribed to group VI (Fig. 60). Limited amount of raw material to work with meant that the beads already in possession were likely to be kept in use as long as possible. However, the chart also includes the unbroken objects classified to the group – approximately half of the objects from group VI were found in a damaged condition. The remaining finds are those for which drilling the secondary, or tertiary perforation, was not necessitated by fracture, but rather by the need to strung them in a particular fashion.

Fig. 60. Temporal fluctuations in numbers of contexts and finds classified to group VI during the Bronze Age.

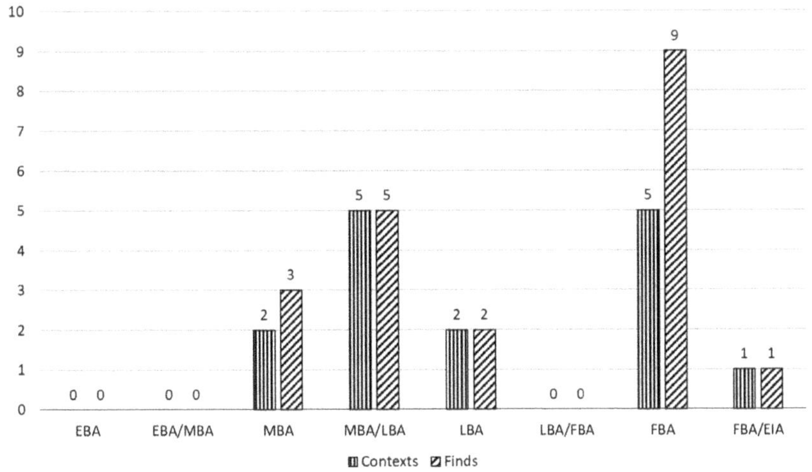

Chronologically, the types of finds from group VI correspond with similar forms from group I. If we consider the finds diachronically by dividing them into those damaged and undamaged (see Tab. 41), then it becomes clear that broken ones are all dated to MBA–LBA (1650–1150 BC). It could mean two things: (1) during MBA–LBA users of amber in the circum-Adriatic zone were more eager to restore broken amber adornments, (2) during FBA (1150–950 BC) amber craftsmen were more skilful in performing complex perforations in amber. The second hypothesis is plausible in light of the fact that FBA processing of amber has been proven at least in Northern Italy. Furthermore, chronological review of finds attributed to groups III–V has shown that the eccentric or V-shaped perforations became definitely more common in the later stages of the Bronze Age (cf. section 3.5–3.7). Nevertheless, recovery of broken beads with attempted secondary perforations from MBA–LBA contexts potentially proves that certain contemporaneous circum-Adriatic communities importing amber were able (had some knowledge and skills) to drill-process the material in order to turn it into jewellery. It seems improbable that those broken pieces, often with unfinished perforations, were streamed into circulation. They are few in comparison with finished and undamaged objects in the entire assemblage of amber finds from the study area. However, one can wonder how much of the production waste has not been preserved or recovered from the archaeological record, that could enrich group VI.

3.8.2. Spatial distribution

Finds classified to group VI come from parts of every macro-region distinguished before. However, objects dated to the earlier MBA–LBA horizon, i.e. those predominantly broken – fragmented, reveal a particular affiliation with Northern Italian provinces of Emilia, western Veneto and the Garda Lake basin (see Tab. 41). There, the types such as VIF121, VIF222, VIF161, VIF162w or VIF162y are matched by numerous discoidal, biconical or lenticular beads belonging to group I. This spatial correspondence of similar forms, along with the fact the abovementioned VIF-types were recovered in a broken condition predominantly from settlements, further renders plausible the hypothesis of local manufacture of amber jewellery by communities of the *Palafitte-Terramare* culture. Even though there are no well-documented places of amber processing ('workshops') in the contexts of MBA–LBA sites in the circum-Adriatic zone, the above observations can be considered as an indirect evidence supporting this notion.

During FBA amber adornments with secondary perforations had a wider distribution, encompassing, in addition to Northern Italy, also north-central Lazio, Lika and Glasinac. However, in the latter two regions, amber products with more complex perforations can be considered incidental.

3.9. Group VII

Group VII is one of the smallest, comprising six artefacts in total (Tab. 42). The common feature characterising them is the presence of multiple perforations, forming a complex structure of apertures and channels. Such objects, popularly called by archaeologists 'spacers' or 'spacer plates', were most probably intended to play a particular function in large and multi-row necklaces of beads.[523] As the name implies, their role was to separate beads inserted on strings into segments. Depending on the number of perforations, a single spacer could be fitted onto two or more rows of beads.

Tab. 42. List of the finds classified to group VII

Context	Type	Quantity	Phase	Region
Olmo di Nogara, grave 122	VIIE814	1	LBA (1300–1150 BC)	Western Veneto
Plemmirio, tomb 10	VIIF222	2	MBA (1650–1300 BC)	Sicily
Rocca di Oratino, phases I–III, superficial layers and sunken structures	VIIF721y	1	LBA (1300–1150 BC)	Molise
Val di Non, precise context unknown	VIIE611	2	MBA/LBA	Trentino-Alto Adige

523 N. Negroni Catacchio et al. distinguish spacers as a separate category of amber finds, consisting of a single type – rectangular forms (2006, 1450–1452, Fig. 5). Extensive overview of amber spacers in the European Neolithic and Bronze Age was published by C. Du Gardin (2003). Amber spacers in the *Mycenaean* culture were divided into four types: 21. Rectangular, 22. Triangular, 23. Eight-shaped, 24. Other by J. Czebreszuk (2011, 92–94). Detailed analysis and detailed typology of amber spacer linked to the *Únětice* culture was published by M. Ernée (2012, 88–93, Obr. 11–12). Amber spacer from the area of Germany were classified to types 22a, 22b and 22c by G. Woltermann (2016a, 51–52, Abb. 18).

There are five objects which, according to previous studies as well as the criteria applied in this work, can be certainly classified as spacers and included in group VII, and one which potentially belongs to this category; however, its poor state of preservation (it is broken) does not allow for an unambiguous classification – VIIF721y (Tab. 42). The most often cited examples of amber spacer from Italy, and for that matter from the entire circum-Adriatic zone, are the two specimens discovered in the bog near Cles in the Val di Non area (cf. Bellintani 2002; Angelini et al. 2003). They have rectangular cross-sections, rectilinear sides and bases (VIIE611), and are internally transected by multiple parallel-straight perforations running along the shorter side (L2), crossed by a concentric perforation along the longer side (L1).[524] One of them additionally reveals traces of two unfinished perforations located parallel to the bore along L1 (Fig. 61:1).

Fig. 61. Examples of the types of amber finds belonging to group VII: 1 – Val di Non, precise context unknown (Angelini et al. 2003, Fig. 2.2); 2 – Plemmirio, tomb 10 (Tusa 1999, Fig. 29:1); 3 – Olmo di Nogara, grave 122 (Salzani 2005, Tav. XIII:B); 4 – Rocca di Oratino, phases I–III, superficial layers and sunken structures (Copat/Danesi 2010, Fig. 4:13).

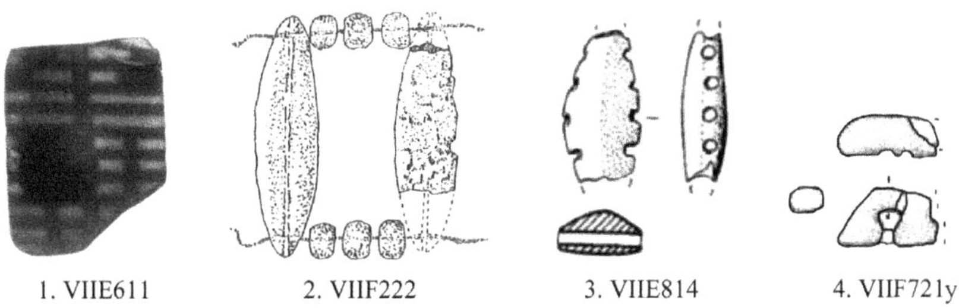

The two objects from Plemmyrion, tomb 10 have more oblong form (VIIF222) and possess fewer perforations: two parallel on the opposite, distal ends and one running perpendicularly along the longer side of the object, connecting the latter two (Fig. 61:2).[525] Remaining two amber spacers were found individually, not in pairs. One of them, although broken at the ends, largely retained its form: polygonal cross-section, rectilinear sides and convex-flat bases (VIIE814), with four parallel perforations, transecting the object along the shorter side (Fig. 61:3).[526] The sixth amber spacer, discovered

524 N. Negroni Catacchio et al. cite the two specimens from Val di Non as examples of type 1 of Italian amber spacers, at the same time highlighting their uniqueness in Italy (2006, 1458). In terms of dimensions, as well as formal and technological aspects, the two spacer plates from Val di Non resemble German analogies from Schwanheim near Frankfurt (Du Gardin 1986, Fig. 7:62) and that of Essingen near Aalen in Wurttemberg (Merhart 1940, Abb. 1:6), corresponding chronologically with spacers of Val di Non (cf. Angelini et al. 2003, 229).

525 Often described in the literature as 'spacer beads' (see Tusa 1999, 490–491; Angelini et al. 2003, 229; Cultraro 2007, 381–382), the two finds were discussed at length by O. Cornaggia Castiglioni and G. Calegari (1978).

526 The artefact was discussed by R. C. De Marinis and L. Salzani, who compared it with other European amber spacer plates and found the closest analogies among specimens related to the *Tumulus* culture of phase BzC (De Marinis/Salzani 2005, 428).

only recently, is preserved only in a fragment with trapezoidal cross-section, convex sides and rectilinear bases (VIIF721y; Fig. 61:4). One can notice a hole which is not fully enclosed due to a missing fragment, and traces of another hole suggesting more elongated original form of the object, potentially with greater number of perforations. Poor state of preservation allows only to hypothesise that the object served as a spacer.

3.9.1. Chronology

All six objects included in group VII are chronologically consistent, indicating MBA–LBA (ca. 1650–1150 BC) as a time span during which amber spacers occurred in the circum-Adriatic zone. It largely corresponds to the occurrence period of amber spacers in other parts of Europe. The two specimens from Val di Non are dated by their resemblance to some of the specimens related to the *Tumulus* culture in Germany (ca. BzB–BzD) (cf. Bellintani 2002, 39; Angelini et al. 2003, 229). Similar chronology pertains to the two spacers from Plemmyrion, tomb 10 which, although lacking precise analogies, are dated basing on the context to phases *Thapsos I–III* (LH IIIA–B; ca. 1420–1270/1250 BC) (Tusa 1999, 490–491; Cultraro 2007, 381). The remaining two amber spacers are slightly younger, both falling into LBA (ca. 1300–1150 BC).

3.9.2. Spatial distribution

Within the scope of the circum-Adriatic zone, amber spacers are known only from the area of Italy. However, all six amber spacers are not concentrated in a single region or macro-region, but they are scattered in several places, quite remote from each other. The two specimens from Val di Non (Trentino-Alto Adige) and the one from Olmo di Nogara, grave 122 (western Veneto) are geographically the closest. Further south, in Molise was found the suspicious object from Rocca di Oratino. The last two spacers come from south-east Sicily (Plemmyrion, tomb 10). This rather wide but sparse distribution, together with analogies in the neighbouring parts of Europe, suggests that these objects are imports which found their way to Italy by different channels.

3.10. Group VIII

Group VIII covers all amber finds from the study area which are lacking a perforation, not counting the finds for which presence of a hole(s) for suspension can be assumed based on broken off fragments (Tab. 43). Absence of perforation can be considered as a sign of unfinished/incomplete level of processing at which those objects were left or, alternatively, of a specific purpose intended for them, i.e. other than beads, buttons or pendants (for instance, as an application). To the first group can definitely be classified some of the finds from Campestrin di Grignano Polesine, where discovery of amber-processing waste and reconstruction of the Tiryns-type beads *chaîne opératoire* allows to consider the finds missing a perforation as semi-finished products (cf. Bellintani et al. 2015a). They include six different forms in average and elongated categories, decorated with a single horizontal rib. Four objects combine oval cross-section with rectilinear sides and bases (VIIIE211a), or rectilinear sides and skewed-flat bases (VIIIA281) or concave sides and rectilinear bases (VIIIE231a) (Fig. 62:1, 3–4). Another four artefacts from the site feature triangular cross-section and either rectilinear or pitched sides (VIIIE418a and VIIIE451a) (Fig. 62:5–6). The last object also has conical profile, but its cross-section is more akin to a trapeze (VIIIA751a) (Fig. 62:2).

Tab. 43. List of the finds classified to group VIII

Context	Type	Quantity	Phase	Region
Campestrin di Grignano Polesine, late BR–initial FBA layers	VIIIA218a	1	LBA/FBA	Western Veneto / northern Po Plain
Campestrin di Grignano Polesine, late BR–initial FBA layers	VIIIA751a	1	LBA/FBA	Western Veneto / northern Po Plain
Campestrin di Grignano Polesine, late BR–initial FBA layers	VIIIE211a	2	LBA/FBA	Western Veneto / northern Po Plain
Campestrin di Grignano Polesine, late BR–initial FBA layers	VIIIE231a	1	LBA/FBA	Western Veneto / northern Po Plain
Campestrin di Grignano Polesine, late BR–initial FBA layers	VIIIE418a	1	LBA/FBA	Western Veneto / northern Po Plain
Campestrin di Grignano Polesine, late BR–initial FBA layers	VIIIE451a	3	LBA/FBA	Western Veneto / northern Po Plain
Frattesina di Fratta Polesine, phases 2–3	VIIIE711c	1	FBA (1150–950 BC)	Western Veneto / northern Po Plain
Island of Krk, precise context unknown	VIIIE211a	1	FBA (1150–950 BC)	Kvarner and Northern Dalmatia
Santa Rosa di Poviglio, *Villaggio Grande*, 'summit' layer – phases 3 and 4	VIIIF154	1	LBA (1300–1150 BC)	Emilia/southern Po Plain
Santa Rosa di Poviglio, *Villaggio Grande*, 'summit' layer – phases 3 and 4	VIIIF154k	1	LBA (1300–1150 BC)	Emilia/southern Po Plain

Single unperforated object was found on the nearby settlement of Frattesina di Fratta Polesine, in the residential layers dated to the 3rd phase of the settlement. It has elongated form with trapezoidal cross-section, and rectilinear sides and bases, decorated with multiple and evenly spaced transverse grooves (VIIIE711c) (Fig. 62:7). Thanks to the latter feature, the find can be considered as a semi-finished bead with the Allumiere-type decoration (Bietti Sestieri et al. 2015, 431).

From the opposite side of the Adriatic originates a piece of amber allegedly found together with several other amber adornments bearing the Tiryns and the Allumiere-type decorations, collectively attributed to a disturbed context (grave?) from the Krk Island (see Blečić Kavur 2014, 171–172). It has roughly oval cross-section, rectilinear sides and bases, and exhibits a partly formed horizontal rib in the lower half (VIIIE211a) (Fig. 62:8). The object has been classified as the Tiryns type (cf. Palavestra 1993, 62–63; Blečić Kavur 2014, 56–57, cat. no. 23), but the question of its perforation remains problematic. One can notice a groove along the longer side of the object; it is difficult to tell from the figures alone[527] whether it is a natural cavity or a trace of perforation that caused the lump to break. Even, if the latter is true and there was an attempt to drill through the object, it is probable that the latter remained in an unfinished form, and the underdeveloped decorative rib seems to confirm this hypothesis.

The last two artefacts included in group VIII, in contrast to those discussed above, were possibly never intended to have a suspension hole. Those are very peculiar looking objects resembling small discs with slightly conical profiles, and one of them is decorated on the dorsal side with five evenly distributed groups of concentric circles (VIIIF154k) (Fig. 62:9). This is the only element decorated in this manner in the entire dataset, so far not registered among Bronze Age amber assemblages from other parts of Europe (Bellintani 2014, 116). Its ornamentation and absence of a perforation suggest that it was used in a different way than as an element of jewellery, perhaps as an application or adornment of some other object. Recurrence of the decorative motif consisting of concentric circles on artefacts made of bone and antler in the *Terramare* culture, e.g. combs and *rotelle*, suggest a local manufacturing of the discussed amber discs (Bellintani 2014, 116; cf. Provenzano 1997).

Fig. 62. Examples of the types of amber finds belonging to group VIII: 1–6 – Campestrin di Grignano Polesine, late BR–initial BF layers (Bellintani et al. 2015a, Fig. 2A:2–5; 2021, CGP 3, 6, 9, 12, 45, 46); 7 – Frattesina di Fratta Polesine, phases 2–3 (Bietti Sestieri et al. 2015, Fig. 2:12); 8 – Island of Krk, precise context unknown (Blečić Kavur 2014, Fig. 21:23); 9 – Santa Rosa di Poviglio, Villaggio Grande, *'summit' layer – phases 3 and 4 (Bellintani 2014, Fig. 4).*

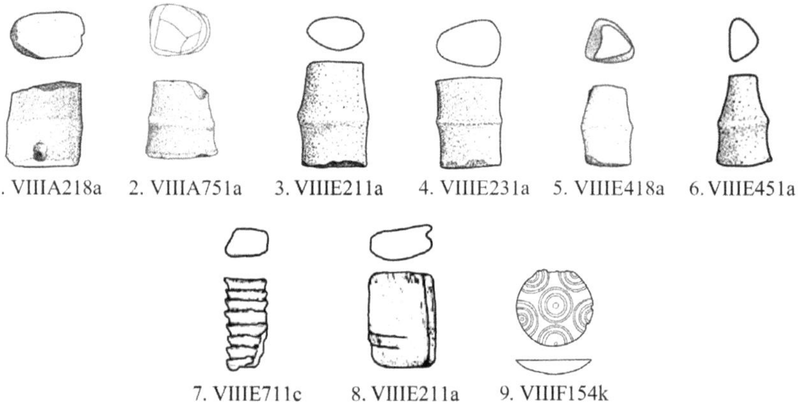

3.10.1. Chronology

All finds assigned to group VIII fall into a period from LBA to FBA (ca. 1300–950 BC) (Tab. 43). This observation gains importance in the light of the fact that most

527 Despite repeated requests, authorities of the Museum of Natural History in Vienna, where the collection of finds from Krk Island is kept, has not issued a permit to inspect and sample the artefact.

of the above-discussed artefacts are considered as semi-finished products – pieces of evidence documenting local processing of amber in the circum-Adriatic zone. It leads to conclusion that full-scale working of amber developed in this part of Europe in the advanced stages of the Bronze Age. However, it is worth remembering some of the finds with hybrid perforations (group VI) dated to MBA (1650–1300 BC). They potentially show that drilling perforations in amber, arguably the most difficult task in the process, was definitely attempted at an earlier with various results, sometimes allowing for production of crude beads or pendants. Moreover, if the local craftsmen could carve out a quite sophisticated decoration on VIIIF154k from Santa Rosa di Poviglio, one can presume that by LBA (1300–1150 BC) they have already mastered shaping of amber lumps into adornments. But it seems that stable/stationary amber manufacturing centres with standardised processing came to being in the circum-Adriatic zone not earlier than the turn of LBA and FBA.

3.10.2. Spatial distribution

Distribution of the finds classified to group VIII is limited to the northern part of the circum-Adriatic zone – area around the head of the Adriatic (*Caput Adriae*) (Tab. 43). Two of the sites which yielded between them ten non-perforated amber artefacts are located in western Veneto, in the lower Po Valley (Campestrin di Grignano Polesine and Frattesina di Fratta Polesine). Another one, which brought two finds from discussed category, is situated in Emilia/southern Po Plain (Santa Rosa di Poviglio). Finally, the last of the listed non-perforated finds can be only vaguely attributed to the Island of Krk in the Kvarner Bay.

3.11. Group IX

The last group distinguished within the framework of discussed typology includes four finds which have only partially completed perforations (Tab. 44). Two of them are decorated with motif 'c' – multiple and evenly spaced transverse grooves; they both have truncated-conical profiles, but one exhibits circular (IXE151c) and the other quadrangular cross-section (IXE551c) (Fig. 63:1–2). The third still is elongated, but has rectilinear sides and bases, and features a single horizontal ridge in the middle of its profile (IXE711a) (Fig. 63:3). The fourth artefact can be described as a low conical button with unfinished or poorly executed V-shaped perforation (IXF154) (Fig. 63:4).

Tab. 44. List of the finds classified to group IX

Context	Type	Quantity	Phase	Region
Campestrin di Grignano Polesine, late BR–initial BF layers	IXE711a	1	LBA/FBA	Western Veneto / northern Po Plain
Campestrin di Grignano Polesine, late BR–initial BF layers	IXF154	1	LBA/FBA	Western Veneto / northern Po Plain
Frattesina di Fratta Polesine, hoard 1 *Tesoretto*	IXE151c	1	FBA (1150–950 BC)	Western Veneto / northern Po Plain
Frattesina di Fratta Polesine, hoard 1 *Tesoretto*	IXE551c	1	FBA (1150–950 BC)	Western Veneto / northern Po Plain

All those objects can be regarded as semi-finished products – interpretation which appears plausible considering that the finds were discovered on two sites known for other evidence of amber processing: Campestrin di Grignano Polesine and Frattesina di Fratta Polesine. However, it is worth remembering that in Northern Italy were documented several cases of amber adornments inserted on bronze pins (cf. De Marinis 2007, Fig. 5). Interestingly, in one of the two decorated artefacts from Frattesina (IXE551c) a fragment of bronze wire/rod was found stuck inside the hole, potentially indicating similar use of the object as a pinhead (cf. Negroni Catacchio 1972, 17). From the hole of the second find (IXE151c), instead, was pulled out a bone fragment possibly belonging to a broken pin (cf. Negroni Catacchio 1972, 17). However, both pieces of bronze and bone could be tips of drills used to perforate the beads which snapped in the process and were left inside the holes, thus leaving the objects unfinished.[528] The latter find (IXE151c) is also partly intersected along D by two orthogonal perforations what also makes it a likely candidate for group VI (finds with hybrid perforation); however, considering a formal similarity to object IXE551c and presence of a pin/drill fragments inside the incomplete perforation, it was decided to classify it to group IX.

Fig. 63. Examples of the types of amber finds belonging to group IX: 1–2 – Frattesina di Fratta Polesine, hoard 1 Tesoretto *(Negroni Catacchio 1972, Figs. 15 and 16); 3–4 – Campestrin di Grignano Polesine, late BR–initial BF layers (Bellintani et al. 2015a, Fig. 2A:6; 2021, CGP 26).*

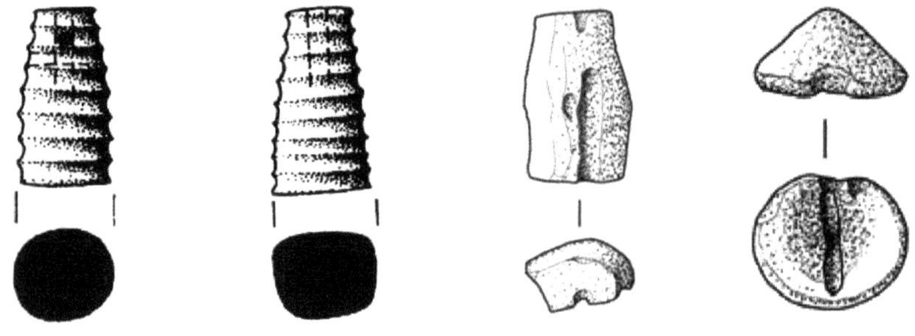

3.11.1. Chronology

All four finds classified to group IX can be dated according to the chronology of their context, i.e. from the turn of LBA and FBA up to the mid FBA (BR 2–BF 2; ca. 1260/1220–1010/1000 BC). This period corresponds to the time, when local amber craftsmen in the circum-Adriatic zone achieved the highest level of technological advancement in the Bronze Age, as also indicated by other groups of finds.

528 For an overview of prehistoric tools used to drill in amber, see Popkiewicz 2012, 2016; Popkiewicz/Czebreszuk 2016.

3.11.2. Spatial distribution

The finds related to group IX come from two closely located sites, situated within western Veneto/northern Po Plain. In the light of gathered evidence, the area can be regarded as an important centre of amber processing in LBA–FBA (ca. 1300–950 BC).

3.12. Typology-based network for reconstruction of interregional circulation of amber

Typological considerations of Bronze Age amber finds from the circum-Adriatic zone helped to define temporal and spatial occurrence of nine categories of artefacts. The goal of this chapter is to determine similarity-measured relationships between regional assemblages of amber finds (divided into groups and types) in four consecutive phases of the Bronze Age (EBA, MBA, LBA and FBA)[529] and use those observations to reconstruct interregional circulation of amber in the study area. It was decided to base this part of the work on the NA methodology and tools.

The analysis was carried out using UCINET 6 software. Creation of the networks was based on similarity measurement between the regions with the *averaged cross-product*, calculated on the basis of types of amber finds co-occurring in regions across a contingency table. The *cross-product* method takes each entry of the row for an object (a column), and multiplies it times the same entry for another object (another column), and then sums the result. Usually this method is used for binary data, because the result is a count of co-occurrence. With binary data each product is '1' only if both objects (columns) share the variable (in this case amber type), and the sum across the objects yields the number of variables (types) in common – a valued measure of strength. Then the resulting sum is averaged – divided by the total number of variables (Σxy/n) giving a mean value (cf. Hanneman/Riddle 2005). The level of similarity on the resulting graphs is expressed through the width of links joining the nodes (regions) – the more two regions have in common, the wider the link is. For each network the *density* – one of cohesion coefficients deriving from the total number of ties divided by the total number of possible ties – was calculated (cf. Hanneman/Riddle 2005). Next the *betweenness* centrality was calculated for the nodes, to assess importance of individual regions for the network's existence. Index of *betweenness* denotes how many geodesic paths in a network pass through a given node, thus indicating how important the latter is for connecting remaining nodes (cf. Freeman 1979; Freeman et al. 1991). Size of a node denotes percentage of the maximum possible *betweenness* in a network – the larger the node is, the greater its centrality. Finally, the nodes were divided into two classes on the basis of *core-periphery* analysis (cf. Borgatti/Everett 1999; Everett 2015). First class – cores – groups the nodes characterised by high *degree* – number of ties with remaining nodes that stems from sharing many variables (types of amber beads) in common. The second class – peripheries – comprises nodes with low *degree* of ties, thus having few variables in common with the others. The networks were subsequently transposed over a map of Europe (shaded relief with coastlines and watercourses) using QGIS 3.26 software.

529 Contexts classified to the intermediate stages (EBA/MBA, MBA/LBA, LBA/FBA and FBA/EIA) were used in both periods to which they pertain, thus occur twice in the analysis.

In a similar way, NA was applied by M. Furholt to measure incidences of coherence among regionally grouped artefacts from the Central European Late Neolithic (Furholt 2009). Another example of NA application in a congruous way is by M. Przybyła, who included sites from the MBA in the Carpathian Basin in a geographically fixed network, measuring tightness of relationships through shared decorative motifs on pottery (Przybyła 2016).

The choice of applying NA in the way explained above is grounded in an assumption that formal similarities among amber artefacts stem from interactions between actors involved in the process of amber circulation, from an object's inception to its deposition. Amber, as a resource not commonly available in Europe, lends itself well to such analytical approach. As with the other categories of handcrafted products that ended up in archaeological context, similarity between two or more amber objects (e.g. beads) potentially means that they have something in common beyond superficial appearance. They could share similar histories by being produced by the same craftsman, or by being copies manufactured by different people but on the basis of a single model, among various other possibilities. Thus, certain features in morphology of amber adornments could be interpreted as trademarks of craftsmen from specific regions, or evidence of use of similar processing techniques ('know-how') which themselves were subject to exchange through contacts. Of course, it is possible that similar inventions and creations took place independently. However, with such hardly available material in the circum-Adriatic zone as amber, this option seems less likely. The given arguments do not merely stem from the *diffusionistic* vision of the material's dissemination from a source (creator – active agent) to a final destination (recipient – passive agent). Although the premise of formal similarity is crucial in the discussed approach, it does not prejudge a way that the similarities between objects occurred, nor it equates them on the semantic level. Each member of a category possibly possesses specific 'biography'/'itinerary' and could have been perceived ('translated') differently (see section 1.2.2). Here, the focus falls predominantly on the statistical side of similarity – the more similarities two regions had in common, the more likely it is that they interacted in one way or another (see section 1.2.1). Character of these interactions revolves around amber circulation, but observations derived from NA do not preclude other relationships among the regions included in it. One certain fact is that the knowledge of amber and its processing in areas devoid of fossil resins (such as the majority of the circum-Adriatic zone) was only starting to be developed in the 2nd millennium BC, and in this early stage, one can hope to see, where it took hold first and where last, which regions were playing roles of important hubs of its redistribution and which were dependent on them, which communities where the most active in its acquisition and which the least, and so on.

In the next step, results of NA are confronted with selected archaeological data from the sites with amber, to discuss the plausibility of using degrees of similarity between assemblages of amber beads and derived coefficients as proxies for interregional relationships in the circum-Adriatic zone during the Bronze Age involving amber transfer. Thus, the figures obtained through NA are just a starting point for further considerations on the level of affiliation between the regions, using other categories of finds, chiefly bronze items.

Several aspects of the applied approach require explanation, before moving on to description of NA results. The detail level of the typology, which is the basis for similarity measurements among the regions in NA, means that only the perfectly matching artefacts were treated by the program as a link. It is argued that this solution provides a certainty that behind a tie joining two regions stands an actual morphological correspondence between amber artefacts these two include. There often is, however, a great deal of resemblance between two objects, despite differences in one aspect of their morphology (e.g. cross-section) resulting in the absence of a link in the analysis. Therefore, the raw outcome of NA based on a very rigid classification of similarities is supplemented in the discussion below with additional comments on the typology and relations between amber finds.

For an overall comprehension of NA, it seems important to answer the question: why entire regions, and not individual sites or contexts, were used as the nodes? Apart from technical issues – fewer nodes allow for easier and more readable presentation of the network, regions as larger and more robust units of the analysis offer much more information as sums of all the constituent findspots. In chapter 2 the provinces of amber's acquisition were distinguished in a way to overlap with culturally defined units, such as cultures, facieses, groups, etc. Since the dawn of archaeology, similarities among material objects, that are foundation for such distinctions, have been commonly interpreted as testimonies of interpersonal and intergroup interactions which in effect led to emergence of culturally coherent communities (Renfrew/Bahn 2007, 107, 163). Direct translation of unified assemblages of material culture into social organisations known from history, such as ethnic groups, tribes or nations, frequently came under a well-argued criticism in the last 70 years of developments in archaeological theory since the end of the Second World War (e.g. Hodder 1986). While the paradigm towards interpretation of material culture has changed and various new concepts have been proposed, it has been always in the centre of archaeological reasoning that material culture is meaningful. It is not the task of this work to polemicize with approaches to and views on material residues of the past that over the years led to distinction of a myriad of archaeological taxa in the study area. The choice was rather to accept and adapt to the existing state by finding a single underlying principle in the previous interpretations of the gathered data that can be applied in the study of amber across such a vast space. The basic rule is that people, willingly or not, interact with other people – a process that often ends up with exchange of information and objects, and in a longer perspective may result in emergence of features unifying some groups and differentiating them from others (cf. Wobst 1977; Hodder 1979). Also, it is one of the basic assumption guiding the applications of NA (e.g. Blake 2014, 71). Furthermore, it can be argued that people belonging to one group (e.g. community) will more likely interact more often and intensely between themselves, than with members of other groups (Blake 2014, 73). This brings us to the answer to the question raised at the start of this paragraph: this work assumes occurrence of interactions (which involved amber exchange) on the local – micro-level, and seeks to investigate circulation of amber on the level of interregional relations.

Furthermore, it is worth explaining the approach to interpretation of the ties resulting from NA. On a purely methodological level, they indicate similarities between regional assemblages of amber finds. Mapping of the network carried out in the following sections does not pretend to indicate the actual routes in the geographical sense via which amber was translocated. Instead, it is aimed at suggesting directed relations between the nodes – regions participating in amber circulation, although their relative locations in the landscape of the circum-Adriatic zone may lend useful clues as to ways by which this process was happening. In reality, it is difficult to escape from questions about the possibilities of transporting resources or processed goods over long distances. Having in mind the lack of stable infrastructure in the modern sense and technological limitations of the means of transport available at the time, it could be quite challenging to move and navigate in the landscape of Bronze Age Europe. Here can be mentioned a few innovations which reached Europe by the 2nd millennium BC, potentially facilitating translocation of amber. With regard to land transport, one can recall introduction of donkey as a pack animal; donkey bones were discovered on several BR–BF sites in Italy, most importantly from the perspective of this work at Coppa Nevigata in northern Apulia (Bökönyi/Siracusano 1987, 207) and Frattesina di Fratta Polesine in western Veneto (De Grossi Mazzorin 2019, 129–130). Although the evidence from the circum-Adriatic zone at hand prevents from recognising this invention as widespread, it does suggest a south-to-north pattern of spread between BR and BF, probably thanks to the contacts with the Aegean (Blake 2014, 62–63). Second, it is worth mentioning the advancements in seafaring, i.e. introduction of sail in the Aegean at around the beginning of the 2nd millennium BC, allowing for longer and uninterrupted journeys (Knappett 2011, 135). Having a substantial evidence of contacts between Italy and the Aegean in the later stages of the Bronze Age, it is highly likely that small, wind-powered ships were the mode of transportation used on the Tyrrhenian, Ionian and Adriatic Seas.

In the recent monographs related to the prehistory and early history of the Mediterranean, authors are devoting a great deal of attention to natural factors that have conditioned movement and connectivity across the sea basin, including directions of the winds and currents as well as different kinds of coastlines (Horden/Purcell 2000, 121–143; Broodbank 2013, 54–62, 72–79). One of the major notions is that sea unites, forming a sort of hinterland for the communities living on its margins. These are so-called 'gateway communities' which maintain contacts with other micro-regions, and its members are engaged in interactions such as exchange or trade. Such communities 'base a considerable part of their economy upon the goods entering and leaving their territories, namely, upon their favourable geographic position' (Palavestra 2007, 353–354). They inhabit 'gateway settlements', through which goods and people are being funnelled in both directions. Core territories, seen in the scope of the Adriatic Basin as main regions of amber's acquisition, 'may be composed of far-flung coastlands whose functional proximity is the product of seaborne connectivity' (Horden/Purcell 2000, 133). Liburnian and Iapodean tribes inhabiting areas of Dalmatia and Lika during the 1st millennium BC, known for their interest in amber, might serve as a good example (Palavestra/Krstić 2006, 48–49).

Thanks to aforementioned studies, we are now aware that in the scope of the Adriatic Sea, different forms of trade could operate independently, yet supplementing each other on occasion. In comparison to often overestimated large-bulk, long-distanced traffic, short-distanced more flexible trade resembling a sort of 'cabotage' most probably had greater significance as it could deliver goods to more peripheral regions (Horden/Purcell 2000, 144–145, 150–151).

On the other hand, trade in the inland zone could operate along the most accessible stripes of terrain which were eventually transferred into actual roads in the Roman times, e.g. Via Egnatia. However before that, people were trying to organise surrounding landscape so that it could serve their purposes better. It is suggested that Iron Age barrows, as an artificial element of the natural landscape, except their funerary function could have played the role of fixed markers in terrain facilitating communication and transport (Palavestra 2007, 353). A. Palavestra notes that 'the most important communication routes, which from the prehistoric times used to connect the Adriatic coast with its hinterland, led to any significant princely grave in the central Balkans that was rich with amber' (Palavestra/Krstić 2006, 55). Same routes were used for caravan trade during the antiquity and Medieval periods (Palavestra/Krstić 2006, 60). It is worth mentioning that similar ideas were applied even earlier by scholars who attempted to show interconnections between various Western Balkan sites with amber, by characterising their geographic position and landscape in immediate vicinity (cf. Todd et al. 1976).

Another important aspect of the applied approach is the use of other data associated with amber finds as a help in reconstruction of the latter's circulation. The following description mainly utilises to that effect objects – artefacts found on the sites, or more specifically in the contexts with amber. Significance of particular categories or types of finds is explained whenever referring to these sources by citing relevant literature. What remains to be discussed is the method of selecting these sources, and their value/usefulness in the adopted reasoning.

As amber finds are the focus of this work, the scope of considerations is largely limited to the sites and contexts from which these finds originate. One way of looking at NA carried out in this place is to perceive the nodes as interlinked actors capable of both receiving/acquiring as well as processing/redistributing, according to the rule 'if amber was deposited in a region (implying its use by local people), then it could have been exchanged'. However, there are areas lacking evidence of amber which, nonetheless, could have participated in this process. Among the data related to amber finds from the studied zone are often evident traces of contacts with such amber-less places. Does it mean that we as archaeologist are missing a part of the picture, or do those non-amber testimonies stem from different channels of resources and goods circulation? Hopefully, new research and excavations will allow to clarify this issue.

The following discussion considers only those non-amber artefact which co-occur with the amber ones; however, there are many other finds regarded as useful indicators of interregional contacts which were found elsewhere. On the other hand, sites with amber reflect various modes of deposition (see chapter 4). There are graves with respect to which one can suppose, with a good deal of probability, a direct link between amber finds and the person buried, as well as amber and other elements of inventory. Similar intentionality in deposition allows to envisage connection between individual components of hoards. But there are also settlements, where finds have often been recovered from vast and thick layers designating long time horizons.[530] In those cases the relationship between amber and other finds seems more loose and indirect; however, one still cannot definitely rule out even a remote connection between them. In the analysis below, the non-amber finds from settlements, cemeteries and hoards are considered alike, even though they are not ascribed the same informative value, what hopefully is sufficiently clarified in the descriptions of their contexts. In some cases analogies for non-amber objects were found outside the exact contexts or sites of amber finds' origin. Still, such counterparts are cited, if considered vital for supplementing the evidence for a relationship between two regions. One can therefore

530 This observation is particularly relevant in relation to the *Terramare* settlements and the *palafitte* dwellings which were largely excavated in the 19th century with now-archaic methods, ascribed often leaving incomplete and imprecise documentation (cf. Bernabò Brea et al. 1997e).

characterise the applied methodology as a ‘middle-of-the-road’ approach: largely rejecting the evidence not related to sites with amber as less likely linked to amber circulation and, at the same time, including the data stemming from the places which revealed presence of the discussed material in a broad way.

Realising that there are numerous possibilities for how two objects could end up in a single context, it is necessary to tackle also the reason for and value of associations between amber and other co-occurring finds. As expressed in the section 1.2.3, the undertaken analysis will hopefully manage to characterise the last stage of life of objects, including the physical travels they went through, prior to being buried in the ground. At the foundation of the applied approach to meet this challenge lie two theoretical concepts: that the co-occurring artefacts share history of use, and that through analysis of various aspects of the objects, i.e. typology and chemical composition, it is possible to determine their provenience. For example, infrared analysis of amber finds can trace back their origins in the macroscale to fossil resin deposits, but their typology, in conjunction with typology and technological markers of coinciding products made from other materials, have a potential to uncover their itineraries at a more regional level. Although the concept of typological convergence as a proof of the same origin and dispersion from a single location goes back to *diffusionistic* approach, the questions about the provenience of objects and inference based on their immediate environment are promoted by the newer *contextual* archaeology (cf. Hodder 1986, 150–187). With the constantly growing pole of data and new developments in scientific methods specifying sources of raw materials, it is increasingly feasible to reconstruct paths of products’ circulations. In doing so, one can rely on several different sources of information: with respect to bronze items, these are i.a. typologies, lead-isotopes analyses and metal processing-related finds, e.g. casting moulds, semi-finished products, etc.

The author by large abstains from making definite claims regarding the character of amber circulation between the regions. Such detailed determination of the human interactions involving amber would need a much more thorough analysis of the data on a micro-level – an objective this work does not undertake. Instead, the description below often resorts to more neutral terms, such as circulation or exchange, not implying a precise form or mode in which amber changed hands. Only occasionally, a more narrow interpretation is suggested. Even if there is a link between two regions stemming from similarity among amber artefacts, it does not necessarily mean that the contexts which yielded those finds are directly related as, e.g. deposits made in a single chain of events by the same people. The conducted analysis is only an attempt to understand the complex problem of amber circulation in the circum-Adriatic zone. It seemed advisable to the author to approach obtained results with cognitive flexibility and leave space for future reconsiderations of the topic.

3.12.1. EBA (2150–1650 BC)

During EBA amber in the circum-Adriatic zone occurred in two regional concentrations: one in Northern Italy – in Trentino and around the Garda Lake, and the second on Sicily (Fig. 64). Only one site is situated outside these two clusters, in Tuscany. Out of 18 contexts with amber considered in this section, 8 can be decisively dated to EBA (2150–1650 BC), while 10 have chronology possibly extending onto early MBA.

The very condensed distribution of amber finds, limited to four regions, results in a very simple network, comprising only four nodes and two links (one node is detached from the rest) (Fig. 65). The network has the averaged *density* value of 0.333, indicating rather weak level of interconnectedness. The linked regions are Trentino-Alto Adige and the Garda Lake basin, and Trentino-Alto Adige and Sicily, while Tuscany remains unconnected. This means that Trentino-Alto Adige is the focal node in the network – the core, concentrating all *betweenness.* The key position of Trentino-Alto Adige in the EBA network is solely due to abundance and typological variety of amber finds from the Molina di Ledro settlement (cf. Rageth 1974, Taf. 87:1–24). One has to remember the issues with dating the layers at the site from which amber finds were recovered (see section 2.2.4). Only this site contains types of amber artefacts which appeared both in Northern and Southern Italy during EBA; however, it has to be emphasised that similarity level is small as can be inferred from the thin links. Trentino-Alto Adige has two types in common with the Garda Lake basin (IF122 and IA122; cf. Fig. 35:18, 45:10) and only one with Sicily (IF128; cf. Fig. 36:1), among the total of six or seven types that each of these provinces yielded.

Fig. 64. Distribution of the sites with amber finds dated to EBA and EBA/MBA in the circum-Adriatic zone.

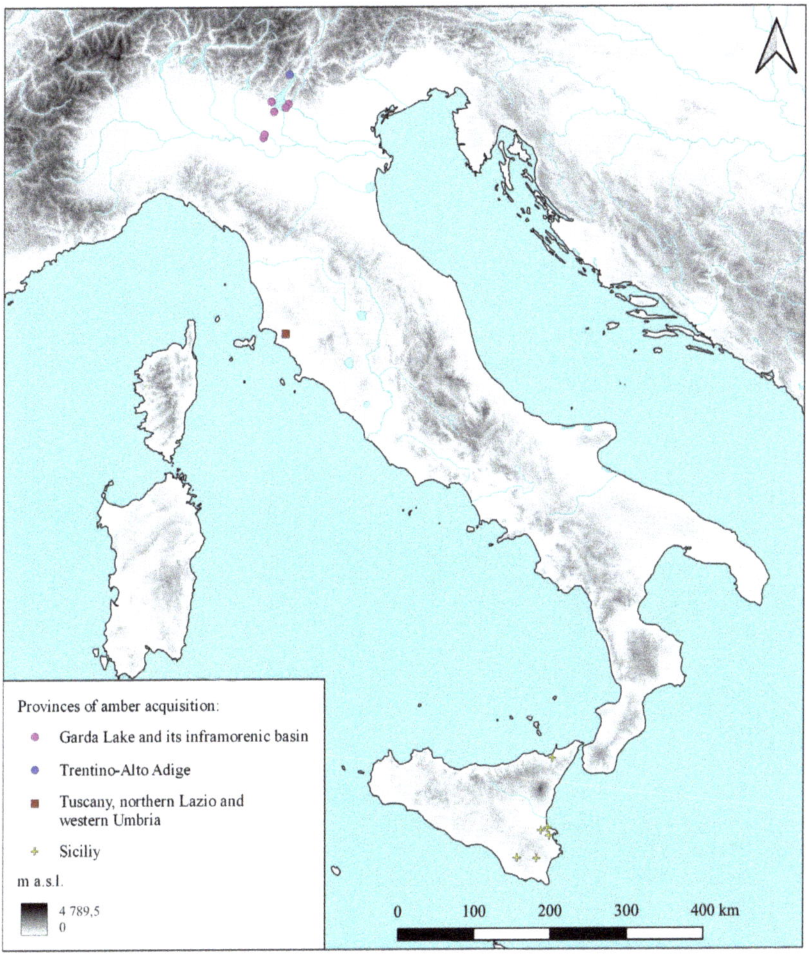

In reality most of the finds from the period resemble discoidal or spheroidal forms, with the differences mainly in the shape of the sides and bases – presumably effects of varying level of craftsmanship. This observation is particularly relevant in the case of Northern Italy, proving that communities inhabiting Trentino and the Garda Lake area during EBA used similar forms of amber jewellery. That the amber finds from the area were distributed within the same network of amber circulation corresponds well with the fact that both provinces at that time appear united under the *Polada* culture (see sections 2.2.4 and 2.2.5). Unification of material culture in the area probably reveals stabilisation of social relations and tightening of communal bonds which facilitated communication and flow of resources. While presence of amber in the earlier – classical stages of the *Polada* culture (BA 1) is uncertain, its occurrence is surely attested for the ultimate phase (BA 2) (1800–1750/1650 BC). Concentration of the sites with amber around the Garda Lake and within its inframorenic basin speaks for non-random spatial pattern in amber's acquisition and use. In this narrow and geographically well-defined area, amber could have easily circulated on a regular basis, although not a grand scale, through various kinds of interactions by communities inhabiting the local pile-dwellings.

Fig. 65. Network analysis of EBA and EBA/MBA contexts with amber grouped into regions.

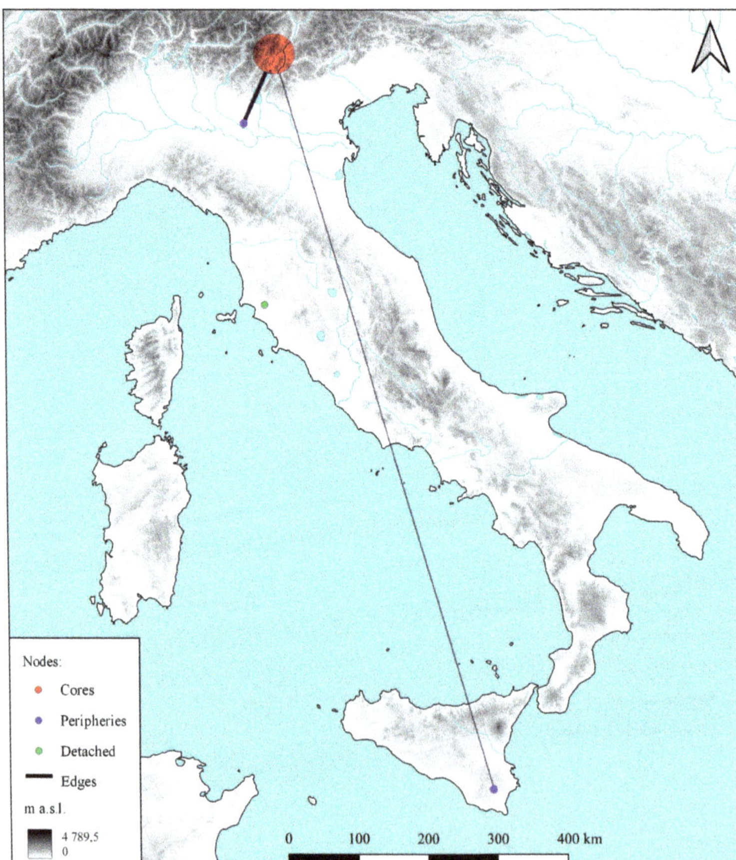

However, some of the finds possess quite specific forms not found elsewhere: e.g. IF611 from Lavagnone, phase 4 (De Marinis 2007, Fig. 2), or IF122z from Grotta dell'Artofago (Fig. 35:22). This last bead found in Tuscany through its morphology resembles numerous artefacts from Northern Italy, however possesses a distinct feature in the form of elongated groove across the perforation's outlet (recess 'z'). This kind of recess may attest to a specific way of drilling the perforation, however might as well be something unwitting. This feature does not appear too often on amber finds from the study area, and stylistically and geographically the closest analogies were found at Lavello, ipogeo 743 (IF224z; Fig. 36:4) and Castione Marchesi, layers 2–3/phases 1–4 (IF262z; Fig. 37:5); both, however, are younger spanning MBA–FBA (ca. 1650–950 BC). It is worth remembering that the context in which the bead from Grotta dell'Artofago was found is not well defined and its chronology ranges from BA 1B to BM 2A (ca. 2100–1500 BC). Still, the cave did not reveal any evidence of long-distance contacts, other than the bead. So determining how or by which way this particular amber artefact came to Tuscany is impossible at the current stage.

Rare types of amber objects were recorded also on Sicily. Tomb 9 at the Castelluccio cemetery, dated to ca. 2200–1800/1700 BC, returned beads in three variants, quite unusual for the period: IF162, IA511 and IE441 (Orsi 1892, Tab. III:15, 18; Massari et al. 1996, Fig. B:5–7). Except for IF162, which is frequently attested in the MBA–LBA period in the northern Adriatic area (cf. section 3.3.1), the above-listed forms remain without direct analogies. Furthermore, amber beads from the tomb were proven to have been made of simetite – fossil resin native to the island (cf. Angelini/Bellintani 2017). Thus, it is allowed to suspect local amber processing in the *Castelluccio* culture, although this was a non-standardised practice, undertaken sporadically, considering the low overall number of amber artefacts and lack of uniformity among them (six types represented by one bead each). In the light of these observations, low *betweenness* centrality and peripheral position of Sicily, as indicated by the network analysis (Fig. 65), should not prejudge the role of the island in amber circulation during EBA (hence it was given the symbol of triangle to distinguish it from the remaining regions).

Interestingly, the only object typologically matching the finds from Northern Italy – Trentino is the bead from Cava Secchiera which turned out to be the only succinite-made item detected among FTIR-analysed EBA amber finds from Sicily (Orsi 1893b, Pl. II; Angelini/Bellintani 2017, Fig. 1:c). Similar morphology might not be enough to confirm Northern Italian provenience of the find. Nor does it offer a clue regarding the way through which it came to Sicily, as any evidence of mutual contacts between the *Castelluccio* and *Polada* cultures, as seen on the basis of the contexts with amber, is elusive.

3.12.2. MBA (1650–1300 BC)

During MBA (ca. 1650–1300 BC) the circum-Adriatic zone witnessed an 'amber boom', visible through a dramatic growth of amber finds and a widening spatial scope of their distribution. In comparison with the preceding phase (EBA), the number of regions in which amber artefacts have been registered increased from 4 to 24 (Fig. 66). Amber's use spread beyond Italy to various parts of Western and Central Balkans. In total there are 130 contexts which pertain to MBA; however, it has to be stressed that 10 of them were classified as EBA/MBA (already included in the preceding EBA network), while 53 chronologically overlap both MBA and LBA. As described in the chronological analysis (chapter 2), these contexts cannot be unambiguously ascribed to a single phase, thus are appearing in more than one network.

Fig. 66. Distribution of the sites with amber finds dated to EBA/MBA, MBA and MBA/LBA in the circum-Adriatic zone.

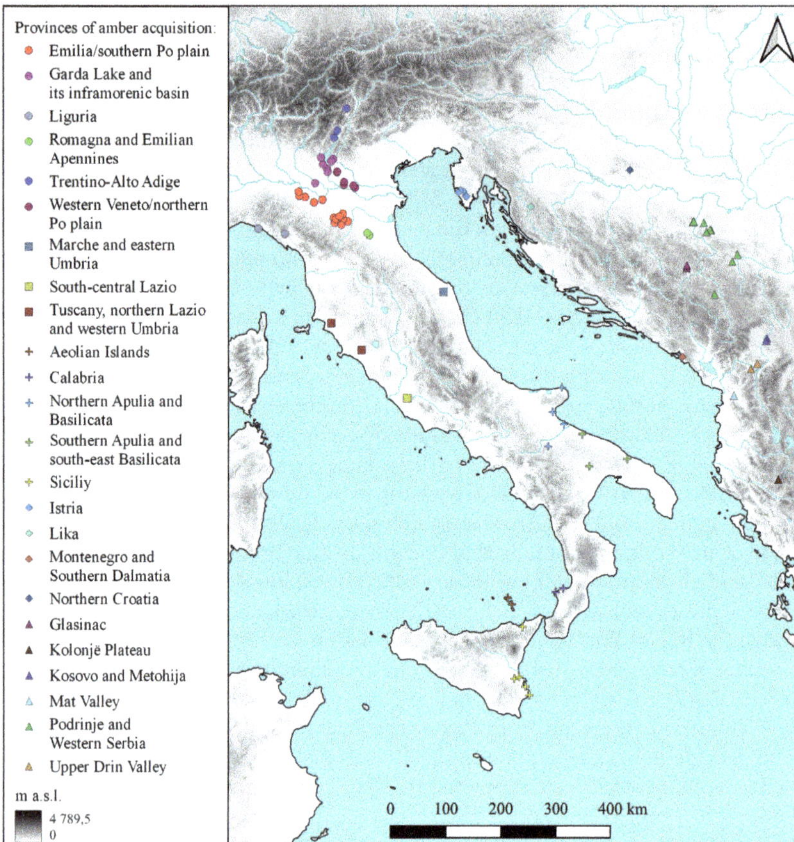

One of the more conspicuous concentrations of sites is visible in the Po Valley, adjacent to the cluster around the Garda Lake evident already in the previous phase. Another, although less dense, group of sites appeared in Apulia and Basilicata. Considerable cluster of sites is still present in the east Sicily but is now accompanied by contexts with amber in the neighbouring Aeolian Islands and Calabria. Furthermore, a complex of sites, quite compact but nonetheless abundant in terms of finds, emerged in Istria. Finally, a significant group of sites with amber burgeoned in Podrinje and Western Serbia. From there other Central Balkan contexts with amber continue south along the West Morava and the Drina Rivers in a somewhat meridional distribution. Some regions located in between above-listed areas also yielded amber finds dated to the period, but usually confined to single sites.

Fig. 67. Network analysis of EBA/MBA, MBA and MBA/LBA contexts with amber grouped into regions.

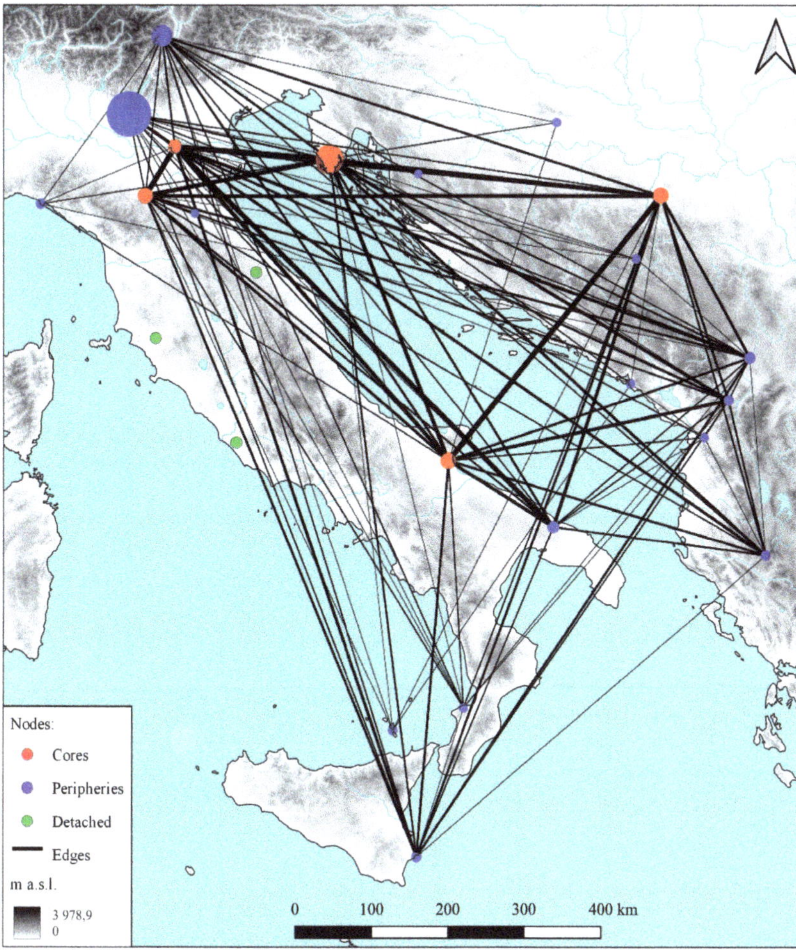

Increased number of MBA provinces of amber's acquisition, in comparison with EBA, resulted in a larger number of nodes (23 in total) and a much more complex network overall (Fig. 67). The network is characterised by a slightly larger level of *density* than the one obtained for EBA graph, amounting to 0.435. This suggests better, but still moderate interconnectedness of nodes. Despite a large number of ties (220 in total) stretching between many nodes, structure of the network is uneven, with certain nodes having fewer and other more connections. This translates into the similarity of amber assemblages in the regions – although the stylistics of amber jewellery during MBA seem rather simple (cf. sections 3.3–3.11), there are certain differences between the regions in terms of typological compositions of amber assemblages and there hardly is one province which documents the whole spectrum of available types. Those differences stem from a number of possible reasons, such as regional preferences towards certain forms of amber jewellery, different suppliers, reliance on manufacturers with varying processing skills and all sorts of other factors that could have had an influence.

It is evident that three nodes situated in Central Italy: Tuscany, northern Lazio and western Umbria, Marche and eastern Umbria and south-central Lazio, are detached from the remaining nodes (they were marked green). It means that, typologically speaking, the MBA amber artefacts found in those regions do not fully match the finds known from other provinces, although they can appear quite similar. On top of that it has to be added that the amber finds from Marche and eastern Umbria (site Ripabianca di Filottrano) could not be classified due to absence of graphic representations of their form; thus, the province was not eligible for network analysis. On the other hand, among the interconnected nodes, it is easy to notice varying thickness of the links which expresses different levels of similarity between pairs of regions. The thickness increases proportionally to the number of shared types (the thinnest links inform that there is only one type in common).

Fig. 68. Tabular summary of core-periphery *analysis of EBA/MBA, MBA and MBA/LBA contexts with amber grouped into regions (presence of a digit at the intersection of a column and a row implies presence of a tie between two regions, while the value gives a number of types they share; calculated with the maximum correlation algorithm described by Borgatti/Everett 1999).*

		14 Is	18 Po	5 Em	4 We	9 No	6 Ro	1 Li	8 So	2 Tr	10 So	11 Ca	12 Ae	13 Si	3 Ga	15 Mo	16 Li	17 No	7 Tu	19 Gl	20 Ko	21 Up	22 Ma	23 Ko
14	Istria		6	8	10	7		1		4	4	1	1	3	3	1		1		1	3	6	1	5
18	Podrinje and Western Serbia	6		7	7	10				4	6	2	1	4	2	1				1	5	4		3
5	Emilia	8	7		10	5				2	3	2	1	4	2	1				1	4	3		2
4	Western Veneto	10	7	10		7		1		3	9	1		5	6					1	6	4	2	5
9	Northern Apulia and Basilicata	7	10	5	7			1		4	6	2		4	2			1			4	6	2	4
6	Romagna and Emilian Apennines														1									
1	Liguria	1			1	1				1	1							1			1			
8	South-central Lazio																							
2	Trentino - Alto Adige	4	4	2	3	4		1			2	2	1	2	2	1		1			2	1		2
10	Southern Apulia and south-east Basilicata	4	6	3	9	6		1		2				2	1						2	2	1	3
11	Calabria	1	2	2	1	2				2				1	1						1	1		
12	Aeolian Islands	1	1	1						1						1								
13	Sicily	3	4	4	5	4				2	2	1			2					1	4	2		1
3	Garda Lake basin	3	2	2	6	2	1			2	1	1		2			1			1	2	1		1
15	Montenegro and southern Dalmatia	1	1	1						1			1											
16	Lika														1									
17	Northern Croatia	1				1		1		1														
7	Tuscany, northern Lazio and western Umbria																							
19	Glasinac	1	1	1	1									1	1						1			
20	Kosovo and Metohija	3	5	4	6	4		1		2	2	1		4	2					1		2		1
21	Upper Drin Valley	6	4	3	4	6				1	2	1		2	1						2		1	3
22	Mat Valley	1			2	2					1											1		1
23	Kolonjë Plateau	5	3	2	5	4				2	3			1	1						1	3	1	

The thickest links are present between the regions classified as cores (red circles), while the thinnest ones predominantly connect peripheral areas (blue circles). The core nodes form a subgroup characterised by high *density* of ties; not only are they fully interconnected between themselves with strong ties suggesting many types in common but they also possess links to most of the nodes defined as peripheries (Fig. 68). Conversely, the peripheral regions are more connected to the core areas and less between themselves. The *density* level among the cores amounts to 7.700 and between the peripheries to 0.405. Cores with their numerical preponderance of amber artefacts over other regions, and bigger representativeness of the typological variety of the finds, can be considered as chief recipients of amber. In total there are five cores in the network: Emilia, western Veneto, Istria, northern Apulia and Basilicata, and Podrinje and Western Serbia. Interestingly, all five macro-regions are represented by at least one core, with the exception of Central Italy which is disconnected from the network as explained above. Only Northern Italy has two cores: Emilia and western Veneto.

Tab. 45. Regions with EBA/MBA, MBA and MBA/LBA contexts with amber ranked according to their coreness *(calculated with the Minres algorithm described by Borgatti/Everett 1999),* betweenness *(calculated with the algorithm described by Freeman 1979) and* degree.

Province:	Coreness:	Betweenness:	Norm. Betweenness:	Degree:	Norm. Degree:
Aeolian Islands	0.027	0.000	0.000	5	0.227
Calabria	0.084	0.000	0.000	10	0.455
Emilia/southern Po plain	0.327	7.224	3.127	15	0.682
Garda Lake basin	0.160	38.057	16.475	15	0.682
Glasinac	0.043	0.000	0.000	7	0.318
Istria	0.364	20.674	8.950	18	0.818
Kolonjë Plateau	0.194	1.333	0.577	12	0.545
Kosovo and Metohija	0.225	2.924	1.266	14	0.636
Liguria	0.039	0.750	0.325	7	0.318
Lika	0.004	0.000	0.000	1	0.045
Mat Valley	0.053	0.000	0.000	6	0.273
Montenegro and southern Dalmatia	0.027	0.000	0.000	5	0.227
Northern Apulia and Basilicata	0.361	7.817	3.384	15	0.682
Northern Croatia	0.021	0.000	0.000	4	0.182
Podrinje and Western Serbia	0.357	7.224	3.127	15	0.682
Romagna and Emilian Apennines	0.004	0.000	0.000	1	0.045
Sicily	0.210	1.057	0.458	13	0.591
South-central Lazio	0.000	0.000	0.000	0	0.000
Southern Apulia and south-east Basilicata	0.265	2.917	1.263	13	0.591
Trentino – Alto Adige	0.184	12.983	5.620	16	0.727
Tuscany, northern Lazio and western Umbria	0.000	0.000	0.000	0	0.000
Upper Drin Valley	0.222	1.783	0.772	13	0.591
Western Veneto/ northern Po plain	0.420	5.257	2.276	15	0.682

All the nodes can also be ranked by their *coreness* – valued indication of closeness to the core of each actor for continuous (ratio-level) *core-periphery* model (cf. Borgatti/ Everett 1999). Among the cores, region with the greatest *coreness* is western Veneto, followed by Istria, northern Apulia and Basilicata, Podrinje and Western Serbia and, finally, Emilia (Tab. 45). In this order they can be regarded as the focal points towards which amber was gravitating. Assuming that communities inhabiting above-listed regions had the power and capacity to obtain considerable quantities of amber in various forms, it seems safe to say that they most probably formed the 'skeleton' of the network through which amber was circulating, and by which it was most probably spreading to the peripheries. Regions with *coreness* ranging from 0.265 to 0.160 (Tab. 45) can be considered as semi-cores as they also yielded considerable quantities of morphologically varied amber artefacts, but with narrow distribution, indicating more restricted access to amber and influence over what was being received. These regions usually comprise fewer contexts with amber than the cores. Remaining nodes are characterised by very low *coreness*, suggesting occasional access to amber, the latter usually limited to one or two types.

Distribution of the types has significance for linking the nodes, especially in relation to peripheral regions which usually posses small and undifferentiated assemblages of amber artefacts. Not many of them has the ability to reach all remaining nodes with one *geodesic* (direct link), and the connection is allowed only through intermediate ones. Thus, certain nodes situated on the shortest path between pairs of indirectly connected regions are said to be in a privileged position. This characteristic can be measured with the *betweenness* centrality which indexes the percentage of all geodesic paths from neighbour to neighbour that pass through a given node (Freeman 1979). Furthermore, values of this centrality measure can be divided by the maximum possible *betweenness* in a network and expressed as a percentage, thus giving the *normalised betweenness* values (Tab. 45). Additionally, one can include in the consideration *degree* centrality, i.e. numbers of ties adjoining a node, which can be also normalised through dividing the value by the maximum possible *degree* the node would have in an ideal situation (i.e. when connected to all other nodes) (Tab. 45).

All the core regions are characterised by slightly increased levels of *betweenness* (as expressed by the sizes of their nodes) (Fig. 67; Tab. 45). However, the two regions with the index reaching the highest values are the Garda Lake basin and Trentino-Alto Adige, both located in the northernmost part of the study area and not classified as cores. They owe their *betweenness* to the links with several remote regions with low *degree* levels, namely Liguria, Romagna, Lika and Northern Croatia, while at the same time being well connected to all the cores and most peripheries. It was already mentioned that the Garda Lake basin and Trentino-Alto Adige are among the nodes with a medium *coreness* levels. Still, their location at the northern border of the circum-Adriatic zone, near the areas vital for the inflow of amber, in connection with a longer tradition of amber import going back to EBA, made them important hubs in the network of amber circulation. Quite probably it is there that the material first arrived and from where it was subsequently redistributed. Similar role could have been played by certain regions with large levels of *betweenness* located further south, especially the cores which were potentially responsible for redistribution of amber in their respective macro-regions.

It may seem an overestimation to decide that the regions in which amber has been found in abundance played the role of redistribution centres; after all, they could have only obtained the material for their own needs – conclusion particularly plausible in cases, where amber finds are known primarily or only from graves. However, it is possible the communities acquiring amber through a range of contacts had the ability of transferring it further, especially to neighbours with whom they kept cultural and economical relations. People interact in various ways, and objects, such as amber, are often a part of those interactions – especially, if they are members of an organised community, sharing same culture, language and territory. To determine where discussed interactions most probably took place, and how all the indexes applied in NA described above (*similarity, coreness, betweenness*) translate into the prehistoric realities, it is necessary to include in the considerations other data related to analysed finds and sites.

It has already been mentioned that Trentino-Alto Adige and the Garda Lake basin possibly played the roles of important intermediaries in amber transfer. They manifest numerous links to various provinces, particularly strong in the case of cores, among them Emilia and western Veneto being the nearest. Many other ties, however, stretch beyond Northern Italy to reach more distant parts, implying that some forms of amber adornments were used over a vast areas.

Focusing on the nearest neighbours adjacent to Trentino-Alto Adige and the Garda Lake, it is a priority to tackle their relations with Emilia and western Veneto. Relationships between these four provinces during MBA were evidently tight, as documented by many similarities in the sphere of material culture, the latter being often referred to as the *Palafitte-Terramare* culture (see sections 2.2.4–2.2.7). Contacts between the provinces can be effectively shown on the basis of numerous shared types of bronze items, found on the sites which yielded amber finds datable to MBA. Bronze objects were chosen as the main basis for further considerations, as similarly to amber, both copper and tin have geographically limited occurrence, their smelting was a profession of skilled artisans and the products had to be channelled into circulation in order to reach metal-poor areas.

In the contexts of interactions within the scope of Northern Italy, the most illustrative products of metallurgy are pins representing types S. Caterina (Fig. 69:A–E), with three rings (Fig. 69:F–K) and Pieve S. Giacomo (Fig. 69:L–M), to name but a few (cf. Carancini 1975). The second category of bronze finds showing interregional similarities are daggers which include types S. Ambrosio, var. B (Fig. 70:A–B), Pieve S. Giacomo, var. B (Fig. 70:C–E), Capurso, var. A (Fig. 70:F–G), Castellaro di Gottolengo, var. B (Fig. 70:H–I), S. Agata (Fig. 70:J–K) and Bacino Marina (Fig. 70:L–M) (cf. Bianco Peroni 1994). To those two numerically preponderant categories can be added less numerous similarities among axes with raised edges of the type Montata dell'Orto or Montale (Fig. 71:A–D) (cf. Carancini 1997, Figs. 223:22–23 and 227:19–22) and some of the early variants of axes with median fins, among the latter the Grenchen type in particular (Fig. 71:E–F) (cf. Abels 1972, 80). Finally, to the common items can be ascribed bronze swords of the type Bigarello (Fig. 71:G–H) (cf. Bianco Peroni 1970, 16–19) and pinheads (*rotelle*) of the type Gazzade (Fig. 71:I–J) (cf. Carancini 1975, 325; 1997, Fig. 234:88–90).

Abovementioned examples are only a portion of similarities shared by Trentino-Alto Adige, the Garda Lake basin, Emilia and western Veneto during MBA. To render a full picture of interactions between these regions, one would have to recall numerous other finds not associated with amber, not to mention a multitude of common features observed on the basis of pottery. Still, this handful of analogies among bronze artefacts proves that there was a circulation of ready-made products, casting forms and metallurgical technology, perhaps carried by specialists in this field – metallurgists practising particular techniques of bronze processing, in a same way as other artisans had processed amber.

Fig. 69. Examples of bronze pins shared by some of the MBA sites with amber in the Garda Lake basin (D, F, G, J, L), western Veneto (E, H, K, M) and Emilia (A–C, I): A – Anzola dell'Emilia (Finotelli et al. 1997a, Fig. 200:20); B – Redù-Pilastro (Malnati/Cardarelli 2003, Fig. 58:9); C – Sant'Ambrogio (Cremaschi 1997, Fig. 179:3); D, G, J, L – Isolone del Mincio (Guerreschi et al. 1985, Tav. XXII:st.8920, st.9111, st.9153, st.9216); E – Scalvinetto, tomb 315 (Salzani 2004, Fig. 4:C); F – Castellaro Lagusello, layer C/C1 (Nava 1982, Fig. 25); H – Olmo di Nogara, grave 154 (Salzani 2005, Tav. XVI:F); I – Gaggio, 3rd and 4th phase of the settlement (Balista et al. 2008, Fig. 24:10); K – Olmo di Nogara, grave 89 (Salzani 2005, Tav. X:A); M – Olmo di Nogara, grave 112 (Salzani 2005, Tav. XII:B).

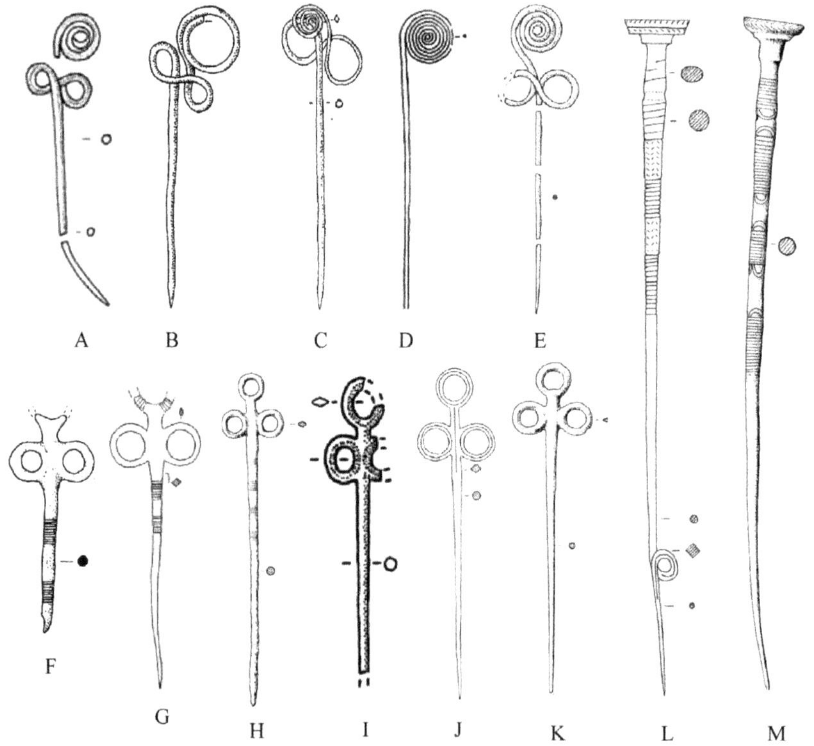

Fig. 70. Examples of bronze daggers shared by some of the MBA sites with amber in the Garda Lake basin (A, G, I, K), western Veneto (B, E, M) and Emilia (C, D, F, H, J, L): A – Castellaro Lagusello, layer C/C1 (Nava 1982, Fig. 11); B – Povegliano Muraiola, phase IV (Salzani 1997b, Tav. 55:5); C – Casinalbo (Bianco Peroni 1994, Tav. 42:669); D, F, J – Montale, phase II (Carancini 1997, Fig. 228:29–30); E – Olmo di Nogara, grave 411 (Salzani 2005, Tav. XXXIV:E); G, I, K, – Isolone del Mincio (Guerreschi et al. 1985, Tav. XX:st9023; Bianco Peroni 1994, Tav. 62:1096); H, L – Redù-Pilastro (Bianco Peroni 1994, Tav. 37:557, 62:1102); M – Finilone Valle (Salzani 1996, Fig. 21:2).

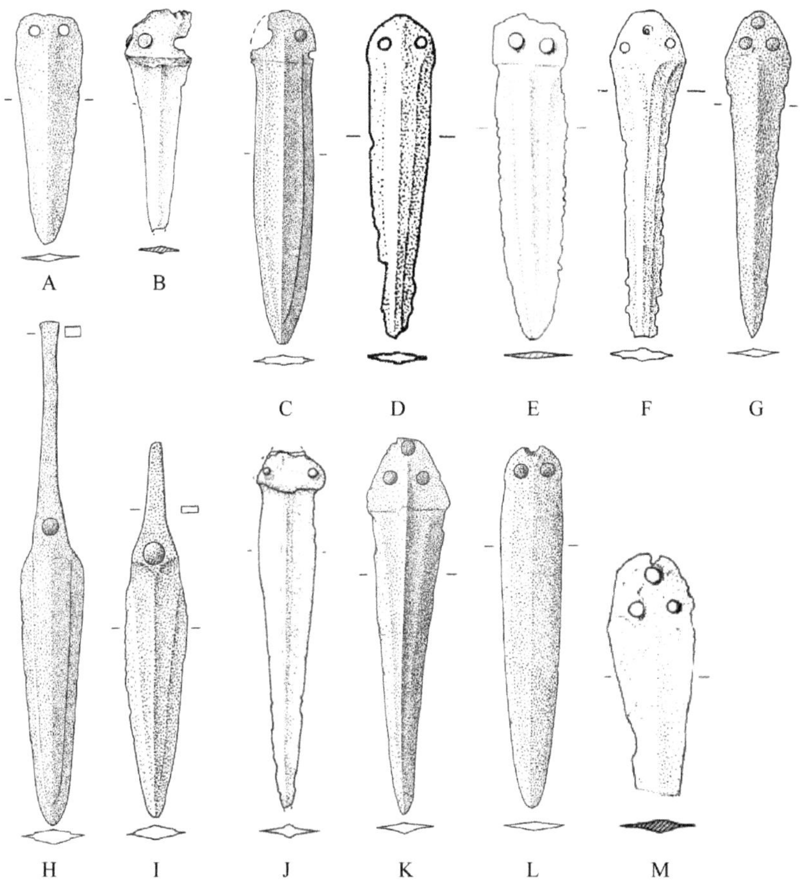

Fig. 71. Examples of bronze axes, swords and pinheads (rotelle) *shared by some of the MBA sites with amber in Trentino-Alto Adige (F, H), the Garda Lake basin (D, J) and Emilia (A–C, E, G, I): A, I – Casinalbo (Carancini 1975, n. 2646; Carancini 1997, Fig. 227:20); B, E – Castione Marchesi, layers 2–3/phases 1–4 (Mutti et al. 1988, Fig. 71:3–4); C – Montale, phase II (Carancini 1997, Fig. 227:19); D, J – Isolone del Mincio (Guerreschi et al. 1985, Tav. XXI:st.9348, XXII:st.9057); F – Fiavè-Carera, horizon 6 (Perini 1987, Tab. III:16); G – Gaggio, 1st phase of the settlement (Balista et al. 2008, Fig. 19:9); H – Molina di Ledro, strata I–III/IV (Rageth 1974, Taf. 26:1).*

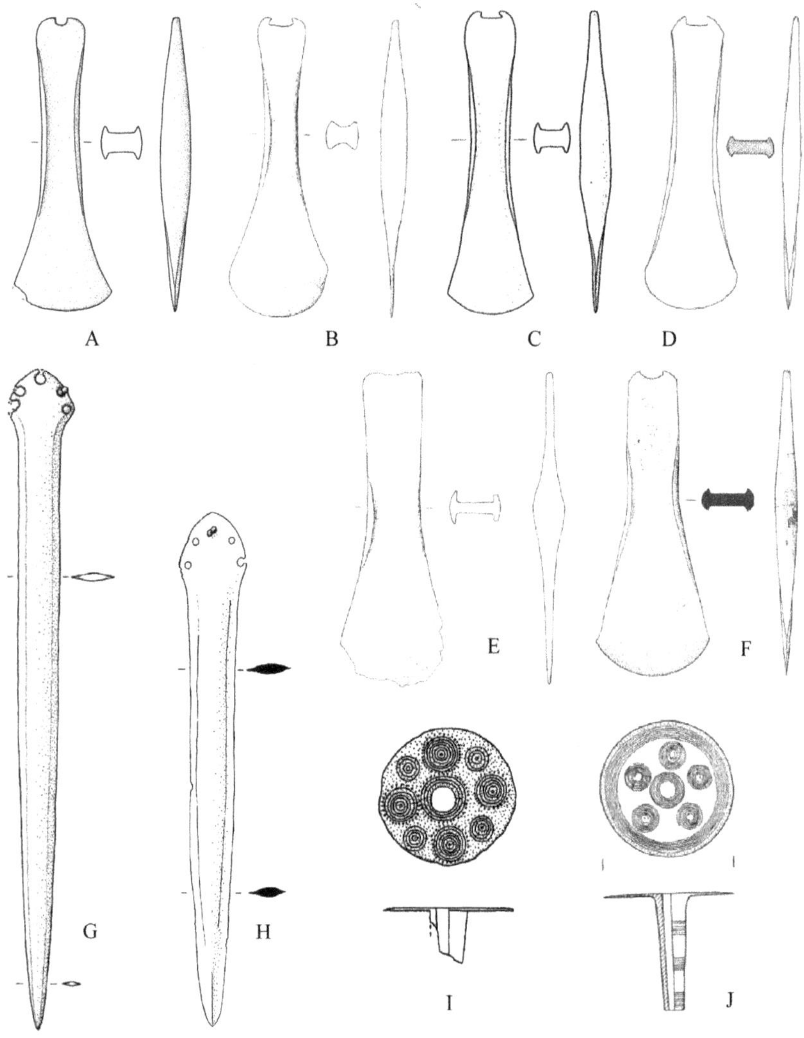

Relations between the domain of the *Palafitte-Terramare* culture and the neighbouring provinces of amber's acquisition, indicated by the results of NA, are less tangible through other categories of data. Amber beads of types IF151 and IA122 found in the Ligurian settlements of Castellaro di Camogli (Fossati/Milanese 1982, Tav. V:89) and Castellaro di Zignago (Mannoni/Tizzoni 1980, Fig. 7:s) have been confirmed in many contexts located further east, especially in the regions characterised by considerable level of *betweenness* (e.g. Trentino-Alto Adige, Emilia, western Veneto, northern Apulia and Basilicata, Istria and Kosovo and Metohija). While relations with more distant among abovementioned provinces are difficult to grasp on the basis of available material, some traces of contacts with closer neighbours seem to be reflected through pottery found on the Ligurian sites. Bead IA122 from Castellaro di Camogli finds most analogies in Trentino,[531] whereas bead IF151 from Castellaro di Zignago in western Veneto[532] (see also sections 3.3.1 and 3.3.2). Analysis of the pottery found at the former sites revealed, apart from evident contacts with the closely situated facieses of *Scamozzina-Alba* and *Viverone* in Lombardy and *Grotta Nuova* in Tuscany, particularly strong associations with the *Terramare* culture of the Po Plain (Delfino 2010, 429–430). Among the ceramic materials from Castellaro di Zignago were observed affinities with the *Palafitte-Terramare* culture of the more advanced stage that, together with the numerically preponderant influences of the *Canegrate* culture of advanced BR, open the possibility of a later – LBA origin of amber materials from the site (Mannoni/Tizzoni 1980, 277–278). Nonetheless, in both cases ceramic indicators point to the east, potentially showing a direction from which amber artefacts were brought to Liguria.

Looking farther to the east of the circum-Adriatic zone, one cannot avoid noticing the strong ties between the Northern Italian and the Western Balkan provinces of amber's acquisition during MBA, i.e. Istria, Lika and Northern Croatia. Majority of amber finds from these regions do not exemplify any particular forms or decorative features which would clearly set them apart from amber assemblages present at the time in other parts of the study area – most of them represent simple, generally discoidal forms. However, there is a notable co-occurrence of finds with biconical profile which can be divided into discoid-lenticular types (e.g. IF162, IF262), and average or flattened carinated beads (e.g. IA161, IF161, IF261) (cf. sections 3.3.1.2 and 3.3.2.2). In Istria examples of discoid-lenticular and flattened biconical beads were found at cemeteries of Mušego, Vrčin and Žamnjak.[533] With respect to biconical beads, the closest parallels come from sites of the *Palafitte-Terramare* culture, both settlements and graves.[534]

531 Molina di Ledro, strata I–III/IV (Rageth 1974, Taf. 92:17–18).

532 Povegliano Muraiola, phase IV (Salzani 1997b, Tav. 56:15); Olmo di Nogara, grave 388 (Salzani 2005, Tav. XXX:F).

533 Bačić 1960, Tab. VI:1; Buršić-Matijašić 1989, Fig. 1:1–2, 10–11; Mihovilić et al. 2009, 55.

534 E.g. Isolone del Mincio, precise context unknown (Negroni Catacchio et al. 1999, Fig. 12:21); Castione Marchesi, layers 2–3/phases 1–4 (Mutti et al. 1988, Fig. 81:6–7, 9); Olmo di Nogara, graves 85 (Salzani 2005, Tav. IX:E–F); Scalvinetto di Legnago, graves 233 and 354 (Salzani 1994, Tav. 5:7–8; 2004, Fig. 4:G).

Furthermore, sites from the same cultural ecumene offer the best counterparts for the Istrian lenticular beads.[535]

There are few other finds from the Istrian contexts with amber which can be cited as testimonies of contacts with Northern Italy. For sure, genetic ties with some of the hillforts in Fruli belonging to the *Castellieri/Gradine* culture can be traced on the basis of pottery; common ceramic vessels, in particular the handles, are well documented in the case of the Vrčin hillfort (Buršić-Matijašić 1997; cf. Cardarelli 1983). Bronze objects are usually poorly represented in the Istrian contexts, and those indicative in terms of cultural affiliation and chronology are very scarce. Other metal objects associated with amber in Istria, i.e. jewellery made of spirally wound wire or tape such as rolls or tubes (*saltaleoni*), rings (*fermatrecce*) and bracelets with spiral endings, or domed buttons with a lug on the inner side, do not provide precise comparisons.[536] They quite commonly appear together with amber in Northern Italian contexts though.[537] The evidence presented here can be supplemented with the clay bread-loaf idol (ger. *brotlaibidol*; it. *oggetto enigmatico*) with an oval outline and five circular, impressed markers with inscribed cross symbols from the Vrčin hillfort (Fig. 72:A). The style of decoration, corresponding with the sign CC1 *coppella con croce, segno base* (cf. Piccoli/Zanini 1999), is present on numerous similar idols from Northern Italian sites, including those which delivered amber artefacts (Fig. 72:B).

535 E.g. Sant'Ambrogio, precise context unknown (Negroni Catacchio 1975b, Fig. 2); Castione Marchesi, layers 2–3/phases 1–4 (Mutti 1988, Fig. 81:1–3); Isolone del Mincio, precise context unknown (Negroni Catacchio et al. 1999, Fig. 11:11–12); Santa Rosa di Poviglio, *Villaggio piccolo* (Bellintani et al. 2004, Fig. 274:2); Olmo di Nogara, graves 112, 172 and 411 (Salzani 2005, Tav. XII:E, XVII:D, XXXIV:G–H).

536 Žamnjak, tumulus (Bačić 1960, Tab. VI:1); Vrčin, graves 7–8, 11–12, tumulus B (Buršić-Matijašić 1989, Figs. 1:4–8, 2, 3:3–8 and 4:9–15); Škicini, tumulus (Buršić-Matijašić 2003, Figs. 6 and 7); Mušego, tumuli 7 and 8 (Mihovilić et al. 2009, 55).

537 Castione Marchesi, layers 2–3/phases 1–4 (Mutti et al. 1988, Fig. 78:13–14); Fiavè-Carera, horizon 6 (Perini 1987, Tab. III:13, 17–18); Scalvinetto, tomb 233 (Salzani 1994, Tav. 5:1–4); Olmo di Nogara, graves 388 and 411 (Salzani 2005, Tav. XXX:G, L, XXXIV:B, I, M).

Fig. 72. Examples of clay bread-loaf idols shared by some of the MBA sites with amber in Istria (A), the Garda Lake basin (B), Northern Croatia (C) and Emilia/southern Po Plain (D): A – Vrčin (Mihovilić et al. 2017, 25, n. 8); B – Isolone del Mincio (Guerreschi et al. 1985, Tav. XVIII:ED821524); C – Alilovci Lipje, trench A/large house (Mavrović Mokos 2019, Fig. 3); D – Gaggio, 1st phase of the settlement (Balista et al. 2008, Fig. 19:8).

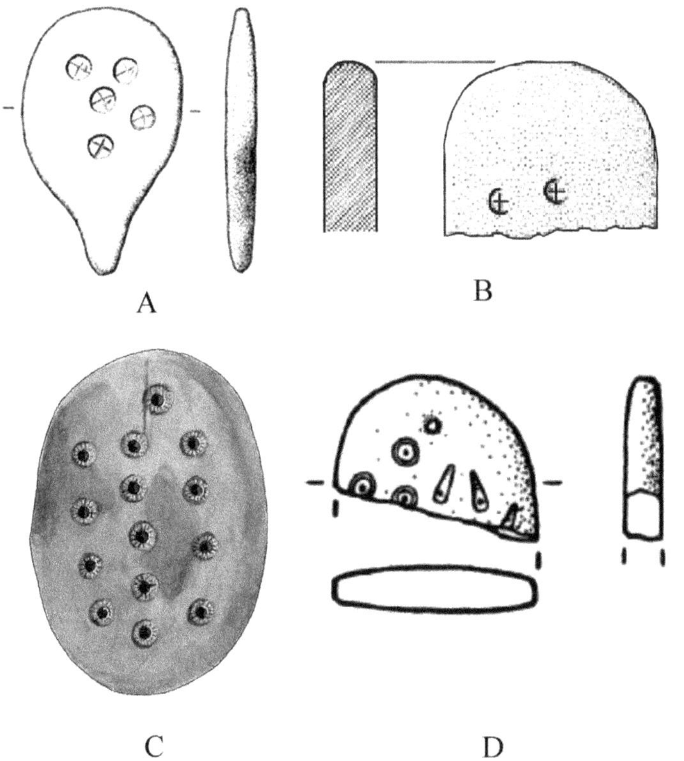

In the middle of the Požega Valley in Northern Croatia, a single amber bead type IA122 was found in the so-called test 'trench E/small house' on the site Alilovci Lipje (Mavrović Mokos/Pavličić 2015, 21, Fig. 16). From the adjacent trench A/large house comes a clay bread-loaf idol (Fig. 72:C). The idol from Alilovci Lipje is rather simply decorated with round impressed markers with central point and radial incisions/notches (marker/sign C6 *con punto centrale e campitura raggiata* according to Piccoli/Zanini 1999), arranged without a particular pattern, thus can be inserted into group 3 in the proposition by S. Carafa (Carafa 2013, 136–139). This group of clay idols is richly represented in Italy, also on some of the sites which yielded amber artefacts (Fig. 72:D). Not counting the collections from two other Istrian hillforts of Monkodonja and Nezakcij, the abovementioned Balkan bread-loaf idols are isolated findings, discovered far from the main concentrations of similar objects (Mihovilić et al. 2017). Thus, the cited idols from Northern Italy may be regarded as the closest counterparts, both in morphological as well as geographical sense (cf. Cupitò/Leonardi 2015, 215;

Cupitò et al. 2015). Still, the Croatian idols are not missing more distant analogies in other parts of Europe either, e.g. Southern Germany (cf. David 2016), Poland (cf. Fogel/Langer 1999) or the Carpathian Basin (cf. Bándi 1974). The latter evidence goes hand in hand with an abundant ceramic material from Alilovci Lipje that strongly manifests influences of the Carpathian Basin cultures, chiefly the *Litzen*-type pottery, and form of the amber bead which has also been documented in Hungary in chronologically equivalent contexts (Sprincz/Beck 1981, 481, Fig. 3:31, Tab. 2).

In the region of Lika, separated from the Adriatic Sea only by the Velebit mountain range, a discoid amber bead with biconical, slightly skewed profile (IF167), similar to the one found in the Garda Lake basin, was discovered in the mass grave no. 4 of the Bezdanjača cave near Prozor.[538] The grave falls into the Ist horizon of the cave (ca. 1400–1250 BC), so is coeval with the heyday of the *Palafitte-Terramare* culture in the Po Plain – a period to which belong many analogies cited above. But similarities end on the amber bead, not counting simple bronze adornments such as domed buttons with lugs or *saltaleoni* also known from Italy[539] which, however, serve as poor cultural and chronological indicators. Remaining materials found inside the cave show southern links to the hillforts of Bosnia and Herzegovina and northern ones with Pannonia (cf. Karavanić 2007, 43–44).

More perspicuous residues of interactions are tangible along the north–south axis, between the regions of *Caput Adriae*, including the Garda Lake basin, Emilia, western Veneto and Istria, on one hand, and Central and Southern Italian provinces of amber's acquisition on the other. Similarities among some of the bronze artefacts are recognisable on the sites with amber from the Po Plain, through Romagna, all the way down to Apulia. Two closely located settlements in Romagna – Monte Castellaccio and San Giuliano di Toscanella Imolese – yielded amber beads which hardly have exact analogies in the circum-Adriatic zone: IF164w and IF124y, respectively (Pacciarelli 1996c, Tav. 1:F; Miari 2007a, Fig. 3:22). While the first of the two types remains without a counterpart, the second one is represented among the materials from Castellaro Lagusello, layer D (Negroni Catacchio et al. 1999, Fig. 10:1). Nonetheless, the morphologically similar specimens (differing only in regularity of the cross-sections or the manner of perforations) were discovered in many contemporaneous contexts in the Garda Lake basin, Emilia, western Veneto and northern Apulia.[540] From the same regions, and in some cases from the same contexts, are known analogies for the bronze items discovered on the two sites from Romagna. To the level/phase 2 at Monte Castellaccio is attributed at least one dagger blade representing the namesake Monte Castellaccio type which was confirmed also at Isolone del Mincio (Fig. 73:A–B) (cf. Bianco Peroni 1994, 88–89). This observation fits well into the notion of a general influx of the *Palafitte-Terramare* influences on the *Farneto-Monte Castellaccio* facies

538 Drechsler-Bižić 1979–1980, Tab. XXV:6; cf. Isolone del Mincio, precise context unknown (Negroni Catacchio et al. 1999, Fig. 12:19).

539 Drechsler-Bižić 1979–1980, Tab. XXV:4–5; cf. Isolone del Mincio, precise context unknown (Guerreschi et al. 1985, Tav. XXIII).

540 Castione Marchesi, layers 2–3/phases 1–4 (Mutti et al. 1988, Fig. 81:11); Olmo di Nogara, graves 422 and 435 (Salzani 2005, Tav. XXXV:C, XXXVI:A); Coppa Nevigata, *Recent Apennine* stage (Cazzella/Moscoloni 1999, Fig. 11:3); Madonna di Loreto, 'Hypogeum of Ivories' (Tunzi Sisto 2006, Fig. 1:1).

during BM 1b–2 (see section 2.2.8). The neighbouring, but slightly later, settlement of San Giuliano di Toscanella Imolese delivered bronze daggers in three different types that can be correlated with specimens from Emilia and northern Apulia, including types Campegine, var. A (Fig. 73:C–D) and Manaccora (Fig. 73:E–H) (cf. Bianco Peroni 1994, 101–108, 146–148). Presence on the site of bronze objects dispersed both to the north and the south of Italy illustrates well the peculiar position of Romagna during BM 3–BR at the junction of the *Terramare* and the *Apennine – Subapennine* cultures (see section 2.2.8). It appears that the two contexts with amber in Romagna have much in common with other Italian regions that together with general similarity of amber artefacts (despite the slight differences in some aspect of their morphology) render plausible the circulation of amber down the Apennine Peninsula along the Adriatic coast.

Fig. 73. Examples of bronze daggers shared by some of the MBA sites with amber in Romagna (A, C, E), Garda Lake basin (B), Emilia (D, H), and northern Apulia (F–G): A – Monte Castellaccio, level/phase 2 (Bianco Peroni 1994, Tav. 45:735); B – Isolone del Mincio, precise context unknown (Bianco Peroni 1994, Tav. 46:743); C, E – San Giuliano di Toscanella, excavations of 1901 (Morico 1996, Tav. 1:3503, 3506); D – Redù-Pilastro (Malnati/Cardarelli 2003, Fig. 58:2); F – Grotta Manaccora, Grotticella Funeraria *– SS layers (Bianco Peroni 1994, Tav. 80:1456); G – Madonna di Loreto, 'Hypogeum of Bronzes' (Peroni 1999a, Fig. 1:2); H – Anzola dell'Emilia (Finotelli et al. 1997a, Fig. 200:18).*

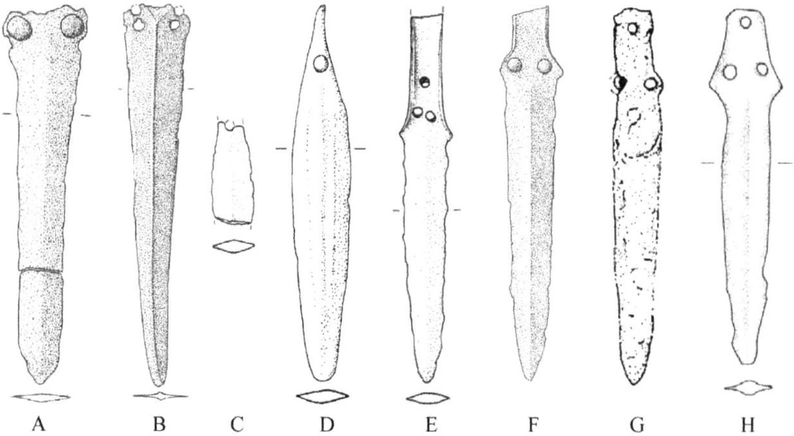

Fig. 74. Examples of bronze pins and daggers shared by some of the MBA sites with amber in northern Apulia and Basilicata (A, D–I, K, L–N, P, T), Trentino-Alto Adige (O), the Garda Lake basin (B, F, R), western Veneto (C, J), and south-central Lazio (S): A – Grotta Manaccora, Grotticella Funeraria *– SS layers (Carancini 1975, Tav. 56:1855); B – Ca' Nova di Cavaion (Rapi 2011, Fig. 4:3); C – Finilone Valle (Salzani 1996, Fig. 22:1); D, E, I, P – Madonna di Loreto, 'Hypogeum of Ivories' (Peroni et al. 2003, Fig. 12:5957, 5813, 5960, 5975); F, R – Isolone del Mincio (Guerreschi et al. 1985, Tav. XX:st.9245, XXII:st.9140); G, H, K, N – Madonna di Loreto, 'Hypogeum of Bronzes' (Tunzi Sisto 1999a, Tav. XVI:827; Peroni 1999a, Figs. 1:3 and 2:3); J – Povegliano Muraiola, phase II (Salzani 1997b, Tav. 55:10); L, M, T – Toppo Daguzzo, tomb 3 (Cipolloni Sampò 1986, Fig. 8:1, 5); O – Fiavè-Carera, horizon 6 (Perini 1987, Tab. III:27); S – Villaggio delle Macine (Angle et al. 2016, Fig. 3).*

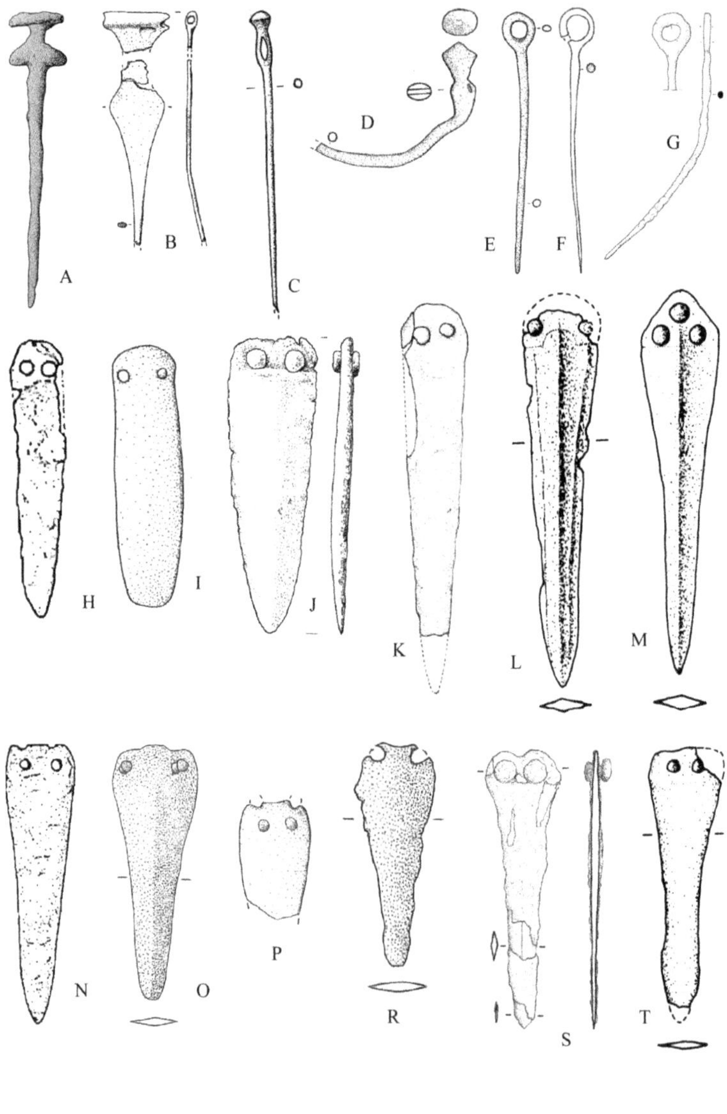

Northern Apulia and Basilicata – the core province of amber's acquisition in Southern Italy during MBA – yielded many more clues of interactions with Northern Italy, the Po Plain in particular. Similarity of amber beads between this region and Emilia as well as western Veneto is clearly shown by the results of NA (Fig. 67). Correspondence among amber adornments is matched by similarity of bronze artefacts co-occurring with them in the contexts. The two most abundant categories of finds demonstrating interactions are, again, bronze pins and daggers. The former include the types Porto di Pacengo (Fig. 74:A–B), Boccatura del Mincio/Cattaragna (Fig. 74:C–D) and with the single ring-head (Fig. 74:E–G) (cf. Carancini 1975; Rapi 2011). Among the daggers are present examples of the types S. Ambrosio in var. A (Fig. 74:H–J) and C (Figs. 70:A–B and 74:K–L), Capurso, var. A (Figs. 70:F–G and 74:M) and Gualdo Tadino (Fig. 70:N–P) (cf. Bianco Peroni 1994).

This spectrum of shared bronze items can be supplemented with swords of the type Castione Marchesi (cf. Bianco Peroni 1970, 16–17).[541] Above-listed similarities between amber-bearing sites show a kind of relationship between the Po Plain and northern Apulia-Basilicata (cf. Peroni 1999a, 217). A valuable comment in this respect is that a recent lead-isotope analysis of swords from Grotta Manaccora and Trinitapoli representing the types Manaccora and Pertosa – forms absent among bronze swords in the Po Plain – showed that they most probably were made from southern Alpine copper, thus adding a new dimension to the reconstructed relationship between Northern and Southern Italy during MBA (cf. Jung et al. 2011; Jung/Mehofer 2013). These contacts might have facilitated the influx of amber to the southern Apennine Peninsula. A potential intermediary role of Romagna has already been mentioned; however, presence of amber in other Central Italian regions demands attention as well.

As described in sections 2.2.9–2.2.12, finding amber materials from Tuscany, Umbria, Marche and Lazio surely dating back to MBA is very problematic, as most of them appeared in mixed contexts, alongside materials chronologically spanning BM-BF. Furthermore, none of amber artefacts from those provinces has adequate analogy in the study area. In south-central Lazio, there is only one context datable to MBA; however, it offers better chances of attaching it to the network. The amber bead from Villaggio delle Macine is also one of its kind by having convex recess around perforation's outlet (IF123y; Fig. 35:25), however can be confronted with other similar beads with hourglass-shaped perforations, i.e. IF121y and IF221y known from several coeval context located in Italy.[542] Enlarged perforations' outlets and similar ovoidal form of the cited beads, together with parallel chronology of the finds (BM 1–2), give certain credence to a possibility of them being made in with similar techniques, maybe by the same person. Villaggio delle Macine also yielded i.a. an Ortucchio type bronze dagger (Fig. 74:S) and a green-blue glass bead.[543] Similar daggers are known from two Italian MBA sites with amber: one in the Garda Lake basin and the other

---

541 Castione Marchesi, layers 2–3/phases 1–4 (Bianco Peroni 1970, Fig. 3.18); Madonna di Loreto, 'Hypogeum of Bronzes' (Tunzi Sisto 1999b, Tav. 2:1139).

542 E.g. Fiavè-Carera, horizon 6 (Perini 1987, Fig. 68:a1, a2, a4); Lipari 'Acropolis', hut Delta XXI (Bernabò Brea/Cavalier 1980, Pl. 98:i); Santa Domenica di Ricadi, tomb 4 (Pacci 1987, Fig. 4:10); Toppo Daguzzo, tomb 3, burial no. 3 (Cipolloni Sampò 1986, Fig. 9:5).

543 Bellintani et al. 2007, Fig. 1; Angle et al. 2016, Fig. 3; cf. Bianco Peroni 1994, 89–91.

in northern Basilicata (Fig. 74:R, T). The glass bead, in turn, upon chemical analysis revealed composition resembling high magnesium glasses (HMG) which were used during MBA only in Southern Italy, and to which were assigned i.a. glass finds from Grotta Manaccora and Coppa Nevigata (Angelini et al. 2001, 2006). The settlement of Villaggio delle Macine apparently combines influences arriving from Northern and Southern Italy at the time of MBA (cf. Bellintani et al. 2007).

Contacts between the core provinces in Northern Italy and other regions in Southern Italy along the lines indicated by NA are much less evident in archaeological sources from MBA contexts with amber. Common types of bronze artefacts are missing and what is left are some remote similarities visible among pottery, e.g. rod-like handle with snail-horn appendices recorded in tomb 4 of Santa Domenica di Ricadi in Calabria that reflects lugs known from Emilia-Romagna (Pacci 1987, 45, Fig. 2:12, Tav. 5). However, what is more surprising is the low interconnectedness among Southern Italian provinces of amber's acquisition during MBA. Some regions do not reveal any similarity between amber artefacts, even though the contexts which yielded the latter bear testimonies of local interactions between the neighbouring communities. For instance, graves of Gallo di Briatico and aforementioned Santa Domenica di Ricadi in Calabria delivered pottery showing affiliations with north-east Sicily – *Rodì-Tindari* facies, and the Aeolian Islands – *Capo Graziano* and *Ausonian I* cultures, respectively (cf. Pacci 1987, 44–45; Pacciarelli 2001, 185).

The key to understanding the influx of amber to Southern Italy seems to be the core area of northern Apulia and Basilicata which is connected to nearly all abovementioned provinces due to similarities in amber artefacts (except for the Aeolian Islands) and, moreover, by some of the finds associated with amber. In the case of Calabria, tomb 4 of Santa Domenica di Ricadi, again, can be cited as an example, where some amber beads (types IF121) and pottery (Pacci 1987, Fig. 2:1, 4:11–12), find analogies in Grotta Manaccora (Baumgärtel 1953, Fig. 3:3, 6; Manfredini 1979, Fig. 6:2). To this evidence can be added small spheroidal and discoidal beads in white, yellow or green glass paste found in the tomb 4 also having counterparts in Apulia (Pacci 1987, Fig. 4:13–18). Moving farther to the south, co-occurrence of similar amber beads (types IA161; IF122, IF221) and swords (and similar daggers) falling into the group Pertosa-Thapsos on Sicily (Plemmyrion, Thapsos)[544] and in northern Apulia and Basilicata (Grotta Manaccora, Toppo Daguzzo, Madonna di Loreto)[545] can be listed here as an evidence of interactions. Furthermore, some of the sites in Apulia and Sicily yielded similar cylindrical and biconical beads made of faience/glassy faience made in the mixed alkali technology with low magnesium and low potassium levels (LMLK; cf. Angelini et al. 2005) which further strengthen the links indicated by similarities among amber beads.[546]

544 Voza 1973, 40, n. 116, Tav. 6:116; 2008, Fig. 12; Jung 2009, Fig. 11:2–3; cf. Bianco Peroni 1970, 24–25; Bietti Sestieri/Lo Schiavo 1976, 165; Bettelli 2006.

545 Bianco Peroni 1970, Tav. 6:39; Cipolloni Sampò 1986, Fig. 9:1–3; Tunzi Sisto 1999b, Tav. 1:1157, 1394, 1496.

546 Masseria Chiancudda, layer of *Punta le Terrare* facies (Cinquepalmi et al. 2006, Fig. 1.A); Plemmirio, tomb 48 (Colini 1904, Fig. 54; Genovese 2015, Fig. 7); Madonna di Loreto, 'Hypogeum of Bronzes' (Tunzi Sisto 1999a, Tav. X:644, 828, 1042, 1769); cf. Bellintani et al. 2006, 1505, Fig. 1; Bellintani 2012, 16, Fig. 1.

Relationship between the northern and southern part of Apulia and Basilicata during MBA was surely a more developed one, taking into consideration geographic closeness and affiliation with the *Protoapennine* culture (see sections 2.2.14 and 2.2.15). Amber beads in both regions also reflect a good degree of resemblance, similarly to certain forms of pottery and jewellery found with them. Analogies are well documented in the case of graves of Grotta Manaccora, Toppo Daguzzo and Madonna di Loreto on one side, and those of Bisceglie-La Chianca and Murgia Timone on the other. As examples one can mention shallow bowls with small handles, concave bases, straight walls and everted rims,[547] or jugs with strap handles slightly protruding over the rim by trapezoidal appendices.[548] Bronze jewellery includes round buttons/plaques with holes at the edges or lugs on the inner sides,[549] and spectacle pendants or spirally coiled bronze adornments (*fermatrecce*).[550] Important component present in graves with amber in northern and southern parts of Apulia and Basilicata are spheroidal glass beads of greenish or blueish colour, classified as the so-called 'plant ash' glass characterised by distinct levels of magnesium (LMG and HMG).[551]

Lack of direct analogies in amber between the Aeolian Islands and the remaining regions of Southern Italy is puzzling; however, it has to be stressed that the materials from Portella di Salina, hut F could not be classified due to absence of figures, thus reducing the basis for statistics to merely a single context. Amber bead found inside hut Delta XXI at the Lipari 'Acropolis' (IF121y; Bernabò Brea/Cavalier 1980, Pl. 98:i), if not for the hourglass-shaped perforation, would match some of the contemporaneous finds from Sicily and northern Apulia (IF121).[552] Moreover, glass and faience beads found in the context with amber on the islands bear superficial resemblance to already mentioned vitreous materials in Southern Italy.[553] Since other commodities evidently

547 Bisceglie, dolmen La Chianca (Cataldo 1995, 268, Fig. XXXVII:7); Madonna di Loreto, 'Hypogeum of Bronzes' (Cataldo 1999, Tav. V:825); Murgia Timone, tomb 1 (Matarese 2018, Tab. 1:T2CR/64).

548 Bisceglie, dolmen La Chianca (Cataldo 1995, 268, Fig. XXXVII:3), Madonna di Loreto, 'Hypogeum of Bronzes' (Peroni 1999a, Fig. 2:5), Grotta Manaccora, *Grotticella Funeraria* – SS layers (Recchia 1993, Fig. 15:3).

549 Bisceglie, dolmen La Chianca (Cataldo 1995, Fig. XXXVII:16a–b); Grotta Manaccora, *Grotticella Funeraria* – SS layers (Recchia 1993, Fig. 27:U); Murgia Timone, tomb 1 (Matarese 2016, Fig. 2:B); Madonna di Loreto, 'Hypogeum of Bronzes' (Tunzi Sisto 1999a, Tav. XVIII, IXX, XX).

550 Grotta Manaccora, *Grotticella Funeraria* – SS layers (Recchia 1993, Fig. 27:R–S); Murgia Timone, tomb 1 (Matarese 2016, Fig. 2:C); Madonna di Loreto, 'Hypogeum of Bronzes' (Tunzi Sisto 1999a, Tav. VIII:1719, IX:1632–1633, XI:1401, 1440, 1717, 1631).

551 Grotta Manaccora, *Grotticella Funeraria* – SS layers (Manfredini 1979, Fig. 6:3); Murgia Timone, tomb 1 (Matarese 2016, Fig. 2:G); Toppo Daguzzo, tomb 3, burial 3 (Cipolloni Sampò 1986, Fig. 9:6); Madonna di Loreto, 'Hypogeum of Bronzes' (Tunzi Sisto 1999a, Tav. X:713, 714, 836, 845, 959, 1306, 1639, 1640, 1768); cf. Angelini et al. 2001; Conte et al. 2019.

552 Grotta Manaccora, *Grotticella Funeraria* – SS layers (Manfredini 1979, Fig. 6:2c); Madonna di Loreto, 'Hypogeum of Bronzes' (Tunzi Sisto 1999a, Tav. X:945); Plemmirio, tomb 48 (Genovese 2015, Fig. 8:c); Thapsos, tomb D (Cultraro 2007, Pl. XCVIII:c9).

553 Lipari 'Acropolis', hut Delta XXI (Bernabò Brea/Cavalier 1980, 516); Portella di Salina, hut F (Bernabò Brea/Cavalier 1968, Tav. LXXXV); cf. footnotes 546 and 551.

reached the Aeolian Islands from Sicily and south parts of the Apennine Peninsula, it cannot be completely ruled out that amber was travelling via the same routes.

Proofs for trans-Adriatic contacts during MBA which possibly involved amber exchange can be found in northern Apulia on one side of the sea and Istria on the other. Considerable similarity between amber assemblages from the two regions, indicated by the NA results (Fig. 67), comes from a number of shared types from IF subgroup (cf. section 3.3.1). Other similarities can be captured through analysis of co-occurring bronze objects. These include two objects from the Vrčin hillfort which were hitherto disregarded as coming out-of-context, typologically insensible, or presumably younger than the Bronze Age. As shown by the recent re-examination of materials from the Vrčin necropolis, their incorporation into the same time span with the remaining materials, i.a. amber beads, seems valid on the basis of Italian analogies, in particular from the southern part of the Apennine Peninsula (cf. Cupitò et al. 2018, 517–518). More specifically, the similarities include a wave-shaped/serpentine bracelet with spiral endings from grave 13 which bears a striking resemblance to a piece of bronze wire found in the Manaccora cave (Fig. 75:A–B). Furthermore, a small dagger with the blade visibly reduced by repeated sharpening and a short, but broad handle, supported by three rivets in triangular arrangement, was classified as the Manaccora type (Fig. 75:C). Daggers of the type were found in several contexts with amber in Italy, including the eponymous cave in Apulia (Fig. 73:E–H).

Fig. 75. Examples of bronze bracelets, daggers and spectacle pendants shared by some of the MBA sites with amber in Istria (A, C, D) and northern Apulia (B, E–F): A, C, D – Vrčin (Buršić-Matijašić 1989, Figs. 1:9, 3:10 and 4:8); B, F – Grotta Manaccora (Baumgärtel 1953, Tab. XI:1, X:4); E – Madonna di Loreto, 'Hypogeum of Bronzes' (Tunzi Sisto 1999a, Tav. XI:1440).

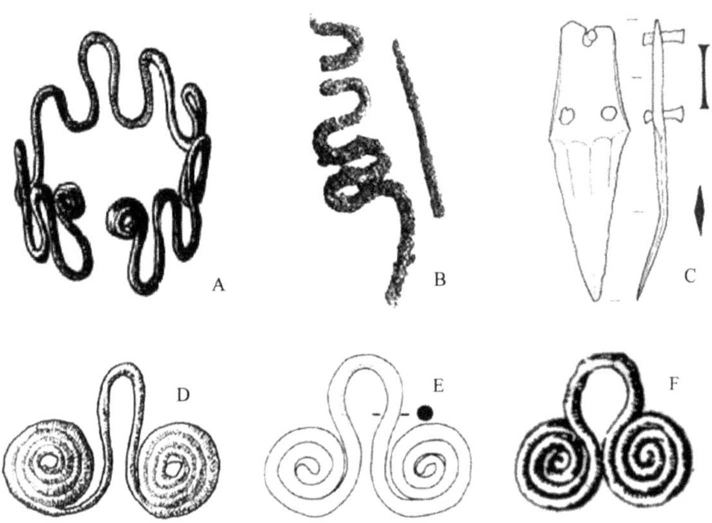

Other metal objects associated with amber in Istria, i.e. jewellery made of spirally wound wire or tape such as rolls/tubes (*saltaleoni*), rings (*fermatrecce*) or domed buttons with a lug on the inner side are also present in various graves in Apulia and Basilicata; however, they do not provide precise comparisons.[554] Better chances of finding direct connections are offered by spectacle pendants with raised connecting parts between the spirals which are attested in discussed regions in MBA (Fig. 75:D–F); however, this form of jewellery stayed in use during the subsequent stages of the Bronze Age and expanded in spatial scope of distribution.

Slowly emerging relationship between Istria and Southern Italy can be reinforced with yet another element: faience bead from Mušego, tumulus 9 (Mihovilić et al. 2009, 56). The bead, as of this moment only mentioned in the literature, has an annular form with wide central hole and double horizontal grooves decorating the lateral surfaces. It resembles some of the faience and glassy-faience beads known from the amber-containing contexts in Southern Italy.[555] Observations on distribution and technology of adornments made of vitreous paste in Italy show faience and glassy-faience beads to be predominantly known from the southern part of the Apennine Peninsula during BM 3–BR (ca. 1450/1400–1150 BC), so around the time of the Mušego tumuli construction (cf. Bellintani et al. 2006, 1505–1512, Tab. I, Fig. 1:14–26). Our knowledge concerning the bead from Mušego, tumulus 9 could benefit from analysis characterising its chemical composition to determine the provenience. Surely its morphology and the fact of occurrence together with amber presents an interesting clue to investigate.

Finally, northern Apulian and Basilicatan amber collections indicate a significant degree of similarity with assemblages from Central Balkan regions (Podrinje and Western Serbia, Kosovo and Metohija, Upper Drin Valley, Mat Valley, Kolonjë Plateau). In contrast, strong ties built on resemblance of amber finds from provinces of amber's acquisition in Northern Italy and Istria, and those in Central Balkans are not matched by presence of other imports suggesting interactions between them. In fact, the possibility of direct contact is only attested by pins with two opposing spirals in the upper part of the stalk from grave 1 at Iglarevo in Metohija that fit the Peschiera type known in abundance from Northern Italy (Fig. 76:A–C) (cf. Carancini 1975, 12–13, 130–133; see footnote 167). Still, the spectacle-head pins have a wide European distribution and are affiliated with various cultures of the late Middle (the *Tumulus* culture) and the early LBA (the *Urnfield* culture); hence, origin of the Iglarevo examples does not necessarily go back to Northern Italy (cf. Vasić 2003, 27, Taf. 9:133, 10:134–135).

More punctual and geographically circumscribed are analogies among bronze items associated with amber in Central Balkans on one hand and northern Apulia on the other. The most obvious element shared by contexts with amber from the opposing sides of the southern Adriatic, already brought up in literature,[556] are spectacle

554 Žamnjak, tumulus (Bačić 1960, Tab. VI:1); Vrčin, cemetery (Buršić-Matijašić 1989, Fig. 1:4–8, 2, 3:3–8, 4:9–15); Škicini, tumulus (Buršić-Matijašić 2003, Figs. 6 and 7); Mušego, cemetery (Mihovilić 2008, 36, 46–47; Mihovilić et al. 2009, 55); Madonna di Loreto, 'Hypogeum of Bronzes' (Tunzi Sisto 1999a, Tav. XIII, IXX, XX); Murgia Timone, tomb 1 (Matrese 2018, 150–153).

555 E.g. Madonna di Loreto, 'Hypogeum of Bronzes' (Tunzi Sisto 1999a, 214, Figs. 41–43, 46, Tav. X) and Portella di Salina, hut F (Bernabò Brea/Cavalier 1968, Tav. LXXXVb).

556 E.g. Bietti Sestieri/Lo Schiavo 1976, 181, Fig. 11:5–6; Vanzetti 1999, 226, Fig. 9; Cazzella/Recchia 2005, 142.

pendants with coiled connecting parts (also interpreted as braid holders). In northern Apulia they appeared in numerous examples inside Grotta Manaccora and 'Hypogeum of Bronzes' of Madonna di Loreto (Fig. 76:D–E). This variant of pendants is widely distributed among MBA sites in the Balkans, including Podrinje and Western Serbia, southern Dalmatia and Montenegro, and Mat Valley (Fig. 76:F–H). Geographically, the closest analogy for the Apulian specimens of spectacle pendants are those from Velika Gruda in Montenegro. However, amber finds from grave 9 of Velika Gruda (IF121y and IF244y), which possibly overlap MBA and LBA, do not have analogies in Apulia, although certain resemblance can be detected among some beads representing types IF121 and IF221y.[557] The most adequate MBA analogies for the amber object of IF121y type from the grave can be found in Istria on one hand,[558] and Podrinje and Western Serbia on the other.[559] Even though there is an evidence of interactions between Montenegro and Istria during MBA gathered outside the discussed contexts,[560] the spectacle pendants directly associated with amber from Velika Gruda, grave 9 testify to contacts with the Balkan interior and, also, with Apulia.

A less commented case is the pin from Grotta Manaccora (Fig. 76:I) which manifests a certain degree of similarity with some pins from Podrinje and Western Serbia, particularly the examples from Banjevac-Jovanin Breg (Fig. 76:J) and Paulje (Fig. 76:K). The cited pins from Serbia are classified to the group of pins with disc-shaped heads with middle thorns (ger. *Plattenkopfnadeln mit Mitteldorn*), and are regarded as a characteristic product of Western Serbian – the *Brezjak* culture's metallurgy conceived under influence of the *Tumulus* culture around BzC/D (Vasić 2003, 48; Filipović 2013, 66–68). Since there are no exact analogies for the pin from Grotta Manaccora in Italy (cf. Carancini 1975, 54, 246), it seems probable that it came as an import from the Balkan hinterland. In this place it is also worth mentioning presence of the Boiu/Sprockhoff Ia type sword in tumulus H at the cemetery of Paulje (Bulatović et al. 2017, T. XVI:24) which by its form resembles swords of the type Manaccora from the eponymous site (Bianco Peroni 1970, 52–55, Taf. 15:110; cf. Cazzella/Recchia 2005, 142).

Thanks to amber, as well as co-occurring bronze artefacts, Central Balkan regions of amber's acquisition are densely interconnected, forming a discrete sub-network (Fig. 67). Moving to discussion of relationships between the regions in the Balkan interior in MBA, a number of bronze items that bind adjacent areas can be brought forward. Starting from the northernmost region – Podrinje and Western Serbia – the most notorious element recurring together with amber are open bronze bracelets made from a bronze bar of semi-circular (D-shaped or C-shaped if hollow) or flattened triangular cross-section, with tapered or widened ('stamp-shaped') ends (Fig. 77:A–F).

557 Madonna di Loreto, 'Hypogeum of Bronzes' (Tunzi Sisto 1999a, Tav. X:945) and Toppo Daguzzo, tomb 3, burial no. 3 (Cipolloni Sampò 1986, Fig. 9:5).
558 Mušego, tumulus 9 (unpublished; Arheološki muzej Istre Pula, inv. No. 329, 330).
559 Savin Lakat, cremated burial (Lazić 2007, T. IV:16).
560 There are certain sites which, due to presence of specific bronze objects, link the area of Montenegro with Istria, and further with the Danubian Basin and the Central Europe in the course of the MBA and subsequent transition to the LBA. For example, from the Spič hoard in the coastal part of Montenegro comes a large assemblage of bronze axes exemplifying a transitional form from axes with pronounced heel (ger. *Absatzbeile*) to axes with median wings/fins (ger. *Lappenbeile*) (Žeravica 1993, 69–70, Taf. 19:236–239; cf. Hänsel et al. 2010, 17–18).

On the basis of terminations' shapes and compositions of engraved ornaments, bracelets from the Čačak area were divided into several types and subtypes (cf. Dmitrović 2016, 164–169, Fig. 113:1–4). Still, the decorative motifs are largely similar across Central Balkans, and ornamental syntaxes rarely indicate strictly regional affiliation or chronology. Thus, the typology cited above can be roughly transposed onto other areas, by approximation of given specimens to examples portrayed on the figures and characterised in the descriptions. The bracelets representing type I-A, distinguished by garland and fish-bladder motifs, were recorded in contexts with amber in Podrinje and Western Serbia, Glasinac, Kosovo and Metohija and Upper Drin Valley (Fig. 77:A–D). Other types do not have the same wide distribution; however, there are certain inter-type similarities like the decorative motif of hatched triangles appearing on some of the bracelets from Podrinje and Glasinac (Fig. 77:E–F).

Fig. 76. Examples of bronze pins and spectacle pendants shared by some of the MBA sites with amber in Kosovo and Metohija, the Garda Lake basin and western Veneto (A–C), and northern Apulia, Montenegro, Western Serbia and the Mat Valley (D–K): A – Iglarevo, grave 1 (Bunguri et al. 2006, Fig. 4.70:f); B – Isolone del Mincio (Guerreschi et al. 1985, Tav. XXII:st.9022); C – Olmo di Nogara, grave 72 (Salzani 2005, Tav. IX:A); D – Madonna di Loreto, 'Hypogeum of Bronzes' (Tunzi Sisto 1999a, Tav. XV:1446); E, I – Grotta Manaccora, Grotticella Funeraria *– SS layers (Baumgärtel 1953, Tab. XII:6; Recchia 1993, Fig. 27:T); F – Velika Gruda, grave 9 (Della Casa 1996a, Fig. 44:17); G – Ravnine, mound 3 (Nikitović/Vasić 2002, Tab. I:2); H – Shtogj, tumulus 2 (Kurti 2017a, Fig. 7:1); J – Banjevac, tumulus 2 (Bulatović et al. 2017, T. IX); K – Karavlaške kuće, mound 1, grave 2 (Kosorić/Krstić 1970, Tab. IV, Fig. 2).*

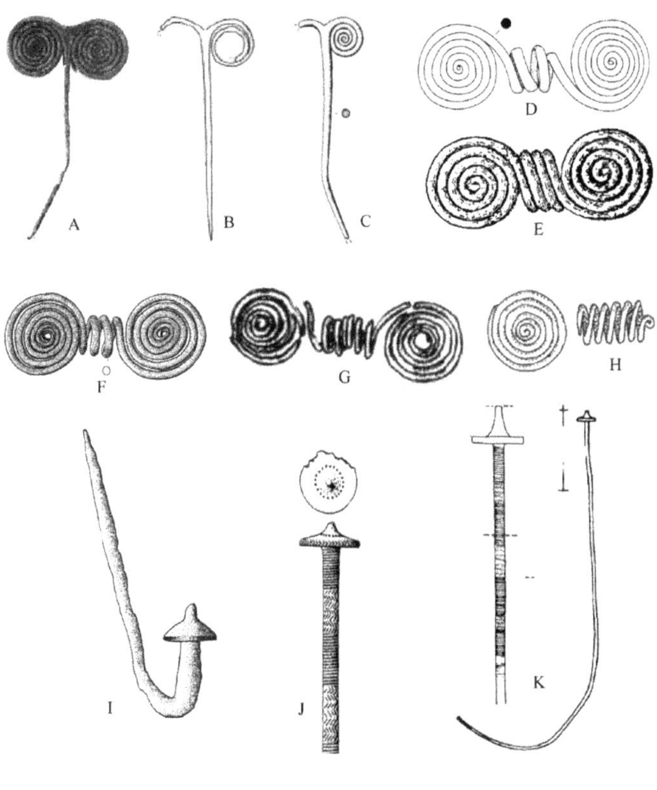

Fig. 77. Examples of bronze bracelets and pins shared by some of the MBA sites with amber in Podrinje and Western Serbia (A, E, H, J–K), Glasinac (B, F), Kosovo and Metohija (C, G, M) and Upper Drin Valley (D, I, L): A – Vranjani, tumulus 1 (Zotović 1985, T. IX:3); B – Gučevo, tumulus IV, grave 1 (1895) (Benac/Čović 1956, Pl. V:1); C, G – Iglarevo, grave 1 (Bunguri et al. 2006, Fig. 4.70:c–d); D, I, L – Çinamak, tumulus 16, grave 19 (Kurti 2017a, Fig. 4:1, 7–8); E, K – Paulje, tumulus K (Bulatović et al. 2017, T. XVII:26, 29); F – Rudine, tumulus I (1894) (Benac/ Čović 1956, Tab. XIV:6); H – Paraćin (Garašanin 1983e, T. CI:I); J – Belotić, tumulus XIX (Garašanin 1983d, Tab. CII:4); M – Boka-Prčevo, tumulus XIV, grave 23 (Tasić 1979, T. III:1).

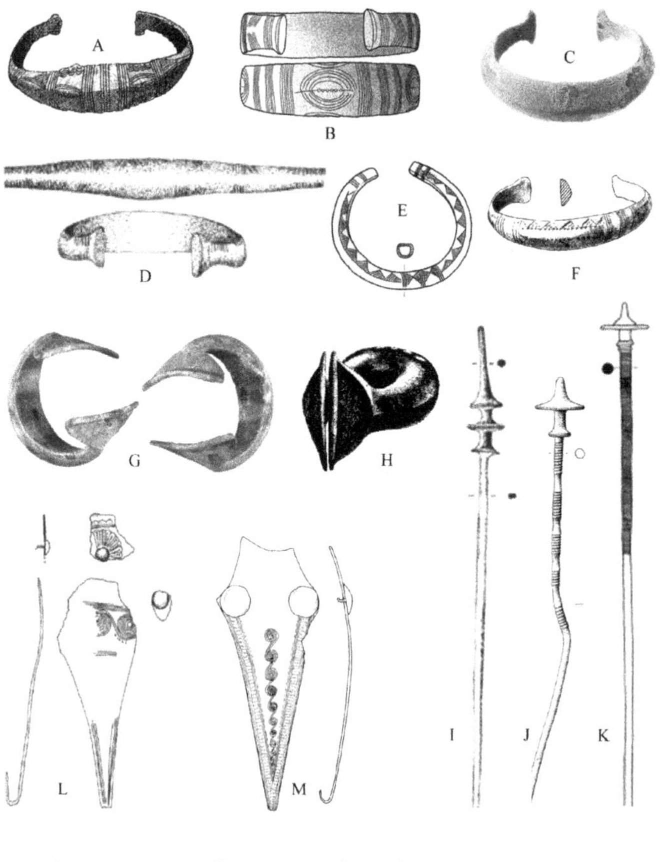

There are other categories of bronze artefacts that connect individual provinces of amber's acquisition in Central Balkans. The two closest regions – Podrinje-Western Serbia and Glasinac – show the tightest cultural affinity by sharing multiple elements of grave inventories including i.a. bronze torcs with rolled terminals[561] and pendants with thorns (ger. *Stachelscheibe*).[562] Grave 1 of Iglarevo, apart from yielding the type

561 Vranjani, tumulus 1 (Zotović 1985, T. IX:8); Borovsko, tumulus II, grave 1 (Benac/ Čović 1956, Tab. XXVII:8).

562 Bela Crkva, mound III, grave 1 (Garašanin/Garašanin 1958, 40, Fig. 15:b); Han-Osovo, tumulus 1, grave 2 (Drechsler-Bižić 1983a, T. XXXIX:6).

I-A bracelet, contained the Juhor-type anklets which by analogies with specimens from the Great Morava Valley – the *Paraćin* culture strengthens the connection with Serbia (Fig. 77:G–H).

Proceeding to the Upper Drin Valley, one of the pins found in Çinamak, tumulus 16, grave 19 draws similarities between the spindle-head pins known from MBA barrows in Podrinje: it has length of 30.5 cm and three perpendicular discs placed on the thickened upper part of the stalk, with the top one extending into a conical thorn (cf. Vasić 2003, 47–48, Taf. 17–18:254–257) (Fig. 77:I–K). The other pin from the same grave, characterised by head in the shape of a lozenge with the upper sides rolled forming a cylindrical cone, can be classified to the group of *Hülsenkopfnadel*, for which the closest parallels are known from Western Serbia.[563] Same grave yielded bronze belt hook for which analogies are also known from Metohija, but the spiral ornaments on both examples point to cultural influences originating from an area situated even further north, in Western Serbia and East Bosnia (cf. Kurti 2017a, 96) (Fig. 77:L–M).

Through its eclectic inventory, Shtogj, tumulus 2, grave 9 symbolises transitional character of Bronze Age culture of the Mat Valley (see section 2.3.13.2). It contained i.a. a leaf-shape spearhead of northern origin (Fig. 78:A) and a sword of Aegean provenience (or inspired by Aegean sword-making) dated circa LH IIIA/early IIIB, being a local variant of the type C swords, or a hybrid of the types A and C (Fig. 78:B). It is worth mentioning that in grave 8 at the Iglarevo cemetery in Metohija were found two *Mycenaean* rapiers of the A-type dated to LH IIB–IIIA (Fig. 78:C), together with a bronze leaf-shaped spearhead quite like the one from Shtogj (Fig. 78:D). Presence of discussed swords and spearheads, rather unique in the Balkans, in the Mat Valley and Metohija certainly creates a link between them. Furthermore, the tumulus of Shtogj yielded a spectacle pendant mentioned earlier (Fig. 76:H) and a pin with disc-head and biconical, perforated swelling of the stalk (Fig. 78:E). Identical pins were found in the coeval tumuli of the Kolonjë Plateau, e.g. Rehovë, graves 81 and 192, and were classified by S. Aliu as the type IIa (Aliu 2012, 300; cf. Kurti 2017a, 107–107) (Fig. 78:F). Worth mentioning is that versions of the nail-head pins with perforated necks are known from several sites in Western Serbia (cf. Vasić 2003, 33, Taf. 11:163, 166; Dmitrović 2016, 176, Fig. 114:3).

Finally, two pins from Rehovë tumulus, grave 186, defined as type VA by S. Aliu and described as of northern, that is Central European origin (Aliu 2007, 240, Pl. LXIIa; 2012, 193–199), are closely related to the Paarstadl type (Fig. 78:G). Pins with cone-shaped heads and thickened, pierced necks, akin to the Paarstadl type, occurred between the 16th and 14th centuries BC in a vast area mainly encompassing the Danube Basin, but significant concentrations have also been noted in southeast Albania and western Macedonia (Pabst 2013, 155–157, Abb. 1). A related form of pin with spherical swelling of the neck (sometimes pierced) appeared in considerable numbers in Kosovo, i.e. in grave 4 at Iglarevo, where it has been associated with the *Iglarevo* and *Brnjica* groups (cf. Vasić 2003, 65–67, Taf. 24:378–383) (Fig. 78:H). Furthermore, at least one pin of the Paarstadl type was discovered in Western Serbia – Čačak area (Fig. 78:I). This evidence supplements well the similarities in amber jewellery between the Kolonjë Plateau and the other Central Balkan provinces of amber's acquisition. The cited Balkan examples of the Paarstadl-type pins seem a proper analogy; however, it has to be mentioned that

563 Kurti 2017a, 88–92, Fig. 4:2; cf. Vasić 2003, 16, Taf. 3:29–46, Taf. 51, Taf. 70:81–82; Dmitrović 2016, 174, Fig. 114:1.

there is a certain level of affinity between pins of the type Paarstadl and those representing the Italian type Boccatura del Mincio/Cattaragna, like the specimen from Madonna di Loreto, 'Hypogeum of Ivories' (Fig. 74:D). Although the latter are by large dated to BR – LBA (see footnote 169), this dating also covers the later part of the chronology of graves from the Rehovë tumulus. Thus, a potential connection between Albania and northern Apulia (also suggested by similarity of amber objects), already suggested some time ago (Vanzetti 1999, 226; Kurti 2017a, 86), should not be disregarded.

Fig. 78. Examples of bronze spearheads, swords and pins shared by some of the MBA sites with amber in the Mat Valley (A–B, E), Kosovo and Metohija (C–D, H), Kolonjë Plateau (F–G) and Western Serbia (I): A–B, E – Shtogj, tumulus 2, grave 9 (Kurti 2017a, Fig. 7:3–4, 7); C–D – Iglarevo, grave 2 (Bunguri 2006, Fig. 3.6); F – Rehovë tumulus, grave 81 (Aliu 2012, II:42, 44); G – Rehovë tumulus, grave 186 (Aliu 2012, Tab. XV:200–201); H – Iglarevo, grave 4 (Bunguri et al. 2006, Fig. 4.69:b); I – Mojsinje, mound V, grave 2 (Dmitrović 2016, Fig. 114:2).

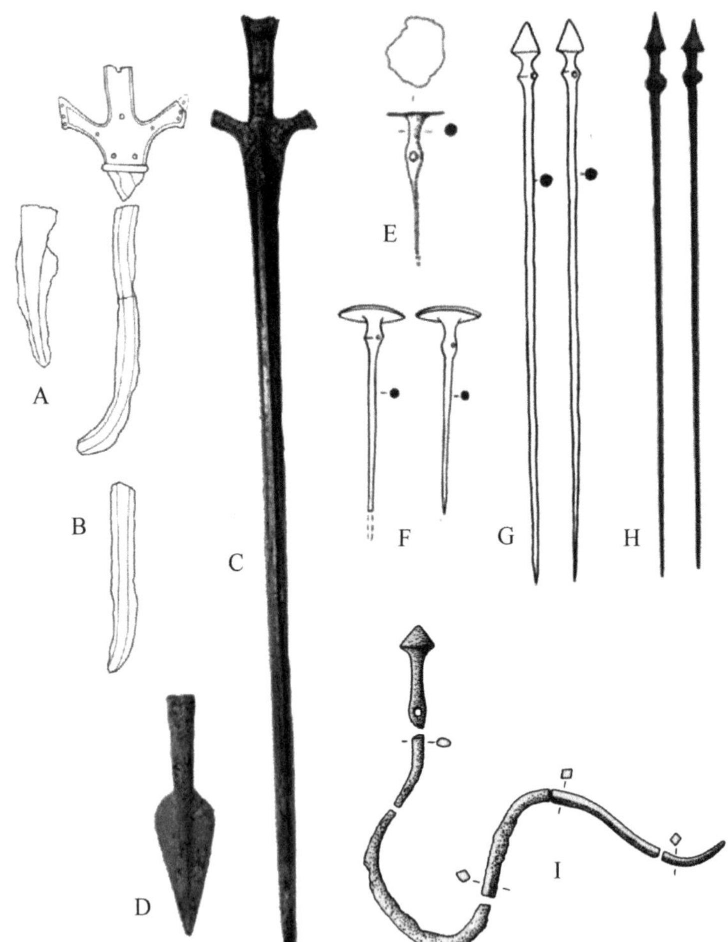

Altogether, distribution of the discussed types of bronze items suggest presence of a developed network of contacts in Central Balkans during MBA which followed similar meridional path like the one that linked Northern Italy with Apulia around the same time. Similarities stemming from distribution of amber, bronze and other categories of finds, if present, allow to reconstruct a chain of interlocked regions and communities by which amber could have circulated in the discussed period (Fig. 79). Two kinds of links can be proposed as the basis for the network: (1) 'probable connections' distinguished between regions characterised by presence of typologically identical amber as well as associated artefacts (or pointing to close cultural affiliation), and (2) 'suspected connections' linking regions with related, but not exactly similar amber and/or associated finds (or pointing to more distant cultural affiliation). The directions, visualised on the figure as arrows symbolise interactions between the regions, here considered as centres of amber's acquisition and redistribution, without dictating the course of amber circulation, although, on the basis of provenience analyses, it seems likely that it mostly proceeded along the north–south axis.

Fig. 79. Reconstruction of amber circulation in the circum-Adriatic zone during MBA.

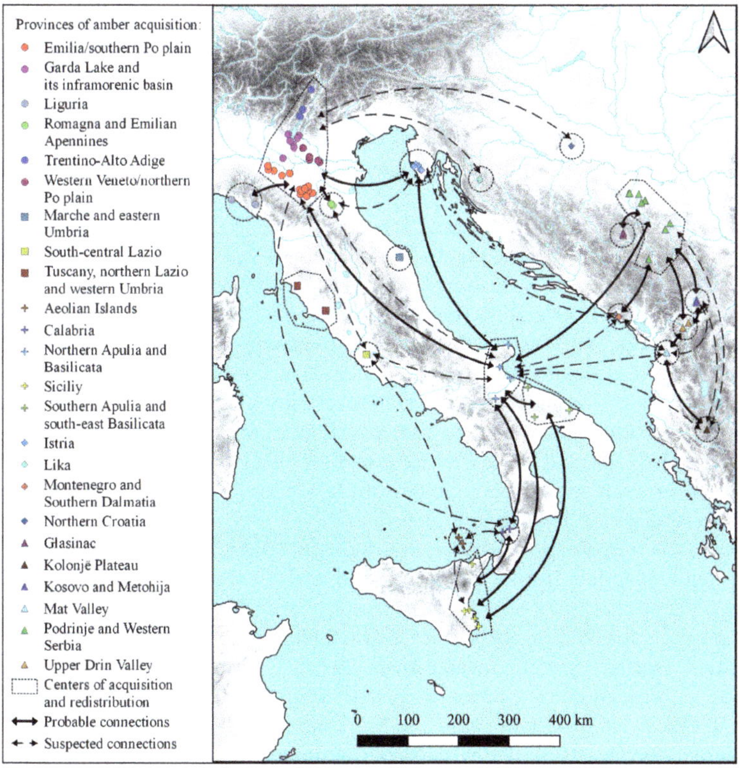

Key positions in the network are taken by the core regions which, apart from the largest assemblages of amber artefacts, delivered the most abundant evidence for interregional contacts. Thus, one can summarise it in a simple principle: the greater the numerosity and complexity of amber collections from a region, the greater the chances of occurrence of related data testifying to long-distance interactions.

Four regions in Northern Italy: Trentino-Alto Adige, the Garda Lake basin, western Veneto and Emilia were amalgamated due to close cultural affinity (the *Palafitte-Terramare* culture) and numerous material evidence of multidirectional (and multidimensional) interactions among them. Together they form the most important Northern Italian centre of amber's acquisition and, likely, redistribution. From there typological compatibility of amber and other, mostly bronze and pottery finds, render circulation of amber plausible in three directions: to the west with Liguria, to the south with Romagna and northern Apulia and Basilicata, and to the east with Istria. Other ties coming out of the *Palafitte-Terramare* culture ecumene seem less clear, however show possibility of amber transfer from Northern Italy to and Lazio, and even further to the Aeolian Islands, Lika and Northern Croatia. Istria despite sharing many similar amber artefacts with other circum-Adriatic regions is lacking solid evidence to link it beyond the most adjacent areas, and even then the data is quite sketchy, also due to state of research. Nonetheless, available finds discovered on the Istrian sites with amber, mainly the Vrčin hillfort, testify to trans-Adriatic contacts with the Po Plain and northern Apulia.

The latter region is of considerable importance for tracing circulation of amber in the southern part of the circum-Adriatic zone. Apart from the aforementioned connections to the north (Po Plain), probably involving contacts with Romagna and Lazio, the region is densely connected with southernmost parts of the Apennine Peninsula (Calabria and southern Apulia and Basilicata) and Sicily, with suspected ties with the Aeolian Islands. The peripheral regions of Southern Italy maintained relations among themselves; however, it seems that with respect to amber, mediation of northern Apulia was a key factor. Furthermore, there are indications of contacts between northern Apulia and some Central Balkan regions which supplement similarities among amber artefacts. The most precise evidence of these interactions was found in Podrinje and Western Serbia (pins), whereas with respect to Montenegro, the Mat Valley and Kolonjë Plateau, the similarities are less specific, coming down to spectacle pendants (braid holders). It is interesting whether above-listed similarities stem from occasional (less likely regular) exchange of amber between northern Apulia and Central Balkans, and what was the direction of this translocation: was it always southward, or did it sometimes follow north in a 'counter-stream' movement as well (e.g. from Apulia to Albania and into Serbia)?

Despite the remarks given above, Podrinje and Western Serbia and other Central Balkan regions appear not so much involved with the rest of the network. Besides certain finds pointing to a possibility of contacts with Apulia, there is not much evidence speaking for relations between Central Balkans and other provinces of amber's acquisition. During MBA the former more likely were integrated in a second, inland chain of amber circulation, proceeding from Podrinje and Western Serbia, up the Drina and West Morava Rivers, down to Metohija and Albania. It is also possible that this sub-network was supplied with amber by intermediaries other than in the case Northern Italian or Western Balkan centres, most likely operating from the Carpathian Basin. Although it is too early at this point to name those exchange partners, but it will not be an exaggeration to point to the Podrinje and Western Serbia as the main recipient of amber, as well as its redistributor in the Central Balkan area.

3.12.3. LBA (1300–1150 BC)

Regional distribution of amber in the circum-Adriatic zone during LBA looks largely similar to the previous period (Fig. 80; cf. Fig. 66). However, it has to be reminded that some of the sites (those defined as MBA/LBA due to prolonged or imprecise chronology) already appearing on the MBA map are still present in the picture. Yet, the continuing occurrence of amber in most parts can be explained by uninterrupted cultural development from MBA to LBA, such as in the Po Plain where the *Terramare* culture entered its final, most mature stage, or in Podrinje, where the same cemeteries and burial rites were used and practised up to the end of BzD.

Fig. 80. Distribution of the sites with amber finds dated to MBA/LBA, LBA and LBA/FBA in the circum-Adriatic zone.

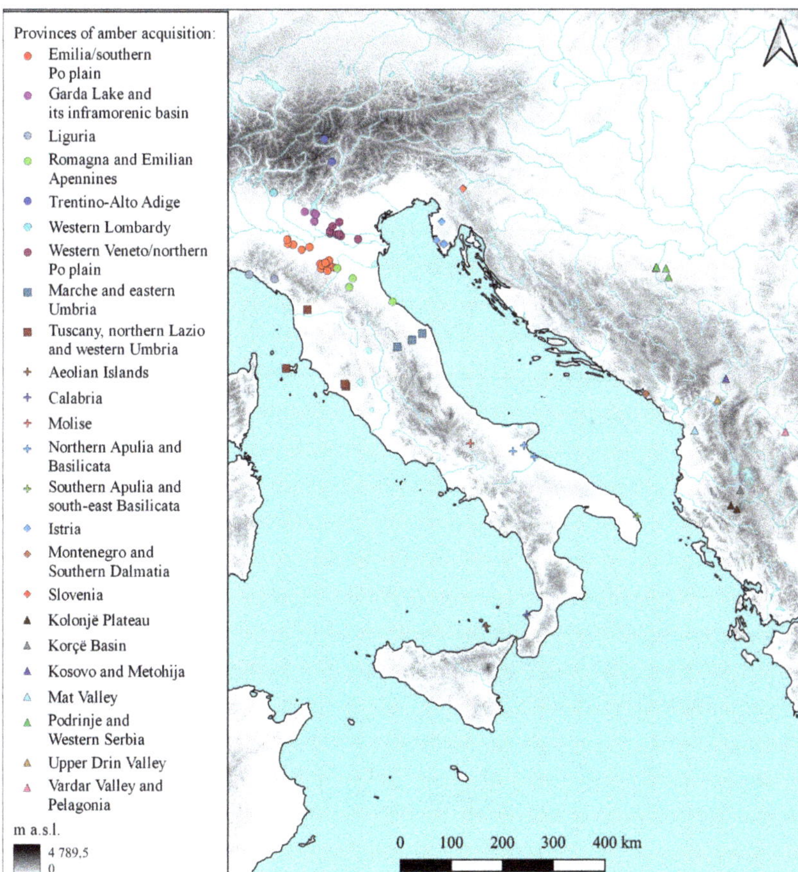

The transition from MBA to LBA is accompanied by various amplitude of fluctuations in the numbers of contexts and finds in the regions (cf. section 2.5). One can spot a visible drop in number of sites with amber, e.g. in Podrinje, and simultaneous increase in, e.g. Marche and eastern Umbria. Despite the steady cultural development, amber disappeared for few centuries from the Glasinac and south-central Lazio. Similar interlude is observable on Sicily, where once very popular and accessible through local resources as well as import, amber seemingly stopped being used for a time span of ca. 250 years. In this case the pause could have been caused by more profound changes in the local culture and society, recognised as the *Pantalica I – Nord* phase/culture. At the same time, amber was introduced for the first time in western Lombardy, Molise, Slovenia and the Vardar Valley.

Fig. 81. Network analysis of MBA/LBA, LBA and LBA/FBA contexts with amber grouped into regions.

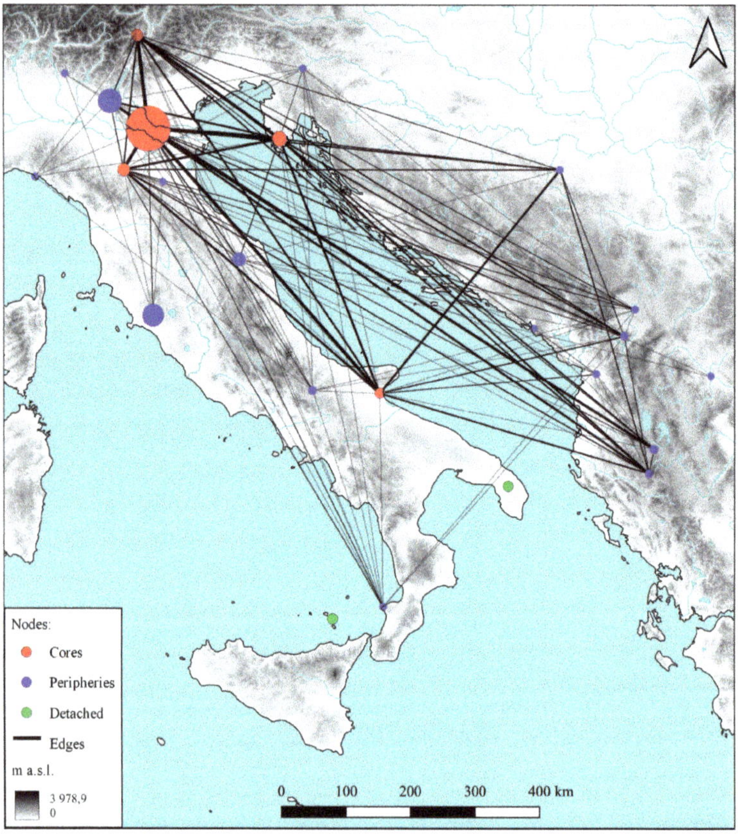

Regardless of the changes in the spatial distribution of amber at the turn of MBA and LBA, the network built upon similarities between regional assemblages of the finds largely retained the previous structure (Fig. 81), and even slightly increased in interconnectedness (average *density* value of 0.451). Emilia, western Veneto, northern Apulia and Basilicata, and Istria still are classified as the 'cores' due to the most numerous and typologically manifold collections of amber artefacts (Fig. 82). As a novelty, Trentino and Alto Adige emerged as a core region, proving that this Alpine region enjoyed access to bulk amber supply and was crucial for its transfer further south, what is indicated by considerable levels of *betweenness* and similarity with other nodes. Another change in this respect is Podrinje and Western Serbia which, together with the reduction of the numbers of contexts and finds, lost the 'core' status. Still, it kept strong ties with provinces on the western shore of the Adriatic, along with regions such as the Upper Drin Valley, Kolonjë Plateau and Korçë Basin (Tab. 46).

Fig. 82. Tabular summary of core-periphery *analysis of MBA/LBA, LBA and LBA/FBA contexts with amber grouped into regions (presence of a digit at the intersection of a column and a row implies presence of a tie between two regions, while the value gives a number of types they share; calculated with the maximum correlation algorithm described by Borgatti/Everett 1999).*

		6 Em	14 Is	3 Tr	11 No	5 We	2 We	7 Ro	8 Tu	4 Ga	10 Mo	1 Li	12 So	13 Ca	9 Ma	15 Mo	16 Sl	17 Po	18 Ko	19 Va	20 Up	21 Ma	22 Ko	23 Ko
6	Emilia		6	4	6	9				2	1			2	2	1	1	3	2		2	1	3	2
14	Istria	6		9	6	11				5	1	1		1	2	1	1	3	1		4	1	5	4
3	Trentino - Alto Adige	4	9		5	10				3	1	1		1	1		1	2	2		4	1	4	2
11	Northern Apulia and Basilicata	6	6	5		9				1	1			1	1		1	6	1		3	2	4	1
5	Western Veneto	9	11	10	9		1		2	8	2	1		2	3	1	1	4	2	1	3	1	6	7
2	Western Lombardy					1			1	1														
7	Romagna and Emilian Apennines								1															
8	Tuscany, northern Lazio and western Umbria					2	1	1		1														
4	Garda Lake basin	2	5	3	1	8	1		1		1	1		1	1		1		1		1		1	4
10	Molise	1	1	1	1	2				1		1		1	1		1		1		1			
1	Liguria		1	1		1				1	1													
12	Southern Apulia and south-east Basilicata																							
13	Calabria	2	1	1	1	2				1	1				1		1		1		1			
9	Marche and eastern Umbria	2	2	1	1	3				1	1			1			1	1	1	1	1		1	1
15	Montenegro and southern Dalmatia	1	1			1																		1
16	Slovenia	1	1	1	1	1				1	1			1	1				1		1			
17	Podrinje and Western Serbia	3	3	2	6	4									1						2		3	
18	Kosovo and Metohija	2	1	2	1	2				1	1			1	1		1				1			
19	Vardar Valley and Pelagonia					1									1									
20	Upper Drin Valley	2	4	4	3	3				1	1			1	1		1	2	1				2	
21	Mat Valley	1	1	1	2	1																	1	1
22	Kolonjë Plateau	3	5	4	4	6				1					1			3			2	1		2
23	Korçë Basin	2	4	2	1	7				4					1	1						1	2	

The core areas are characterised by high levels of *betweenness* and *degree* centrality (Tab. 46). However, after western Veneto which has the highest *betweenness* among all the nodes, the second and third place with respect to this centrality measure is taken, respectively, by the Garda Lake basin, and Tuscany, northern Lazio and western Umbria. They owe their high *betweenness* to links with peripheral regions situated at the outskirts of the circum-Adriatic zone, such as Liguria, western Lombardy, Romagna and Emilian Apennines, reaching as far south as the Vardar Valley and Pelagonia. Important role in this respect is played also by Marche and eastern Umbria. It is evident that the core areas, despite having yielded the largest and the most diverse amber collections, sometimes lack the few types that would allow them to be connected with some of the minor nodes. By LBA the major provinces like western Veneto and Emilia enjoyed the steady influx of amber and presumably began to manufacture amber jewellery themselves, however still relied on intermediaries to deliver their products to some of the locations. Nevertheless, it appears that in the later stages of the 2nd millennium BC, the centre of gravity in the amber circulation network became firmly anchored in the north-western side of the Adriatic. At the same time amber movement in Central Balkans began to decline. Also Southern Italy as a whole apparently shows less involvement in amber exchange in LBA. As mentioned before, northern Apulia and Basilicata retained their core position;

however, overall the amount of amber at that stage is lower in this macro-region than before. The single find dated to the period from Lipari 'Acropolis', hut Beta I-tomb III at the Aeolian Islands could not be typologically determined due to absence of figures, making the region not eligible for the analysis. Southern Apulia delivered only a single find at that stage (bead IE218a from Roca Vecchia, 'Cult Center') which does not have an analogy in the circum-Adriatic zone, resulting in the region being detached from rest of the LBA network. Only Molise and Calabria are the only other provinces in Southern Italy with significant *degree* levels (Tab. 46).

Tab. 46. Regions with MBA/LBA, LBA and LBA/FBA contexts with amber ranked according to their coreness *(calculated with the Minres algorithm described by Borgatti/Everett 1999),* betweenness *(calculated with the algorithm described by Freeman 1979) and* degree.

Province:	Coreness:	Betweenness:	Norm. Betweenness:	Degree:	Norm. Degree:
Calabria	0.085	0.000	0.000	11	0.500
Emilia	0.309	6.955	3.011	16	0.727
Garda Lake basin	0.217	22.596	9.782	16	0.727
Istria	0.407	10.021	4.338	17	0.773
Kolonjë Plateau	0.210	1.012	0.438	11	0.500
Korçë Basin	0.164	1.786	0.773	10	0.455
Kosovo and Metohija	0.092	0.000	0.000	11	0.500
Liguria	0.033	0.000	0.000	5	0.227
Marche and eastern Umbria	0.125	8.536	3.695	15	0.682
Mat Valley	0.053	0.000	0.000	7	0.318
Molise	0.085	1.400	0.606	12	0.545
Montenegro and southern Dalmatia	0.026	0.000	0.000	4	0.182
Northern Apulia and Basilicata	0.315	3.455	1.496	15	0.682
Podrinje and Western Serbia	0.158	0.000	0.000	8	0.364
Romagna and Emilian Apennines	0.007	0.000	0.000	1	0.045
Slovenia	0.072	0.000	0.000	11	0.500
Southern Apulia and south-east Basilicata	0.000	0.000	0.000	0	0.000
Trentino – Alto Adige	0.335	6.021	2.607	16	0.727
Tuscany, northern Lazio and western Umbria	0.033	20.000	8.658	4	0.182
Upper Drin Valley	0.171	1.196	0.518	13	0.591
Vardar Valley and Pelagonia	0.013	0.000	0.000	2	0.091
Western Lombardy	0.020	0.000	0.000	3	0.136
Western Veneto	0.552	51.021	22.087	20	0.909

Fig. 83. Examples of bronze pins shared by some of the LBA sites with amber in the Garda Lake basin (A, D, G, J, M), western Veneto (B, E, H, K, N) and Emilia (C, F, I, L, O): A, D, G, J, M – Peschiera – Imboccatura del Mincio (Aspes 1980, Fig. 41:1–2, 4, 7, 9); B – Olmo di Nogara, grave 190 (Salzani 2005, Tav. XVIII:B–C); C, F – Santa Rosa di Poviglio, Villaggio Grande, *'heaps of ash' layer – phase 2 (Bernabò Brea/Cardarelli 1997, Fig. 186:1, 10); E – Olmo di Nogara, grave 243 (Salzani 2005, Tav. XXI:B); H – Olmo di Nogara, grave 208 (Salzani 2005, Tav. XX:B); I – Beneceto, advanced Recent Bronze Age (BR2) layers (Bernabò Brea et al. 2008, Fig. 13:12); K – Castello del Tartaro, tomb 300 (Salzani 2011, Fig. 5:A1); L, O – Santa Rosa di Poviglio,* Villaggio Grande, *'summit' layer – phases 3 and 4 (Bernabò Brea/Cardarelli 1997, Fig. 189:12, 14); N – Olmo di Nogara, grave 252 (Salzani 2005, Tav. XXII:C).*

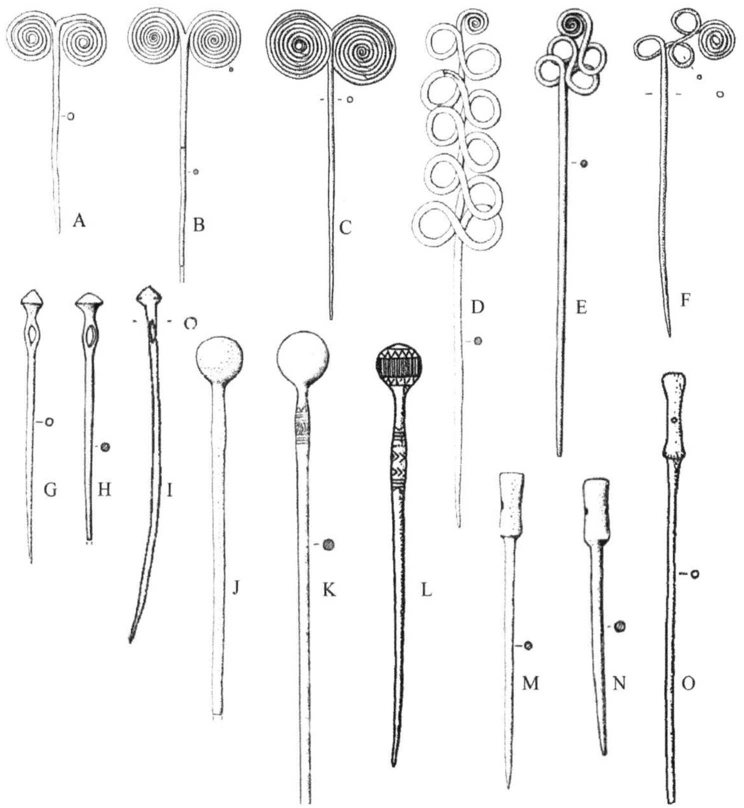

As the *Terramare* culture continued its existence in the Po Plain during LBA, so did the two centres of amber's acquisition identified with the culture's regional groupings inhabiting northern (western Veneto) and southern (Emilia) sides of the lower Po. The numerosity and typological diversity of amber assemblages suggest sustained inflow of the material to these provinces, adding yet another proof confirming the notion of uninterrupted cultural development in this part of Italy between MBA and LBA (see sections 2.2.6 and 2.2.7). Closely connected in cultural terms to the *Terramare* phenomenon still was the Garda Lake basin and its pile-dwelling settlements (see section 2.2.5). Here, amber also retained a significant representation in LBA, resulting in a considerable level of *coreness*. Furthermore, *betweenness* index informs that the province potentially played the role of a redistribution centre, as it did in MBA.

The close cultural affinity and the continuing access to bulk supply of amber went hand in hand with similarity of amber artefacts in the three discussed regions: very close between western Veneto on one hand and Emilia or Garda Lake basin on the other, and slightly more remote between the latter two. This similarity is reflected in typological congruity of many finds co-occurring with amber in these regions, again, mainly bronze items. The most representative in this context are pins, e.g. of the types Peschiera (Fig. 83:A–C), Bacino Marina (Fig. 83:D–F), Boccatura del Mincio/Cattaragna (Fig. 83:G–I), Franzine (Fig. 83:J–L) or Ca del Lago (Fig. 83:M–O) (cf. Carancini 1975).

Bronze pins like those listed above, aside from pottery, are the category that neatly captures the connections between *palafitte* of the Garda Lake basin, cemeteries of western Veneto and *terramare* of Emilia. However, one has to realise the specificity of bronze items which, due to limitedness of metal ores, were subject to circulation from the birth of metallurgy. Towards the end of the Bronze Age, metallurgical industries in various parts of the circum-Adriatic zone were becoming increasingly unified in the formal sense. A novelty among bronze-made elements of costume linking some of the sites with amber in Northern Italy, mostly those dated to the advanced phase of LBA (BR 2), were violin-bow fibulae with twisted/corrugated arch (Fig. 84:A–B), also denominated as the Peschiera-type fibulae due to their particular occurrence around the Garda Lake (cf. Eles Masi 1986, 2–4). Similar fibulae are found in many areas, far beyond the borders of Italy (cf. Teržan 2007; Pabst 2014), showing how products, models (i.e. casting moulds) and probably metallurgists travelled over long distances, much like in the case of amber. Other types of finds common among LBA contexts with amber in Northern Italy are, e.g. daggers of the type Redù, var. A (Fig. 84:C–D) (cf. Bianco Peroni 1994, 143–145), sickles of the type Castions di Strada (Fig. 84:E–G) (cf. Carancini/Peroni 1999, 56, Tav. 27:14), axes with median fins/wings of the type Pertosa (Fig. 84:H–I) (cf. Carancini 1997, Fig. 224:12) and bone or bronze pinheads (*rotelle*) of the types Gazzade (Fig. 84:J–K) and Narce (Fig. 84:L–N) (cf. Carancini 1975, 78, 254, 325; 1997, Fig. 234:88–90, 92–93). This multitude of shared types of artefacts hints to well-established relationships between the Garda Lake basin, western Veneto and Emilia, facilitating circulation of amber in Northern Italy during LBA.

Fig. 84 Examples of fibulae, daggers, sickles, axes and pinheads shared by some of the LBA sites with amber in the Garda Lake basin (A, C, E, H, J, L), western Veneto (B, E, F, M) and Emilia (D, G, I, K, N): A, C, H, J, L – Peschiera – Imboccatura del Mincio (Aspes 1980, Figs. 40:1, 43:4, 7 and 44:6, 9); B – Sabbionara, sector II, US 27 (Salzani 1990–1991a, Fig. 21:18); D, N – Redù-Pilastro (Bianco Peroni 1994, Tav. 78:1413; Malnati/Cardarelli 2003, Fig. 58:10); E – Isolone del Mincio (Guerreschi et al. 1985, Tav. XXIII:st.9151); F, M – Finilone Valle (Salzani 1996, Figs. 17:7 and 23:1); G – Anzola dell'Emilia (Desantis et al. 2011, 26, Fig. 2); I – Castellazzo di Fontanellato (Mutti 1993, Fig. 56:4); Castione Marchesi, layers 2–3/phases 1–4 (Mutti et al. 1988, Fig. 101:1).

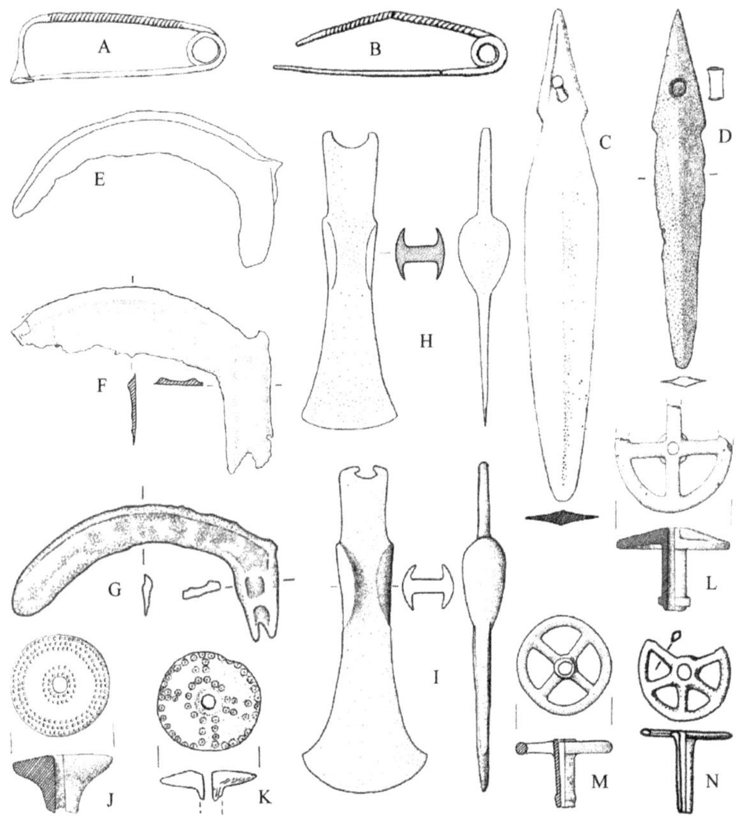

Relations between this main centre of amber's acquisition in LBA and other regions in Northern Italy appear less explicit, when looking at the data derived from the contexts containing amber. There are two sites with amber in Trentino-Alto Adige datable to the period, both of them being interpreted as cultic places – votive troves, thus involving highly selective deposition of objects. Both of them produced sets of amber artefacts greatly corresponding with the forms found to the south, particularly in western Veneto. In case of the Val di Non bog near Cles, the exact context of amber finds remains undetermined; hence, they cannot be associated with other finds known from the area (cf. Marzatico/Tecchiati 1998, 28–29; 2002, 50). The high-altitude site of Maso Finale, aside from amber, delivered several glass beads, e.g. barrel shaped/fusiform in dark blue-black past with spiral decoration whitish inlay, which are comparable with some of the vitreous materials found at LBA sites with amber in western Veneto and Emilia, e.g. from the settlement of Santa Rosa di Poviglio – *Villaggio Grande* and the necropolis of Franzine (Putzer 2012, Abb. 10:36–38, 41; cf. Bellintani et al. 2006, Fig. 2:15–16). Although the glass beads from Maso Finale have not been subjected to analysis of chemical composition, their surface colour is similar to the surface weathering typical of high magnesium brown glass (HMBG), thus matching the cited analogies (Bellintani 2014, 120, fn. 7; Bellintani/Gonzato 2017, 178). The cultural bond between Trentino and the Garda Lake basin or the Po Plain seems to loosen during LBA, with fewer similarities noticeable especially among pottery which testifies to separate identity of communities inhabiting the former area (the *Dos Gustinaci* facies, later to morph into the *Luco/Laugen* culture) (cf. Marzatico/Tecchiati 1998, 29–30; De Marinis 1997, 412; see section 2.2.4). Still, this does not completely rule out the possibility of amber coming to the Po Plain through the mountains of Trentino-Alto Adige, as it most probably happened during MBA.

Attribution of amber finds from Costa Cavalina cave in western Lombardy to LBA is based on the generally recognised introduction of the Tiryns type around the 12th century BC, however cannot be verified, as apart from a ceramic bowl, reportedly resembling BR pottery forms (Negroni Catacchio 1999, 245), the cave did not contain any other chronologically indicative materials. By their form, the amber beads from Costa Cavallina (IA231a, IE211a and IE231a) find exact analogies in the Garda Lake basin, western Veneto and Tuscany.[564] Analogy from western Veneto (IE211a) is of particular importance, as it comes from Fondo Paviani – site with well-documented occupation at the time of BR–BF transition. The latter settlement is situated in the same province as Campestrin di Grignano Polesine – a contemporaneous site for which production of beads with the Tiryns-type decoration is attested. The settlement of Fondo Paviani also returned a torc made of thick, richly decorated bronze rod with one biconical and one folded end, which finds the closest parallels in BR 2 contexts of the *Canegrate* culture situated in Lombardy (cf. Cupitò 2011b). Presence of similar amber and bronze finds such as those from Costa Cavallina and Fondo Paviani implies contacts between Lombardy and Veneto around the advanced stage of LBA.

When it comes to Liguria, the sites already active during MBA (Castellaro di Camogli and Castellaro di Zignago) delivered evidence of continuous occupation in LBA, entailing prolonged chronology of amber finds. Considering the latter phase,

564 Poggiani Keller 1989, Fig. 98:1–3; cf. Peschiera – Imboccatura del Mincio, precise context unknown (De Marinis 2007, Fig. 5); Fondo Paviani, advanced BR–early BF layers (Fasani/Salzani 1976, Fig. 2:3–4); Ponte San Pietro Valle, tomb 1 (Negroni Catacchio 1999, Fig. 3:34).

the closest counterparts for the finds in geographic sense are known from the Garda Lake basin and western Veneto. The most precise seems to be the analogies for IF151 bead from Castellaro di Zignago, zone S, layer G, found among BR-dated graves from western Veneto.[565] This observation corresponds well with the already mentioned conclusions reached upon analysis of pottery from the Ligurian site which point to relationships with the *Canegrate and* the *Palafitte-Terramare* cultures in their advanced phases of development (Mannoni/Tizzoni 1980, 277–278).

Fig. 85. Examples of fibulae and pinheads shared by some of the LBA sites with amber in Tuscany, northern Lazio and western Umbria (A–D), the Garda Lake basin (cf. Fig. 84:A, J), western Veneto (cf. Fig. 84:B, M) and Emilia (cf. Fig. 84:K, N): A – Fossa Nera di Porcari (Andreotti/Zanini 1995–1996, Fig. 2:4); B–D – Scarceta, sector D – casa-laboratorio *no. 137 (Poggiani Keller 1999, Figs. 96:4–5 and 98:1).*

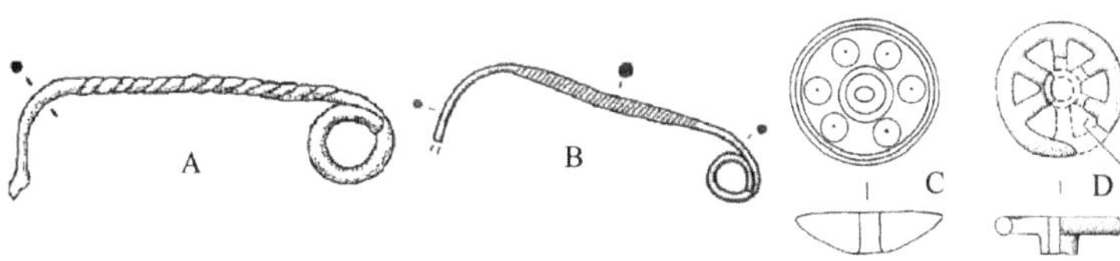

Moving counterclockwise around the Garda Lake and the Po Plain, the next region to discuss – Tuscany, northern Lazio and western Umbria (northern Etruria) – provides a richer data for determining cultural links with other provinces. The amber finds from northern Etruria morphologically correspond with artefacts from four other regions: western Lombardy, the Garda Lake basin, Romagna and, to the biggest *degree*, western Veneto (Fig. 81). Four contexts from the region, all dated to LBA/FBA: tomb 1 of Ponte San Pietro Valle, 'house-laboratory' of Scarceta, Omo Masso and Fossa Nera di Porcari, yielded in total seven amber beads decorated with the Tiryns-type motif (IA111a, IE111a, IA231a, IE251a, IE261a and IE451a). Among them IE111a and IE451a morphologically correspond with some of the artefacts from the Garda Lake basin and western Veneto.[566] This clue is supported by presence on the same Tuscan sites of other elements co-occurring with amber having the *Palafitte-Terramare* cultural provenience, i.e. Fossa Nera di Porcari delivered pottery represented in such contexts

565 Mannoni/Tizzoni 1980, Fig. 7:s; cf. Castello del Tartaro, tomb 166 (Salzani 2011, Fig. 3:B5); Olmo di Nogara, grave 358 (Salzani 2005, Tav. XXIX:C); Olmo di Nogara, grave 441 (Salzani 2005, Tav. XXXVI:A); Scalvinetto, tomb 282 (Salzani 2004, Fig. 3:B).

566 IE111a: Omo Masso, precise context unknown (Negroni Catacchio 2014, Fig. 8); cf. Ponte San Marco, BR–BF horizon (Ruggiero 1994, Fig. 89); IE451a: Fossa Nera di Porcari, precise context unknown (Andreotti/Zanini 1995–1996, Fig. 2:13); cf. Campestrin di Grignano Polesine, late BR–initial FBA layers (Bellintani et al. 2015a, Fig. 2A:1).

as the 'summit' layer (BR 2) of Santa Rosa di Poviglio – *Villaggio Grande* (Andreotti/Zanini 1995–1996, 320, Fig. 3:3; cf. Bernabò Brea et al. 1997d, Fig. 188:2). To categories of finds shared by Tuscany and western Veneto can be included i.a. violin-bow fibulae with twisted arches (Fig. 85:A–B; cf. Fig. 84:A–B) and pinheads of the types Gazzade and Narce (Fig. 85:C–D; cf. Fig. 84:J–K, L–N).

Finds from Tuscany allow the node of Romagna and Emilian Apennines to be connected with the network. The sites of Omo Masso and Borgo Panigale share beads of type IE261a.[567] Unfortunately, both beads come from uncertain contexts, and there are no associated finds which could be confronted to find if there is a connection between the sites beyond morphological similarity of amber finds. Another amber artefact decorated with the Tiryns motif (IA211a) from Romagna and Emilian Apennines was found at Monte Battaglia (Fig. 45:3). Its closest counterparts would be IE211a beads found in western Veneto and western Lombardy (cf. Fig. 49:3).[568] Furthermore, bead of the type IF124y from San Giuliano di Toscanella Imolese also has the best analogies in the area of western Veneto, where specimens classified as IF124 were discovered in several graves.[569]

567 Cf. Catarsi Dall'Aglio 1976, Fig. 6:16; Negroni Catacchio 2014, Fig. 8. It is worth mentioning that the lost amber bead from the FBA (?) layers of Coppa Nevigata from the profile resembles IE261a judging by its profile (Belardelli 2004, Fig. 31:219), hence could be considered an analogy upon typological verification.

568 Fondo Paviani, advanced BR–early BF layers (Fasani/Salzani 1976, Fig. 2:3–4); Costa Cavallina, cave burial (Poggiani Keller 1989, Fig. 98:3).

569 Castello del Tartaro, tomb 350 (Salzani 2011, Fig. 5:B3–4); Olmo di Nogara, graves 190 and 441 (Salzani 2005, Tav. XVIII:E, XXXVI:C).

Fig. 86. Examples of pins, pinheads, daggers, fibulae, axes and sickles shared by some of the LBA sites with amber in Romagna (A–C, E–G, I, K, N, P, S), the Garda Lake basin (J, L, R) western Veneto (cf. Fig. 83:A) and Emilia (D, H, M, O, T): A–B, E, G, I, K – Riccione (Bermond Montanari et al. 1992, Fig. 17:1–3, 5–6, 10); C, N, P – Monte Battaglia, hoard (Bermond Montanari 1997, Fig. 128:1, 129:3, 130:7); D – Sant'Ambrogio (Cremaschi 1997, Fig. 179:4); F – San Giuliano di Toscanella Imolese (Bianco Peroni 1994, Tav. 79:1443); H – Castellazzo di Fontanellato (Bianco Peroni 1994, Tav. 83:1500); J, L, R – Peschiera – Imboccatura del Mincio (Aspes 1980, Fig. 42:13; Eles Masi 1986, Tav. 1:1; Bianco Peroni 1994, Tav. 89:1585); M – Redù-Pilastro (Malnati/Cardarelli 2003, Fig. 58:11); O, T – Castione Marchesi, layers 2–3/phases 1–4 (Mutti et al. 1988, Fig. 72:4, 76:3); S – Borgo Panigale (Catarsi Dall'Aglio 1976, Fig. 6:6).

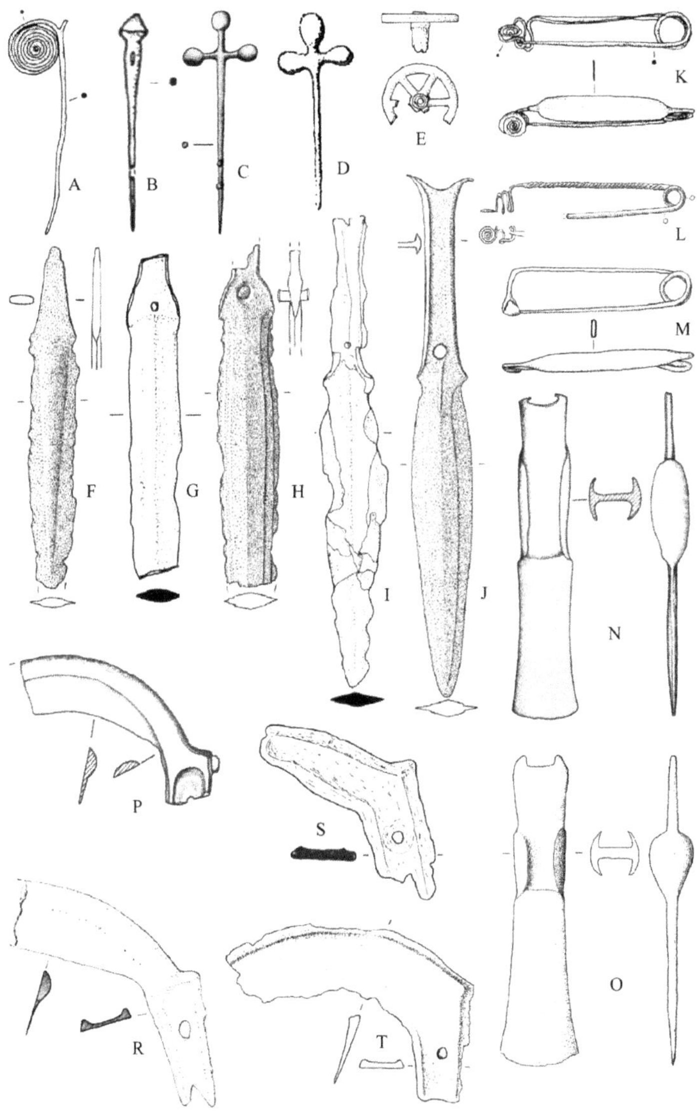

Abovementioned observations on similarities between amber beads can be supplemented with numerous types of bronze finds common for Romagna and the adjacent centre of amber's acquisition in the Garda Lake basin and the Po Plain. They include pins of the types Peschiera (Fig. 86:A; cf. Fig. 83:A–C), Boccatura del Mincio/Cattaragna (Fig. 86:B, cf. Fig. 83:G–I) and Cogolara (Fig. 86:C–D), and the Narce-type pinheads (Fig. 86:E, cf. Fig. 84:L–N) (cf. Carancini 1975). There are at least three types of daggers occurring with amber in those regions: Redù, var. A (Fig. 86:F; cf. Fig. 84:C–D), Pertosa, var. B (Fig. 86:G–H) and Bertarina (Fig. 86:I–J) (cf. Bianco Peroni 1994). To interregional contacts can be credited examples of fibulae with twisted/corrugated arches,[570] but also one violin-bow fibula of specific kind which seems to combine distinct features present in fibulae from some of the LBA sites with amber in the Garda Lake basin and Emilia (Fig. 86:K): the spiral bracket present on some specimens from Peschiera – Imoccatura del Mincio (Fig. 86:L) and widened 'foliated' arc resembling the fibula from Redù-Pilastro (Fig. 86:M). Other common elements include the Pertosa-type axes (Fig. 86:N–O), and sickles of the types Merlara (Fig. 86:P–R) as well as some placeable between the types Castions di Strada and Muscoli (Fig. 86:S–T) (cf. Carancini 1997, Fig. 224:12; Carancini/Peroni 1999, 56, Tav. 27:14, 20). Following the above given analogies, it seem plausible that amber reached Romagna through contacts with the areas directly to the north – the Garda Lake basin and the Po Plain.

Many of the abovementioned types of finds were also registered on sites with amber in Marche and eastern Umbria – a region which has evident links in terms of amber morphology with the Garda Lake basin, western Veneto, Emilia and Romagna. One can mention the Boccatura del Mincio/Cattaragna-type pins (Fig. 87:A; cf. Fig. 83:G–I, 86:B) and the Narce-type pinheads (Fig. 87:B; cf. Figs. 84:L–N and 86:E) (cf. Carancini 1975). Furthermore, similarities include three variants of violin-bow fibulae: with twisted/corrugated arch (Fig. 87:C; cf. Fig. 84:A–B), with two nodules on the arch (Fig. 87:D–E) and with two nodules and foliated arch (Fig. 87:F–G) (cf. Eles Masi 1986). The most numerous category of analogies are bronze daggers covering the types Pertosa, var. B (Fig. 87:H; cf. Fig. 86:G–H), Bertarina (Fig. 87:I; cf. Fig. 86:I–J)[571] and Glisente (Fig. 87:J–K), also including specific variant of the latter type described as *daghe* (Fig. 87:L–M) (cf. Bianco Peroni 1994). It is worth mentioning that some ceramic vessels from Moscosi di Cingoli, i.e. biconical vases decorated with bosses, *cuppelle* and grooves, combined in complex syntaxes, have analogies at sites in Emilia such as the settlement of Santa Rosa di Poviglio – *Villaggio Grande* or the cemetery of Casinalbo.[572] Moreover, the settlements of Santa Rosa di Poviglio – *Villaggio Grande* and Moscosi di Cingoli share almost identical barrel-like glass beads in brown paste, decorated with incrusted spirals in white, produced of HMBG – variety of vitreous materials so far not attested outside Northern and Central Italy (Fig. 87:N–O) (cf. Bellintani et al. 2006, 1507, 1525; Bellintani 2012, 20). The compositional and formal similarities between

570 San Giuliano di Toscanella Imolese (Pettazzoni 1916, Fig. 32:1–2; cf. Fig. 84:A–B).

571 It is worth noting that a casting mould for the Beratrina-type daggers was discovered in the layer US19b at the settlement of Fondo Paviani, dated to BR 2 (Marcon/Mazzoli 2015, Fig. 1:A). Results of petrographic analysis shows that the casting mould was made of talc schist rock, obtained from the central-western Alps, specifically from the Valchiavenna area (Marcon/Mazzoli 2015, 845).

572 Sabbatini/Silvestrini 2005, 648, Fig. 3; cf. Bernabò Brea et al. 1997d, Fig. 216; Cardarelli 1997, Figs. 396 and 398.

LBA assemblages from sites with amber in Marche, eastern Umbria, Romagna, the Po Plain and the Garda Lake basin are remarkable. It seems, therefore, justified to extend the links indicating contacts from the Po Plain over Romagna to Umbria and Marche.

Fig. 87. Examples of pins, pinheads, fibulae and daggers shared by some of the LBA sites with amber in Marche and eastern Umbria (A–D, F, H–J, L, N) and the Garda Lake basin (E, G, K, M), western Veneto and Emilia (O): A, B, F, H–J, N – Moscosi di Cingoli, phase 4 (Silvestrini/Pignocchi 1999, Fig. 9:9; Sabbatini/Silvestrini 2005, Fig. 4:1–2, 4, 10, 14, 17); C–D, L – Gualdo Tadino, hoard (Peroni 1963b, Tav. 3:6, 4:27); E, G – Peschiera (Eles Masi 1986, Tav. 3:53, 60); K – Ponte San Marco, BR–BF horizon (Poggiani Keller 1994, Fig. 80:1); M – Isolone del Mincio (Bianco Peroni 1994, Tav. 75:1360); O – Santa Rosa di Poviglio, Villaggio Grande, *'summit' layer – phases 3 and 4 (Bellintani/Biavati 1997, Fig. 354:6).*

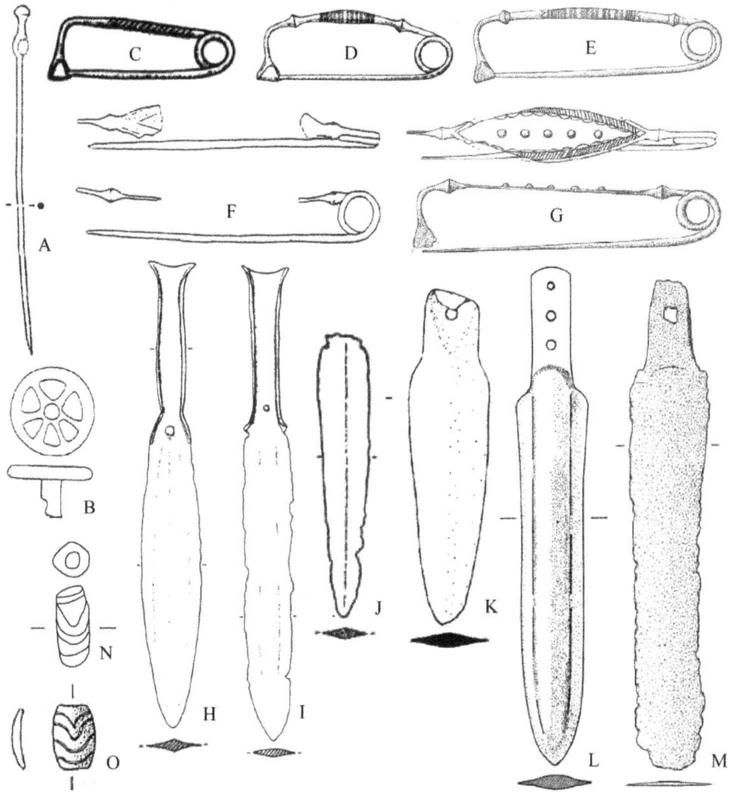

The interregional ties built upon similarity of amber beads and associated categories of finds reach further south of the Apennine Peninsula, to Molise and Apulia. Northern Apulia and Basilicata – a core region rich in amber artefacts in LBA – delivered particularly abundant evidence of contacts with the areas to the north, i.e. the Garda Lake basin, western Veneto and Emilia, just like it did in the previous phase (MBA).[573] The hypogea of Trinitapoli yielded examples of the Cattaragna (Fig. 74:D; cf. Fig. 83:G–I), the latter also found in Romagna (cf. Fig. 86:B) and Marche cf. (Fig. 87:A), and Ca del Lago types of pins (Fig. 88:A; cf. Fig. 83:M–O) (cf. Carancini 1975). Furthermore, among shared items should be also highlighted a sickle attributable to the type Muscoli (Fig. 88:B–C) and a spearhead of the type Fucino (Fig. 88:D–E) from Coppa Nevigata (cf. Carancini/Peroni 1999, 56, 58, Tav. 27:20, 28:45).[574]

It is also worth mentioning two pins with potentially northern origin: one coming from the *Subapennine* (BR) layers of the settlement of Coppa Nevigata classified to type Biandronno, var. A (Belardelli 2004, 98, Fig. 34:1; cf. Carancini 1975, 45, 222), and one with onion-shaped head and the neck decorated with oblique lines found beside the burial of female adorned with a necklace of amber beads in the sector C2 of the 'Hypogeum of Bronzes' at Madonna di Loreto which resembles types Monza or Sover (Tunzi Sisto 1999b, 215, Tav. XVI:1361; cf. Carancini 1975, 45–46, 222–224).[575] Finally, glass is of importance for tracing LBA contacts between northern Apulia and the Po Plain; the LMLK glassy-faience beads in lenticular form with radial incisions on the sides were found in the 'Hypogeum of Bronzes' of Madonna di Loreto, as well as on several sites in Emilia and western Veneto, including the settlement of Santa Rosa di Poviglio – *Villaggio Grande* and the cemetery of Franzine Nuove[576] (Fig. 88:F–G).

Second, northern Apulia shares some of the types in amber and bronze with Romagna as well as Marche and eastern Umbria. Among the latter, apart from the already mentioned Cattaragna-type pins, can be listed the Torre Castelluccia-type daggers in variants A (Fig. 88:H–I) and B (Fig. 88:J–K) (Bianco Peroni 1994, 122–130), and axes with short or medium-short, but wide-folded median wings (Fig. 88:L–N) (cf. Carancini 1997, Fig. 224:6–7; Carancini/Peroni 1999, Tav. 27:11). The similarities also include bone 'handles' with incised concentric circles.[577]

573 Cf. Peroni 1999a, 217; Cazzella 2009, 161; Recchia 2009, 230.

574 Casting mould designed for producing spearheads similar to the type Fucino was found alongside the cited example from Casinalbo (cf. Säflund 1939, Tav. 71:2).

575 A. M. Tunzi Sisto, quoting one of the examples from L. Carancini's catalogue, proposes different attribution of the pin to the type Sover (Tunzi Sisto 1999a, 271; cf. Carancini 1975, T. 54:1767). However, style of decoration of the neck as well as slightly biconical shape of the head of the specimen from Madonna di Loreto prompts to reconsider the latter classification.

576 Franzine Nuove, grave 481 (Aspes 1980, Fig. 1:1); cf. Bellintani et al. 2006, 1505–1507, 1524–1525.

577 Catarsi Dall'Aglio 1976, Fig. 6:9; Silvestrini/Pignocchi 1999, Fig. 9:8; Belardelli 2004, Fig. 31:213.

Fig. 88. Examples of pins, sickles, spearheads, glassy-faience beads, daggers and axes shared by some of the LBA sites with amber in northern Apulia (A, C–D, F, H, K–L), the Garda Lake basin (B), Emilia (E, G, J), Marche and eastern Umbria (I, N), and Romagna (M): A, F – Madonna di Loreto, 'Hypogeum of Ivories' (Tunzi Sisto 1999a, Tav. X:1769, XVI:198); B – Isolone del Mincio (Guerreschi et al. 1985, Tav. XXIII:st.9151); C–D, H, K–L – Coppa Nevigata, BR–BF stage (Bianco Peroni 1994, 65:1160; Belardelli 2004, Fig. 34:8, 11, 14–15); E – Casinalbo (Carancini 1997, Fig. 233:76); G – Santa Rosa di Poviglio, Villaggio Grande, *'summit' layer – phases 3 and 4 (Bellintani 2012, Fig. 1:15); I, N – Moscosi di Cingoli, phase 4 (Silvestrini/Pignocchi 1999, Fig. 7:7, 9:2); J – Beneceto, BR 2 layers (Bernabò Brea et al. 2008, Fig. 13:16); M – Borgo Panigale, 1957 campaign, precise context unknown (Catarsi Dall'Aglio 1976, Fig. 6:1).*

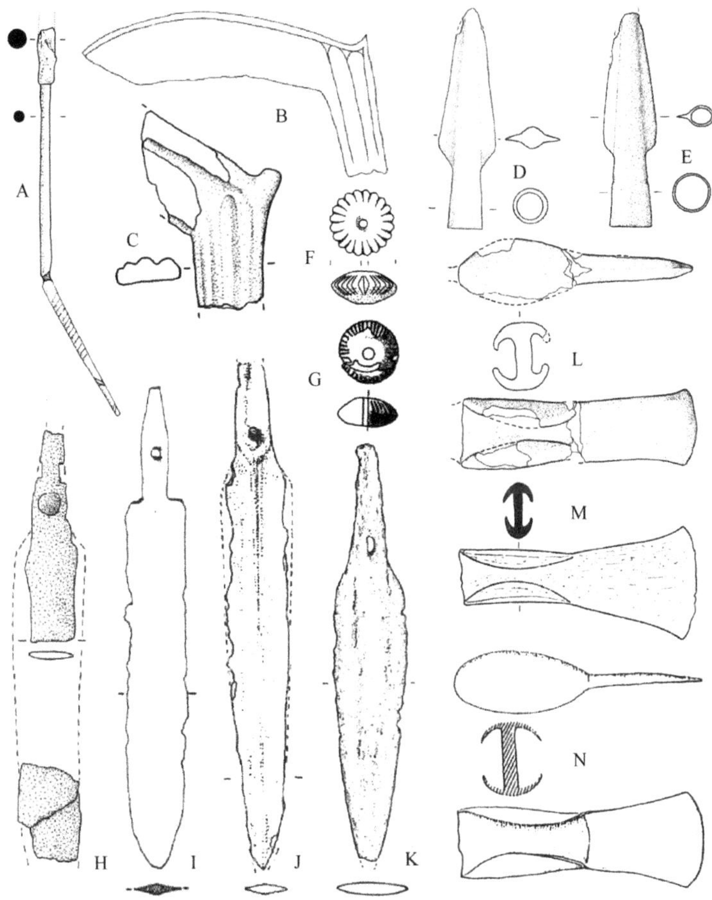

Fig. 89. Examples of spearheads, knives and pinheads shared by some of the LBA sites with amber in northern Apulia (A, D, F–G), Molise (B, E), southern Apulia (H), the Garda Lake basin, western Veneto (C) and Emilia: A, D, F–G – Coppa Nevigata, BR–BF stage (Belardelli 2004, Fig. 35:19; Cazzella et al. 2010, Fig. 56; Cazzella/Recchia 2016, Fig. 2:1–2); B, E – Rocca di Oratino, phases I–III (Copat/Danesi 2010, Fig. 4:6–7); C – Fondo Paviani, advanced BR – early BF layers (Fasani/Salzani 1975a, Tav. X:6); H – Roca Vecchia, Sector IX, US 11349, Phase II (Pagliara et al. 2008, Fig. 16:4).

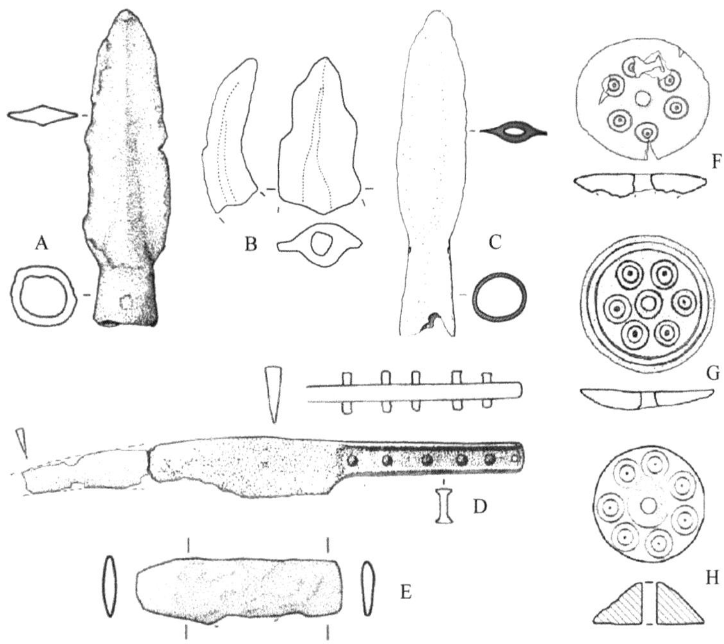

Third, certain types of artefacts, some of them having genetic links with the Garda Lake basin and the Po Plain, link northern Apulian sites with amber with those in the neighbouring regions, i.e. Molise and southern Apulia. In the first case, these are the Pila de Brancon-type spearheads (Fig. 89:A–C) (cf. Carancini/Peroni 1999, Tav. 27:16) and the Scoglio del Tonno-type knives (Fig. 89:D–E) found i.a. at Coppa Nevigata and possibly at Rocca di Oratino (cf. Bianco Peroni 1976, 14–15).[578] In the second case, one can mention the Gazzade-type pinheads from Coppa Nevigata and Roca Vecchia (Fig. 89:F–H; cf. Fig. 84:J–K), the latter site also notable for yielding a violin-bow fibula and the Baierdorf-type knife in BR 2 layers (Pagliara et al.

578 Due to bad state of preservation, the spearhead and the knife blade from Rocca di Oratino cannot be securely determined in terms of typology; however, the abovementioned classification is suggested by authors of the survey at the site (cf. Copat/Danesi 2010).

2008, 267, Fig. 15:B). Both these objects are reported from Campestrin di Grignano Polesine (Bellintani et al. 2015a, 420), while casting mould for the Baierdorf-type knives was discovered at Fondo Paviani (Fasani/Salzani 1976, Fig. 2:9). These observations support association between southern Apulia and western Veneto built upon co-occurrence of not identical, but morphologically similar amber beads with the Tiryns-type decoration, which also include semi-finished forms lending a clue about their processing place.[579]

While distribution of the finds of northern provenience all the way along the Adriatic side of the Apennine Peninsula from the Po Plain to Apulia seems to testify to presence of a developed interregional network of interactions, the relationships with provinces of amber's acquisition on the Tyrrhenian side of Italy during LBA seem vague. In Calabria the sole context with amber datable to this phase – Santa Domenica di Ricadi, tomb 4 – has already been described in terms of discernible testimonies of outer influences among the finds in the previous section. Thus, it remains only to emphasise the links visible on the basis of pottery, mainly the snail-horn appendices, with the *Ausonian I* culture on the one hand, and the *Subapennine* culture on the other, with similarities pointing to sites such as i.a. Lipari 'Acropolis' and Coppa Nevigata (cf. Pacci 1987, 46, Fig. 2:1, 12). Ties built upon similarities in morphology of amber beads reach far to the north with the strongest links pointing to Emilia and western Veneto. Resemblance between the Calabrian amber beads and the contemporaneous finds from the Aeolian Islands – Lipari 'Acropolis', hut Beta I-tomb III cannot be determined, because amber bead from the latter context lacks illustration. To this evidence can be added a badly preserved bronze pin with cylindrical stem surmounted by a small oblong head, decorated on the neck with horizontal grooves from tomb 4 Santa Domenica di Ricadi (Fig. 90:A) which, in the opinion of M. Pacci, has analogies in Northern Italy among the pins with small poppy-head (Pacci 1987, 47; cf. Carancini 1975, 50, 236, Taf. 53) – type also attested on some of the LBA sites with amber in Emilia (Fig. 90:B).

579 IE218a: Roca Vecchia, 'Cult Center', Sector IX, US 11349, Phase II (Guglielmino 2021, Fig. 8); IE211a: Fondo Paviani, advanced BR–early BF layers (Fasani/Salzani 1976, Fig. 2:3–4); VIIIA218a: Campestrin di Grignano Polesine, late Recent–initial FBA layers (Bellintani et al. 2021, CGP 46).

Fig. 90. Examples of pins shared by some of the LBA sites with amber in Calabria (A), Emilia (B): A – Santa Domenica di Ricadi, tomb 4 (Pacci 1987, Fig. 4:3); B – Beneceto, BR2 layers (Bernabò Brea et al. 2008, Fig. 13:10).

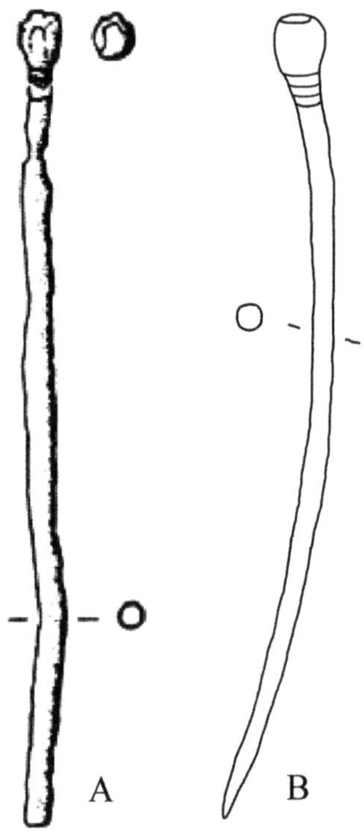

Trans-Adriatic circulation of amber during LBA seems probable in the light of similarities among amber finds from the Italian and the Balkan regions, supplemented with occurrence of few objects interpretable as 'imports'. The strongest ties resulting from NA connect the core provinces of Istria on the east side of the sea, with western Veneto, Emilia and northern Apulia on the west side. Due to issues with dating, or prolonged chronology, some of the Istrian sites with amber and related finds can, again, be brought forward as analogies in the LBA discussion, even though they have partly been used in the MBA analysis (cf. section 3.12.2.). Still, there is one context in Istria securely dated to LBA – the sepulchral cave of Laganiši which yielded a dagger of the Tenno type (Fig. 91:A). Daggers of the Tenno type chronologically fall into BR, while their distribution is limited to Northern Italy, and includes the Peschiera del Garda – Imboccatura del Mincio site (Fig. 91:B) (cf. Bianco Peroni 1994, 156–157). To the group of shared metal objects should be added, as in the previous phase, the spectacle pendants with raised connecting parts which in LBA appeared beyond Istria (Fig. 91:C) and northern Apulia (Fig. 91:D), with one specimen known from a site with amber in Romagna – the settlement of Riccione (Fig. 91:E). Amber finds from the latter region in general do not resemble exactly those from Istria; however, certain similarity is observable between the bead from San Giuliano di Toscanella Imolese (IF124y) and the finds from Mušego and Vrčin.[580]

Fig. 91. Examples of bronze daggers and spectacle pendants shared by some of the LBA sites with amber in Istria (A, C), the Garda Lake basin (B), northern Apulia (D, H), Romagna (E), Montenegro (F), the Mat Valley (G), western Veneto (I): A – Laganiši, sepulchral cave (Mihovilić 2008, 45, Fig. 36), B – Peschiera – Imboccatura del Mincio (Bianco Peroni 1994, Tav. 87:1557), C – Vrčin, tumulus B (Buršić-Matijašić 1989, Fig. 3:1), D, H – Madonna di Loreto, 'Hypogeum of Bronzes' (Tunzi Sisto 1999a, Tav. XI:1440, XV:1398), E – Riccione (Bermond Montanari et al. 1992, Fig. 17:4), F – Velika Gruda, grave 9 (Della Casa 1996a, Fig. 44:16); G – Shtogj, tumulus 9 (Kurti 2017a, Fig. 7:1); I – Scalvinetto di Legnago, grave 282 (Salzani 2004, Fig. 3:A).

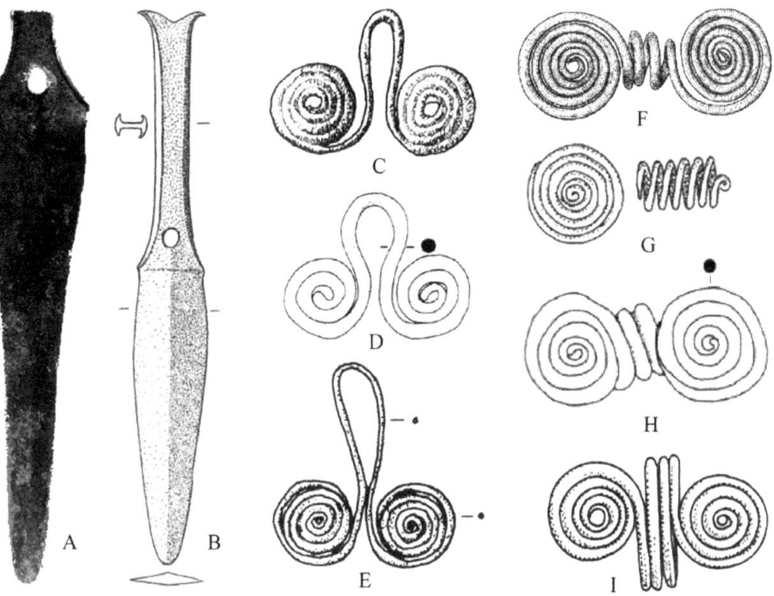

Spectacle pendants (or braid holders) with coiled connecting parts, same as those from grave 9 of Velika Gruda in Montenegro (Fig. 91:F), or grave 2, tumulus 9 of Shtogj in the Mat Valley (Fig. 91:G), come in handy as indicators of LBA contacts across the lower Adriatic. In Italy contemporaneous pendants are known in abundance from the hypogea of Trinitapoli in northern Apulia (Fig. 91:H) and, in at least one specimen, from the necropolis of Scalvinetto in western Veneto (Fig. 91:I). Convergences in bronze jewellery are followed by similarities in amber adornments; some amber beads (i.e. IF121y) from Scalvinetto, tomb 282 – the western Venetian context which yielded the abovementioned spectacle pendant – correspond typologically with the finds from grave 9 of Velika Gruda (Salzani 2004, Fig. 3:B; cf. Della

580 Morico 1996, 350, Tav. 1:D; cf. Buršić-Matijašić 1989, Fig. 1:2; Mihovilić et al. 2009, 55.

Casa 1996a, 44:18). Beads representing types IF223, on the other hand, were found in the 'Hypogeum of Bronzes' of Madonna di Loreto (Tunzi Sisto 1999a, Tav. X:716) and Shtogj, tumulus 2, grave 9 (Kurti 2017b, Pl. XCVIIIf:4). The latter context also shares beads of the types IF228 with western Veneto.[581] In the face of limited links between Montenegro and the Mat Valley and the other Balkan provinces (Fig. 81), one can hypothesise some form of relationship between the two regions and the above-listed parts of Italy during LBA.[582]

With respect to other Italian and Balkan regions, bronze pins offer the best chances of pinpointing possible directions of the trans-Adriatic interactions during LBA. Among the rich collection of pins recovered from the Ljubljanica River in Slovenia, there is one with amber adornment inserted onto the upper part of the stalk (Fig. 92:A). Form of the amber pinhead, classified as IF121, finds numerous analogies in the LBA contexts in Italy. However, the overall design incorporating amber adornment on a bronze pin, so far, has only been registered in the Garda Lake basin, i.e. at the *palafitte* of Maraschina and Peschiera del Garda (Fig. 92:B–C). According to Slovenian researchers, the bronze part of the Slovenian pin with amber head resembles some of the Italian pins with globular heads and enlarged necks (Turk/Gaspari 2009, 70; cf. Carancini 1975, Tav. 47:1436–1454). Furthermore, another pin from the same collection of pins from the Ljubljanica River, i.e. one with a club-shaped head and horizontally ribbed neck, possesses analogies in Northern Italy, some of them coming from sites with amber in the Garda Lake basin and Emilia.[583]

Although during LBA Podrinje and Western Serbia lost the status of a core region, there are still present some artefacts potentially testifying to long-distance relationships with the Apennine Peninsula. Especially prominent is the similarity level between amber beads from Podrinje and northern Apulia (Fig. 81), replicating the same strong bond already seen in MBA (Fig. 67). This link can be further strengthened by bringing forward several pins that these two regions have in common. One of them comes from Grotta Manaccora – context dated to MBA, however possessing several elements of inventory potentially of later origin (cf. Recchia 1993, 1999). The pin has profiled, biconical head and swollen neck decorated with horizontal bands of incised lines (Fig. 92:D). G. Carancini places it among the pins with flattened biconical heads and enlarged necks, found at sites in the vicinity of Peschiera del Garda (Imboccatura del Mincio and Bacino Marina) and dated to BR (Carancini 1975, 196, Tav. 44:1347–1351). Still, the object from Grotta Manaccora can be correlated with some pins from the Paulje necropolis in Podrinje, among them two from mound A dated to LBA and one found outside the graves (Fig. 92:E) (Bulatović et al. 2017, T. XIII:3–4; cf. Vasić 2003, 55). Decoration pattern on the necks (vertical bands of incised lines or thin grooves) enables to ascribe the pins to the Deinsdorf type – a large category of pins widespread during LBA in the areas encompassed by the late *Tumulus* and the

581 Kurti 2017b, Pl. XCVIIIf:5; cf. Olmo di Nogara, graves 190, 243 and 252 (Salzani 2005, Tav. XVIII:D, XXI:F, XXII:D).

582 Acceptable only if the strong relations of Montenegro and, particularly, Mat Valley with the Balkan interior visible during MBA would turn out to be unsubstantiated or chronologically inconsistent (cf. section 3.12.2.).

583 Turk/Gaspari 2009, 70; cf. Carancini 1975, 184–189, Pl. 42:1275, 43:1295–1296; Aspes 1980, Fig. 41:14; Bernabò Brea et al. 2008, Fig. 13:7.

early *Urnfield* cultures.[584] Furthermore, the earlier mentioned Monza/Sover-type pin from the 'Hypogeum of Bronzes' of Madonna di Loreto (Fig. 92:G) finds strikingly similar counterpart in grave 1, mound 1 of Karavlaške kuće (grave 2 from the same mound contained amber) on the Bosnian side of Podrinje (Fig. 92:F). This particular variant of bronze pins is not specific for southern Apennine Peninsula or Central Balkans, as morphologically it replicates pins with slightly flattened ball-like heads (ger. *Kugelkopfnadeln*) and swollen or corrugated necks, generally attributed to the early *Urnfield* culture in the circum-Alpine zone and the Carpathian Basin.[585] However, distant southward localisation of the discussed examples of pins, away from the main centres of the *Urnfield* culture, together with their association with amber is noteworthy. Presence of similar amber adornments and bronze pins, although the latter may not belong to exactly the same types, allows to suspect continuation of interactions between northern Apulia and Podrinje during LBA.

Regions located further south in the Balkan hinterland, i.e. Kosovo and Metohija, the Upper Drin Valley, the Korçë Basin and the Kolonjë Plateau, despite manifesting ties with Italian regions based upon similarities among amber artefacts, are missing any clear evidence of contacts with the latter. The only common element that comes to mind are the earlier mentioned pins with conical heads and swollen, pierced necks from the tumuli on the Kolonjë Plateau (Fig. 78:G). It has been argued in the literature that these pins, along with specimens classified as the Boccatura del Mincio/Cattaragna type particularly widespread in Northern Italy (cf. Fig. 83:G–I), belong to a single family of pins epitomised by the Paarstadl type (Aliu 2012, 300; Pabst 2013, 155–156, Abb. 1). Although the Italian and the Albanian pins differ in details, a general design seems to be consistent, providing some credibility to the hypothetical relationship between the upper-west and lower-east Adriatic. One more object to be mentioned in the context of possible Albano–Apulian relations is the spearhead from BR–BF layers of Coppa Nevigata, attributed to the type Pazhok quite widespread in southern Albania during LBA (Belardelli 2004, 34:13; cf. Bodinaku 1995, Fig. 5:1–3). Pattern of spatial distribution of the cited finds belonging to the same categories and types certainly allows to suspect interactions between the opposite shores of the lower Adriatic in the discussed phase. There is also a hint at possible relations between south-east Albania and northern Italy is offered by the comparison of glass jewellery from Barç, tumulus 1, grave 47, comprising cylindrical beads decorated with spiral incrustation of whitish enamel (Andrea 1985, Tab. VI:3–4), which generally resemble vitreous materials from western Veneto (Fig. 87:N–O); however, analysis of chemical composition of the Albanian finds is required before the connection can be proved.

584 Cf. Novotná 1980, 101–102; Gedl 1983, 24–27; Říhovský 1983, 15–16.

585 Carancini 1975, 45–46, 222–224; Vasić 2003, 53; cf. fn. 583.

Fig. 92. Examples of bronze pins shared by some of the LBA sites with amber in Slovenia (A), the Garda Lake basin (B–C), northern Apulia (D, G) and Podrinje (E–F): A – Ljubljanica River (after T. Lauko, National Museum of Slovenia); B – Maraschina (De Marinis 2007, Fig. 5); C – Peschiera – Imboccatura del Mincio (De Marinis 2007, Fig. 5); D – Grotta Manaccora, Grotticella Funeraria *– SS layers (Carancini 1975, Tav. 44:1350); E – Paulje, cemetery (Bulatović et al. 2017, T. XXVI:133); F – Karavlaške kuće, mound 1, grave 2 (Kosorić/Krstić 1970, Tab. IV:1); G – Madonna di Loreto, 'Hypogeum of Bronzes' (Tunzi Sisto 1999a, Tav. XVI:1361).*

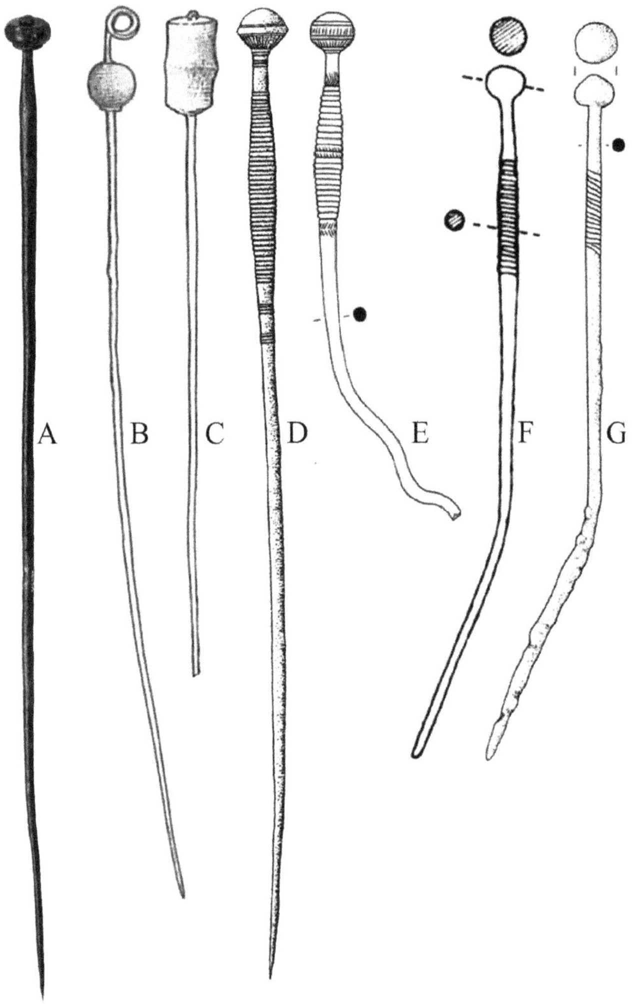

Inner Balkan relationships between provinces of amber's acquisition during LBA appear definitely less distinct, than in the previous phase. Istria still lacks more direct connection with the Balkan hinterland beyond the superficial similarity among amber adornments. The only meaningful object is the axe with short median wings, narrow, semi-circular cutting blade and rounded heel with trapezoid recess, found in the Laganiši sepulchral cave and determined as the type Freudenberg (Fig. 93:A) (cf. Mihovilić 2008, 45). Comparable type of axe (Fig. 93:B), defined as the type Gmunden (Žeravica 1993, 71, Taf. 19:242) comes from mound IV of Karavlaške kuće (Bosnian Podrinje), neighbouring mound 1 which yielded contemporaneous amber artefacts (cf. Kosorić/Krstić 1970). There are visible differences between the axes in the shape of heel and cutting edge, but nonetheless they definitely fall into the same family of axes with median wings. Both the types Freudenberg and Gmunden represent the forms characteristic for LBA, however were not particular to the Balkans, as they most probably have their origins in the ecumene of the early *Urnfield* culture in the circum-Alpine zone: Austria, Bavaria, Bohemia and the Carpathian Basin (cf. Mayer 1977, 93–95; Žeravica 1993, 70–71). Thus, rather than evidence of contacts between Istria and Podrinje, they should be perceived as testimonies of influences of the *Urnfield* complex in the Balkans.

Fig. 93. Examples of axes shared by some of the LBA sites with amber in Istria (A) and Podrinje (B): A – Laganiši, sepulchral cave (Mihovilić 2008, 45, Fig. 36); B – Karavlaške kuće, mound IV, grave 1 (Kosorić/Krstić 1970, Tab. V:3).

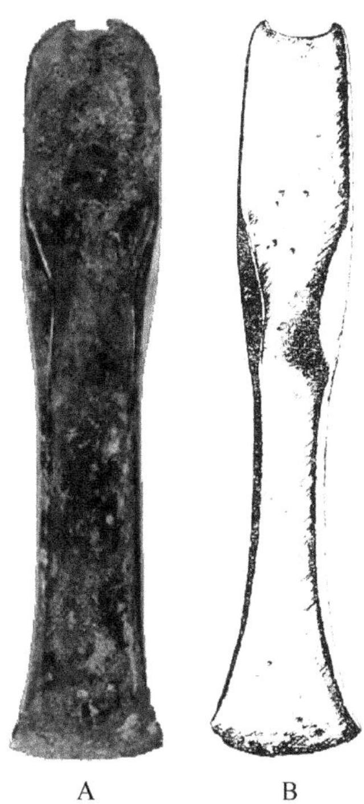

A B

With the decrease in number of amber artefacts and contexts, the MBA network of amber circulation in Central Balkans appears to have started to disintegrate already in LBA. In Metohija and the Upper Drin Valley, some context with prolonged chronology, already used in the MBA network, possibly pertain to LBA; however, no new amber finds emerged in those regions during the latter phase. Between Podrinje and areas to the south of Central Balkans, the bond is still visible through the distribution of the open bracelets with stamp-shaped terminals (type I) which includes Metohija and the Upper Drin Valley (cf. Fig. 77:A–F), in the latter case strengthened by similarities among amber finds. At the turn of MBA and LBA, bracelets in Podrinje and Western Serbia became more diverse in terms of decorations, evolving into types I-B, I-C and I-D; however, the basic morphology had been kept.[586] Apart from Serbia, examples of types ID can be found among certain amber-containing contexts in Metohija[587] and the Upper Drin Valley.[588] However, no such bracelets, nor any other kind of finds with similar provenience accompany amber finds further south beyond the northernmost Albania.

On the other hand, the remaining Balkan provinces of amber's acquisition to the south seem to have been well interconnected during LBA. As explained earlier, Montenegro and the Mat Valley, although not sharing exactly the same amber artefacts, have in common other elements of grave inventory, namely the spectacle pendants with coiled connecting parts (Fig. 91:F–G) – a feature that links both regions with certain parts of Italy. The tumulus 2 of Shtogj also yielded the previously discussed pins wit nail heads and swollen, perforated necks which form a link with the Kolonjë Plateau (Fig. 78:E–F), specifically the tumulus of Rehovë which, moreover, contained similar amber beads (IF223) (cf. Kurti 2017b, Pl. XCVIIIf:4, XCIX:g).

Closely located regions of the Kolonjë Plateau and the Korçë Basin are culturally related, with similarities, in addition to amber beads, encompassing a number of finds' categories, including pottery. One can mention the cup with S-shaped profile and a single handle protruding above the rim, finished with horn-like or saddle-like appendix found in grave 186 of the Rehovë tumulus, or the *kantharos* (double-handle beaker) from grave 201 of the same tumulus which have analogies i.a. among the pottery discovered at the Maliq settlement (phase IIId) in the Korçë Basin.[589] More importantly, the tumuli of Luaras and Rehovë in the Kolonjë Plateau, and the tumulus 1 of Barç in the Korçë Basin are the only contexts in Albania (aside from Shtikë also in the Kolonjë Plateau) which contained carnelian beads, thus proving that they were involved in the circulation of this rare material during LBA (cf. Kurti 2017b).

At the moment unclear is the connection between the Korçë Basin and the Kolonjë Plateau with the territory of the North Macedonia. The amber finds from the cemetery of Dimov Grob in the Vardar Valley are quite unlike the artefacts from the tumuli of Barç, Luaras or Rehovë, as they exhibit elongated or average fusiform beads (IE121 and IA121; cf. Videski 2004, Figs. 33 and 34, T. XLI-XLII). These forms have more in

586 E.g. Belotić, tumulus X, grave 3 (Garašanin/Garašanin 1958, Fig. 3:a; cf. Dmitrović 2016, 166–168).
587 Boka-Prčevo, tumulus XIV, grave 19 (Tasić 1979, T. IV:3–4).
588 Krumë, tumulus 1, grave 1 (Jubani 1982, Tab. III:1).
589 Aliu 2012, 280–281, 285, Tab. XV:195, XVII:212; cf. Prendi 1977–1978, Tab. VIII:3, 5; 1982, Abb. 7:1.

common with the concurrent barrel-shaped beads (type 111 by J. Czebreszuk) from some of the *Mycenaean* sites in Greece (cf. Czebreszuk 2011, 88, Tab. 8, III:30–33). The closest specimens from south-east Albania would be IE261 and IE841 from Barç, tumulus 2, grave 2 (Kurti 2017b, Pl. XCVIII:e). This feeble link can be reinforced by a handful of other finds shared by contexts from those regions. These are, again, pins with swollen and pierced necks (cf. Aliu 2007, Pl. LXIV:a; Videski 2007a, Pl. LV:b), and the double-handle beakers with hemispherical bodies (Aliu 2012, Tab. XVII:212; Videski 2007a, Pl. LIV:a). At the first sight, the specimens from Albania and North Macedonia manifest certain differences. However, in the recent years, strong arguments have been brought forward for ascribing these regional variants of the Paarstadl-type pins and the double-handle beakers to a wider phenomena, linked to translocation of people and materials across the Balkans (cf. Bulatović 2011, 124–127, Fig. 1; Pabst 2013, 156–158, Abb. 1). To conclude, even though the morphology of amber beads does not testify to close connection between the areas of south-east Albania and the North Macedonia, both were involved in the network(s) of contacts which, if not amber, facilitated circulation of other goods and ideas during LBA.

Fig. 94. Reconstruction of amber circulation in the circum-Adriatic zone during LBA.

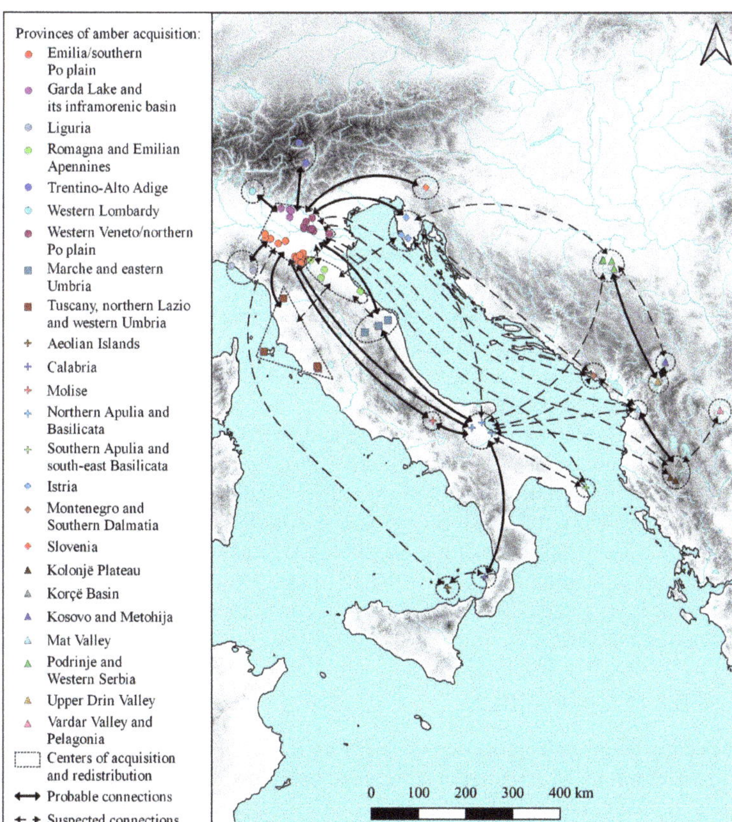

Overall, the structure of interregional relationships between the provinces of amber's acquisition in the circum-Adriatic zone during LBA is uneven (Fig. 94). On the one hand, mainland part of Italy abounds with evidence substantiating a wide network of interactions involving circulation of amber. The ties seem to follow a similar pattern to the one observed in MBA, however with addition of links to several peripheral regions, especially in Central Italy. On the picture, it gives the Italian network more robustness and interconnectivity. Hence, it is possible to attribute it with a certain stability and spatial expansion of contacts established already in the previous phase. Nature of those interactions is difficult to concretize, but co-occurrence with amber of numerous objects falling into same categories and types over a vast area lends some credibility to the idea of exchange of materials, products or services, e.g. by travelling merchants or craftsmen. From the Garda Lake basin and the Po Valley, through Romagna, Tuscany, Umbria and Marche, to Molise and Apulia, amber is associated with items crossing cultural and geographic boundaries, so it is plausible that Apennine Peninsula was the main meridional axis of amber translocation during LBA. Moreover, through mediation of the core regions, amber occasionally reached the outskirts of the circum-Adriatic zone (e.g. the Aeolian Islands, Calabria, Liguria, western Lombardy and Slovenia).

The Balkans, on the other hand, are weakly interconnected on the whole. There are certain elements found together with amber that allow to suspect interactions between some regions; however, there is no solid evidence to envisage these contacts as intense and stable as in the case of Italy. Further functioning of the MBA network in Central Balkans in the discussed period is not supported by available data, and continuation of a large-scale supply of the area in amber is not substantiated by appearance of new contexts surely dated to LBA. On the map, the Balkan area appears as if partitioned into two smaller networks: one to the north (Podrinje and Western Serbia, Kosovo and Metohija, and Upper Drin Valley) and second one to the south (Montenegro, Mat Valley, Korçë Basin, Kolonjë Plateau and Vardar Valley), whereas Istria and Slovenia seem to gravitate towards the Northern Italian centre.

In the light of these observations, it is increasingly probable that amber was brought to the southernmost Balkans by mediation of Southern Italy, possibly involving maritime voyages. Trans-Adriatic contacts which included amber translocations previously were largely limited to interactions between the core regions: mainly Istria on the one side, and the Garda Lake basin – Po Plain and northern Apulia on the other. With LBA there are several finds allowing to include also the southern Balkan peripheries of amber's acquisition in those channels of amber circulation. Thus, after the Central Balkan MBA 'amber route' was broken, amber could still reach Albania thanks to trans-Adriatic relationships with Italy. Similar connection could exist between northern Apulia and Podrinje, as in the previous phase. However, those trans-Adriatic links at the present stage of research rely on modest material evidence and need to be investigated further.

3.12.4. FBA (1150–950 BC)

Last two centuries of the 2nd millennium BC brought significant changes on the map of amber distribution in the circum-Adriatic zone (Fig. 95). Previously dense grouping of contexts in the Po Plain underwent a marked rarefication, with western Veneto being the only region which retained considerable representation of finds (cf. section 2.5.1).

Amber completely disappeared from Emilia and became barely present in the Garda Lake basin. In the remaining regions of Northern Italy, amber finds relevant to FBA were found on no more than three sites each, although some of them pertain also to final LBA – BR 2 (those with LBA/FBA chronology). In most of these areas, amber was known already in the earlier phases of the Bronze Age, but there is, notably, one new province of amber's acquisition – Piedmont. As mentioned previously, during FBA amber became more widespread in Central Italy, reaching new heights in the numbers of contexts and finds (cf. section 2.5.2). The material returned to south-central Lazio, but towards the end of the phase found a new group of recipients in north-central Lazio as well. In Southern Italy, on the other hand, amber became more sparsely distributed and considerably fewer in terms of the finds-per-context ratio. Still, the same provinces participated in its acquisition as before, with Sicily re-emerging as a recipient after a cessation of amber's use during LBA.

Fig. 95. Distribution of the sites with amber finds dated to LBA/FBA, FBA and FBA/EIA in the circum-Adriatic zone.

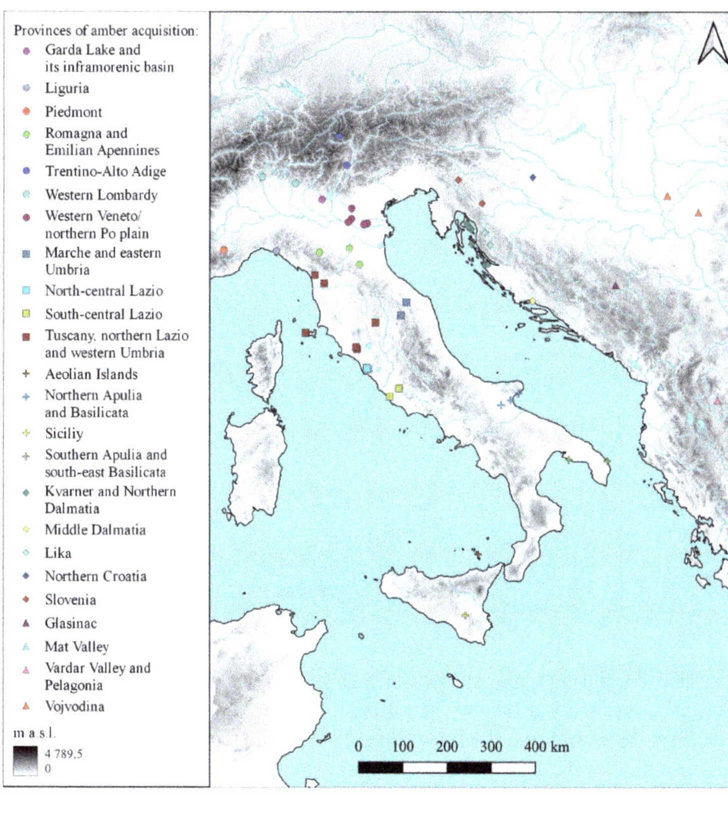

By the end of the 2nd millennium BC, amber almost completely ceased to be imported to Central Balkans, and its acquisition largely shifted to the east Adriatic coastal zone. Here, too, a major change occurred around the transition from LBA to FBA: instead of

Istria, amber began to pour into Kvarner and northern Dalmatia. Some regions in the vicinity started or continued to acquire amber as well, albeit in smaller quantities and stylistic variety, i.e. Slovenia, Lika, Northern Croatia or middle Dalmatia. Elsewhere in the Balkans, distribution of amber is limited to singular contexts which yielded few artefacts. The only exception is Vojvodina – the extreme north-east part of the study area which began to acquire amber for the first time in FBA and right away managed to accumulate substantial amount of this material.

Fig. 96. Network analysis of LBA/FBA, FBA and FBA/EIA contexts with amber grouped into regions.

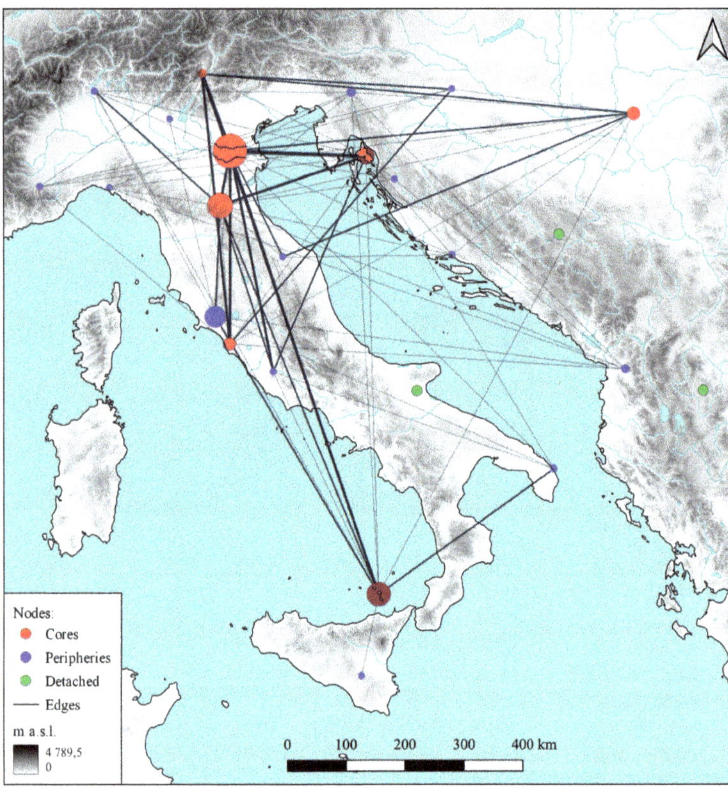

The network resulting from analysis of regional assemblages of amber dated to FBA (Fig. 96) has a markedly different structure, than the ones shown while discussing MBA and LBA. The number of ties and interconnectivity among the nodes is definitely lower than in the previous two phases, what is clearly expressed by the *density* value of 0.261. The network's centre of gravity is still located on the upper Adriatic, but no longer the ties densely cross its lower part, as they did in the previous phases. The eastern side of the lower Adriatic became almost devoid of amber, while the western side turned considerably poorer than before, leading to the loss of *coreness* and *betweenness* by northern Apulia which becomes detached from the rest of the nodes (Figs. 96–97;

Tab. 47). Instead, the southern tip of the meridional axis of the network shifted towards the Tyrrhenian Sea, having foothold on the Aeolian Islands. This region gained status of a core in the FBA network and is characterised by strong level of similarity with western Veneto, the latter province still playing the role of a major core.

Fig. 97. Tabular summary of core-periphery *analysis of LBA/FBA, FBA and FBA/EIA contexts with amber grouped into regions (presence of a digit at the intersection of a column and a row implies presence of a tie between two regions, while the value gives a number of types they share; calculated with the maximum correlation algorithm described by Borgatti/Everett 1999).*

		7 R	14 A	21 V	4 T	6 W	10 N	16 K	5 G	8 T	3 W	11 S	12 N	1 L	2 P	15 S	13 S	17 M	18 L	19 S	20 N	9 M	22 G	23 V	24 M
7	Romagna and Emilian Apennines		2	1		3	1	3		2	2	2		1				1	1	1		1			1
14	Aeolian Islands	2		1	3	2		1		2	1	1				1	2			1					
21	Vojvodina	1	1		2	2	1								1			1		1	1	2			
4	Trentino - Alto Adige		3	2		4	3	1							1						2				
6	Western Veneto	3	2	2	4		3	4		1	1	3			1		1	1	1	1	1	2			
10	North-central Lazio	1		1	3	3		1		1					1						2				1
16	Kvarner and northern Dalmatia	3	1		1	4	1				1	2		1			1	1		1					1
5	Garda Lake basin									1															
8	Tuscany, northern Lazio and western Umbria	2	2			1	1		1		1						1								1
3	Western Lombardy	2	1			1		1		1										1					
11	South-central Lazio	2	1			3		2									1								
12	Northern Apulia and Basilicata																								
1	Liguria	1						1												1					1
2	Piedmont			1	1	1	1														1				
15	Sicily		1																						
13	Southern Apulia and south-east Basilicata		2			1		1		1		1													
17	Middle Dalmatia	1		1		1		1														1			
18	Lika	1				1																			
19	Slovenia	1	1	1		1		1			1			1											1
20	Northern Croatia			1	2	1	2								1										
9	Marche and eastern Umbria	1		2		2												1							
22	Glasinac																								
23	Vardar Valley and Pelagonia																								
24	Mat Valley	1					1	1		1				1						1					

Furthermore, while the Adriatic side of Italy (east of the Apennines) seems to have been vital for amber circulation down the peninsula during MBA and LBA, there is much evidence that in FBA this role was taken over by provinces on the Tyrrhenian coast. That is suggested i.a. by high *coreness* and *betweenness* of Tuscany, northern Lazio and western Umbria, and north-central Lazio (Fig. 97; Tab. 47), as well as the general increase of amber in this part of Italy. The former region derives its *betweenness* from presence of several forms – mostly beads decorated with the Tiryns-type motif – which it shares with a number of other provinces. The latter region, although characterised by lower *betweenness*, has typologically richer assemblage of amber artefacts which places it among the cores, with the strongest links towards western Veneto and Trentino-Alto Adige. This last region, known for amber's acquisition already since EBA, in this phase provided a large and diverse set of amber finds which also gives it a prominent position. On the same meridional axis, another FBA core region is Romagna and Emilian Apennines. This last province, together with western Veneto, has the strongest ties to Kvarner and northern Dalmatia – also a core region and the main recipient of amber on the east coast of the Adriatic. While Istria ceased to function in this capacity, Kvarner and northern Dalmatia immediately emerged as a sort of its successor. The last core in the FBA network is Vojvodina which indicates similarities in morphology of amber artefacts with many regions; however, these do not have the same high level as in the case of the remaining cores.

Tab. 47. Regions with LBA/FBA, FBA and FBA/EIA contexts with amber ranked according to their coreness *(calculated with the Minres algorithm described by Borgatti/Everett 1999),* betweenness *(calculated with the algorithm described by Freeman 1979) and* degree.

Province:	**Coreness:**	**Betweenness:**	**Norm. Betweenness:**	**Degree:**	**Norm. Degree:**
Aeolian Islands	0.300	25.312	10.005	11	0.478
Garda Lake basin	0.018	0.000	0.000	1	0.043
Glasinac	0.000	0.000	0.000	0	0.000
Kvarner and northern Dalmatia	0.318	14.967	5.916	12	0.522
Liguria	0.071	0.000	0.000	4	0.174
Lika	0.035	0.000	0.000	2	0.087
Marche and eastern Umbria	0.106	0.000	0.000	4	0.174
Mat Valley	0.106	2.441	0.965	6	0.261
Middle Dalmatia	0.088	0.476	0.188	5	0.217
North-central Lazio	0.247	8.829	3.490	9	0.391
Northern Apulia and Basilicata	0.000	0.000	0.000	0	0.000
Northern Croatia	0.124	0.000	0.000	5	0.217
Piedmont	0.088	0.000	0.000	5	0.217
Romagna and Emilian Apennines	0.389	27.479	10.861	14	0.609
Sicily	0.018	0.000	0.000	1	0.043
Slovenia	0.141	3.954	1.563	8	0.348
South-central Lazio	0.159	0.200	0.079	5	0.217
Southern Apulia and south-east Basilicata	0.106	0.786	0.311	5	0.217
Trentino – Alto Adige	0.283	2.698	1.067	7	0.304
Tuscany, northern Lazio and western Umbria	0.177	21.817	8.623	8	0.348
Vardar Valley and Pelagonia	0.000	0.000	0.000	0	0.000
Vojvodina	0.230	9.594	3.792	10	0.435
Western Lombardy	0.124	0.686	0.271	6	0.261
Western Veneto	0.548	40.762	16.111	16	0.696

The peripheral provinces are weakly interconnected what translates into considerably lower level of average *density* characterising the network (0.261) in comparison with the earlier phases. Two of the nodes in the eastern part of the study area (Glasinac in Bosnia and Pelagonia in North Macedonia) are even detached from the rest of the network. This shows how differentiated amber artefacts became during FBA, leading to emergence of certain regionalisms partly touched upon already in sections 3.3–3.11. This regional specificities stem from, more than anything else, changes in the stylistics of amber adornments which began to be more commonly ornamented with a series of decorative motifs widely known as the Tiryns and the Allumiere types.

Previous considerations on the chronological differentiation of elongated beads (subgroup IE) – the category of finds exhibiting the greatest presence of ornamentations – proved a certain gradualness in introduction of decorative motifs (Fig. 52). In this sense, evolution of amber adornments in FBA indicates much more dynamism, than in the previous phases, and allows for a more detailed assessment of amber circulation. Reconstruction of the development of decorative styles on amber beads is feasible on the basis of Italian contexts which, in comparison with the Balkan ones, are usually characterised by a finer chronological adjustment.

As explained previously (cf. section 3.3), ornament based on a single median rib (variant 'a') can be said, with a considerable probability, to have first been applied in LBA, as reliably indicated by the find from Roca Vecchia, 'Cult Center' (Sector IX, US 11349, Phase II), but became more common during the early to mid-FBA (BF 1–2; ca. 1150–1000 BC). There is, however, noticeable decrease in number of artefacts manifesting this decoration during the late stage of FBA – phase BF 3 (ca. 1030/1010–950/910 BC). First use of the decoration involving parallel, transverse grooves (var. 'c') on a bead's surface can be dated as early as BF 1 (ca. 1130–1000 BC), as shown i.a. by Roca Vecchia, 'Hut-temple' (Sector IX, US 11322, Phase VI), whereas broader adoption of the motif took place in the subsequent BF 2 (ca. 1100–1030/1010 BC). Other decorative motifs utilising grooves (var. 'd'–'i') are present on artefacts coming predominantly from Italian contexts dated to BF 3 (ca. 1030/1010–950/910 BC). More precisely dated Balkan contexts with such decorated amber beads tend to confirm the above-described trend, but unfortunately many more can be only generally dated to FBA. One should not forget that on the east Adriatic coast (Kvarner and Lika) were also recorded other variants of the grooved decoration, practically unregistered in Italy (var. 'b' and 'f'), indicating partly independent local amber processing during FBA.

The presented 'micro-chronology' of stylistics of amber ornamentation is based on previously evaluated dating of contexts (cf. chapter 2). These contexts and amber artefacts can, thus, be divided into two horizons within FBA, adding more depth to the network analysis. The first one covers BF 1–2, and involves circulation of the amber artefacts adorned with motifs 'a' and 'c', in addition to those undecorated. The second one includes a different set of ornaments, almost without var. 'a' and with a lower contribution of var 'c', but extended by the new motifs (var. 'b', 'd'–'i'). The proposed division of contexts has, in some cases, implications for the assessment of the nodes, i.e. their roles in the overall circulation of amber in the circum-Adriatic zone. Detailed chronological considerations need to supplement evaluation of the relationships between the regions on the basis of amber and associated data.

Occurrence in FBA of many more amber artefacts specific in terms of form, perforation and decoration, than during the preceding phases, allows to more effectively trace circulation of the material in the circum-Adriatic zone. At the same time, however, using bronze products to that effect becomes more complicated. FBA is a period when basic forms – designs of utensils from a number of categories, e.g. fibulae or pins, became widespread in the Adriatic Basin and further in the Mediterranean, providing grounds for distinguishing a metallurgic *koiné*.[590] Seemingly, this gives an opportunity to observe circulation of materials in a wider scale, than before. However, adaptation of these designs in various areas often involved adding specific features which lead to creation of numerous regional variants. Potentially, this indicates circulation of not complete objects, as in the previous phases, but of raw materials which were then processed locally, perhaps under external inspiration, with the necessary technological know-how already in place.

A way to disentangle the issue of FBA metallurgic *koiné* is to rely on provenience analyses involving searching for trace elements in samples extracted from the finds and comparing them with outcrops of these materials in different regions. In the case of copper used for bronze production, the most common method currently in use is the lead isotope (^{204}Pb, ^{206}Pb, ^{207}Pb and ^{208}Pb) ratio comparisons (cf. Radivojević et al. 2018); however, it will take some time before an extensive database encompassing the circum-Adriatic zone is created. Useful information can be also derived from analyses of glass beads – another category of finds often accompanying amber in the FBA contexts. Analysis of composition of vitreous materials allows to classify them within technological groups having specific spatial and chronological characteristics, whereas comparison of strontium and neodymium isotopes (^{87}Sr/^{86}Sr and ^{143}Nd/^{144}Nd) between the samples of finds and those obtained from sand and ashed plants deposits can help to pinpoint probable production places (Henderson 1988; Henderson et al. 2015).

Processing of all three materials (amber, bronze and glass) is well documented on FBA settlements in western Veneto. Significance of Campestrin di Grignano Polesine for production of amber adornments decorated with the Tiryns-type ornament (motif 'a') has already been mentioned in the discussion devoted to LBA. The site most probably continued to function as amber-processing centre in the early FBA – BF 1. The settlement of Fondo Paviani, which was active around the turn of LBA and FBA (BR 1/2 – BF 1–2), produced evidence of glass working comprising processing waste composed of a series of fused red-blue annular beads, and a small block probably belonging to an ingot which upon analysis revealed LMHK composition with copper (Cu) and cobalt (Co) serving as colourants (Cupitò et al. 2015b). Furthermore, the site yielded casting moulds and pieces of slag-documenting metallurgic activities, with the lead-isotopes ratios determined on finished items as well as ingots pointing to copper ores of south-east Alps, specifically volcanogenic massive sulphides deposits (VMS) of the Valsugana area (Vicenzutto et al. 2015). It has to be added that a single amber lump from the site might be seen as evidence of local processing of this material, however

590 For references to the idea of the FBA *koiné* in the Adriatic Basin and the objects, particularly metal, included in it, see e.g. Carancini/Peroni 1999; Teržan 2007; Borgna/Càssola Guida 2009; Borgna 2013; Jung/Mehofer 2013; Iacono 2013; Blečić Kavur 2014; Cazzella/Recchia 2018.

only an indirect one and not clearly defined in terms of chronology (LBA–FBA) (cf. Strafella et al. 2015).

Finally, a rich collection of data referring to various crafts, including metal, bone, antler, ivory, glass and presumably amber, comes from the settlement of Frattesina di Fratta Polesine. It can be said about Frattesina that bronze casting was performed there on an industrial scale (Pearce 2019, 341). This is confirmed by collections of at least 100 casting moulds,[591] as well as four 'founder's hoards' (Bellintani 2000, 65). The recent lead-isotope analyses which took into consideration 14 bronze objects from Frattesina,[592] mostly recovered from hoard no. 2, showed that the copper used to make these artefacts does not originate from either Etruria or Sardinia, and that it most likely comes from a single mining district (Jung et al. 2011). Together with new findings regarding copper mining in Trentino during FBA, the available evidence supports earlier hypotheses that southern Alpine copper formed the bulk or raw material processed at Frattesina.[593]

Potentially even bigger is the importance of Frattesina for production and circulation of vitreous materials in FBA. Until lately the finds from the site such as crucibles containing red, green, blue and white glass, ingots of glass (fitting the crucibles), chunks/lumps of raw furnace glass and malformed beads allowed to suspect that glass was reheated and worked on the site (cf. Towle et al. 2001); however, no tangible archaeological evidence for the fusion of glass has been found there. The dominant colour of the Frattesina glasses is translucent turquoise, but opaque red, opaque white, translucent cobalt blue and translucent green glasses have also been found, whereas their chemical composition matches the mixed alkali glass of LMHK group – the earliest indigenous European glass invented at the end of the 2nd millennium BC, around the time when single alkali – sodium glasses of the Near East provenience became more rare in Europe.[594] Recent isotope analysis of five samples of raw glass from Frattesina using thermal ion mass spectrometry showed that three have Nd and Sr rations typical to silica and calcium of sands from the Po delta, one is a mixture of glasses containing contrasting isotope compositions, and another one has lower Sr isotope signature corresponding with ultrapotassic volcanic rocks in the Rome province (Henderson et al. 2015, 6–7; Henderson 2019, 325–326). The evidence indicates Frattesina and its immediate vicinity to be one of the two production centres of LMHK-mixed alkali glasses, the second one potentially located in Lazio, as well as a 'processing plant', where glasses of more than one origin were remelted, mixed and moulded into new adornments.

591 The largest portion was published in Le Fèvre-Lehöerff 1994, 171, 187–212, Figs. 4–8, 11–12, 19, 21–22, 24, 26, 28–29, 31–32, 37–38.

592 These are three Matrei-type knives, four pick-ingots, two plano-convex ingots, a socketed shovel, a Ponte San Giovanni–type winged axe and an anvil (Jung et al. 2011, Tab. 23.1).

593 Pearce 2019, 342; cf. Pearce 2000; Silvestri et al. 2015; Giardino/Guida 2018; Villa/Giardino 2018.

594 Cf. Henderson 1988, 2019; Angelini et al. 2004; Bellintani et al. 2006; Bellintani 2012; Henderson et al. 2015.

At last, it is possible that Frattesina also engaged in amber processing (cf. Bellintani 2014, 117; Bietti Sestieri et al. 2015, 431). The relevant sources include amber objects decorated with transverse grooves – decorative motif 'c', lacking perforations (group VIII; cf. section 3.10.), or with unfinished borings (group IX; cf. section 3.11.). Same as in the case of LMHK glasses, the rank of Frattesina as an important hub in amber circulation and a potential amber-processing centre is attested by a large concentration of beads with matching ornaments.

Significance of western Veneto for circulation of amber during the first stage of FBA (BF 1–2) is clearly recognisable through contexts located within regions characterised by high *coreness* and *betweenness* values immediately to the south: Romagna and Emilian Apennines, and Tuscany. The settlement of Borgo Panigale near Bologna revealed considerable evidence, discussed in the previous section (3.12.3.), attesting to occupation of the site during LBA (BR). However, there are proofs that it remained inhabited during early stages of FBA (BF 1) as well, i.a. pinhead in the shape of a wheel with four spokes, doubled at the ends (Fig. 98:A) and head of a pin probably belonging to the Sover type (Fig. 98:C) (cf. Carancini 1975, 52–53, 242–243), both having analogies at BF 1 sites with amber in western Veneto (Fig. 98:B, D–E). This evidence, together with resemblance between some of the non-identical, yet contemporaneous amber finds, e.g. IA211a from Monte Battaglia and IE211a from Fondo Paviani, advanced BR – early BF layers (cf. Figs. 45:3 and 49:3), means that the amber beads with the Tiryns-type motif could have been brought to Romagna from the northern side of the Po. Even though the cited examples of amber beads are slightly different in terms of lateral surfaces or dimensions ratio, it is still possible that they were made in the same place, as showed by slight morphological divergencies in objects from the workshop of Campestrin di Grignano Polesine presumably produced by the same artisan: VIIIA218a and VIIIE211a (Fig. 62:1, 3).

Fig. 98. Examples of pinheads, pins, torcs shared by some of the FBA sites with amber in Romagna and Emilian Apennines (A, C, J), western Veneto (B, D–E, K, N, R), Tuscany (F, I, M, P) and the Garda Lake basin (G–H, L, O): A, C – Borgo Panigale (Catarsi Dall'Aglio 1976, Fig. 6:2–3); B, K, P – Narde II, 'Ustrinum', US 1009 (Colonna et al. 2010, Tav. 56:11, 14–15); D–E, N – Mariconda di Melara, upper layer (Salzani 1973, Tav. XVIII:10, XIX:6–7); F, M, O – Scarceta, sector D – casa-laboratorio *no. 137 (Poggiani Keller 1999, Fig. 98:1, 5, 12); G–H, L, N – Ponte San Marco, BR–BF horizon (Ruggiero 1994, Fig. 78:5–6, 79:1, 4); I – Fossa Nera di Porcari (Andreotti/Zanini 1995–1996, Fig. 2:15); J – Monte Battaglia, hoard (Bermond Montanari 1997, Fig. 130:14); S – Frattesina di Fratta Polesine, phase 2 (Bietti Sestieri et al. 2015, Fig. 1:11).*

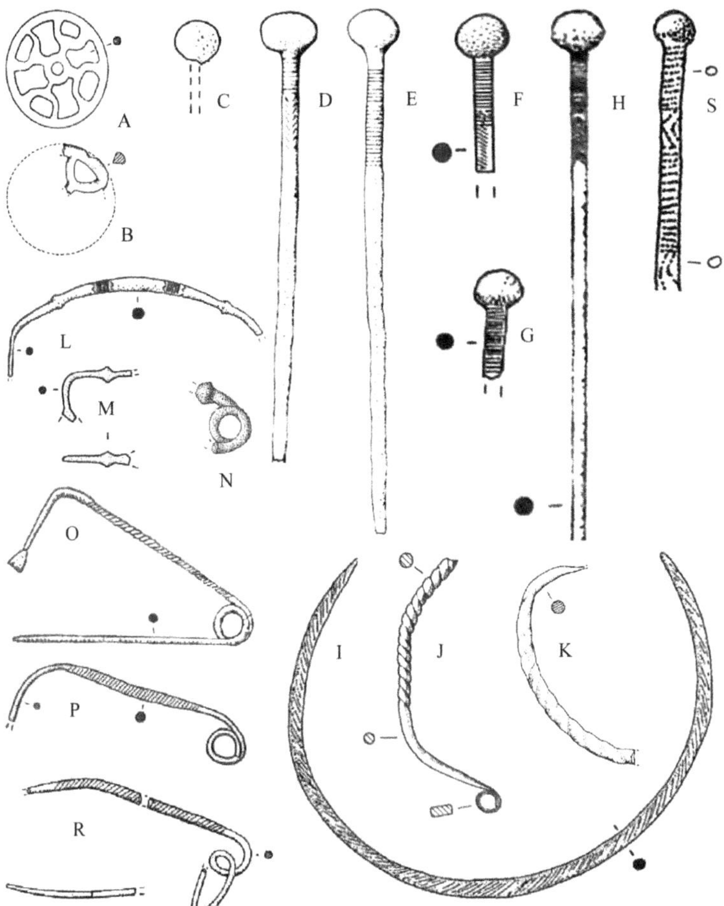

The Sover-type pins appeared together with amber decorated with the Tiryns-type motif (IE251a) also in Tuscany, precisely on the settlement of Scarceta, in the sector D – 'house-laboratory' (Fig. 98:F). The same site, interestingly, yielded two beads in clay apparently imitating their amber counterparts with the Tiryns-type decorations (Poggiani Keller 1999, Fig. 88:3–4). This finding, in conjunction with other data such as products in different stages of finishing, tools and processing waste, contribute to a specific function of the structure no. 137 'house-laboratory', potentially also as a small-scale amber workshop oriented towards production of the Tiryns-type decorated beads (cf. Poggiani Keller 1999, 123–124). Moreover, same context abounds with small annular glass beads made of monochrome paste in various shades of blue (Poggiani Keller 1999, Fig. 104:4–31) – type of vitreous products that can be correlated with glass production in western Veneto, as explained above, and which will turn up again and again in relation to other sites. Connection between early FBA contexts with amber in western Veneto, Romagna and Tuscany, beyond the Tiryns-ornamented amber beads, is somewhat supported by co-occurrence in the relevant contexts of the bronze torcs with decoration imitating twisted bronze rod ('false corrugation') (Fig. 98:I–K) – objects typical for the LBA and the EIA in the area covered by the *Urnfield* culture.[595]

The discussed interregional links built around distribution of the Tiryns-type decorated amber beads can be extended in the north-west direction, to include the Garda Lake basin and western Lombardy. In addition to IE111a bead (Fig. 49:1), which has precise analogies in Tuscany,[596] the BR–BF layers at Ponte San Marco yielded i.a. the Sover-type pins (Fig. 98:G–H) and a cylindrical bead made of light-blue vitreous paste (Ruggiero 1994, 133). These materials only strengthen the evidence of contacts along the same lines (between the Garda Lake basin, western Veneto and Tuscany) pertaining to late LBA-early FBA (BR 2/BF 1), e.g. the fibulae with two nodules (Fig. 98:L–M) or twisted arches (Fig. 98:N–P) (cf. Eles Masi 1986, 2–4, 11–12; Savella 2015, 7–11, Tav. 72). The amber beads from Costa Cavallina in western Lombardy (IE211a, IE231a, IA231a; Poggiani Keller 1989, Fig. 98:1–3), due to the Tiryns-type decoration and overall morphological similarity, certainly form a link with western Veneto, Romagna and Tuscany.[597] Unfortunately, absence of materials at the site specifically belonging to FBA hinders more detailed discussion, beyond what has already been said in section 3.12.3.

Stylistic similarities with amber and glass beads from Maso Finale provide an argument for a link between Trentino-Alto Adige and western Veneto around BF 1–2. They have been partly discussed in the previous section, as the site belongs to the group of contexts with LBA/FBA chronology, and in fact matches well with some of the LBA amber and glass finds from western Veneto and Emilia (cf. section 3.12.3). Still, the connection based on amber is equally firm during FBA, and some of the vitreous finds from Maso Finale also fit the general picture of glass industry of the period. Namely, these are the flattened-spherical/annular beads without decoration, made of azure-greenish blue glass, and the fusiform beads with spiral decoration (Fig. 99:A–C) – forms continuously produced in the first part of FBA (BF 1–2) and distributed beyond Northern Italy (Putzer 2012, Abb. 10:39–41; cf. Bellintani 2012, 21,

595 For an overview of chronology and spatial distribution see Vasić 2010, 42–43.

596 Omo Masso, precise context unknown (Negroni Catacchio 2014, Fig. 8).

597 E.g. IE211a – Fondo Paviani, advanced BR – early BF layers (Fasani/Salzani 1976, Fig. 2:3–4); IE231a – Bismantova – Campo Pianelli, grave XXXI (Catarsi/Dall'Aglio 1978, Fig. XXI:1); IA231a – Ponte San Pietro Valle, tomb 1 (Negroni Catacchio 1999, Fig. 3:34).

Fig. 3:1, 7, 12–14, 17, 19, 30, 4). Since the chemical analyses are lacking, it cannot be ruled out that at least some specimens from Maso Finale were produced in LMHK technology, as were some of the similar beads found on other sites with amber.[598] Even if all the beads were made in HMBG technology as suspected (Bellintani 2014, 120, fn. 7; Bellintani/Gonzato 2017, 178), it would not counter the possibility of FBA chronology of the discussed materials, as both types of glass in similar fusiform and annular shapes were discovered on the site Salorno – Cava Girardi, dated to *Luco A*/ BF 1–2/HaA1–A2 (ca. 1200–1060 BC) and also located in the Trentino-Alto Adige area (cf. Bellintani 2012, 20; 2014, 124).

Fig. 99. Examples of glass beads decorated with inlayed spiral and 'eyes' ornaments shared by some of the contexts with amber in western Veneto (A–B, F–H), Trentino-Alto Adige (C), the Aeolian Islands (D, J), western Lombardy (E) Emilian Apennines (I) and northern Apulia (K): A, F–G – Frattesina di Fratta Polesine, phases 1–3 (Bellintani 1997, Fig. 9; Bietti Sestieri et al. 2015, Fig. 2:14); B – Mariconda di Melara, upper layer (Salzani 1973, Tav. XVIII:16–20); C – Maso Finale, cultural layer (Putzer 2012, Abb. 10:41); D, J – Piazza Monfalcone, tomb 31 (Bellintani 2012, Fig. 3:30–31); E – Isolino Virginia, central part, superficial layers (De Marinis 2009c, Fig. 5:4–5); H – Narde II, grave 75 (Colonna et al. 2010, Tav. 13:B15); I – Bismantova – Campo Pianelli, grave XXXI (Catarsi/Dall'Aglio 1978, Fig. XXI:2); K – Monte Saraceno, tomb 84/IV (Nava/Preite 1995, Tav. XIII:24).

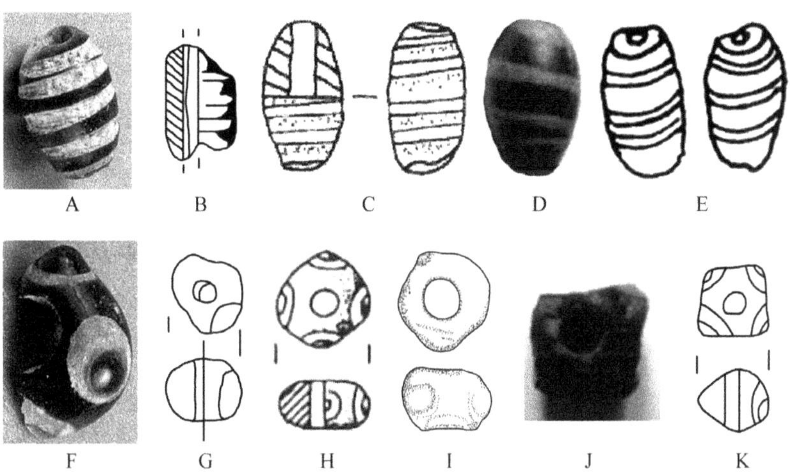

Amber-related interactions between western Veneto on one side and the Garda Lake basin, Trentino-Alto Adige, Romagna and Emilian Apennines and Tuscany are readable in relation to BF 2. From the same layer at Ponte San Marco that produced the Tiryns-type decorated amber bead (US 427) comes an undecorated simple bow

598 Bismantova – Campo Pianelli, grave XXXI (Catarsi/Dall'Aglio 1978, Fig. XXI:2); Fondo Paviani, advanced BR–early BF layers (Cupitò et al. 2015b, Fig. 1:1, 4); Mariconda di Melara, upper layer (Salzani 1973, Tav. XVIII:18–19); Piazza Monfalcone, tomb 31 (Bernabò Brea/Cavalier 1960, Tav. XLIII:4).

(IE121 and IE261) and violin-bow fibulae with corrugated or nodule-decorated arches (Fig. 87:C–D), still find good comparisons at this stage in areas to the north, particularly in western Veneto (Fig. 98:N, R).[603] The fibulae are also known from some contexts with amber in Tuscany and the Garda Lake basin (Fig. 98:L–M, O–P); however, those regions do not share similarities with Marche and eastern Umbria in amber. Influx of amber to Marche and eastern Umbria from western Veneto is plausible also in the light of finds from the settlement of Monte Croce Guardia: fragment of an amber bead from the foundation ditch of the western wall of the hut 1 (layer US4), and annular glass bead made of blue paste and fragment of a bone comb of the type Frattesina from a large waste pit found on the back of the hut 1 (layer US5) (Cardarelli et al. 2017, 354). A. Cardarelli compares the amber artefact to elongated biconical bead, such as the ones from the hoard of Gualdo Tadino or tomb XXXI of Bismantova (Cardarelli et al. 2017, n. 36, Fig. 23:4; cf. Fig. 50:20–21). However, in the view of its triangular cross-section, it seems that other analogies should be considered (e.g. Fig. 49:22, 50:2, 15). Nonetheless, due to fragmentary state of the object, its precise typology cannot be safely determined. The bone comb, on the other hand, can be related to a rich collection of similar items recovered from the hoard 1 *Tesoretto* of Frattesina representing type 2, ver. C (Fig. 100:I–J; cf. Bettelli/Damiani 2005). Similar provenience is indicated by the glass bead which, although not analysed in terms of chemical composition, has numerous analogies at Frattesina (Cardarelli et al. 2017, fn. 37, Fig. 23:5; cf. Bellintani/Stefan 2009, Tab. 1:1).

Besides the settlement of Colle della Mola, which yielded typologically undefined bead, the evidence of amber during FBA (specifically phases BF 1–2) in south-central Lazio comes from the grave of Ardea – Campo del Fico. Amber artefacts from this context have the most precise analogies with some of the finds coming from western Veneto, as shown by the network analysis (Fig. 96). However, remaining materials from the context are not easily put into the narrative about contacts between the two provinces. Among them is a simple bow fibula with two ribbed nodules and slightly thickened middle section of the arc decorated with a complex pattern including 'herringbone' and zig-zag motifs (Fig. 100:K). Although the two decorative motifs are not foreign to FBA fibulae from western Veneto, and coexist chiefly in variants with twisted upper part of the foot, e.g. arch of a broken fibula from Frattesina (Fig. 100:L), they are not encountered in the version with nodules (cf. Eles Masi 1986, 20–25, Tav. 7:134–149, 9:179–180, 10:197–198). Another clue as to the provenience of amber from Ardea is the decoration on the neck of biconical vase from the grave, consisting of applied metal (tin) strips oriented obliquely and perpendicularly, forming a horizontal 'ladder' mesh or pattern (although it is hard to reconstruct it completely, due to the state of preservation) (Delpino 1987, Fig. 1, 16). According to F. Delpino, this particular decorative motif applied with metal was unknown in Italy at the time of FBA, but appears on numerous examples from lake-dwellings of eastern France and Switzerland dated to HaA2–HaB2 (*Rhin-Suisse-France orientale Urnfield* culture), from where it could have been borrowed (Delpino 1987, 13, with further literature in footnote 12). Furthermore, the motif of engraved, hanging semi-circles ('festoons') present on the vase from Ardea below the neck is rarely attested on vessels in *Protovillanovan* contexts in southern Etruria or Lazio; however, it appears frequently at sites in Veneto,

603 Analogies in amber include i.a. Fondo Paviani, advanced BR–early BF layers (Strafella et al. 2015, Fig. 1:D) and Narde II, grave 75 (Colonna et al. 2010, Tav. 13:B11).

i.a. on pottery from Frattesina, hence was possibly brought from the transpadane area (Delpino 1987, 13–14; cf. Fasani/Salzani 1975b, 296, 315, no. 144).

Also the Aeolian Islands appear to have been entwined with western Veneto in contacts involving amber and glass translocation. The beads made in both materials, found on the cemetery of Piazza Monfalcone on Lipari, especially those from grave 31, reflect a considerable level of similarity with finds from a number of contexts in western Veneto. They include amber beads with the Tiryns (e.g. IE211a) and the Allumiere (e.g. IE221c) types of decorative motifs, as well as annular glass beads in various shades of white, blue and green, and barrel-shaped/fusiform ones decorated with spirals or 'eyes' motif (it. *occhi*) (Fig. 99:A–B, D, F–G, J).[604] Chemical analysis of seven out of the total of 700 glass beads recovered from grave 31 of Piazza Monfalcone found two to be made of high magnesium plant ash glass (HMG), while another four of mixed alkali glass (LMHK) (Conte et al. 2019, Tab. 5). The latter are likely to have been imported to Lipari from the glass-production centre of Frattesina (Conte et al. 2019, 1833). Some interregional affinity is also exemplified by the simple bow fibulae with corrugated-twisted middle part of the arches co-occurring with amber in grave 18 at Piazza Monfalcone and at Frattesina (Fig. 101:A–B). This kind of fibulae was generally more common for Southern Italy during FBA; however, several examples are also reported from the northern part.[605] Given the formal as well as structural similarities among amber and associated finds, it seems probable that some interaction between the Aeolian Islands and western Veneto occurred in the period. Amber collection from Lipari bears some resemblance to assemblages from Tuscany, Romagna and Emilian Apennines; however, these most likely stem from similar relationships between the latter and the Po Plain. Noticeably fewer and weaker links adjoin those regions to Apulia.

Southern Apulia yielded the earliest secure context with an amber bead decorated with the Allumiere-type motif – IE221c from Roca Vecchia, 'Hut-temple' (Sector IX, US 11322, Phase VI) dated to BF 1 (ca. 1130–1100 BC). So, the presence of the Allumiere type in Apulia slightly precedes its occurrence in the other parts of the circum-Adriatic zone, generally falling into BF 2 (1100/1060–1010/1000 BC), although the difference is minimal. Regardless of the dating, the specimen from Roca finds counterparts in the Aeolian Islands, south-central Lazio and western Veneto, among the Italian provinces.[606] The Roca Vecchia settlement offers other evidence useful in tracing contacts between southern Apulia and Northern Italy; however, it is not directly connected with the context of the abovementioned amber bead. From the 'hoard of gold' stored in an empty posthole of the 'Hut-temple' (layers US 10333 and 10334) were recovered numerous and manifold objects chronologically frameable between BR and BF

---

604 Bernabò Brea/Cavalier 1960, Tav. XLIII:1, 4; cf. Bellintani/Stefan 2009, 83; Bellintani 2012, 21, Fig. 3:29–31.

605 F. Lo Schiavo considers solely example no. 96 given by P. von Eles Masi as the direct counterpart to her type 26 (Lo Schiavo 2010, 110, fn. 37; cf. Eles Masi 1986, Tav. 5:96). But there are several other analogies that can be brought forward, including the two fibulae from Frattesina which, however, differ in the cross-section of the foot which in the former is square, as opposed to circular one (e.g. Eles Masi 1986, Tav. 4:75–75A, 6:111–112).

606 Piazza Monfalcone, tomb 31 (Bernabò Brea/Cavalier 1960, Tav. XLIII:1); Ardea – Campo del Fico, pit-grave (Delpino 1987, Fig. 18:8–9); Narde II, grave 75 (Colonna et al. 2010, Tav. 13:B4).

2/3 (ca. 1300–950 BC), including simple bow fibulae with ribbed nodules on the arc, simple bow fibulae and torcs decorated with torsion (Maggiulli G. 2009, Fig. 2:1.4–1.5, 1.10) and numerous glass beads of which 20 analysed belong to mixed alkali glasses (LMHK) some with high levels of potassium (Conte et al. 2019, Tab. 2). These materials, although not contextually related to the IE221c amber bead from the site, suggest relations with at least two Italian regions of amber's acquisition located to the north along the links resulting from NA: western Veneto and south-central Lazio.

Fig. 101. Examples of bronze fibulae, pins and bone (E – ivory?) spacers shared by some of the FBA sites with amber in western Veneto (A, J), the Aeolian Islands (B, D), southern Apulia (C, G), northern Apulia (E, H–I, K) and Tuscany (F): A, J – Frattesina di Fratta Polesine, phases 2–3 (Eles Masi 1986, Tav. 6:111; Bietti Sestieri et al. 2015, Fig. 1D:12); B – Piazza Monfalcone, tomb 18 (Lo Schiavo 2010, Tav. 12:88); C, G – Torre Castelluccia, ripostiglio *(Gorgoglione et al. 1993, Tav. LVIII:1–2; Lo Schiavo 2010, Tav. 11:83); D – Piazza Monfalcone, tomb 34 (Lo Schiavo 2010, Tav. 11:80); E – Coppa Nevigata (Cazzella/Recchia 2016, Fig. 2:5); F – Scarceta, sector D –* casa-laboratorio *no. 137 (Poggiani Keller 1999, Fig. 96:2); H – Monte Saraceno, tomb 86/IV (Lo Schiavo 2010, Tav. 26:229); I, K – Monte Saraceno, tomb 84/IV (Nava/Preite 1995, Tav. XIII:4; Lo Schiavo 2010, Tav.12:86B).*

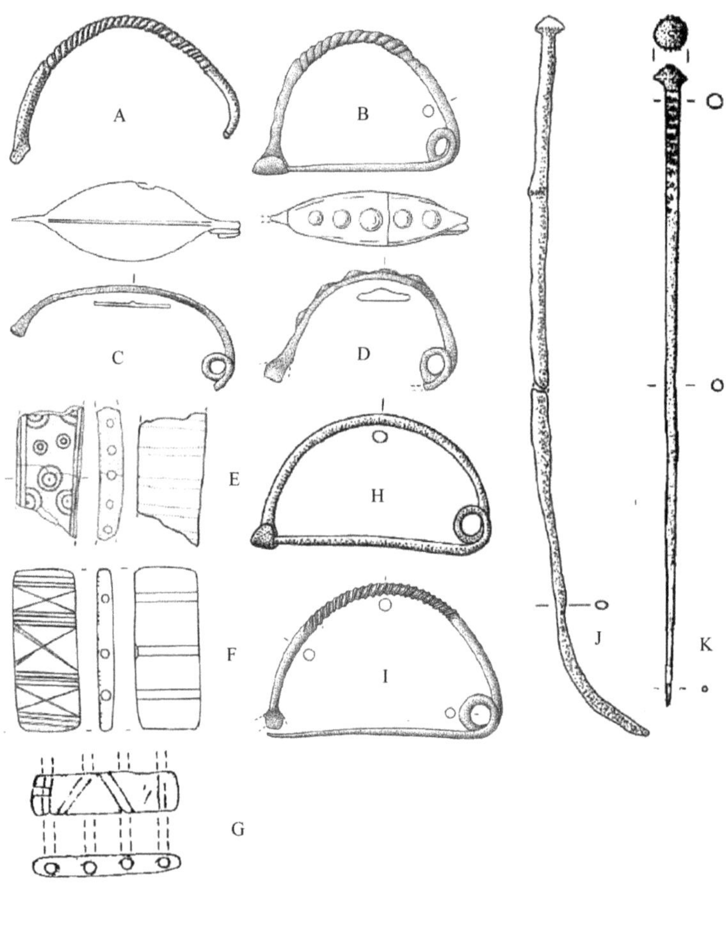

Connection between southern Apulia and the more closely located Aeolian Islands is tangible upon consideration of the inventory from the hoard/lumber room (*ripostiglio*) in the hut 7 of Torre Castelluccia dated to BF 1–2. The simple bow fibula with foliated and decorated arc has analogy i.a. in tomb 34 of Piazza Monfalcone also containing amber; both are classified to type 22 by F. Lo Schiavo (Lo Schiavo 2010, 105–106, nos. 80 and 83), although differ in decoration of the bow (Fig. 101:C–D). This link corresponds with co-occurrence of beads type IA218a in both regions (Fig. 45:5). Also the pottery found in the layers US2-3-4 of the hut 7 of Torre Castelluccia reveals affinity with the *Ausonian II* culture (Gorgoglione et al. 1993, 87). It should be added that also glass beads from Torre Castelluccia were determined as belonging to mixed alkali glasses with high levels of potassium (LMHK) (Conte et al. 2019, Tab. 2).

More difficult to interpret is the position of northern Apulia in the network. NA results in no links connecting this region with the other provinces. One can suspect some kind of relation, though, from the presence of amber beads decorated with the Tiryns-type motif in northern Apulia. The bead from BF layers of Coppa Nevigata, judging by the available figures, features elongated form and double-pitched profile, thus could be with some probability assigned to type IE161a or IE261a (cf. Belardelli 2004, Fig. 31:219). If that is the case, then there are three possible analogies for the object: one each from Romagna, Tuscany and the Aeolian Islands (cf. Fig. 50:25).[607] Overview of LBA-related materials from Coppa Nevigata (the discussed bead possibly originates from BR 2/*Subapennine* layers) allows to suspect that it arrived in Apulia via contacts with Northern Italy, along the axis Po Plain-Romagna-Marche (cf. section 3.12.3.). With respect to FBA clues potentially helpful in determining provenience of the bead from Coppa Nevigata are very scant among materials from the site; one can mention the Fucino type bronze spearhead (Bellardelli 2004, Fig. 34:12; cf. Carancini/Peroni 1999, 58, Tav. 28:45). Nor does the other Tiryns-decorated bead from northern Apulia – the one from Capitanata (IE221aw) – can be used to include the region in the network, as it lacks precise analogies and its context is unknown (cf. section 2.2.14). One is, thus, forced to rely only on the decoration of the beads from Coppa Nevigata and Capitanata in seeking potential partners in circulation of amber. Nonetheless, it is interesting to note the presence (although not exclusive) of bone spacers in some of the contexts with Tiryns-decorated amber beads in Apulia, both northern and southern, and Tuscany (Fig. 101:E–G; cf. Cazzella/Recchia 2016, 364).

Northern Apulian site more opulent in FBA materials – the cemetery of Monte Saraceno – does not indicate a great deal of similarity with amber assemblages from other regions either. Amber finds from the site, by large, lack decorative motifs and do not match other artefacts in terms of morphology, partly due to their bad state of preservation. Finds such as the IE822 bead from tomb 84/IV can be compared with, e.g. IE824 from the roughly contemporaneous grave 4, trench D (1960) at the Allumiere – Poggio La Pozza cemetery or IE828 from Fondo Paviani, but difference in shape of the bases makes the typological match impossible.[608]

---

607 Borgo Panigale, 1957 campaign, precise context unknown (Catarsi Dall'Aglio 1976, Fig. 6:16); Omo Masso, precise context unknown (Negroni Catacchio 2014, Fig. 8); Piazza Monfalcone, tomb 31 (Bernabò Brea/Cavalier 1960, Tav. XLIII:1).

608 Cf. Nava/Preite 1995, Tav. XIII:28; Peroni 1960, Fig. 11:15; Strafella et al. 2015, Fig. 1:E.

The only object found at Monte Saraceno which exhibits decoration is IVE411j from tomb 86/IV (Fig. 57:6). The manner of decorating amber with criss-crossing grooves or incisions resulting in a mesh/net pattern is rather unique in the scale of the entire Bronze Age in the circum-Adriatic zone. The only other amber artefact ornamented with intersecting lines – IIIF224j – comes from Allumiere – Poggio La Pozza, grave 1, trench F (1960) (Fig. 56:6). However, it differs morphologically from the Monte Saraceno pendant, and it is clear that the decorative motif on the two artefacts, although similar in the general design, is not identical. The pattern on IVE411j is much denser than in IIIF224j. Interestingly, if not for the two vertical grooves/incisions, the ornament on the former object could be classified as decorative motif 'c' – multiple and evenly spaced transverse grooves. In that case the closest analogy, albeit with different perforation, would be VIE411c from grave 75 of Narde II necropolis (Fig. 59:1). The BF 3 (ca. 1030–950 BC) chronology of the object stands in agreement with some of the materials from the tomb 86/IV at Monte Saraceno, i.e. the undecorated simple bow fibula type 35 (Lo Schiavo 2010, 126, no. 229; Fig. 101:H). The ornament on IVE411j raises questions regarding the pendant's history: was it originally designed the way it is by a craftsman, thus making it an outstanding case in the analysed collection, or is it a derivation from the decorative motif 'c', only altered at the later stages? If we were to agree with the first interpretation, then the pendant's origin should be sought in provinces with well-attested use, or processing of amber adornments with the Allumiere-type ornament.

From the chronological standpoint, the finds from Monte Saraceno can be dated to BF 2–3 (1100–950 BC) at the earliest; thus, the proposed analogies should come from the same time span. Amber bead from tomb 84/IV of Monte Saraceno (IE822) can be as old as BF 2, as indicated by presence of a simple bow fibula with twisted middle part of the arch (Fig. 101:I). By its morphology, the fibula can be assigned to the same category (Lo Schiavo's type 27) with specimens from Piazza Monfalcone and Frattesina (Fig. 101:A–B; cf. Lo Schiavo 2010, 109, no. 86B). Furthermore, all the tombs with amber from Monte Saraceno contained pins of the Torri d'Arcugnano type, widespread in Northern Italy, and present in substantial numbers in layers pertaining to phase 2 of Frattesina (Fig. 101:J–K; cf. Carancini 1975, 46, 226). Finally, the tombs yielded a rich collection of annular glass beads in various shades of blue, some decorated with the 'eyes' motif (it. *occhi*) which, as mentioned before, have very good analogies among the finds from Piazza Monfalcone and Frattesina (Fig. 99:F, H, J–K), although absence of chemical analyses hampers a full identification of their origin (Nava/Preite 1995, Tav. XIII:24, XIV:43–44; Preite 2001, Fig. 7:12–14). So, despite incompatibility among amber artefacts, there are signs of relationship between northern Apulia, the Aeolian Islands and western Veneto during advanced stages of FBA (BF 2–3).

The last phase of FBA (BF 3; ca. 1030/1010–950/910 BC) saw a substantial reshuffling on the map of amber's acquisition in Central and Southern Italy. With respect to regions in the west part of Central Italy, chronological overview of the contexts (cf. section 2.5.2) allows to suspect that they functioned as amber recipients at slightly different times during FBA: Tuscany, northern Lazio and western Umbria, as well as south-central Lazio in BF 1–2 (but with some sites in northern Tuscany dated up to BF 3a), while north-central Lazio in BF 3. The latter region is famously known for beads decorated with the Allumiere-type decoration in variants 'c', 'd' and 'e', however yielded even larger quantity of non-decorated amber artefacts which, altogether, result in connections with a number of other regions. The strongest ties link north-central

Lazio with western Veneto and Trentino-Alto Adige. These links stem from similarities among the non-decorated objects with biconical profile (IF161 and IE161),[609] conical buttons perforated perpendicularly (IIIF154)[610] and discoidal beads with hybrid perforation (VIF121).[611] While the biconical beads or those with hybrid perforation are not region-specific, the conical buttons are quite characteristic, for they essentially occurred during FBA only in the discussed three regions (cf. section 3.5), although one should note the slight chronological differences between the contexts. Furthermore, both western Veneto and north-central Lazio share quite similar forms exhibiting decorative motifs var. 'c' and, to a lesser extent, 'e' which were found in contemporaneous contexts.[612] On the other hand, western Veneto lacks any evidence of application of var. 'd' and, conversely, north-central Lazio so far has not revealed amber beads decorated with var. 'g', 'h' or 'i', which during BF 3 became quite widespread in the northern part of the circum-Adriatic zone, and western Veneto in particular. Remaining data, as witnessed on the basis of the contexts with amber, cannot help to decisively determine the relationship between the two. Traces of contacts between the discussed regions are confined to the fragment of a simple fibula with corrugated arc (imitating series of nodules) found in Allumiere – Poggio La Pozza, grave 4, trench D (1960) (Fig. 102:A) which belongs to a type sporadically registered in the area of the *Tolfa-Allumiere* group, but widely distributed in the *Protovillanovan* culture of the Po Plain (Peroni 1960, 359; cf. Eles Masi 1986, Tav. 8:169–174, 9:175–177), including two specimens from an amber-rich grave 437 of Desmontà (Fig. 102:B). One can also mention a bronze wheel-shaped pinhead with six spokes from Allumiere – Poggio La Pozza, grave 2, trench F (1960) (Peroni 1960, Fig. 11:13) – a rather common item in FBA – for which, nonetheless, a relevant casting mould was found at Frattesina (Bellintani 1973, Tab. VI:10). Finally, it is worth mentioning that 'about thirty or forty' beads of blue glass paste were reportedly found at Poggio della Capanna together with amber, however became lost afterwards (Fugazzola Delpino 1992, 286). Above-described evidence fits well into the notion of western Veneto, Frattesina in particular, as an FBA centre of amber, glass and bronze processing. However, in the absence of chemical analyses, Northern Italian provenience of discussed material from north-central Lazio can only be hypothesised. Regarding amber, there are undeniable similarities between some Northern Italian finds and those from north-central Lazio. However, certain decorative motifs typical only for the latter region possibly indicate a local processing of amber in the area in BF 3.

---

609 Allumiere – Poggio La Pozza, grave 4, trench D (1960) and grave 1, trench F (1960) (Peroni 1960, Fig. 11:8, 17); Frattesina di Fratta Polesine, hoard 1 *Tesoretto* (Negroni Catacchio 1972, Fig. 2); Poggio della Capanna, grave (Fugazzola Delpino 1992, Fig. 10:8); Maso Finale, cultural layer (Putzer 2012, Abb. 10:34–35).

610 Allumiere – Poggio La Pozza, grave 4, trench D (1960) (Peroni 1960, Fig. 11:16); Fondo Paviani, advanced BR–early BF layers (Strafella et al. 2015, Fig. 1:F); Maso Finale, cultural layer (Putzer 2012, Abb. 10:25).

611 Allumiere – Poggio La Pozza, grave 2, trench F (1960) (Peroni 1960, Fig. 11:23); Frattesina di Fratta Polesine, phases 2–3 (Bietti Sestieri et al. 2015, Fig. 2:13).

612 E.g. IE421c – Allumiere – Poggio La Pozza, grave 2, trench F (1960) (Peroni 1960, Fig. 11:17a–b); IE451c – Narde II, grave 75 (Colonna et al. 2010, Tav. 13:B3); IE211e – Allumiere – Poggio La Pozza, grave 2, trench F (1960) (Peroni 1960, Fig. 11:18); IE111e – Narde II, grave 75 (Colonna et al. 2010, Tav. 13:B10).

Fig. 102. Examples of fibulae, daggers (pendants), pins, combs, pinheads, torcs and bracelets shared by some of the FBA sites with amber in north-central Lazio (A), western Veneto (B), Sicily (C, E), the Aeolian Islands (D), Liguria (F, K), Piedmont (G–I) and western Lombardy (J): A – Allumiere – Poggio La Pozza, grave 4, trench D (1960) (Peroni 1960, Fig. 11:14); B – Desmontà, grave 437 (Salzani 2013, Tav. XXXII:5); C – Madonna del Piano, tomb 11 (Orsi 1905, Fig. 6); D – Piazza Monfalcone, tomb 31 (Bernabò Brea/Cavalier 1960, Tav. XLI:2a); E – Madonna del Piano, tomb 46 (Bernabò Brea et al. 1969, Fig. 22:g); F, K – Castellaro di Uscio, horizon 3 (Del Lucchese 1990, Fig. 110:2–3); G–I – Monte Cavanero, hoard (Rubat Borel 2009b, Fig. 47:2, 49:8, 67:5); J – Isolino Virginia, superficial layers (De Marinis 2009c, Fig. 5:2).

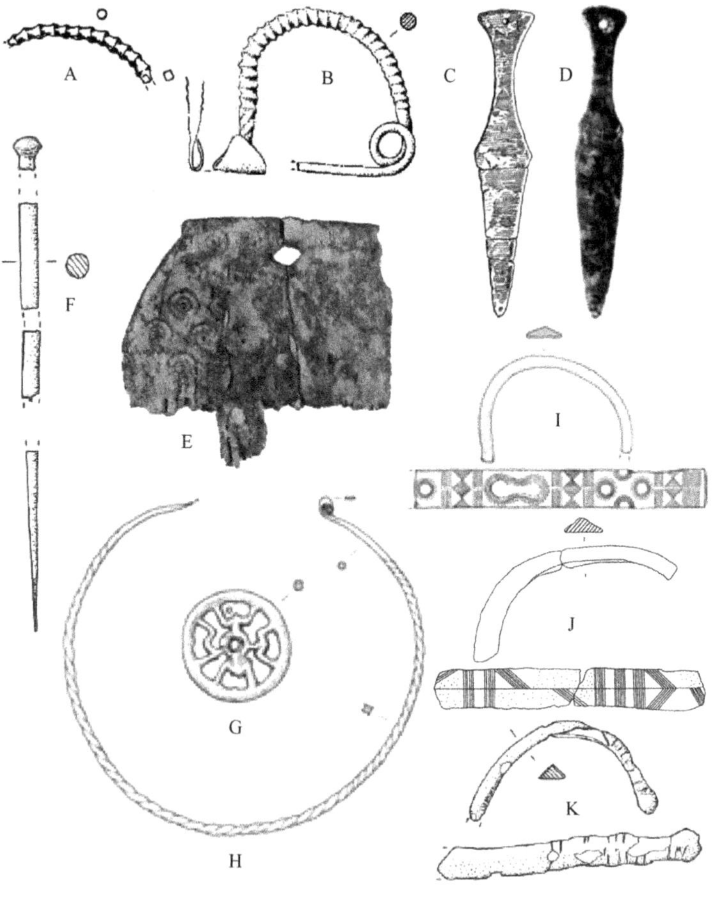

Situation of Sicily in the FBA network of amber circulation is puzzling. The only link the region shares with other nodes is with the nearby Aeolian Islands. This shows a low level of similarity between Sicilian FBA amber finds and those in the remaining parts of the circum-Adriatic zone. However, it also results from a low number of typologically classified amber artefacts from Sicily which mostly lack appropriate figures. This basically restricts the basis for building the similarity ties to a single object from tomb 46 of the Madonna del Piano cemetery, representing type IE241 (Bernabò Brea et al. 1969, Fig. 22:b). The bead finds the closest analogy among the large assemblage of amber adornments from tomb 31 at Piazza Monfalcone, Lipari (Bernabò Brea/Cavalier 1960, Tav. XLIII:1). The link between Sicily and the Aeolian Islands might not be surprising as the two regions are closely located, and their cultural development throughout FBA was partially convergent, as seen on the basis of the Madonna del Piano necropolis (cf. section 2.2.18). Some connection derived from contexts with amber situated in the two regions is evident when comparing miniature daggers (pendants?) with fused handles (Fig. 102:C–D). Problem is, however, that there is a chronological discrepancy between the tombs with amber at the Piazza Monfalcone necropolis and those from Madonna del Piano: the former are dated to phase *Ausonian IIa*/BF 1–2 (ca. 1130/1100–1035/1010 BC), while the latter to *Pantalica II – Cassibile*/BF 3-PF IA (1035/1010–910 BC). Some Sicilian graves attributed to the earlier horizon of *Pantalica II – Cassibile* (FBA A) can be suspected partly contemporaneous with those of Piazza Monfalcone dated to *Ausonian IIb*. However, tomb 46 of Madonna del Piano, dated to subphase 2b of the *Pantalica II – Cassibile* stage (ca. 960/950–910 BC), is at least 50 years younger than the latest moment at which Piazza Monfalcone ambers could have been deposited. One could suspect that the Sicilian bead was kept as a heirloom, thus appearing in a younger context. Another complication is the provenience of amber samples from the Madonna del Pianno which, unfortunately, cannot be linked with particular beads from the tombs. The analysis showed that some of them match the succinite, whereas one proves the use of simetite for production of FBA amber jewellery (cf. Angelini/Bellintani 2017). This observation allows a hypothesis that FBA communities inhabiting Sicily sourced and processed amber locally; thus, Sicilian artefacts can differ from those in other parts of the circum-Adriatic zone. Nevertheless, presence of succinite means that at least part of amber was imported from the north. This connection can be associated with a comb from tomb 46 decorated with concentric circles (Fig. 102:E; cf. Fig. 100:J). The Sicilian comb was made of ivory, as were some of those found in Frattesina di Fratta Polesine, hoard 1 *Tesoretto*. Although, the discussed specimens come from chronologically discrepant contexts and differ in morphology resulting in different typology (cf. Bettelli/Damiani 2005), the exotic material and ornamentation pattern certainly links them. It is worth adding here that processing of elephant ivory is documented at Frattesina (Bietti Sestieri/de Grossi Mazzorin 1995, 2005). So, it cannot be ruled out that Sicily and western Veneto were integrated into the same network of materials and products exchange during FBA.

During the last phase of FBA (BF 3), amber kept reaching north-west part of Italy in small quantities. The simple flattened discoidal beads (IF121 and IF121y; Maggi 1990, Fig. 111:4–5) found in the layers of horizon 3 at Castellaro di Uscio in Liguria have the geographically closest analogies among the materials from grave XXXI of Bismantova in Emilian Apennines (Catarsi/Dall'Aglio 1978, Figs. XIII:6 and XXI:1). Morphology of the Ligurian beads itself is not of much help in determining their provenience, but some of the other finds from Castellaro di Uscio pertinent to the same phase can be used as auxiliary evidence to a certain degree. Bronze pin found in the same layer with one of the amber beads is said to resemble the type Torri d'Arcugnano or Terlago; however, due to fragmentary state, it is impossible to specify the typological attribution (Del Lucchese 1990, 170; Fig. 102:F). Nonetheless, both types belong to the family of pins with slightly biconical nail-heads which characteristically appeared in Northern Italy (Lombardy, Trentino and Veneto) during advanced and late stages of FBA (BF 2–3), e.g. the already mentioned specimens from phase 2 of Frattesina (Fig. 101:J), thus can be linked with the *Protovillanovan* culture (cf. Carancini 1975, 226–231; Salzani 1978, 142). Furthermore, to horizon 3 of Castellaro di Uscio can be cautiously linked three spheroidal beads made of turquoise-blue and dark-blue glass paste, similar to those found in a number of contexts with amber in Northern Italy (Maggi 1990, 176–177, Fig. 111). While these indications are not very specific and include chronologically discrepant contexts, they allow to see a faint link between Liguria and western Veneto.

In western Lombardy the single amber find datable to BF 3 comes from Isolino Virginia, and is characterised by oblique, regular grooves applied on elongated cylindrical bead (IE211g; Fig. 49:7). The most akin counterpart is the IE221g bead from Desmontà, grave 437 in western Veneto (Fig. 49:17), whereas more remote analogies were found in the Balkans – IE458g (Fig. 50:18). The bronze knife found in the same stratum as the amber bead from Isolino Virginia was classified as the type Bismantova, var. C – a variety that has been confirmed only in the extreme north of Italy, thus probably reflecting local production (Bianco Peroni 1976, 61, Taf. 34:283, 61:D). On the other hand, two blue, barrel-shaped glass beads with spiral decoration of white filaments, found in another trench than amber but still pertaining to the same horizon, represent type of vitreous adornments present in some of the already described contexts with amber, most importantly Frattesina (Fig. 99:A–E). Although the two beads have not been checked for provenience, hitherto conducted analyses of similar objects allow to suspect western Venetian production.[613] It is worth mentioning in this place that elongated amber beads decorated with the Allumiere-type ornament (also in the oblique-grooves version) were found at the sites Montlingerberg and Hauterive-Champréveyres in Switzerland, on the latter associated with LMHK barrel shaped with spiral decoration (Jennings 2014, Fig. 12:b–c, Bellintani 2012, Fig. 3:12–14). Thus, the finding of Isolino Virginia could be an important link in a transalpine circulation of amber and glass during FBA.

In FBA numbers of hoards with amber increased as a part of the pan-European phenomenon of hoarding at the turn of the 2nd and 1st millennia BC. To group of the already discussed hoards of Frattesina *Tesoretto*, Gualdo Tadino or Torre Castelluccia can be added the deposit from Monte Cavanero near Chiusa di Pesio in Piedmont. Chronology and cultural attribution of its content has partly been discussed in section 2.2.2. It is time now to name some of the materials which could possibly point to the route the ambers of Monte Cavanero went through before being buried at the site. Amber beads from the hoard find analogies in Trentino-Alto Adige, western Veneto and north-central Lazio among the Italian provinces. One of the beads has flattened double-pitched profile (IF161) – a form rather common in FBA, while the other one is more specific and can be characterised as a flattened spheroidal pendant (IIF225; Fig. 55:3). While the biconical bead has numerous analogies in i.a. Trentino-Alto Adige, western Veneto and north-central Lazio, the pendant is unique in the scope of FBA in the circum-Adriatic zone. Fortunately, more specific comparisons are offered by the bronze items from the Monte Cavanero site.

613 Cf. Bellintani 2012, 21; Cupitò et al. 2015b; Conte et al. 2019.

The oldest bronze finds from the hoard have the *Terramare*/Po Plain provenience and date back to MBA–LBA, but many more finds are datable to FBA. Some of them can be used to verify the links built on the similarities among amber finds. The most precise analogies can be presented in relation to the pinhead with four M-shaped motifs resting on a ring-shaped ferrule (Fig. 102:G), which would be i.a. the specimens from Narde II, *Ustrinum* and Borgo Panigale (Fig. 98:A–B; cf. Rubat Borel 2009b, 99). Both at Narde and Monte Cavanero, these pinheads coincide with corrugated torcs (Fig. 98:K, 102:H). The Monte Cavanero hoard contained also open bracelets representing the Zerba type, indicating more local north-west Italian background, with some of the specimens, although not belonging to the same variety, coming from sites with amber in western Lombardy and Liguria (Fig. 102:I–K; cf. Paltineri/Rubat Borel 2010). More meaningful are the results of glass analysis which indicate LMHK type for the annular beads in blue-turquoise colour found in the hoard (Angelini 2009). Providing that the amber beads from Monte Cavanero have FBA, or even MBA/LBA chronology, this short overview of the data suggest that they could have been brought from north-east Italy, western Veneto in particular. Still, the Monte Cavanero hoard yielded a substantial number of bronze artefacts with cultural affiliation going beyond the boundaries of the circum-Adriatic zone. A fitting example is the pendants of the Réallon type from the namesake hoard in the French Alps (ca. 950–800 BC) which also contained three amber and two vitreous beads (Rubat Borel 2009b, 80–82, 103, Fig. 56; cf. Hausmann 1996–1997). Thus, it cannot be excluded that amber finds have a provenience external to the study area.

Another FBA hoard with amber – Debeli Vrh in Slovenia represents an intriguing case of a bead with the Tiryns-type decoration (IE211a) which may have travelled a long and complex route before it was eventually deposited. Apart from the above-mentioned object, the hoard yielded 12 other amber beads (Hirschbäck-Merhar 1984, Tab. IX:17). However, these do not reflect any clearly defined forms which might correspond with assemblages from the other circum-Adriatic provinces of amber's acquisition, with the exclusion of Vojvodina (IE841; cf. Rašajski 1988: Figs. 58–60). FBA contexts with amber in the study area, especially the Italian part, rarely include decorated beads alongside irregular and unadorned objects, the Apulian collections from Monte Saraceno and Torre Castelluccia being notable exceptions. In Northern Italy assemblages reflect a great deal of uniformity, testifying to a growing prowess in amber processing. Is it possible then that IE211a bead from Debeli Vrh, for which there are numerous analogies in Italy accompanied by evidence of amber processing,[614] was added to amber artefacts coming from a different area?

The bulk of assorted materials associated with the beads in the Debeli Vrh hoard, mostly scrap metal from bronze weapons, tools and jewellery, are classifiable to the IInd horizon of hoards in the Balkans (ca. 1200–1050 BC) and bear a strong connection to the *Urnfield* culture of the eastern Alpine zone, Pannonia and the Carpathian Basin (Turk 1996, 110–115). Nonetheless, several elements of the hoard's inventory point to Italian origin, and find analogies in some LBA–FBA context with amber on the Apennine Peninsula. The most prominent among them is a well-preserved

614 Frattesina di Fratta Polesine, hoard 1 *Tesoretto* (Negroni Catacchio 1972, Fig. 18); Fondo Paviani, advanced BR–early BF layers (Fasani/Salzani 1976, Fig. 2:3–4); Costa Cavallina, cave burial (Poggiani Keller 1989, Fig. 98:3); Piazza Monfalcone, tomb 31 (Bernabò Brea/Cavalier 1960, Tav. XLIII:1).

With respect to trans-Adriatic contacts seen in the light of decorated amber beads, it is vital to consider FBA contexts from Lika. Relatively modest collection of amber artefacts from the region datable to the discussed period comprises, nonetheless, forms and ornaments which can be related to some of the finds from western Veneto and Emilian Apennines. Bead representing type IE451c coming from Kompolje, grave 293 (1903–1905) has direct analogies in both the abovementioned Italian provinces.[619] The connection is strengthened by occurrence in Lika of beads decorated with the typical Tiryns-type motif all coming from the same grave at the Kompolje cemetery: IA411a (Fig. 45:9), IE748a (Fig. 50:10) and VIF218a (Fig. 59:8), although in terms of morphology, they differ slightly from Northern Italian finds possessing the same decoration. In the same manner, certain indirect analogies, based on the presence of the decorative motifs 'a' and 'c', for the beads from Lika can be found in Kvarner and northern Dalmatia. More importantly, however, the similarities between the two regions comprise some objects adorned with the motifs 'b' (Figs. 45:4 and 59:4) and 'f' (Fig. 49:6, 9). First of them has not been registered in the remaining parts of the circum-Adriatic zone so far. Variant 'f' appears on few finds from Italy, but they are morphologically remote from the specimens from Kvarner and Lika. Unfortunately, due to poorly equipped context, the listed amber beads from Lika cannot be correlated with other objects which could verify signalised links with western Veneto, Emilian Apennines or Kvarner.

On the other hand, connections between Kvarner and northern Dalmatia, and Italy are reflected in some of the bronze objects. The adduced graves from Privlaka and Vrsi share pairs of ribbed-cuff bracelets with rolled-up ends and open bracelets with double-pitch (triangular or V-shaped) cross-sections (Batović 1983, Tab. XLIV:1–6, 13–18) – objects prevalent in various parts of the *Urnfield* culture ecumene, thus in the context of coastal Croatia considered as manifestations of the continental metallurgical *koiné* (cf. Blečić Kavur 2014, 84–87). Such bracelets were not that common on the Apennine Peninsula or southern Alpine zone; however, some counterparts are known from the territory of Daunia (northern Apulia), and the *Protogolasecca* culture in Lombardy and Piedmont, in the latter case co-occurring with amber i.a. in the hoard of Monte Cavanero, Chiusa di Pesio (Blečić Kavur 2014, Figs. 46 and 49; Rubat Borel 2009b, Figs. 47:1–2 and 48:1–5). While the double-pitch bracelets from the hoard include geometric ornaments – feature lacking on the specimens from Croatia – and were assigned to the type Zerba, Buco del Diavolo variety (Rubat Borel 2009b, 64–65), the ribbed ones provide quite punctual analogy in relation to those from Privlaka and Vrsi (Blečić Kavur 2014, 85) (Fig. 104:A–B).

Better comparisons between the opposite coasts of the Adriatic are offered by the simple fibulae with two knobs on the arc, present in i.a. grave 89 of Vrsi (Fig. 104:C). This category of fibulae firmly stands as one of the hallmarks of the Mediterranean *koiné*, with the distribution range encompassing the entire length of the Apennine Peninsula.[620] Although the fibulae from Vrsi have been ascribed to the 'Liburnian' variant – a local variety recognised by small size, decoration with parallel lines, triangles and zig-zag or 'herringbone' motifs in metopes, and large, round knobs,[621] their form

619 Narde II, grave 75 (Colonna et al. 2010, Tav. 13:B3); Bismantova – Campo Pianelli, grave XXXI (Catarsi/Dall'Aglio 1978, Fig. XXI:1).

620 Teržan 2007, 162; Blečić Kavur 2014, 56–60; cf. Eles Masi 1986; Lo Schiavo 2010; Savella 2015.

621 Blečić Kavur 2014, 49–50, Fig. 22; cf. Glogović 1987, 80–81; 2003, 8–11.

mirrors the basic concept replicated on several Italian sites with amber. One can simply invoke the content of grave XXXI of Bismantova, where two fibulae with characteristic double swellings, and decoration with 'herringbone' and linear motifs (Fig. 100:C) were found associated with 18 amber beads including type IF121 present also at Vrsi (cf. Batović 1983, Tab. XLIV:12; Catarsi/Dall'Aglio 1978, Fig. XXI:1). However, the similarities are not limited to Northern Italy but continue southwards; in Lazio the grave of Ardea – Campo del Fico yielded a fibula with thickened middle section of the arch, decorated with a complex pattern including 'herringbone' motif, hatched diamonds and triangles, fitted between two nodules with four ribs each (Fig. 100:K), whereas in northern Apulia, grave 84/IV of Monte Saraceno produced another variant characterised not only by double/ribbed nodules but also exhibiting 'herringbone' motifs and linear incisions, classified by F. Lo Schiavo as type 20, variant A – the 'Gargano' variety (Lo Schiavo 2010, 102–103; Fig. 104:D). Still, the grave XXXI of Bismantova presents the best analogies both in terms of morphology of the fibulae as well as amber beads, and also for the fact that it contained annular glass bead in greenish-blue/turquoise colour, same as those from Privlaka, grave 87 (Batović 1983, 315).

Fig. 104. Examples of bronze bracelets, fibulae, tweezers and pendants shared by some of the FBA sites with amber in Kvarner and northern Dalmatia (A, C, E, G–H), Piedmont (B), western Veneto (F) and northern Apulia (D, I): A – Privlaka, grave 87 (Batović 1983, Tab. XLIV:15); B – Monte Cavanero, hoard (Rubat Borel 2009b, Fig. 48:1); C – Vrsi, grave 89 (Batović 1983, Tab. XLIV:7–8); D, I – Monte Saraceno, tomb 84/IV (Nava/Preite 1995, Tav. XIII:2, 7); E, G–H – Island of Krk, precise context unknown (Blečić Kavur 2014, 171, Fig. 21:18–19); F – Narde I, grave 221 (Salzani 1990–1991b, Fig. 4:3).

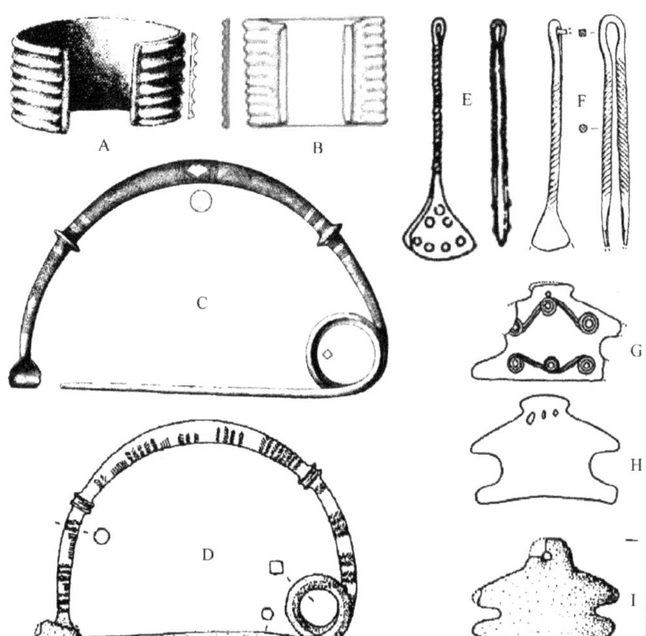

Some punctual analogies from Italy can be brought forward with respect to some of the finds from the Krk Island. The tweezers ascribed to the presumed grave on the island are made from thin metal sheet, twisted in the handle-part, with triangularly shaped clips, and are decorated with seven incised circles (Fig. 104:E). They have direct comparisons among the finds from the Frattesina necropolises, for instance, in grave 221 from Narde I (Fig. 104:F), where it was hung on a small bow fibula typical of BF 3 (Salzani 1990–1991b, 126; cf. Blečić Kavur 2014, 139). Same context on Krk produced four trapezoidal-anthropomorphic pendants made of flat bronze sheet with four opposing protrusions on the sides, and one on the top which has a narrow perforation for suspension; moreover, one of them is decorated (Fig. 104:G–H). Similar item, albeit belonging to a different variety was discovered in the already mentioned grave 84/IV from Monte Saraceno in northern Apulia (Fig. 104:I). Nevertheless, objects representing the types II (Monte Saraceno) and III (Krk) are considered closely related to the *Dobova-Velika Gorica* group of the *Urnfield* culture (Blečić Kavur 2014, 69, Fig. 33). The anthropomorphic pendants from Monte Saraceno (also known from graves 22/IV, 77/IV and 15/VI) form an isolated group on the Apennine Peninsula, and are interpreted as imports from the east Adriatic coast (Nava/Preite 1995, 91). This would be the third, after the bracelets and fibulae, category of bronze items held in common by Kvarner and northern Dalmatia, and northern Apulia. This observation is interesting, given the dissimilarity between amber finds from the two regions. The abovementioned metal objects could, perhaps, testify to yet another kind of interaction, not involving amber circulation. Still, the possibility of amber exchange between them should not be completely rejected, despite the lack of morphological congruity between the artefacts.

More evidence of trans-Adriatic contacts is revealed by materials retrieved from disturbed tumuli at the cemetery of Garica on the Krk Island. Collection of materials from the site includes a flattened spheroidal amber bead (IF122; Blečić Kavur 2014, Tab. 1:171) which itself is not a precise indicator of provenience; however, metal artefacts accompanying it have that quality. The large decorated arched fibula (Fig. 105:A), while having no good parallels on the eastern side of the Adriatic, finds good analogies particularly in the territory of Picenum (nowadays the region of Marche) dated to BF 2–3, e.g. on the cemetery of Pianello di Genga (Savella 2015, Tav. 12:82, 13:86; Blečić Kavur 2014, 56). Numerous simple-arch fibulae with similar decoration of alternating parallel lines and the 'herringbone' motif are known from contexts with amber in Northern (Fig. 100:C, L), Central (Fig. 100:E–F) and Southern Italy (Fig. 105:B); however, all of them are lacking a foot with embossed dots. The latter are nonetheless present in many other fibulae from northern part of the peninsula, the best examples being those from Casalecchio and Coste del Marano (Bietti Sestieri 1973, Fig. 9:10, 19:6, 20:5–6; Savella 2015, Tav. 33:235). Discussed fibulae chronologically can be placed in BF 2–3, while the specimen from Garica, in accordance to the proposition of M. Blečić Kavur, can be regarded as an import from the Apennine Peninsula (Blečić Kavur 2014, 56).

Fig. 105. Examples of bronze fibulae and razors shared by some of the FBA sites with amber in Kvarner and northern Dalmatia (A, C, E, G), northern Apulia and Basilicata (B, D), western Veneto (F) and Emilian Apennines (H): A, C, E, G – Garica (Blečić Kavur 2014, Tab. 1:160–162, 168); B – Lavello (Lo Schiavo 2010, Tav. 14:106); D – Monte Saraceno, tomb 86/IV (Nava/Preite 1995, Tav. XIV:8); F – Narde I, grave 39 (Salzani 1990–1991b, Fig. 23:7); H – Bismantova – Campo Pianelli, grave I (Catarsi/Dall'Aglio 1978, Fig. VII:5).

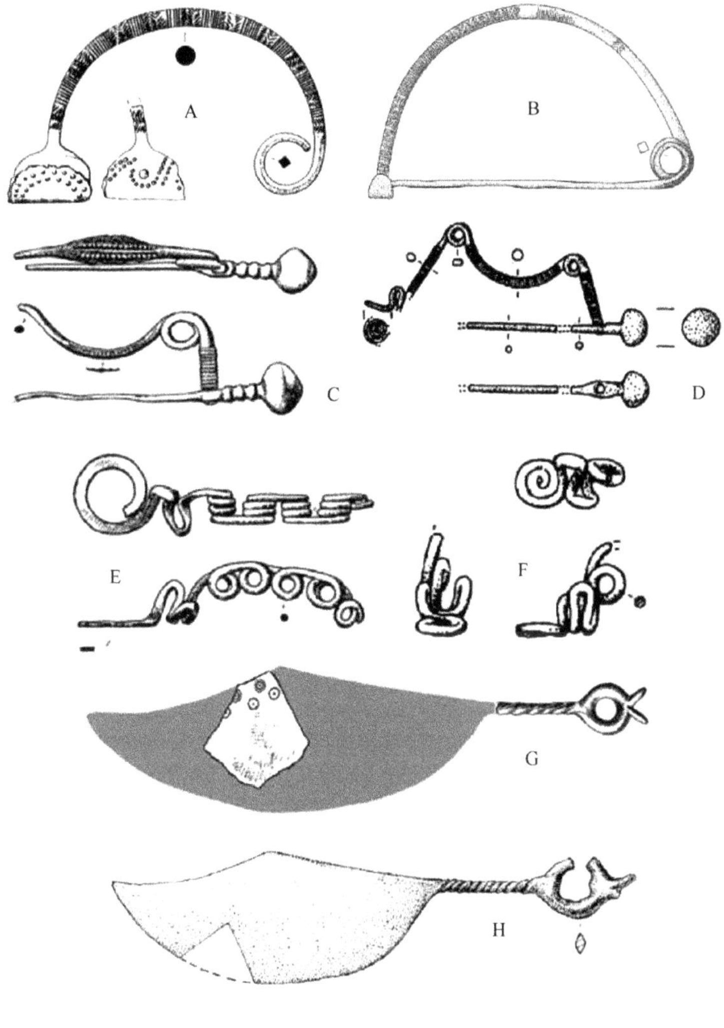

Furthermore, Garica yielded examples of two other types of fibulae which can be found among contemporaneous Italian sites containing amber. The two-part serpentine fibula with a saddle-shaped bow (Fig. 105:C) – an example of the Adriatic *koiné* bronzes – has recently been defined as, most probably, a late product of a local, east Adriatic workshop, succeeding earlier variants interpreted as Italic imports, under

Fig. 106. Examples of bronze fibulae, bracelets, torcs and pendants shared by some of the sites with amber in middle Dalmatia (A–C, E–G), Kvarner and northern Dalmatia (C) and the Mat Valley (H): A–C, E–G – Trcela-Vranjic (Marović 1960, Figs. 1:1, 3, 2:1, 5, and 3:1–2); D – Vidasi (Glogović 1991, T. 1:1); H – Perlat, tumulus 2, grave 3 (Bietti-Sestieri/Lo Schiavo 1976, Fig. 17:1).

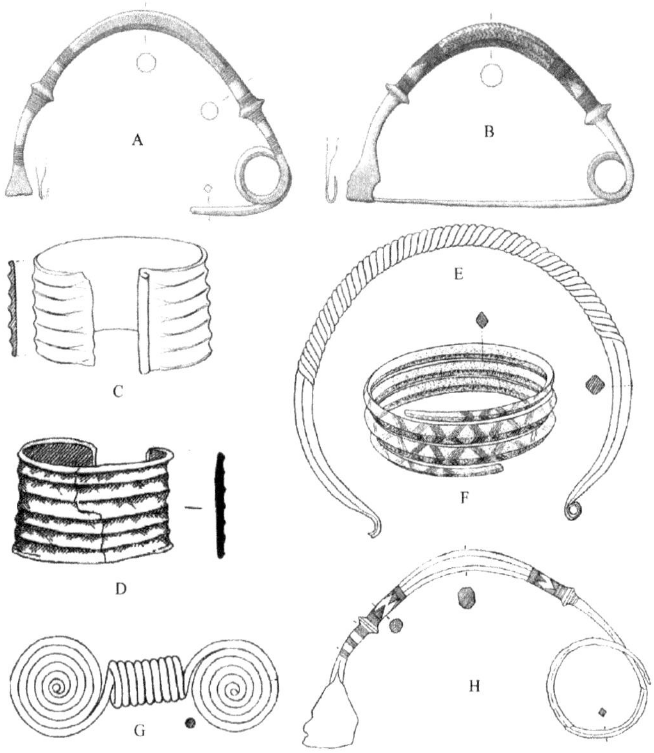

Other finds from Trcela-Vranjic that have both specific spatial and chronological attribution are ribbed-cuff bracelets, and imitated twisted torque (Fig. 114:C, E). The former objects are typical elements of jewellery encountered in contemporary sites in Dalmatia, Istria and Northern Croatia; bracelets of this kind appeared together with amber in several, already discussed contexts in Kvarner and northern Dalmatia,[625] although it is important to remember about their incidental occurrence in northern Apulia and Piedmont (Figs. 104:A–B and 106:D).

625 Privlaka, grave 87, Vrsi, grave 89 (Batović 1983, T. XLIV:13–15); Vidasi (Glogović 1991, T. 1:1); Garica (Blečić Kavur 2014, T. 1:164–165). Consequently, the ribbed-cuff bracelets are chronologically comparable with the arched fibulae with two knobs on the bow and pins of the Šula type (Blečić-Kavur 2014, 84–85). Aside from Vranjic, ribbed-cuff bracelets are known from eight other localities in Middle Dalmatia, e.g. Balina glavica (Batović 1983, 339, Sl. 21:11).

Imitated twisted torques, on the other hand, are considered as objects rather untypical for the eastern Adriatic coast, presumably imported from the Balkan interior, where they can be found in numerous contexts (cf. Vasić 2010, 38–42). The torque from Trcela-Vranjic is made of massive bronze rod with rhombic cross-section, seemingly corrugated in its middle section (form of decoration performed by engraving), with ends rolled up, therefore more akin to younger versions dated to HaA2/B1 (ca. 1100–950 BC). The closest analogies in middle Dalmatia are known from Balina Glavica (Batović 1983, 339, Sl. 21:7), while further north-west in Lika, they were found in EIA/*Iapodean* graves on the cemetery of Kompolje (Drechsler-Bižić 1961, Tab. XXI:2; 1968, Tab. III:30). On the other hand, despite generic similarity, massiveness and rhombic cross-section sets the torc from Trcela-Vranjic apart from some of the already invoked specimens from Italy which also come from slightly older contexts (Fig. 98:I–K). The only tentative link between middle Dalmatia and western Veneto are five disc-shaped glass-paste beads of yellowish-brown tinge (Marović 1960, Fig. 1:2) – rare kind of vitreous artefacts considering the form and colour, even among materials from Frattesina (cf. Bellintani/Stefan 2009, Tab. 1).

Gathered evidence from Trcela-Vranjic points to, apart from the northern section of the east Adriatic coast, the links with the Balkan interior, particularly Bosnia and Herzegovina, but also the Sava-Drava interfluve. Among the remaining materials from Trcela-Vranjic are four bracelets made of bronze tape with rhombus cross-section which is slightly tapered at the ends, decorated on the outer side with the 'herringbone' motif (Fig. 106:F) and spectacle pendants with coiled intermediate part (Fig. 106:G). In Bosnia and Herzegovina similar combinations of the Golinjevo-type fibulae, corrugated torcs, spirally coiled bracelets and spectacle pendants appear in the context of graves and hoards dated to the 4th horizon of hoards/HaB1 (ca. 1060–950 BC) (Gavranović 2011, Abb. 256). In the Sava-Drava interfluve, massive corrugated torcs and bracelets with rhombic cross-sections and decoration consisting of geometric motifs (oblique lines, triangles, rhombuses) appear in the *Dobova-Velika Gorica* group (Gabrovec 1983, T. X:18–21, 23). Still, the abovementioned parts of the Balkans so far have not provided evidence of amber during FBA.

Furthermore, Trcela-Vranjic can be related to the occurrence of amber in the Mat Valley in Albania – the site of Perlat, despite no apparent link in the network analysis. Perlat, tumulus 2, grave 3 is the only context in Albania which yielded amber artefact comparable with the Allumiere type in the oblique-grooves variant (IE458g; Fig. 50:18). Oblique grooves are not encountered on FBA amber beads in the Balkans, while they are occasionally occurring in northern Italy, probably revealing the birthplace of this ornament.[626] Interestingly, there also is one artefact exhibiting such ornament from Romania.[627] Apart from the decorated bead, Perlat also yielded finds classified as IE221 and IF121 – more common types of amber adornments not only found predominantly in Italy[628] but also recorded in several FBA contexts in Western

626 IE221m – Desmontà, grave 437 (Salzani 2013, Tav. XXXII:35); IE211m – Isolino Virginia (De Marinis 2009b, Fig. 5:3).

627 Dridu settlement (Boroffka 2001, Abb. 3:21).

628 IE221 – Allumiere – Poggio La Pozza, grave 2, trench F (1960) (Peroni 1960, Fig. 11:20), Fossa Nera di Porcari, precise context unknown (Andreotti/Zanini 1995–1996, Fig. 2:14); IF121 – Castellaro di Uscio, horizon 3 (Maggi 1990, Fig. 111:5), Bismantova – Campo Pianelli, grave XXXI (Catarsi/Dall'Aglio 1978, Fig. XXI:1).

Balkans.[629] Helpful factor in tracing a potential route through which amber jewellery came to Albania in FBA might be the fibula found in Perlat, tumulus 2, grave 3. As explained in section 2.3.13.2, the grave from Perlat contained an early variant of the Golinjevo-type fibula, closely resembling the specimens from Trcela-Vranjic in middle Dalmatia (Fig. 106:A–B, H). Of course, it is possible that the fibula and amber jewellery reached the Mat Valley independently from different locations – hypothesis strengthened by the apparent lack of similarity between amber finds from Perlat and Trcela-Vranjic. Nonetheless, evidence gathered from two sites seems to indicate an NW–SE axis of contacts along the east Adriatic coast. One could imagine a chain of exchange between the regions leading from Caput Adriae, through Dalmatia to Albania, which facilitated circulation of goods. If that was the case in FBA, amber beads and bronze fibulae could have arrived at Perlat at different times or circumstances, however via the same route.

Contexts with amber in the Balkan hinterland are linked with the other regions predominantly by beads with double-pitched profile, either in flattened (IF161) or elongated variant (IE161 and IE261). Elongated biconical beads, like the one from Vidasi, were particularly widespread in domain of the *Urnfield* culture, just as flattened biconical beads were common in the areas influenced by the preceding *Tumulus* culture.[630] They were discovered in three Balkan contexts with amber, where bronze objects related to the *Urnfield* metallurgical tradition have also been recognised: Križevci-Ciglana settlement in Northern Croatia (Homen 1982, Tab. II. 7), Majdan hoard in Serbian Banat (Rašajski 1988, Fig. 61) and the already discussed grave of Trcela-Vranjic in middle Dalmatia (Marović 1960, Fig. 2:3–4). Interestingly, in Italy these beads were known since LBA and have not been found associated with explicit evidence pointing to the *Urnfield* influence, although one can infer their 'external' origin from the absence of evidence testifying to their local manufacturing in the circum-Adriatic zone at the time, as well as striking similarity to the finds recorded on the northern side of the Alps. An indication of the transalpine circulation of elongated biconical beads could be the deposit of such artefacts discovered in Maso Finale – Finail Valley in Trentino-Alto Adige (Putzer 2012, Abb. 10:34–35). Vojvodina yielded also artefacts in cylindrical (IF111)[631] and spheroidal (IA121)[632] forms which, however, due to broad chronology and dispersion do not have the same quality as indicators of interregional contacts as the biconical beads.

The region of Northern Croatia, represented by the site of Križevci-Ciglana, delivered biconical amber beads identical to those from i.a. Trentino-Alto Adige, western Veneto or north-central Lazio, however manifests different cultural connotations. Ceramic and bronze artefacts recovered from the site are related to the *Urnfield* culture on the territory of Drava-Sava interfluve. While the pottery has affiliation with the local groups belonging to the *Urnfield* complex, chiefly the *Virovitica II/Zagreb* group (Homen 1982, Tab. I: 5–5a;

629 IF121 – Ljubljanica River, precise context unknown (Turk/Gaspari 2009, Fig. 69), Island of Krk, precise context unknown (Lo Schiavo 1970, 425; Fig. XXIII:11), Vrsi, grave 89 (Batović 1983, T. XLIV:12).

630 E.g. Sprincz/Beck 1981, Fig. 5:1–15, Tab. 2; Boroffka 2001, Abb. 4:7–8, 12, 14, 21–22; Chvojka 2017, Fig. 6; Gogâltan 2016, Fig. 7; Stuchlík 2017, Abb. 7–8; Woltermann 2016a, 43, 170, 175, Abb. 168.

631 Majdan, hoard (Rašajski 1988, Fig. 63).

632 Banatski Dvor, site 22, object 38 (Hristov 2017, T. 16:23).

Dizdar et al. 2011, 78–79), morphology and ornamentation of the bronze pins have more of an interregional relevance. Pins with profiled/ribbed biconical heads, biconical/onion-shaped heads and club-shaped heads (Fig. 107:A–C) are all frequent in the context of the 1st, 2nd and 3rd phase of the *Urnfield* culture in Northern Croatia,[633] however also occur in significant numbers in Transdanubia, where they are typical mostly for the earlier phase of the local *Urnfield* culture (BzD–HaA2).[634] Yet, the club-shaped pins also appeared in Northern Italy, occasionally in LBA contexts with amber: Peschiera – Imboccatura del Mincio (Aspes 1980, Fig. 41:14) and Beneceto, BR 2 layers (Bernabò Brea et al. 2008, Fig. 13:7). Similar pin was recovered from the Ljubljanica River and is ascribed to the same period as the pin with amber head (Turk/Gaspari 2009, Fig. 69). When it comes to the pin with biconical/onion-shaped heads, it is possible to discern certain affinity with the Sover type, particularly apparent when comparing decorative patterns on the pins from Križevci and Frattesina (cf. Figs. 98:S and 107:B). This relationship potentially documents some links in metallurgy and amber, if the biconical beads from the latter site are considered.[635] To this evidence can be added few glass beads made of blue paste found at Križevci-Ciglana which by their form match some of the materials from Frattesina, although their chemical composition is unknown (Homen 1982, Tab. II:6; Bellintani/Stefan 2009, Tab. 1:1).

Furthermore, the pins with club-shaped and ribbed biconical heads are thought to document east–west communication to the south of the Carpathian Basin (Ložnjak Dizdar 2014, 241). Their distribution follows the Drava-Sava interfluve up to Vojvodina (Vasić 2003, 85–86, Taf. 63:A). Similar interregional status can be applied to elongated biconical amber beads which seem to be incorporated into amber jewellery at similar date, i.e. after 1200 BC, in the Carpathian Basin. Within the latter area, elongated biconical beads occurred on three sites located to the south of the Lake Balaton in Transdanubia, dated to the *Kurd* horizon of hoards, and at least on three more in Romania belonging to similar period.[636] The coevality and co-occurrence of the aforementioned pins and amber beads is potentially a sign of them being transferred along similar routes.

The Romanian contexts with amber beads from that period may have more relevance for the deposit of 45 amber beads in the hoard of Majdan near Vršac – the richest FBA amber assemblage discovered so far in the Balkans. Similarities between the beads from Majdan and those from contemporaneous sites in Romania do not concern only the biconical beads but include also those with oblong-polyhedral form (IE841; Fig. 50:11).[637] A short distance separating Romanian deposits containing such beads from the hoard of Majdan in the western part of Banat proves that the latter is, in fact, the most south-western extension of significant amber finds concentration in Transylvania and Banat. This hypothesis can be reinforced with some of the bronze objects included in the Majdan hoard: situla of the Hajdúböszörmény type,[638] spectacle

---

633 Cf. Vinski Gasparini 1983c, 652–662, T. LXXXVI:13–14, LXXXVII:15, LXXXIX:10; XCIII:11.

634 Cf. Říhovský 1983, 26–27, 36, 44, Taf. 4:44, 10:191–192, 13–14, 17–21.

635 IE261 and IF161 – Frattesina di Fratta Polesine, hoard 1 *Tesoretto* (Negroni Catacchio 1972, Figs. 2, 6).

636 Sprincz/Beck 1981, 483, Fig. 5:12–13, 26, 34, 8, Tab. 2; Boroffka 2001, 399, Abb. 4:7–8, 12, 14, 17, 19–22; Gogâltan 2016, 153–156, Figs. 7–8, 9–10.

637 Rašajski 1988, Figs. 58–60; cf. Boroffka 2001, Abb. 3:22, 24, 26, 4:9, 11, 16; Gogâltan 2016, Fig. 11.

638 Rašajski 1988, 25–26, Figs. 51–54.

Glasinac – Vrlazje, tumulus IV, grave 2 (1894) – yielded much more informative materials, i.a. late variant of the Golinjevo-type fibula dated to phase IIIc2 and a spectacle fibula of the type Suseni or Rudovci dated between phases IIIc–IVb (Fig. 108:A–B).[641] These finds have typically Central Balkan affiliation, and could be associated with the older versions of the Golinjevo and spectacle fibulae, like those from Trcela-Vranjic (Fig. 106:A–B) and Majdan (Rašajski 1988, Figs. 3 and 108:C). Moreover, the amber bead from Vrlazje, tumulus IV, grave 2 (1894) (VIF714; Fig. 59:15) represents quite a unique form, so far unknown in other regions in the Bronze Age, and at best can be interpreted as a peculiar product created as a result of modifying another bead or a raw lump of amber, in order to adapt it to personal needs.

The node symbolising Pelagonia in the network analysis is unconnected to any other region, despite the former yielding amber beads seemingly in uncomplicated forms (IA121 and IA811; Kitanovski 1960, Fig. 14:a–b). Close, albeit not identical, analogies for the both beads can be found in many FBA contexts within the circum-Adriatic zone.[642] Still, the assemblage from Popadin Dol possesses a specific character, and cannot be linked directly to any other province of amber's acquisition. This results from a specific variant of simple bow fibulae with two nodules on the arc found together with amber in the grave (Fig. 108:D–E). Chronology of the fibulae was discussed earlier (cf. section 2.3.12). By the fact that they exhibit two nodules it is possible to include them in the large group of similar fibulae widespread in the Mediterranean during FBA (cf. Teržan 2007, Pl. XXXVI:b; Blečić Kavur 2014, Fig. 18). Also, the 'herringbone' motif adorning the arcs of the discussed fibulae is not alien to the other specimens presented so far (Figs. 107:C–F and 112:C–D). Despite those features, the fibulae from Popadin Dol, along with other examples from Pelagonia, are usually classified as a separate variant – a distinction to which contributes the remote concentration of these fibulae with respect to other findspots. The closest and most frequently invoked analogy is the fibula from Vergina, Central Macedonia (Fig. 108:F; cf. Vasić 2003, 46). However, in relation to the fibulae from Popadin Dol, the example from Vergina has a different foot – larger and decorated with embossed dots, thus perhaps should be considered as yet another variant. Aside from bracelets, which for the lack of ornamentation are not precise contacts indicators, the grave of Popadin Dol contained a single annular glass bead made of dark-blue paste (Kitanovski 1960, Fig. 14:c). By the formal similarity, the bead can be related to many abovementioned objects from the Balkans (Križevci, Privlaka) and Italy (i.a. Bismantova, Frattesina, Piazza Monfalcone), however as with other Balkan glass finds absence of chemical analyses hinders determination of its provenience. To conclude, there are several indications of relationship between Pelagonia and some other provinces of amber's acquisition, specifically the areas where fibulae with two nodules on the arc coincide with blue glass beads: western Veneto, Emilian Apennines, northern Apulia, Kvarner

641 Benac/Čović 1956, 10, Pl. IV:3, 5; Cwaliński/Pravidur forthcoming, Fig. 5:2; cf. Bader 1983, 56–60, Taf. 11–13; Vasić 1999, 29–36, Taf. 8–11; Pabst 2008, 624, Abb. 11:2, 12:2; Gavranović 2011, 34, Abb. 1.

642 IA122 – Maso Finale, cultural layer (Putzer 2012, Abb. 10:28); Piazza Monfalcone, tomb 31 (Bernabò Brea/Cavalier 1960, 119, Tav. XLIII:1); Banatski Dvor, site 22, object 38 (Hristov 2017, T. 16:23); IA222 – Narde I, grave 169 (Salzani 1990–1991b, Fig. 37:6); Narde II, grave 75 (Colonna et al. 2010, Tav. 13:B9); IF811 – Campestrin di Grignano Polesine, late BR–initial BF layers (Bellintani et al. 2015a, Fig. 2A:8).

and northern Dalmatia. Still, for now these links should be regarded only as cautious suggestions.

Fig. 108. Examples of bronze fibulae shared by some of the sites with amber in the Glasinac (A–B), Vojvodina (C), and Pelagonia (D–E): A–B – Vrlazje, tumulus IV, grave 2 (1894) (Cwaliński/Pravidur forthcoming, Fig. 6:1–2); C – Majdan, hoard (Rašajski 1988, Fig. 3); D–E – Popadin Dol, grave 1 (Kitanovski 1960, Figs. 2 and 4); F – Vergina, mound E/II (Pare 1998, Abb. 21:1).

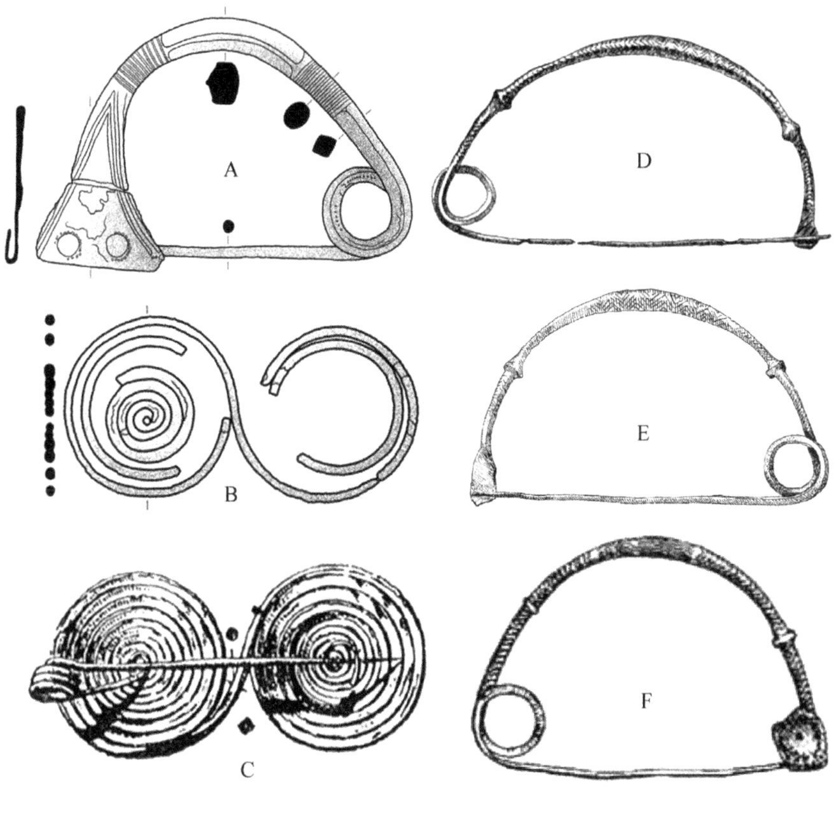

Summary of the considerations on amber circulation in the circum-Adriatic zone during FBA can be divided into two parts. The first one describes finds and contexts dated roughly to the first part of the period (BF 1–2, HaA1–B1a, ca. 1200/1150–1030/1010 BC). The second one covers the latter part (phases BF 3, HaB1b–B2, ca. 1030/1010–910 BC). One has to keep in mind, though, that many contexts cannot be strictly attributed to only one of the stages distinguished above. Thus, they will appear as sites on both the maps presenting reconstruction of amber circulation in the circum-Adriatic zone during FBA.

Fig. 109. Reconstruction of amber circulation in the circum-Adriatic zone during the first stage of FBA (BF 1–2, HaA1–B1a, ca. 1200/1150–1030/1010 BC).

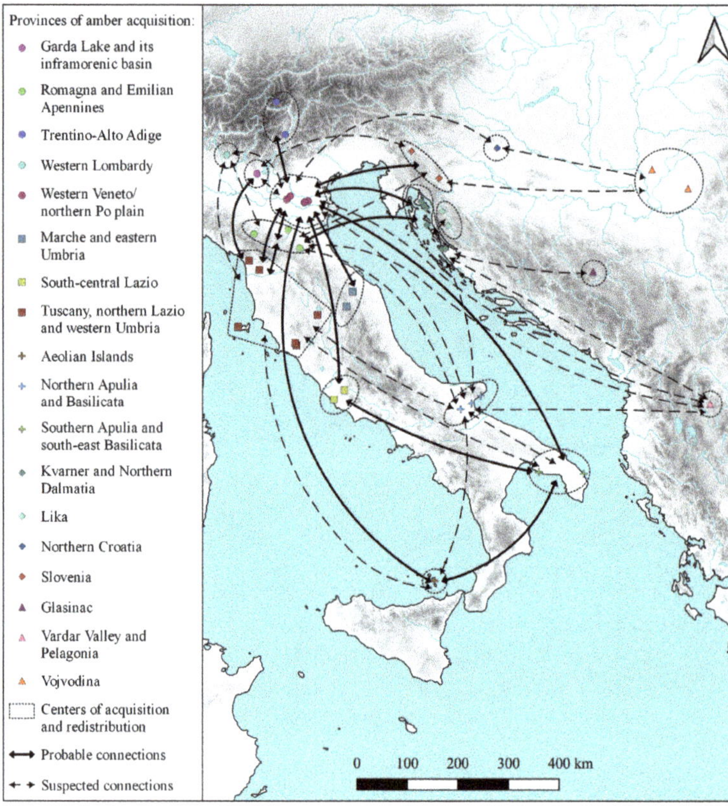

At that earlier stage, there are many attestations of amber in both Italy and the Balkans (Fig. 109). There is a plenty of evidence for interregional relationships which potentially drove the dissemination of amber in the northern part of the study area. The focal point in that respect is western Veneto which continued to be a major recipient of amber as in the previous phases, however in FBA became also an important centre of its processing. In relation to many regions, associations of amber beads adorned with the Tiryns- and the Allumiere-type motifs in their classic versions (var. a and c) with particular types of metal and glass objects allow to trace origins of those materials back to western Veneto, where their local manufacturing is well documented. Those links span the entire length of the Apennine Peninsula, from the Alps to southern Apulia and the Aeolian Islands. At the current state of research, there is enough evidence to support the contacts which contributed to the influx of amber from the Po Plain, across Romagna and Emilian Apennines, to Tuscany, Lazio and further south along the Tyrrhenian coast to Lipari. Less clear are the testimonies of interactions with the Adriatic side of Italy, particularly in relation to northern Apulia. Nonetheless, also here it seems plausible to suspect the transfer of amber from

Northern Italy. Even if these were not direct contacts, amber could reach Apulia via certain other regions of Central (Tuscany and Lazio) and Southern Italy (the Aeolian Islands) for which there are certain attestations. Still, it has to be investigated further: why some amber assemblages (like the one from Monte Saraceno in northern Apulia) differ to such an extent from remaining collections, despite good indications of contacts seen among the co-occurring finds? Is it because some links are still missing from the available evidence, or perhaps one has to presume a local manufacturing of amber jewellery (as in the case of the Scarceta settlement in Tuscany) resulting in slightly different forms? An interesting kind of relationships linked western Veneto with Trentino-Alto Adige; it could have involved southward inflow of processed (and maybe also raw) amber in stylistic typical for Central Europe – the *Urnfield* culture (e.g. elongated biconical beads), as well as northward redistribution of beads with the Allumiere-type decoration.

Clues testifying to amber-related interactions with Northern Italy can be found in Western Balkans. They involved import of Italian bronze and, presumably, amber products (e.g. Debeli Vrh in Slovenia, Garica in Kvarner), as well as adaptation of foreign types by local artisans, resulting in creation of specifically regional variants. In the case of bronze imports, typological indicators mainly point to Northern and Central Italy, less likely Southern Italy, and include typical FBA innovations i.a. in fibulae, but also comprise older – LBA types (e.g. sickles from Debeli Vrh). However, when it comes to amber, the most precise analogies for the Slovenian, Kvarnerian and northern Dalmatian finds were recorded in western Veneto, Romagna and Emilian Apennines, and also the Garda Lake basin if the overall design of amber pinheads (i.e. the one from Ljubljanica River deposit) known from LBA is considered. On the other hand, the flow of goods was not unidirectional; garment accessories, i.e. bracelets and pendants of Western Balkan provenience were identified in northern Apulia, i.a. on the cemetery of Monte Saraceno, whereas other artefacts, e.g. certain types of serpentine fibulae are possibly products of interregional contacts. A specific case of an object with a mixed background and vast distribution is the simple fibulae with two nodules on the arc, appearing in a number of provinces of amber's acquisition, yet often representing different varieties. Similarly, the Tiryns and Allumiere types of ornaments underwent local adaptation leading to creation of a number of variants. Thus, it can be said about amber-related interactions, especially in the northern part of the Adriatic, that they changed in the course of the Bronze Age from passive reception of ready-made products or preconceived forms to more proactive posture in setting the stylistics of amber jewellery. Unfortunately, a finer adjustment of chronology of the contexts with amber in Kvarner and northern Dalmatia is infeasible at present. However, it can be inferred from similarities with Italian amber finds that a true growth of the region as a centre of amber's acquisition took place in the advanced and late stages of FBA (HaA2–B1; ca. 1100–950 BC), so roughly contemporary to BF 2–3. From there, amber could be streamed inland, i.e. to Lika (possibly also participating in amber processing in the same fashion as Kvarner), and perhaps also Glasinac, if the Tiryns-decorated beads from the area are considered. The Kvarner Bay could have provided the latter two regions with amber as a gateway to the Adriatic. Still, imprecise chronology of the contexts and absence of other data linkable to amber from Glasinac and Lika prevents from making more concrete conclusions.

Discussed phase witnessed intensified inflow of amber to the north-west Italy (western Lombardy, Liguria and Piedmont). There are only hints at possible relations of the three regions with western Veneto, either derived from similarity of amber finds or presence of other objects having counterparts in the latter. As mentioned before, the bead decorated with oblique grooves (Allumiere var. m) found at Isolino Virginia in western Lombardy forms a neat link in presumed dispersion of such amber adornments into the Alps, specifically Switzerland. In Liguria combination of glass finds and bronze pins of the Torri d'Arcugnago type from Castellaro di Uscio seems to point at western Veneto, despite apparent similarity in amber. Finally, in Piedmont specifically perforated amber pendants, similar to those from north-central Lazio and western Veneto, were found in the hoard of Monte Cavanero. The hoard yielded several other finds, datable between LBA and FBA, which potentially have origins in the post-*Terramare* area, or even further east in Dalmatia (i.e. ribbed-cuff bracelets). It might be possible, as in the case of the Debeli Vrh hoard, that a part of the hoard was assembled in the lower Po Plain – a region maintaining contacts with the east Adriatic shore, before being completed with other materials of north-west Italian/south-east French provenience prior to its deposition at the start of the EIA. Question remains: where and when exactly amber beads from the hoard were collected? Their similarity to some Italian finds also indicates a link to western Veneto. On the top of that, contexts with amber in all three north-west Italian regions share the Zerba-type bracelets tying them to the *Protogolasecca* culture and a local network of contacts.

At the end of the Bronze Age in Western Balkans, a route down the east Adriatic coast came into prominence. It is evidenced by the amber and bronze finds from Trcela-Vranjic in middle Dalmatia and Perlat in the Mat Valley. Together with finds from Kvarner and northern Dalmatia, these artefacts build a picture of culturally interlinked areas close to the Adriatic littoral which could have channelled amber along the NW–SE axis. Dalmatia also benefitted from the contacts with the Balkan hinterland, although it seems unlikely that amber reached it that way. There still is the unsolved issue of the Tiryns-decorated beads from burial mounds at Vrlazje on the Glasinac plateau, for which Kvarner and northern Dalmatia, nonetheless, offers the closest analogies. Some of the finds from Trcela-Vranjic, such as the corrugated torc and spectacle pendants, can be related to the FBA culture of Bosnia and Herzegovina, i.a. Glasinac, while the Golinjevo-type fibulae (e.g. those from Trcela-Vranjic and Vrlazje) in different variants have a clearly confined distribution between the Upper Drina and the Adriatic coast. Still, there are no similarities between amber beads from the Glasinac and middle Dalmatia, hence only the suspected connection. Typologically related spectacle fibulae are the basis for a link between the Glasinac and Vojvodina; however, also here the absence of more clear indicators of contacts leaves suspected influx of amber from the latter area to the former in the sphere of hypotheses.

4. Material and symbolic status of amber in the Bronze Age societies of the circum-Adriatic zone

This chapter deals with issues revolving around material and symbolic status of amber in the eyes of the circum-Adriatic societies utilising it during the 2nd millennium BC. The term 'material status' is understood here as the totality of aspects related to the functionality and utility value of amber for the said communities. In turn, the 'symbolic status' explains the non-utilitarian features of amber which had significance in the sphere of religious beliefs and cultural values. Despite the apparent opposition, both sides of amber were entwined warranting the role and use of the substance (e.g. in rituals which utilised both the physical as well as supernatural properties of amber).

These issues are investigated, first of all, through analysis of the forms of amber's deposition in the study area. Second, the work attempts to disentangle some of the meaning of amber through determination of its social attribution, i.e. sex, age and social identity or status of its users.[643] Third, the issues are approached through comparison with other elements of material culture co-occurring with amber in the contexts. Finally, observations inferred from the abovementioned stages of analysis were subjected to discussion in the light of previous views on the status of amber for prehistoric communities.

4.1. Forms of amber's deposition

Basis for the following considerations is the assumption that different kinds of contexts in which amber was found possibly testify to various aspects of its material and symbolic status. In fact, the collected data provides information on several ways in which amber artefacts were buried in the ground. Three major types of deposition could be distinguished: (1) in residential contexts (settlements), (2) in sepulchral contexts (graves) and (3) in hoards (considered also as votive offerings).[644] The numbers of contexts representing them are not equal and vary considerably across the circum-Adriatic zone (Fig. 111). To present these quantitative differences, contexts

643 After M. Parker Pearson, 'social status' can be defined as 'formed out of three dimensions: one's political, kinship and gender entitlements, one's lifestyle (as a totality of cultural practices such as dress, speech, outlook, bodily dispositions and tastes that organize our perceptions of the social), and one's economic class (in the Marxist sense of people's relationship to the ownership or control of the means of production)' (Parker Pearson 1999, 83).

644 There is one site with amber for which the type of deposition could not be determined – Capitanata in northern Apulia. For that reason, it was left out from further considerations.

were grouped basing on their location in the five macro-regions, already used in the discussion on their chronology (section 2.5): Northern Italy, Central Italy, Southern Italy, Western Balkans and Central Balkans (Fig. 112). In this way, more general tendencies in the manners of amber deposition can be captured.[645]

Out of the total 84 findings of amber in settlements, the vast majority (56) were discovered in Northern Italy. In the remaining macro-regions, these numbers are significantly lower, respectively amounting to 15, 9, 3 and 1. The largest group of contexts is formed by graves (172 in total), and this category is usually dominating in all five macro-regions. The only difference is Central Italy, where there almost is twice fewer graves (8) than settlements (15) among contexts with amber. Also, in Northern Italy there notably is a balanced proportion of settlements (56) and graves (58), with a slight advantage of the latter. However, in Southern Italy, Western Balkans and Central Balkans, the graves visibly predominate over the remaining types of contexts (39, 25 and 42, respectively). The least represented kind of amber deposition is that in hoards and votive offerings. Most frequently amber has been recorded in such contexts in Northern Italy (5 times in total). Two hoards/votive offerings are known from Western Balkans, specifically from the present-day territory of Slovenia. Each of the remaining macro-regions yielded a single case of hoard/votive offering with amber.

Fig. 111. Cases of the three types of amber deposition in the study area measured by numbers.

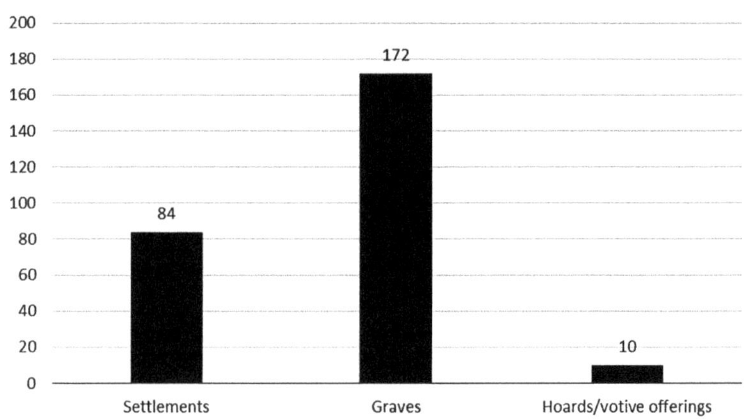

645 The exact numerical distribution of the three types of deposition can be deduced from the summaric tables of contexts with amber for individual regions (sections 2.2.1–2.3.13).

Fig. 112. Cases of the three types of amber deposition in the five macro-regions of the circum-Adriatic zone measured by numbers.

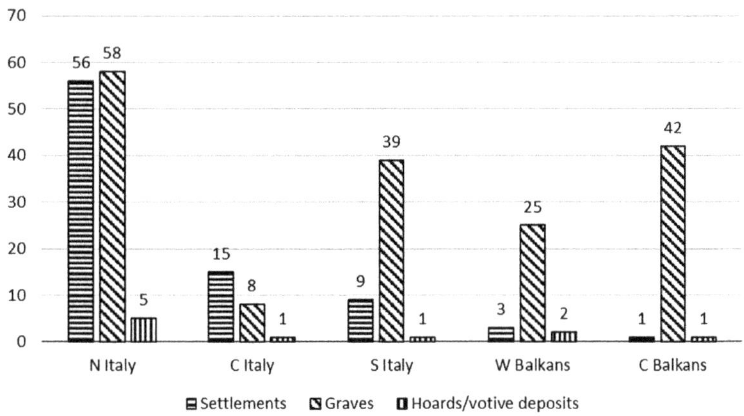

Each of the three modes of deposition is highly selective, meaning that there are all sorts of factors deciding why and how an object found its way into the ground, from a human decision-making to purely accidental reasons. Furthermore, deposition can be seen as a process during which different strategies of objects manipulation and treatment come to play, leaving them in various states, from apparently unchanged to completely altered. On the top of that, there are natural and anthropogenic post-depositional agents influencing preservation of the finds; thus, one can expect to find only a portion of the actual amount of amber used in the past. While deposition of amber in graves and hoards/votive offerings can be regarded as 'intentionally selective', in the case of settlements this process can be more complicated, involving both the 'intentional' as well as the 'unintentional'/'involuntary' side of human behaviour. This is merely a useful simplification, serving only to signalise basic differences in the discussed categories of archaeological record, but does not explain the peculiarities of each individual act of deposition. It is argued that one can, nonetheless, discern circumstances and motives behind some cases of amber's deposition in the study area by carefully examining available data relevant to the context and the finds themselves.

4.1.1. Settlements

Findings of amber from settlements contribute a large portion to the entire collection of finds. They potentially offer a great deal of information on amber's function and status in the circum-Adriatic communities of the Bronze Age. However, due to variety of activities that took place within the settlements, it is not easy to propose a coherent picture of what actually the deposits of amber on those sites stand for.

4.1.1.1. Theoretical considerations

By accepting their largely residential function, settlements more than anything else are supposed to reveal a domestic – 'daily' use of amber, as opposed to sites with presumed ritual or ceremonial function, e.g. cemeteries and hoards. However, the division into sacred and profane can be misleading, because both spheres of life most likely and used them as constituents and relays of identity, thus becoming 'inalienable' (cf. Mauss 1990; Weiner 1992). Through the cycles of their use, they acquire a life of their own (cf. Kopytoff 1986), and their biographies are closely tied to those of their owners, evolving in meaning from utilities to symbols, e.g. weapons and armours wielded by Homeric heroes as exponents of their heroic deeds (cf. Crielaard 2003). Burying these objects with their owner could have been a culturally accepted and ritually prescribed way of dealing with such possessions.

However, the Bronze Age grave inventories are often limited to few recurring types of artefacts, albeit differing in terms of morphology – typology, which hardly represent all spectrum of possible activities the deceased took on during his/her lifetime (Blischke 2002, 35). Thus, either we are dealing here with societies where possessions, especially dress and garments, were highly codified by normative tradition – this precisely is the opinion of K. Kristiansen who states that personal identities were being firmly institutionalised by the MBA in Northern Europe (Kristiansen 2012, 386), or the burials represent only a handful selection of items required or allowed by the burial rite. Furthermore, it is possible that grave goods were procured or made specifically for the occasion of the burial. Some of them reproduce symbols linked to the cults prevailing at the time (e.g. pins and pendants with solar motifs). The canonical nature and uniform combination of grave inventories, instead of presenting the entire, often multifaceted character, are more likely to show the image of a person's former 'self' relevant only 'in the face of death' from religious standpoint. Thus, the objects in a grave can be regarded as having specific, death-related field of meaning which is adjusted by conscious and planned actions by the participants of burial ceremony (Blischke 2002, 38).

The role of actors attending a funeral is highlighted in many modern analytical approaches to burials. Yet, most of them reflect a belief that grave goods are meaningful and had served a community to create an image of a person upon his/her death. Most likely it was a cultural idealisation rather than the true representation of an individual in life (Fontijn 2002, 218). It can be argued that the personal view of one's personhood/identity is not as important during the burial as much as the image of an individual preserved in the mind of a community. M. Parker Pearson stresses that status is not a static characteristic given for the lifetime 'to be reflected in mode of burial and associated grave goods but a panoply of practices which are historically situated and open to manipulation' (Parker Pearson 1999, 84). J. Brück goes even further to state that grave goods are representation of a relational construction of the deceased's 'self' – personhood which is built and dissembled through circulation and deposition of possessions contributing to enduring interpersonal ties (Brück 2006).

It may seem difficult to decide whether amber jewellery found in graves was a possession of the deceased or a gift from the mourners. However, I would argue that there is a way to work around this problem. The act of depositing an object beside body in grave creates a direct link between the two. It is based on the notion shared by a community organising a funeral that a given object or material (e.g. amber) 'fits' the identity or status of the buried person, and is 'right' according to the cultural rules, stemming from social realities or religious beliefs. It might be a possession important for the deceased, or a thing which was not his/her, but somehow reminds the mourners of the person. The choice of how a grave will be arranged and equipped lies predominantly within the competences of mourners; however, it is done according the certain rules and through collective negotiation, resulting in a burial that reflects the status of a deceased inherent in the collective consciousness. While the social structure to certain extent defined the roles and position of the groups, particular identities could be subtly accentuated through selection and manipulation of body and inventory, at the same time respecting religious norms and burial rite principles. As J. Sofaer notes, 'the body of a person lies in intimate contact with artefacts in a grave. The grave constructs and restricts, forcing the person and objects into association' (Sofaer 2006, 50). Once an object and a body are deposited in the grave, they are inextricably tied with a bond that will last forever, or as long as the grave will stay intact.[650]

For the aims of the following analysis, the question who possessed or who added a certain object to the burial can be regarded irrelevant. One can argue that in this way the considerations are limited merely to a simple overview of burial ritual and a place of amber within it. However as amber artefacts have predominantly been found in graves, it is a sign that this was a common practice and culturally constituted role of amber to be buried with certain types of people, and through decoding some of the characteristics of the deceased, we can determine social attribution of amber.

D. Fontijn concludes the analysis of Bronze Age grave deposits in the southern Netherlands by saying that although it is difficult to explain the meaning or a value of every bronze item found in the grave (e.g. gold ornaments, copper daggers), one can risk the statement that body ornaments and daggers served as the paraphernalia of a specific personal identity (Fontijn 2002, 218). Amber beads certainly qualify for the category of body ornaments and as such could be used to express deceased personhood and status. By being visually attractive objects, they lend themselves for display and for that reason it can be assumed that they were worn as jewellery. However, it would be too simplistic to treat amber beads or pendants as adornments serving just to add glamour or grandeur to the costume. Within the socially constituted system of signs and symbols, amber jewellery, just as any other element of material culture, could be used as a relay for communicating specific information.

Beads in various cultures are used to send signals directed at a particular audience who are capable of interpreting the meaning correctly because they share similar social traditions (Choyke/Bar-Yosef Mayer 2017, 15). For instance, in Western Ethiopia women of Mao community traditionally wear many necklaces made of thick amber beads which are crucial in their constitution as persons: necklaces are worn from a very early age, and they are only taken away in traumatic occasions, such as severe illnesses or mourning (González-Ruibal 2012, 75).

Dress is a non-verbal signal of a person's stage in life: infancy, childhood, adolescence, marriage, parenthood, old age, widowhood, death and, finally, burial or disposition of the dead body. Thus, personal ornaments and dress are an important unspoken system of visual communication, via symbols, concerning where and how a person 'fits in' to a social setting (Wright 2012, 423). Furthermore, objects are often being ascribed magical and healing properties which can be integrated into dress. Beads are widely used as amulets in connection with averting illness, aiding in pregnancy and other such practices (Wright 2012, 424; cf. Beck 1976). Amber beads are no different in that respect; their potential use in that capacity by prehistoric communities is extensively discussed in archaeological literature (e.g. Gestoso Singer 2016, 253–256).

650 One should consider here the cases of graves' rearrangement and removal of the burials or grave goods. This may be seen as deliberate alteration or deconstruction of a deceased afterlife image. See Blischke (2002, 39–50) on the deliberate re-openings and re-furnishing of the graves in the Carpathian Bronze Age.

Fig. 115. Fluctuations in the numbers of inhumed and cremated burials with amber in the circum-Adriatic zone during the Bronze Age.

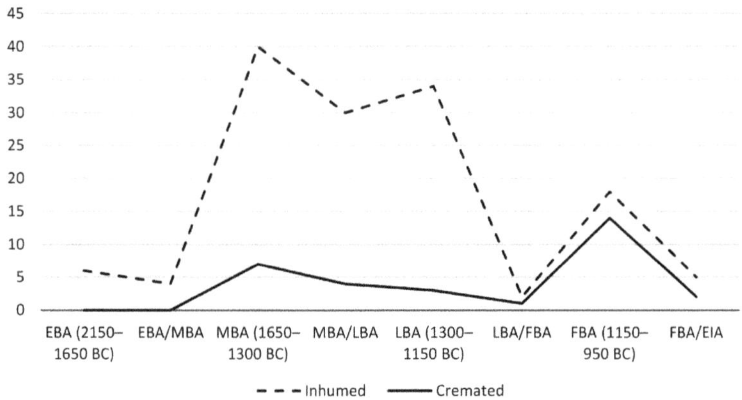

From the chronological perspective, the inhumed burials with amber were more numerous than those cremated in the study area in every stage of the Bronze Age (Fig. 115). Even during LBA and FBA, when cremation became the leading burial rite in much of Europe – the *Urnfield* period, amber was still more often deposited together with unburnt bodies. This observation can possibly be biased/distorted by the loss of finds due to incineration on the pyre. However, it is important to stress that the practice of burning corpses of the dead does not necessarily lead to combustion of all the grave goods, as there is a considerable portion of cremated burials in cases of which amber and other elements of inventory have been preserved. The funerary customs practised, e.g. in Podrinje and Western Serbia or north-central Lazio apparently involved adding the offerings to graves only after cremation of a body was complete. Conversely, the case of grave 115 at Bovolone, which yielded a scorched, but not completely burnt piece of amber (Angelini 2010a), suggests that in Veneto and Emilia during MBA–LBA amber was placed on the pyre along with other grave goods prior to the cremation and consequently was combusted in the process. This conclusion follows suit the previous observations on the general absence of grave goods in the *Terramare* culture urn graves. Fire traces on the few fragments still preserved imply that all elements of inventory that survived the cremation were subsequently separated from the bones and deposited away from the urns.[651] This custom, although practised on more strict terms in MBA than LBA when limited numbers and categories of grave goods started to be inserted into urns, can potentially explain absence of amber in cremated burials in the *Terramare* area and the quantitative discrepancy between deposits of amber in the settlement and the cemeteries of Emilia. While it could have been barred from such graves by religious

651 Cardarelli/Tirabassi 1997; De Marinis/Salzani 1997; Cardarelli 2014, 857–861; 2015, 175.

norms, it is more likely that as an easily combustible material, amber usually did not last through the cremation process, and the lump from Bovolone, grave 115 is a luckily preserved piece which escaped this fate. Thus, one can assume that in parts of the *Terramare* ecumene practicing cremation amber was more numerous than the statistics based on the current state of knowledge show. On the other hand, in western Veneto during FBA grave goods were commonly added to the burials after cremation of the body; thus, one can expect that the few burials with amber known from that period present a picture not far from the truth.

Differences in the burial rites and in the treatment of grave offerings show changing attitudes towards the deceased. Those shifts in funerary customs had an impact on the treatment of amber within the sphere of ritual. According to one mode of burial, body of the deceased was dressed in the posthumous outfit and placed unburnt in the grave, as way to preserve person's corporeal form and identity. In the second funerary tradition, body was cremated, but grave goods were nonetheless deposited in grave. The third tradition ordered combustion – destruction of dead man's body and all material exponents of one's status or personhood, and separation of their remains from the ashes. Without entering into discussion about the religious foundations of the rituals of inhumation and cremation, it should be noted that amber was present in both of them. Although its place in the funerary rituals was probably justified by religious and social customs, amber itself did not dictate the mode of burial – it was treated in the same way as other elements of grave inventories.

Similarly, there was no particular grave form reserved for the burials with amber. The overview of grave types recorded in the circum-Adriatic zone in relation to amber finds mentions almost every grave form used in Europe during the Bronze Age. Grave form was dependant on the social relations and religious norms; however, to some degree it must have been influenced by local geomorphology – e.g. it is difficult to imagine construction of underground hypogea in the wet and sandy Po Plain.

Fig. 116. Distribution of graves with amber in sector C of the Olmo di Nogara cemetery: black dashed sub-parallel lines – gullies of the Bronze Age road; dark grey line – BM 2 graves; black line – BM 3 graves; light grey lines – BR graves; black mirrors of Venus – adult females; white shields of Mars – adult males; black and white circles – unsexed adults; crosses – subadults; triangles – sex and age not determined; black squares – cremation graves; grey squares – sword-bearing males (after Vanzetti 2010, Fig. 6). Burials with amber are coloured red accordingly.

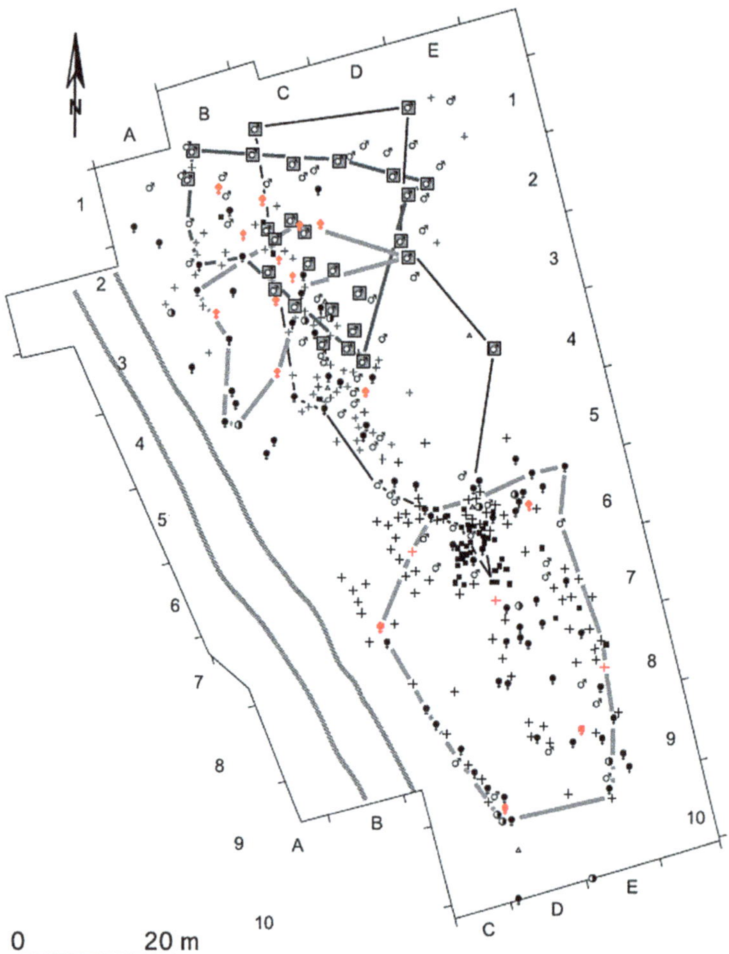

Fig. 117. Distribution of graves with amber in sector C1 of Madonna di Loreto, 'Hypogeum of Bronzes'. Hatched-grey ovals (grave inventories containing ornaments); hatched-grey ovals filled red (grave inventories containing amber beads); black triangles (bronze swords); grey triangles (bronze knives); white triangles (bronze daggers); grey arrows (bronze arrowheads) (after Peroni 1999b, Fig. 1).

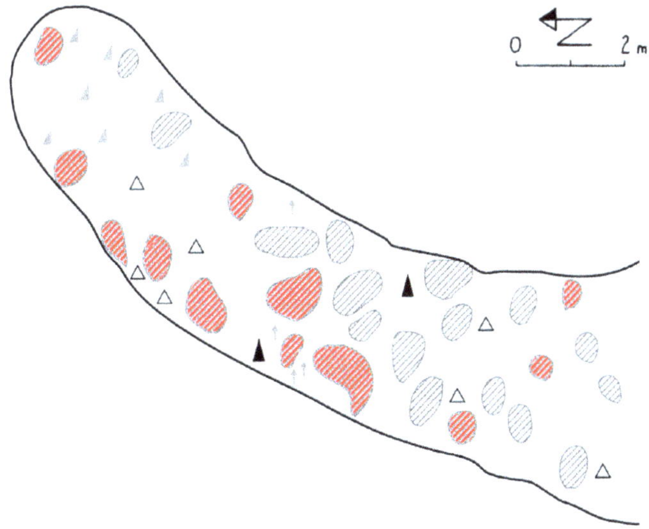

One can presume that if amber served as a denominator of specific social status, graves containing it could be spatially arranged in clusters or occupy isolated parts of a burial ground. Yet, none of the analysed cemeteries revealed any specific rules for situating graves with amber. In many cases they are too few to infer any meaningful distribution pattern from their location. Nevertheless, even with respect to larger necropolises with substantial numbers of burials equipped with amber, no discernible arrangements could be spotted. Main effort in the analysis of cemeteries was focused on two sites with well-defined spatial limits which yielded considerable numbers of burials with and without amber artefacts: Olmo di Nogara in western Veneto and 'Hypogeum of Bronzes' at Madonna di Loreto in northern Apulia. Both have been subject of numerous studies devoted to determination of social structure of the buried communities.[652] These considerations paid much attention to the internal structure of the necropolises, i.e. distribution of burials with specific inventories across space and time. As a result, scholars proposed several types of community organisation ranging in complexity from kinship-based, largely egalitarian societies with equal access to the burial ground and rites of deposition, to early stratified societies in the chiefdom model with a prominent role or warriors and their wives, for whom specific parts of the cemeteries were reserved. While these hypotheses will be commented in a greater

652 E.g. Peroni 1999b; Recchia 1999; Tunzi Sisto et al. 2003; Salzani 2005, 292–305; Cupitò/Leonardi 2005; Vanzetti 2010.

Fig. 119. Cases of coincidence with amber of various forms of bronze weapons and tools in the five macro-regions of the circum-Adriatic zone.

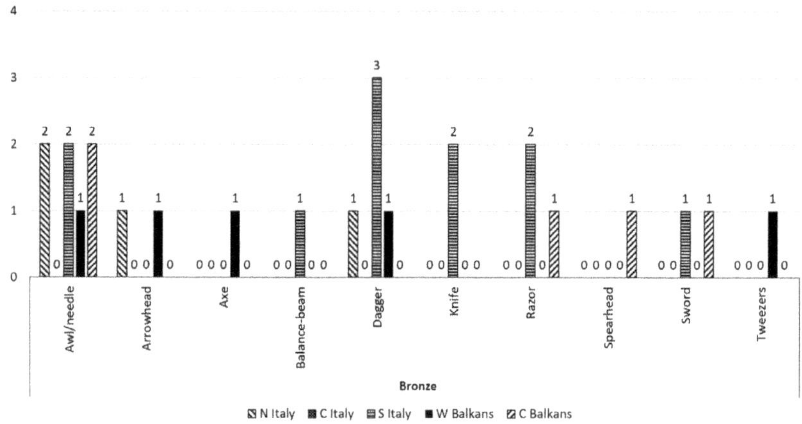

Bronze weapons and tools are far less common in relation to the graves with amber than the jewellery (Fig. 119). The only category that can be considered 'typical' for the sepulchral contexts analysed here are awls or needles which have been registered in all macro-regions, with the exception of Central Italy. Also daggers seem to be more frequent than the remaining forms of bronze armaments; however, they are largely limited to Southern Italy, as are knives and razors. Still, none of them exceeds the quantitative threshold of three contexts. Other kinds of weaponry and tools are restricted to singular cases of coincidence with amber per region.

Fig. 120. Cases of coincidence with amber of adornments and tools made of various organic and non-organic materials in the five macro-regions of the circum-Adriatic zone.

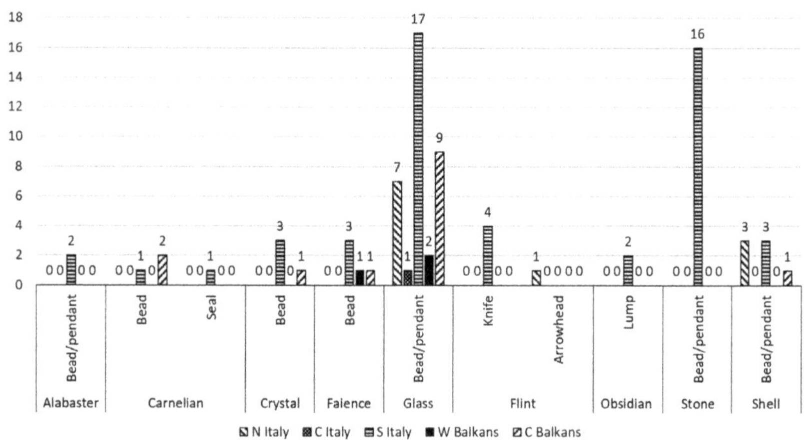

There is plenty of non-bronze adornments, mainly beads and pendants, known from the graves with amber in the circum-Adriatic zone (Fig. 120). Vast majority of them is made of glass which happens to be a particularly frequent element of the Southern Italian grave inventories, but appears quite often in the remaining macro-regions as well. Also faience beads have been spotted together with amber in graves on several occasions, however only in Southern Italy and the Balkans. Another material used for manufacturing pendants which were later deposited with amber in tombs are shells of marine molluscs (mainly *Cardiidae*, *Dentaliidae* and *Glycymerididae* families). Pierced shells were found chiefly in the Northern and Southern Italian contexts, but there is also one assemblage of such pendants coming from Central Balkans. Adornments produced of non-organic materials such as various kinds of rock and minerals have distribution limited exclusively to Southern Italy. Especially Sicilian cemeteries yielded rich and varied collection of beads and pendants made of alabaster, carnelian, crystal, sedimentary rock (stone) and, occasionally, lumps of obsidian. Furthermore, tombs of the early *Castelluccio* culture delivered sets of flint knives, whereas one EBA grave in Northern Italy (Sorbara, grave 54) delivered single flint arrowhead. Flint objects have not been registered in later sepulchral contexts with amber.

Fig. 121. Cases of coincidence with amber of adornments, tools and vessels made of various organic and non-organic materials in the five macro-regions of the circum-Adriatic zone.

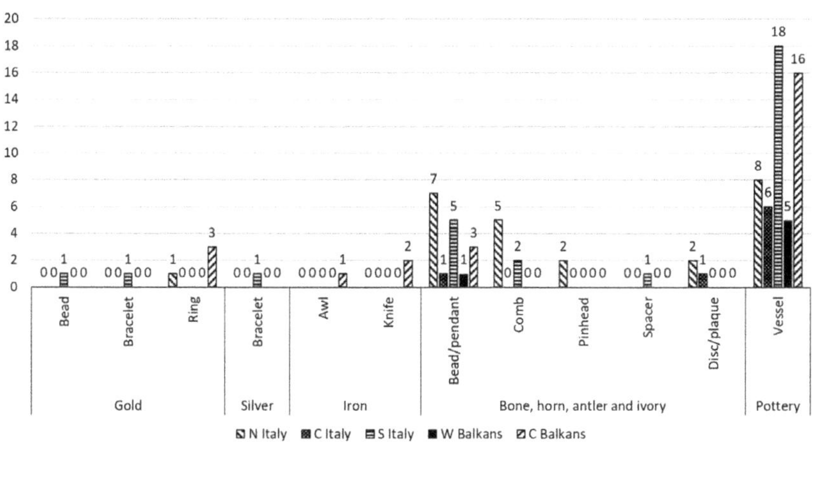

Apart from bronze, presence of items produced of three other metals has been documented in the circum-Adriatic graves with amber: gold, silver and iron (Fig. 121). They are very rare and limited to singular occurrences. Gold and silver bracelets are known solely from Southern Italy, as is a single gold bead. Gold rings, on the other hand, were found once in Northern Italy and thrice in Central Balkans. Irons tools, i.e. awl and two knives come from the Glasinac barrows which, however, are dated broadly to the turn of FBA and EIA. Much more frequent are adornments and tools made of animal cartilage tissue, i.e. bone, horn, antler and ivory. These are mostly beads which co-occur with amber in graves from all five macro-regions. In Northern and Southern Italy, one can witness the custom of adding combs to the burials. Other elements falling into this category are pinheads, spacers and discs/plaques.

Finally, one should highlight the presence of pottery in many graves with amber. Ceramic vessels, however, played a different role to the previously described items. In many cremated burials ceramic vessels were used as containers (urns) for ashes and bones remaining after incineration of bodies. This explains presence of vessels in Northern and Central Italy, as well as Central Balkans, where most of the cremated burials with amber have been recovered. Otherwise, pottery deposited in graves most probably contained food and drink fetched for the deceased's journey into the underworld, or as an offering to the deities. It is also likely that broken pieces of pottery remain after funerary feasts held to commemorate and bid farewell to the buried (Parker Pearson 1999, 10–11; Kristiansen/Larsson 2005, 245). This concept seems particularly plausible in relation to Southern Italy, where the rock-embedded tombs (hypogea) are often large enough to accommodate numerous burials, but also mourners whose gatherings might have a character of celebrations – hypothesis already proposed for some of them (e.g. Murgia Timone – Matarese 2018, 208–209).

Fig. 122. Results of correspondence analysis of variables symbolising different types of grave goods co-occurring with amber (1st and 2nd eigenvectors). Variables circumscribed with dashed circle symbolise items commonly included with amber in grave inventories. Variables with underlined labels symbolise items rarely occurring with amber in grave inventories.

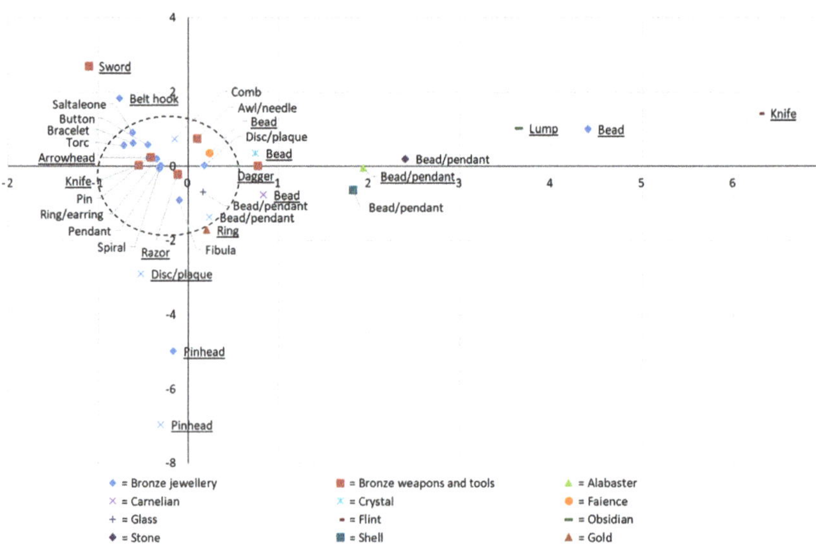

Above-described artefacts appeared with amber in various configurations. Combinations of grave goods can bring us closer to determining the status of people buried with amber. To measure associations between non-amber elements of inventory from the graves, it was decided to use correspondence analysis (CA). Results of the analysis were visualised on a plot against the 1st and the 2nd eigenvectors (Fig. 122), accounting for the highest percentage of the total decomposition of chi-square statistic, created using CAPCA 2.1 software (Madsen 2007).[654]

Distribution of variables on the chart has four-partite structure: it comprises the main concentration of features around the intersection of the axes (1st and 2nd eigenvectors), with smaller clusters branching out along the axes to the right, top and bottom. The main grouping of variables (circumscribed with dashed circles) mostly includes items commonly occurring with amber in graves (discovered in more than the arbitrarily adopted threshold of five contexts – cf. Figs. 118–121). Their close position with respect to the intersection of the eigenvectors – point 0 – means that they are characterised by low inertia, i.e. they are not bound by a specific pattern of correspondence. In reality, they were discovered in various configurations; however, relative proximity between certain variables indicates some of the more frequent associations: e.g. adjacency between bronze *saltaleoni*, bracelets and torcs shows a set commonly occurring with amber mostly in Central Balkans, while glass or bone beads beads/pendants and bronze fibulae form another combination frequent in the FBA graves in Northern and Southern Italy, as well as Western Balkans. Slightly more specific in terms of chronology and geography are bone combs, bronze awls/needles, discs/plaques and faience beads which appeared in fewer graves with amber, mostly those from Northern or Southern Italy dated to MBA–LBA. So the grave inventories differ to some extent across space and time (cf. Fig. 123), especially if one would take into consideration also typology of the objects; however, in their core they combine several classes of bronze ornaments – elements of dress.

Outside the main cluster are located objects rarely coinciding with amber in graves (registered in fewer than six contexts in total). Some of them still may be regarded as typical additions to the burials with amber, but only within the borders of certain regions and time frames of specific periods: e.g. mineral beads and flint knives in Sicily during EBA–MBA, or bronze and bone pinheads in western Veneto and north-central Lazio during FBA. Some of them potentially testify to a specific status of people buried with them; here the most likely indicator of status is sword – symbol of warriorhood which is typically positioned in contrast to more female-related objects not linked to the sphere of combat, e.g. jewellery. Yet, it has to be stressed that two burials in which amber was found beside bronze swords (Shtogj, tumulus 2, grave 9 and Madonna di Loreto, 'Hypogeum of Bronzes', burial of the 'Chief') also yielded elements of garment, i.e. bronze pin in the first case, and buttons, rings and belt hook in the second. Hence, the symbol of sword is located quite close to the main concentration of variables on the chart. Even inside the dashed circle, one can spot variables which are considered as armaments and tools or, otherwise, objects from the male domain: bronze arrowheads,

654 The category of pottery was omitted from the analysis for the reasons explained above. Parameters set for the correspondence measurement assumed the entry threshold of at least two incidences of variables in objects. This resulted in rejection of some of the objects and variables which did not meet this requirement.

daggers, knives and razors. These objects are in fact rare among analysed inventories, and most of them (all razors and arrowheads, and some of the daggers) come from mass graves where separation of grave goods into individual burial outfits was impossible. Still, the association between daggers or knives and amber is undeniable in at least three cases of tombs with singular interments (Olmo di Nogara grave 411, Madonna del Piano, graves 46 and 211), although it has to be mentioned that these were rather small-sized variants.

In the end it can be concluded that amber, by large, was deposited as a grave good with items lacking connotations with the sphere of war/combat (weapons) or specific activities related to craft/production (tools). It was found most often in combination with items considered, on a functional level, as elements of dress – jewellery. This observation matches the commonly assumed role of amber beads and pendants as personal adornments – additions to the costume. However, on a semantic level, amber artefacts could have played a much more deep and complex role, as argued in the theoretical foreword to this chapter (cf. section 4.1.2.1). It can be posited that there was a particular norm, stemming from perceived meaning of amber which barred it from being buried with weaponry/tools in some cultures, or allowed such deposition only in specific cases in other communities. Perhaps the circum-Adriatic communities of the Bronze Age saw amber as unfitting to the warrior identity, but compatible with other, non-militant social roles, involving also beads and pendants of various materials as similar relays of personal status. Supplementing above considerations with the information about sex, age and social status of the people buried with amber can help to better address this issue.

4.1.2.4. Social attribution: sex, age, identity and status

Only a portion of the rich collection of burials with amber has been a subject of anthropological analysis. Among the collected data, there is a much greater number of sex determinations relying on the traditional division into gender-related categories of objects included in grave inventories (i.e. weapons to men, jewellery to women). Although such approach may sound archaic, relevancy of this intuitive categorisation often finds confirmation in the light of anthropological analyses. Nevertheless, it was decided to focus on the cases for which anthropological expertise has actually been done.

Observations on sex and age of the skeletal remains of persons buried with amber are available in relation to 24 cemeteries (Tab. 48). They differ in terms of detail and some of them might require verification due to old date of the analysis. Presence in the dataset of numerous mass graves means that the bones were usually described collectively, without specifying which skeleton revealed particular features. Often, due to severe mixing of the human remains coming from a grave or cemetery, it was possible only to generally characterise the entire buried population. On the other hand, newer researches offer better resolution of the anthropological analyses, not only allowing for detailed characterisation of each burial in terms of sex and age but sometimes also extended by palaeopathological studies. These more comprehensive and detailed analyses facilitate determination of the social attribution of amber in graves. Results of anthropological studies given in Tab. 48 are supplemented with abridged descriptions of amber objects and other artefacts found in the analysed graves.

Tab. 48. Abridged description of results of anthropological analyses and grave goods related to the Bronze Age burials with amber in the circum-Adriatic zone

Name	Region	Phase	Anthropological description	Amber artefacts	Other artefacts	References
Gallo di Briatico, tomb	Calabria	MBA (1650–1300 BC)	Female individual (?)	1 bead	1 bronze awl, 1 silver bracelet, 1 crystal bead, 1 stone bead, 1 carnelian seal, pottery	Pacciarelli 2001
Sorbara, tomb 54	Garda Lake basin	EBA (2150–1650 BC)	Adult individual	2 beads	1 flint arrowhead, pottery	Baioni 2000; De Marinis/Valzolgher 2013
Krmedski Novi Grad, grave 1	Istria	MBA (1650–1300 BC)	Two adults and an infant	1 bead	2 bronze earrings	Bačić 1957; Palavestra 1993
Laganiši, sepulchral cave	Istria	LBA (1300–1150 BC)	Twelve individuals: three adult females and one adult man, and eight infants under 10 y.o.	2 beads	1 bronze arrowhead, 1 axe, 1 bracelet, 1 dagger, 1 ring, 6 saltaleoni	Mihovilić 2008
Mušego, tumulus 6	Istria	MBA/LBA	At least three adult individuals (including one male) and an infant.	6 beads	none	Mihovilić et al. 2009, 2011
Mušego, tumulus 7	Istria	MBA/LBA	Between four and six individuals: two males and a female, all aged 30–40 y.o., and between one and three infants.	14 beads	1 bronze ring, 4 saltaleoni, 3 spirals	Mihovilić et al. 2009, 2011
Vrčin, cemetery	Istria	MBA/LBA	Among remains of deceased interred in over a dozen of graves, presence of 11 males, 11 females and 12 children was ascertained.	77 beads	2 bronze bracelets, 1 pendant, 55 rings, 19 saltaleoni	Buršić-Matijašić 1989; Palavestra 1993

(Continued)

Tab. 48. Continued

Name	Region	Phase	Anthropological description	Amber artefacts	Other artefacts	References
Žamnjak, tumulus	Istria	MBA (1650–1300 BC)	Three individuals (two adults and an infant).	2 beads	2 bronze saltaleoni	Bačić 1960; Palavestra 1993
Boka-Prčevo, tumulus XIV, grave 19	Kosovo and Metohija	MBA/LBA	One adult individual and an infant	40 beads	2 bronze bracelet, 2 saltaleoni, 1 belt hook, 3 buttons	Tasić 1979
Iglarevo, grave 1	Kosovo and Metohija	MBA (1650–1300 BC)	Female individual	160 beads and pendants	2 bronze anklets, 3 bracelets, 1 pin, 2 saltaleoni, 1 glass bead, 1 crystal bead	Palavestra 1993, 1997; Bunguri 2006
Iglarevo, grave 4	Kosovo and Metohija	MBA (1650–1300 BC)	Infant	1 bead	none	Palavestra 1993, 1997
Trcela kod Vranjica, grave	Middle Dalmatia	FBA (1150–950 BC)	Three individuals: two adults and an infant	3 beads	8 bronze bracelets, 2 fibulae, 6 pendants, 18 rings, 1 torc, 5 glass beads	Marović 1960; Palavestra 1993
Velika Gruda, grave 9	Montenegro and Southern Dalmatia	MBA/LBA	Seven individuals: one adult aged above 40 y.o., one juvenile aged between 12 and 17 y.o. and 5 other adults.	2 beads	3 bronze buttons, 2 pendants, 5 saltaleoni, 1 spiral	Della Casa 1996a
Monte Saraceno, tomb 84/IV	Northern Apulia and Basilicata	FBA (1150–950 BC)	Eight individuals: two adult males aged 20–30 y.o., one mature female; one infant aged 1–2 y.o., one infant aged 4–6 y.o., one juvenile aged 10–15 y.o. and two other subadults.	2 beads	3 bronze fibulae, 3 pins, 7 pendants, 1 ring, 29 buttons, 3 spirals, 1 razor, 1 disc, 4 glass beads, 1 stone bead	Nava/Preite 1995
Monte Saraceno, tomb 43c/IV	Northern Apulia and Basilicata	FBA (1150–950 BC)	Six individuals: 1 adult male, 1 adult female, 1 adult n.d., 1 perinatal subadult and 2 infants aged 1–3 y.o.	1 bead	2 bronze spirals, 1 pin, 1 pendant, 6 buttons, 1 razor, 3 glass beads, 2 bronze fibulae	Preite 2001

Name	Region	Phase	Anthropological description	Amber artefacts	Other artefacts	References
Toppo Daguzzo, tomb 3, burial 3	Northern Apulia and Basilicata	MBA (1650–1300 BC)	Female individual	1 bead	1 glass bead	Cipolloni Sampò 1986
Madonna di Loreto, 'Hypogeum of Bronzes'	Northern Apulia and Basilicata	MBA/LBA	Burial of the *Capo* ('Boss') – adult male individual discovered in the sector C1.	2 beads	16 bronze buttons, 3 rings, 1 sword, 1 belt hook	Tunzi Sisto 1999c
Madonna di Loreto, 'Hypogeum of Bronzes'	Northern Apulia and Basilicata	MBA/LBA	Burial of the *Signora delle ambre* ('Lady of ambers') – adult female individual discovered in the sector C1.	5+ beads	1 faience bead, 1 glass bead, 2 bronze saltaleoni, 3 rings and reels, 1 pin, 1 pendant	Tunzi Sisto 1999c
Madonna di Loreto, 'Hypogeum of Bronzes'	Northern Apulia and Basilicata	MBA/LBA	Burial of a female individual in the sector C2.	2 beads	2 glass beads, 1 saltaleoni, 1 ring, 1 pin	Tunzi Sisto 1999c
Madonna di Loreto, 'Hypogeum of Bronzes'	Northern Apulia and Basilicata	MBA/LBA	Burial of a female individual in the corridor A.	1 bead	3 bronze pendants, 21 glass beads, 1 ring, 1 pendant	Tunzi Sisto 1999c
Madonna di Loreto, 'Hypogeum of Bronzes'	Northern Apulia and Basilicata	MBA/LBA	Burial of an 'Unarmed male individual'	1 bead	Bronze rings, buttons and pendants	Tunzi Sisto 1999c
Paulje, tumulus A	Podrinje and Western Serbia	LBA (1300–1150 BC)	Two individuals: adult male aged about 40 y.o. and an infant aged about 10 y.o.	1 bead	7 bronze bracelets, 6 buttons, 6 earrings, 2 pins, 7 saltaleoni	Bulatović et al. 2017

(Continued)

Tab. 48. Continued

Name	Region	Phase	Anthropological description	Amber artefacts	Other artefacts	References
Paulje, tumulus K	Podrinje and Western Serbia	MBA (1650–1300 BC)	Infant 5–10 y.o.	6 beads	6 bronze bracelets, 15 pendants, 2 pins, 1 torc	Bulatović et al. 2017
Vranjani, tumulus 1	Podrinje and Western Serbia	MBA (1650–1300 BC)	Female individual	1 bead	2 bronze bracelets, 2 buttons, 5 pendants, 2 saltaleoni, 1 torc	Zotović 1985; Palavestra 1993
Castelluccio, tomb 9	Sicily	EBA (2150–1650 BC)	Skeletal remains of at least 28 individuals, including 1 infant.	4 beads	16 flint knives, 1 obsidian lump, 3 stone pendant, 1 bronze bead, pottery	Orsi 1892
Bisceglie, dolmen La Chianca	Southern Apulia and south-east Basilicata	MBA (1650–1300 BC)	Skeletal remains of 13 individuals including one adult female and child (boy in the age of 10–12 y.o.).	5 beads	1 bronze plaque	Cataldo 1995
Popadin Dol, grave 1	Vardar Valley and Pelagonia	FBA (1150–950 BC)	Adult individual	2 beads	8 bronze bracelets, 2 fibulae, 1 ring, 1 spiral, 1 glass bead	Kitanovski 1960; Palavestra 1993
Dimov Grob, grave 54	Vardar Valley and Pelagonia	LBA (1300–1150 BC)	Mature female individual (ca. 40 y.o.)	1 bead	2 glass beads, 2 gold earrings, 2 bronze pins, 14 bronze pendants, 25 shell pendants, 2 bone beads, pottery	Videski 2004; Papazovska 2017
Dimov Grob, grave 73	Vardar Valley and Pelagonia	LBA (1300–1150 BC)	Mature female individual (40–45 y.o.)	2 beads	3 glass beads, 2 bone beads, 2 bronze earrings, 2 bronze pins, pottery	Videski 2004; Papazovska 2017

Name	Region	Phase	Anthropological description	Amber artefacts	Other artefacts	References
Dimov Grob, grave 120	Vardar Valley and Pelagonia	LBA (1300–1150 BC)	Young female individual (ca. 10 y.o.)	1 bead	2 faience beads, 2 glass beads, 2 bone beads, 2 bronze pendants, 1 bronze earring	Videski 2004; Papazovska 2017
Franzine Nuove, grave 12	Western Veneto	LBA (1300–1150 BC)	Female individual	2 beads	2 bronze pins, 2 bronze earrings	Aspes 1973; Corrain et al. 1984; David-Elbiali 2010
Franzine Nuove, grave 123	Western Veneto	MBA/LBA	Infant	1 bead	1 bronze earring, 1 gold ring	Aspes 1973; Corrain et al. 1984; David-Elbiali 2010
Franzine Nuove, grave 159	Western Veneto	LBA (1300–1150 BC)	Female individual	8 beads	2 bronze pins	Aspes 1973; Corrain et al. 1984; David-Elbiali 2010
Franzine Nuove, grave 523	Western Veneto	LBA (1300–1150 BC)	Adult female individual (25–40 y.o.)	3 beads	2 bronze pins, 1 glass bead	Aspes 1973; Corrain et al. 1984; David-Elbiali 2010
Narde I, grave 169	Western Veneto	FBA (1150–950 BC)	Adult female individual	4 beads	1 bone pinhead, 2 bronze fibulae, pottery	Salzani 1990–1991b; Cardarelli et al. 2015
Narde I, grave 281	Western Veneto	FBA (1150–950 BC)	Infant (1–3 y.o.)	1 bead	1 bronze fibula, 1 bone pendant, 15 bone beads, 11 glass beads, pottery	Salzani 1990–1991b; Cardarelli et al. 2015
Narde II, grave 75	Western Veneto	FBA (1150–950 BC)	Mature male individual (30–50 y.o.)	24 beads and pendants	1 horn pinhead, 1 bronze pinhead, 12 bone beads, 6 glass beads, pottery	Colonna et al. 2010; Cardarelli et al. 2015
Olmo di Nogara, grave 18	Western Veneto	LBA (1300–1150 BC)	Female adult (25–40 y.o.)	2 beads	2 bronze pins	Salzani 2005; David-Elbiali 2010; Canci et al. 2015

(Continued)

Tab. 48. Continued

Name	Region	Phase	Anthropological description	Amber artefacts	Other artefacts	References
Olmo di Nogara, grave 388	Western Veneto	MBA (1650–1300 BC)	Female infant (7–12 y.o.)	2 beads	2 horn combs, 3 bronze pins, 1 ring, 1 saltaleoni, 1 pendant	Salzani 2005; David-Elbiali 2010; Canci et al. 2015
Olmo di Nogara, grave 390	Western Veneto	MBA (1650–1300 BC)	Senile female individual (>60 y.o.)	2 beads	3 bronze pins, 2 saltaleoni,	Salzani 2005; David-Elbiali 2010; Canci et al. 2015
Olmo di Nogara, grave 411	Western Veneto	MBA (1650–1300 BC)	Probably male individual – advanced adult with features of senility (Salzani 2005, 59), or senile female (>60 y.o. – David-Elbiali 2010, 241, Tab. VI), or senile female (50–60 y.o. – Canci et al. 2015, 330).	2 beads	pottery, 2 bronze rings, 3 pins, 1 dagger, 1 saltaleoni, 1 horn comb	Salzani 2005; David-Elbiali 2010; Canci et al. 2015
Olmo di Nogara, grave 422	Western Veneto	MBA (1650–1300 BC)	Female individual deceased in adult (35–40 y.o. – Salzani 2005, 65), or mature (40–60 y.o. – David-Elbiali 2010, 241, Tab. VI), or senile age (50–60 y.o. – Canci et al. 2015, 329).	1 bead	2 bronze pins	Salzani 2005; David-Elbiali 2010; Canci et al. 2015
Olmo di Nogara, grave 435	Western Veneto	MBA (1650–1300 BC)	Mature female individual, at least 50 years old (40–60 y.o.)	3 beads	3 bronze pins	Salzani 2005; David-Elbiali 2010; Canci et al. 2015
Olmo di Nogara, grave 441	Western Veneto	LBA (1300–1150 BC)	Adult female individual (25–40 y.o.)	2 beads	1 horn comb	Salzani 2005; David-Elbiali 2010; Canci et al. 2015

Name	Region	Phase	Anthropological description	Amber artefacts	Other artefacts	References
Olmo di Nogara, grave 493	Western Veneto	MBA (1650–1300 BC)	Juvenile female individual (18–25 y.o.)	2 beads	3 bronze pins, 1 ring	Salzani 2005; David-Elbiali 2010; Canci et al. 2015
Olmo di Nogara, grave 30	Western Veneto	LBA (1300–1150 BC)	Adult individual – Salzani 2005, 124, or adult female (25–40 y.o. – David-Elbiali 2010, Tab. VI), or juvenile female (18–25 y.o. – Canci et al. 2015, 332).	1 bead	2 bronze pins, 1 ring	Salzani 2005; David-Elbiali 2010; Canci et al. 2015
Olmo di Nogara, grave 32	Western Veneto	MBA (1650–1300 BC)	Adult female (40–50 y.o. – Salzani 2005, 125), or mature (40–60 y.o. – David-Elbiali 2010, Tab. VI), or young adult (25–35 y.o. – Canci et al. 2015, 334).	2 beads	3 bronze pins	Salzani 2005; David-Elbiali 2010; Canci et al. 2015
Olmo di Nogara, grave 85	Western Veneto	MBA (1650–1300 BC)	Senile individual with male features (Salzani 2005, 150), or senile female individual (>50 or >60 y.o. – David-Elbiali 2010, Tab. VI; Canci et al. 2015, 334).	2 beads	3 bronze pins, 1 horn comb	Salzani 2005; David-Elbiali 2010; Canci et al. 2015
Olmo di Nogara, grave 89	Western Veneto	MBA (1650–1300 BC)	Mature female individual (40–60 y.o.)	2 beads	3 bronze pins	Salzani 2005; David-Elbiali 2010; Canci et al. 2015
Olmo di Nogara, grave 100	Western Veneto	MBA (1650–1300 BC)	Senile individual with male features (Salzani 2005, 159), or probably adult female (David-Elbiali 2010, Tab. VI), or mature female (35–50 y.o. – Canci et al. 2015, 330).	2 beads	3 bronze pins	Salzani 2005; David-Elbiali 2010; Canci et al. 2015

(Continued)

Tab. 48. Continued

Name	Region	Phase	Anthropological description	Amber artefacts	Other artefacts	References
Olmo di Nogara, grave 112	Western Veneto	MBA (1650–1300 BC)	Adult female individual (ca. 40 y.o. – Salzani 2005, 164), or mature female (40–60 y.o. – David-Elbiali 2010, Tab. VI), or young adult female (20–30 y.o. – Canci et al. 2015, 329).	2 beads	3 bronze pins, 1 horn pinhead	Salzani 2005; David-Elbiali 2010; Canci et al. 2015
Olmo di Nogara, grave 122	Western Veneto	LBA (1300–1150 BC)	Young adult female (18–25 y.o.)	2 beads and 1 spacer	1 bronze pin	Salzani 2005; David-Elbiali 2010; Canci et al. 2015
Olmo di Nogara, grave 154	Western Veneto	MBA (1650–1300 BC)	Young adult female (18–25 y.o.)	4 beads	3 bronze pins	Salzani 2005; David-Elbiali 2010; Canci et al. 2015
Olmo di Nogara, grave 155	Western Veneto	MBA (1650–1300 BC)	Young adult female (18–30 y.o.)	1 bead	2 bronze pins	Salzani 2005; David-Elbiali 2010; Canci et al. 2015
Olmo di Nogara, grave 172	Western Veneto	MBA (1650–1300 BC)	Mature female (40–60 y.o. – David-Elbiali 2010, Tab. VI), or adult female (25–35 y.o. – Canci et al. 2015, 329).	2 beads	3 bronze pins, 1 bone button	Salzani 2005; David-Elbiali 2010; Canci et al. 2015
Olmo di Nogara, grave 190	Western Veneto	LBA (1300–1150 BC)	Mature female (40–60 y.o. – David-Elbiali 2010, Tab. VI), or adult female (25–35 y.o. – Canci et al. 2015, 332).	9 beads	2 bronze pins	Salzani 2005; David-Elbiali 2010; Canci et al. 2015
Olmo di Nogara, grave 208	Western Veneto	LBA (1300–1150 BC)	Senile female (>60 y.o. – David-Elbiali 2010, Tab. VI; 50–60 y.o. – Canci et al. 2015, 332).	1 bead	1 bronze pin, 1 saltaleoni	Salzani 2005; David-Elbiali 2010; Canci et al. 2015

Name	Region	Phase	Anthropological description	Amber artefacts	Other artefacts	References
Olmo di Nogara, grave 243	Western Veneto	LBA (1300–1150 BC)	Infant (0–6 y.o.)	5 beads	2 bronze pendants, 2 pins, 1 needle, 1 button, 5 saltaleoni, 1 horn comb	Salzani 2005; David-Elbiali 2010; Canci et al. 2015
Olmo di Nogara, grave 252	Western Veneto	LBA (1300–1150 BC)	Mature female (40–60 y.o. – David-Elbiali 2010, Tab. VI), or adult female (30–40 y.o. – Canci et al. 2015, 332).	1 bead	2 bronze pins	Salzani 2005; David-Elbiali 2010; Canci et al. 2015
Olmo di Nogara, grave 307	Western Veneto	LBA (1300–1150 BC)	Infant (0–6 y.o.)	1 bead	2 bronze pins, 1 pendant, 1 bone bead	Salzani 2005; David-Elbiali 2010; Canci et al. 2015
Olmo di Nogara, grave 353	Western Veneto	LBA (1300–1150 BC)	Mature female (40–60 y.o. – David-Elbiali 2010, Tab. VI), or senile female (>50 y.o. – Canci et al. 2015, 332).	1 bead	3 bronze pins, 2 rings	Salzani 2005; David-Elbiali 2010; Canci et al. 2015
Olmo di Nogara, grave 358	Western Veneto	LBA (1300–1150 BC)	Adult female (25–40 y.o.)	2 beads	2 bronze pins	Salzani 2005; David-Elbiali 2010; Canci et al. 2015
Olmo di Nogara, grave 365	Western Veneto	LBA (1300–1150 BC)	Infant (0–6 y.o.)	4 beads	1 bronze ring, 1 pendant	Salzani 2005; David-Elbiali 2010; Canci et al. 2015
Desmontà, grave 437	Western Veneto	FBA (1150–950 BC)	Female adult individual (20–40 y.o.).	6 beads and pendants	8 bronze fibulae, 2 glass beads	Salzani 2013

The primary observation stemming from the anthropological overview of people buried with amber is the domination of female and, to a lesser extent, children among the skeletal remains. There are only three cases of men that were evidently given amber as a part of grave inventory: Narde II, grave 75 (Colonna et al. 2010, 211), Madonna di Loreto, 'Hypogeum of Bronzes', burials of the 'Chief' (it. *Capo*) in the sector C1 and the 'Unarmed male individual' (Tunzi Sisto 1999c, 212).[655] Aside from the three instances listed above, skeletons of males have also been registered in graves with multiple burials, with respect to which there is no chance of ascribing amber to an individual. Publications of these collective tombs are often limited to the estimated number of buried individuals, while leaving sex of the adults undefined. Nevertheless, in many such contexts, it was possible to identify the remains of both women and men. Furthermore, it is worth noting that the body of at least one infant has been identified in majority of them. On the other hand, sex and age of the people buried with amber is quite clear with respect to graves with single interments. They include almost indifferently burials of women in various ages, and infants. There are only few single-interment graves in which the skeletons lack determination of sex in addition to age (Sorbara, tomb 54 and Popadin Dol, grave 1). Thus, one can assume with considerable dose of probability that amber in the mass graves was attached to female and infant burials (girls?) as well.

With reference to collective burials, it is worth noting that representatives of both sexes were usually found together and it was only once that a child burial was accompanied by a single male skeleton (Paulje, tumulus A). Apart from that, there is also the context of Boka-Prčevo, tumulus XIV, grave 19, where an infant was found in company of a single adult, although sex of the latter individual remains undetermined. In several Balkan graves, there is a recurring configuration of burials with two adults and a child (Krmedski Novi Grad, grave 1, Trcela kod Vranjica, grave, Žamnjak, tumulus), or 3–4 adults and several children (Laganiši, sepulchral cave, Mušego, tumulus 6, Mušego, tumulus 7). These observations suggest that we are probably dealing here with family tombs, although confirmation of such hypothesis would require biological determination of the degree of kinship between the burials. Existence of family ties between the burials has already been proposed for some of the Southern Italian tombs (hypogea) with multiple burials, e.g. Toppo Daguzzo, tomb 3, inside which one female individual equipped with a single amber bead (burial no. 3) was accompanied by a group of 10 other individuals (6 men, 3 women and 1 child) which altogether can be considered as members of a lineage (Cipolloni Sampò 1986, 5; cf. Recchia 1999). Similar interpretation was proposed with respect to Monte Saraceno, tomb 84/IV, comprising burials of eight individuals of both sexes and in various age (Pacciarelli 2012, 223).

655 Formerly skeletons from three graves with amber at the cemetery of Olmo di Nogara (no. 85, 100 and 411) were considered as males, although in all these cases the sex markers were not unambiguous (cf. Salzani 2005, 59, 150, 159, 427). Recent analyses reviewed those determinations, concluding that they all belonged to female individuals, rather than male ones (David-Elbiali 2010, Tab. VI; Canci et al. 2015, 330, 334).

While there clearly seems to be some regularity in sepulchral deposition of amber according to the sex of the deceased, age appears to have played no part as a criterion. Amber artefacts occur with all age classes, from the young infants (*Infans I* – ca. 0–7 years of age) up to senile persons (*Senilis* – above 55 years of age). Moreover, there is no apparent difference in the inventories of burials belonging to infants or women in particular age classes. They all possess sets comprising mostly between 1 and 10 amber beads or pendants. There are several burials of women more lavishly outfitted with amber than the others. Here, one can mention grave 1 at the Iglarevo I cemetery which yielded astonishing assemblage of 160 amber beads and pendants, or Boka-Prčevo, tumulus XIV, grave 19 containing 40 beads, both located in Kosovo and Metohija region. Higher level of differentiation between the burials is evident on the basis of the remaining, non-amber categories of grave goods. As shown by the CA (Figs. 122 and 123), grave inventories are largely similar across the circum-Adriatic zone and mainly comprise various elements of bronze jewellery, sometimes supplemented with adornments made of other materials (e.g. glass or bone). There are, however, certain categories of items, i.e. weapons and tools, that rarely appear in association with amber, while products of particular materials (e.g. obsidian or gold) have limited spatial occurrence, contributing to a regional character of grave inventories. Earlier recognised combinations of grave goods (Fig. 122) can now be correlated with the data related to the anthropological analyses (Fig. 123) as a starting point for the discussion on identity and status of the people buried with amber.

The plot resulting from the CA of objects (Fig. 123) shows that the vast majority of graves falls into the area around the intersection of the eigenvectors, predominantly on the left side of the 2nd eigenvector (vertical axis), coinciding with the main cluster of variables symbolising items regularly co-occurring with amber (Fig. 122). The concentration of objects in this place includes contexts from all regions of amber's acquisition. Here are also located many female and infant burials listed in Tab. 48. As noticed above, sex and age of the deceased were not interdependent with the amount of amber deposited in graves. The CA shows that females and infants possessed largely compatible inventories, involving same categories of jewellery. Yet, some graves possess richer assemblages than the other, i.e. they include larger number of amber and other items, sometimes combining them with products made of other rare ('exotic') materials. These differences potentially stem from various social statuses of the people buried in those graves, and often prompted scholars to consider the most outstanding burial outfits as markers of great prestige and prominent social position of their owners.

Fig. 123. Results of correspondence analysis of object symbolising graves with amber divided into regions (1st and 2nd eigenvectors). Burials discussed in the text are labelled.

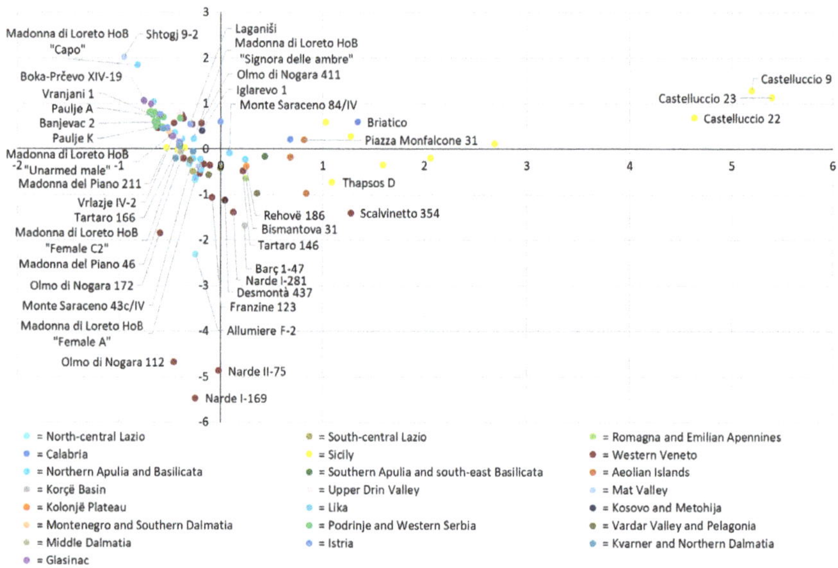

Quantitative discrepancies with respect to amber artefacts are most obvious when looking at the contexts, such as Iglarevo, grave 1, Banjevac, tumulus 2, Boka-Prčevo, tumulus XIV, grave 19, Castello del Tartaro, grave 166, Allumiere – Poggio La Pozza, grave 2, trench F (1960), Franzine Nuove, grave 481 (in descending order). In all of them, amber beads most probably formed sumptuous necklaces, although it is not always certain if they were actually worn by the deceased at the time of burial, or were inserted into the grave in the final stage of funeral (e.g. as in the case of cremated burials, where amber was added after cremation of the body). Interestingly, those graves do not show any striking qualitative differences in composition of the remaining grave goods – they include the most common elements of jewellery in the dataset, placing them on the chart among other graves poorer in amber (except for Allumiere – Poggio La Pozza, grave 2, trench F). Furthermore, with respect to the three richest graves among those above listed, it should be mentioned that amber necklaces found there were largely made of crudely cut and rudimentarily worked beads, requiring fewer time to process than some of the artefacts in other tombs. One can, thus, pose a question whether the number of amber artefacts itself is a good indicator of wealth and prestige.

Presence of amber, even in much smaller numbers than in the cases mentioned above, has often been considered as a marker of prominent social standing. One of the most often cited examples of such approach in the circum-Adriatic zone are the burials from Madonna di Loreto, 'Hypogeum of Bronzes' in Apulia, and the cemeteries of western Veneto, most importantly Olmo di Nogara. In both the cases, grave inventories can be compiled with results of anthropological analyses (sex and age of the deceased), and compared with other features related to the cemeteries (chronology and distribution of depositions). Scholars dealing with those sites from the standpoint of social structure of the buried populations invariably came to conclusion that amber, together with elements of bronze jewellery (mainly pins), formed characteristic attires of women belonging to the highest social strata ('elites'), within communities ('chiefdoms') centred around armed ('militant') males – warriors, with whom those females were affiliated as wives within extended families (e.g. clans, lineages).[656] Occasionally, some of the more extensive cemeteries were subject to statistical analyses with methods explicitly aimed at quantifying the differences among grave inventories (e.g. rank-size approach) and, on that basis, distinguishing horizontal social groups (lineages) and vertical classes (hierarchy) of the particular communities (e.g. Vanzetti 2010; Cardarelli et al. 2015). The notions of prestige and elitism have also been applied in relation to some of the more lavishly equipped burials from other cemeteries, combining amber and bronze jewellery with adornments attributed with high value due to their rarity: i.a. gold rings, shells of tropical molluscs (cowrie), or beads made of glass or semi-precious stones such as carnelian (e.g. Bismantova – Campo Pianelli, grave XXXI, Piazza Monfalcone, grave 31, Rehovë, grave 186; cf. Negroni Catacchio 2009; Kurti 2017b). Motivation to emphasise high social rank is even greater, when a tomb contains objects of foreign provenience pointing to contacts with more developed civilisations, here mainly the *Mycenaean* culture; presence of Aegean pottery and some ornaments has been used to explain identities and to accentuate statuses of people buried in contexts such as Thapsos, tomb D, Barç, tumulus 1, grave 47 or Dimov Grob, grave 54.[657] While it is not the author's intention to question other scholars' opinions, often backed by solid analyses of the relevant data, it is argued that certain ways of amber appropriation in the circum-Adriatic zone have been overlooked in the search for status and identity. Here will be presented several, rather obvious regional peculiarities in the funerary/sepulchral usage of amber; however, the whole topic certainly deserves more attention.

In the 'Hypogeum of Bronzes' at Madonna di Loreto in Apulia, painstakingly conducted excavations revealed numerous female burials adorned with amber and other jewellery, i.a. glass or faience beads, bronze pins, pendants, rings and *saltaleoni*. Distribution of the paraphernalia against the bodies is not always ideally preserved; however, by their proximity to different parts of the skeletons, it is possible to reconstruct the costumes worn by the deceased at the time of funeral. Consequently, the beads, pendants and rings were hanged around the head and neck, the pins were attached to the upper torso, whereas the discs, plaques and buttons would be placed at the abdomen's height. Although in the qualitative sense these burials do not differ significantly from each other, they can be divided into poorer and richer deposits in quantitative terms. Thus, alongside 'standard' female burials with amber, such as those discovered in corridors C2 and A, there is the more sumptuously equipped burial of the 'Lady of ambers' (it. *Signora delle ambre*) at the entrance to sector C1 (Fig. 124). Still, each cited burial with amber is distinguished by its richness from average graves in the hypogeum (cf. Tunzi Sisto 1999c, 214–215). Interestingly enough, excavators

656 Peroni 1999b; Tunzi Sisto et al. 2003; Cupitò/Leonardi 2005; Vanzetti 2010; Cardarelli 2015.

657 Bodinaku 1995; Alberti 2006; Bejko 2007; Videski 2007a, 2007b; Kurti 2017b; Papazovska 2017.

also noted the presence of male burial(s) in the tomb outfitted in the same manner, e.g. the 'Unarmed male' (Tunzi Sisto 1999c, 212). If not for the anthropological determination of sex, this burial could have been regarded as yet another female, judging only by the composition of grave goods (cf. Fig. 123). Question is whether we are dealing here with social acquiescence to dressing a part of the male population with female-prescribed outfits, or should those burials be regarded as occasional deviations from a generally accepted norm. These 'womanly dressed' male burials could be a sign of inter-sex transferability of socially determined gender roles,[658] e.g. some men's enactment of woman-related activities or, looking from the other end of the spectrum, adoption of identity which did not revolve around stereotypically male-oriented occupations.

Fig. 124. Schematic drawing of the burial of the 'Lady of ambers' in Madonna di Loreto, Hypogeum of Bronzes (Vanzetti 1999, Fig. 1A).

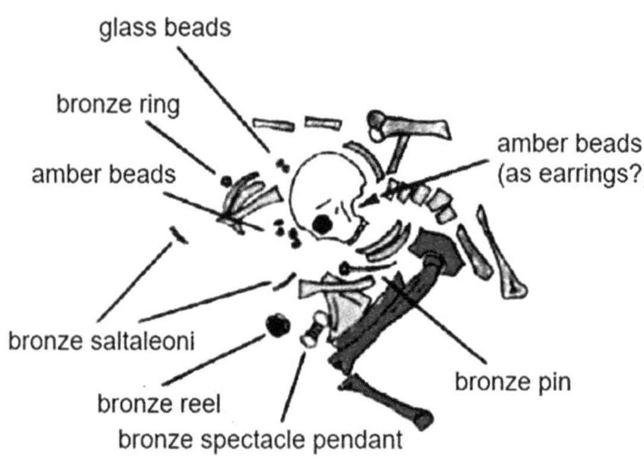

Same hypothesis can be applied in relation to grave 75 of the Narde II cemetery, situated in the lower left part of the CA chart (Fig. 123). The context is specific due to composition of its inventory, comprising i.a. bone/horn pinheads, which make it stand apart from the main concentration of objects, as well as male sex of the deceased. While the combination of objects recovered from this grave differs from most of the burials with amber, it is still quite similar to the inventory of Narde I, grave 169 which turned out to belong to an adult female individual, or Narde I, grave 281, assigned to an infant (1–3 y.o.). Description of the anthropological analysis of cremated remains from the urn of grave 75 does not leave a room for doubt (Colonna et al. 2010, 211); however, it might be worthwhile to re-study them for certainty. It is worth noting that the grave also differs from the regular post-humous female outfits by the sheer abundance of amber artefacts which comprise 24 beads and pendants.

658 For an overview of the discussion on 'gender' in archaeology and its application in the context of European Bronze Age, see e.g. Bergerbrant 2007.

The image of unarmed ('non-militant') males contrasts with other individuals of the same sex equipped with swords and buried on cemeteries such as Olmo di Nogara or Madonna di Loreto. Sector C2 of the 'Hypogeum of Bronzes' yielded a number of such burials, now attributed to warriors interred with members of their families which include females dressed in amber (i.a. the 'Lady of ambers') (cf. Peroni 1999b). Yet, even among those distinctly 'male' inventories one can find amber beads – few of them were found in the burial of the 'Chief' (it. *Capo*) belonging to an adult male individual (Tunzi Sisto 1999c, 212–213). The sword certainly makes a difference when comparing this assemblage with other grave sets incorporating amber, and contributes to the slightly offset position of the burial on the CA plot (Fig. 123). Nevertheless, it also contained some of the items more commonly registered in the female graves (e.g. bronze rings and buttons); hence, the distance of the symbol from the main cluster of objects on the chart is not so big. In the upper left corner of the diagram is located another grave combining bronze sword with amber jewellery: Shtogj, tumulus 2, grave 9 from the Mat Valley. While there is no information regarding sex or age of the person buried in this context, it is likely that this is yet another male burial with amber. How can we explain those rare cases in which amber was deposited with armed males, apparently against the prevailing customs?

J. Maran approaches the early *Mycenaean* burial of a presumed male individual from Kakovatos, tholos A, dressed in full 'warrior' paraphernalia, including bronze sword and scabbard, as well as roughly 500 amber objects, by using the idea of individual acts of amber appropriation (Maran 2013). The scholar revised the old interpretation of the amber finds as elements of a necklace, instead opting for their use as *appliqués* to a leather shoulder belt, arranged from reworked amber jewellery or partly manufactured in Greece from imported raw amber. In functional terms, the belt was used to suspend the scabbard of the sword, whereas on a symbolic level the amber attachments could have played a magically protective role as talismans aiding the warrior in combat. One can propose a similar interpretation with respect to the two sword-bearer's burials with amber in Apulia and the Mat Valley. Although such an elaborate reconstruction of the mode in which amber elements were fitted into the outfits cannot be conducted in the two given cases, it is still plausible that they served as a kind of amulets. Their magical role could have been borrowed from female dress, itself normally lacking any 'active' means of defence, thus probably entrusting their well-being to tokens of supernatural powers and possibly using amber beads and pendants in that capacity.

Alternatively, amber beads found a new meaning and usage in the hands of people with whom amber was not popularly associated. It might be suspected that the male burials equipped with amber signify deliberate and ostentatious breaking of the rules governing amber's use. Those few 'warrior' male burials (i.e. containing bronze swords) may be an expression of the appropriation of amber by force (e.g. using one's physical strength or social position). Yet, these isolated cases do not break, but confirm the rule regarding the use of amber in the circum-Adriatic zone, visible on the basis of the collected anthropological data.

General picture shows amber as relevant to the domain opposite to violence, combat and warriorhood. Its connotations with items related to dress and decoration, mostly belonging to females, are clearly visible. Male sex in the burials with amber is more often disguised than explicitly accentuated, when looking at the grave inventories from archaeological perspective. But that does not necessarily mean that it was the mourners' intention to hide the true identity of the deceased. Personhood or status could be accentuated in delicate ways, unobservable or not yet understandable to us. It is the matter of recognising position of individual *parures* on the scale from particular to standardised and common identities, or from non-restrictive to the normative and uniform funerary customs and practices.

Vast majority of burials show that amber jewellery was added to graves in the form of beads and pendants strung as necklaces, or perhaps also bracelets. Very often tombs yielded only single or few beads which might have been hanged independently or in conjecture with beads made of other materials. In some fortunate cases, these sets of amber adornments were discovered around the necks of the deceased. It is easy to imagine this as the most popular way of wearing amber in the real life. Yet, there is also one alternative way of composing amber into posthumous outfits, visible among analysed graves. It combined bronze pins with amber beads, but not as pinheads as in the cases of finds from Peschiera del Garda or Ljubljanica River discussed earlier (see sections 2.2.5 and 2.3.6), but as clasps inserted onto the pointy ends of pins, presumably to fasten a cloth or a shroud. The manner of attaching amber beads to bronze pins is best documented in relation to cemeteries of western Veneto: predominantly at Olmo di Nogara (De Marinis/Salzani 2005, 427–428), however occurs also in some cases at Castello del Tartaro (graves 300 and 350; Salzani 2011, Fig. 5:A–B), Franzine di Nuove (graves 12 and 159; Aspes 1973, Fig. 1:1–4) and Scalvinetto di Legnago (graves 373 and 376; Salzani 2004, Fig. 5:A–D). At Olmo di Nogara amber elements used as clasps are often larger than average beads, and were regularly discovered on breasts of the deceased near collarbones, with the pins being oriented upside down in relation to the body (Fig. 125). It cannot be ruled out that this practice was also used elsewhere in the circum-Adriatic zone; however, absence of graphic presentations of graves does not allow to verify this supposition. Whether such arrangement of grave goods emits the costume worn on the daily basis is up for discussion. Similar disposition of bronze pins is evident among graves of the *Tumulus* culture in Germany, although they lack amber beads at their tips (Wels-Weyrauch 1989, 128, Fig. 8A). Based upon observation on burials with well-preserved textiles from Northern Germany and Denmark, S. Bergerbrant and K. Kristiansen argue that pins served to fasten the outerwear – cloaks which could have been later utilised as shrouds (Bergerbrant 2007, 50, 59, 81, 83, 85; Kristiansen 2013, 763). Yet, these pins are devoid of any clamps. Reconstruction of a dress with inclusion of amber-clenched pins has not been attempted so far, so it is difficult to say whether such wearing style would be practical or convenient.

Fig. 125. Schematic drawing of the burial in grave 172 at the Olmo di Nogara cemetery with two bronze pins with amber clasps placed on breasts of the deceased (Salzani 2005, Fig. 409).

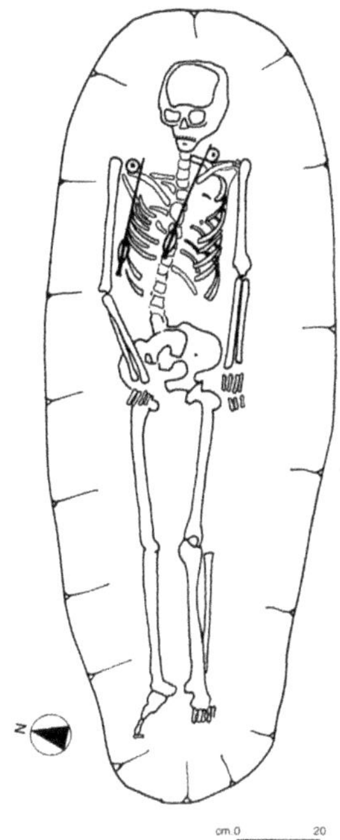

At the same time, the cemeteries of western Veneto delivered also burials presenting more common use of amber in the form of necklaces. In the previous studies, the burials outfitted with amber in the region have been attributed to women of prominent social standing at the apex of hierarchy, within societies at first (BM 2–3) resembling conical clan model, and later (BR) gradually morphing into a 'nobles-clientele' structure.[659] Yet, while the studies on the development of social structure paid a lot of attention to the male burials – representations of warriors, much less consideration was given to the female burials. It is interesting whether the two modes of amber deposition in graves (as clasp and as necklaces) simply illustrate two casually adopted manners of attaching amber to dress or stems from more formalised rules dictating dress code appropriate to person's status. The uniformity of burial inventories seems to indicate quite rigid norms in this regard. Apart from the pins with numbers predominantly varying from one to three specimens, graves at the western Venetian cemeteries contained mostly bronze rings, *saltaleoni*, and bone/horn combs, sometimes supplemented with bone beads (e.g. Olmo di Nogara, grave 172) or pinheads (e.g. Olmo di Nogara, grave 112),

659 Cupitò/Leonardi 2005; Vanzetti 2010; Cardarelli 2015.

shell pendants (e.g. Scalvinetto, grave 354), glass beads (e.g. Castello del Tartaro, grave 146) or gold rings (e.g. Franzine, grave 123). Burials containing such 'extras' are usually slightly divergent from the main concentration of objects on the CA chart (Figs. 122–123). Thus, there seems to have been a standard in the burial outfits ('a bare minimum') which included bronze pins, most probably needed to fasten the cloak/shroud, to which were added more personal items, possibly to express personhood of a buried individual. Still, all of them found acceptation among the community, and were incorporated into funerary customs as appropriate elements of women and children's attire. The place of amber in this context is twofold: on the one hand, it could play a functional role by not letting the cloak/shroud to slip off the pins and unfold, but on the other hand, it still adorned the person, either as a necklace or as separate beads. Whether the manner of amber deposition as a clasp to a pin or as a necklace had deeper social significance/reasoning is still to be determined. One can also presume the symbolic meaning of amber as a talisman/amulet protecting the deceased and fitting within religious beliefs of a society. Similarly unified rules related to the dress/posthumous outfit of the dead are observable in other communities, where amber was frequently used in that capacity, e.g. in north-central Lazio, Podrinje and Western Serbia, Istria, or, to a lesser degree, in northern Apulia. This regularity is well represented on the CA results, where graves from the abovementioned regions tend to be clustered together (Fig. 123). Only in singular cases, one can notice derogations from the prevailing rules of dress and burial.

One of the strikingly different inventories containing amber in western Veneto is grave 411 at Olmo di Nogara. The burial, belonging to senile female individual, in addition to commonly appearing grave goods: bronze pins, rings, needle and horn comb, comprised two objects absent from other graves with amber at the site. These are a small bronze dagger of the Pieve S. Giacomo type and a clay bowl decorated on the bottom with concentric *cuppelle* and pitched grooves, forming together a star symbol. Should the inclusion of the dagger be considered as an example of the deviation from the normally respected rules of burial? At Olmo di Nogara daggers were also found with the burials of children twice (De Marinis/Salzani 2005, 406). Although quite rare, they sometimes co-occur with amber in other regions of amber's acquisition; however, connecting them to specific individuals is largely impossible. Daggers in female burials are known from many cultures, e.g. Scandinavian graves during periods II and III of the *Nordic* Bronze Age (Bergerbrant 2007, 100). They are usually interpreted as signs of more active role of women in warfare, or as defensive weapons intended to grant the women more protection. However, daggers could have been used for a number of different purposes which were not directly linked to the sphere of combat. In the case of grave 411, attention is drawn by the small size of the blade and its association with the bowl (Canci et al. 2015, 332). Perhaps the women from grave 411, who died in old age, enjoyed a great esteem among people and played an important part in the life of the community, thus was granted extra grave goods linked to her status. The dagger and the bowl decorated with the distinct motifs might be the paraphernalia of a person, who was responsible for spiritual guidance (priestess?) and ritual performance (dagger and bowl for preparing sacrifice as offerings?). These are only guesses, but the tomb surely deserves to be set apart from other graves with amber from the site.

During FBA compositions of grave inventories in western Veneto changed (e.g. pins were replaced by fibulae and pinheads), however still kept a considerable degree of uniformity, as indicated by graves with amber from Narde and Desmontà. This observation finds reflection in the results of CA (Fig. 123). Question is whether the social attribution of amber stayed intact. By the sole fact of fewer burials with amber, as well as occasional admission of males to this material, one can presume that the group of its users narrowed down, perhaps to a small but more elite social strata, at the same time allowing for slightly more personal – individually decided forms of its appropriation. Graves from the Narde I necropolis have been hitherto put into hierarchical order basing on their inventories (Cardarelli et al. 2015). As a result, grave 169 was classified to group F3 comprising rich assemblages of female burials, while grave 281 was ascribed to group I4 comprising rich inventories of children graves. Similar analysis has not been carried out in relation to the Narde II cemetery; however, just by a simple comparison, grave 75 can be placed in the same if not the wealthier class of burials as tomb 169. The above observations are difficult to reconcile with conclusions of the recently published strontium isotope-based analysis of burials from the both cemeteries of Narde (Cavazzuti et al. 2019). Authors of the paper claim that 'the Narde II sector may be characterised as representing a local, kinship-based population, while the more privileged social groups of Narde I show a greater degree of mobility – and therefore tend to incorporate more non-local adults. Through a system of alliances, the nascent elite of Narde I, which was less reliant on traditional kinship ties than the population represented by Narde II, extended its influence over wider territories' (Cavazzuti et al. 2019, 641). If the burials of Narde II comprise the indigenous part of the community, clinging to tradition and showing less social differentiation or wealth, then the tomb 75 has to be seen as an exception, especially in the light of anthropological determinations of sex which show that western Veneto from MBA up to FBA deposited amber exclusively in female and children burials. Grave 75, thus, may be considered as a proof of loosening the previously tight rules concerning dress and posthumous outfits.

The process of changing the social attribution of amber by opening it to different sepulchral contexts of deposition is also apparent in other 'core' regions of amber's acquisition towards the later phases of the Bronze Age. In northern Apulia the transition from MBA through LBA to FBA is usually seen as accompanied by changes in the social structure of communities inhabiting the region. The development lead from the social groups relying on kinship (extended families – clans or lineages) with the powerful position of male warriors to increasingly hierarchical society in which the traditionally respected divisions based on blood-ties were renounced in favour of a structure promoting political authority and economic control in the hands of certain individuals or 'classes' (a nascent elite). Although the dynamics and scale of this transformation are debated, the sole factor of change is generally recognised by scholars dealing with the subject.[660] Evidence of amber in sepulchral contexts in the region during FBA comes from the Monte Saraceno necropolis.[661] The basic idea of interring multiple burials in rock-embedded tombs remained largely intact since MBA; however,

660 Cf. Peroni 1999b; Cazzella 2010; Pacciarelli 2012; Cardarelli et al. 2015; Cazzella/Recchia 2015.

661 Nava/Preite 1995; Nava 2001; Preite 2001.

one can suspect a narrowing of admittance to this form of burial to a certain group of people. The tombs of Monte Saraceno delivered great richness of grave goods; thus, it might be that only the highest-ranking members of society and their immediate families were allowed to be buried there. Amber found in those mass graves cannot be surely attributed to specific individuals; however, its social availability certainly was more limited than in the preceding periods.

In Istria during the Bronze Age amber shows a relatively stable manner of deposition which perhaps testifies to its well-grounded status in the local culture. Graves with amber either had form of a tumulus or a cist, and usually housed multiple burials. While grave inventories do not seem diverse and rarely include anything more than simple bronze jewellery (rings, *saltaleoni*, pendants, etc.), this apparent poverty might result from grave robberies. The recent-most burial with amber on the peninsula (the funerary cave of Laganiši) differs considerably from the remaining tombs not only for the place of deposition (cave) but also due to more complex inventory, comprising i.a. bronze dagger and axe. Again, the level of association between these two items and the two amber beads from the context cannot be estimated, because of the state in which the burial was discovered. Nonetheless, the overall combination of grave goods may signify certain evolution of the burial rite and the position of amber within it. In fact between LBA and FBA, the cultural landscape of Istria underwent profound changes with most of the hillforts (cr. *gradine*) being abandoned and inhumation being replaced by cremation (Mihovilić 2013a, 2013b). Perhaps in those changing conditions, amber lost its previous significance and ceased to be imported, or found a completely new manner of deposition, e.g. the one which involved burning of garments on the pyre with the body, as sometimes indicated by traces of intentional damage on other grave goods (Mihovilić 2013a, 65). When acquisition of amber shifted to Kvarner and Dalmatia during FBA, the character of amber deposition in graves was already different. The severe lack of data in relation to the majority of amber-containing burials on the Croatian part of the Adriatic coast prevents from making decisive conclusion with respect to the social attribution of amber or its status in the local society. However, judging by the composition of grave goods, one can suspect a more individualised approach to amber, less restricted by the norms of deposition and, at the same time, limited to certain groups or identities (persons with a prominent status?).

Also in Podrinje and Western Serbia, the last phase of amber's presence there (LBA) saw a decrease in the amount of material both in number of contexts and artefacts per burial. Throughout the entire period of amber's acquisition in the area, the grave inventories remained largely standardised, comprising bronze pins, bracelets, pendants, *saltaleoni* and rings. Although little is known about social structure of the local society during the Bronze Age, one can discern markers of incipient social stratification (Dmitrović/Ljuština 2010). Moreover, it seems justified to suspect re-evaluation of amber's status and social attribution in the region around MBA/LBA. During MBA amber jewellery is found only in relation to females (e.g. Paulje, tumulus K and Vranjani, tumulus 1), while in LBA amber appears in a double burial comprising an adult male and an infant (Paulje A). Still, the number of anthropologically analysed burials is very low, so any hypothesis advanced in relation to social attribution of amber in Podrinje and Western Serbia has to be taken with a grain of salt.

In the neighbouring Glasinac area, certain long-term development in the appropriation of amber is discernible as well. The few Bronze Age burials with amber show compositions of grave goods largely similar to those observed in the tumuli of Podrinje and Western Serbia. The change came around the transition from the Bronze to Iron Age, when the burial outfits started to embrace more categories of items, i.e. not only ornaments but also weapons and tools.[662] The latest tombs with amber from the Glasinac illustrate this metamorphosis of the burial rite; the best example is Vrlazje, tumulus IV, grave 2 (1894) which, apart from bronze fibulae, contained an iron knife. Unfortunately, none of the analysed graves from the region was determined in terms of sex and age of the deceased, and it is unlikely they ever will, considering the state of preservation of materials from the Glasinac tumuli. Nonetheless, the new (Iron Age) rules of deposition of amber and other grave goods can be read as a sign of changing attitudes towards certain materials and objects, and their place in the local culture.

The rules of deposition must have been less rigid in the areas where amber was not frequent and occurred in short periods, i.e. where it was not commonly adopted by social norms and incorporated into the routinely worn costumes. This observation corresponds mostly in relation to the southern regions of amber's acquisition in the circum-Adriatic zone, on the CA plot located predominantly on the right side of the vertical axis (2nd eigenvector) (Fig. 123). In south-east Albania (Kolonjë Plateau and Korçë Basin), amber was found on several occasions in association with gold rings and/or carnelian beads (Barç, tumulus 1, grave 47; Luaras tumulus, grave 46; Rehovë tumulus, grave 186). Although anthropological analyses are missing, L. Bejko considers amber and glass beads as well as gold rings among the typical markers of female burials in the tumuli of Barç and Rehovë (Bejko 2007); however, it is unlikely that they were seen that way by the local Bronze Age community. There is only a handful of graves equipped with these items, in comparison with hundreds of burials lacking them (cf. Andrea 1985; Aliu 2012). Furthermore, the former incorporate mostly items of foreign provenience, rarely supplemented with locally made products, mostly pottery. In the more recent publications, carnelian and amber beads, together with gold spirals and rings, are considered as luxury items and tokens of prestige which appeared in Albania in the 'second wave' of imports from the Aegean (Kurti 2017b). This interpretation is plausible in light of the rich evidence pointing to the contacts between indigenous southern Albanian Bronze Age cultures and the *Mycenaeans* (cf. Bejko 1994; Bodinaku 1995). However, it is worth paying attention to the composition of grave inventories which contain amber. Barç, tumulus 1, grave 47, aside from glass and amber beads, and gold rings, contained a stirrup-jar of the late *Mycenaean* provenience (LH IIIC-*Submyceanean*) (Andrea 1985, Tab. VI). Barç, tumulus 2, grave 2, on the other hand, possessed a *kantharos* (double handled beaker) and two bronze pins which suggest relationship with the northern Balkans (Andrea 1985, Tab. XVII). Similar ceramic vessel was found in Rehovë tumulus, grave 201, whereas grave 186 from the same mound, in addition to amber and carnelian, yielded two Paarstadl-type pins (Aliu 2012, Tab. XV, XVII). Rather than an unilateral link with the Aegean, the cited burials present a mix of both northern and southern influences. It is difficult to say whether we are dealing

662 Cf. Benac/Čović 1956, 1957; Čović 1983c, 1987.

here with elites or nobility, as except for the 'exotic' artefacts in their inventories, these grave are not distinguished by anything special – they are constructed just as the other tombs located in the same grave mounds. It is tempting to see these burials as belonging to foreigners; however, it seems more likely that they signify persons, who were involved in interregional contacts, and through communing with the external societies built their own identity. Thus, rather than directly copying *Mycenaean* practices in terms of the use of imports, the communities inhabiting Albanian territory had their own way of adapting amber and other foreign goods to local burial and presumably day-to-day customs.

These considerations should also include three graves with amber at the Dimov Grob cemetery in the Vardar Valley (nos. 54, 73 and 120) which are interpreted as evidence of the Aegean/*Mycenaean* influence over North Macedonia during LBA. The burials belong to female individuals which could have a specific status among the population of the *Ulanci* group. Tomb 54 is especially worth mentioning due to its rich and varied inventory, comprising bronze pins, gold rings, bronze, glass and amber beads, shell pendants and a mix of Aegean-inspired (matt-painted alabastron) and locally made pottery (Videski 2004, 31–32, 78, T. XLI-XLII, Figs. 33 and 34). On the top of that, the burial contained a belt consisting of 14 axe-shaped pendants (*labryses*) divided from each other by five glass and bronze beads (Videski 2007a, T. LV:e). The object has clear connotation with the Aegean, where *labryses* – double axes had a cult status; thus, the tomb 54 at Dimov Grob has been interpreted as a burial of priestess (Videski 2007a, 213). Presence of the *labrys* belts in North Macedonia adds a 'spiritual/religious' link with the Aegean which seems to fit the presumed magical symbolic of amber beads found in the same context.

A specific case is the one of amber on Sicily. Many Sicilian graves, especially in the early stages of the *Castelluccio* culture, combined amber with various local materials, such as alabaster, obsidian or flint, unavailable in the other regions. This results in their remote position of the right side of the vertical axis (2nd eigenvector) on the CA plot (Fig. 123). Among them, the most characteristic are inventories of graves no. 9, 22 and 23 from the eponymous Castelluccio cemetery. All three tombs contained i.a. flint knives, although it is difficult to say whether they should always be correlated with amber finds. From the description of grave 9 published by P. Orsi, it can be inferred that the context included at least 28 individuals, among them one infant (Orsi 1892, 18). The author provided quite detailed characterisation of the latter burial and its components which clearly states that it included flint knives and several beads/pendants. It should be stressed that amber beads from the tomb were determined as simetite – local Sicilian amber, hence opening a room for discussion about the potential differences in appropriation of succinite and simetite. Perhaps the latter substance had other connotations in the semantic field which determined its role in the burial rite of the *Castelluccio* culture differently from the Italian mainland. It seems that at least before the arrival of succinite in Sicily in more advanced stages of the Bronze Age, simetite was utilised by the local Sicilian communities as yet another interesting mineral to create adornments. Thus amber falls into plethora of various non-organic materials (minerals and sedimentary rock) which were used by the *Castelluccian* society to manufacture beads and pendants, also seen suitable as grave offerings.

During MBA Sicilian communities of the *Thapsos* culture fully embraced imported amber – succinite which replaced simetite in graves. Succinite beads, pendants and even spacers, apart from still recurring adornments made of locally obtained materials (e.g. beads of mineral and sedimentary rock), were deposited together with other products of foreign provenience, mostly pointing to contacts with Aegean – *Mycenaean* culture. The tombs still had a form of underground hypogea and communal character (mass interments), but the burial outfits became less uniform, as if to accentuate identities or statuses of buried individuals or entire groups of people. The differences in wealth between the *Thapsian* tombs are well illustrated in the case of the eponymous cemetery. Statistical analysis, aimed at determining hierarchy among the graves basing on the wealth indexes ascribed to various categories of grave goods, shows that amber appeared in the richest tombs used in phases 1, and 3 (tomb D), as well as the more modest ones of the phase 2 (tomb 61), although outcome of the analysis in the latter case raises reservations due to the possible robbery of the tomb (Alberti 2006). More sophisticated grave inventories are also evident with respect to the tombs with amber at the Plemmirio necropolis. Thus, it seems plausible to suspect that with the change of its source (simetite to succinite), amber in Sicily also evolved in terms of social attribution by finding a narrower group of consumers.

Presence of crystal, obsidian and stone items resulted in convergent position of the MBA–LBA tombs from Calabria – region neighbouring with Sicily and having access to similar resources. In the context of the Calabrian burials with amber, it is worth mentioning the tomb of Gallo di Briatico (Pacciarelli 2001, 185–187). Inhumed remains of the deceased, presumably a female individual (very scarce dental pieces), were placed inside a large vessel (*dolium*) covered with a lid. The burial was equipped with a bronze awl, a silver-wire spiral bracelet, a large bead or pinhead of rock crystal, an amber bead, a greenstone ophiolitic bead and a seal of carnelian (Pacciarelli 2001, Fig. 109). Almost all finds from the context can be considered as rare/exotic/luxury items; however, the seal attracts the most attention. It has almost rectangular outline and lenticular cross-section with through hole, decorated on one side with carved 'V'-motif filled with small semi-circles and with a pair of diverging beams engraved at the base. Most likely it is a product of Aegean provenience (Greek mainland rather than Crete), where seals were used for administrative purposes as proof-marks of authority, however could have had a double meaning as talismans (cf. Pacciarelli 2001, 187; Jung 2013, 245). The seal certainly forms a link with the Aegean, but the overall picture of the Briatico tomb (rest of the inventory and form of the burial – inhumation in pithos) seems to prove indigenous southern Apennine descent of the person buried in it (Jung 2013, 245). It is argued that the burial of Briatico complements the picture of Southern Italian elite tombs of MBA, such as those of northern Apulia (Pacciarelli 2001, 187).

With its re-appearance on Sicily during FBA, the status of amber in the local society and various spheres of its life (e.g. burial ritual) was re-appraised once more. Amber beads are present at the Madonna del Piano cemetery in graves with singular interments, either buried in pits or inside *pithoi*, by the composition of inventories ascribed to female individuals (Bernabò Brea et al. 1969, 226). Their potential meaning as an indicator of prestige was put forward by some scholars (Nicoletti 2006, 174). In comparison with the *Castelluccian* or *Thapsian* tombs, Madonna del Piano delivers the picture of increasingly rich and 'personalised' burial outfits which comprise objects related more to the costume (bronze fibulae, pendants, bracelets and ring, plus ivory combs), in similar vain to FBA inventories from the Italian mainland. Flint knives known from EBA were replaced by bronze knives what shows continuation of the burial custom in some respect. With the absence of anthropological determination of sex and age of the burials at Madonna de Piano, it is difficult to say anything more about the identity or status of the deceased buried with amber at the site.

The above discussion on the social attribution of amber, its status and forms of cultural adaptation and appropriation may seem selective, because out of necessity it was limited to some of the most characteristic examples. Graves not included here or mentioned briefly still deserve attention, as their reconsideration may bring to light previously unnoticed facts about relationship between amber and circum-Adriatic communities using it in the Bronze Age. Still, a number of observations was obtained through this overview which is summarised below.

4.1.2.5. Amber in the context of cemeteries

The discussed graves offer a good deal of important information on the role of amber in funerary practices in the study area. With a certain dose of probability, the attained observations can be used to reconstruct amber's status also in the everyday life of the circum-Adriatic communities of the Bronze Age. While amber on its own was not a criterion which decided the placement of a burial within a cemetery, or selection of a particular grave form, it has often been found in parts of necropolises or tombs reserved for specific social groups. Across the entire study area, there is evidence of an almost unanimous association of amber with women and children. Such uniform social attribution attests to a common approach to amber by the communities inhabiting both sides of the Adriatic during the 2nd millennium BC. These people must have recognised certain qualities inherent in the material, both in the physical as well as symbolic sense. Similarities in the forms of amber's adoption and, subsequently, adaptation to local cultures seem to confirm a specific and legible semantic content of amber, the notion of which spread along with the succinite inflow from the north. This 'travelling package of meaning' helped to incorporate amber into the temporal as well as post-mortem domains over a vast area.

Amber's primary functions, at least from the archaeological perspective, were connected with dress – a concept first conceived for the world of living and later transposed onto the funerary customs. Amber was worn both as an adornment/emblem – a category which can emit a number of potential messages, among them most obviously a social position, and a spiritual token – amulet or talisman. Its status apparently corresponded with religious beliefs widespread at the time; thus, amber found its way into burial practices. It is widely accepted that the sun played an important part in the Bronze Age cosmology and religion.[663] Between ca. 1600 and 1000 BC, Europe witnessed adoption of a more or less uniform system of beliefs which can be characterised as 'solar cult' that permeated from the Near East (Kristiansen/Larsson 2005, 288–294; Goldhahn 2013, 251–253). In short, the solar cults worshipped in different parts of the Europe have many aspects in common; the most important ones are day/night dichotomy and voyage of the sun across the sky as foundations of the Bronze Age cosmology which guided the life of local societies. In ideographic form, and presumably oral tradition, the abovementioned

663 E.g. Green 1991; Kaul 2005; Kristiansen/Larsson 2005; Goldhahn 2013.

phenomena were visualised/described as 'the sun-goddess, who drives the sun over the sky in a chariot drawn by horses, or she can be represented by a sun-disc as well. At dawn the sun goddess is captured and held prisoner, but her twin brothers come to her rescue so the sun can rise again in the morning' (Kristiansen/Larsson 2005, 297). There is plenty of material evidence of the solar cult from all across Europe (cf. Kristiansen/Larsson 2005, 294–319). J. Czebreszuk writes extensively about the 'solar' symbolism of amber in the *Mycenaean* culture (Czebreszuk 2011, 164–166). He observes that for prehistoric people absence of the sun in the night was explained by its temporary descent into the underground somewhere in the north, so the realm from which amber (succinite) came as well. In reference to the myth of the sun's daily journey, pieces of amber could be interpreted as congealed tears of sun-goddess wept while being imprisoned. Specific properties of amber also played a part in consolidating its 'divine' origin; despite being obtained from the ground, it was different from other stones: translucent, light, warm and easily burnt at the same time producing characteristic smell. Thanks to them, amber potentially had a number of applications in religious practices. Furthermore, K. Kristiansen and T. Larsson note that 'the Delphic Apollo had strong northern links with the solar deity of the Baltic, from which amber came' (Kristiansen/Larsson 2005, 44). It can be inferred from these observations that amber could have been associated with the cult of sun also among the communities of the circum-Adriatic zone.

Yet, while its visual attractiveness and compelling semantics helped amber to be eventually adopted by numerous communities, the process of its assimilation ran sometimes differently depending on the region. Even in the societies which tended to have a standardised place for amber in the social sphere, there are cases of a more individualised approach to using the substance. Visualisation of the CA results for objects also revealed which graves correspond with items rarely or untypically occurring with amber. It is interesting to highlight that among them are burials characterised by specific anthropological features, not always in compliance with the prevailing picture of the social group/part of a population commonly buried with amber. Depositing amber in non-female or non-infant burials, in association with previously unprecedented items, shows that the material's social status was discursive to a degree. Some of these burials reveal images of specific identity which precisely may be why amber was treated there differently to the 'standard' way. In the course of the Bronze Age, amber's status could undergo re-appraisals in the face of changing social structure, not to count more individualised acts of its appropriation, and amber as a substance rarely enjoyed longer periods of stability in terms of social attribution. It is also interesting whether succinite and simetite had similar connotations in Sicily. Use of simetite-made beads and pendants precedes occurrence of succinite on Sicily; thus, it raises a question if both substances were regarded as fundamentally different or fell into the same semantic field.

In the end it can be argued that amber artefacts deposited in graves were important attributes of the people they were buried with. Amber use as an element of costume was sanctioned by the social rules regarding the dress code relevant to particular members of society, predominantly woman, but also children (girls?). The level of formalisation differed among the communities: it seemed to be more explicit and rigid in regions with longer histories of amber's usage and greater access to this substance. If dress was a culturally sanctioned exponent of one's identity or status ascribed for a lifetime or its certain stages, then it seems reasonable to treat amber jewellery found in the graves as possessions of the deceased, connected with them through long and tight association in the meaning of 'inalienable' possessions (Weiner 1992). The rules of burial incorporated certain norms regarding the posthumous outfit, and amber certainly was allowed to be taken by the deceased into the afterlife, most probably because it had a death-related field of meaning (Blischke 2002). The symbolic connection of amber with the sphere of magic and spiritual powers was not only activated at the time of the funeral but more likely served as an important quality in the daily use of amber. Earlier it was argued that the post-mortem outfits were devised through adaptation of the notion of dress worn by the living. Same process of transposition from the sphere of life to the sphere of death included amber jewellery. It is possible that before being deposited with the burial, amber adornments were rearranged on a deceased body prior to the interment, e.g. as in the case of amber clasps on pins in some of the western Venetian graves. In regions where deposition of amber objects in graves appears to have been a common and standardised habit, the items included in the burial inventories might be regarded as culturally prescribed, indispensable paraphernalia of the dead which needed to be provided by the mourners from the deceased's personal possessions or on their own expense. Thus, the beads found in the burials could be treated as objects with at least twofold function: to serve as a relay of the buried persons' identity and as devotional articles linked with religious beliefs of a community. Amber could have been put inside the graves as a gift to deities which would secure the safe passage of the buried person to the underworld. In the same way it could represent the membership of the deceased to the sun cult. Regardless of who owned or donated them, selection of the grave goods was in the end targeted at reflecting an idealised image of a society member. Many personal possession which the person acquired in the course of his/her life were discarded or could not be brought in to the grave, but amber along with few other categories of objects were allowed into the afterlife.

In addressing the issue of amber's status among the Bronze Age communities of the circum-Adriatic zone, one cannot avoid the question of its presumed high value and association with the higher classes of society as an indicator of prestige and wealth. Studies regarding the social structure in the study area are not offering a single, clear or indisputably accepted model of society; however, most of them point to incipient social hierarchisation and departure from the previous, traditionally family-oriented organisation. Dynamics of this process differ from region to region, and it is difficult to state whether amber was exclusively a property of elites, because it would mean that many communities possessed highly developed social stratification already at the start of the Bronze Age. Nevertheless, sepulchral contexts in which amber has been discovered often reveal special features distinguishing them from other graves.

Amber itself was not readily available material in the circum-Adriatic zone, and its procurement required certain knowledge (concerning sources of the material or the routes and centres of its circulation), power (to organise an expedition to bring it or to acquire it in a different way, e.g. through gifts exchange) and wealth (resources and commodities which could be traded in exchange for amber). Furthermore, association of amber in graves with other rare and exotic materials seem to underline even more the capabilities of these people or social groups to import and acquire foreign goods. That this ability was not in the hands of the entire community can be illustrated by the levels of amber's commonness in the graves of the more spacious cemeteries (Tab. 49). In relation to the cemeteries where exact number of grave with amber can be confronted with the overall number of excavated graves, the level of amber's commonness, expressed as a percentage, rarely exceeds 5 %. Only with respect to the necropolis of Piazza Monfalcone in Lipari and the tumuli complexes of Podrinje (Padine-Ročević and Paulje)[664] the coefficient is higher, varying between 10 % and 20 %. The very limited dissemination of amber within the circum-Adriatic communities is most striking in the case of western Venetian cemeteries of MBA, LBA and FBA. These vast necropolises comprised several hundreds of graves each, but the number of graves containing amber represents only a small fraction.

Tab. 49. Coefficients of amber's commonness on cemeteries with single-interment graves

Cemetery	**Region**	**Phase**	**No. of graves total (excavated)**	**No. of graves with amber**	**Percentage of graves with amber**
Piazza Monfalcone	Aeolian Islands	FBA (1150–950 BC)	57	6	10.53 %
Iglarevo I	Kosovo and Metohija	MBA (1650–1300 BC)	39	2	5.13 %
Padine-Ročević complex (Jezero, Karavlaške kuće, Šundinovača)	Podrinje and Western Serbia	MBA–LBA (1650–1150 BC)	57	10	17.54 %
Paulje	Podrinje and Western Serbia	MBA–LBA (1650–1150 BC)	20	3	15 %
Bismantova – Campo Pianelli	Romagna and Emilian Apennines	FBA (1150–950 BC)	49	1	2 %

(*Continued*)

664 Calculating the percentage of excavated graves with respect to the two cemeteries is burdened with some difficulty. The tumuli rarely contained just a single burial, yet very often the individual graves inside the mound were not clearly delimited, or could not be recognised due to a poor state of preservation of bones. Furthermore, graves sometimes lacked inventories or suitable chronological indicators among them; thus, some of the burials could be later than the Bronze Age. In case of the Padine-Ročević complex, it was possible to sum up all the burials in the mounds, but regarding Paulje only the overall number of excavated mound was ascertained.

Tab. 49. Continued

Cemetery	Region	Phase	No. of graves total (excavated)	No. of graves with amber	Percentage of graves with amber
Dimov Grob	Vardar Valley and Pelagonia	LBA (1300–1150 BC)	124	3	2.42 %
Castello del Tartaro	Western Veneto	LBA (1300–1150 BC)	396	5	1.26 %
Desmontà	Western Veneto	FBA (1150–950 BC)	441	1	0.23 %
Franzine Nuove	Western Veneto	MBA–LBA (1650–1150 BC)	582	10	1.72 %
Olmo di Nogara	Western Veneto	MBA–LBA (1650–1150 BC)	517	26	5 %
Narde I and II	Western Veneto	FBA (1150–950 BC)	~900	3	0.3 %
Scalvinetto di Legnago	Western Veneto	MBA–LBA (1650–1150 BC)	535	6	1.12 %

Tab. 50. Coefficients of amber's commonness in tombs with multiple burials

Grave	Region	Phase	No. of burials (total)	No. of amber artefacts (total)	Object-per-person ratio
Laganiši, sepulchral cave	Istria	LBA (1300–1150 BC)	12	2	0.17
Velika Gruda, grave 9	Montenegro and Southern Dalmatia	MBA/LBA	7	2	0.29
Madonna di Loreto, 'Hypogeum of Bronzes'	Northern Apulia and Basilicata	MBA–LBA (1650–1150 BC)	200	205	1.025
Grotta Manaccora, *Grotticella Funeraria*	Northern Apulia and Basilicata	MBA (1650–1300 BC)	26	4	0.15
Toppo Daguzzo, tomb 3	Northern Apulia and Basilicata	MBA (1650–1300 BC)	11	1	0.09

Tab. 50. Continued

Grave	Region	Phase	No. of burials (total)	No. of amber artefacts (total)	Object-per-person ratio
Monte Saraceno, tomb 84/IV	Northern Apulia and Basilicata	FBA (1150–950 BC)	8	2	0.25
Monte Saraceno, tomb 43c/IV	Northern Apulia and Basilicata	FBA (1150–950 BC)	6	1	0.17
Castelluccio, tomb 9	Sicily	EBA (2150–1650 BC)	28	4	0.14
Castelluccio, tomb 22	Sicily	EBA (2150–1650 BC)	26	1	0.04
Castelluccio, tomb 23	Sicily	EBA (2150–1650 BC)	20	2	0.1
Plemmirio, tomb 48	Sicily	MBA (1650–1300 BC)	50	15	0.3
Thapsos, tomb 61	Sicily	MBA (1650–1300 BC)	19	1	0.05
Thapsos, tomb D	Sicily	MBA (1650–1300 BC)	49	4	0.08
Bisceglie, dolmen La Chianca	Southern Apulia and south-east Basilicata	MBA (1650–1300 BC)	13	5	0.38
Murgia Timone, tomb 1	Southern Apulia and south-east Basilicata	MBA (1650–1300 BC)	54	24	0.44

It is more difficult to estimate commonness of amber with respect to mass graves such as the Southern Italian hypogea, where it is not always possible to assign the artefacts to individual burials. Here one can, nonetheless, calculate the ratio of amber object-per-person by dividing the overall quantity of amber finds by the total number of burials (Tab. 50). It is clear that the coefficient is almost always lower than 1, meaning that there rarely is a case in which each burial would be given at least one piece of amber. In fact the only tomb where this ratio is above 1 is the 'Hypogeum of Bronzes' of Madonna di Loreto (1.025), but even there the distribution of amber was not uniform (cf. Vanzetti 1999, Fig. 2B). In the remaining contexts, this proportion is much lower. Hence, even in the case of the mass graves, one can suspect that amber was a rare grave offering, connected predominantly to individuals, rather than entire population buried in a tomb. Combining the observations stemming from Tab. 44 and

45 with conclusions of the previous chapter (4.1.2.4.), it seems reasonable to say that amber was attributed to certain members of society as markers of their status and identity in the representative sense, and as amulets/talismans to protect them from evil and misfortune in the religious one, because their lives apparently mattered or those people were particularly concerned with those issues. Whether those individuals belonged to 'elite' or, in other words, high-ranking groups in terms of wealth, power or birth right is difficult to say, but they must have enjoyed quite a distinct social position, enabling them to acquire or be given such rare products as amber jewellery. From here emerges a picture of amber as a quite luxurious material which only some people could afford.

4.1.3. Hoards/votive offerings

Hoards and votive deposits with amber in the circum-Adriatic zone form the smallest category of finds with respect to the type of deposition. Nevertheless, they are all quite different, making each one of them a specific case in terms of localisation, context and content. For those reasons they do not lend themselves easily to interpretation, and require thorough analysis in order to understand why they were deposited, and what was the sense behind adding amber to them.

4.1.3.1. Theoretical considerations

Hoards are one of the most debated issues in European archaeology. It is noticeable how the paradigm of hoards analysis underwent a major change over the years. Nowadays, interpretations increasingly depart from the nomothetic explanations of hoarding as a practice forced upon people by the desire to hide the valuables in times of trouble, or in the face of imminent danger (cf. Brunn 1968, 232). Also, many of the previous hypotheses considering hoards as deposits of scrap metal awaiting to be recycled, or otherwise as properties of travelling metallurgists and merchants (cf. Brunn 1968, 231), have been extensively criticised. The chief argument raised by the critics is that in all such approaches it is implied that these deposits were meant to be recovered from the ground, but the circumstances prevented this from happening.[665] Such implicitly assumed rule is difficult to defend when confronted with the periodicity of hoarding phenomena in Europe which would mean that at some stages people were more able to retrieve their hidden possessions, while at other times they were less likely to do it, as if prone to forgetfulness.[666] One may argue, following Reinecke's *Katastrophentheorie*, that ethnic conflicts and migrations led to drastic changes on the map of Bronze Age Europe, resulting in founders of the hoards dead or forced to flee, leaving behind great masses of belongings (cf. Harding 2000, 355; Fontijn 2002, 14). However, in many regions sites contemporaneous with the nearby hoards do not show any signs of unrest or crisis whatsoever, but instead bear evidence of thriving development. Implausibility of hoards' interpretations as personal valuables or commodity stocks is evident in the light of the observations showing that many of them consist of broken/fragmented (often purposefully as

665 For an overview of the archaeological approaches to hoards, see Fontijn 2002, 13–22.
666 Pauli 1985, 196; Needham 2001, 275–298; Bradley 2005, 146–147.

analyses show) objects.[667] On the other hand, the traditional interpretations of these 'metal scrap hoards' as concentrations of broken (useless – devalued by default) intended for remelting and recasting does not always make sense, as they sometimes include non-metal finds, e.g. amber as in the cases discussed here (cf. Dietrich 2014, 472–473). Some of the recently proposed alternative analytical approaches to hoards containing entire series of similar objects in complete or/and fragmented form focus on calculating their weight as the means of measuring value of metal in pre-monetary societies ('weighted currency economy');[668] however, standardised weight systems are difficult to prove for larger regions (Hansen 1994, 378–379). While one cannot rule out that one of the above-presented interpretations is probable to explain reasons behind deposition of some hoards, it has to be preceded by a careful idiographic analysis.

As a counterweight to the approaches trying to rationalise hoarding in modern 'common-sense' categories of minimising loss and maximising gain, there are archaeological works putting more emphasis on the cultic/religious aspect of depositing wealth. The definition of hoards as ritual deposits motivated by beliefs/religion was proposed quite early (e.g. Worsaae 1867), and was intensively developed by the German school of archaeology.[669] To put it simply, such interpretations can be characterised as steered by the conviction that the deliberate deposits of objects were not intended for people to be recovered, but were meant to remain buried forever. Sometimes a bit more specific term 'votive offerings' is used, implying that an object or collection of objects was deposited as a gift or sacrifice to the gods or supernatural powers, in the hope of receiving something in return, i.e. good fortune, blessing, forgiveness, etc. (cf. Bradley 1990; Hänsel 1997). Such interpretation has usually been reserved for remote places from which retrieving the objects would be impossible or hardly feasible, such as swamps, rivers or lakes. Places like these have magical connotations derived from legends and myths, e.g. the story of the King Arthur ordering the sword Excalibur to be thrown into the enchanted lake (or, according to a different version, being granted the sword by Lady of the Lake; cf. Mallory 1997). However, it has to be remembered that long-term transformations in the landscape caused by the climate could dramatically change the environment in which a hoard was found (Dietrich 2014, 468). Furthermore, the mystical aura ascribed to wet areas may not reflect what prehistoric people thought about them (cf. Bradley 2013, 124–125).

Over the years archaeologists debated the profane and sacred character of prehistoric hoards, and various criteria of differentiating between the two have been proposed (cf. Fontijn 2002, 15–17, Tab. 2.3). One of the most commented attempts at establishing general rules of identifying 'ritual' and 'non-ritual' hoards is by J. Levy (1982). In the wake of many previously used approaches, she distinguished four categories of hoards: 'ritual hoards' situated in unrecoverable contexts, away from the settlements (e.g. wet environment or under a mound), recognisable also due to the presence of residues associated with ceremonial practices: animal bones, broken pottery, traces of fire; 'personal hoards' comprising varied, but not damaged objects, usable after the recovery; 'craftsmen hoards' composed of metal scrap intended to be

---

667 For the latest contributions to understanding the phenomenon of fragmentation in hoards, see Dietrich 2014; Brück 2016; Hansen 2016.

668 E.g. Sommerfeld 1994; Pare 1999; Turk 2001.

669 E.g. Brunn 1968; Kubach 1985; Hansen 1991, 1994.

re-processed by smiths; 'merchants hoards' assembled from series of unused, freshly made objects belonging to one type, ready to be distributed. The latter three categories are found in easily retrievable places, close to the settlements, and can be defined as 'profane' (cf. Fontijn 2002, 17–18, Tab. 2.4; Gori 2014, 274).

The conclusions at which Levy arrived have since been criticised by a number of scholars. R. Bradley stresses that the categorisation of hoards was adjusted to the rich archaeological record in Denmark, and sees the limitations of applying Levy's criteria in a larger and more diverse spatial context (Bradley 2005, 146). D. Fontijn complains about Levy's overreliance on ethnographic and historical analogies, repeating the old assumptions regarding the meaning of hoards' location and composition, as well as inconsistency of arguments related to the rules of differentiating between 'ritual' and 'non-ritual' hoards (Fontijn 2002, 18). To the arguments *contra* any attempts of dividing hoards into ritual and profane on the basis of their location can be added the observation by S. Hansen that content of the hoards deposited in wet places does not differ much from those situated in dry places. While rejecting dichotomy between the two categories as merely epistemological, Hansen is in favour of interpreting hoards as votive offerings (Hansen 1991; after Harding 2000, 362, fn. 21). The significance of spatial context for interpretation of hoards should not be disregarded; however, it has to be determined through idiographic analysis of the relationships between hoards, other sites in the area and key features of the landscape (cf. Fontijn 2002, 259–272; Maciejewski 2015, 2017). Also, the possibility of a hoard being deposited for reasons other than ritual can be a plausible explanation, but only when circumstances of a discovery and context of a finding is thoroughly analysed (Harding 2000, 365–368).

In contrast to deterministic perception of hoards as either profane or sacred, some scholars promote alternative perspectives on hoarding as the phenomenon ritual in practice, but having a complex nature involving religious, social and/or economical aspects. The common ground for such interpretations is the conviction that hoarding had a deeper ideological motivation and symbolic meaning. There are several major contributions to the topic which attempt to characterise rationale behind deposition of hoards.

In the approach of K. Kristiansen, hoarding is perceived as an economy-driven practice, involving ritual 'consumption of goods' (Kristiansen 1978). The basis in which the author builds this interpretation is the assumption that bronzes – the main component of hoards – possessed the status of prestige symbols among Danish Bronze Age communities which, however, was dependant on the quantity of bronze in circulation. Thus to upkeep their value, the bronzes had to be buried in the ground in order to create scarcity (cf. Fontijn 2002, 18; Brück/Fontijn 2013, 199, 202). Another, more cross-cultural approach was formulated by R. Bradley who sees hoarding as manifestation of competition among the LBA communities of Britain, and uses analogies taken from ethnography, i.a. *potlatch* ceremony of the north-west-coast Kwakiutl native Americans, to build his case (Bradley 1990). The 'competitive consumption' involves bronzes exchange as prolonged process of gift-giving and receiving, aimed at building position in the intergroup relationships which could also involve ceremonial destruction of things; the process ends when one of the participating sides offers the ultimate sacrifice to the gods as a means to reciprocate a gift (or 'pay off the debt'), at the same time maximising the prestige (Mauss 1990, 22–25; cf. Gregory 1980). An important contribution to the discussion is that by J. Brück, who perceives hoards as one of the stages in 'lives' of artefacts they contain. Thus, from the analytical standpoint, they have to be integrated in the larger narrative beginning with ore procurement, melting casting, use and recycling (Brück 2016). The 'lives' of things are tightly connected to the histories of individuals, and their contextual transition and material transformations ought to be treated as liminal stages (rites of passage) in the process of construction, transformation and dissolution of personhood (cf. Battaglia 1990). Concurrently, Brück proposes to entirely drop the category of ritual as an analytical tool, as it implies existence of the 'non-ritual', whereas one should consider material and spiritual spheres seamlessly intertwining in the lives of prehistoric people (Brück 1999, 325; cf. Fontijn 2002, 21). The approach of D. Fontijn (2002), elaborated on the basis of finds from the southern Netherlands, in several ways follows the one of Brück. Fontijn argues that the massive deposition of valuables in later European prehistory, especially during the Bronze Age, is without ethnographic counterpart, and therefore demands a theorising that is historically specific (Fontijn 2002, 20). His integrative approach calls for accepting the unique rationale ('sacrificial mentality') forcing the Bronze Age people to dispose property that archaeologists might not be able to ever understand, but can get a grasp of its social and economic implications (Fontijn 2002, 275). The 'sacrificial economy' of the Bronze Age apparently demanded a part of the objects imported from abroad to be renounced and buried. It can be inferred from Fontijn's opinion that hoarding has a direct relation with the gift exchange as one of its manifestations; here the act of giving away was the most important part of the process, but from the performative aspect was embedded in a specific social and ritual context. Objects, depending on their form, to some extent were governed by the laws of deposition; however, their meaning and value was fluid, building up and transgressing the boundary between gift and commodity at different moments during the history of their use. The deposition of things in hoards can be perceived as culmination of their entire cultural biographies; it prevented the object from playing any role in future exchange histories (Fontijn 2002, 217). As for the selectiveness in depositing various categories of objects in different contexts or configurations, the author interprets it as a system of 'keeping things apart', intended on 'resolving ideological and political tensions stemming from different (sometimes conflicting) ideas and values that every society has' (Fontijn 2002, 279).

Each of the above-described interpretations of hoards in Bronze Age Europe has its proponents, as well as critics for whom they fall short of explaining the complexity of the discussed phenomenon. However, they certainly are innovative and offer a new look on some of the hoards already known for a long time. In the following, the available evidence of amber-containing hoards discovered in the circum-Adriatic zone will be reviewed and confronted with these approaches in order to determine which one of them reflects the specificities of given cases in the fullest manner.

4.1.3.2. Amber in the context of hoards/votive offerings

As mentioned at the beginning of this chapter, the hoards/votive offerings with amber in the circum-Adriatic zone might be few and sparse, but each of them is quite distinct. Despite apparent peculiarities differentiating these contexts, there are several characteristics in common for all of them. First of all, it has to be stressed that none of the hoards/votive deposits discussed here consists purely of amber; every one of them contained at least one category of non-amber finds. Furthermore, amber does not even form a majority of finds in terms of quantity in these contexts; the number of amber artefacts in these hoards vary from 1 (Monte Battaglia, Ljubljanica River) to 45 (Majdan). Consequently, all of the hoards can be described as of mixed composition. In each case, but one, the hoards or contexts in which amber was deposited yielded bronze artefacts. But can we interpret amber in hoards according to same principles as metal? I would argue that the studies in this regard can be of some use.

The hoards/votive offerings were spotted in various locations: most were located in secluded places, remote from other traces of Bronze Age habitation, usually in mountainous or hilly landscape; however, there are two such contexts detected within settlements (Frattesina di Fratta Polesine, hoard 1 *Tesoretto* and Torre Castelluccia, *ripostiglio*). The hoards/votive offerings deposited in uplands were by large found on the summits or slopes, rarely in the valleys, but there are also two cases of 'aquatic' deposits: one in a river (Ljubljanica River) and another in a marsh (Val di Non). The 'settlement' hoards were placed in the vicinity or inside the houses, sometimes in adjacent compartments. Using the criteria proposed by J. Levy (1982), one could try to divide the hoards with amber into the 'ritual' and 'non-ritual' ones. The location, being the chief criterion, would mean that the hoards outside the settlements fall into the 'ritual' category, whereas the two hoards known from settlements can be considered in utilitarian terms, either as the 'personal', 'craftsmen' or 'merchants' hoards.

Fig. 126. Non-amber content of the hoard Tesoretto *of Frattesina di Fratta Polesine (photography by Michele Baldo, Museo archaeologico nazionale di Fratta Polesine, with the permission of Italian Ministry of Culture).*

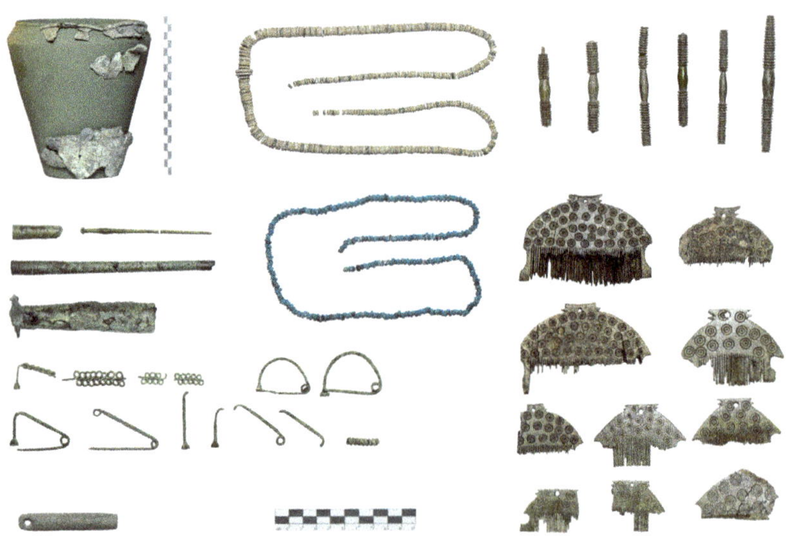

The deposit denominated as the hoard 1 *Tesoretto* from Frattesina contained 38 amber beads, 1 stone pendant, 3 ivory batons, 10 ivory combs, 374 bone beads, numerous annular beads of pale-blue glass paste and various bronze elements, including several complete and fragmented bronze fibulae as well as fragments of bronze *situla*

assigned to the type Kurd (Fig. 126).[670] The circumstances of discovery (the materials were reportedly recovered from a clod of earth after ploughing of the field on which the site is located; cf. Bellintani/Peretto 1972, 32; Bellintani 2000, 65) raise question about character of the original context, i.e. were all these objects deliberately and simultaneously deposited in one place? Still, the finding is usually referenced to in archaeological literature as a hoard – interpretation reinforced by the fact that from the same area of the Frattesina settlement also come the three other similar deposits, commonly known as the 'smelters' hoards (cf. Bellintani 2000, 65). The *Tesoretto* hoard does not fall neatly into any of the categories of 'non-ritual' hoards coined by Levy. It is difficult to treat it as the 'merchants' hoards, because the finds wear traces of being used and are broken in many cases. Furthermore, with respect to some amber artefacts, absence of perforation prevents from considering them as finished products, whereas presence of bronze and bone tips in two of them can be a sign of failed attempts at drilling, making the objects unsuitable for use (see sections 3.10. and 3.11.). Thus, it is unlikely that the finds from *Tesoretto* were stored in the ground in anticipation of the right moment for their recovery and distribution which, contrary to their owners' expectations, never came. Also, interpretation as a 'craftsmen' hoard does not appear plausible in the view of Levy's definition, as the finding largely contained non-metal finds which can hardly be recycled. While it can be speculated that some of the damaged objects (e.g. ivory combs or bronze fibulae) were intended to be turned into new forms, most of the bone, glass and amber beads were good enough to be worn. Finally, identification of the hoard as a 'personal' deposit seems the most acceptable, but Levy emphasises that one of the criteria for such deposits is the unspoiled state of the objects enabling them to be used upon recovery. As explained above, many of the objects from *Tesoretto* do not meet this requirement.

Same problems arise when trying to define character of the hoard *ripostiglio* from Torre Castelluccia. The hoard was found inside the compartment – house no. 7, in a small stone enclosure concealing two ceramic vessels and collection of assorted objects, i.a. bronze bracelets, torcs, fibulae, razor, dagger, glass bead, stone pendant, bone spacer, ornamented elephant tusk, and six amber beads (Gorgoglione et al. 1993). Space between the stones revealed traces of fire seen mainly as reddened soil, while some of the finds were broken or damaged. Question remains whether deliberately or due to post-depositional factors. The setting in which the hoard was found prompts various interpretations. One of them is that the context is in fact a cremated burial; however, it cannot be proved as no bones were discovered at the spot (Gorgoglione et al. 1993, 84). Classification of the finding as the 'merchants' hoard, according to Levy's definition, is less acceptable than in the case of Frattesina's *Tesoretto*, as the hoard from Torre Castelluccia did not include such numerous series of similar products. Furthermore, its interpretation as the 'craftsmens' or 'smelters' hoard does not withstand the argument that the hoard comprises a selection of various materials, also non-metal objects, which is relatively small to obtain substantial amount of material to produce new objects. At last, it is possible that the hoard consisted of personal belongings – interpretation even more plausible considering its location inside a house. However, traces of combustion and state of preservation of the objects suggest also an alternative solution, seeing the Torre Castelluccia hoard as a ritual deposit. Earlier it

670 Bellintani/Peretto 1972; Negroni Catacchio 1972; Bellintani 2000, 65.

was shown that the amber finds from the 'Cult Center' and 'Hut-temple' structures at the nearby settlement of Roca Vecchia can be integrated into the narrative describing those buildings as places of ceremonies and rituals, where food was consumed and votive offerings were deposited (cf. section 4.1.1). Amber at Torre Castelluccia was found in direct association with multiple objects concentrated in a single, clearly delimited space, hence its classification as a hoard. However, the reasons behind burying amber in those two sites in southern Apulia possibly stem from religious practices or social ceremonies involving offerings. Both sites revealed traces of fire and broken pottery, regarded by Levy as auxiliary indicators of ritual practices. From the standpoint of Levy's classification, the only difference between them is presence of food remains at Roca Vecchia reinforcing the latter's ritual character.

It is clearly visible that the mixed compositions of the two hoards described so far stand in contrast with the purely metal content of the vast majority Bronze Age hoards known from Europe, rendering previously applied strict classifications, such as the one by Levy, unsuitable. Also the location criterion does not seem to be of essential importance for distinguishing the 'non-ritual' hoards, as shown by the given examples from Roca Vecchia and Torre Castelluccia. Therefore, it seems necessary to test relevance for other theoretical approaches towards the hoards.

The remaining eight hoards with amber were all discovered in areas outside the settlements; therefore from the perspective of older concepts regarding hoards, they can be considered as ritual deposits. Moving aside the debate on correct criteria of distinguishing ritual deposits, the 'ritual' function has already been proposed for three of them. Two of the latter qualify as 'aquatic' deposits – votive offerings submerged in wet areas: collection of amber artefacts, including two spacers, from Val di Non bog near Cles (Angelini et al. 2003, 228–229) and the pin with amber head from Ljubljanica River (Turk/Gaspari 2009, 70). Although being amber these are exceptional finds, and they fit into the general trend in depositing wealth in water reservoirs, observable in both sub-Alpine zone, including Trentino-Alto Adige and Slovenia, around the later part of the Bronze Age (cf. Marzatico/Tecchiati 1998, 30; Turk/Gaspari 2009, 68–69). The second hoard from Trentino-Alto Adige – the one discovered in the Finail pit (Maso Finale) high in the Alps (2460 m a.s.l.) can be ascribed to another category of votive offerings, involving sacrificial combustion of gifts, known as *Brandopferplätze*, loosely translated to 'votive bonfires', widespread in the latter half of the Bronze Age in the Alpine area (Putzer 2012, 162–163; cf. Marzatico/Tecchiati 1998, 30, 39). The meaning of the aquatic hoards as well as votive bonfires is usually interpreted as a peculiar form of sacrificing 'gifts to the gods' linked to the religious beliefs of local populations (Marzatico/Tecchiati 1998, 43; Turk/Gaspari 2009, 68). What is specific about them is that they rarely include amber.[671] The circumstances of finding the amber assemblage in Val di Non are unclear, and it cannot be ruled out that the beads were a part of a larger deposit (see section 2.2.4). With respect to the Ljubljanica River find, it is notable that the amber was inserted on a bronze pin, therefore posing question whether it was the pin that was considered as an eligible gift, or the amber pinhead, or both? Finally, the amber beads revealed at the *Brandopferplatz* of Maso Finale were

671 A. Putzer notes only one other *Brandopferplatz* that yielded amber – St Walburg/Ulten in South Tirol (Putzer 2012, 162) – which unfortunately could not be included in this analysis in time before publication of the book.

found not concentrated in a single place, like in most of the hoards, but scattered all over the site. A. Putzer writes that 'the role played by the numerous beads in the ritual has to be left open' (Putzer 2012, 164). Assuming that the function of the site was based on sacrificial burning of votive offerings, it seems relevant to highlight amber's flammability, especially when powdered, giving a bright flame. Perhaps the sense of using amber at Maso Finale was to burn it by throwing pieces to the fires inside the four pits uncovered at the site, all of them filled with charcoals and ash, and the amber beads found in the 'cultural layer' is what has been left after such rituals? These divagations are suited to the sites with a specific function like Maso Finale. The greater problem is to find an answer why and to what extent amber was considered eligible to be deposited as a votive offering/hoard?

It is tempting to explain deposition of amber as a practice aimed at preserving its value as a luxury, rare and prestigious material, in line with K. Kristiansen's concept regarding bronze deposition (Kristiansen 1978), especially in the light of the chronology of the hoards with amber which falls into LBA–FBA. At that point amber had been present in the circum-Adriatic zone for some time, and in some parts was becoming increasingly available and abundant (e.g. western Veneto). However, small number of hoards with amber refrains from seeing its economically reasoned deposition as a mass phenomenon. Moreover, the hoards were found by large in provinces relatively poor in amber (e.g. Piedmont, Trentino-Alto Adige, Marche and eastern Umbria, southern Apulia, Slovenia and Vojvodina). Hoarding amber was not particularly widespread in the main provinces of amber's acquisition during LBA and FBA (e.g. western Veneto, Emilia, the Garda Lake basin, Romagna and Emilian Apennines, Istria or Kvarner and northern Dalmatia). As mentioned earlier, there are no hoards consisting purely of amber, nor there are hoards where amber forms a substantial percentage of the entire content. Amber rather appears as accompanying addition to the prevalent bronze objects. Therefore, the discussed hypothesis favoured by Kristiansen does not seem to reflect the idea behind depositing amber in hoards.

Another alternative is to perceive the hoards as deposits of the personal belongings of the deceased which were banned from being added to graves by tradition or socio-religious norms, or which could not be deposited together with body of the deceased due to some other reasons (cf. section 4.1.2). This interpretation of the hoards was advocated by some of the German scholars, who coined the term *Totenschätze* loosely meaning the 'treasures of the dead' (Hundt 1955; Torbrügge 1985). One can recall application of this notion in the context of river deposits of bronze pins – objects often co-occurring with amber in graves – in the southern Netherlands (cf. Fontijn 2002, 240–241), and use it as an analogy to explain deposition of the pin with amber head in the Ljubljanica River.

The hoard whose content entirely fits the general 'model' of grave inventory observed in the FBA burials with amber is the one from Frattesina *Tesoretto.* The context yielded objects known from numerous graves analysed so far: i.a. glass beads, bone beads, bronze fibulae. It might be puzzling why such a grave-like assemblage was not accompanied by skeletal remains. Surely, one can come up with a number of possible answers; however, there is one hypothesis which seems worth considering in the light of the modern discussion on the hoards' meaning. By accepting two laws: that the objects deposited in graves are linked to the identity of buried person, and that objects are marked with meanings arising from histories of their use by individuals, one can imagine that there were certain cultural norms regulating relationships between people and objects, i.e. how the latter are to be used as a medium of expression for the ones 'self'. From the analysis of burials with amber, it follows that amber was, by large, reserved for women and, far less commonly, men in the age varying from infant to senile, lacking items described as tool and weapons in their inventories. While it is difficult to determine other identity features of the persons buried with amber, it may be assumed that somehow it was right for them to be associated with this material. Deposits of jewellery like the hoard *Tesoretto* of Frattesina can, therefore, be interpreted as manifestations of the rites of passage, involving dissolution of one's former personhood and entering a new social identity potentially linked with marital status, e.g. marriage, widowhood, etc. (cf. Brück/Fontijn 2013, 208). After D. Fontijn, we can consider intentional deposits of personal belongings as 'a practice in which the paraphernalia of a social role were laid down. They were not handed over to others to start a new life of circulation. Neither were they physically associated with the remains of what was once a living person, thus creating an indissoluble link between the deceased and the statuses associated with the imagery. Similar to the argument about weapons, we can imagine that depositing ornaments in natural places may be a way to deal with personal identities that should be temporary, ambiguous ones, related to special roles' (Fontijn 2002, 239). This, of course, is only one of the probable explanations; however, to some extent it lends itself to application in contexts of three other hoards with amber in the circum-Adriatic zone: Gualdo Tadino in Umbria (Peroni 1963b), Monte Cavanero in Piedmont (Venturino Gambari 2009) and Majdan in Vojvodina (Rašajski 1988).

All the three hoards listed above comprised mostly unbroken elements of jewellery, not unlike the ones from graves with amber: i.a. bronze pins, fibulae, various pendants, bracelets, torcs, golden discs, glass beads. They are supplemented by few other items with some connotations to burials with amber: awls, combs, tweezers, belt-hooks. Furthermore, the hoards of Monte Cavanero (Fig. 127) and Majdan included bronze vessels of several kinds. The hoards' contents reveal a considerable level of 'eclecticism' regarding cultural provenience of finds it comprised, not in the sense of exoticism of materials from which they were manufactured (although amber, gold and glass qualify as such), but in terms of their morphology (see section 3.12.4.). As in the case of Frattesina *Tesoretto*, the complex and genetically varied inventories from Gualdo Tadino, Monte Cavanero and Majdan seemingly attest to the 'high status' of people who deposited them or to whom they belonged – possibly females 'dressed in internationality' (cf. Fontijn 2002, 243). One can recall the idea of travelling wives, whose status was about to change when married and admitted into a new community.[672] As opposed to many burials of women whose foreignness is recognisable thanks to the burial outfit, in the cases discussed here we are possibly dealing with renouncing the previous cultural/social affiliation by burying costume (dowry?) which serves as an identity indicator. In the context of Umbria, Piedmont and Vojvodina, where it was very rare and certainly is not attested in graves, amber could be viewed by the local communities as alien to their culture and not fitting to the image of feminine attire, what possibly explains its absence in burials.

672 Cf. Jackenhövel 1991; Kristiansen/Larsson 2005, 232; Bergerbrant 2007, 118–129.

Fig. 127. Content of the hoard of Monte Cavanero (Venturino Gambari/Ferrero 2013, Fig. 3, with the permission of Soprintendenza per i Beni Archeologici del Piemonte.).

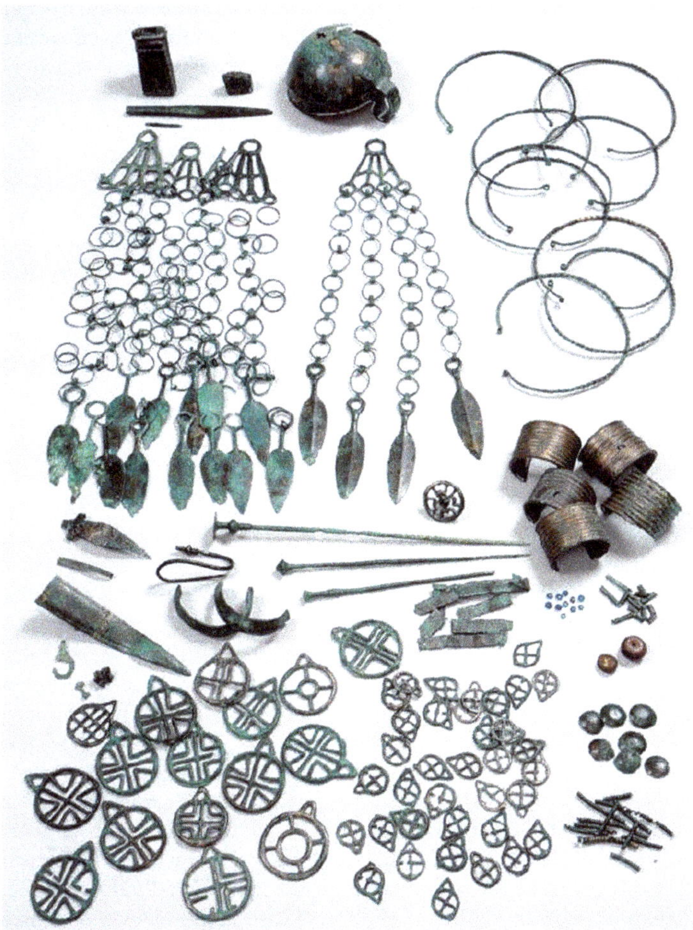

Still, there are some features in the three above-discussed hoards that stand in opposition to their interpretation as renounced personal belongings. First of all, each of them yielded few elements uncommon for burial *parures* including amber: tools such as chisels (Gualdo Tadino and Monte Cavanero), hammers and casting moulds (Monte Cavanero), axes and sickles (Majdan), or even weapons such as daggers (Gualdo Tadino) and sword blades (Monte Cavanero and Majdan). Presence of the bronze hammer and stone-casting mould even make the hoard of Monte Cavanero eligible for the definition of a 'metallurgist's/smith's/smelter's' storage in the light of the traditional classifications of hoards. Another problem with this one is the broad chronological range of the finds, spanning several centuries (cf. section 2.2.2). While those seemingly 'unfitting' objects are vastly outnumbered by jewellery in the discussed hoards, they still prompt to seek a different, hopefully more suitable interpretation.

Applying the above-discussed concept to the remaining hoards with amber in the circum-Adriatic zone (i.e. Debeli Vrh and Monte Battaglia) is even more difficult. The hoard of Debeli Vrh in Slovenia (Hirschbäck-Merhar 1984) contained a massive collection of fragmented as well as complete bronze weapons and tools, including predominantly socketed axes (15 pieces) and sickles with tongue-grip (22), supplemented with several fragments of swords, daggers, knives and spearheads. They vastly outnumber elements of jewellery found in the hoard which mainly comprise broken fragments of bracelets, fibulae and a torc, together with few pieces of bronze belonging to unidentifiable objects. Also the hoard of Monte Battaglia in Romagna-Emilian Apennines (Bermond Montanari 1997) yielded bronze axes and sickles (two pieces each), as well as small portions of blades, which match in numbers the ornaments (bronze pins, bracelets and torc) coming from the deposit. Objects such as bronze swords, axes, sickles, razors, knives or spearheads are very rarely registered in the burials with amber (see section 4.1.2.3). As noted by D. Fontijn, 'weaponry was an ambiguous category in material culture, and that martial identities were temporary ones, constructed and deconstructed in ritualised circumstances' (Fontijn 2002, 221). Deposition of bronze armaments, swords in particular, outside the settlements and cemeteries, was a common phenomenon in many parts of Europe during the Bronze Age. Weapons, especially those wearing marks of use, could have particular kinds of life-paths (by dealing a lethal blow in battle) which forever branded them. Such connotation could have been at odds with the cultural values and social identities, and prevented them from being buried in graves. For getting rid of such objects in an 'honourable' manner, specific places were chosen, which had a special meaning in the consciousness of people, often far away from living areas to avoid 'crossing paths' with such items again. In this way swords were removed from the society, and martial ('warrior') identities, that they are linked to, were temporarily renounced (Brück/Fontijn 2013, 209). One type of depositing weapons could have been hoards; T. Mörtz identified a particular group of the 'scrap metal hoards' as ritual offerings after successfully conducted war-like events (Mörtz 2013, 59, after Hansen 2016, 199).

The hoard of Debeli Vrh contains axes and several other weapons, thus qualifies best for this category. However, it also comprises items usually not associated with warfare, i.e. tools such as sickles, or jewellery. Also amber stands in contrast with such attribution. It is worth noting that amber beads from Debeli Vrh were found inside one of the axe's socket, giving a way for its interpretation as a 'hoard within a hoard' – specific cultic phenomenon noted in several cases in Europe (cf. Dietrich 2014). Another problematic issue connected with Debeli Vrh and Monte Battaglia is that in those hoards the non-amber objects often bear marks of intentional destruction. This includes weapons and tools as well as ornaments which in grave inventories comprising amber normally reveal relatively good state of preservation. This leads us to the phenomenon of 'fragmentation' encountered in many Bronze Age hoards which itself poses many questions and has been discussed extensively in the passed years.

The recent investigations into this phenomenon emphasise the symbolic nature of fragmenting – dividing objects as means to enter them into another meaningful stage in the semantic continuum through which every thing passes during its lifetime (Fontijn 2002, 26; Brück/Fontijn 2013, 212). It could have been the way to deal with personal objects possessing a meaning and value to their owners/users which were required to be sacrificed as a votive offering or a gift. According to the *pars pro toto* idea, upon breaking one part of the object stayed in the hands of the giver, while the other one was deposited, thus retaining a material link between the donor and the offering, while adding to it a metaphorical connection between the donor and the receiver (Hansen 2016, 200). Even though amber finds coming from the hoards are not fragmented – chopped or broken, it is still possible that they were subject to partitioning, e.g. by dividing necklaces and leaving a portion for the hoard, whilst keeping the reminder. Assuming the religious/cultic dimension of hoarding, the fragmented material could have been divided into a divine and human share (Nebelsick 2000, 169–170). S. Hansen wonders if the fragmentation of objects in hoards 'reflects a trend towards expanded participation in the "Salvation" (ger. *Heil*), which came with the expansion of the burial custom of urn graves from the 13th century BC onwards' (Hansen 2016, 200). If one insists on the interpretation of the hoards as 'graveless grave goods' (*Totenschätze*), then the fragmentation of objects can be explained through certain ethnographic analogies. For instance, in communities of the Tiwi Aborigines, personal property is treated as an objectification of immaterial memories about the deceased, hence is subjected to deliberate ritual destruction by mourners and become *taboo*, because it is considered too painful to be kept alive (Venbrux 2007). When using this interpretation, it is hard to overcome the observation that weapons and tools in general were not deposited as burial inventories together with amber. Nonetheless, it has to be noted that the same act of property destruction is used by the Tiwi Aborigines to sever relations between the people as 'management of emotions' (Venbrux 2007). The above considerations lead to an assertion that hoarding amber in connection with other non-amber artefacts should be perceived as a specific, ritualised practice embedded in beliefs and social realities of the Bronze Age communities in the circum-Adriatic zone.

There are reasons to believe that practice of depositing large amounts of bronze in the ground had to do with the religion and cosmology professed by Bronze Age people. Explanation of hoarding as a ritual of offering gifts to the supernatural powers was already brought into discussion in relation to the aquatic deposits and *Brandopferplätze* of Ljubljanica River, Val di Non and Maso Finale, as well as the in-house hoard of Torre Castelluccia. Now it is time to check if a similar interpretation can be applied in the case of the large hoards/votive offerings with amber located outside settlements or wet areas.

In as many as four hoards with amber, there is a recurring feature of the sun symbol among the bronze finds, visible as a combination of circles with or without dots and crosses.[673] The solar motif appears on the large ornamented bronze discs

673 Decorations comprising circles; concentric circles with or without a dot in the centre; circles divided into four, six or eight equal segments; and circular motifs filled with various decorative spiral elements are generally considered to be sun symbols by archaeologists working in the contexts of the Bronze Age (Pásztor 2015, 1341; cf. Fredell 2003). The cross-circle (sun cross) is posited as being a universal Bronze Age sign that originates from the moving four-spoke wheel and symbolizes the orb of the sun with the four cardinal points (Pásztor 2015, 1346; cf. Larsson 1999).

('shields') from Gualdo Tadino, Debeli Vrh and Majdan in the form of engraved or embossed concentric circles, sometimes arranged in more complex patterns (Fig. 128:A, C–D). Here it is worth noting that a 'solar shield' very similar to the one from Gualdo Tadino was recently discovered in the so-called 'hoard of gold' inside the Roca Vecchia's 'Hut-temple' (Fig. 128:B) – context also known for revealing amber beads possibly deposited during religious ceremony. It is important to take the material from which the discs were manufactured into consideration. The shields from Gualdo Tadino and Roca Vecchia were made of gold, whereas those from Debeli Vrh and Majdan from bronze (in the former case determined precisely as the white bronze – cf. Hirschbäck-Merhar 1984, 97). This observation shows that the discs stand apart from the remaining objects in the hoard not only due to their specific form, but also the source material. The lavishly ornamented golden discs from Gualdo Tadino and Roca Vecchia are definitely unique within the scope of the circum-Adriatic zone, however they share similarities with many shields known from Europe,[674] most notably the Trundholm sun chariot (Kristiansen/Larsson 2005, Fig. 133) and the golden sheets found at Delos and Cephalonia (Bettelli 2012, Fig. 2:3–6). The simpler shields from Debeli Vrh and Majdan are characteristic for the Carpathian Basin, while their distribution in the Balkans encompasses mostly the Sava-Drava interfluve and falls into the IInd horizon of the *Urnfield* culture hoards.[675] These objects are universally treated as religious symbols or liturgical paraphernalia belonging to priests and priestesses connected with the cult of sun personified by the sun-goddess (or the 'sun maiden') and her twin brothers ('divine twins') which became widespread in Europe in the second half of the 2nd millennium BC contemporary to similar cults in the *Mycenaean* Greece and the Near East – Hittite Kingdom (Kristiansen/Larsson 2005, 251–319).

674 De Angelis et al. 2007, 119; De Angelis/Gori 2017, 370; cf. David 2003.

675 Cf. Vinski-Gasparini 1973, T. 46:21, T. 53:38, T. 58:6; 1983c, T. XCIII:15, 17, 19; Rašajski 1988, 23.

Fig. 128. Discs and pendants with solar/wheel motifs: A – Gulado Tadino, hoard (Peroni 1963b, Fig. 5:1–2); B – Roca Vecchia, 'Hut-temple' (Scarano/Maggiulli 2014, Fig. 6:b); C – Majdan, hoard (Rašajski 1988, Fig. 14, 30); D, E – Debeli Vrh, hoard (Hirschbäck-Merhar 1984, T. VIII:1); F – Monte Cavanero, hoard (Rubat Borel 2009b, Figs. 53 and 54).

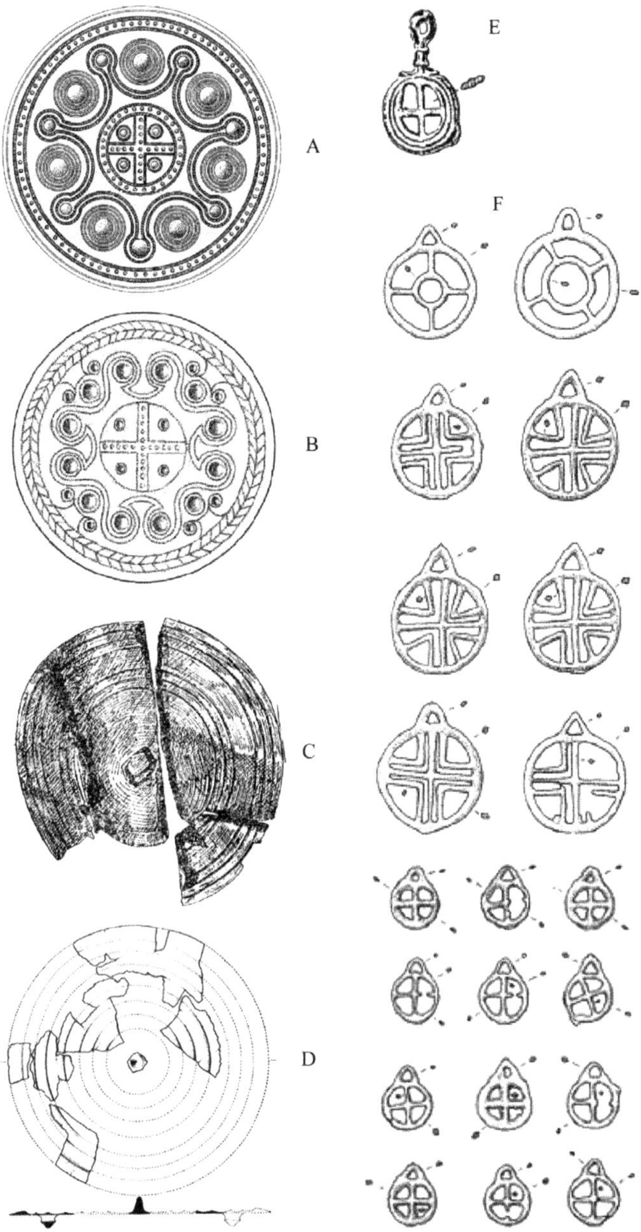

The wheel/sun iconography was reproduced on different types of bronze objects from this period such as pendants, pinheads, armour and belts, and also appears on objects made of tusk, bone, clay, and even wood and cloth (De Angelis/Gori 2017, 363). The solar motif is also apparent on the pendants from Majdan and Monte Cavanero, consisting of crosses inscribed in circles (Fig. 128:E–F). The pendant from Majdan falls into the type Bingula Divoš distributed predominantly in the Sava-Drava-Danube confluence (cf. Pabst 2008, Abb. 8:2), whereas the pendants from Monte Cavanero represent numerous types defined by F. Holste (1939) and G. Kossack (1954), with good analogies in the hoards of Villethierry in France and Kanalski Vrh I in Slovenia (Rubat Borel 2009b, 72–79, Figs. 51–54; cf. De Angelis/Gori 2017). It is interesting to note that the pendants from Monte Cavanero upon inspection did not reveal any traces of use, but instead possessed casting imperfections as if they were just finished, hence the conclusion that they were made with the intention to be buried in the hoard (De Angelis/Gori 2017, 368–370). Overall, the hoards present us with the evidence of deliberate deposition of objects with deep religious/cultic symbolic and meaning, rendering the ritual interpretation of hoarding plausible. Thus amber, if available, could have been used in votive offerings connected to the solar cult alongside other, more commonly sacrificed items.

So far we have seen that hoards with amber rarely consist of uniform selection of goods, reflecting the 'model' grave inventories in which amber was found. The objects co-occurring with amber in hoards represent various functional categories, all of them possessing different meanings, from the strictly utilitarian belonging to the sphere of mundane life, to deeply symbolic connotations which resurfaced when the objects were used in rituals and ceremonies. They all have varied social attribution, ranging from jewellery (e.g. amber beads) usually ascribed to women (but not always as shown by anthropological analyses), weapons and tools typically seen as men's domain, to rare and peculiar items possibly handled as insignia of political or spiritual sovereignty. Moreover, all of the hoards contained various categories of objects made of exotic materials/substances (extraneous for the region of deposition), or made in a manner (morphology and decoration) foreign to the region of deposition. From the above stems the image of hoards as reification of communal identity and interpersonal relationships vs graves which can be seen as reification of personal identity.

There was no universal law obliging people to deposit amber in the sacrificial ritual of making offerings. In comparison with the widespread practice of hoarding in the study area which is well documented in relation to the advanced stages of the Bronze Age (LBA and FBA), hoards with amber seem few and far between. Certainly, the limited availability of this material had an impact on the rate of amber deposition in hoards. Nevertheless, judging by the relatively low number of hoards with amber in relation to numerous bronze hoards registered in the circum-Adriatic zone, it can be inferred that amber was not commonly considered as a regular component of the deposits. Perhaps it has to do with amber's varying status or norms governing its deposition as an offering.

Amber artefacts, prior to the deposition in a hoard, had a firm connection to the individuals as personal belongings, although presumably regulated by cultural norms of costume with respect to social status. Amber went to hoards, only if it was considered a valid and valuable contribution in a given circumstances. Upon its deposition in such collective deposits, value of amber was socially reconstituted as a part of the communal offering/gift, although it retained a link to its owner-giver, which could be manifested by dividing the necklaces in two parts: one which would be buried and the other which will stay with the donor. Here, the notion of hoards as a gift exchange seem the most suitable to interpret why and how amber was sometimes included in such deposits.

It is a common perception that hoarding practice was a highly codified, complicated and repetitive type of ritual (De Angelis/Gori 2017, 363). However, I would argue that there were various rationales behind the hoards, from the performative side disguised in a ritual, but with clear goals to be achieved. It is difficult to pinpoint in each case what was the reason of hoarding amber. Nevertheless, a few plausible explanations can be given.

Amber itself was an import in the circum-Adriatic zone and as such it was often buried in hoards together with other materials of foreign, or mixed local and foreign provenience. Here we go back to the opinion expressed by D. Fontijn that hoarding in the eyes and practice of Bronze Age people could have been a meaningful and right way to deal with imported objects (Fontijn 2002, 275). Amber, thus, could change hands in a communal gift-giving between representatives of culturally different communities at the geographical boundaries (e.g. hills and mountains – Monte Cavanero, Monte Battaglia, Gualdo Tadino, Debeli Vrh, rivers and marshes – Ljubljanica, Val di Non), or at the *entrepôt* (e.g. Frattesina). It was a peculiar manner of gift-giving/exchange which involved burying a part (either in form of fragments, or in form of selected artefact, e.g. amber beads), in turn receiving a bulk of the material which would be actively used by the locals. In this place it is worth reminding the discussion about the exchange and trade outlined in section 1.2.1. Interactions involving goods transfer (including amber) could have had various other forms which did not leave recognisable residues, be it exchange or trade. Rather than simply acquiring commodities, hoarding as a specific form of gift-giving was more about the exchange in which other gains were at stake. Attaining them apparently required renouncing some material wealth. Amber sometimes took part in such interactions and could have change hands on those occasions.

From one perspective, deposition of a hoard was an aftermath of exchange between communities – an act possibly obliging the participants to offer a part of the acquired materials to the deities ('the sun goddess') out of gratitude or/and in supplication for prosperity and favour. As noted by M. Mauss, gift exchange does not only take place between men but also between humans and supernatural powers (Mauss 1990, 39–49). Thus, each intentional deposit – hoard – can be interpreted as an offering, gift, act of exchange between people, or between people and the world around them. Consequently, some of the offerings could have been precisely destined for religious ceremonies. On such specific occasions, the deposits were supplemented with objects with concrete religious symbolism (e.g. solar shields) – tangible expressions of intentions, whereas amber, aside from playing a passive role of an offering, could have been actively manipulated in the rituals (e.g. by being burnt – Maso Finale). Some goods obtained from afar could have been produced to be sacrificed immediately upon exchange (e.g. wheel pendants from Monte Cavanero), or used for some time until they were eventually considered as 'ready' to be buried.

It is important to emphasise the visual aspect of hoarding. Arguably, the act of renouncing wealth was more important than the actual commodity/utility value of objects being deposited. From the performative side, hoarding ritual was intended to have an impact on the spectators/participants. Hoarding can be understood as an arena for reconstruction of social structure and positions of the actors within it, for power is a product of interpersonal relationships and interactions, and must be constantly renegotiated in the face of shifting social and material conditions (Brück/Fontijn 2013, 204). A. Weiner distinguishes two types of transactions that are fundamental to social and cultural reproduction: those that has to do with maintaining status quo, or stasis, and those that has to do with developing status quo by giving gifts (Weiner 1992). Exchange can have a social, economic or religious dimension, but in every variant things are sacrificed in hope of certain rewards, be it maintaining status (prestige) of the giver by material manifestations of power or being granted a blessing from the gods.

It can be concluded that hoarding was a ritual with a deeper meaning, extending beyond mere exchange of goods. This approach seems to be the most universal and acceptable as it is very capacious in terms of actual events it could encompass: e.g. commemoration of the dead involving deposition/burying of their belongings, rites of passage for community members or religious ceremonies involving votive offerings. In the context of analysed hoards, the ritual was possibly conducted on the occasions of i.a. demonstrations of power aimed at enhancing social or political positions, ending conflicts, making alliances, religious ceremonies involving votive offerings or social events involving admitting new members to a community (e.g. weddings). During the rituals participants, headed by the main actor of this event, be it a bride, a chief, a warrior or a priest, deposited portions of their belongings followed by others in attendance, who added their own gifts in the effect creating a varied assemblage of objects which was at the end covered with earth.

4.2. Amber as a 'sémiophore' – proposition of a new interpretative model

To conclude the considerations commenced in chapter 4, I will first recapitulate the observations on its functional-utilitarian aspects and then move on to its symbolic meaning linked with social and religious spheres. Amber beads have been usually interpreted as finished goods used as jewellery, presumably as elements of larger, composite necklaces or bracelets, perhaps in some cases also sawn into the fabric of a garment, or attached as decorative additions to costume accessories (e.g. pins or fibulae) (cf. Negroni Catacchio 2011a, 58). The notion of amber as an adornment finds confirmation in the analysed data from the circum-Adriatic zone. Amber was worn there during the Bronze Age as a part of dress, either on daily basis or, otherwise, in circumstances when it was required to be displayed (i.e. during specific events). Redesigned and socially accepted version of the costume 'of the living' was used as the posthumous outfit; thus, amber went into graves. Other forms of amber's use are less documented or remain speculative. Amber could have been also used in medicine, as its therapeutic properties were already noted by ancient scholars (cf. Negroni Catacchio 2009, 193). Apart from attractive looks and presumed healing properties, amber's significance stemmed presumably from its spiritual/divine connotations, particularly with the solar cult (cf. Czebreszuk 2011, 165). Its role resided in the religious sphere of life, i.e. as a devotional symbol ostentatiously worn or deposited in ground (i.e. in graves during funerals, or hoards during religious ceremonies), as a talisman providing magical powers (e.g. blessing, protection, strength to its bearers) or as a liturgical item used in rituals (e.g. burnt as incense). Here, we have already touched upon the issue of amber's status.

Amber was used precisely because it had a capacity of emitting certain meanings important from the perspective of prehistoric inhabitants of the circum-Adriatic zone. The message it was carrying was facilitated in reception by amber's inherent properties (translucence, warm colour, propensity for inclusions, distinct smell when burnt, lightness, etc.). However, the meaning ascribed to amber was coined by people who tried to rationalise and interpret those apparent qualities as signs of supernatural powers and divine origin. Thus, the special cultural status of amber was built upon its natural traits. It is probable that its rarity, distant origin and inherent qualities contributed to amber's status as an attribute of certain social group whose privilege was owning and wearing objects made from it. The latter could serve as a tangible sign communicating affiliation of an individual with such groups, along with other information which archaeologist are yet unable to recognise. Potential identities of individuals buried with amber seem to be discernible in some cases (e.g. marital status – married woman, widow; social role – priestess, noble, non-militant members of a society). However, proving some unified identity of amber users is difficult, considering age and, to a lesser degree, sex variations among skeletal remains, as well as the slight differences among grave inventories. It is quite likely that amber's material and symbolic status, unambiguous and defined as it was, was sometimes adapted by individuals beyond the existing social norms or customary patterns; rare cases of such behaviour are visible in the circum-Adriatic zone (e.g. burials of armed males adorned with amber). Still, it is clear that amber was predominantly an attribute of women and children (girls?), but rarely of men. Even the male burials with amber often lack very specific paraphernalia of warriorhood. The latter appeared with amber also in several hoards, but even these are sporadic. Perhaps its status was irreconcilable with the sphere of warfare, but instead matched certain features/ideas personified by femininity. Maybe women and children, usually appearing in sepulchral contexts as peaceful and defenceless, were perceived as society members exceptionally deserving this gift. However, this right was limited only to a very small portion of population. Statistics on the commonness of amber in cemeteries support the idea of amber as an exclusive possession of high social strata ('elites'), who could afford and import it, and then use it as an emblem of their position/rank. Yet, this interpretation is quite simplistic and skips many peculiar forms of amber adaptation/appropriation.

To better understand its status, amber should be perceived in more individualised terms: certain properties and symbolism it carried appealed to people, who eventually adopted it as an element of costume. Thus amber became their personal property and tied its history to their biographies. Usually amber objects ended their lifecycle in graves, but on several occasions found their way to hoards which symbolise complex and varied events which were grounded in the ritual of gift-giving and depositing votive offerings. In the course of hoarding, amber was renounced by an individual (lost its specific social attribution) and became a part of collectively deposited assemblage (gift, offering), albeit with a link to the donor. Amber artefacts appearing in graves might not have been the deceased's personal property, although such solution is probable, but they still belonged to someone and, thus, had a history of use. That amber items (jewellery) were used, and not only made for specific occasions like funerals, can be inferred from the findings of amber on settlements. Depositing personal belongings as inalienable goods fits well into the context of graves and hoards with amber. However, gift-giving as a way to create an invisible but long-lasting bond between the people, or people and things, events, spirits or gods, through renouncing and depositing a part of property is also a sound concept in this regard.

As for the often assumed prestige and luxuriousness of amber among communities, it would be an overstatement to label it as an international symbol of wealthy, upper classes of society, who all treated it alike. Amber was relatively costly in comparison with other locally, available materials as its import from the far-away lands must have involved lots of effort. But that does not mean that every person who could afford it desired it. One should not forget that there are regions and contexts (e.g. graves) which revealed evidence of considerable 'richness', in the material sense, and highly developed social structure, but somehow amber is absent there, e.g. Campania (cf. Pacciarelli 2001, 85–87), or the *Vatin* culture (cf. Ljuština 2011). If people in those places clearly had wealth and capacity to import other precious and rare materials, why they were not interested in amber? It is easy to say that they might not have a taste for it, but it seems more plausible that amber with its symbolic and meaning did not fit the culture of those people, or did not meet their needs. Thus, amber should not be universally regarded as a possession or emblem of elites, but its prestigious value during the Bronze Age can be accepted in the light of its inaccessibility and rarity, both in chorographic and social aspects.

Fig. 129. Pomian's model of the 'sémiophore' (Jackson 2006, Fig. 1).

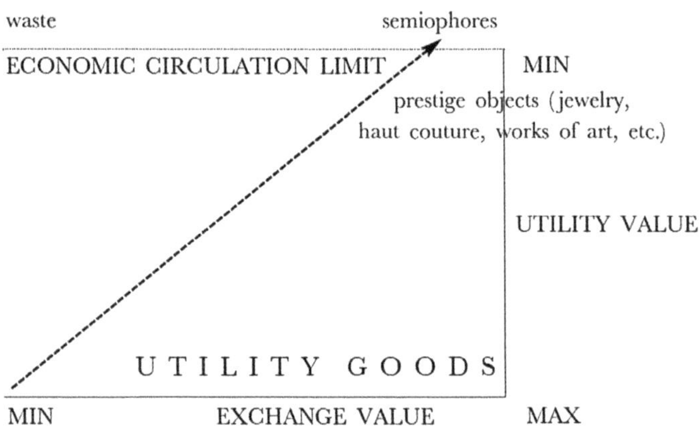

In the author's opinion, the status of amber can be effectively described with the concept of 'sémiophores' invented by K. Pomian. It may be argued that this idea has a capacity to embrace all the previously cited theories concerning circulation and deposition of things. In his book devoted to the phenomenon of art and curiosities collectors of the 16th–17th century in Europe, Pomian fully presented the idea of 'sémiophores' and how they relate to collections (Pomian 1996). Sémiophores (literally 'meaning carriers') are objects recognised in a given community as, first and foremost, conveyors of meanings. Thus, they are produced or exhibited in such a way as to attract the gaze at the same time excluding any other function, or while maintaining their utilitarian function. According to Pomian's original concept, their utility value, however, is limited or entirely non-existent; the more an object is loaded with meaning, the less its usefulness and vice versa (Pomian 1996, 48; Domańska 2016, 350). For the object to become a sémiophore, it has to increase its 'exchange value' at the expense of the 'utility value' – a process which can eventually lead to their exclusion from a common economic circulation (Fig. 129). A simple example might be nails – widespread tools with high utility value but low exchange value. However, upon being placed in a specific context, subsequently fixed as significant in history/memory, a nail can become filled with meaning and marked with specific connotation which would result in abandoning its original function and turning into a sémiophore, e.g. 'holy nails' – relics claimed to be used in Christ crucifixion and venerated in the Christianity. The value of sémiophores is determined by the attraction of people to gather and possess them; the fact of having them adds prestige to their owners (Pomian 1996, 23).

In reality, there are also items that seem to have both utilitarian and sémiophore qualities. They position themselves at various levels on the scale between useless and highly practical objects, and between those meaningless and those saturated with meaning. A good example might be vintage instruments (e.g. guitars) of famous musicians – they are revered by their fans who long to have them; however, staggering prices achieved by these things at auctions stop many from buying them, so only the richest can afford them. Still, these instruments can be played, thus fulfilling their utilitarian role; however, they are more likely to be kept unused at display, to preserve their value and, at the same time, be tokens of prestige and adoration. The visibility is one of the most important features and prerequisites of sémiophores.

No object is essentially a sémiophore. It is about the role that an observer (interpreter) assigns to the object. Sémiophores reveal their meaning when they are made available to viewers (Pomian 1996, 48). Both utility and meaning assume the existence of an observer, because they are only relations that individuals or groups maintain through objects with their visible or invisible environment (Pomian 1996, 47). Sémiophores, as objects primarily carrying a meaning, are objects representing the invisible sphere – they mediate between the viewers in the observable world and concepts, images or events embedded in the sphere of memory, imagination, spirituality, etc. (Pomian 1996, 37). E. Domańska considers sémiophores as material actors, agents who introduce changes in the environment: 'in this way, they also become an interesting example of relational items whose status and meaning are determined by a changing place in networks, including collections' (Domańska 2016, 349). In this way the concept of sémiophores corresponds with Actor-Network-Theory by B. Latour (2005) described earlier in this work (see section 1.2.2).

Sémiophores are often included and set for display in collections. The 'collection' according to Pomian is 'any set of natural objects or products of human activity, temporarily or permanently kept outside the area of economic activities, subjected to special care in a closed place, adapted for this purpose and exposed for viewing' (Pomian 1996, 22). The essence of collection is to ensure a link between what is invisible and visible (Pomian 1996, 13), by being viewed/received by some actual or virtual viewers (Pomian 1996, 26–28). The basic illustration of this concept are art collections exhibited in galleries, museums, etc. and kept outside the common commodity market (they are only sold at auction houses). However, Pomian stresses that grave goods, votive offerings and outstanding gifts (e.g. royal) known from ancient times can also be interpreted as collections (Pomian 1996, 25–30). Recognising equipment of a grave or hoard as collection seems justified, because from our point of view, it is not important to consider their recipients as living, dead, ghosts or gods. It is important to allow virtual viewers, existing in death, different from our view in time or space, which is implicit in the very act of placing objects in the graves or offering them in temples (Pomian 1996, 36).

The content and nature of a collection depends on the status of collectors, the place they occupy in the hierarchy of power, prestige, education and wealth (Pomian 1996, 14). Pomian also sees people of high standing in social hierarchy as sémiophores – representatives of the invisible sphere, e.g. authority from god – chiefs and priests (Pomian 1996, 49–50). Thus, objects can become sémiophores in the hands of certain people as insignia of their power and social position.

Amber lends itself reasonably well to the definition of sémiophore. Although predominantly used as an adornment, amber's utility value was quite small as shown by its use in prehistory and history. It certainly was limited by the physical properties of amber. Beads and pendants are the most obvious example of putting amber to use, but jewellery is a quite specific category of items. It is not located on the same level of utility as, e.g. tools; jewellery's role lays within its emotive capabilities when viewed, thus adornments fulfil the visibility criterion for sémiophores (the requirement of being exposed to view) by default so to speak (cf. section 4.1.2.1). The looks attracting attention might only be an invitation to start delving into an object's meaning. Was it only amber's attractive appearance that dictated its function and causing it to serve primarily as decoration? Amber could have been utilised as a medicine, although such function remains speculative with respect to the Bronze Age. Same as in the case of its burning as incense or wearing as amulet, its assumed healing properties derived from belief rather than actual effects. Even today amber is used in the largely same way, despite the omnipresent technological progress. One cannot rule out few more applications of amber in the past, but they are not indicated by the analysed data. Thus, with its restricted usefulness and substantial popularity associated with certain implied meanings, amber oscillates somewhere between rather low utility value and considerably high exchange value.

Amber beads or necklaces found in grave inventories or hoards can be regarded as collections or their parts. In the individual aspect, amber artefacts can be sémiophores conveying identity and status of the person they were signifying (adorning or buried with). In a broader sense, connected to its religious/spiritual symbolism, amber could be perceived as a link with the sphere beyond the real – observable world, a tangible proof of the existence of supernatural powers. Its connotations with the sun symbolism – cult have already been discussed. It also seems fitting here to recall histories about the Hyperboreans, recorded in the Greek literature of the Archaic and Classical periods and known to this day, thanks to Hecataeus of Abdera and his work *On the Hyperboreans* from the 4th century BC. They were considered to be the happy people of the followers of Apollo, the solar god (Winiarczyk 2006). According to the tradition dating back to the 7th century BC, these peoples gave their gifts to the oracle of Delos every year, among them amber, through two Hyperborean priestesses (Bouzek 2007). This story finds confirmation in archaeological discoveries that led to the finding of the graves of two women from the Bronze Age near the Artemis temple in Delia (Winiarczyk 2006, 43). The hyperborean myth has a lot of significance for interpreting the relationship between the north and south Europe in the context of amber translocation, already in the period of the *Mycenaean* culture (Czebreszuk 2011, 167). Also Pomian refers to the myth, describing the gifts sent by the Hyperboreans as collections representing people living in the far-away and hardly accessible, if not legendary lands (Pomian 1996, 36). Thus, through its rarity and remote origin, amber's role as a sémiophore is revealed; the artefacts made of this material granted its users/viewers connection with the distant, mystical region in the north, where it came from. Amber's inherent capacity for attracting curiosity, transferring messages and evoking emotions already in the raw form gave it a considerable potential as a sémiophore. Once it was ascribed a meaning, it could easily emit it, thanks to its natural properties. If objects accumulate meaning throughout their lives – histories of use, then the longer and more intricate are their biographies/itineraries, the more significance and, thus, exchange value they possess (cf. section 1.2.2). Intentional deposition of amber in the ground, either in graves or hoards, could have been a way to preserve its meaning and high exchange value as a sémiophore, which it could lose upon entering or (re-entering) commodity exchange (cf. section 4.1.3.1).

It may be difficult at first to comprehend how amber entered into circulation and moved across large distances as a sémiophore. As Pomian notes, exchange value of objects has its basis in their meaning (Pomian 1996, 48). It can therefore be assumed that it is in the interest of the collection owners (individuals or communities) to keep sémiophores outside the area of utility and economic circulation, because only then they can retain their significance and prestigious value. Furthermore, Pomian notes that in some cultures sémiophores are not exchangeable for things; their commodification is barred by tradition and social norms (Pomian 1996, 48). On the other hand, the more an object is loaded with meaning, the greater its value, which is henceforth measured by the number of things it can be exchanged for (Pomian 1996, 49). Pomian notices the paradox of his definition of sémiophore as an object of high exchange value, but kept out of the economic cycle. He responds to that by saying that sémiophore can conditionally enter economic cycle only if it is 'considered possible'; the object then gains a 'guise of utility' and moves back into the area of economic circulation (Pomian 1996, 49; cf. Fig. 129). Nonetheless, Pomian did not specify the mechanism by which sémiophores are or can be commodified.

A suitable explanation of this problem can be borrowed from M. Bloch's and J. Parry's idea about relationship between gift and commodity (Bloch/Parry 1989). Throughout their lives objects can become subject to the short-term sphere of exchange (i.e. commodity exchange – trade) as well as to the long-term sphere of exchange (i.e. exchange of personified gifts between people, and between people and the supernatural) (Bloch/Parry 1989; cf. Fontijn 2002, 31, 254). Thus, objects can be commodities or things at one stage in their cultural biography and gifts possessing specialised meanings at another. Using examples from early historic societies and anthropological analogies, Bloch and Parry are able to list various procedures, that every society has, which allow the objects to move between the short-term and long-term spheres of exchange. The mechanism allowing the objects to move from one to the other sphere works by moral evaluation: 'While the long-term cycle is always positively associated with the central precepts of morality, the short-term order tends to be morally undetermined since it concerns individual purposes which are largely irrelevant to the long-term order. If, however, that which is obtained in the short-term individualistic cycle is converted to serve the reproduction of the long-term cycle, then it becomes morally positive' (Bloch/Parry 1989, 26).[676]

To translate the above considerations into the context of sémiophores, there are two main, legalised ways in which they can circulate among people (apart from other less commonly practised and often illegal practices, e.g. robbery of art). The first one is through gift-giving (long-term sphere of exchange) which arguably is a more personalised form of passing an object considered as a sémiophore. The second one is by commodification of a sémiophore and streaming it into a wider circulation, however requires moral/ethical re-evaluation (e.g. most people think long and intensively before selling something to which they are emotionally attached or what has a great meaning for them, be it a family heirloom or work of art). Sometimes commodification of these materials can be the only possible way to acquire them in the first place (especially if they need to be imported from afar, as in the case of amber), so later they can play their role in the long-term exchange again, e.g. as important regulators of communal life and social structure. Each act of commodification of a sémiophore involves the risk of losing that original meaning and opening it to new definitions upon entering a different social context. Nevertheless, this does not mean that the loss of meaning and value is inevitable; observations from the circum-Adriatic zone, i.e. similarities in use and deposition of amber, tend to show that the status of amber was consistent by large; however, its social attribution and forms of appropriation varied slightly in space and time.

Decoding amber's status and appropriation in the regionally varying cultural, social and religious realities of the Bronze Age requires approaching almost each case of its occurrence individually. Nevertheless, basing on the observations stemming from the hitherto conducted overview, a simplified model showing history of its introduction, incorporation and eventual renunciation by the circum-Adriatic societies can be proposed. The model depicted here is inspired and replicates some of the aspects of the diagram elaborated by P. Jackson, who used sémiophores to develop his idea of information economics, especially information as a measure of uncertainty (Fig. 130). In order to analyse this phenomenon, Jackson proposed to replace the utility value and exchange value used in Pomian's theory with concepts of entropy[677] and redundancy,[678] which he writes about, referring to the creator of information theory C. E. Shannon (1948).

Transmittance of a message 'becomes costly in terms of information economy, if the number of allowed elements of a set (its redundancy) increases in spite of its

676 D. Fontijn describes this transformation on the example of bronze axes which, apart from their strictly utilitarian function, played a double role in the Bronze Age exchange: as a commodity money and a source of raw material (Fontijn 2002, 251). Through specific biographies they moved in status from tools to commodities to gifts in various order, while the two possible transformation mechanisms could be re-shaping the objects (e.g. bronze re-melting and re-casting) or *pars pro toto* sacrifice (e.g. offering a part of newly acquired goods to the deities) (Fontijn 2002, 254).

677 In information theory 'entropy' can be interpreted as the uncertainty of occurrence of a given elementary event in the next moment. If any event in the set of events occurs with a probability equal to 1, then the entropy of the system is 0, because it is known in advance what will happen – there is no uncertainty.

678 In information theory, redundancy refers to an excess of information/meanings that a message contains in relation to the quantity necessary to convey its content.

relative entropy' (Jackson 2006, 21). In other words, if the message become too loaded with information/meaning, it might be difficult and time-consuming to decode it, especially if the recipients are not prepared or equipped to do it. On the other hand, 'if a process is perfectly predictable and patterned (no entropy), on the other hand, it has no information value' (Jackson 2006, 221). In Jackson's model the curve 'a' signifying increase of information economy reaches its climax – perfect information economy – at the lowest redundancy and moderate entropy, meaning that an object and the meaning(s) it carries are both deemed useful and easily readable by society or its part, but important rather in utilitarian terms – low meaning (Fig. 130). Prestige (curve 'b'), on the other hand, raises to the maximum at the highest redundancy and moderate entropy, thus inducing things (including symbols and paraphernalia) into sémiophores, emitting costly and increasingly ambiguous information that can be decoded only with certain knowledge. The model tends to depart from Pomian's very strict division of objects into utilities and sémiophores, and shows how they are transformed in the continuum between minima and maxima of exchange and utility values. Jackson gives a number of examples illustrating this process which are not going to be described here. Nevertheless, it is important to note his remarks on applicability of the model. Jackson warns that the graph presented on Fig. 130 is only a simplification that 'will give a wrong impression if read as a representation of actual facts'; however, 'if different but coalescing processes and forces are to be represented synoptically, certain optical inaccuracies are inevitable' (Jackson 2006, 222).

Fig. 130. Jackson's model of information economy (Jackson 2006, Fig. 2): a – information economy increase curve; b – prestige increase curve; c – technical utility increase; d – information/uncertainty increase; e – ambiguity increase; f – technical utility decrease; A – redundant patterning, isochrestic choice; B – prestige (charisma, prestige objects, rhetoric, ritual actions and paraphernalia, symbolism, etc.); C – redundant noise (infinite set equiprobability); D – insignificant utility, ergonomics, nutrition, technical expediency, tools, etc.; E – maximum information economy (significant utility, binary systems, radio communication, stenography, etc.); F – inertia (exhaustion, nondegradables, pollution, wear, etc.).

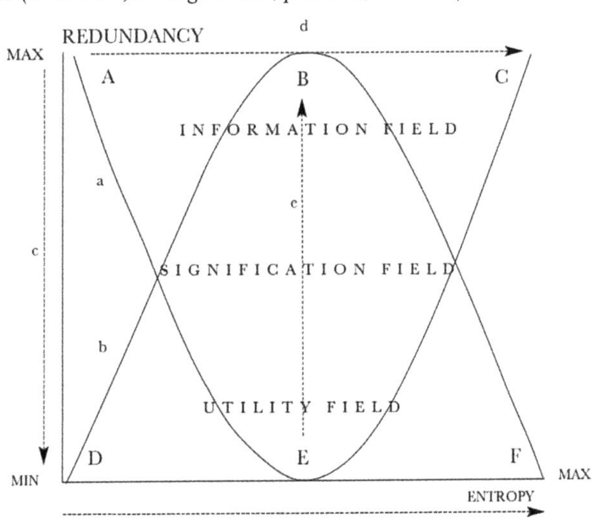

With that in mind, it was decided to reconfigure the model in keeping with Pomian's and Jackson's basic assumptions, to build a graph illustrating a generalised pattern of amber's incorporation into culture by circum-Adriatic societies (Fig. 131). Trajectory of this process is illustrated by two curves with six segments (a–f) in relation to six dimensions: dissemination level in society, time, entropy, redundancy, utility value and exchange value. Some of the areas – liminal stages in the life of objects were also borrowed from Jackson's model of information economy (cf. Fig. 130). Like in Jackson's case, the presented graph also requires a comment regarding its applicability. The dimensions are not scaled and lack precisely determined minima and maxima. They are only to give impression about the general trends of increase and decrease in relation to each other and to the curves signifying the processes of amber adoption. In fact, the latter should be placed somewhere in the middle of the redundancy, utility and exchange values: it may approach the extremities of scales, but cannot be measured in absolute numbers. Furthermore, at low levels of redundancy and with the continuously raising entropy, the curve may seem to drop infinitely, at the same time indicating increase in utility value. Thus the graph should be seen as an illustration of one in the series of similar episodes, each ending with the reset (zeroing) of entropy. Quoting Jackson, 'the entropy beyond the so-called "information field" should thus be considered qualitatively different from the entropy within the so-called "utility field". In the latter case, entropy merely implies the successive but inevitable disintegration of bodies and physical artefacts, not the entropy of stochastic processes' (Jackson 2006, 222). The model also is not precisely defined in terms of time; the timescale at the base of the graph can equally reflect long phenomena lasting several centuries (or archaeological periods/phases), as well as short-term events – histories of singular amber objects. It is argued that the presented model can be applied to single amber artefacts, their entire assemblages in regions or cultures, short phases and longer periods, because they are all governed by similar principles. Still, the model requires contextualising with respect to particular regions of the circum-Adriatic zone and stages of the Bronze Age. In summary, the general history of amber introduction, adoption and renunciation can be described as follows.

Fig. 131. Trajectories of amber incorporation into societies.

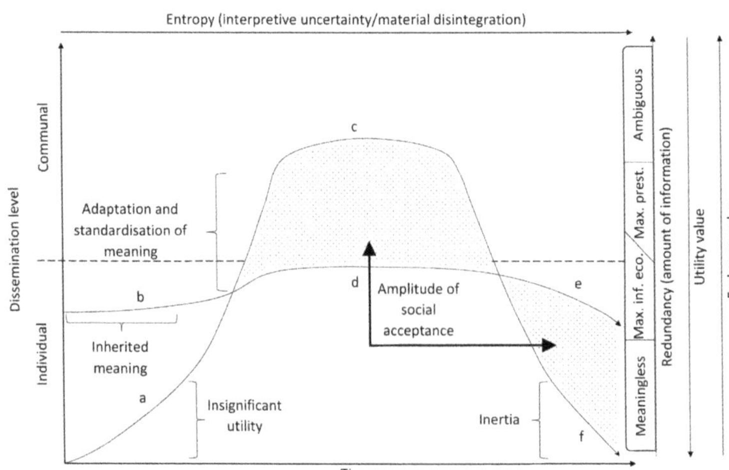

The introduction of amber to a community always proceeded through individuals, who came into contact with the material first. People possibly took interest in amber due to its utilitarian aspect, i.e. its attractive looks and potential as an adornment (segment 'a' on Fig. 131). Thus, initially amber could have been nothing more than a curious trinket without any deeper meaning or place in a society. This explanation is particularly plausible for the regions, where amber occurs naturally, e.g. Baltic shore or Sicily. On the other hand, amber might have been desired and imported precisely because of its status among the communities already using it (sémiophore status – high exchange value), the knowledge of which has spread beyond the geographical boundaries of a region in which the given fossil resin occurs (b). In this variant, apart from its physical qualities apparent to spectators, amber inherited meaning and symbolism assigned to it by the people who earlier had contact with the substance, and integrated it into their culture. This semantic content might have been adopted by the people new to amber. Both ways (a and b), those were always individuals who decided its introduction and they were the only ones who at first comprehended its meaning and function. That is particularly evident when comparing the numbers of the earliest contexts or artefacts of amber in regions of the circum-Adriatic zone. On its own amber was silent; certain visible traits certainly drew people's attention, but its symbolism was yet to be ascribed (a), or might pass unnoticed or be misinterpreted (b). Transmission of this content could not have taken place without an active agency of individuals, who were interested in disseminating this material. Wider adoption of amber required those persons to have a considerable influence over a community.

After the introduction stage, amber could gain a wider recognition and be adopted by a community or social group (c), or remain a substance valued only by individuals, not bound to any group by status or identity (d). If it did not meet a wider recognition, as in the latter case (d), amber was perceived by the public mainly in utilitarian terms (as jewellery) with fairly insignificant and narrowly understood meaning/symbolism. Hence at that level it was not overload with information in the eyes of beholders, although for its users/owners, it could have a personal value (as 'inalienable' good). Still, its impact as a sémiophore was rather low. Here we can mention a number of the lone cases of amber's sepulchral deposition within some of the circum-Adriatic regions that appear as exceptions from locally prevailing burial norms. Usage of amber in those instances could have been substantiated by its inherited meaning derived from previous social context. However, the failure to increase its redundancy – exchange value (converted to things for which it can be exchanged) meant that it remained sparse in those areas.

Only when the meaning of amber and custom of its use were internalised by a portion of population, especially a specific social group, it began to act as a significant element of gifts (long-term exchange) or trade (short-term exchange). Wider incorporation of amber into culture was accompanied by adaptation and standardisation of meaning in the local cultural context (c). At certain point in the raise of dissemination level from individual to communal, amber started to enjoy maximum information economy – its symbolism/meaning was acknowledged by a wider group or community, but was not banal/trivial on the one hand, nor too complex on the other. The process of its incorporation into culture can be explained with the concepts of 'organic' and 'intentional' hybridity (Werbner 1997). In contrast to 'organic hybridity' meaning adoption of ideas, words and objects in unreflected way, 'intentional hybridity' can be understood as conscious adaptation of ideas and objects borrowed from the outside by social actors to distance themselves from other groups within a given society (Maran 2012, 64). Upon its introduction to a new social, cultural or religious context, amber was prone to reappraisal in the semantic sense, and could be imposed a different meaning. Furthermore, new attributes might be added to older meanings, causing an increase in redundancy – process more dynamic when occurring at the communal/group level. It eventually led to decrease in information economy, however gained in terms of prestige. This process can be seen as reflectance of 'intentional hybridity' in amber's incorporation, and is somewhat confirmed by the narrow social attribution amber among the circum-Adriatic communities, which boiled down to a relatively small group of women and children, and only occasionally men. Upon adaptation and standardisation of its meaning, amber fully turned into a sémiophore – its widely recognised status, although now not fully comprehended by the entire population, was more important than its utility value (what translated into growing exchange value). Along with growing information redundancy, amber at some point garnered maximum prestige as a symbol and paraphernalia of social standing, identity, religion, etc. From then on it seems reasonable to presume that amber was subject to various norms which regulated its use and deposition. This is visible in cases of communities possessing substantial quantities of amber and indicating normalised social attribution of amber (e.g. western Veneto, Emilia, or Podrinje and Western Serbia during MBA–LBA). Nevertheless, even within communities where amber had precisely defined meaning and social attribution, happened cases of its more individualised appropriation; one can mention here male burials with amber in the Hypogeum of Bronzes at Trinitapoli. The variation in scale and form of amber's adaptation is meant to be illustrated by the amplitude of social acceptance on Fig. 131.

Even if amber and its meaning have arrived from afar and nestled in human consciousness, they were still open to renegotiation in the face of changing social conditions. Thus, the role and status of amber was never given once and for all. For it to fulfil its function and transmit a specific message, it had to be continuously imported and displayed, while assuring that the social structure and status are maintained. If amber was to sustain its status, then its users (or people who it was ascribed to) had to make sure that it does not become more redundant which would eventually lead to an uncontrollable growth of ambiguity. Thus, its accessibility had to remain limited. High ambiguity in connection with common availability of the material (after it was commodified) may cause the material to lose its prestigiousness and become something of ordinary (e.g. works of art which became over-present in the popular culture to an extent that they are considered tacky or vulgar). This problem tended to grow in time with increasing entropy, when status of amber could encounter various challenges, and the material itself faced the imminent disintegration. As mentioned above, the influence of entropy could have a material dimension: wear and tear of objects leading to their reduction to waste, and unforeseen events such as retention/obstruction in import. But entropy can also be recognised as questioning the previously obtained status in the face of changing social or economical conditions. One could mention in this place widespread emergence of the *Urnfield* culture during LBA which in some regions (e.g. Istria, or Podrinje and Western Serbia) accompanied amber's withdrawal from use. Breaks in amber's import due to various reasons (e.g. wars, economical crises, etc.) could have been another factor in raising the entropy and undermining amber's once well-established status.

These factors at some point could render amber's status untenable. In the course of social transformations, amber could lose its previous meaning what would inevitably lead to growth of the utility value at the cost of exchange value. In such situations amber fell back in dissemination level to the domain of individuals (e), or was gone from an area because it became inert – was meaningless. Situation depicted by segment 'e' is reflected by FBA contexts from western Veneto which no longer show specific social attribution and widespread use known from the preceding phases. Otherwise, amber ceased to be used temporarily because even its utility value was not considered significant enough to incline people to import it (f). On the other hand, high entropy could trigger a new approach to amber which might result in its comeback to a wider public, although in different social conditions.

The periodic popularity of amber and changes in its status and social attribution are well documented with respect to historic times. After the first stage of its international 'career' in the late prehistory and antiquity, amber was largely 'forgotten' during Medieval. At that time its use was mainly limited to production of devotional items (e.g. crucifixes) and common objects used almost exclusively in the areas, where amber was readily available (circum-Baltic zone). The revival of amber, which took place during the 16th–18th century AD, saw increase in export of amber products manufactured in ever growing number of forms – usually luxury commodities (furniture and art) intended for royalty, aristocracy or bourgeoisie (cf. Christ 2018). Decrease in its significance came again in the late 19th–20th century AD, when amber became identified with cheap and often tasteless jewellery (cf. Sobecka 2016). Only recently there has been a renewal in interest in amber; it finds new applications and the material is slowly gaining in status, but already in a different social context. Conclusions derived from observations of amber's history during modern times confirm the changing status and social attribution of amber in the long-term perspective, although one can risk a statement that at the core of people's interest in this material always reside the unique physical properties and its supposed relationship with the sphere of magic, spirituality and supernatural.

5. Amber in the 'macro'-perspective: circum-Adriatic zone and neighbouring regions

Knowing the highly localised geographic distribution of fossil-resin sources as well as the European-wide dispersion of amber-made products during the Bronze Age, amber finds from the circum-Adriatic zone cannot be considered in isolation. On the one hand, communities from the regions under study by large had to import amber from the outside. Localisation of the largest European deposit of amber in the circum-Baltic zone suggests that this was the go-to area for craftsmen and merchants ('amber mongers') in the past, as it is in the present. Thus, to properly delineate the circum-Adriatic branch of the amber route, it seems necessary to determine first which parts of Northern Europe provided amber and which ones mediated in its transfer to the south.

On the other hand, abundance of amber in Greece during the Bronze Age prompts to consider the role of the circum-Adriatic zone, itself placed in-between Northern Europe and the Aegean, in the process of meridional circulation of the material. It is right to expect that if the Italian and the Balkan regions of amber's acquisition took part in the transfer of amber further south and east, there should be some traces of contacts preserved on the sites and in the contexts included within the scope of this study. Furthermore, to evaluate to what degree the circum-Adriatic zone provided the Aegean with amber, it would be helpful to compare amber artefacts from the two areas quantitatively and qualitatively. Outcome of such analysis may then be confronted with the hitherto posited concepts about the nature of interactions between Northern and Southern Europe during the 2nd millennium BC.

5.1. Imported or locally produced? Transfer of amber between Northern Europe and the circum-Adriatic zone

In chapter 4 it was described what drove people of the circum-Adriatic zone to obtain amber and how their attitudes to the material were changing during the Bronze Age. Nevertheless, the question about the exact realisation of amber's import has so far been left untackled. There is a whole series of issues connected with this problem. In the beginning, it can be asked: from where was amber imported?

That the amber used by circum-Adriatic communities during the Bronze Age came, in the dominant part, from the north of Europe has been credibly proven by chemical analyses of the finds, chiefly IR spectroscopy (cf. section 2.4). Unfortunately, the scientific methodology currently used cannot differentiate between samples obtained from various deposits of succinite. Thus, the source areas of amber utilised in prehistory have to be specified using other criteria, such as typology of amber artefacts and co-occurring finds.

The question remains, however, whether amber was transported in unprocessed form (raw lumps or prefabricated elements) or as already finished products. If the former is true, one would expect to find evidence of local processing in the study area, e.g. production waste, tools, blocks of raw materials, semi-finished products, etc. Furthermore, it is also necessary to consider the character of processing, i.e. was it performed by specialists (skilled practicians manufacturing specific forms of objects in series) or non-specialists (common craftsmen having basic knowledge in treatment of various materials)? Was it stationary (performed in a dedicated place – workshop) or mobile (artisans travelling with their tools and materials from place to place)? As it occurs, one question leads to another.

Amber is a rather soft material, so any likely candidate with a set of basic tools commonly used in the Bronze Age (e.g. blades, drills, scrappers made of flint, bone or metal) could engage in its processing, if first managed to get hold of the raw lump. However, it takes time and practice to master the process, especially perforation drilling which, in the case of more sophisticated forms, involved the use of lathe. To go through the learning curve required a steady access to raw material, which was not easy for most inhabitants of the circum-Adriatic zone, taking into account the necessity of importing it from afar.

G. Woltermann (2016a, 207), in part following P. Wielowiejski (1997), applies three criteria allowing for interpretation of a finding or findings group as a 'work place/station':[679]

1. Evidence of a material (amber) in different stages of production (raw material, waste, semi-finished products, finished products, repaired items);
2. Evidence of one or several abovementioned production stages in significant quantities which surpass the assumed personal needs, or indicate prolonged (multi-phase) use;
3. Evidence of a handicraft device or a tool specified for a material (or a production step), the spatial positioning and socialization of which attests to its use at the place of its deposition;

Points 1 and 2 are considered indispensable for distinguishing a 'work place/station', and can optionally substitute for point 3. By using these criteria, one may try to verify where and when it is possible to speak of indigenous amber processing in the circum-Adriatic zone.

Already at the beginning of this work, it was explained that there could have been a number of different ways (events, actions, interactions, transactions, etc. – cf. section 1.2.1) through which amber changed hands, eventually ending up in the circum-Adriatic zone. One of the more likely mechanisms of its translocation were exchanges (amber as a gift) and trade (amber as a commodity). Chapter 4 explains how materials (resources) and items (products) transgress the boundary between gift and commodity, or inalienable and alienable. If amber took part in such transactions, it is interesting to consider for what it was exchanged/traded. Moving further along this line of thinking, it can be asked whether we can recognise those things (resources, products, goods etc.) in the material evidence available to archaeologists, and do they say anything about the potential exchange participants or trade partners cooperating

679 By the term 'work place/station' G. Woltermann understands aggregated finds testifying to processing by a single person or group of people, at the same time advising its use over the term 'workshop' which is 'too closely linked to the idea of a building' (Woltermann 2016a, 207).

with the communities of the circum-Adriatic zone. Answering such questions requires an in-depth overview of the literature and finds related to the circum-Adriatic zone in the entire part of the European continent which was involved in amber circulation, from the latter's sources all the way to places of its deposition. This is a challenge surpassing the scale of this work. Nevertheless, it is argued that acts of exchange/trade, especially if occurring frequently in a region between same two parties, could leave by-products of such transactions. These 'by-products' could be imports (materials and objects of foreign origin) or external influences on metallurgy or other crafts, discernible among finds from the circum-Adriatic regions of amber's acquisition, or even context of its deposition. Thus, through an overview of literature and assembled data related to the regions, sites and context with amber in the study area, i.e. provenience and typology of co-occurring finds, it is attempted to pinpoint areas, cultures and communities in Europe providing the circum-Adriatic zone with amber.

Only a small portion of the typologically analysed amber finds from the circum-Adriatic zone – 25 artefacts in total – can be securely dated to EBA (ca. 2150–1650 BC). To this group of the earliest finds can be added a second one, almost as large – 21 artefacts, comprising objects chronologically overlapping EBA and MBA. There are several commonalities between the two. First, amber finds from the both groups are typologically homogeneous to a large extent: they only include forms perforated concentrically along H, with the dominance of flattened beads, followed by the average ones and then the elongated ones. Predominantly, the artefacts represent discoidal (IF121, IF122) or spheroidal beads (IA122) which are widely attested in other regions of Europe at the time, including Hungary (Sprincz/Beck 1981, 478, 481, Tab. 2), France (Du Gardin 1986, 547), Romania (Boroffka 2001, Abb. 3:1–12), Poland (Bukowski 2002, Fig. 8:c, 11:b, 14:2, and 15:d), Bohemia (Erneé 2012, 82, Obr. 7:A) and Germany (Woltermann 2016a, 40–41, 143, 147). Most of the remaining types can be regarded as variants of the abovementioned discoidal and spheroidal beads, distinguished due to irregularities in their forms (e.g. IF128, IA125), which are not recognised as individual types by other authors. However, it needs to be stressed that even the rare EBA–EBA/MBA types, which appeared only in single cases within the study area (e.g. IA111, IA511, IF611), find good analogies in the areas to the north, particularly in the large assemblages of amber adornments belonging to the *Únětice* culture in Poland (Bukowski 2002, Fig. 14:3 and 16:c), Bohemia (Erneé 2012, Obr. 7–8) and Germany (Woltermann 2016a, 143, 147, Abb. 92, 94). This stylistic similarity between the circum-Adriatic and Northern-Central European finds, in addition to complete absence of any evidence of local amber processing in the former zone, lends a strong argument for the extraneous production of the analysed amber artefacts. The only clue regarding the local processing of fossil resin is the simetite origin of the earliest amber finds on Sicily related to the *Castelluccio* culture. So far no site pertaining to amber 'work place/station' according to the definition cited above has been found on the island, but distribution of simetite in EBA meant that its processing must have taken place locally and mainly for the internal use. It means that other parts of Italy had to satisfy the demand for amber by importing succinite. Although, in comparison with the Late Neolithic, the testimonies of amber working during EBA in Northern Europe – along the southern coasts of the North and Baltic Seas – are much fewer, there still are some convincing findings. From the Dutch peninsula of Noord-Holland amber beads, semi-finished products and production waste are known from the sites of the *Hilversum* culture, while five graves and one hoard dated to the period in Northern

Germany returned lumps of raw amber (Woltermann 2016a, 215).[680] Amber blanks are also mentioned from EBA-dated graves related to the *Únětice* or *Iwno* cultures in north-central Poland, and it is possible that the *Rzucewo* culture amber workshops at the Vistula estuary were still operating at that time.[681]

It is important to note that the vast majority of EBA and EBA/MBA amber finds (39) in the circum-Adriatic zone originate from the two northernmost Italian provinces: Trentino-Alto Adige and the Garda Lake basin, situated at short distance from the amber-rich regions of Europe during EBA: Southern and Eastern Germany, Bohemia and the Danube Basin. It also has to be underlined that at the current stage of research, we are at disposal of IR spectra of six artefacts dated to EBA from sites in Trentino-Alto Adige (Molina di Ledro) and the Garda Lake basin (Lagazzi di Vho), all indicating succinite (cf. Angelini/Bellintani 2005; Instituto Donegani – Novara, unpublished results). Location of the Northern Italian amber-bearing sites within Alpine valleys (Trentino-Alto Adige), or at their extensions to the south (Garda Lake basin) shows the most likely routes for amber to arrive in the area.

Fig. 132. Some of the EBA bronze and clay objects from Northern Italy with analogies in Central Europe: A–D – Molina di Ledro (Rageth 1974, Taf. 17:4, 19:6, 24:1, 87:3); E – Ca' Nova di Cavaion (Aspes 1987a, Fig. 3:5); F – Lavagnone (De Marinis 2001, Fig. 59).

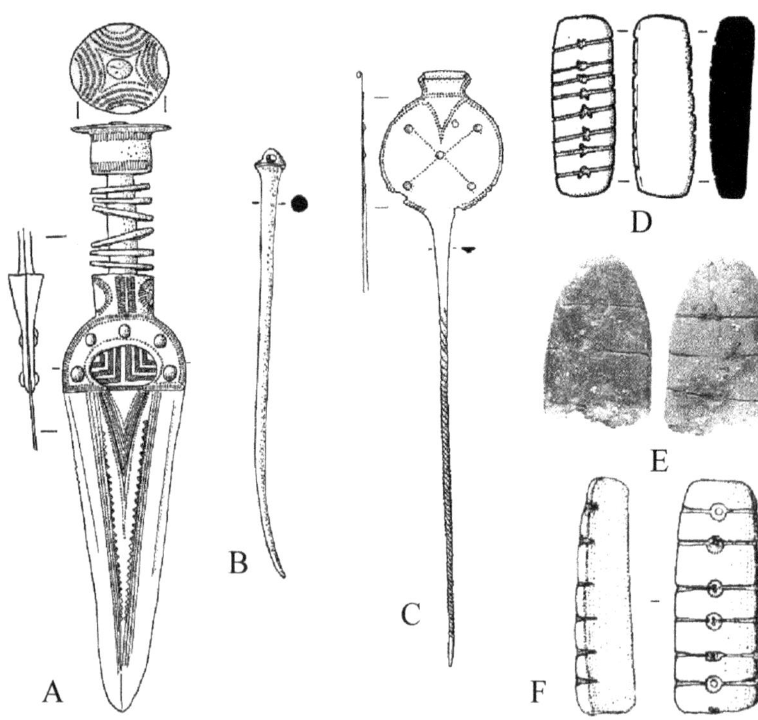

680 See also Gogâltan 2016, 148 for evidence of amber processing around 1900–1850 BC at the site of Pecica in south-west Romania (Arad county).

681 Bukowski 2002, 31, 35; cf. Mazurowski 2014; Butrimas et al. 2018.

To further substantiate the relationships between Central Europe and Northern Italy, one can cite a handful of analogies among the non-amber finds. To begin with, the latter area has revealed evidence of interactions with the Danube Basin in the form of pottery. Some settlements in western Veneto, most of all the pile-dwelling of Canàr, yielded a distinct ceramic component of the *Wieselburg-Gata* ancestry, also enriched by types that can be placed side by side with the *Untervöbling*, *Kisapostag* and *Vatya* elements (Cupitò/Leonardi 2015, 206–207). These are not isolated findings, as EBA pottery of Danubian provenience has also been noted in the Garda Lake basin (De Marinis 1997, 406). Contacts with amber-using neighbours to the north are equally documented by some of the bronze and clay finds discovered on the sites with amber in the latter region. It is fitting to mention in this context i.a. the full-grip bronze daggers and the pin with a small lug (eyelet) at the top of the stalk (ger. *Ösenkopfnadel*) from Molina di Ledro, proving contacts with the classic *Únětice* phase (Fig. 132:A–B).[682] From the same site come bronze pins with twisted stalk and flat, disc-like head (ger. *Scheibenkopfnadel*; Fig. 132:C) with wide distribution range stretching from the south of France, over the Alps, to upper Rhine and Danube Basins, to Bohemia, Moravia and Hungary.[683] If not imported, those bronze objects could have been made under foreign influence. Finally, it is worth mentioning the impressed and fired clay tablets (it. *tavoli enigmatici*, ger. *brotlaibidole*) which appeared on several Northern Italian sites with amber dated to the turn of EBA and MBA (Fig. 132:D–F).[684] Contemporaneous objects belonging to this category have a wide distribution, spanning the entire circum-Alpine zone, with two large concentrations: first in the Garda Lake basin and the second in the middle Danube Basin, where the abovementioned Italian specimens find most numerous and precise analogies.[685] It is worth reminding that in the past they have been interpreted i.a. as tallies for the amber trade (Bándi 1974, 251).

List of similarities between Northern Italy and Central Europe/Danube Basin should be supplemented by observations related to metallurgy. It has been noticed that Italian bronzes in the initial phases of EBA used reduced copper from fahlore minerals. Since the reduction process was not developed enough, considerable percentages of As, Sb, Ag and Ni remained which, when combined, produced the effect of an alloy increasing the hardness of copper. The same phenomenon occurred with respect to BzA1 in Central Europe, among the *Adlerberg*, *Singen*, *Straubing* and *Únětice* cultures (De Marinis 1997, 406; cf. Radivojević et al. 2018). Moreover, presence of amber in Trentino-Alto Adige and the Garda Lake basin already in EBA can be explained by closeness to important sources of raw materials such as the copper of Trentino, the flint of Lessini and Baldo, as well as also available on the morainic hills porphyry and granite for millstones (De Marinis 1997, 416).

682 Rageth 1974, 208, 232, Taf. 19:6, 24:1, 25:1, Karte 1; cf. De Marinis 1997, 406; David-Elbiali/David 2009, 320, 330–334, Fig. 3:H–J; Kneisel 2012, 213. Fig. 3.

683 Rageth 1974, 98–100, 208, 219, 223, 234, Taf. 17:1–4, Karte 4; cf. Carancini 1997, 92, Fig. 222:69.

684 Rageth 1974, Taf. 87:3; Aspes 1987a, Fig. 3:5; De Marinis 2001, Fig. 59.

685 Cf. Bándi 1974; Carafa 2013; Battisti 2015; David 2016.

Fig. 133. Sites with amber in Northern-Central Europe (dots) ca. 2100–1750 BC and reconstruction of amber's inflow and redistribution (arrows) to and from the Únětice *culture in Bohemia (grey colour) (Ernée 2017, Abb. 4).*

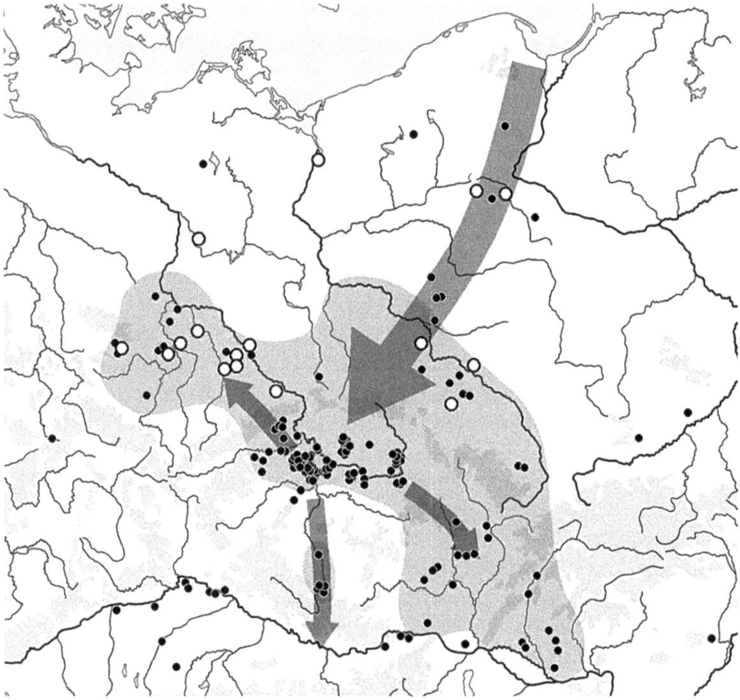

To conclude, it can be said about the Garda Lake basin and Trentino-Alto Adige area that they most probably owe supply in amber to the access to strategic resources and the favourable location in the European trade network of EBA. Arguably, the most important trading partners at this stage were the transalpine cultures such as the *Straubing* and *Únětice*, and some of the units from the Carpathian Basin (e.g. *Vatya*), all of them known for using amber. Especially the *Únětice* culture of Bohemia emerges from the hitherto conducted studies as the most important nodal point and arranger of the European amber circulation during EBA (Ernée 2012, 2013, 2017). Following M. Ernée, it seems safe to say that amber, either in raw or processed state, was brought from Eastern Pomerania (Pomerelia – Gdańsk Bay) to Central Bohemia, and from there it was redistributed in three major directions: towards Saxony, upper Danube Basin and Tisa-middle Danube Basin (Fig. 133) – the areas whose contacts with Northern Italy are proven by the abovementioned finds (Fig. 132). In contrast to Central Bohemia, however, all these regions lack evidence of technically more developed and stylistically more sophisticated use of amber (the *Únětice* culture of Bohemia is known for i.a. spacer plates and amber incrustations), as well as pale in quantitative comparison with widespread distribution and richness of amber exhibited by the Czech cemeteries.[686] So while amber shows all signs of being incorporated into the *Únětice* culture in Bohemia at the communal level, in Northern Italy during EBA, it was still sporadic and limited to individuals (cf. section 4.2). From this it can be inferred that amber was unlikely the main object of trade, but rather its by-product or an item casually exchanged from time to time by representatives of communities that, nonetheless, kept contacts.

During MBA (1650–1300 BC) and LBA (1300–1150 BC), amber jewellery in the circum-Adriatic zone continued to replicate forms known in other regions. Next to flattened- and average-sized beads in discoidal and spheroidal forms, in more significant numbers, cylindrical beads (IA111, IA2111, IF111, IF211) started to appear, what corresponds with their quantitative increase in Northern Europe. From the area of Southern German regional groups of the *Tumulus* culture 3315 amber artefacts resemble disc-like forms, while those akin to cylinder number 27 objects (Woltermann 2016a, 167). The latter are known in 24 examples from Northern Germany dated to the IInd period of the *Nordic* Bronze Age (Woltermann 2016a, 154). Although in fewer numbers, MBA specimens of flattened and average cylindrical beads have been recognised in Bohemia (Chvojka 2017, Fig. 3:6–7), Moravia (Stuchlík 2017, Abb. 5:1–11) and Hungary (Sprincz/Beck 1981, 477, 481, Tab. 2).

Furthermore, flattened conical beads with circular or oval cross-sections, and truncated or convex tops (IF151, IF251, IF154, IF254), which were introduced in the circum-Adriatic zone around EBA/MBA and continued to be used there until the end of LBA, are also documented, albeit in fewer quantities, during those stages in Southern Germany (Woltermann 2016a, 168). They have stayed in use there until the *Urnfield* period. In other countries of Central Europe such as Poland, Bohemia and Hungary, occurrence of conical amber beads is also linked with the development of local manifestations of the *Tumulus* culture.[687]

The greatest novelty of the period, however, are the beads with double-pitched profiles (IF161, IF261, IA161, IA261) which were introduced on a large scale during the same period all across Europe. Few of such beads are known from the EBA contexts in Northern Germany: there are eight flattened biconical beads from sites Fallingbostel and Vorwohlde, dated to the I[st] period of the *Nordic* Bronze Age (ca. BzA2–BzB1), and only one in the entire German part of the *Únětice* culture – Dieskau hoard (Woltermann 2016a, 142–143, 147, Abb. 88; 2016b, Fig. 3:D–E). A single bead from grave 68 of the Battonya cemetery (Békés County, Hungary) attributable to the *Maros/Mureş II* culture (ca. 2000–1600/1500 BC) was classified to type IXA comprising flattened biconical beads by E. Sprincz and C. W. Beck. However, by the author's own admission, it 'is rather crudely made and suggests minimal working to take advantage of the natural form of the raw amber', and judging by its images it lacks the clearly outlined double-pitched profile (Sprincz/Beck 1981, 482, Fig. 4:29; cf. Gogâltan 2016, 149, Fig. 4).

686 Ernée 2013, Abb. 2–3; cf. Sprincz/Beck 1981; Woltermann 2016a, 147–152.
687 Sprincz/Beck 1981, 483, Fig. 5:19, Tab. 2; Bukowski 2002, Figs. 39, 40:30–31 and 44:1–9; Chvojka 2017, Figs. 2, 3:3, 8, 11.

Fig. 134. Contexts with amber finds dated to the BzB–C and IInd Nordic *Bronze Age period in Germany; points symbolising findspots are scaled according to quantity of finds and the most important sites are labelled (after Woltermann 2014, Abb. 2).*

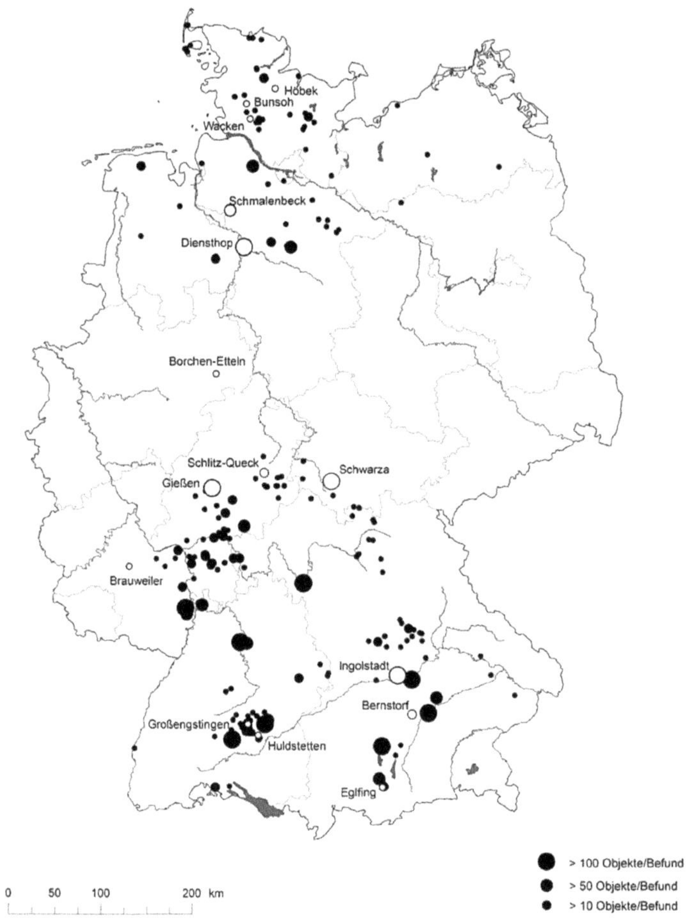

It is clear that biconical amber beads became widespread in Central Europe only with the emergence of the *Tumulus* culture around the 16th century BC (Czebreszuk 2011, 51, Fig. 13:1–3; Woltermann 2016b, 386). From contexts of the *Tumulus* culture, regional groups in Southern Germany alone, 419 biconical amber beads corresponding with abovementioned types are known (Woltermann 2016a, 167, Abb. 114–115). In France they began to be used not before MBA, and their largest concentration has been recognised in the area of the *Hagenau* group of the *Tumulus* culture – the Appenwihr cemetery (Du Gardin 1986, 549, Fig. 2:1–2, 11). This novelty could have been transferred from Southern Germany further east through Bohemia, where presence of biconical beads is not attested in the otherwise plentiful amber collections of the *Únětice* culture, but has been detected in barrows of the *Bohemian Palatinate Tumulus* culture (Chvojka 2017, Figs. 2:6, 3:13–14, and 4:3–4; cf. Ernée 2012). Afterwards, it could have been passed on to the Carpathian Basin, where biconical amber beads are truly present starting from the period following the *Koszider* horizon (after ca. 1600/1500 BC), and reach climax at the time of the *Kurd* horizon, so around 1200–1100 BC (Sprincz/Beck 1981, 482–483, Fig. 5:4–15, 22–42, Tab. 2). MBA is also the period when lenticular beads (IF162 and IF262) became more popular in many European areas: e.g. Germany (Woltermann 2016a, 41), Hungary (Sprincz/Beck 1981, 481) or France (Du Garding 1986, 549), in sync with similar phenomenon in the circum-Adriatic zone.

Abovementioned forms of amber beads (discoidal, conical and biconical) had rather broad distribution range within the circum-Adriatic zone during MBA – a period during which amber adornments started to be more available in this part of Europe. One can easily discern a dramatic growth in the number of sites with amber, as well as the spatial scope of its occurrence. The increase of amber is especially prominent in the northern part of the study area, along the border with the Alps and the Carpathian Basin. On the other hand, amber spacers, very characteristic for MBA in general and the *Tumulus* culture in particular, are significantly less represented in the circum-Adriatic zone. Comparing with six objects assigned to group VII comprising amber spacers and spacer-like artefacts, there are 159 spacers known from Germany alone (Woltermann 2016a, 169). Some of them are quite similar to the Northern Italian specimens in terms of perforation patterns (see section 3.9). A considerable number of amber spacers come from east and south-east French sites dated to MBA; suffice to mention the rich assemblage from the Hasard cave in the lower Rhône Valley, or from mound 53 at Oberfeld cemetery in the Hagenau forest which seem most similar to the Northern Italian examples, in particular the finds from Olmo di Nogara, tomb 121 (cf. Du Gardin 1986, Figs. 5:1–5 and 6:2–11). Thus, despite being much lower in quantity, the spacers in the circum-Adriatic zone, especially its northern part, find good analogies in the neighbouring countries, further confirming the interregional circulation of amber during MBA.

Quantitative comparison shows that even if combined, the regional assemblages from the circum-Adriatic zone are lower than collections of amber from the *Tumulus* groups in Southern Germany (Fig. 134). This discrepancy is quite striking, if one juxtaposes the total number of amber finds from MBA (and MBA/LBA) core regions of amber's acquisition in the former area, e.g. western Veneto – 90, Istria – 124, Podrinje and Western Serbia – 233 or even northern Apulia and Basilicata – 612, with some of the richest contexts from the latter territory, e.g. the Ingolstadt hoard which comprises 2800–2900 amber beads alone (Woltermann 2016a, 164–165). Still, when put against other provinces influenced by the *Tumulus* culture, e.g. West Bohemia, where circa 144 artefact pertaining to MBA were found (Chvojka 2017, 107), or Moravia where this sum amounts to 48 beads (Stuchlík 2017, 116), the individual regions of amber's acquisition in the study area do not appear that inferior in terms of sheer numbers of finds. One can forward a hypothesis that during MBA, amber became a commonly recognised symbol and attribute of certain social groups (females with distinct social position?) in the core regions of Central Europe and the circum-Adriatic zone influenced by the *Tumulus* culture (cf. Woltermann 2016a, 196). This picture contrasts with amber's status in Northern Europe (e.g. Jutland or Vistula estuary) which in the Late Neolithic as well as the Bronze Age used to be associated with both males and females, and worn according to gender-based rules quite different than those observed to the south (cf. Siemen 2016; Woltermann 2016a, 194–206). Some authors go as far as to say that during the Bronze Age amber in the Nordic zone somehow lost its social significance and became merely a commodity traded with other regions.[688]

From MBA and the first half of LBA, there still is no solid evidence of amber processing in the circum-Adriatic zone, according to the earlier described criteria applied by G. Woltermann. In Northern Europe, on the other hand, there are contemporaneous attestations of amber working on sites such as Bjerre (Thisted commune) in north-west Jutland, where a settlement dated to the IInd–IIIrd period of the *Nordic* Bronze Age has been uncovered. Inside and around a large three-nave house with a floor plan of 8.5x25 m were collected over a hundred pieces of raw amber, thus interpretation of the building as an amber workshop or a hut belonging to amber gatherers (Woltermann 2014, 79, Abb. 10; 2016a, 217, Abb. 164). Absence of finished products (completed amber beads) at Bjerre and other similar Northern European sites suggests that final processing of the material took place elsewhere, presumably further south, where a variety of MBA–LBA forms is well documented (Woltermann 2014, 83). However, as R. F. Mazurowski notes, a complete operational chain of production will never be available at the amber working sites, as the finished ornaments were eventually taken away (Mazurowski 1984, 18). Amber blanks found in several Northern German graves, belonging almost without exception to wealthy males, indicate a connection between the social status and identity of the buried, and the economic importance of extraction of raw materials. A prominent addition to these graves are bronze awls, suggesting the use of these tools in amber processing (Woltermann 2014, 81–82, 215). In Southern Germany, graves with raw amber, not to mention settlements with such finds, are rare but constitute important points in the debate (Woltermann 2016a, 216). Suffice to mention the hillfort of Bernstorf in the Upper Bavaria which, aside from six hardly processed amber beads and 34 blanks, yielded also the infamous amber seal with carved inscriptions allegedly in Linear B – a subject of much controversy (Woltermann 2016a, 165–167; cf. Harding/Hughes-Brock 2017). Overall, taking into account the richness of amber in the *Tumulus* culture in Central Europe, there is strikingly little evidence of amber processing in its ecumene. All these observations led G. Woltermann to propose a model of amber circulation during MBA which assumes collecting and pre-processing of raw material (including a removal of weathering crust and damaged areas) at the source (Baltic coast), and afterwards its transport inland, where the final processing was carried out by travelling craftsmen (Woltermann 2016a, 256). This arguably involved more fine-tool shaping of the material and less manual 'hard-work' (i.a. decortication and chopping/cutting), thus produced little waste which, assuming mobile character of the industry, consequently is difficult to trace on archaeological sites.

The import from southern Scandinavia and Northern Germany – the most likely areas supplying the rest of Europe in amber at the time – was taking place along several major German rivers, among them the Weser having the crucial importance, through the *Lüneburg* group into the *Fulda-Werra* group and the burial mound groups in the *Main-Neckar* region, ending in Bavaria (Woltermann 2016a, 158). During MBA the importance of Oder and Warta rivers and their interfluve for the inflow of amber

688 Cf. Bukowski 1998, 371; Jennings 2014, 25; Earle et al. 2015, 647.

decreased, as illustrated by the significantly lower numbers of finds in comparison with EBA, especially with respect to Pomerania (cf. Bukowski 2002, 61–63, Map 2). During late EBA and early stages of MBA (until ca. 1500 BC), amber found considerable popularity in some cultures of the Carpathian Basin, particularly the *Otomani-Füzesabony* and *Vatya* (Jaeger 2017). Although Bohemia and the Carpathian Basin still showed an interest in amber during MBA, in terms of numbers the import decreased and its focus shifted to the west, towards the *Tumulus* groups in Southern Germany (cf. Chvojka 2017, 107–111, Fig. 1; Gogâltan 2016, 151–152, Fig. 6). It is there that amber was most probably subjected to the final processing by local specialists, although this is inferred not from evidence of workshops, which is still lacking, but indirectly from the abundance and technological advancement of the finds. They include biconical beads and spacer plates indicating forms and manual dexterity that was unknown in Northern Germany at the time (Woltermann 2014, 83; 2016a, 259). Thus, it seems plausible that biconical beads, spacers and other amber artefacts from the circum-Adriatic zone which find analogies to the north were imported. Italy and Western Balkans most likely obtained amber via Alpine passes through contacts with the *Tumulus* groups of Germany (Bellintani 2014, 2017), whereas Central Balkans gained access to this material probably through a chain of intermediaries which included the *Middle Danubian Tumulus* culture and the *Tapé* group by route leading along the Danube and Tisa (Palavestra 1993, 16; Ljuština 2019, 98).[689]

This hypothesis is to some extent reinforced by the presence of non-amber imports and influences in the study area that reveal genetic links with Central European cultures. Among the earliest examples are pins with obliquely perforated globular or biconical heads which appear in a number of Northern Italian contexts dated to BM 1–2, e.g. Lavagnone 5–6; Fiavè 6; Isolone del Mincio (Fig. 135:A–D) and find numerous analogies to the north of the Alps around BzA2c–BzB early (De Marinis 1999, 44–48, Fig. 16–19; cf. David-Elbiali/David 2009, Figs. 5:F–I and 6:E). Transalpine contacts in early MBA are substantiated by the bronze diadems from Ledro which have the most precise counterparts on the cemetery of Pitten in Lower Austria (Fig. 135:G; Rageth 1974, 112–116; Urban 1993, 248). Same connection is proven even more clearly by appearance of the Sauerbrunn-Boiu swords in Northern Italy, especially at Olmo di Nogara (Fig. 135:E–F; De Marinis 1999, 64, 75–80, Figs. 34–36; cf. De Marinis/Salzani 2005, 393–395, 397–403, Figs. 3, 6–9). The northward links in Northern Italian metallurgy continued in more advanced stages of MBA, as illustrated by axes with median wings or daggers with trapezoidal heels (De Marinis 1999, 63; cf. David-Elbiali/David 2009, Figs. 7:A–B, D, F and 8:A–D). While some similarities between Northern Italian and Danubian pottery dated to MBA do exist, they are quite remote and at best indicate weak connection between the two areas (cf. Urban 1993, 259–262).

---

689 It is worth considering the possibility of amber coming to Central Balkans during MBA and LBA from the south-west via Southern Italy, as indicated not only by similarities among amber and certain bronze artefacts but also by some other factors, i.a. absence of amber in transitive areas immediately to the north of Podrinje and Western Serbia, i.e. in Vojvodina (cf. section 3.12.2.).

Fig. 135. Some of the MBA bronze objects from Northern Italy with analogies in Central Europe: A–B – Fiavè-Carera, horizon 6 (Perini 1987, Tab. II:9, III:28); C – Isolone del Mincio (Guerreschi et al. 1985, Tav. XXII:st9081); D – Lavagnone, phase 5–6 (De Marinis 1999, Fig. 16:3); E–F – Olmo di Nogara, graves 475 and 201 (De Marinis/Salzani 2005, Figs. 3:1 and 8:4); G – Molina di Ledro (Rageth 1974, Taf. 22).

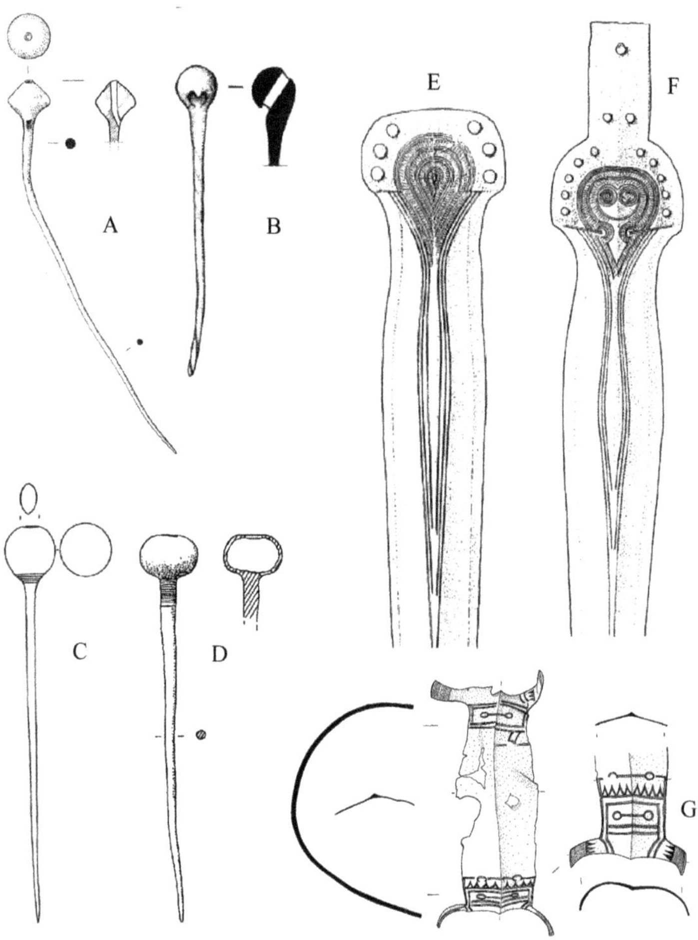

Also the Balkan part of the circum-Adriatic zone is rich in attestations of contacts with the cultures of Central Europe at this stage. The earliest testimonies come in the form of pottery found together with amber at Alilovci Lipje in Northern Croatia. This ceramic material strongly manifests influences of Carpathian Basin cultures, including the *Litzen*-type pottery (Mavrović Mokos/Pavličić 2015, 16–21, 23–24, Figs. 12–13, 18). Bronze objects co-occurring with amber in West Serbian and East Bosnian contexts (including Glasinac), such as pins with spindle-, nail- and stamp-heads (Fig. 136:A–C); open bracelets with expanded or thinned ends (Fig. 136:D–G); heart-shape and lunate pendants (Fig. 136:H–I); or spiral armbands (Fig. 136:J) by their form and decoration are related to similar objects from the Carpathian Basin. The pins were assigned to phases BzC–BzD of the Central European periodisation, and despite certain regional features, their form generally indicates the *Tumulus* culture influences.[690] The open bracelets made from a bronze bar of semi-circular (D-shaped) or a flattened triangular cross-section, with tapered and later widened (stamp-shaped) ends and various engraved decorative motifs, occurred in Glasinac, Podrinje and Western Serbia in various phases of advanced and late MBA (BzB late/C–BzD). Although considered as products of local metallurgy,[691] they bear striking similarity to the contemporaneous bracelets from the Carpathian Basin and the Central Europe.[692] The heart-shape pendants can be fitted into variants 6 and 7 by Hänsel, dated to phases MDII–III of the Bronze Age in the Danubian Basin (BzB–BzC1) and widespread in Central Europe during the *Tumulus* period.[693] Also lunate pendants can be found on terrains included in the *Tumulus* culture during the MBA; in Hungary they were used from the *Koszider* until *Forró* and *Ópályi* horizons,[694] while in western Poland they are known from phases BzB–BzC, however in some cases continued to be used up to phase BzD.[695] Bracelets made of spirally wound wire were very common in Central Europe, where

---

690 Vasić 2003, 31–32, 38–39, 48, Tab. 11:155, 12:182, 17:254, 70; Filipović 2013, 59, 66–68, Tab. 1. Spindle-head pins (ger. *Spindlekopfnadeln*), similar in form, appear during the MBA in a geographically limited scope, covering terrains influenced by the Polish variant of the *Tumulus* culture (also known as the *Vorlausitzerkultur*): western Poland, Lusatia and Saxony in Germany (Gedl 1975, Tab. II:12–14, X:7–8, XI:10, 13; 1983, 59–66, Taf. 15–19:242–243), and parts of Bohemia, Moravia and Slovakia under influence of *Middle Danubian Tumulus* culture (Říhovský 1979, 37–42, Taf. 7:105–117, 8:118–120; Novotná 1980, 72–77, Taf. 17–19:435–439). Stamp-head pins (ger. *Petschaftkopfnadeln*) are characteristic for the *Tumulus* culture during the phases BzB late and BzC, that is the classical phase of the culture, however in some cases were continuously used until the end of the MBA/early stages of the *Urnfield* culture (Říhovský 1979, 62–63, Taf. 11:222–226, 12:227–240; 1983, 14–15, Taf. 5:63–64; Hochstetter 1980, Taf. 42:6; Beilage 1; Gedl 1983, 70–71, Taf. 20:272–275, 60). Nail-head pins with perforated necks (ger. *Lochhalsnadeln mit nagelkopf*), similar in ornamentation pattern to the examples from Western Serbia, can be found among pins from Bohemia (Čujanová-Jílková 1970, Taf. 2:1), Slovakia (Novotná 1980, Taf. 8:304) and Western Hungary (Říhovský 1983, Taf. 3:35–36).

691 Kosorić/Krstić 1970, 31; Nikitović/Vasić 2002, 29; Dmitrović 2016, 166–168.

692 Cf. Willvonseder 1937, Figs. 7:7 and 8:18; Mozsolics 1967, Abb. 21:1–2, Taf. 60:7, 63:3–6, 69:5, 53:5a-b; 1973, Taf. 4:3a, 134:6–8; Hänsel 1968, 92–94, Beilage 4:18; Gedl 1975, Tab. XI:11, XII:1–3, XIII:15, XLII-XLIII; Čujanová-Jílková 1978, Fig. 106:37; Hochstetter 1980, Beilage 1:42–45, 48–52; Stuchlík 1993, Figs. 172:18–21, 177:11, 13 and 179:4–7.

693 Hänsel 1968, 117–118; Beilage 4:55, 5:45; cf. Čujanová-Jílková 1978, Figs. 106:38 and 129:18; Wels-Weyrauch 1978, 63–65, Tab. 16; Hochstetter 1980, Tab. 80:5, 83:16; Stuchlík 1993, Fig. 177:2–3.

694 Mozszolics 1967, 87, Taf. 19:1a-b, 2a-b, 28:27–30; 29:7–12, 70:4–6; 1973, Taf. 151:17–19, 173:23–24.

695 Gedl 1975, Tab. III:6, XVIII:27–28, XXXVIII:1–5, Tab. Chron.

they appeared as early as EBA, and stayed in use until the *Urnfield* culture; however, specimens with spiral terminations were considerably popular in the *Tumulus* culture.[696] Influences from the same cultural milieu are evident in pottery coming from the West Serbian tumuli: these are mainly small amphorae with cylindrical necks and expanded bellies, adorned with warts or bulges and grooves (Fig. 136:K; cf. Filipović 2013, 63–64). Although they are definitely most visible with respect to Podrinje and Western Serbia, influences of Central Europe (*Tumulus* culture in particular) on the local metallurgy are exhibited by pins, bracelets, pendants, etc. also in the Balkan regions of amber's acquisition further south.

Fig. 136. Some of the MBA bronze and clay objects from the Balkans with analogies in Central Europe: A, C – Banjevac, tumulus 2 (Bulatović et al. 2017, T. IX); B – Belotić, mound VII, grave 1 (Garašanin/Garašanin 1962, Fig. 6a–b); D – Ravnine, mound 3 (Nikitović/ Vasić 2002, Tab. III:8); E – Vranjani, tumulus 1 (Zotović 1985, T. IX:3); F – Rudine, tumulus I (1894) (Benac/Ćović 1956, Tab. XIV:6); G–K – Paulje, tumulus K (Bulatović et al. 2017, T. XVII:26, XIX:30–31, 36–45, 47); L – Bezdanjača, mass grave 9 (Drechsler-Bižić 1979–1980, Tab. XXVI:2); M – Krmedski Novi Grad, grave 1 (Batović 1983, Tab. XLII:6); N – Vrčin (Buršić-Matijašić 1989, Fig. 3:1).

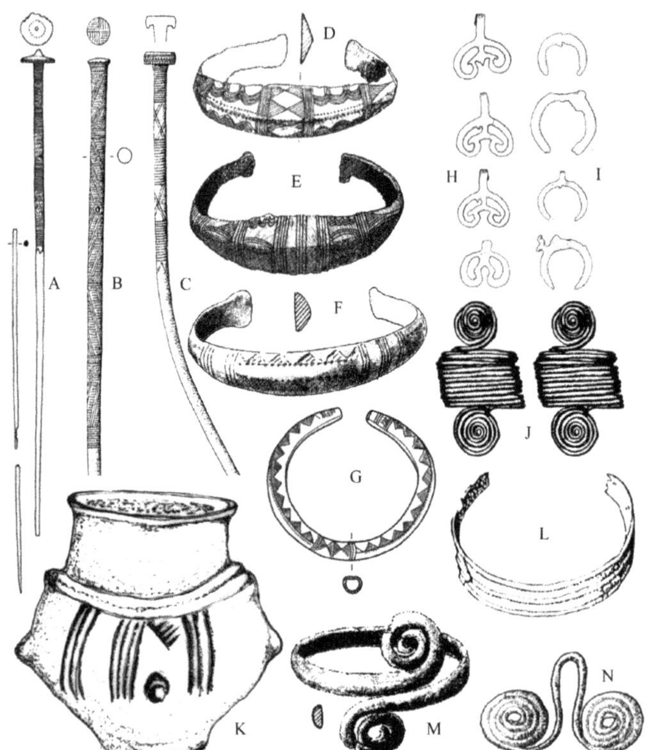

696 Dmitrović 2016, 172; cf. Gedl 1975, Tab. IV:1, IX:2–3; XLIII:8–9; Čujanová-Jílková 1978, Fig. 129:16; Hochstetter 1980, Beilage 1:36.

These links are less evident with respect to MBA sites with amber in Western Balkans. They are at best documented in Lika, i.a. by some of the bronzes belonging to the first horizon of Bezdanjača cave, e.g. an open bracelet with expanded ends, decorated by parallel ribs (Fig. 136:L) which has analogies in the *Tumulus* culture of the Central Europe (BzB/C1) and the Carpathian Basin (BzB2/C1–BzD).[697] These observations contribute to the general picture of Northern Croatia and Lika as a transitional zone between the Adriatic coast, Pannonian plain and Danube Basin, not only in terms of geography but also due to a mixture of various cultural traditions.[698]

Such punctual analogies are missing from the Istrian contexts with amber. Some similarities in material culture between Istria and the Central Europe, or the Carpathian Basin can be traced on the basis of simple, wire-made elements of jewellery: bracelets with spiral discs at the ends[699] and spectacle pendants (Fig. 136:M–N).[700] However, these are quite generic forms, therefore cannot serve as good influence indicators. Still, the evidence of contacts between the circum-Adriatic zone and the parts of Europe situated to the north is quite compelling. But does it mean that all amber jewellery was imported in a finished form to the circum-Adriatic zone?

G. Woltermann observes that despite a considerable level of formal unification of amber products during MBA, demonstrating increasing 'professionalisation' in jewellery production, the workmanship of artefacts is strikingly heterogeneous, showing that the processing was carried out both by specialised artisans and relatively inexperienced craftsmen (Woltermann 2014, 83–84). While the morphological similarities potentially testify to Southern German provenience of many finds (i.e. biconical beads and spacers), it cannot be excluded that amber adornments were occasionally copied, with varying level of success, by craftsmen from other areas. Among the collected finds, there are many examples of irregularities, e.g. skewed-flat bases or convex-straight sides. Furthermore, it is possible that some of the finds with less regular forms – especially those with polygonal cross-sections found in abundance in Central Balkans and Southern Italy – were made by chopping larger pieces of amber – raw lumps, but also quite possibly ready-made products, and turning them into smaller but more numerous beads by simply piercing them and leaving the surfaces unpolished. Such rudimentary processing could have been easily done by average craftsmen equipped with right tools. Still, no certain proofs of MBA amber working in the circum-Adriatic zone have been discovered, so one can only speculate it basing on indirect clues. On the other hand, examples of amber pieces with attempted secondary perforations (group VI) prove that the locals were capable of or at least tried to 'recycle' broken beads. Thus, it can be suspected that they had technology and tools to do so, however might not be experienced enough or have skills necessary to be successful in handling amber.

If amber was a commodity imported from Northern Europe either in raw or finished form, it is interesting what the circum-Adriatic communities could offer for trade. Significant increase in number and spatial distribution of amber during MBA indicates

697 Hänsel 1968, 102, Taf. 42:7; Gedl 1975, XIX:12, XXVI:8; Čujanová-Jílková 1978, Fig. 106:13; Hochstetter 1980, Beilage 1:46; Stuchlík 1993, Fig. 172:3.

698 Cf. Karavanić 2007; Ložnjak Dizdar 2007, 2011.

699 Batović 1983, 290, Sl. 19:11, Tab. XLII:6; cf. Hochstetter 1980, Beilage 1:40; Gedl 1975, Tab. XIX:1–2, 7–9, 15, XX:8–10, XXI:5–6, Tab. Chron.; Čujanová-Jílková 1978, Fig. 129:10–11.

700 Buršić-Matijašić 1989, Fig. 3:1; cf. Wels-Weyrauch 1978, 77–86, Tab. 20; Batović 1983, 292, Sl. 19:9.

that it found a more widespread reception. Perhaps certain parts of Southern Europe at this stage began providing some of the in-demand materials and consequently took more prominent positions in the European network of goods circulation. Or maybe this was the moment when amber finally became commonly considered as a commodity worth exchanging for other goods.

The increase of amber in the circum-Adriatic zone during MBA corresponds with the spread of bronze products in the Nordic zone – an area lacking sources of copper or tin which might have been exploited at the time due to technological problems. It has already been suggested that amber was exchanged for raw metal ore, especially copper from Central and Southern Europe (Ling et al. 2014; Earle et al. 2015). Recently conducted lead-isotope analyses proved some of the bronze artefacts from Sweden to be made of copper from regions coinciding with amber finds, i.a. North Tirol in Austria (Ling et al. 2014). However, the picture for a while was unclear, whether some or any of the copper sources from the circum-Adriatic zone played a part in supplying other parts of Europe during the Bronze Age (cf. Bellintani 2014, 117–118). More importantly, a new study focused on the provenience of copper used to produce 118 swords dated to ca. 1600–1100 BC and collected from the vast area between Northern Italy and Scandinavia proved the south-east Alpine copper ores to be the main source of metal between 1500 and 1300 BC, and the dominant supplier between 1300 and 1100 BC (Ling et al. 2019). These observations are consistent with earlier research indicating that copper extraction in the eastern Trentino (in the Valsugana and nearby areas) saw revival during MBA and reached a climax in LBA–FBA, supplying most of Italy in the material.[701]

Fig. 137. Model of MBA trade system with amber being supplied by Jutland in exchange for copper from several European sources (i.a. Trentino) by mediation of the Tumulus *groups in Southern Germany (Ling et al. 2019, Fig. 21).*

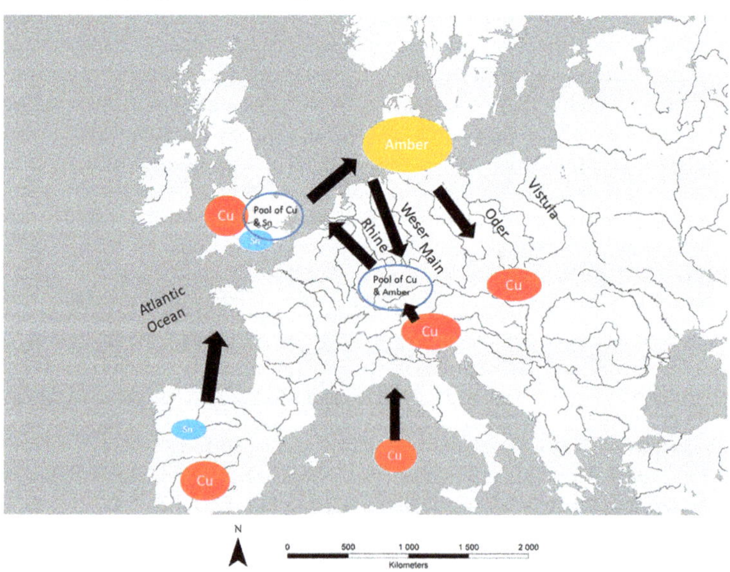

701 Bellintani 2014, 118; cf. Cierny 2008; Marzatico et al. 2010; Jung et al. 2011; Silvestri et al. 2015.

Interestingly, the study proved lack of dependency between a source of copper and typology of swords, thus it is likely that raw copper (in ingots) was circulated and turned into actual objects only at the final destination. Authors of the study integrated its outcome with observations on the distribution of amber finds in Germany to argue that the Weser and Rhine rivers were the most commonly used waterways – routes for the translocation of copper and amber during MBA. In their model the *Tumulus* groups from Southern Germany played the crucial part as intermediaries in the circulation of copper from the Alpine zone, tin from British Isles and amber from Scandinavia (Fig. 137). The second most used copper source was the Slovak Carpathians; the analysis showed that Northern German full-hilted swords dated to the IInd period of the *Nordic* Bronze Age were predominantly casted from Slovakian copper (Ling et al. 2019, Fig. 12a, 18). The resources were most likely transported from the Carpathian Basin along the Elbe and Oder rivers which could also explain presence of amber in Bohemia, Moravia and Slovakia during MBA. From the perspective of this work, the most significant conclusion stemming from the above-discussed study is that by MBA Northern Italy became one of the main hubs for circulation of goods inside the pan-European network of trade. However, how can we rationalise the abundance of amber below the areas of copper mining, i.e. in the Garda Lake basin and the Po Plain, and at the same time its relatively moderate quantity in Trentino-Alto Adige?

P. Bellintani explains this apparent inconsistency in the model described above by adhering to the concept of 'unequal exchange' employed by S. Shennan in his study of earlier Bronze Age north Alpine smelting sites (Bellintani 2014, 119; cf. Shennan 2010). Accordingly, the Alpine communities kept a close control of the metal extraction, while the trade was in the hands of the Po Plain communities, who got the greater benefit. The latter had an important advantage of developed agriculture and crafts (e.g. metallurgy and textile production) which made other regions dependent on them, and allowed to enter trade with high-quality products not just raw materials. Thus, amber first reached the main centres of goods production and distribution on the macro-regional scale ('markets') and only later was channelled on the micro-level through local networks.

It is worth considering the location of other regional concentrations of amber within the circum-Adriatic zone with respect to chorography of various raw materials. The core provinces of amber's acquisition sometimes occupy places rich in natural resources, such as sea salt along the shores of northern Adriatic, particularly Istria (Montagnari Kokelj 2007), and tin in Western Serbia (Huska et al. 2014; Mason et al. 2016, 2020). Although so far there are only hints of using these resources in the Bronze Age and the true scale of exploitation or export range of these materials remains unknown, their potential importance for explaining participation of Istria and Podrinje-Western Serbian in the long-distance trade should not be overlooked. Solving those problems requires more research to be conducted, but at this moment one can argue that these core areas of amber's acquisition on their own had some valuable resources for trade to offer. It is debatable whether amber was an equivalent of copper in this pan-European trade network (cf. Kristiansen/Suchowska-Ducke 2015, 369; Radivojević et al. 2018, 163–164); after all there is less amber finds than the metal ones, and amber had a much smaller utility and much narrower social attribution. Tin arguably possessed greater importance in transaction, as it was a necessary ingredient of bronze metallurgy. Furthermore, there were other materials that could have been provided by Northern Europe, but which unfortunately are invisible in the archaeological data, e.g. furs, hides, timber, slaves, etc. (Earle et al. 2015, 635). Still, having a valuable resource for trade meant that it was much easier to obtain amber, even if the latter was not the main purpose of engaging in transactions.

To conclude the considerations on cooperation between the circum-Adriatic zone and Northern and Central Europe during MBA, it seems right to say that by this stage circulation of amber was becoming increasingly 'normalised', its processing more proficient and distribution much wider. Probably, the production was concentrated in the areas occupied by the various *Tumulus* groups that had an easier access to the sources of Baltic amber. Individual craftsman or their groups from certain areas were able to manufacture entire series of products in standardised form, what can be shown on the basis of biconical beads or spacer plates. These adornments were likely exported in batches, as shown by the hoard of Ingolstadt (cf. Woltermann 2016a, 250–254). Perhaps this process was handled by travelling artisans or merchants specialising in amber, who operated from the Baltic coast, along the rivers of European lowlands, up to the Alpine zone and the Carpathian Basin. However, the quality of objects is varied, suggesting that amber working was engaged in also by the less-proficient craftsmen. It seems likely that those 'amateurs' had a considerable share in this process, being responsible for the heterogenic image of amber typology in the circum-Adriatic zone during MBA. By that time amber was beginning to be transported to the northern fringes of the study area 'in bulk'. It is symptomatic that the main regions (cores) of amber's acquisition are placed in the vicinity of sources of strategically important resources (copper, tin, salt), therefore could actively participate in the pan-European trade network. This does not preclude obtaining amber in different ways, e.g. through occasional exchange, either casual or more ceremonial. But it does suggest that within the core areas amber was commoditised to a certain degree; thus, trade might supersede other forms of interactions as the major way of amber circulation. From the core amber could be exchanged via local networks further south the circum-Adriatic zone. While there are no solid proofs for amber processing in the circum-Adriatic zone during MBA, imported adornments could have been subjected to secondary processing (chopping/dividing into smaller pieces and/or perforating new holes for suspension).

Available evidence attests to continuity in the technology and stylistics of amber working in Europe at the turn from MBA to LBA. The palette of amber forms largely remained unchanged, and still comprised predominantly flattened or average discoidal, biconical and cylindrical beads. Types corresponding to IF161, IA161 and related variants are particularly represented (71 artefacts) in Southern German contexts dated to BzD and attributed to the late *Tumulus* – early *Urnfield* culture (Woltermann 2016a, 169–170), but even more so in Northern Germany, where sites falling into IIIrd period of the *Nordic* Bronze Age yielded 205 such objects which equals 35.3 % of the total number of amber finds from this phase (Woltermann 2016a, 172). From LBA onwards biconical beads were increasingly present in Poland (Bukowski 2002, 87) and Hungary (Sprincz/Beck 1981, 482–484, Tab. 2). Discoidal and flattened cylindrical beads are present in substantial numbers as in the preceding periods in all regions of amber's acquisition neighbouring with the circum-Adriatic zone.[702] A notable similarity between Southern German and circum-Adriatic amber assemblages is appearance of conical beads perforated concentrically along D/L (IIIF154, IIIF254, IIIA254), which interestingly are absent in other regions (Woltermann 2016a, 170, 173). A significant novelty

702 Sprincz/Beck 1981, Tab. 2; Bukowski 2002, 77; Woltermann 2016a, 170–173.

in the repertoire of amber adornments in the circum-Adriatic zone during LBA was introduction of many elongated versions of cylindrical, barrel-shaped (fusiform) or biconical beads (e.g. IE121, IE221 and IE161). This change is also reflected in contemporaneous amber collections from countries to the north.[703] Although the elongated beads in LBA were still numerically inferior in comparison with the flattened ones, they became more numerous in the next phase – FBA.

As the cultural map of the study area remained largely untouched at the turn from MBA to LBA, so did persist the earlier connections between regions of amber's acquisition in the circum-Adriatic zone and Northern as well as Central Europe. The demographic and economic growth evident in the milieu of the *Palafitte-Terramare* culture in Northern Italy was matched by a steady amber import, as well as copper export which increasingly dominated Northern European market (Ling et al. 2019). As in the preceding phase (MBA) amber imported from the north first reached the core regions in the northern part of the circum-Adriatic zone, and from there was redistributed to the provinces further south. This situation of continuous cultural development and amber influx lasted for the better part of LBA (LBA 1; until ca. 1200 BC). The contacts with the late *Tumulus* culture – nascent *Urnfield* culture are confirmed by a collection of artefacts, although numerically inferior in comparison with MBA. With respect to Italy, it is worth mentioning the presence of flange-hilted swords (ger. *Griffzungenschwerter*) belonging to the Naue II type, also known as the types Sprockhoff IIa – Nenzingen – Reutlingen – Cetona according to other typologies.[704] Examples of such swords, specifically of the types Cetona, were found on the sites with amber in Northern Italy: e.g. Olmo di Nogara, grave 41 (De Marinis 1999, Fig. 36:11). Apart from signalising a change in combat style (from thrusting to slashing), the widespread distribution of these swords has been interpreted as a sign of long-distance mobility of people, i.e. warbands or mercenary warriors (Suchowska-Ducke 2015; Kristiansen/Suchowska-Ducke 2015). Other Italian finds that have links to the early *Urnfield* culture are the Baierdorf-type bronze knives possessing wide distribution range from the southern Baltic coast to the circum-Alpine zone (Bianco Peroni 1976, 13–14; cf. section 3.12.3). Finally, it is necessary to mention violin-bow fibulae, abundantly represented on the LBA Italian sites with amber which at the time became one of the most popular elements of costume from the Danube Basin to the Aegean (Pabst 2014, 2018). Further east, contacts between the Balkans and Central Europe are testified by, e.g. Freudenberg/Gmunden-type axes from the Laganiši cave in Istria and the Karavlaške kuće cemetery in Podrinje (Fig. 93; cf. Pászthory/Mayer 1998, 93–98). Northward connection is reinforced by occurrence of numerous pins which appear in the areas influenced by the late *Tumulus* and early *Urnfield* cultures, including northern fringes of the circum-Adriatic zone. Among them can be singled-out pins of the Deinsdorf type (Fig. 92:D–E), pins with club-shaped heads (ger. *Keulenkopfnadeln*) (Fig. 107:C), or the ones with spherical heads and incised/grooved necks (Fig. 92:F–G), which all have good analogies in the Carpathian Basin (Říhovský 1983, 15–16, 36; Novotná 1980, 101–102) and even further north, e.g. in Poland (Gedl 1983, 24–26, Taf. 5:82–101, 6:103–116). To this evidence can be added occasionally occurring vessels reproducing some of the forms and ornaments typical for the late *Tumulus* culture (Filipović 2013, 64, Tab. III:1).

703 Sprincz/Beck 1981, Tab. 2; Bukowski 2002, 74–75, Fig. 55; Woltermann 2016a, 42–43, 170, 172, 252–254, Tab. 29.

704 Cf. Sprockhoff 1931; Cowen 1955; Catling 1956; Bianco Peroni 1970; Schauer 1971.

During FBA amber fishing and possibly also processing was still taking place on Jutland, as indicated by the finding of numerous pieces of raw amber found together with other artefacts datable to Vth period of the *Nordic* Bronze Age at the Bulbjerg and Troldting settlements in Thisted commune (Bukowski 2002, 83; Woltermann 2016a, 217). Amber blanks were also frequently found inside grave inventories dated to the IVth and Vth periods in Northern Germany (Woltermann 2016a, 178). That amber was transported south in unprocessed form can be proven on the basis of blanks found on the settlements of Thalmassing in the Upper Palatinate, and Bad Buchau in Baden-Württemberg (Woltermann 2016a, 175). Raw lumps of amber were also found in graves dated to the IVth–Vth periods of the *Nordic* Bronze Age at the cemetery of Siemirowice (Lębork commune) in Eastern Pomerania (Bukowski 2002, 76). Findings like those listed above indicate that the area of amber collecting was broader in FBA than in the two preceding phases of the Bronze Age (MBA and LBA).

Inflow of ready-made amber products to Southern Europe can be proven on the basis of the elongated biconical beads (IE161, IE261). While this form occurred earlier in the *Tumulus* culture milieu of Southern Germany (Woltermann 2016a, 167, Tab. 29), it became widespread during the advanced stages of the Bronze Age, FBA in particular. Significant increase in numbers of those beads is noted in the case of Southern Germany, while in smaller quantities they are known from the northern part of the country (Woltermann 2016a, 175, 178). Elongated biconical beads became a staple amber adornment in parts of Central Europe influenced by the *Urnfield* culture: Bohemia, Moravia, southern Poland, Hungary and Romania.[707] Excellent processing quality of these beads, some of which appear to be almost turned, prompted some scholars to suggest technological development in amber working in LBA–FBA possibly related to technical innovations such as the advent of lathe (Bukowski 2002, 87; Woltermann 2016a, 174). Beads in this form could be imported to Italy from the north, or inspired by contemporaneous amber jewellery of the *Urnfield* culture; it is characteristic that elongated biconical beads in the circum-Adriatic zone appeared mostly along its northern edge – in the area, where the *Urnfield* influences were the strongest.

Yet aside the elongated biconical beads, amber jewellery in the circum-Adriatic zone, in terms of stylistic, became more independent from the rest of Europe during FBA. Not only amber was locally processed in some parts of the study area (chiefly Northern Italy – western Veneto), but also new, unprecedented features (i.e. decorations) were invented there, while other concepts were revived. This is exemplified, i.a. by pendants – amber artefacts perforated eccentrically along D/L (group IV; cf. section 3.6), buttons with V- and Y-shaped perforations (group V; cf. section 3.7), some of the beads with hybrid perforations (group VI; cf. section 3.8), but most of all by the decorated beads (motifs 'a'–'i'). Frequent application of secondary or more complex drilling, and carved or incised decorations shows that circum-Adriatic artisans were increasingly proficient in this craft. Ornaments customarily classified as the Tiryns and Allumiere type are the prime examples of indigenous designs which gained a considerable popularity in the area, becoming prominent symbols of amber processing of the

707 Sprincz/Beck 1981, 483, Figs. 5:12–13, 26, 34 and 8, Tab. 2; Boroffka 2001, 399, Abb. 4:7–8, 12, 14, 17, 19–22; Chvojka 2017, Fig. 6; Gogâltan 2016, 153–156, Figs. 7–8, 9–10; Stuchlík 2017, Abb. 7–8; Purowski 2018, Figs. 2–3, 5.

period. While the former were certainly manufactured at Campestrin, the latter might have been produced at Frattesina, as can be inferred from the pieces with unfinished drilling or entirely missing a perforation (cf. sections 3.10. and 3.11.).

By HaA2/HaB1 the cultural links between the circum-Adriatic zone and the *Urnfield* culture to the north were tightened and some of the non-amber objects found on analysed sites testify to this. Contexts which yielded those objects are concentrated at the northern fringes of the study area. Similarities are clearly manifested by bronze pins with spherical or slightly biconical ('onion-shaped') heads and decorated necks. They appeared in a number of Italian contexts with amber as the type Sover (Fig. 98:C–H, S; Salzani 1990–1991b, Fig. 59:2), and a few of the Balkan ones (Fig. 107:B; Glogović 1991, T. 1:4); however their analogies have been registered further north, e.g. in Transdanubia (Říhovský 1983, Taf. 17–21). Similar distribution pattern is revealed by the pins with slightly biconical, profiled heads like the one from Križevci-Ciglana (Fig. 107:A; cf. Říhovský 1983, 26–27, Taf. 10:191–192; Blecić-Kavur 2014, 72–75; Fig. 38). Widespread occurrence of twisted or imitated twisted bronze torcs in the circum-Adriatic zone should not be surprising, as it fits into a common phenomenon of increased incidence of such torcs in Europe from the beginning of the *Urnfield* culture at about BzD/HaA1 (Vasić 2010, 43). Swords, again, provide much help as precise indicators of contacts; one can mention here Italian swords of the Allerona type from the Narde necropolis (tombs 168 and 227) which can be correlated with many other European swords belonging to the Naue II type (Salzani 1989a, Figs. 12:5 and 16:3; cf. Bianco Peroni 1970, 66–71). Furthermore, pieces of the sword found in the hoard of Debeli Vrh are cognate with the type Reutlingen, variant Staro Topolje (Hirschbäck-Merhar 1984, T. 7:1; cf. Harding 1995, 38–39, T. 13:93). Numerous analogies on the northern side of Alps can be found with respect to the Matrei-type knives from Frattesina (Salzani 1989b, 67, Fig. 1:3; cf. Bianco Peroni 1976, 16). Another element with such broad distribution range is razor of the Fontanella/Oblekovice type (Fig. 105:G–H).[708] There are many other bronze finds which show a broad spatial affinity with the units related to the *Urnfield* culture, for instance, socketed axes or sickles with tangs. Away from the centre of the circum-Adriatic zone extraneous influences were even stronger, suggesting cultural affiliation of such remote provinces of amber's acquisition with neighbouring regions outside the study area. As examples can be brought the hoards of Monte Cavanero in Piedmont, which included objects linking it to south-east France (e.g. pendants of the Réallon type), and Majdan in Vojvodina, the latter with composition replicating many hoards of the *Kurd*, *Gyermely* and *Hajduböszörményi* horizons in the eastern Carpathian Basin (e.g. situla of the eponymous Hajduböszörményi type) (cf. section 3.12.4.).

However, it is not always certain whether we are dealing here with imports from the north or exports from the south. It is possible that some of the bronze items mentioned above were conceived and produced in the circum-Adriatic zone. This option seems plausible regarding the Allerona-type swords from Narde or the Matrei-type knives from Frattesina, proven to be made of copper most likely coming from southern Alpine ores (Jung et al. 2011). On the other hand, products of Northern Italian provenience such as amber beads of the Tiryns and Allumiere types and glass beads made in LMHK technology reached areas far to the north (Fig. 138). Notable examples

708 Jockenhövel 1971, 205–207; Bianco Peroni 1979, 59; Weber 1996, 234–240.

of far-flung amber findings of the circum-Adriatic provenience are the lake-dwellings of Montlingerberg and Hauterive-Champréveyres in Switzerland, where decorated amber beads representing both the Tiryns (var. 'a') and Allumiere (var. 'c' and 'g') types were discovered (Jennings 2014, Fig. 12:b–c, Bellintani 2012, Fig. 3:12–14). Significant concentration of the Tiryns- and Allumiere-ornamented beads has been recorded on Sardinia (i.a. elongated cylindrical beads decorated with var. 'f' from Romanzesu), where amber could have been sent in exchange for copper.[709] Sardinia also deserves attention due to evidence of local amber processing, as well as presence of non-succinite fossil resins among the analysed finds which elevate the island to the rank of a production and redistribution centre, possibly using other sources or routes of raw amber influx (Bellintani 2014, 124; cf. Angelini 2012).

Fig. 138. Distribution of some of the products genetically linked with Italy during FBA and beginning of the EIA (12th–10th centuries BC) (Bellintani 2014, Fig. 6).

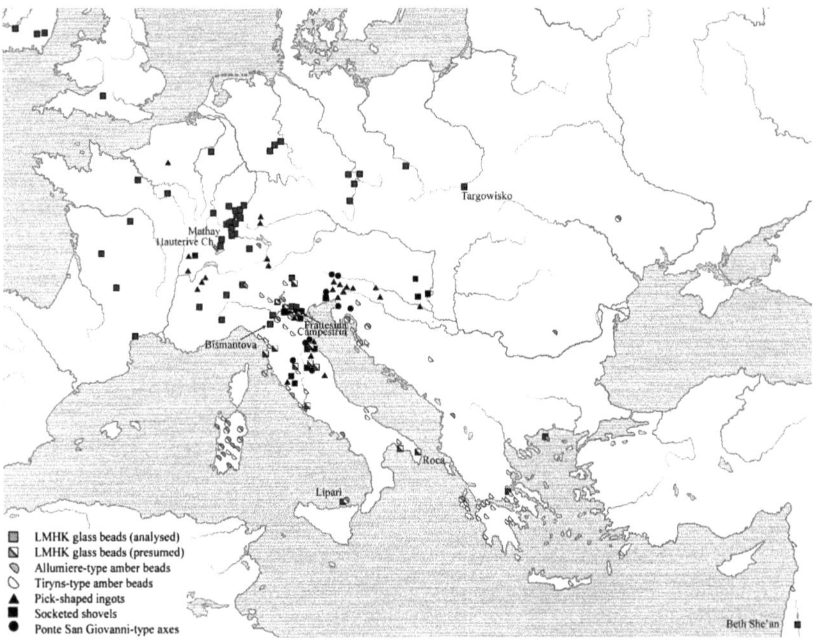

To the north of the Danube, there are few isolated sites on which amber beads representing one of the discussed types were discovered. Conspicuous assemblage of beads featuring the Tiryns- and Allumiere-type decorations (var. 'a' and 'c') was found at the cemetery of Hordeevka (barrows 31 and 38) in Ukraine, dated to ca. 12th century BC.[710] Single specimen of the Allumiere type (var. 'g') comes from LBA

709 Cf. Lo Schiavo 1982; Negroni Catacchio 1999; Bellintani et al. 2012; Jung et al. 2011; Ling et al. 2014.

710 Berezanskaja/Kločko 1998; Ślusarska-Michalik 2003; cf. Harding 2007, 50.

(HaA1) hoard recovered at the locality Dridu in Romania (Boroffka 2001, 404, Abb. 3:21; Negroni Catacchio 2014, 8, Fig. 15). Recently, a group of previously unknown Tiryns- and Allumiere-ornamented beads (var. 'a' and 'c') from middle Bohemia (cemeteries of Drahelčice and Únětice) has been published (Ernée 2017, 126, Abb. 12). Thus, one can speak of export of finished (processed) goods from the northern regions of the circum-Adriatic zone (most likely western Veneto, and possibly Kvarner and northern Dalmatia) to other parts of Europe.

Basing on the gathered information, one can reconstruct several ways of amber's inflow into the circum-Adriatic zone during FBA. The main one, well documented by the finds from Italy and Austria, which transported both processed and raw amber, went through the Alpine passes, along Adige, to western Veneto. However, it seems justified to assume yet another, more easterly route which led through Carinthia and/or Styria to Slovenia, and further east to the Croatian Littoral zone. The latter area could also obtain amber, thanks to mediation of the *Urnfield* groups in Transdanubia and Northern Croatia, as indicated by the findings of biconical amber beads in Vidasi, Križevci and Hungarian hoards of the *Kurd* horizon (cf. Sprincz/Beck 1981, Fig. 8). Less frequented route led into Romania (cf. Gogâltan 2016, Fig. 12), from where amber reached Banat (i.a. the hoard of Majdan and settlement of Banatski Dvor). Finally, it is possible that amber reached north-west fringes of the circum-Adriatic zone by the way which followed through western Alps (Switzerland and France) – this route can be inferred from the context of Monte Cavanero in Piedmont and is supported by other evidence of interactions between the *Rhin-Suisse-France orientale Urnfield* culture and the *Protogolasecca* culture (cf. David-Elbiali 2014).

On the other hand, finished goods (i.a. processed amber and glass) were exported to the north from Northern Italy. These are mainly the Tiryns- and Allumiere-decorated amber beads and mixed alkali (LMHK) glass beads (cf. Jennings 2014). Apart from them, one can suspect export to the north of V-perforated buttons, popular and possibly manufactured at the time in the northern regions of the circum-Adriatic zone (cf. sections 3.7 and 3.11) and found i.a. at Montlingerberg in Switzerland (Jennings 2014, 28, Fig. 12:c) as well as Montpezat and Tharaux in Occitany (Du Gardin 1986, 553, Fig. 2:8–9). Yet, from the sparse findings of amber products of northern Adriatic provenience, it can be inferred that they were not regularly traded; more plausible seems to be their occasional exchange. Rarity of the Tiryns- and Allumiere-type decorations on the finds from beyond the Alps poses questions. In the light of quite narrow social attribution of amber in the circum-Adriatic zone during FBA, it can be deduced that those specifically shaped amber adornments were tightly knit into culture and identity of their users, and circulated through a small although occasionally far-reaching network. Perhaps they were not subjected to a more casual trade to preserve this status, or perhaps their meaning could not be translated into social realities of communities to the north (cf. Jennings 2014, 28). The limited occurrence of both the Tiryns- and Allumiere-ornamented beads in Central Europe contrasts with their much more widespread use in the Mediterranean (i.e. Sardinia and Greece). These beads represent the clearest manifestation of amber translocation to Southern Europe via the circum-Adriatic zone during FBA; however, this process started much earlier.

5.2. Intermediary or consumer? Amber in contacts between the circum-Adriatic zone and Greece

Presence of amber artefacts in the *Mycenaean* culture is a part of the more general subject of contacts between Greece and the rest of Europe during the Bronze Age. Pioneering investigations into the provenience of amber finds from the *Mycenaean* contexts conducted by C. W. Beck and his collaborators showed that they were predominantly made of succinite.[711] This raised a question from where exactly and how amber reached the Aegean. The issue has been debated in numerous publications by a constantly growing group of scholars.[712] Additionally, this question is linked to the appearance of amber objects in other areas of the Eastern Mediterranean, such as Crete, Levant and Egypt.[713] Various authors saw the Central Mediterranean and the circum-Adriatic zone as crucial areas for the flow of goods, amber among them, in and out of the Aegean; however, they characterised importance of constituent regions, i.e. Sicily and the Aeolian Islands, the Apennine Peninsula, the Balkans, differently. Here it is attempted to reconstruct the spatial dynamics of amber influx into the Aegean (and with respect to the Eastern Mediterranean) via the circum-Adriatic zone.

Amber made its first appearance in Greece around MH III–LH I (ca. 1800–1635/1600 BC, so roughly around the turn of EBA and MBA), but already on a grand scale. The earliest contexts with amber are concentrated at three sites: Mycenae, Pylos and Peristeria on Peloponnese, and yielded around 1,560 artefacts between them, although the lion's share (1,290 finds) comes from the Shaft Grave IV at Mycenae (Harding/Hughes-Brock 1974, Tab. I; Maran 2013, 148). About the same time, amber began to appear in more substantial numbers in the circum-Adriatic zone; however, it was still largely confined to the initial regions of its acquisition in Trentino and the Garda Lake basin in the north, and Sicily in the south. Still, the quantities of amber from the LH I-contemporaneous contexts in the study area, even if combined, are far lower than the corresponding quotas from the Aegean cited above. The Shaft Graves of Mycenae, in particular, are astonishing in terms of quantity and quality of inventories. The list of finds comprises 'ostrich eggs from Nubia sent through Egypt and Crete, lapis lazuli from Mesopotamia, alabaster and faience from Crete, raw ivory from Syria, silver from Anatolia, and amber from Prussia brought down the Adriatic or out of Odessa across the north Aegean' (Vermeule 1964, 89). The above description is interesting as it assumes the 'Prussian' (south-east Baltic) origin of amber. Such statement may be considered as going too far; even though the provenience analyses with IRS credibly proved Baltic (succinite) attribution of the majority of sampled artefacts from the Shaft Graves of Mycenae (Beck et al. 1972),[714] the exact provenience of amber in geographic

711 Beck 1966b; Beck et al. 1968, 1970, 1972; Beck/Beck 1995.

712 I.a. Harding/Hughes-Brock 1974; Bouzek 1993, 2007; Hughes-Brock 1985, 1993, 1999; Maran 2004; Czebreszuk 2007.

713 Cf. Hood 1993; Gestoso Singer 2008, 2016; Cultraro 2011.

714 According to the study, spectra of several samples were identified by computer program as walchowite (type of fossil resin occurring in Moravia; cf. section 1.3.3.4); however, upon reconsideration the differences in the spectra have been attributed to contamination during conservation treatment in the museum (Beck et al. 1972, 362–364, 367, 373).

sense is not readable through chemical analysis. At the moment the only solution to trace the routes via which amber reached the Aegean is through comparison of artefacts' morphology.

The amber finds of the *Mycenaean* culture have received a detailed typological study relatively recently (Czebreszuk 2011). From an overview of morphology of the finds assigned to the Shaft Graves period, specifically those dated to MH III–LH I, it occurs that they cover most of the types present at the time in the circum-Adriatic zone (cf. Czebreszuk 2011, 114–116). The beads can be typologically encapsulated within four leading types: spheroidal (Czebreszuk's type 101 corresponding to IA122), flattened spheroidal (type 102 – IF121, IF221, IF122, IF222), lenticular (type 103 – IF162, IF262) and annular (type 108 – discoidal or flattened spheroidal beads with expanded perforation outlets 'w', 'x', and 'y'), each of them represented by at least several dozens of pieces in the Shaft Graves of Mycenae (Czebreszuk 2011, Tab. 1–2, 7). Apart from those, there are types known from singular examples at this early stage: elongated cylindrical (type 107 – IE111), fusiform/barrel shaped (type 111 – IE121, IE221) or flattened biconical (type 105 – IF161, IF261), the latter two also marking their first appearance during EBA/MBA (Czebreszuk 2011, 116, Tab. 4, 6, 8). However, there is one important category of finds which differentiates the Aegean and the circum-Adriatic amber assemblages – spacer plates famously known from *Mycenaean* graves and almost absent in the study area, certainly at such an early stage. It does not seem necessary to delve into discussion about the genetic and chronological relationship between the *Mycenaean* amber spacers and those from other parts of Europe, i.e. England – the *Wessex* culture, Germany – the *Tumulus* culture or Bohemia – the *Únětice* culture.[715] A necessary observation to mention is that the closest analogies to those technologically advanced and stylistically sophisticated products are separated from Greece by hundreds of kilometres. At the current state of debate, it seems that the specimens from the British Isles are morphologically the closest counterparts for the Aegean spacers, although there still are many unsolved questions related to nature of contacts between the *Mycenaean* and *Wessex* culture, or the mode of amber's transportation. While the spacers originating in the *Tumulus* culture milieu can be rejected as potential inspiration on the chronological, as well as partly stylistic grounds, one should not neglect the option that the whole idea of amber spacers was born in the milieu of the *Únětice* culture. From there it gradually spread, but only to select parts of Europe where complex amber adornments found appreciation. It should also be considered possible that amber spacers were manufactured locally in the Aegean; H. Hughes-Brock on the basis of tablets with inscriptions in linear B *re-di-na-to-mo*, translated to 'resin-cutter', suggests that specialised craftsmen worked with amber in Greece during the Late Helladic period (Hughes-Brock 1985, 259). Yet, so far no Bronze Age amber workshop has been detected in Greece what might be surprising taking into account the multitude of evidence convincingly documenting processing of other exotic materials in the Mycenaean culture (Czebreszuk 2011, 151–153).

Another problematic issue is the vast disparity in numbers between the *Mycenaean* LH I amber collection and the one from the circum-Adriatic zone dated to EBA and

715 The debate was summarised i.a. by Harding 1984, 76–79; Du Gardin 2003; Czebreszuk 2011, 116–123.

EBA/MBA. The latter amounts to no more than 60 artefacts what pales in comparison with over 1,500 finds from *Mycenaean* contexts (Harding/Hughes-Brock 1974, Tab. I). In truth, the disproportion is evident with respect to most regions of amber's acquisition in Europe during EBA; the only area surpassing the richness of early *Mycenaean* amber is the Bohemian *Únětice* culture where approximately 4,000 amber items have been found (Ernée 2012).

This does not mean that the circum-Adriatic zone was disconnected or played no part in the import of amber to the Aegean during the first half of the 2nd millennium BC. After all amber, somehow, reached the *Mycenaean* tombs on the Peloponnese, and it is more than likely that the route of its translocation partly crossed the area under consideration. In fact parts of Southern Italy have produced a growing evidence of contacts with the *Mycenaeans* beginning from MH III/LH I, so the time when amber began reaching Greece. The main source of information about those interactions is imported *Mycenaean* pottery which in respect to LH I–II is predominantly concentrated in the Tyrrhenian Basin: on the Phlegraean and Aeolian islands as well as Sicily (Bettelli 2002, 251). Presence of LH I pottery was surely confirmed at four sites: (1) Punta Mezzogiorno settlement on the Vivara Islet, (2) Capo Graziano settlement on the Filicudi Island, (3) Lipari 'Acropolis' settlement on the Lipari Island and (4) Monte Grande settlement near Agrigento on Sicily.[716] These findings, often cited as the earliest evidence of the *Mycenaeans*' interest in their western 'neighbours', can be supplemented with few other objects documenting this relationship, coming from the *Castelluccian* sites with amber on Sicily.

Fig. 139. Probable Aegean imports from Southern Italy dated to the early Mycenaean *period (MH–LH II): A – Monte Sallia, tomb 9 (Orsi 1923, Tav. II:11); B – Castelluccio, tomb 22 (Orsi 1892, Tab. V:7); C – Villaggio delle Macine, MBA layers (Bellintani et al. 2007, Fig. 1); D–E – Gallo di Briatico, tomb (Pacciarelli 2001, Fig. 109:4, 8).*

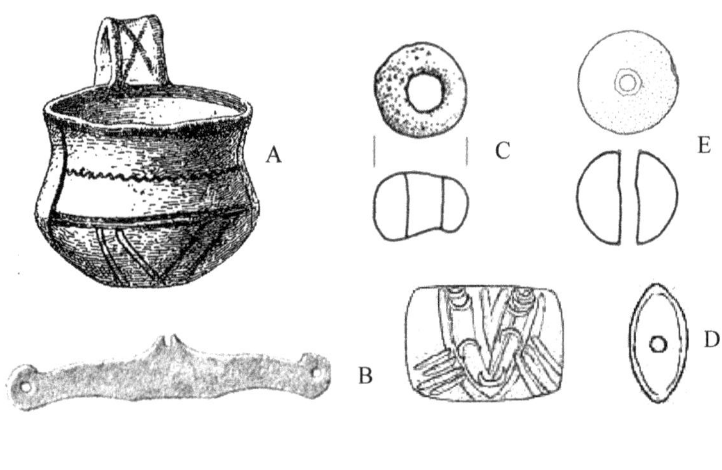

716 Alberti/Bettelli 2005, 548–549; Bettelli 2011, 110, Fig. 1B; Jung 2017, 623–624.

Inside tomb 9 at Monte Sallia was discovered a black matt-painted carinated cup with high-swung handle, decorated with zig-zags and lines (Fig. 139:A). W. Taylour thought the cup, in terms of technology, morphology and ornamentation, is similar to the MH Minyan ware (literally based on a Minyan prototype) from sites such as Eleusis, Zygouries and Korakou (Taylour 1958, 54–56). M. Bietti Sestieri lists the vessel as an evidence of contacts between the Aegean and Sicily before the developed *Mycenaean* culture (Bietti Sestieri 1988, 32, Fig. 10). Furthermore, from tomb 22 of the Castelluccio necropolis was recovered an object made of bronze sheet, with sharply cut edges, interpreted as a balance beam (Fig. 139:B). Similar bronze-scales are attested on several LH I–II sites in Greece (cf. Crispino/Cultraro 2015, 212–213). That Greece was involved in the broader network of amber circulation at the dawn of the *Mycenaean* culture is evidenced by provenience analyses indicating mostly Baltic amber (succinite) and morphological similarity of the earliest amber finds with their counterparts to the north, including Italy (cf. Czebreszuk 2011). However, it should be mentioned that Sicilian amber (simetite) was detected among the fragments of amber bead(s) from Pylos – Vayenas (V1) tholos, containing MH–LH II/III pottery (Beck 1966b, 208; Beck/Hartnett 1993, 42). This proves that in the early-middle stages of the Bronze Age, simetite or objects made of it were also exported outside Sicily to a certain degree. Succinite also reached Sicily during EBA, as documented, i.e. by the spectrum of the bead from tomb 10 of Cava Secchiera (cf. Angelini/Bellintani 2017), thus presenting archaeologist with an interesting case of two types of fossil resins being utilised roughly at the same time by certain Mediterranean communities.

All evidences described above support the idea of amber import to the Aegean around the 17th century BC via the Central Mediterranean, thanks to mediation of the communities inhabiting islands and coast of the Tyrrhenian and Ionian Seas, some of which utilised amber themselves, i.e. the *Castelluccio* culture. The four above-listed findspots of LH I pottery have been used by some authors as the basis for reconstruction of the *Mycenaean* maritime trade network which enabled amber's inflow into the Aegean (Maran 2004, Abb. 1; Cazzella/Recchia 2009, Figs. 1 and 148). It is speculated that the interest of the *Mycenaeans* in the Central Mediterranean was moved by the search for metal, given that eastern routes and markets were still under Cretan control (Bettelli 2002, 251). Others stressed that rather than metal those were other materials, such as sulphur and alum which occur on Sicily, that the *Mycenaeans* were interested in (Jones et al. 2014, 448–449). In fact the metal analyses using lead isotopes so far have not revealed Sardinian or southern Alpine copper among the indigenous Aegean bronze products (cf. Jung/Mehofer 2013). Truth be told, all the potential reasons for the *Mycenaean* presence in the Central Mediterranean are guesses at best, as so far no solid evidence support exploitation of the local resource, be it copper, sulphur or alum, at this stage of the Bronze Age (cf. Cazzella/Recchia 2018, 12–14).

Fig. 140. Mycenaean *sites with amber and findspots of Helladic pottery in the Central Mediterranean dated to LH I: (1) Vivara, (2) Filicudi, (3) Lipari, (4) Monte Grande, (5) Peristeria, (6) Pylos, (7) Mycenae (Maran 2004, Abb. 1).*

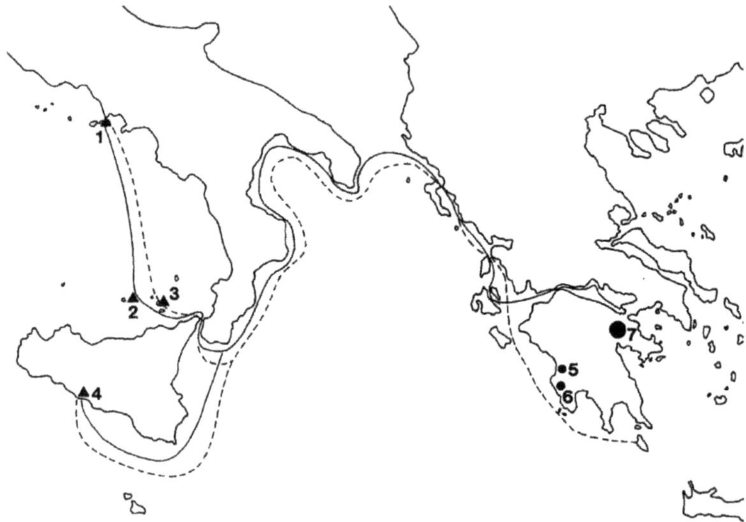

Even despite some obvious problems such as quantitative and qualitative discrepancies between amber assemblages from the early *Mycenaean* (MH III–LH I) and EBA/MBA circum-Adriatic zone, the inflow of amber to the Aegean via islands of the Central Mediterranean remains the most plausible hypothesis (Fig. 140). It is supported not only by the evidence of contacts between communities inhabiting both areas but also by absence of amber anywhere nearer the Aegean in this period. The lack of more sophisticated amber adornments such as spacers or large discs in the Italian EBA contexts can be explained by the absence of highly stratified social structure which demanded the elites to manifest their position with more magnificent paraphernalia – the function which is attributed to the outstanding pectoral necklaces with spacers in the *Mycenaean* and *Wessex* cultures (cf. Beck/Shennan 1991, 77–89, 133–137; Maran 2013, 148–149). It can be further inferred that the communities inhabiting Southern Italy during EBA and at the beginning of MBA played mainly the function of intermediaries, using amber for their own purposes to a limited extent and occasionally providing the Aegean with the locally occurring fossil resins (simetite).

Islands and coasts of the Tyrrhenian Sea, in the opinion of some scholars, acted as key nodes in the exchange/trade network used by the *Mycenaeans* in the first half of the 2nd millennium BC (Cazzella/Recchia 2009, 31; Bettelli 2015, 208–209). Scholars are usually shying away from stating the long-term presence of *Mycenaean* sailors, merchants or settlers in the areas beyond the Aegean during the Shaft Graves period. It seems plausible that amber reached the Aegean as a concomitant of trade in tin with Atlantic Europe, i.e. Brittany and Cornwall (Maran 2013, 149), although the idea of all the LH amber being brought to Greece in a single maritime journey made by a lone vessel, as proposed by Harding and Hughes-Brock (1974, 159), certainly works on the imagination. R. Jung and M. Pacciarelli, partly following R. Peroni, explain the interaction between Southern Italian natives and the *Mycenaean* sailors coming to the shores of Italy through exchange of goods and their purposefully uneven redistribution among the communities (Jung/Pacciarelli 2017, 190–191). In exchange for the locally produced goods' surplus, the ruling circles obtained other products or resources of foreign provenience, some of which were delivered to a community in order to meet current needs, while others were kept from wider access in order to maintain or increase personal economic and political power. Here, the authors point to the *Mycenaean* pottery (comprising not only storage vessels, that one would expect when thinking of maritime transport of goods, but tableware – Bettelli 2015, 209) as one of the key items used to legitimise the high social position of local elites. Yet, similar interpretation can be used to explain slow emergence of amber adornments in the EBA Italian context. The *Mycenaean* demand on amber possibly prompted local individuals to seek adornments made of it for themselves (especially those made of succinite consequently superseding simetite). By copying the custom of wearing amber jewellery, they might have hoped to enhance their own position.

It is difficult to determine where to locate the preceding section of the 'Aegean-bound amber route' of the early 2nd millennium BC. Some authors suggest that it leads along the Rhône and Rhine rivers to the North Sea (Gerloff 1993, 85; Czebreszuk 2011, 158–159), but it is also possible that it went through the Garda Lake basin and the Alpine passes of Trentino and Alto Adige (cf. section 3.12.1.). Problem with reconstruction of this route stems from the fact that the evidence of interactions with the Aegean basically ends in the Tyrrhenian Sea to reappear on the British Isles, thus creating a vast gap (cf. Maran 2004, 55; 2013, 149–150). Yet, there is one important finding from Switzerland – Zürich-Mozartstrasse which delivered an amber disc wrapped in gold foil (Maran 2004, 56) – a quite unique find with analogies known from Wilsford, grave 8 and Preshute, grave Ia in England on the one hand (Harding/Hughes-Brock 1974, 156; Harding 1984, 79–82), and Knossos-Isopata – the so-called 'Tomb of the Double Axes' on Crete (Harding/Hughes-Brock 1974, 157; Czebreszuk 2011, 95) on the other, although the latter is dated to a later period (LM IIIA1).[717] The amber finds from Swiss settlements such as Zürich-Mozartstrasse and Koblach-Kadel (the latter known for an amber spacer; cf. Stahl 2006, 140, 147–148; Maran 2004, 156) are located not too far from the amber-bearing contexts of the Garda Lake basin and Trentino, and there are proofs of contacts between the *Rhône* culture of western Alpine zone and Northern Italy during EBA (cf. David-Elbiali/David 2009; David-Elbiali 2014). However, the evidence associated with Northern Italian amber finds dated to EBA points more clearly in the north and north-east directions, rather than north-west, regarding the potential sources of fossil resin (cf. section 5.1). Is it possible that there were two routes of amber translocation passing through or intersecting in the Alpine zone during EBA: one leading from

---

717 H. Hughes-Brock solves the chronological discrepancy by interpreting the disc from Knossos as an 'heirloom' which arrived on the Greek mainland around LH I together with other amber objects from the *Wessex* culture, but stayed in use for a longer time, before being eventually deposited on Crete when the island became settled by the *Mycenaeans* (Hughes-Brock 1993, 219).

the British Isles and the second from the Baltic shores? The exact route and mode of transportation of amber which ended up in LH I tombs on the Peloponnese will have to remain a mystery for now. Nevertheless, the link between the Aegean and southern parts of the circum-Adriatic zone seems firm and currently offers the best basis for reconstructing EBA amber inflow to Greece.

The 'Tyrrhenian route' was probably used until the end of LH II. Continuous frequentation of the Tyrrhenian area by the *Mycenaeans* aimed at obtaining amber is even more evident, when the early MBA contexts are added to the picture. Area of amber's acquisition in Italy during BM 1, although rare and sparse, stretched from the Garda Lake basin and along the Tyrrhenian coast, from Tuscany (Grotta dell'Artofago), through south-central Lazio (Villaggio delle Macine) and Calabria (Gallo di Briatico), to the Aeolian Islands (Lipari 'Acropolis') and eastern Sicily (Longane and Cava Cana Barbara). It shifted towards the Adriatic coast only during BM 2. It is worth mentioning that some of the sites listed above, aside from amber, yielded other testimonies of interactions with the Aegean. At Villaggo delle Macine was discovered a sub-spherical bead made of green-blue glass with HMG composition typical for the Eastern Mediterranean vitreous materials (Bellintani et al. 2007; Fig. 139:C). Hut Delta XXI at the Lipari 'Acropolis' returned a similar glass bead made of bluish paste (however not analysed chemically as of now) as well as a fragment of undefined vessel of *Mycenaean* origin dated to LH II (Bernabò Brea/Cavalier 1980, 256; cf. Wijngaarden 2002, 384, cat. No. 247). Finally, the tomb of Gallo di Briatico yielded a carnelian seal (Fig. 139:D) which has been attributed *Minoan* origin and LM IB–LM IIIA–B chronology (Pacciarelli 2001, 187), or *Mycenaean* provenience and LH I–II dating (Jung 2013, 245). Furthermore, the same grave contained a globular bead made of crystal with close analogies in mainland Greece (Jung 2013, 245; Fig. 139:E).

Abovementioned evidence of contacts with the Aegean is supported by the continuing appearance of *Mycenaean* pottery in contexts situated on the Aeolian and Phlegraean islands, as well as on few isolated sites in Calabria, albeit in fewer quantities (Punta Zambrone and Petrosa Cave; Cazzella/Recchia 2009, 32). At Lipari Helladic pottery reached the peak in quantitative terms during LH I–II, then diminished sharply in LH IIIA to remain more or less stable up to LH IIIB, and then underwent further reduction during LH IIIC – time when the *Mycenaean* presence in the Aeolian Islands was at the finish (Bettelli 2002, 254; cf. Wijngaarden 2002, Tab. 15.3). Majority of the Aegean pottery from the ancient-most phase of the *Mycenaean* presence (including MH and LH I–II ceramics as well as Minyan ware and burnished ware) in the Central Mediterranean seems to have come from the mainland Greece, specifically the Peloponnese, but some fabrics have *Minoan* features (Wijngaarden 2002, 204; Bettelli 2015, 209).

This observation corresponds well with distribution of LH II amber finds in the Aegean which are largely confined to two main centres: the Argolid and Messenia with singular findings in Laconia and Beotia (Harding/Hughes-Brock 1974, 147, Fig. 1; Czebreszuk 2011, Fig. 22). The stylistics of amber adornments used by the *Mycenaeans* had not changed since the previous phase; the beads have similar morphology and spacers continued to be worn until at least LH IIA (cf. Czebreszuk 2011). The most remarkable finds come from the three Kakovatos *tholoi*, where over 510 amber artefacts were discovered, comprising, i.a. lenticular, annular and discoidal beads as well as five spacers. Nevertheless, there is a notable drop in numbers of amber finds between

LH I and II, with those pertaining to the latter period amounting to no more than 1,000 finds, even if the contexts with unsecured (LH I or II) chronology are added. A general trend is that from LH II there was a steady decline in numbers of amber finds corresponding to the successive stages of the *Mycenaean* culture, although at the same time amber was becoming more widespread, appearing in increasingly more contexts spreading outside the Peloponnese (Harding/Hughes-Brock 1974, Tab. 2; Czebreszuk 2011, Tab. 12–13).

The classical period, also known as the palatial period (LH IIIA–B, ca. 1420–1220/1210 BC, roughly coeval with MBA 3–LBA 1), was the time of great flourishing of the *Mycenaean* culture which expanded beyond the Peloponnese and intensified the relations with the Central Mediterranean (Bettelli 2002, 251). Amber was more widespread then in Greece, than in the preceding period; however, its quantities continued to decline, as mentioned earlier (cf. Harding/Hughes-Brock 1974, Tab. I; Czebreszuk 2011, Tab. 12–13). Apart from the Peloponnese, objects made of fossil resin started to appear in more substantial numbers in central Greece, and since LH IIIB also further north-east in Aeotlia and Epirus (cf. Harding/Hughes-Brock 1974, 149–150, Figs. 2 and 3; Czebreszuk 2011, 123, Fig. 23). To the south amber reached Crete for the first time; it is highly likely that it was brought there by the *Mycenaeans*, when they moved to the island in LM IIIA after the final collapse of the *Minoan* political system (Harding/Hughes-Brock 1974, 150; Cultraro 2011, 228–232).[718] Singular sites with amber, possibly linked with the classical *Mycenaean* culture, were recorded also far to the east and north-east, in Macedonia, Kos and Rhodes (Czebreszuk 2011, Fig. 23).

With respect to morphology of amber artefacts, the classical period produced beads in most of types distinguished on the basis of evidence available from the Aegean; however, spacers and end-pieces no longer appear, nor do the lavish crescent necklaces (Czebreszuk 2011, 123). In combination with a drastic decline of objects-per-context ratio, it appears that amber was worn in the form of simpler adornments (more modest necklaces or bracelets), or as a part of necklaces incorporating beads of various materials (Czebreszuk 2011, 123). Typologically the beads do not differ significantly from the Shaft-Grave period. Among them the most common are spheroidal (type 101 corresponding to IA122) and flattened spheroidal forms (type 102 – IF121, IF221, IF122, IF222) (cf. Czebreszuk 2011, Tab. 1–2). Very numerous (ca. 100 specimens) are lenticular beads (type 103 – IF162, IF262), although majority of them come from two tombs with uncertain dating (LH IIA–IIIA1): Dendra, tomb 10 in Focis and Routsi, tholos 2 in Messenia (Czebreszuk 2011, Tab. 3). Nonetheless, their broad dating matches the chronology of similar beads in the circum-Adriatic zone (cf. section 3.3.1). Worth noting is the fact that lenticular beads in both areas were almost non-existent by FBA (LH IIIC). Significantly fewer than in the preceding period are annular beads (type 108; cf. Czebreszuk 2011, Tab. 7). In the classical period there was an increase in the number of flattened biconical beads in Greece (type 105 – IF161, IF261); however, they are still much fewer than in the circum-Adriatic zone; the respective ratio is 10 to circa

---

718 There is only one Cretan context with amber safely (?) dated to the pre-LM III period: the tomb of Arvi near Amira attributed to LM I or II (Cultraro 2011, 228). For the discussion about the absence of amber in the *Minoan* culture, see Czebreszuk 2011, 105–108.

111 specimens (cf. Czebreszuk 2011, Tab. 4). Yet, it is interesting to note that flattened biconical beads are more frequent in the northern part of the circum-Adriatic zone than in the southern part (cf. section 3.3.1.2), what possibly explains their rarity in the Aegean. Other kinds of beads appeared incidentally in the Aegean during the classical period; however, one should note the adherence of majority of elongated cylindrical (type 107 – IE111, IE211) and barrel-shaped (type 111 – IE121, IE221) beads to LH III period, signalising the trend towards longer forms, much like in the circum-Adriatic zone at the time (cf. Czebreszuk 2011, Tab. 6, 8; section 3.3.3.1).

The classical period saw reorientation of contacts network in the Central Mediterranean which was now focused more on the Ionian and lower Adriatic basins; the Tyrrhenian area noted a decrease in the *Mycenaean* involvement in trade, although towards LH IIIB Sardinia began to more actively participate in contacts with Greece (Bettelli 2002, 251; Cazzella/Recchia 2018, 15). As before, also in this phase the main source of information on the relations between the Aegean and Italy is *Mycenaean* pottery.

During LH IIIA and IIIB ceramic imports from the Aegean still reached Southern Italy, and similar to the earlier phases, the analytical results indicate mainly the Peloponnese as the region of provenance (Bettelli 2011, 111–112; Jung/Pacciarelli 2017, 190). However this process was subject to temporal fluctuations. Thus, during LH IIIA the Aeolian Islands and Sicily were still the main recipient of imported *Mycenaean* pottery, whereas in the subsequent phases, its destination gradually shifted to the Ionian arc and southern Apulia.[719] The particularly rich collections of Aegean pottery come from sites with amber on the Aeolian Islands (Portella di Salina, hut F), as well as Sicily (Thapsos, tombs 61 and D), to which can be added imports of other materials of similar provenance (present also i.a. in Plemmirio, tomb 48). The assemblage from Salina comprises nine fragments of *Mycenaean* pottery dated to LH IIB–IIIA (Bernabò Brea/Cavalier 1968, 166; Wijngaarden 2002, 227, 329, cat. I, no. 323; Fig. 141:A) and 56 beads of carnelian in various shapes which can be compared with numerous analogies known from Greece, Cyprus and the Near East (Matarese et al. 2015; Jung/Pacciarelli 2017, 190; Fig. 141:B). Similar items were discovered on the Thapsos cemetery; tomb 61 contained i.a. a small piriform jar of *Mycenaean* provenience (described by P. Orsi as amphora), decorated with painted brown-red horizontal bands on white background and dated to LH IIIA2–IIIB (Orsi 1895, Fig. 49; Wijngaarden 2002, cat. IX:16). In addition, it yielded at least nine faience or glassy-faience beads, by the look resembling vitreous materials from Southern Italian sites defined as produced in LMLK (low magnesium and low potassium) technology and plausibly manufactured in the Aegean (Angelini et al. 2005; Fig. 141:C). Tomb D at the same necropolis yielded one of the most astounding assemblages, comprising i.a. a jug with a flattened globular body and wide cylindrical neck, and a deep rounded bowl with horizontal handle, both dated to LH IIIA2–B and possessing good analogies in the mainland Greece (Alberti 2004, 128–134; Fig. 141:E–F). Furthermore, tomb D delivered a collection of seven carnelian and 48 vitreous beads (Voza 1973, 40, nos. 105–115; cf. Matarese et al. 2015). Finally, tombs 48 and 49 of Plemmirio revealed presence of considerable

719 Cf. Wijngaarden 2002; Jones et al. 2005; Jung/Pacciarelli 2017.

assemblages of faience or glassy-faience beads (134 and 14, respectively) which, again, can be compared with LMLK beads from Southern Italy (Fig. 141:D). Last but not least, tomb 48 produced an ivory comb with rectangular shape, decorated with carved spiral band motif in the middle and short notches at the edges, which is assumed to be an import of *Mycenaean* origin (Colini 1904, 255; Bettelli/Damiani 2005, 20; Fig. 141:G). Thanks to the above-listed finds, it appears probable that the communities inhabiting the Aeolian Islands and south-east Sicily around LH IIIA–B procured amber for the *Mycenaeans*, but also consumed it themselves in a manner that seems to be derived from the Aegean (occurrence of amber in graves together with imported items; cf. section 4.1.2.4). It should be underlined that Thapsos (the entire complex on the eponymous peninsula comprising the necropolis and the adjacent settlement) has been described as a *Mycenaean* emporium (Bettelli 2011, 117).

Fig. 141. Probable Aegean imports from the Aeolian Islands and Sicily dated to the classical Mycenaean *period (LH IIIA–B): A–B – Portella di Salina, hut F (Bernabò Brea/Cavalier 1968, Tav. LXXXIV:11–17); C – Thapsos, tomb 61 (Bellintani et al. 2006, Fig. 1:19, 24); D, F – Plemmirio, tomb 48 (Colini 1904, Fig. 54; Bettelli/Damiani 2005, Fig. 1A:2; Genovese 2015, Fig. 7:u); E–F – Thapsos, tomb D (Alberti 2004, Tav. IX:6, 11).*

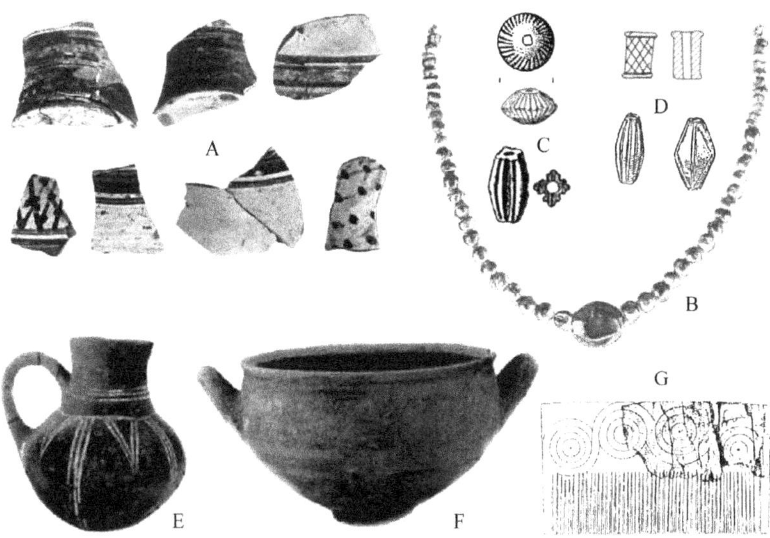

It is also worth noting that from LH IIIA Cyprus began to more actively participate in the Central Mediterranean maritime trade. Cypriot pithoi were discovered at Portella di Salina, while other types of Cypriot wares such as Base Ring II vessels and White Shaved ware jug come from two tombs at the Thapsos cemetery together with *Mycenaean* ceramics (Wijngaarden 2002, 235, Fig. 16.2; Bettelli 2015, 209, 214). These findings correspond well with the fact that LH/LC IIIA saw the first amber import on Cyprus – Enkomi, tomb 67 (Harding/Hughes-Brock 1974, 169, Tab. I).

While during LH IIIA–B south-east Sicily and the Aeolian Islands were important destinations for the *Mycenaean* sailors/merchants, who brought with them genuine Aegean pottery, the period also saw the first local Southern Italian imitations of those vessels. Analyses conducted on the Aegean pottery from Sicily and the Plain of Sybaris (Calabria) proved distinct differences in technology, as well as functionality of the vessels (Jones et al. 2005). It seems that nearly 100 % of *Mycenaean* pottery on Sicily are imports, while the same fact can be attributed only to 25 % of analysed containers from Sybaris. The rest, i.e. 75 % are locally made products, imitating Aegean vessels. On time scale the import of *Mycenaean* pottery to Sicily started in LH I–II, while local products occurred only in the last phase of interactions (LH IIIB–C) when the influx of Aegean pottery was almost extinguished. Completely different picture is visible in the case of the Sybaris Plain, where local products were gradually replacing imports since LH IIIA, eventually surpassing imported ware almost five times around LH IIIB–C (Jones et al. 2005, Pl. CXIX:b). This trend can be connected with amber influx to Southern Italy. On Sicily the decrease in imported *Mycenaean* pottery around LH IIIB corresponds with cessation of amber import to the island, whereas the major influx of amber in the southern Apennine Peninsula in the later stage of MBA and LBA coincides with the gradual increase of genuine as well as imitated Aegean pottery in the area.

The questions of who and why they started to manufacture Aegean-like (or the so-called *Italo-Mycenaean*) pottery in Southern Italy are still debated. Various scholars envisage the *Mycenaean* craftsmen working in foreign lands, while others lean towards the idea of local potters borrowing the technology (introduction of potter's wheel) and stylistics from the Aegean, and combining them with the local *Protoapennine* and *Apennine* impasto tradition.[720] Moreover, apart from the copied wheel-made *Mycenaean* vessels (that is the *Italo-Mycenaean* pottery), Italian potters also manufactured Pseudominyan pottery[721] and *pithoi*. The *Mycenaean* potters, drawn by the prospects of opening new markets for their products, could come to Italy and establish their workshops (Bettelli 2011, 116). In the opposite vision, local craftsmen were prompted to adopt the wheel-based technology and produce their own adaptations of foreign ceramics, in the face of limited access to luxury goods and instability of the Central Mediterranean exchange network (Jung/Pacciarelli 2017, 191).

720 Cf. Bettelli 2002, 257; 2011, 114; Jones et al. 2005, 543; Cazzella/Recchia 2009, 36; 2018, 25.

721 This particular type of pottery production recalls the Minyan ware of MH period, still occurring in Greece at the beginning of LH. Pseudominyan pottery became widespread along the Adriatic and Ionian coasts around LH III, particularly in the Plain of Sybaris with shapes which reproduce both *Mycenaean* and indigenous models, such as carinated cups and bowls (Bettelli 2002, 258).

Fig. 142. Italo-Mycenaean *pottery and vitreous beads of potential Aegean provenience from Apulia dated to the classical* Mycenaean *period (LH IIIA–B): A – Coppa Nevigata,* Recent Apennine *stage (Vagnetti et al. 2012, Fig. 3); B – Coppa Nevigata,* Early Subapennine *stage (Vagnetti et al. 2012, Fig. 5); C – Madonna di Loreto, 'Hypogeum of Bronzes' (Tunzi Sisto 1999a, Tav. X:812, 828); D – Masseria Chiancudda, layer of* Punta le Terrare *facies (Cinquepalmi/Recchia 2010, Fig. 9).*

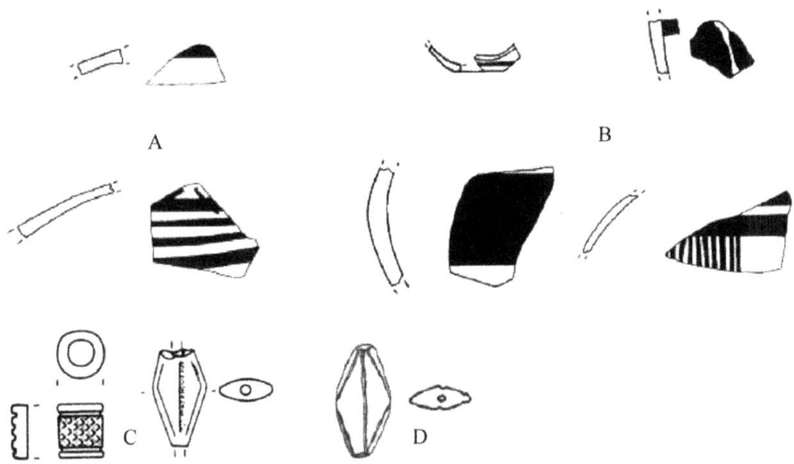

The main workshops producing *Italo-Mycenaean* pottery operated from LH IIIA in coastal settlements such as Scoglio del Tonno, Broglio di Trebisacce, Torre Mordillo and Roca Vecchia, while still receiving imports from the Aegean. It is argued that those settlements acted as 'gateways' between the hinterland and the sea, facilitating transfer of goods, among them amber (Cazzella/Recchia 2009, 33). Other exchanged resources could be foodstuffs such as olive oil, wine and figs, or more refined commodities such as purple dye which was produced from *Murex* snails at Coppa Nevigata (Cazzella/Recchia 2009, 34). It is interesting that from all the abovementioned sites, only Coppa Nevigata delivered evidence of amber earlier than FBA (LH IIIC). Thus, the settlement assumes the key position in the influx of this material to the classical *Mycenaean* culture at least from the turn of MBA and LBA. The ongoing research at the site revealed still growing collection of *Mycenaean* and *Italo-Mycenaean* sherds, bulk of which falls into the LBA 2 (BR 2; recent *Subapennine*) and corresponds with LH IIIC (Cazzella/Recchia 2018, 20–21; cf. Vagnetti et al. 2012). The overwhelming majority of these vessels were locally made, although this conclusion refers mainly to the recent *Subapennine* layers (Jones et al. 2014, 139–142). Still, the horizons pertaining to the recent *Apennine* and early *Subapennine* stages, so those which delivered the earliest objects in amber from the site, produced some Aegean-like pottery (9 and 8 shards, respectively; Cazzella/Recchia 2018, 20; Fig. 142:A–B).

Also the location of the Coppa Nevigata settlement in northern Apulia gives credibility to the idea of the site serving as a major outpost in amber circulation. During later MBA and LBA the region abounds with amber-rich sepulchral context, with some of the largest assemblages of artefacts known from the study area. The overall collections of amber from the hypogea of Madonna di Loreto are on par with those from LH II *Mycenaean* tombs, and easily surpass the amber outfits found in those dated to LH III (cf. Harding/Hughes-Brock 1974, Tab. I). The Hypogeum of Bronzes, apart from the already discussed LMLK glassy-faience beads in lenticular form with radial incisions (Fig. 88:F), yielded other vitreous materials in the same form, and most probably same composition, as those from Sicily and Salina with analogies in the Aegean (Fig. 142:C; cf. Bellintani et al. 2006, 1505, 1508–1509, Figs. 1 and 3; Bellintani 2012, 16, Fig. 1). Although the rock-embedded tombs of Apulia are lacking the more explicit *Mycenaean* imports like their Sicilian counterparts, they reflect a mode of amber adaptation closer to the one observable in the Aegean. In both areas amber seems to have been elevated to the communal level of acceptance as an element with concrete symbolic value, used to adorn and possibly protect (as a talisman) members of upper social strata of both sexes, even though its status was established slightly earlier in the mainland Greece, and in Apulia the crescent pectoral necklaces combining tens of beads are absent (cf. section 4.1.2.4; Maran 2004, 2013). Going back to vitreous materials, it is important to mention that an LMLK glassy-faience pearl, in the same shape as the one from Trinitapoli, was discovered at the inland settlement of Masseria Chiancudda (Fig. 142:D), what supports opinions of some authors about the importance of this site for the interregional exchange (Cazzella/Recchia 2009, 33).

Fig. 143. Exchange interactions in the Central Mediterranean during the later part of MBA and early LBA (LH IIIA–B; ca. 1420–1220/1210 BC). Sites important for tracing Italian – Aegean contacts are numbered; arrows indicate the possible directions of the contacts, not real seaways (Cazzella/Recchia 2009, Fig. 3).

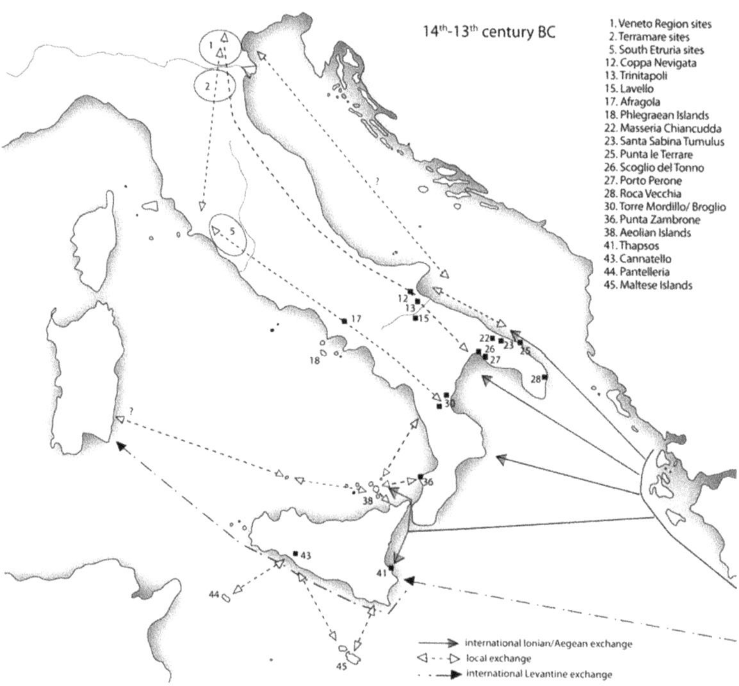

Available data allows for formulation of a hypothesis according to which amber transfer, along with exchange of other goods valued by the *Mycenaeans*, moved from the Tyrrhenian area to the lower Adriatic side by LH IIIB. In essence this change reflects the decreasing 'interest' of the *Mycenaean* sailors/merchants in Sicily and the Aeolian Islands, and assumption of a more prominent role of Apulia in relationships with the Aegean, manifested by the heyday of fortified settlements in the region and local production of goods, including *Mycenaean*-inspired pottery (Fig. 143). It seems probable that by this time communities inhabiting Southern Italy not only acquired the dominant position in amber procurement for the Aegean (at first possibly Sicily and the Aeolian Islands, then Apulia) but also began to consume it themselves to a higher degree.

Evidence of contacts with the Aegean, probably involving amber exchange, is present also when looking to the eastern part of the circum-Adriatic zone. LH III signalises the influx of *Mycenaean* imports to the territory of southern Albania (cf. section 2.3.13). These are chiefly recognised as bronzes, with lesser involvement of pottery. The latter began to appear in LH IIIB and continued to reach Albania until the *Submycenaean* period; however, genuine imports are few (e.g. sherds found in LBA layers of Sovjan 5c; cf. Krapf 2018, 117, 120). One also encounters locally produced ceramics decorated with matt-painted ornaments derived from the Aegean (cf. Krapf 2014, 2018). Nevertheless, these innovations in the pottery-making fall into the end of LBA (late LH IIIB and early LH IIIC); thus, the main source of information about the contacts between Albania and Greece are the non-ceramic finds.

There are several artefacts from the Albanian contexts with amber which, with a good deal of probability, can be considered as imports from the Aegean. Those falling into the period corresponding with LH IIIA–B are the bronze rapier from Shtogj, tumulus 2, grave 9 defined as type A or C (or hybrid of both) in the typology of Sandars (Fig. 78:B), and carnelian beads from grave 186 in the tumulus of Rehovë and grave 46 from the Luaras mound (Kurti 2017b, Pl. XCVIII:c). To this evidence can be added the A-type rapier from grave 2 of Iglarevo in Metohija (Fig. 78:C).

It is symptomatic that appearance of amber in Albania coincides with widening of amber distribution to the northern Greece, beyond the core area of the *Mycenaean* culture. The LH III-dated amber finds were registered in Epirus, Thessaly and Macedonia (Fig. 144). Similarly to the Albanian contexts, the Epirote finds come from cist graves and were found in association with *Mycenaean* bronzes though not much pottery. Harding and Hughes-Brock note that both the tombs and bronzes, i.e. spearhead types, in terms of form appear to be of northern origin and this concurs with the direction of traffic in amber (Harding/Hughes-Brock 1974, 150). One may also wonder if also the swords of the Aegean/Epirote provenience were transferred through the Mat Valley, and across the White Drim and Black Drim confluence, to Kosovo in exchange for amber (cf. Bodinaku 1995, 263; Kurti 2017b, 288–289). Regardless of exact reasons, it appears that a new route for amber influx into the Aegean was opened around LH IIIA–B which led from Central Balkans along the east side of the Adriatic.

Fig. 144. Findings of amber in northern Greece beyond the spatial limit of the Mycenaean *culture: 27. Elafotopos, 30. Gonnos, 44. Kalmbaki-Ioannina, 61. Mazaraki Zitsas, 180. Efira, 195. Spathas/Agios Dimitrios, 206. Aiani, 207. Phaia, Petra Siderokastrou, 208. Agnadero, 209. Ano Komi (Czebreszuk 2011, Fig. 20).*

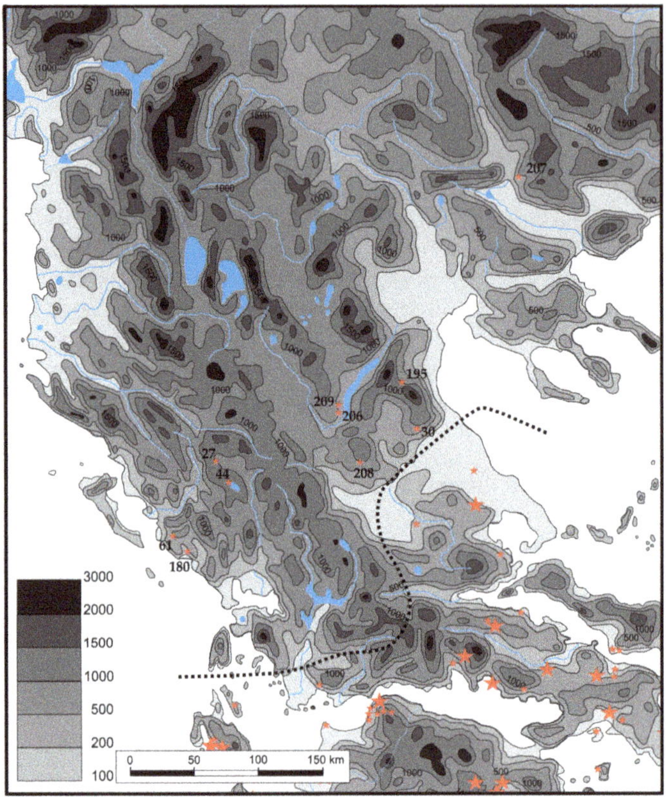

About the same time, Baltic amber reached the outskirts of the Eastern Mediterranean, i.e. Syria. This statement finds a confirmation in the recently conducted provenience analysis of beads and a unique vessel in the form of lion found in Quatna, the latter object most probably produced locally (Mukherjee et al. 2008). Authors of the study, using Py-GC-MS and FTIR, were able to show that these objects, dated to ca. 1340 BC, were made of succinite. Furthermore, 14 beads made of fossil resin were discovered in the royal palace of Ugarit, where they were found together with '*Mycenaean* objects', while a 'large quantity of biconical beads' is reported from the tombs in the adjacent city along with *Mycenaean* pottery (Gestoso Singer 2016, 261, 263). Artefacts made of resin were discovered in the older contexts (dated between 2400 and 1800 BC) in the Near East, e.g. at Assur and Tell Asmar in Mesopotamia (Iraq); however, the finds from the latter were deemed copal from East Africa following provenience analysis (Gestoso Singer 2016, 257; cf. Meyer et al. 1991). These observations cast doubt on the provenience of such early finds in the Near East, especially in the light of absence of succinite-made objects in the Mediterranean at that stage. In Egypt objects made of resin – scarabs – began to occur in the Middle Kingdom at the latest; however, their provenience is unknown. Fossil resin began to be used more frequently during the New Kingdom (Gestoso Singer 2016, 259). Still, it is found in small quantities, mainly as pendants interpreted as amulets, almost always in tombs of the ruling class. However, a good deal of raw lumps and pieces of amber came to light as well; they have been interpreted i.a. as incense or substance used in mummification (cf. Hood 1993; Niwiński 1999). The most famous collection of amber artefacts comes from the tomb of Tutankhamun (ca. 1353/1349–1338/1331 BC; for the dating cf. Bronk Ramsey et al. 2010) and comprises two scarabs and a necklace of 55 beads. Authors such as S. Hood and S. Sherratt argued that morphologically elements from the necklace resemble biconical beads most famously present in the German *Tumulus* culture (cf. section 5.1), but occurring also in considerable number in the circum-Adriatic zone (Hood 1993, 230–232; Sherratt 1995, 203). Nevertheless, in the absence of provenience analyses, the origin of Egyptian finds can only be speculated basing on the literary sources (cf. Niwiński 1999; Gestoso Singer 2016, 259–261).

Role of the circum-Adriatic zone in the inflow of amber to the Near East most probably can be envisaged as a link in the chain of exchange which led through the Aegean. The cargo found on board the shipwreck of Uluburun sheds light on the maritime traffic of goods in the Mediterranean during LH IIIA–B – the time when amber seems to have spread from Sicily to the Levant. Found near Kaş on the south coast of Turkey and dated to the late 14th century BC, the ship carried i.a. a load of beads made of glass, agate, carnelian, quartz, faience, ostrich eggshell and amber (Bass 1991; Pulak 1998). The amber adornments comprised more than 40 beads made of succinite (Pulak 1998, 218). It is debatable whether the adornments were intended for trade or were personal possessions of the ship's crew. Nevertheless, finding such as this one proves that Baltic amber was moved along the west-east axis from Greece to the Levant, probably to be traded or exchanged, e.g. as a royal gift (cf. Pulak 2005). A certain link between Italy and itinerary of the Uluburun ship is a sword of the Pertosa-Thapsos type found among other armaments (Bettelli 2015, 209, Fig. 3:6). It is worth reminding here the swords belonging to the same type were found in some of the Apulian and Basilicatan tombs with amber, i.e. Toppo Daguzzo, tomb 3 and Lavello, hypogeum 743 (cf. section 2.2.14). Thus, a possibility that the amber found on the ship was obtained in harbours of Southern Italy should be taken into consideration.

End of the classical *Mycenaean* period, signalised by the abandonment of many palaces and the fall of political-bureaucratic system connected to them, opened a new chapter in the Aegean-Central Mediterranean relationships. Despite the apparent crisis, amber kept arriving in the Aegean and its use is attested by at least 29 findspots, although the artefacts-per-contexts ratio in LH IIIC reached the lowest level since the beginning of the *Mycenaean* culture (Harding/Hughes-Brock 1974, Tab. 2; Czebreszuk 2011, Tab. 12–13). Spatial structure of amber's acquisition in Greece underwent further changes during the post-classical (or the post-palatial) period. Judging by the location of deposition contexts, amber was utilised in the northern parts of the Peloponnese (Achaea and Argolid) and central Greece (Phocis and Attica); however, it also reached the islands of Cephalonia (Diakata and Metaxata), Thasos (Tziganadika and Vrisoudes), Rhodes (Ialysos) and Crete (Dictean cave) (Harding/Hughes-Brock 1974, Tab. I, Fig. 3; Czebreszuk 2011, 125–126, Fig. 24).

In comparison with the circum-Adriatic zone, morphology of *Mycenaean* amber artefacts dated to LH IIIC seems less complex. Yet, a significant change can be observed at the time; spheroidal (type 101 corresponding to IA122) and flattened spheroidal (type 102 – IF121, IF221, IF122, IF222), lenticular (type 103 – IF162, IF262) and annular (type 108 – discoidal or flattened spheroidal beads with expanded perforation outlets 'w', 'x', and 'y') forms are almost completely gone by this phase (cf. Czebreszuk 2011, Tab. 1–3, 7). Also, the flattened biconical beads (type 105 – IF161, IF261), few as they were during the classical *Mycenaean* period, are almost totally absent (cf. Czebreszuk 2011, Tab. 4). Concurrently grew the quantity of elongated beads, especially the fusiform/barrel-shaped ones (type 111 – IE121, IE221), however they mostly come from a single site – Metaxata on Cephalonia (Czebreszuk 2011, Tab. 8). The most important innovation in the repertoire of *Mycenaean* amber adornments are beads decorated with the Tiryns-type ornament (type 113 – beads decorated with motif 'a'). The oldest bead with the Tiryns-type decoration from Greece – the Napflio tomb – is dated to LH IIIA2–B (Harding/Hughes-Brock 1974, 164). Still, the bulk of these objects fall into LH IIIC, with the largest collection coming from the abovementioned cemetery of Metaxata on Cephalonia (cf. Czebreszuk 2011, Tab. 9). Recent review of the morphology of amber beads decorated with the Tiryns-type motif has shown that Greek specimens from Dictean cave, Ialysos, Metaxata and Tiryns can be classified to the same types together with majority of the finds from Italy (Bellintani et al. 2021). The most famous context of deposition for the Tiryns-decorated beads is the hoard from the eponymous citadel in Argolid, where they were found threaded on spokes of golden wheels (Karo 1930). There has been much discussion about the context of this finding; it is largely suspected that it is a treasure trove of grave robber (Hughes-Brock 1993, 221). If the object is genuine, a surprising similarity links it with gold-wire objects shaped into figures-of-eight from Hradec Králové (Bohemia) discovered in context of the *Urnfield* culture.[722] Nonetheless, R. Jung elaborated on the apparent connection between the design of the Tiryns wheels and the solar motifs found on the Italian gold discs (Jung 2007, 234–240; cf. Fig. 128:A–B). This interpretation can be neatly combined with the notion of the Tiryns-type beads production in Northern Italy, consequently strengthening the argument for translocation of amber to the Aegean via the circum-Adriatic zone.

On the other hand, other forms of amber artefacts characteristic for the study area in FBA are very rare in Greece. Beads decorated with the Allumiere ornament are known just from two sites: Metaxata on Cephalonia and Vrisoudes on Thasos (Czebreszuk 2011, 88). Specimens from those two places depicted in the literature can be ascribed to var. 'e' of decorative motifs in the typology applied for the circum-Adriatic zone, attested among the finds from western Veneto, Kvarner and northern Dalmatia and north-central Lazio (cf. Czebreszuk 2011, Tab. III:28, VII:11; section 3.3.3.2). Elongated biconical beads are much less numerous than barrel-shaped or flattened biconical ones in the *Mycenaean* culture. J. Czebreszuk in his typology classifies the former to the group of amygdaloid beads – type 106, counting in total 10 specimens (Czebreszuk 2011, 79). However, it is not a uniform group, as it also comprises beads with bi-convex profile and polyhedral cross-section, sometimes decorated and, thus, interpreted as seals. The only genuine example of elongated biconical beads in the *Mycenaean* Greece,

722 Pleiner/Rybová 1978, Pl. 94; cf. Jung 2007; Maran 2013, 160.

comparable with the circum-Adriatic specimens, comes from the LH IIIB horizon of the lower citadel of Tiryns (Czebreszuk 2011, Tab. 5:7, Pl. VII:6). It can be inferred from this overview that only a small, morphologically limited portion of amber beads was brought to Greece during LH IIIC, or otherwise that *Mycenaeans* were highly selective in accepting the products offered to them. Still, the beads with Tiryns-type motif present a strong proof for amber-related contacts between the communities of the Aegean and those inhabiting the circum-Adriatic zone – a relationship reinforced by other kinds of archaeological evidence discussed below.

Overall it is possible to consider LH IIIA as the moment of maximum expansion of the *Mycenaean* core towards the Central Mediterranean (Iacono 2013, 61). In the subsequent phases, that is at the turn of LH IIIB and LH IIIC, the direction of relationships changed, with the Central Mediterranean assuming more active posture in arranging local trade or exchange, and perhaps even undertaking travels to the Aegean either peacefully or militantly (Jung/Mehofer 2013; Jung/Pacciarelli 2017, 191). While there are voices arguing that the collapse of the classical *Mycenaean* culture at the end of LH IIIB (late 13th century BC) was caused by an armed invasion from Italy (possibly by tribes included in ancient written sources among the 'Sea Peoples'),[723] it appears that these events did not disrupt or completely break the trade routes, and the relations continued for over another century (Bettelli 2015, 216). It looks as if contacts with the previously established centres in Apulia (Coppa Nevigata, Scoglio del Tonno and Roca Vecchia) and Calabria (Broglio di Trebisacce and Torre Mordillo) were maintained, and production of *Italo-Mycenaean* pottery flourished at those sites in LH IIIC, so during LBA 2–FBA 2. Yet, this network of contacts shows signs of reconfiguration at the time.

During LH IIIC, in place of the previous trans-Ionian connection with Sicily and the Aeolian Islands, the contacts network facilitating trade and exchange with the Aegean moved up the Adriatic towards the Po Plain. There already are some traces of interactions, perhaps based on the goods exchange between the two areas in the earlier phases of the Bronze Age. L. Rhamstorf points to the openwork faience beads (also known as lantern-shaped or wheel-shaped beads) as tangible evidence of participation of the *Terramare* communities in some sort of goods exchange with the *Mycenaeans* (Rhamstorf 2005). Similar beads appear between the 14th (LH IIIA) and 12th century BC (LH IIIC) in various parts of the Eastern Mediterranean basin (Greece, Cyprus, Anatolia, Levant) (cf. Rhamstorf 2005, 664–665, Pl. CLXXI:b). Their occurrence often corresponds with findings of amber in Greece (chamber tombs of Nauplio, Mycenae, Ialysos and palaces of Thebes, Tiryns and Mycenae), Cyprus (cemetery of Enkomi) and Syria (town and cemetery of Alalakh).[724] As it happens, two openwork faience beads were also found together with amber on two sites in the Po Plain both dated to LBA: Santa Rosa di Poviglio, *Villaggio Grande*, 'heaps of ash' layer – phase 2, and Franzine Nuove, grave 523 (Fig. 145:A–B). Apart from the vitreous materials, there are scarce sherds of *Mycenaean* pottery found in the eastern Po Valley, i.e. fragment of the closed-shape vessel from Crosare di Bovolone (Bettelli et al. 2017, 165; cf. Jones et al. 2014, 219, Fig. 4:91). Moreover, a case has been made that the stone weights from the *Terramare* settlements were in some way related to Aegean ponderal system; similar

723 Cardarelli 2009; Kristiansen 2017; Jung/Pacciarelli 2017, 191; cf. Sandars 1985; Cline 2015.

724 Cf. Rhamstorf 2005; Czebreszuk 2011; Gestoso Singer 2016.

weights appear on sites which produced *Mycenaean* pottery in the Po Plain as well as Apulia (Cardarelli et al. 2004; Iacomo 2013, 67).

Fig. 145. Openwork faience beads, Mycenaean *imported pottery and* Italo-Mycenaean *locally produced pottery from Northern Italy dated to LBA–FBA (LH IIIB–C): A – Santa Rosa di Poviglio,* Villaggio Grande, *'heaps of ash' layer – phase 2 (Bellintani/Biavati 1997, Fig. 354:1); B – Franzine Nuove, grave 523 (Aspes 1997, Fig. 420:3); C–E – Fondo Paviani, advanced BR–early BF layers (Jones et al. 2002, Fig. 2:2; Bettelli et al. 2017, Pl. LIV:6, 14); F–G – Frattesina di Fratta Polesine, phases 1–2 (?) (Jones et al. 2002, Fig. 3:1–2).*

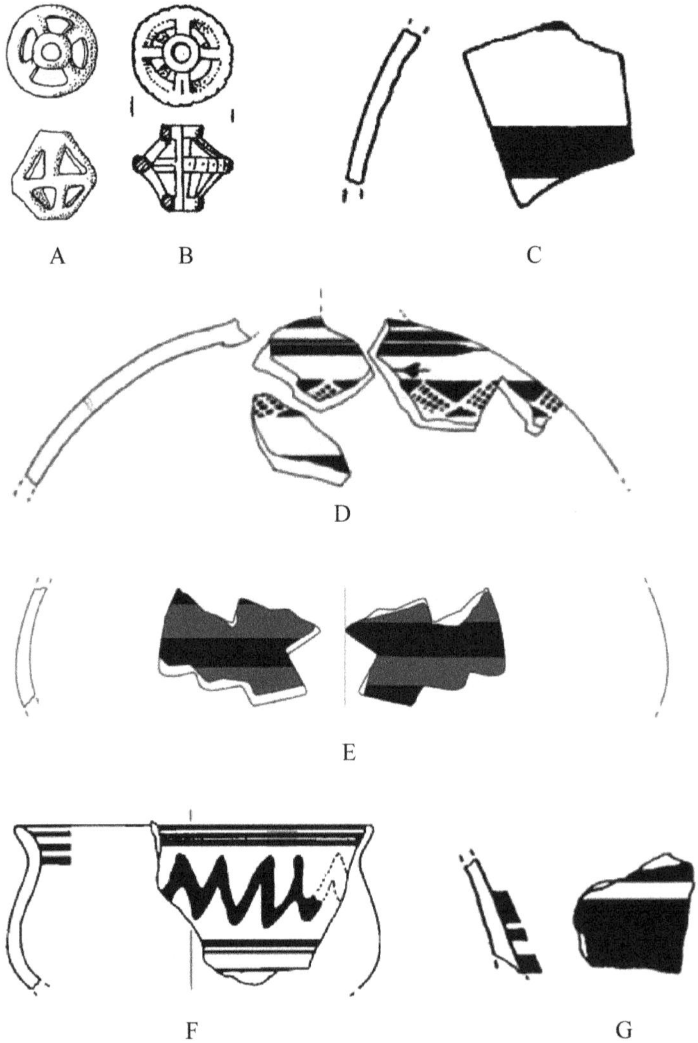

Although it is possible that a stable relationship between the *Mycenaean* Greece and the *Terramare* Italy was initiated at the turn of the 14th and 13th century BC, bulk of evidence confirms that it intensified in the 12th century BC, so at the time of LH IIIC (LBA 2/FBA 1).[725] This process is attested by a greater presence of *Mycenaean* pottery in Northern Italy, findings of which come from seven embanked settlements in the Po Plain: Fondo Paviani, Fabbrica dei Soci, Castello del Tartaro, Frattesina, Montagnana, Terranegra di Legnago and Lovara di Villabartolomea (Jones et al. 2002; Bettelli et al. 2017, Pl. LIII:a). Two of them are of importance for the current considerations as they also provided amber finds: Fondo Paviani and Frattesina.

Archaeometric analyses of *Mycenaean*-type sherds from Fondo Paviani showed that only few can be considered as actually imported from the Aegean, specifically the Peloponnese (Fig. 145:C; cf. Jones et al. 2002, 233, 242 254; Bettelli et al. 2017, 166). Far more pieces belong to locally made *Italo-Mycenaean* pottery which can be divided into two main classes: (1) *Mycenaean* pottery with lustrous monochrome paint (Fig. 145:D) and (2) black-and-red bichrome decorated ware (Fig. 145:E; cf. Bettelli et al. 2017, 167). All fragments seem to pertain to LH IIIC; however, it is not always possible to differentiate them into the consecutive internal stages of this period (Bettelli et al. 2017, 167). While the first class can be said to have been inspired by ceramics imported from the Aegean, the second one appears to have more in common with the Cypriot and Levantine pottery-making of the period (Bettelli et al. 2017, 168–169). Still, vessels belonging to both of them were produced locally from the sedimentary materials obtained from the Po Valley (cf. Jones et al. 2002, 2014).

Collection of *Italo-Mycenaean* pottery from Frattesina di Fratta Polesina is much smaller, comprising only two sherds: one belonging to a cup with S-shaped profile decorated with horizontal straight or zig-zag bands (Fig. 145:F) and a sherd from an open vessel with slightly convex wall also ornamented with horizontal bands of various thickness (Fig. 145:G). Although both fragments seem to fall into advanced/late LH IIIC and most probably were manufactured locally, their stratigraphic position can be attributed only broadly to the 1st and 2nd phase of the settlement (cf. Jones et al. 2002).

Presence of ceramic imports as well as copies of vessel from the Aegean raises questions about the potential presence of *Mycenaean* potters in Northern Italy which, unfortunately, cannot be answered at the moment (Bettelli et al. 2017, 166). Nevertheless, a transfer of technology (introduction of potter's wheel) and derivation of shapes and decorations from the Eastern Mediterranean seems clear. It can be posited that the lower Po Plain, first involved in the *Valli Grandi Veronesi* polity with Fondo Paviani at its centre and subsequently taken over by Frattesina, was a key node in the Mediterranean trade network during the final stages of the Bronze Age (cf. Fig. 146). Its role in this capacity facilitated inflow of other exotic materials such as ivory and ostrich shells (cf. sections 2.2.6 and 3.12.4).

At the current stage of research, it is difficult to pinpoint a single material which prompted *Mycenaean* and possibly other Eastern Mediterranean (Levantine) merchants to seek trade partners so far to the north of the Adriatic. It is unlikely that it was the southern Alpine copper – the key component of exchange between Northern Italian communities and their neighbours to the north. Analyses conducted up to now do not show presence of Italian copper in indigenous *Mycenaean* bronzes locally

725 Cf. Jones et al. 2002, 2005; Cupitò M. 2011a; Bettelli et al. 2017.

manufactured in Greece. The latter appears only in LH IIIC items characterised as imports (cf. Jung/Mehofer 2013; Cazzella/Recchia 2018, 22). Many authors underline the importance of amber in emergence of relationship between the Po Plain and the Aegean.[726] The hypothesis seems plausible especially in the light of local production of the Tiryns- and Allumiere-ornamented beads, from where these products could have been dispersed to the Aegean.

Fig. 146. Exchange interactions in the Central Mediterranean during the later part of LBA and early FBA (LH IIIC–SM; ca. 1220/1210–1070/1040 BC). Sites important for tracing Italian – Aegean contacts are numbered; arrows indicate the possible directions of the contact, not real seaways (Cazzella/Recchia 2009, Fig. 4).

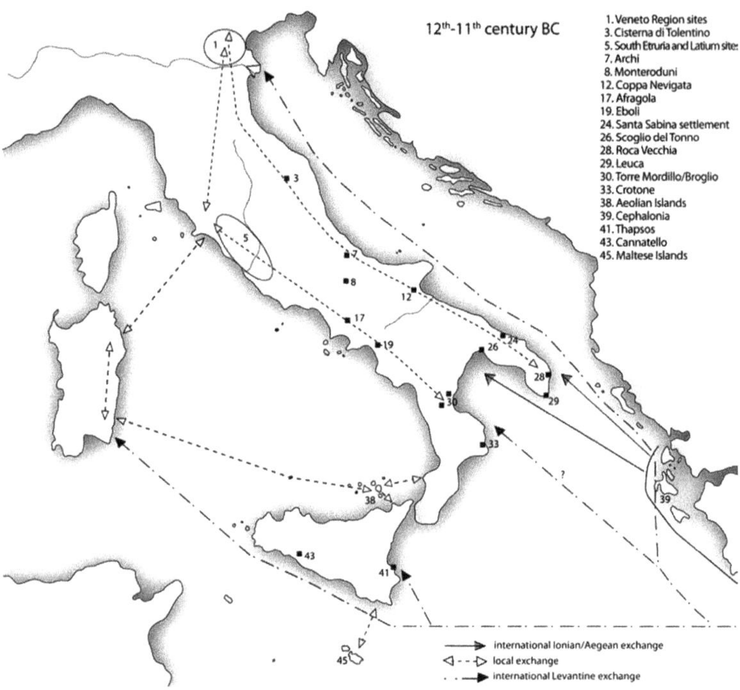

In the discussed period *Mycenaean* influences also reached Central Italy. Locally produced copies of Aegean ceramic vessels dated to LH IIIB/C were discovered at Cisterna di Tolentino in Marche (cf. Vagnetti et al. 2006). Further south *Italo-Mycenaean* pottery was produced i.a. at Coppa Nevigata and Roca Vecchia, but during LBA 2 its production accelerated in comparison with the previous phase (MBA 3–LBA 1). An impressive collection of ceramics has been amassed throughout the years of research at Coppa Nevigata (see Vagnetti et al. 2012 for the latest overview). The recent *Subapennine* (LBA 2) layers at Coppa Nevigata are the richest in terms of

726 E.g. Cazzella/Recchia 2009, 37; Bellintani 2014, 120; Bettelli et al. 2017, 169.

Italo-Mycenaean pottery, amounting to at least 58 fragments mostly referring to LH IIIC (cf. Vagnetti 2012, 424–425). Although the pottery cannot be precisely correlated with the amber finds attributed to LBA (*Subapennine*) and FBA horizons of the site, it is likely that they belong to a roughly contemporaneous period.

Fig. 147. Italo-Mycenaean *locally produced pottery,* Mycenaean *imported pottery and bronze objects with Aegean provenience from Southern Italy and the Balkans dated to LBA–FBA (LH IIIB–C): A – Roca Vecchia, 'Cult Center' (Sector IX, US 11349, Phase II) (Pagliara et al. 2008, Fig. 14:II.1–2); B – Roca Vecchia, 'Hut-temple' (Sector IX, US 11322, Phase VI) (Pagliara et al. 2008, Fig. 14:VI:1); C – Barç, tumulus 1, grave 162 (Bodinaku 1995, Fig. 3:4); D – Barç, tumulus 1, grave 47 (Bodinaku 1995, Fig. 1:7); E – Dimov Grob, grave 54 (Videski 2007a, Pl. LV:e).*

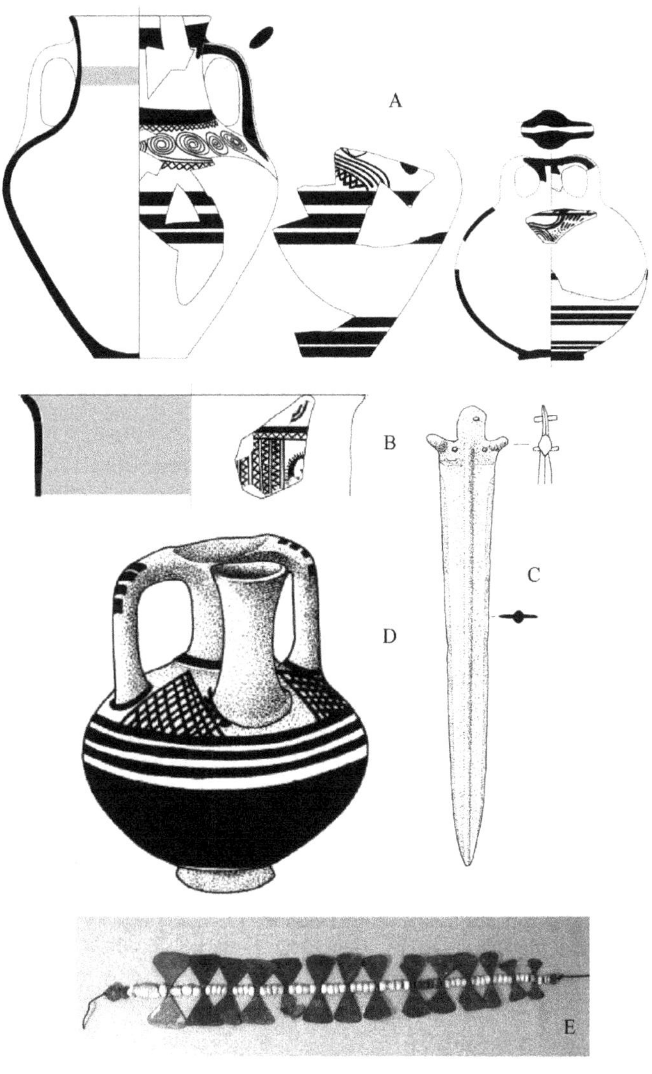

Italo-Mycenaean pottery in the context with amber artefacts is known from Roca Vecchia. In sector IX, layer US 11349, corresponding to the IInd phase of the hut 'Cult Center', alongside a bead with the Tiryns-type decoration were found fragments of numerous vessels, mainly stirrup jars, dated to LH IIIB–C early (Fig. 147:A; cf. Pagliara et al. 2008, 267). Layer US 11322 in the same sector, attributed to the VIth phase of the 'Hut-temple', apart from Allumiere-decorated beads, produced a *skyphos* (deep bowl) decorated with a dense composition of vertical and horizontal zig-zag and straight lines, dated to LH IIIC middle and possibly imported from the north-eastern Peloponnese (Fig. 147:B; Pagliara et al. 2008, 263). Presence during LH IIIC–SM of similar types of amber objects in Northern and Southern Italy, and the Aegean together with other kinds of evidence (predominantly pottery), gives credence to the idea of a trade route along the western Adriatic coast, operated by 'gateway' settlements funnelling goods from the inland to the sea and possibly governed by regional elites (nascent ruling class), who – thanks to surpluses of 'strategic' resources – could afford luxury goods such as amber and high-quality tableware.

Relations between the Aegean and the neighbouring Balkan regions during LH IIIB–C are also visible with respect to amber. Continuing interactions between south-east Albania and Greece can be observed on the basis of graves from the tumuli of Barç. The finds of *Mycenaean* origin comprise i.a. a dagger of type C or G of according to the typology by Sandars from tumulus 1, grave 162 (Fig. 147:C; cf. Bodinaku 1995, 265) and an imported stirrup-jar dated to LH IIIC from tumulus 1, grave 47 (Fig. 147:D; cf. section 2.3.13.4). By the turn of the 13th and 12th century BC, production of Aegean-inspired pottery was taken up also in the Vardar Valley by the communities of the *Ulanci* group (cf. Videski 2007a; Mitrevski 2007). Locally manufactured vessel imitating Aegean models, i.a. alabastra, occur, for instance, in the graves with amber at the Dimov Grob cemetery (cf. section 2.3.12). However, the most striking evidence of contacts with the Aegean comes in the form of the labrys belts, i.e. the one from tomb 54 at the abovementioned cemetery (Fig. 147:E; cf. section 2.3.12). Occurrence of amber together with objects of probable *Mycenaean* provenience in Albania and North Macedonia not only coincides with the beginning of amber import to northern Greece around LH IIIB but also endures during LH IIIC along some of the most prominent centres of amber's acquisition in Greece in their vicinity, i.e. the islands of Cephalonia and Thasos (cf. Czebreszuk 2011, 124–125, Fig. 24).[727]

As mentioned before, LH IIIC and the *Submycenaean* period saw a shift in the relationship between Italy and the Aegean, with the former assuming more active role in arranging the Central Mediterranean network. While many authors tend to see those contacts in terms of exchange and trade, there are some that point to other forms of interactions, including hostile confrontations, and even go as far as to suggest presence of Italic merchants, craftsmen and warbands (mercenaries) on Greek soil.[728]

---

727 At the same time (BzD; LH IIIC), withdrawal of amber from most of the Central Balkan provinces poses a question whether resin-made objects could be imported to Albania and Macedonia from Greece (after being first supplied by Southern Italy). Similarities among amber beads and clearly Aegean/Mycenaean origin of some of the associated artefacts give certain credibility to this hypothesis (cf. Cwaliński 2020, 155, 162, Fig. 11).

728 Cf. Jung/Mehofer 2013; Bettelli 2015; Jung/Pacciarelli 2017.

Reversal in the direction of those interactions and their possible relocation to Greece is indicated by occurrence in the latter area of ceramic and bronze objects of Italian provenience, not to mention other products such as amber and glass beads. Among the products most often cited as clear manifestation of Italic influence over the Aegean in the final centuries of the 2nd millennium BC is the Handmade Burnished Ware (HBW, also known as 'Barbarian Ware')[729] and flange-hilted bronze swords and daggers related to the *Urnfield* metallurgy (Fig. 148).[730] It is interesting to note the cases when objects belonging to those categories were found together with amber in Greece (cf. Czebreszuk 2011, Catalogue; Iacono 2013, Appendix).

Fig. 148. Distribution of handmade burnished ware and Urnfield-*type bronze objects in the Aegean (Iacono 2013, Fig. 5.2).*

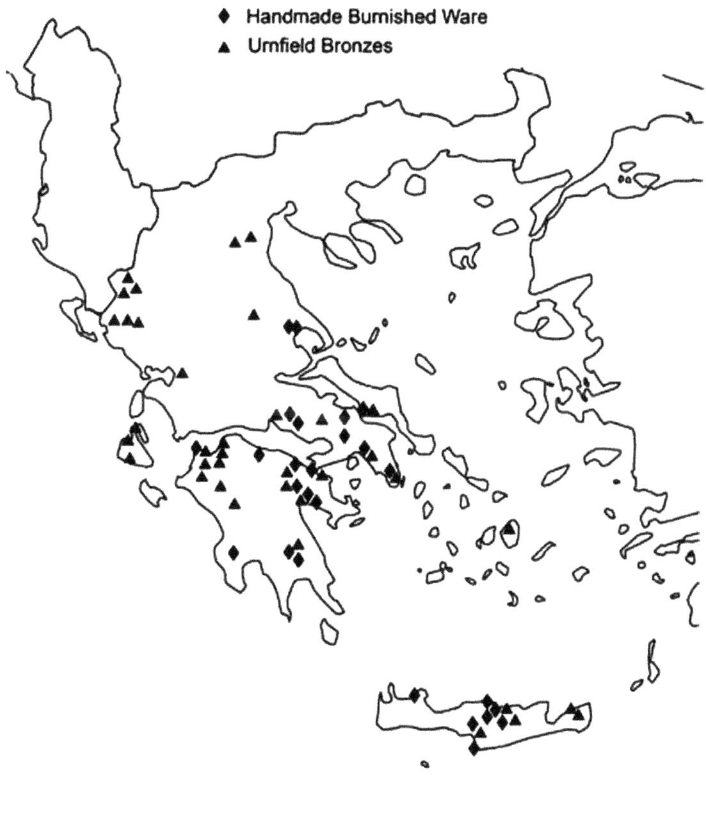

729 HBW is a term used to described vessels made in the impasto technique, modelled by hand and with polished surfaces which often resemble indigenous Italian models of LBA/FBA (Bettelli 2002, 258).

730 Cf. Bettelli 2002, 2011, 2015; Iacono 2013.

Some of the most prominent examples of co-occurrence between Tiryns-type amber beads and bronzes of *Urnfield* design are tomb Gamma at Metaxata (Cephalonia), where the former were found with fragments of at least one violin-bow fibula (Souyoudzoglou-Haywood 1999, 43), a grave from Patras (Klauss) necropolis (Achaea) comprising i.a. a razor of the Scoglio del Tonno type (Cultraro 2006, 1545), or the Dictean cave (Crete) where a bead with the Tiryns-type ornament was discovered inside a cavity which also yielded a Peschiera-type dagger (Cultraro 2011, 232). Worth mentioning is also the votive deposit from Delos which included two amber beads and discs of gold sheet bearing similar decorations to the discs from Gualdo Tadino and Roca Vecchia (Bettelli 2011, 119; 2012, Fig. 2:3–6). Moreover, some tombs at the cemetery of Elateia (Beotia) contained in their inventories amber beads as well as LMHK glass beads which, although not exactly the same as those from Frattesina, could have been brought from the Central Mediterranean (cf. Nikita/Henderson 2006; Czebreszuk 2011, 188–190). Finally, there is a different kind of evidence suggesting presence of Italian migrants in Greece. Latest DNA analysis of faunal remains shows Italian pig haplotype replacing a genetic haplotype of Near Eastern descent in palatial Tiryns in LH IIIC, suggesting a mobility of people and livestock between various parts of the Mediterranean (Mieri et al. 2019).

After the *Submycenaean* period, which already saw further decrease in amount of amber brought to Greece, amber disappeared from the Aegean for several centuries. It is no coincidence that in the course of the middle part of FBA (BF 2) the use of the Adriatic route seems to have come to a halt (Bettelli 2015, 220). The relationship between Italy and the Aegean became looser and did not include amber translocation, even when the material was abundant in certain parts of Italy during the EIA, such as Romagna, Etruria or Picenum.[731] The *Mycenaeans* were soon replaced by Cypriot, Levantine and Phoenician merchants, who once more reorganised the Mediterranean trade routes with special role ascribed this time to Sardinia (see e.g. Nijboer 2008; Bettelli 2015, 221–223). Position of amber in these new interregional relations, however, has yet to be assessed.

Amber-based relation between the circum-Adriatic zone and the *Mycenaean* Greece thus ended after going through approximately three main stages, roughly coeval with the pre-palatial (or shaft grave), palatial (or classical) and post-palatial (or post-classical) periods. It is possible to describe the character of amber circulation between those two areas first (during pre-palatial period) as yet to be specified participation of Southern Italian communities in procurement of amber to the Aegean, with limited exploitation of local fossil resins (simetite), and then also extended by the use of foreign amber (succinite) possibly induced by the contacts with *Mycenaeans.* The second stage (classical period) brought intensification and stabilisation of contacts arranged into two main routes of amber translocation: western – Ionian-Tyrrhenian (earlier) and north-western – Adriatic (later), with possibly a third one passing through Central Balkans. Southern Italian communities began to consume amber on larger scale which locally achieved a communally recognised status and specific, possibly high-class attribution, whereas in Greece amber became more common albeit less prestigious, as noted by J. Czebreszuk (2011, 168). Finally, during the third stage (post-classical period), it is possible that the Aegean became completely dependent

731 Cf. Bonfante 2009; Negroni Catacchio 2011a, 2011b.

in terms of amber import from Northern Italian centres, which by that time began to process the material; one can even risk a conclusion that the status and symbolic connotations of amber, but not the ways of its appropriation (cf. Cultraro 2006), became unified across the circum-Adriatic zone and the Aegean as a result of more direct and closer contacts between the people.

5.3. Circum-Adriatic zone in the European network of amber circulation – some general observations

To conclude chapter 5, observations on amber's inflow and outflow to and from the circum-Adriatic zone presented in the two preceding sections are confronted with the general view of the Bronze Age Europe from the perspective of the World System Theory/Analysis (WST/WSA), already explained briefly in section 1.2.4. Over the years WST/WSA has been criticised in many aspects. Among the main accusations are overreliance on the 'core-periphery' model, simplification of relationships between regions/societies into passive-exploited and active-exploiting, perceiving the Bronze Age Eurasia as a centrally governed system (by elites – ruling classes) with marginal role of individuals, or direct application of the modern economic categories such as 'markets', 'trade' or 'supply and demand' onto the Bronze Age realities.

Notwithstanding the criticism, archaeologists from many generations have been drawn towards the WST/WSA, and the Bronze Age Mediterranean proved to be a particularly fertile ground for applying this approach. It may be argued that the biggest advantage offered by the WST/WSA is the ready-made (although not always precisely defined or unanimously understood) set of terms and concepts – a paradigm – which allows to describe the archaeological record from a larger perspective as an interplay between various regions, communities or political entities (cf. Chase-Dunn/Anderson 2005; Hall et al. 2011). Thanks to a string of publications that were released in the last 20 years, some of which have been brought up in the two previous sections, different parts of Europe and Near East can be included in a Bronze Age world system (if there ever was a single one), stretching along the meridional axis from the British Isles and Scandinavia to the Mediterranean.

Perhaps the most commonly recognised vision of the European Bronze Age world system is the one penned by K. Kristiansen and his various collaborators: M. Rowlands, T. Larsson and T. Earle among others.[732] From their works emerges a picture of the 2nd millennium BC Europe with societies at an incipient level of hierarchisation, governed by elites which legitimise their rule with religion and various manifestations of military power. These communities differed in terms of economy and wealth that they produced, mostly depending on the access to key resources, such as food (e.g. grain, meat, salt, etc.), metals (copper, tin and gold) and other important commodities (e.g. wool, hides, furs, timber, etc.). Among them are also certain rare and prestigious materials, i.a. amber which in the opinion of Kristiansen and other like-minded scholars was the main export commodity and source of wealth of the southern Scandinavian communities.[733] Control over resources determined the position of a community in

732 E.g. Kristiansen/Rowlands 1998; Kristiansen/Larsson 2005; Earle/Kristiansen 2010.
733 Cf. Kristiansen/Rowlands 1998; Kristiansen/Larsson 2005; Earle et al. 2015; Ling et al. 2014, 2019.

the interregional network of contacts and goods' circulation. Consequently, some cultural units are described as cores in the system (e.g. *Otomani-Füzesabony* culture in the Carpathian Basin), while many more fall into the category of peripheries. Yet, this picture was not static; in the course of the 2nd millennium BC, Europe experienced climate changes, political unrests and other events that varied in magnitude from catastrophic to mild effects. They affected balance of power and resulted in shifts in the network (e.g. decline of the Carpathian Basin tells cultures after 1600 BC and rise to a prominent position of the *Tumulus* groups). Kristiansen sees the changes as occurring periodically with two alternating modes of social and economic organisation: one based upon sedentary lifestyle, focused around centres of craft (chiefly metal production) localised in settlements (often fortified) and subsistence provided by agriculture, and the second, more decentralised one, with husbandry as the leading mode of economy and smaller residential units – hamlets (Kristiansen 1998, 300, 303, Fig. 11.2). While the first one is said to have occurred in more stable and prosperous times, the latter is characteristic for periods of shifting balance between cores and peripheries – an 'interval' between collapse of old and rise of new centres.

Northern and Central European cores are envisaged as involved in indirect relations with the cores to the south – secondary states of the *Minoan* Crete and the *Mycenaean* Greek mainland, and through them with various kingdoms/civilisations of the Near East (Parkinson/Galaty 2007, 2010). This relationship was first and foremost based upon circulation of goods (amber again assumes an important role in this context), but included also transfer of know-how and ideology. According to Kristiansen, manifestations of the latter are visible in defensive and sepulchral architecture, technologies and stylistics in which metal objects were manufactured, as well as widespread adoption of religious motifs – i.e. solar cult (cf. Kristiansen/Larsson 2005, 142–250). While Central and Northern Europe were mostly recipients of ready-made objects and predefined concepts, they supplied raw materials, but in the later stages also products of their own. The gradation of the cores is multileveled: the key players in the Central-Northern European network were to a certain extent dependant (often described as 'semiperipheral') from the Mediterranean ones, while the latter were peripheral or even subsidiary with respect to Asiatic centres. Recent studies show that also in the Eastern Mediterranean the system was fluid: various regions competed over dominance in the basin with some regions gaining an upper hand but not for too long, e.g. the *Mycenaean* Greece taking over the *Minoan* Crete as a dominant power in the Aegean in LH/LM III phase, only to lose it at the end of the period (cf. Berg 1999; Papadimitriou/Kriga 2013). This short summary may not reflect all the nuances of the Eurasian world system of the Bronze Age as claimed by the authors writing in this current. Nonetheless, several key aspects for understanding this concept can be listed: (a) core-periphery relations between regions and communities, (b) existence of a network of contacts enabling the flow of information and resources, (c) importance of strategic materials, namely metals and certain 'luxury' items (e.g. amber) for development of cores, (d) stratification of societies with elites striving to take control, (e) formation of the incipient market economy, (f) mechanisms of interaction based predominantly on trade and warfare.

There are several problems with the model advocated by Kristiansen and his collaborators. Namely it portrays development of the Bronze Age in Europe in an unilinear way, basing on the strong belief in repeatability of historical cycles. This approach often neglects individual trajectories and rhythms of cultural development in individual regions, especially on the micro-level. Second, in the pursuit of finding general similarities among archaeological cultures over vast areas, many particularities which differentiate them are omitted. Also the main forces of change are almost always identified with the unequal relationship between cores and peripheries, with the Near East, Cyprus, Crete and Aegean placed in the role of the former and with less developed Northern European areas on the other end of the spectrum.

Although many archaeologists remain sceptical about the plausibility of existence of a Bronze Age world system as in the vision persistently advocated by the likes of Kristiansen, the results of provenience analyses of materials such as copper and amber increasingly show a world where materials were translocated over large distances. The flow of resources arguably resulted from and further strengthened the contacts between communities inhabiting different parts of Europe. It is too simplistic to claim that this developed interregional connectivity resulted from a trade network arranged by elites seated in the remote palaces of the south. This process most likely was initiated from the bottom-up by individuals seeking access to increasingly larger and varied supply of resources. Not even trade was the only type of interaction that caused the goods to flow. As discussed in section 1.2.1, there were various other forms of moving objects in space, and specifying them requires analysing each case of occurrence of a foreign object individually. Still, it is not an exaggeration to say that exchange and trade were among the main ones.

Once the general framework of the European Bronze Age world system was laid down by Kristiansen (1998), the subsequent works helped to refine it theoretically and supplement it with new data, and still continue to do so. Also, there are many new studies conducted by various authors, investigating social inequalities among the Bronze Age communities that show complexity of this aspect (e.g. Kienlin/Zimmermann 2012; Cardarelli et al. 2014). Furthermore, project such as *Forging Identities* helps to understand mobility in the Bronze Age Europe (Suchowska-Ducke et al. 2015). Finally, the results of new analyses of lead isotopes have shown how dynamic was the circulation in metal supply, with the multiple sources being exploited (Ling et al. 2014, 2019). Similar observation can be said about the inflow of Baltic amber; throughout the 2nd millennium BC, various cultural groups of people from different parts of the Baltic and North Seas took over the role of amber's main supplier before ceding it to another community.[734] An attempt was made below to shortly describe the circum-Adriatic zone's situation within the European Bronze Age world system from the perspective of amber circulation.

The circum-Adriatic zone emerges from the recently conducted studies, including the one presented here, as an important region for the long-range transfer of goods and people or, as WST/WSA proponents would put it, for functioning of the European Bronze Age world system. However, it did not assume this status immediately at the beginning of the period. In comparison with the late 3rd millennium BC, when the *Cetina* phenomenon connected opposite shores of the Adriatic with the Aegean,[735] interactions at the start of the 2nd millennium BC seem meagre. By large EBA brought atomisation of many Central and Southern Italian and Western Balkan cultures, and the contemporaneous archaeological data do not show much evidence for

734 Cf. Bukowski 2002; Czebreszuk 2007, 2009; Szmyt/Czebreszuk 2017; Woltermann 2016a.
735 Cf. Della Casa 1995; Maran 1998, 2007; Heyd 2007.

far-reaching stable networks of contacts and exchange. At the same time, that is in MH period, the Aegean experienced stoppage and even regress in many aspects, causing a several-century delay in the development into 'state'-like organism in comparison with Crete (cf. Parkinson/Galaty 2007).

Only at the end of the period, from 1800 BC (BzA2; MH III) participation of the areas around the Adriatic in long-distance exchange seems to have been resumed. Dating problems set aside, only few cultural units from the circum-Adriatic zone show presence of amber during EBA. The earliest evidence from the Sicilian *Castelluccio* culture shows limited exploitation of the indigenous resources of fossil resin (simetite), largely for the local use. Baltic amber (succinite) appears slightly later than simetite; however, when it does it is usually accompanied by testimonies of contacts with foreign cultural units (outside the circum-Adriatic zone). Still, the quantities of imported amber are small in comparison with the *Únětice* culture or the early *Mycenaean* culture. The findings of amber in the *Polada* culture are a far-cry from the abundant assemblages known from the later *Palafitte-Terramare* sites. With respect to EBA there are, so far, no attestations of exportation of southern Alpine copper, showing that the heyday of Northern Italian Bronze Age was yet to come. Also Sicily at this stage does not reveal a strong evidence of close relations with the *Mycenaeans*, although it can be suspected that it somehow was involved in the transit of amber (both succinite and simetite) to the Aegean. All in all, it appears that any activity related to amber translocation in the circum-Adriatic zone during EBA was limited to Northern and Southern Italy. It does not mean that the Balkans were detached from the European world system. There were highly developed (by the European standards of the day) cultural units, such as the *Vatin* culture in northern Serbia and Vojvodina, which show involvement in interactions with the Carpathian Basin; however, for some reasons they did not encompass amber exchange. Still, it is probably true that during EBA the circum-Adriatic regions were situated in the peripheries of amber circulation. This constatation is reinforced by a low level of social acceptation of amber among Italian communities of EBA which, themselves, do not manifest many markers of social stratification at that stage. Perhaps it is right to say that in this phase the circum-Adriatic zone had a more intermediary than consumptive role with respect to amber circulation.

The whole study area started to gain significance in the European world system from MBA onwards, by FBA becoming the leading trail for the communication between Central Europe and the Aegean. Amber and copper can be used as proxies for reconstructing the region's position in the European Bronze Age world system. The raise of amber in terms of numbers (availability) and contexts (commonness) between MBA and LBA is mirrored by the increase in southern Alpine copper's share among European bronze objects (Ling et al. 2014, 2019). Evidence of amber often comes along with other testimonies of contacts between various cultural units to the north (predominantly the *Tumulus* and *Urnfield* cultures) and south (the *Mycenaean* culture) of the circum-Adriatic zone. It is worth noting that amber and other imports occur in the largest numbers in provinces characterised by features considered as indicative of a developed Bronze Age culture: sepulchral markers of incipient social hierarchisation; sedentary lifestyle concentrated in large, often defended settlements; large and diversified subsistence base (food production), access and utilisation of natural resources for local consumption and export. Still, apart from the *Palafitte-Terramare* and *Protovillanovano Padano* cultures, it is difficult to pinpoint units which meet each one of them. A pivotal role in the creation and success of these collective

Intentions and outcome – recapitulation

The aim of this section is to confront intentions behind this work with the outcome, by summarising the results in relation to the goals. To remind what the author has been hoping to achieve, it is necessary to move back to section 1.1, where the goals of this thesis are stated.

First of all, it was meant to reconstruct the dynamics of amber's inflow into the circum-Adriatic zone during the 2nd millennium BC, and describe the communities (archaeological cultures) participating in this process. Second, it was going to determine what varieties of fossil resins (types of amber) were available to and used by those societies, and if succinite (Baltic amber) was dominant among them. Third, by using typology of amber finds and other relevant data, it was hoped to reconstruct the ways of amber circulation within the circum-Adriatic zone. Next, the study was aimed at tackling the material and symbolic status of amber for communities of the circum-Adriatic zone, including its functionality (i.e. modes of using amber), non-utilitarian aspects (i.e. place in beliefs and rituals) and social attribution (i.e. connotations with age, sex, identity and social standing of amber's users). The final goal was to characterise the role of the circum-Adriatic zone and communities inhabiting it within the long-distance contacts network, encompassing Europe during the Bronze Age. This last objective also includes naming the most likely suppliers of amber and other trade partners in its translocation to the north and south of the study area. By simply trying to recapitulate the main accomplishments of the consecutive chapters, I will let the reader decide whether they meet the goals listed above.

Chapter 2 plays a double role: first, it presents the current state of knowledge about amber finds from the study area and reviews it in terms of distribution, chronology and cultural attribution, and second, it orders the database, so it can help answering the questions about the dynamics of amber's inflow and circulation. Consequently, the entire collection of amber artefacts and contexts which yielded them were divided according to geographical and cultural criteria into five micro-regions: (1) Northern Italy, (2) Central Italy, (3) Southern Italy, (4) Western Balkans and (5) Central Balkans (Figs. 11–14). Each one of them encompasses several regions, called 'provinces of amber's acquisition'; overall 34 such provinces were distinguished within the circum-Adriatic zone. They were all characterised from the standpoint of dynamics of amber's inflow, with the results of those considerations compiled in diagrams (Figs. 20, 23, 26, 29). Numbers of amber artefacts and contexts of amber deposition separated into four phases of the Bronze Age (Early, Middle, Late and Final) served to quantify the coefficients of amber's commonness (measured with the total number of contexts in a region) and availability (measured with the total number of finds from a region) (Figs. 21–22, 24–25, 27–28, 30–33).

Several main observations can be inferred from the above-described overview. In the entire circum-Adriatic zone, amber began to be used first in Southern Italy, precisely on Sicily; the oldest amber finds from the island are dated already to the turn of the 3rd and 2nd millennia BC and are ascribed to the *Castelluccio* culture. Slightly later, i.e. around the 19th–18th century BC, amber arrived in Northern Italy, specifically to Trentino-Alto Adige province and the Garda Lake basin, at that time under the influence of the *Polada* culture. Still, its availability and commonness were low in the both macro-regions.

Most of the circum-Adriatic regions, Italian and Balkan alike, greeted amber in the MBA (ca. 16th–14th century BC). This was also a time, when amber reached the highest levels of commonness and availability. In this period the leading centres of amber's acquisition emerged which were to maintain that role in the following phase as well – the LBA (ca. 13th–first half of the 12th century BC). These were the regions influenced by the *Palafitte-Terramare* culture in Northern Italy (provinces of the Garda Lake, western Veneto and Emilia), northern Apulia concurrently inhabited by the *Apennine* culture, Istrian peninsula dominated by the *Castellieri-Gradine* culture, and the Drina Basin at the Bosnian-Serbian border forming the core of the *Brezjak* culture. Other significant provinces of amber's acquisition during the Middle and LBA were the Aeolian Islands (*Milazzese* and *Ausonian I* cultures) and Sicily (*Thapsos* culture).

Turn of the LBA and FBA (12th/11th century BC) brought considerable changes on the map of amber's acquisition in the circum-Adriatic zone. Many of the regions listed above were affected by economic crises and ensuing social upheavals (or other events which enforced sudden cultural changes) which in consequence led to termination of amber import to those areas. The only previous centre of amber's acquisition which adapted to new conditions and continued to acquire amber in bulk was western Veneto, during the FBA influenced by the *Protovillanovan* culture. In this period the first stable work-stations specialised in amber processing on a large scale were established in this province (Frattesina di Fratta Polesine and Campestrin di Grignano Polesine). Import of amber was sustained also, albeit in lower quantities, in northern Apulia and the Aeolian Islands. Place of the old centres of amber's acquisition and redistribution was taken by other regions, such Tuscany and north-central Lazio in Italy, and northern Dalmatia and the Kvarner Bay in Western Balkans. Although those regions still had an access to considerable quantities of amber (what is indicated by occasionally rich assemblages recovered from local contexts), the commonness of the material decreased in general. On the other hand, Central Balkans largely stopped importing amber in the FBA.

With regard to the provenience analyses, it should be noted that few of the amber artefacts turned out to be made of fossil resins locally found in the circum-Adriatic zone. This observation basically applies only to Sicily, which is recognised as the source of simetite. It has been observed that simetite deposits were known and exploited to a limited extent by Sicilian communities since at least the turn of the LCA and EBA. On the other hand, the investigated finds dated to the same period and collected in Northern Italy were classified as succinite originating from Northern Europe. This type of fossil resin became the leading, if not the only, amber used in the circum-Adriatic zone in the MBA and LBA. Only with respect to the FBA, artefacts made of simetite reappeared in Sicily, what may be related to the cultural changes in the research area at the end of the 2nd millennium BC and disruption of the previous succinite circulation network.

The latter was reconstructed for each of the four Bronze Age phases using network analysis. The basis for conducted statistics was the measure of similarity between typologically classified, regional assemblages of amber finds. Classification system was based on a number of criteria such as the location and the course of perforation, the height-to-width ratio, the shape of the cross-section, the shape of the longitudinal section (profile), and the presence and form of decorations (Fig. 34). Typological analysis showed that the vast majority of items can be regarded as ornaments (finished products, with a small proportion of semi-finished products) and led to the distinction of nine functional groups differing mainly in the way objects were hung. Each group includes a number of variants that differ in size, form and, optionally, decoration.

Before starting the network analysis, each group was characterised in terms of chronology and distribution of individual types included in them. This has resulted in interesting observations regarding regional stylistic preferences and trends in amber processing (where this craft is confirmed), which then helped to determine which provinces maintained contacts and exchanged amber. One should mention, e.g., the custom of wearing perforated but poorly shaped (amorphous) beads in the Central Balkans during the MBA and LBA, which to some extent was also adopted in Southern Italy (i.e. northern Apulia). Another important observation is the practice of decorating elongated amber beads with ornaments resembling single transverse ribs (decorative motif 'Tiryns') or parallel grooves (decorative motifs 'Allumiere'), which was most likely born in the northern part of the Adriatic Basin (western Veneto) at the turn of the LBA and FBA, and from there it spread south and north. Thanks to the chronological analysis performed with seriation, it was discerned that in the early stage of the FBA (ca. 1150–1050 BC), the ornamental motifs were quite simple and similar in most regions (variants 'a', 'c' and 'e'). In turn, in the final period (ca. 1050–950/900 BC), their more complicated versions (variants 'b', 'd', 'f', 'g', 'h' and 'i') appeared, which tend to be much more regionalised. This allowed for the distinction of regional ornamental styles, potentially reflecting the local traditions of amber processing: Northern Italian (including the provinces of western Veneto and eastern Lombardy), Central Italian (including northern Lazio) and Eastern Adriatic (including Kvarner, northern Dalmatia and Lika).

The network analysis, involving quantification of similarities between the provinces based on coinciding types of amber artefacts, was supplemented with observations on other categories of shared finds, predominantly pottery and bronzes. In the EBA one can speak of two areas of amber's acquisition within the circum-Adriatic zone: Sicily on the one hand, and the Garda Lake basin and Trentino-Alto Adige on the other (Figs. 64–65). In the light of gathered data, it appears that they obtained amber from two different sources: in the case of Sicily (*Castelluccio* culture) that was locally available simetite, replaced by succinite at the end of the period, while in the north Baltic amber from the start was the only imported fossil resin. Cooperation between the two areas with respect to circulation of amber during the EBA is unlikely at the current stage of research, but should not be completely ruled out.

During the MBA many interregional connections, resulting in a wider circulation of amber, occurred in the circum-Adriatic zone (Figs. 66–67, 79). One should mention in this context the contacts between the *Palafitte-Terramare* culture in Northern Italy (Garda Lake basin, western Veneto and Emilia), and the *Protoapennine* and then the *Apennine* culture in Southern Italy (chiefly Apulia). These cultural units formed key centres of amber's acquisition and redistribution in Italy, as revealed by the *core-periphery* analysis and the *betweenness* coefficient (Fig. 68, Tab. 45). In much lower amounts, possibly as a 'side-effect' of trans-Apennine exchange, amber reached the regions situated in-between the Po Plain and Apulia: i.e. Romagna, Marche, Tuscany and Lazio. From Apulia amber was supplied further to Calabria, the Aeolian Islands and Sicily. It is probable that Istrian peninsula (*Castellieri-Gradine* culture) established contacts with Northern and Southern Italy, facilitating trans-Adriatic flow of amber.

Separate network of amber circulation emerged during the MBA in Central Balkans. It followed from Podrinje and Western Serbia, along the Drina River, to the Glasinac area in eastern Bosnia, and further across Kosovo and Metohija to Albania. While similarities among certain amber and bronze finds suggest some relationships between Central Balkans and Southern Italy, this topic requires much more investigations before such interactions can be confirmed and their character defined. Despite some changes involving cessation of amber influx into several peripheral regions (e.g. Sicily) and emergence of several new provinces of amber's acquisition (e.g. Vardar Valley and Pelagonia in the North Macedonia), the territorial structure of amber circulation remained stable during the LBA (Figs. 80–82, 94, Tab. 46). Only the FBA brought substantial alterations in that respect (Figs. 95–97, Tab. 47).

The pre-eminent role in import, redistribution and processing of amber during that period in Northern Italy was taken over by the local group of the *Protovillanovan* culture in western Veneto (the so-called *Protovillanovano Padano* of the lower Po Plain). Concurrently, a more prominent position in consumption and mediation in amber exchange was adopted by the *Protovillanovan* communities inhabiting Romagna, Tuscany and Lazio. In Southern Italy initially the Aeolian Islands (*Ausonian II* culture) and southern Apulia came to the forefront of amber's recipients, at the same time surpassing northern Apulia which noted a decline in import of amber. Together the abovementioned provinces created a well-interconnected network of amber circulation covering Italy (Fig. 109). Yet, it seems that it was short-lived, as in the later part of the FBA, it underwent rarefication (leaving only western Veneto and north-central Lazio as significant users of amber), whereas its southern section dissolved (with exception of northern Apulia and, perhaps, Sicily) (Fig. 110). On the other side of the Adriatic, amber's main recipient and redistribution node (and potentially also processing centre) became the Kvarner Bay and northern Dalmatia. From there adornments made of fossil resin were brought to the neighbouring provinces. The coastal part of Croatia maintained relations with the opposing shore of the Adriatic what resulted in appearance of similar decorative styles in amber jewellery. In Central Balkans, in turn, a significant destination in amber circulation was the community of the *Belegiš II-Gava* culture in eastern Vojvodina (Banat) which probably kept contacts with culturally related provinces in Northern Croatia, Slovenia and perhaps also eastern Bosnia (Glasinac area).

Regarding amber's status, it is possible to bring forward a series of observations. Typological analysis of the finds alone showed that these are predominantly ornaments in the form of beads, pendants, buttons and, much less often, spacers. Furthermore, amber could be used in a way leaving no trace in the archaeological record, e.g. as an incense or food ingredient; however, those hypotheses still require verification. Although amber adornments mainly come from graves (Fig. 111), anthropological analogies as well as studies into Bronze Age dress permit to suspect that they were not used solely as burial inventories but were also worn during the lifetime. Disposition of grave goods occasionally allows to reconstruct the ways the amber jewellery was arranged into attire: mostly it was composed in necklaces, rarely bracelets, sometimes including adornments made from other materials (e.g. faience, glass, carnelian). From Northern Italy and Slovenia is known a custom of inserting perforated amber beads on pin-stalks, used to fasten robes.

On the basis of anthropological analyses of burials, it was possible to determine that amber, predominantly, was the domain of women and children (girls?) (Tab. 48).

Most of the times amber adornments belonged to graves of lavishly dressed women deceased in juvenile, mature or adult age, who apart from jewellery did not possess items regarded as indicators of social roles. However, in few cases one may infer the identity of these female individuals as persons fulfilling responsibilities connected with religious practices or rituals, i.e. priestesses. From the entire circum-Adriatic zone are known only a few burials of men equipped with amber. Interestingly, by the content of inventories, many of them resemble burials of women, and only anthropological markers allow to tell them apart (Figs. 122–123). Two male burials with amber also contain elements commonly associated with warfare and political power – bronze swords – therefore, they potentially reflect specific appropriation of amber. Calculating the number of graves equipped with amber by the total number of burials in cemeteries revealed that amber was a grave gift used rather rarely, mostly in relation to small social groups characterised by richly and diversely furnished graves (especially in the later stages of the Bronze Age) (Tabs. 49–50). Thus, one can confirm traditionally accepted hypothesis about amber as an attribute of persons with high social standing, or an emblem of small groups of people linked by status or identity.

Amber beads and pendants, aside from the visual – decorative aspect, possibly also had symbolic meaning: not only as emblems of social rank or manifestations of wealth but also as magical elements – amulets/talismans protecting from evil, misfortune, etc., or as devotional objects expressing adherence to a specific set of beliefs, i.e. solar cult particularly popular during the LBA and FBA in Europe. Amber occasionally was buried as a part of hoards/votive offerings sacrificed in various intentions, probably during ceremonies involving participation of larger groups of people. Among hoards with amber can be distinguished those likely having a religious significance or motivation, judging by the presence of other objects considered as symbols related to solar cult, e.g. metal discs decorated with circle, cross and/or sun motifs (Fig. 128).

During the Bronze Age, amber enjoyed periods of popularity among certain communities of the circum-Adriatic zone, which incorporated it into their culture to a varying degree. This process can be explained using the 'sémiophore' concept by K. Pomian, simultaneously applying the model of information economy by P. Jackson derived from the former as well as information theory (Figs. 129–130). Upon accepting amber as a 'sémiophore', adaptation of the model in the realities of the Bronze Age circum-Adriatic zone gives the opportunity to track several different ways of amber appropriation that characterise individual provinces (Fig. 131). First of all, they differ in the degree of social acceptance of amber: from isolated cases of use by individuals, to adoption by entire social groups. Besides, amber's social history has always followed certain fixed trajectories, which began when individual representatives of a given community became acquainted with amber, which was followed by the rejection or very limited appreciation of amber, or, after a period of adaptation and standardisation of meaning, incorporation into the canon of culture. After some time, amber usually lost its status and in order to regain it, it had to go through the entire cycle described above.

Amber, although it was not commonly used in the entire circum-Adriatic zone, could serve as a commodity in trade for other resources. Archaeological studies in this respect most often point to copper from southern Alpine ores as the most likely equivalent of amber (Fig. 137). Recent metal analyses have shown that southern Alpine deposits became one of the main sources of copper used by metallurgists in Northern Germany and Scandinavia from the MBA onwards. It coincided with the development of the *Palafitte-Terramare* culture and the intensification of amber import to Northern

Italy. There are indications that the presence of significant amounts of amber in other key provinces, usually located in the northern part of the circum-Adriatic zone, should be explained by their growing economic importance due to access to strategic raw materials: sea salt in the case of Istria, and later Dalmatia and Kvarner, and tin in the case of Podrinje and Western Serbia. These materials probably meant that the regions in question were included in the far-reaching exchange network, thanks to which Baltic amber could reach so far south. The main trading partners in this process were the communities of Central and Northern Europe also showing a demand for bronze and amber. In the EBA it was predominantly the *Únětice* culture, in the MBA – the *Tumulus* culture, and in the LBA and FBA – the *Urnfield* culture.

Apart from using amber for their own needs, the communities of the circum-Adriatic zone participated in its translocation further south, i.e. to Greece. During the EBA the inflow of Baltic amber to the Aegean took place through Sicily, what had an impact on the local communities as it gave them an opportunity to learn about foreign amber (they already knew simetite). Contacts with the *Mycenaean* culture are evidenced by the presence of pottery and several other items of Aegean provenance on the island (Fig. 139). Still, succinite did not find a widespread use on Sicily at that stage. These contacts intensified in the MBA, but with time the function of amber exchange intermediary began to be taken over by the region of Apulia (Fig. 143). Concurrently, i.e. in the MBA and LBA, amber could also find its way to Greece through the Central Balkan communities, which is documented by the presence of *Mycenaean* bronzes on some Albanian, North Macedonian and Kosovar sites. This coincided with the expansion of the *Mycenaean* culture influences, as well as the consumption of amber, towards the north, into Epirus, Thessaly and Macedonia (Fig. 144). At the end of the *Mycenaean* culture, Italic communities have adopted a more active attitude in arranging contacts with Greece. It is likely that during this phase even Northern Italy directly interacted with the Aegean, allowing the further flow of amber to the south (Fig. 146). This stage heralds a much more prominent role in the processing and circulation of amber that Italian and Balkan societies were to take over during the Iron Age.

Bibliography

Abels 1972: B. U. Abels, *Die Randleistenbeile in Baden-Württemberg, dem Elsaß, der Franche-Comté und der Schweiz.* Prähistorische Bronzefunde IX/4. München 1972.

Adams 1992: R. M. Adams, Anthropological Perspectives on Ancient Trade. *Current Anthropology* 33 (1992), pp. 141–160.

Albanese Procelli 1992: R. M. Albanese Procelli, La necropoli di Madonna del Piano presso Grammichele: osservazioni sul rituale funerario. *Kokalos* XXXVIII (1992), pp. 33–68.

Albanese Procelli 1994: R. M. Albanese Procelli, Considerazioni sulla necropoli di Madonna del Piano di Grammichele (Catania). In: P. Castaldi, G. Maetzke (eds.), *La presenza etrusca nella Campania meridionale. Atti delle giornate di studio Salerno-Pontecagnano, 16–18 novembre 1990.* Istituto Nazionale di Studi Etruschi. Quaderni di «Studi Etruschi», vol. 28. Firenze 1994, pp. 153–169.

Albanese Procelli/Lo Schiavo 2004: R. M. Albanese Procelli, F. Lo Schiavo, La comunità di Madonna del Piano presso Grammichele (Catania): rapporti con l'area calabra. In: I.I.P.P. (eds.), *Preistoria e protostoria della Calabria. Atti XXXVII Riunione Scientifica dell'Istituto Italiano di Preistoria e Protostoria (Scalea, Papasidero, Praia a Mare, Tortora, 29 sett.-4 ott. 2002).* Firenze 2004, pp. 403–420.

Alberti 2004: G. Alberti, Contributo alla seriazione delle necropoli siracusane. In: V. La Rosa (ed.), *Le presenze Micenee nel territorio Siracusano.* Padova 2004, pp. 99–170.

Alberti 2006: G. Alberti, Per una 'gerarchia sociale' a Thapsos: analisi contestuale delle evidenze funerarie e segni di stratificazione. *Rivista di Scienze Preistoriche* LVI (2006), pp. 369–427.

Alberti 2007: G. Alberti, Minima Thapsiana. Riflessioni sulla cronologia dell'abitato di Thapsos. *Rivista di Scienze Preistoriche* LVII (2007), pp. 363–376.

Alberti 2013: G. Alberti, A. Bayesian, ^{14}C chronology of Early and Middle Bronze Age in Sicily. Towards an independent absolute dating. *Journal of Archaeological Science* 40 (2013), pp. 2502–2514.

Alberti 2015: G. Alberti, The aid of Bayesian radiocarbon modeling in assessing the chronology of Middle Bronze Age Sicily at the site level. A case study. *Journal of Archaeological Science: Reports* 2 (2015), pp. 246–256.

Alberti/Bettelli 2005: L. Alberti, M. Bettelli, Contextual Problems of Mycenaean Pottery in Italy. In: R. Laffineur, E. Greco (eds.), *Emporia. Aegeans in the Ceantral and Eastern Mediterranean.* Aegaeum 25. Liège 2005, pp. 547–559.

Alessandri 2013: L. Alessandri, *Latium Vetus in the Bronze Age and Early Iron Age.* BAR International Series 2565. Oxford 2013.

Aliu 2004: S. Aliu, *Tuma e Luarasit.* Tiranë 2004.

Aliu 2007: S. Aliu, Les bijoux de la période du Bronze Récent et de la période du Fer Ancien dans la région de Kolonja. In: I. Galanaki, H. Tomas, Y. Galanakis, R. Laffineur (eds.), *Between the Aegean and Baltic Seas. Prehistory across Borders.* Aegaeum 27. Liège 2007, pp. 239–247.

Aliu 2012: S. Aliu, *Tuma e Rehovës*. Korçë 2012.

Álvarez-Fernández et al. 2005: E. Álvarez-Fernández, E. Peñalver, X. Declòs, La presencia de ámbar en los yacimientos prehistóricos (del Paleolítico Superior a la Edad del Bronce) de la Cornisa Cantábrica y sus fuentes de aprovisionamiento. *Zephyrus* 58 (2005), pp. 147–170.

Anderson et al. 1992: K. B. Anderson, R. E. Winans, R. E. Botto, The nature and fate of natural resins in the geosphere—II. Identification, classification and nomenclature of resinites. *Organic Geochemistry* 18/6 (1992), pp. 829-841.

Anderson 1997: K. B. Anderson, The Nature and Fate of Natural Resins in the Geosphere – VII. A Radiocarbon (14C) Age Scale for Description of Immature Natural Resins: An Invitation to Scientific Debate. *Organic Geochemistry* 25 (1997), pp. 251–253.

Anderson/Crelling 1995: K. B. Anderson, J. C. Crelling, *Amber, Resinite, and Fossil Resins*. Washington 1995.

Andrea 1985: Zh. Andrea, *Kultura ilire e tumave në pellgun e Korçës*. Tiranë.

Andreotti/Zanini 1995–1996: A. Andreotti, A. Zanini, L'insediamento di Fossa Nera di Porcari (Lucca). *Rivista di Scienze Preistoriche* XLVII (1995–1996), pp. 291–330.

Andreotti/Ciampoltrini 2013: A. Andreotti, G. Ciampoltrini, Fossa Nera di Porcari. Un insediamento nella piana dell'Auser intorno al 1200 a.C. In: G. Ciampoltrini (ed.), *Da Fossa Nera di Porcari a Monte Formino di Palaia. La 'Crisi del 1200 A.C.' tra valle del Serchio e Valdarno*. Pisa 2013, pp. 35–58.

Angelini 2009: I. Angelini, Indagini archeometriche dei vaghi in vetro. In: M. V. Gambari (ed.), *Il ripostiglio del Monte Cavanero di Chiusa di Pesio (Cuneo)*. Chiusa di Pesio 2009, pp. 185–192.

Angelini 2010a: I. Angelini, Indagini spettroscopiche sul seperto si natura incognita proveniente dalla tomba 115 di necropoli di Bovolone (VR). In: L. Salzani (ed.), *La necropoli dell'età del Bronzo di Bovolone*. Memorie del Museo Civico di Storia Naturale di Verona – 2. Serie, Sezione Scienze dell'Uomo 10. Verona 2010, pp. 205–211.

Angelini 2010b: I. Angelini, Indagini archeometriche degli ornamenti in ambra provenienti dalla necropoli di Narde. In: L. Salzani, C. Colonna (eds.), *La Fragilità dell'Urna. I recenti scavi a Narde Necropoli di Frattesina (XII-IX sec. a.C.)*. Exhibition catalogue. Rovigo 2010, pp. 135–145.

Angelini 2012: I. Angelini, Ambre protostoriche della Sardegna: indagini archeometriche. In: I.I.P.P. (eds.), *La preistoria e la protostoria della Sardegna. Atti della XLIV Riunione Scientifica dell'Istituto Italiano di Preistoria e Protostoria (Cagliari, Barumini, Sassari 23–28 novembre 2009)*. Firenze 2012, pp. 1151–1161.

Angelini 2013: I. Angelini, Le ambre dalla tomba 347 della necropoli di Desmontà. In: L. Salzani (ed.), *La necropoli di Desmontà (Veronella – Albaredo d'Adige. Verona) – Scavi 1982–2011*. Documenti di Archeologia 56. Mantova 2013, pp. 203–211.

Angelini 2020: I. Angelini, *Report of the Cooperation Agreement between Adam Mickiewicz (University in Poznań, Poland) and Ivana Angelini (INSTM – Research Unit of Padova, University of Padova, Italy) for Analysis of Archaeological Amber*

Finds. Consorzio Interuniversitario Nazionale per la Scienza e Tecnologia dei Materiali. Padova 2020 (unpublished report).

Angelini/Bellintani 2005: I. Angelini, P. Bellintani, Archaeological Ambers from Northern Italy: An FTIR-DRIFT Study of Provenance by Comparison with the Geological Amber Database. *Archaeometry* 47/2 (2005), pp. 441–454.

Angelini/Bellintani 2017: I. Angelini, P. Bellintani, The Use of Different Amber Sources in Italy during the Bronze Age: New Archaeometric Data. *Archaeological and Anthropological Sciences* 9/4 (2017), pp. 673–684.

Angelini et al. 2001: I. Angelini, G. Artioli, P. Bellintani, V. Diella, A. Polla, G. Recchia, G. Residori, Materiali vetrosi da Grotta Manaccora e Coppa Nevigata: inquadramento archeologico, archeometrico nell'ambito dell'età del Bronzo italiana. In: C. Piccioli, F. Sogliani (eds.), *Atti II Convegno Multidisciplinare 'Il vetro in Italia meridionale e insulare', Napoli 5–7 dicembre 2001.* Napoli 2001, pp. 127–138.

Angelini et al. 2003: I. Angelini, G. Artioli, P. Bellintani, Progetto 'Ambre e materiali vetrosi protostorici della Valle dell'Adige nel quadro delle coeve attestazioni dell'Italia del nord'. Primi risultati sulle ambre. *Preistoria Alpina* 39 (2003), pp. 227–241.

Angelini et al. 2004: I. Angelini, G. Artioli, P. Bellintani, V. Diella, M. Gemmi, A. Polla, A. Rossi, Chemical Analyses of Bronze Age glasses from Frattesina di Rovigo, Northern Italy. *Journal of Archaeological Science* 31 (2004), pp. 1175–1184.

Angelini et al. 2005: I. Angelini, G. Artioli, P. Bellintani, A. Polla, Protohistoric Vitreous Materials of Italy: From Early Faience to Final Bronze Age Glasses. In: M. D. Nenna (ed.), *Annales du 16e Congrès de l'Association Internationale pour l'Histoire du verre.* Nottingham 2005, pp. 32–36.

Angelini et al. 2006: I. Angelini, M. Angle, G. Artioli, P. Bellintani, F. Lugli, N. Martinelli, A. Polla, A. Tagliacozzo, A. Zarattini, Il villaggio delle Macine (Castelgandolfo, Roma). In: G. Ghini (ed.), *Lazio e Sabina 3. Atti del Convegno 'Terzo Incontro di Studi sul Lazio e la Sabina', Roma, 18–20 novembre 2004.* Lavori e Studi della Soprintendenza per i Beni Archeologici del Lazio 3. Roma 2006, pp. 157–168.

Angelini et al. forthcoming: I. Angelini, M. Cwaliński, P. Bellintani, G. Recchia, Provenience of Bronze Age Amber Finds from Italy: New Results (in preparation).

Angle/Guidi 2007: M. Angle, A. Guidi, L'antica e media età del bronzo nel Lazio meridionale. In: I.I.P.P. (eds.), *Strategie di insediamento fra Lazio e Campania in età preistorica e protostorica. Atti della XL Riunione Scientifica dell'Istituto Italiano di Preistoria e Protostoria (Roma, Napoli, Pompei, 30 novembre–3 dicembre 2005).* Roma 2007, pp. 147–178.

Angle et al. 2016: M. Angle, P. Cerino, G. Granata, D. Mancini, R. Malinconico, N. Tomei, Il sito su impalcato ligneo del Villaggio delle macine a Castel Gandolfo (Roma). In: E. Calandra, G. Ghini, Z. Mari (eds.), *Lazio e Sabina 10. Atti del Convegno 'Decimo Incontro di Studi sul Lazio e la Sabina', Roma, 4–6 giugno 2013.* Lavori e Studi della Soprintendenza per i Beni Archeologici del Lazio 10. Roma 2016, pp. 315–318.

Appadurai 1986: A. Appadurai (ed.), *The Social Life of Things: Commodities in Cultural Perspective.* Cambridge 1986.

Aranguren/Perazzi 2004: B. Aranguren, P. Perazzi, La Grotta dell'Artofago (Gavorrano – GR). In: N. Negroni Catacchio (ed.), *Atti del VI incontro di studi di 'Preistoria e*

Protostoria in Etruria' (Pitigliano, Valentano 13–15 settembre 2002), vol. II. Milano 2004, pp. 473–482.

Ardesia 2013–2014: V. Ardesia, La cultura Di Rodì-Tindari-Vallelunga in Sicilia: origini, diffusione e cronologia alla luce dei recenti studi. Parte 1. *Ipotesi di Preistoria* 6 (2013–2014), pp. 35–98.

Ardesia/Cultraro 2011: V. Ardesia, M. Cultraro, Le fasi recenti dell'Eneolitico e la transizione al Bronzo Antico 1 nella Sicilia centro-occidentale: alcune considerazioni. In: I.I.P.P. (eds.), *L'età del rame in Italia. Atti del XLIII Riunione Scientifica dell'Istituto Italiano di Preistoria e Protostoria (Bologna, 26–29 novembre 2008)*. Firenze 2011, pp. 121–127.

Ardesia et al. 2006: V. Ardesia, M. Cattani, F. Nicoletti, M. Secondo, S. Tusa, Gli scavi nell'abitato dell'età del Bronzo di Mursia (Pantelleria, TP). Relazione preliminare delle campagne 2001–2005. *Rivista di Scienze Preistoriche* LVI (2006), pp. 293–367.

Argentina et al. 2002: F. Argentina, P. Desantis, F. Finotelli, F. Schneider, Il sito dell'Età del Bronzo di Anzola dell'Emilia. In: T. Ravasio (ed.), *Archeologia in Valle del Samoggia: studi e ricerche sul popolamento antico. Atti del convegno (Bazzano, Rocca dei Bentivoglio, 3 maggio 2001)*. Quaderni della Rocca 9. Bazzano 2002, pp. 39–80.

Aspes 1972: A. Aspes, Materiali inediti dell'Antica età del Bronzo da Cattaragna (Brescia) conservati al Museo Civico di Storia Naturale di Verona. *Natura Bresciana: Annali Museo Civico Storia Naturale di Brescia* 9 (1972), pp. 13–34.

Aspes 1973: A. Aspes, La necropoli dell'età del bronzo di Franzine nuove di Villabartolomea (Verona): risultati delle ricerche 1968–1972. In: I.I.P.P. (eds.), *Atti della XV Riunione Scientifica dell'Istituto Italiano di Preistoria e Protostoria (Verona-Trento, 27–29 ottobre 1972)*. Firenze 1973, pp. 117–125.

Aspes 1976: A. Aspes, Franzine Nuove di Villabartolomea. In: L. Salzani (ed.), *3000 anni fa a Verona. Dalla fine dell'età del Bronzo all'arrivo dei Romani nel territorio veronese. Exhibition catalogue*. Verona 1976, pp. 9–13.

Aspes 1980: A. Aspes, Peschiera. In: A. Aspes (ed.), *La preistoria del Lago di Garda*. Exhibition catalogue. Museo Civico di Storia Naturale di Verona. Verona 1980, pp. 99–105.

Aspes 1987a: A. Aspes, Ca' Nova di Cavaion. In: A. Aspes, L. Salzani (eds.), *Prima della storia: inediti di 10 anni di ricerche a Verona*. Exhibition catalogue. Museo Civico di Storia Naturale di Verona. Verona 1987, pp. 86–91.

Aspes 1987b: A. Aspes, La Quercia. In: A. Aspes, L. Salzani (eds.), *Prima della storia: inediti di 10 anni di ricerche a Verona*. Exhibition catalogue. Museo Civico di Storia Naturale di Verona. Verona 1987, pp. 92–94.

Aspes 1987c: A. Aspes, Franzine Nuove di Villabartolomea. In: A. Aspes, L. Salzani (eds.), *Prima della storia: inediti di 10 anni di ricerche a Verona*. Exhibition catalogue. Museo Civico di Storia Naturale di Verona. Verona 1987, pp. 99–102.

Aspes 1992: A. Aspes, La Quercia di Lazise. Le ricerche e i materiali. In: Neri Pozza (ed.), *C'era una volta Lazise*. Exhibition catalogue. Museo Civico di Storia Naturale di Verona. Verona 1992, pp. 72–86.

Aspes 1997: A. Aspes, Franzine Nuove di Villabartolomea (VR). In: M. Bernabò Brea, A. Cardarelli, M. Cremaschi (eds.), *Le Terramare. La più antica civiltà padana*. Exhibition catalogue. Modena 1997, pp. 717–719.

Aspes/Fasani 1967: A. Aspes, L. Fasani, Aspetti e problemi dell'età del bronzo media e recente nell'anfiteatro morenico del Garda. In: I.I.P.P. (eds.), *Atti della XI e XII Riunione Scientifica dell'Istituto Italiano di Preistoria e Protostoria (Firenze 11–12.2.1967, Sicilia, Palermo-Lipari 22–26.10.1967).* Firenze 1967, pp. 223–228.

Aspes/Fasani 1968: A. Aspes, L. Fasani, Necropoli dell'eta del Bronzo a Franzine Nuove di Villabartolomea (VR) (Scavi 1968). *Memorie del Museo Civico di Storia Naturale di Verona* XVI (1968), pp. 455–482.

Aspes/Fasani 1974: A. Aspes, L. Fasani, Aspetti e problemi dell'antica età del Bronzo in Italia settentrionale. *Preistoria Alpina* 10 (1974), pp. 79–83.

Aspes et al. 1991–1992: A. Aspes, M. Bernabò Brea, E. Bianchin Citton, A. Cardarelli, R. C. De Marinis, L. Fasani, F. Marzatico, L. Salzani, Articolazioni chronologiche e definizione di elementi culturali, II. Area Alpina e Padana centro-orientali. In: D. Cocchi Genick (ed.), *L'età del bronzo in Italia nei secoli dal XVI al XIV a.C. Atti del congresso Viareggio, 26–30 ottobre 1989.* Rassegna di Archeologia 10. Firenze 1991–1992, pp. 39–55.

Aston 2012–2013: D. Aston, Radiocarbon, wine jars and New Kingdom chronology. *Ägypten und Levante* 22–23 (2012–2013), pp. 289–315.

Bačić 1957: B. Bačić, Novi grobovi iz brončanog doba u Istri. *Vjesnik za arheologiju i historiju Dalmatinsku* LVI-LIX (1954–1957), pp. 18–20.

Bačić 1960: B. Bačić, Tumuli iz brončanog doba na Maklavunu i Žamnjaku u južnoj Istri. *Jadranski Zbornik* IV (1960), pp. 197–210.

Bader 1983: T. Bader, *Die Fibeln in Rumänien.* Prähistorische Bronzefunde XIV/6. München 1983.

Bagolan/Vanzetti 1997: M. Bagolan, A. Vanzetti, Bassa Veronese: siti dell'età del Bronzo Medio, alcuni dei quali finiscono agli inizi del Bronzo recente. In: M. Bernabò Brea, A. Cardarelli, M. Cremaschi (eds.), *Le Terramare. La più antica civiltà padana.* Exhibition catalogue. Modena 1997, pp. 356–357.

Bagolan/Leonardi 2000: M. Bagolan, G. Leonardi, Il Bronzo finale nel Veneto. In: M. Harari, M. Pearce (eds.), *Il Protovillanoviano al di qua e al di là dell'Appennino. Atti della Giornata di studio (Pavia, 17 giugno 1995).* Como 2000, pp. 15–46.

Baioni 2000: M. Baioni, La necropoli dell'Antica Età del Bronzo di Sorbara (Asola – MN). *Quaderni di Archeologia del Mantovano* 2 (2000), pp. 41–89.

Baioni et al. 2007a: M. Baioni, G. Bocchio, C. Mangani, Il Lucone di Polpenazze: storia delle ricerche e nuove prospettive. *Annali Benacensi* XIII-IV (2007), pp. 83–102.

Baioni et al. 2007b: M. Baioni, L. Ferrari, F. Leandri, *Il monumento naturale de 'I Lagazzi' di Piadena. Nucleo territoriale n. 14.* Cremona 2007.

Bakarić 2017: L. Bakarić, Japodi. In: L. Bakarić (ed.), *Japodi – Zaboravljeni Gorštaci.* Arheološki muzej u Zagrebu, Katalozi i monografije XI. Zagreb 2017, pp. 21–61.

Bakarić et al. 2006: L. Bakarić, B. Križ, M. Šoufek, *Prehistoric amber and glass from Prozor in Lika and Novo Mesto in Dolenjska.* Arheološki muzej u Zagrebu. Exhibition catalogue. Zagreb 2006.

Baldelli et al. 2005: G. Baldelli, G. Bergonzi, A. Cardarelli, I. Damiani, N. Lucentini, Le Marche dall'antica alla recente età del bronzo. In: I.I.P.P. (eds.), *Preistoria e Protostoria delle Marche. Atti della XXXVIII riunione scientifica dell'Istituto Italiano di Preistoria e Protostoria (Portonovo, Abbadia di Fiastra, 1–5 ottobre 2003).* Firenze 2005, pp. 539–579.

Baldo et al. 2015–2018: M. Baldo, C. Balista, P. Bellintani, Frattesina di Fratta Polesine: estensione, infrastrutture, definizione di aree funzionali ed evoluzione paleo-idrografica del territorio. Metodologie 'a basso impatto' e risultati delle indagini sul campo – anni 2014–2016. *Padusa LI-LIV*, nuova serie (2015–2018), pp. 7–70.

Balista/De Guio 1997: C. Balista, A. De Guio, Ambiente ed insediamenti dell'età del bronzo nelle Valli Grandi Veronesi. In: M. Bernabò Brea, A. Cardarelli, M. Cremaschi (eds.), *Le Terramare. La più antica civiltà padana.* Exhibition catalogue. Modena 1997, pp. 137–160.

Balista et al. 2008: C. Balista, F. Bondavalli, A. Cardarelli, D. Labate, C. Mazzoni, G. Steffe, Dati preliminari sullo scavo della Terramara di Gaggio di Castelfranco Emilia (Modena): scavi 2001–2004. In: M. Bernabò Brea, R. Valloni (eds.), *Archeologia ad alta velocità in Emilia. Indagini archeologiche lungo il tracciato ferroviario.* Quaderni di Archeologia dell'Emilia Romagna 22 (2008), pp. 113–138.

Bándi 1974: G. Bándi, Über den Ursprung und die historischen Beziehungen der Tonstempel der bronzezeitlichen Gruppen: Madarovče und Polada. *Preistoria Alpina* 10 (1974), pp. 237–252.

Banerjee et al. 1999: A. Banerjee, V. Ghiurca, B. Langer, M. Wilhelm, Determination of the Provenance of Two Archaeological Amber Beads from Romania by FTIR- and Solid-State-Carbon-13 NMR Spectroscopy. *Archäologisches Correspondenzblatt* 29 (1999), pp. 593–606.

Barbarić 2009: V. Barbarić, Late Bronze Age in Dalmatia: State of Research. In: E. Borgna, P. Càssola Guida (eds.), *From the Aegean to the Adriatic: social organisations, modes of exchange and interaction in the post-palatial times (12th–11th c. BC).* Studi e Ricerche di Protostoria Mediterranea 8. Roma 2009, pp. 311–323.

Barbaro 2013: B. Barbaro, *Insediamenti, aree funerarie ed entità territoriali in Etruria Meridionale nel Bronzo Finale.* Grandi Contesti e Problemi della Protostoria Italiana 14. All'Insegna del Giglio. Roma 2013.

Barfield 1971: L. H. Barfield, *Northern Italy before Rome.* London 1971.

Barfield 1994: L. H. Barfield, The Bronze Age of Northern Italy: Recent Work and Social Interpretation. In: C. Mathers, S. Stoddart, J. R. Colliss (eds.), *Development and Decline in the Mediterranean Bronze Age.* Sheffield 1994, pp. 129–144.

Barker/Stoddart 1994: G. Barker, S. Stoddart, The Bronze Age of Central Italy: c. 2000–900 BC. In: C. Mathers, S. Stoddart, J. R. Colliss (eds.), *Development and Decline in the Mediterranean Bronze Age.* Sheffield 1994, pp. 145–165.

Bass 1991: G. F. Bass, Evidence of Trade from Bronze Age Shipwrecks. In: N. H. Gale (ed.), *Bronze Age Trade in the Mediterranean.* Jonsered 1991, pp. 69–82.

Bastianelli 1939: S. Bastianelli, Allumiere – Rinvenimento di tombe arcaiche. *Atti della Reale Accademia Nazionale dei Lincei. Serie 6. Notizie degli scavi di antichita* 15 (1939), pp. 45–58.

Bátora 2000: J. Bátora, *Das Gräberfeld von Jelšovce/Slowakei. Ein Beitrag zur Frühbronzezeit im nordwestlichen Karpatenbecken.* Prähistorische Archäologie Südosteuropa 16. Kiel 2000.

Bátora 2013: J. Bátora, The Settlement Structure Problem and the End of the Fortified Settlements from the Final Period of Early Bronze Age in Slovakia. In: H. Meller, F. Bertemes, H. R. Bork, R. Risch (eds.), *1600 – Kultureller Umbruch im Schatten des Thera-Ausbruchs? 4. Mitteldeutscher Archäologentag vom 14. Bis 16. Oktober 2011 in Halle (Saale).* Tagungen des Landesmuseums für Vorgeschichte Halle 9. Halle (Saale) 2013, pp. 374–386.

Batović 1959: Š. Batović, Iz ranog željeznog doba Liburnije. *Diadora: glasilo Arheološkog muzeja u Zadru* 1 (1959), pp. 37–82.

Batović 1960: Š. Batović, Predmeti osebujnih oblika s *područja* Liburna. *Radovi Instituta Jugoslavenske Akademije Znanosti i Umjetnosti u Zadru* VI-VII (1960), pp. 393–425.

Batović 1980: Š. Batović, L'eta del bronzo recente sulla costa orientale dell'Adriatico. *GODIŠNJAK – Centar za Balkanološka Ispitivanja* XVIII (1980), pp. 21–62.

Batović 1983: Š. Batović, Kasno brončano doba na istočnom Jadranskom primorju. In: A. Benac (ed.), *Praistorija Jugoslavenskih Zemalja. IV knjiga: Bronzano Doba.* Sarajevo 1983, pp. 271–373.

Batović 1987: Š. Batović, Liburnska grupa. In: A. Benac (ed.), *Praistorija Jugoslavenskih Zemalja. V knjiga: Željezno doba.* Sarajevo 1987, pp. 339–390.

Battaglia 1943: R. Battaglia, La palafitta del lago di Ledro nel Trentino. *Memorie del Museo di Storia Naturale* 7 (1943), pp. 1–64.

Battaglia 1990: D. Battaglia, *On the Bones of the Serpent: Person, Memory and Mortality in Sabarl Island Society.* Chicago 1990.

Battisti 2015: M. Battisti, Le tavolette enigmatiche della Vallagarina (Rovereto, TN) nel contesto europeo. *Annali Museo Civico Rovereto* 30 (2014), pp. 3–26.

Baumgärtel 1953: E. Baumgärtel, The Cave of Manaccora (Monte Gargano). *Papers of the British School at Rome* 21 (1953), pp. 1–31.

Bazzocchi et al. 2009: M. Bazzocchi, M. Cattani, F. Debandi, Il popolamento della Romagna e delle aree limitrofe nella antica e media età del bronzo. Introduzione al quadro conoscitivo. *Ipotesi di preistoria* 2/1 (2009), pp. 192–196.

Bebber/Buonopane 1982: A. E. Bebber, A. Buonopane, Cattaragna. In: A. Aspes (ed.), *Palafitte: mito e realtà.* Exhibition catalogue. Verona 1982, pp. 69–72.

Beck 1928: H. C. Beck, Classification and Nomenclature of Beads and Pendant. *Archaeologia* 77 (1928), pp. 1–76.

Beck 1976: H. C. Beck, The Magical Properties of Beads. *Bead Journal* 2/4 (1976), pp. 32–39.

Beck 1966a: C. W. Beck, Bemerkungen zur infrarotspektroskopischen Herkunftsbestimmung von Bernstein. *Jahrbuch des Römisch-Germanisches Zentralmuseum Mainz* 13 (1966), pp. 292–295.

Beck 1966b: C. W. Beck, Analysis and Provenience of Minoan and Mycenaean Amber I. *Greek, Roman, Byzantine Studies* 7 (1966), pp. 191–211.

Beck 1970: C. W. Beck, The Origin of Amber Artifacts. In: R. R. Holloway (ed.), *Satrianum.* Providence (RI) 1970, pp. 119–120, 135.

Beck 1971: C. W. Beck, Amber from the Chalcolithic Cemetery of Laterza. *Origini* 5 (1971), pp. 301–305.

Beck 1986: C. W. Beck, Spectroscopic Investigations of Amber. *Applied Spectroscopy Review* 22 (1986), pp. 57–200.

Beck 1996: C. W. Beck, Bernstein. In: P. Della Casa (ed.), *Velika Gruda II. Die bronzezeitliche Nekropole Velika Gruda (Opš. Kotor, Montenegro). Fundgruppen der mittleren und späten Bronzezeit zwischen Adria und Donau.* Universitätsforschungen zur Prähistorischen Archäologie 33. Bonn 1996, pp. 135–136.

Beck 1999: C. W. Beck, The Chemistry of Amber. *Estudios Del Museo De Ciencias Naturales De Alava* 14, Num. Espec. 2 (1999), pp. 33–48.

Beck/Dušek 1969: C. W. Beck, M. Dušek, Die Herkunft des Bernsteins vom trakischen Gräberfeld von Chotin. *Slovenská Archeologia* 18 (1969), pp. 247–258.

Beck/Liu 1973: C. W. Beck, T. Liu, Provenience Analysis of Yugoslavian Amber Artifacts. *Zbornik radova Narodnog muzeja u Beogradu* 7 (1973), pp. 133–142.

Beck/Liu 1974: C. W. Beck, T. Liu, The Origin of Archaeological Amber Artifacts from Yugoslavia. *Bulletin de l'Academie Serbe des Sciences et des Arts, Classe des Sciences Mathematiques et Naturelles* N.S. 13 (1974), pp. 115–118.

Beck/Liu 1976: C. W. Beck, T. Liu, Origine de l'ambre des grottes Hasard et Prevel. *Gallia Préhistoire* 19 (1976), pp. 201–207.

Beck/Sprincz 1981: C. W. Beck, E. Sprincz, A Szegedi Mora Ferenc Muzeum Bronzkori Borostyanko Gyongyeinek Eredete. *Archaeolgia Ertesitö* 108 (1981), pp. 206–210.

Beck/Shennan 1991: C. W. Beck, S. Shennan, *Amber in Prehistoric Britain.* Oxford 1991.

Beck/Hartnett 1993: C. W. Beck, H. E. Hartnett, Sicilian Amber. In: C. W. Beck, J. Bouzek (eds.), *Amber in Archaeology. Proceedings of the IInd International Conference on Amber in Archaeology, Liblice 1990.* Praha 1993, pp. 36–47.

Beck/Beck 1995: C. W. Beck, L. Y. Beck, Analysis and Provenience of Minoan and Mycenaean Amber, V. Pylos and Messenia. *Greek, Roman, Byzantine Studies* 36/2 (1995), pp. 119–135.

Beck/Marková 1998: C. W. Beck, K. Marková, Finds of Amber in the Carpathian Basin in the Bronze Age. In: N. Negroni Catacchio, C. W. Beck (eds.), *Workshop 7. L'ambra in archaeologia. Atti del XIII Congress, Volume 6, Tome I, Forli-Italia, 1996, 8–14 settembre.* Forlì 1998, pp. 409–413.

Beck et al. 1964: C. W. Beck, E. Wilbur, S. Meret, Infrared Spectra and the Origin of Amber. *Nature* 201 (1964), pp. 256–257.

Beck et al. 1965: C. W. Beck, E. Wilbur, S. Meret, D. Kossove, K. Kermani, The Infrared Spectra of Amber and the Identification of Baltic Amber. *Archaeometry* 8 (1965), pp. 96–109.

Beck et al. 1968: C. W. Beck, G. C. Southard, A. B. Adams, Analysis and Provenience of Minoan and Mycenaean Amber, II. Tiryns. *Greek, Roman, Byzantine Studies* 9 (1968), pp. 5–19.

Beck et al. 1970: C. W. Beck, C. A. Fellows, A. B. Adams, Analysis and Provenience of Minoan and Mycenaean Amber, III. Kakovatos. *Greek, Roman, Byzantine Studies* 11 (1970), pp. 5–22.

Beck et al. 1971: C. W. Beck, A. B. Adams, G. C. Southard, C. A. Fellows, Determination of the Origin of Greek Amber Artefacts by Computer-Classification of Infrared Spectra. In: R. H. Brill (ed.), *Science and Archaeology.* Cambridge 1971, pp. 235–240.

Beck et al. 1972: C. W. Beck, G. C. Southard, A. B. Adams, Analysis and Provenience of Minoan and Mycenaean Amber, IV. Mycenae. *Greek, Roman, Byzantine Studies* 13 (1972), pp. 359–385.

Beck et al. 2003: C. W. Beck, E. C. Stout, K. M. Wovkulich, The Chemistry of Sicilian Amber (Simetite). In: C. W. Beck, I. B. Loze, J. M. Todd (eds.), *Amber in Archaeology. Proceedings of the Fourth International Conference on Amber in Archaeology, Talsi 2001.* Riga 2003, pp. 17–33.

Beck et al. 2009: C. W. Beck, E. C. Stout, S. H. Buck, Provenience Analysis of Romanian Amber Artifacts by Infrared Spectroscopy. In: A. Palavestra, C. W. Beck, J. M. Todd (eds.), *Amber in Archaeology: Proceedings of the Fifth International Conference on Amber in Archaeology.* Beograd 2009, pp. 20–29.

Becker et al. 1989: B. Becker, K. D. Jäger, D. Kaufmann, T. Litt, Dendrochronologische Datierungen von Eichenhölzern aus den frühbronzezeitlichen Hügelgräbern bei Helmsdorf und Leubingen (Aunjetitzer Kultur) und an bronzezeitlichen Flußeichen bei Merseburg. *Jahresschrift für mitteldeutsche Vorgeschichte* 72 (1989), pp. 299–312.

Befu 1977: H. Befu, Social Exchange. *Annual Review of Anthropology* 6 (1977), pp. 255–281.

Bejko 1994: L. Bejko, Some Problems of the Middle and Late Bronze Age in Southern Albania. *Institute of Archaeology Bulletin* 31 (1994), pp. 105–126.

Bejko 2007: L. Bejko, Group Identity and Cultural Exchange in the Late Bronze and Early Iron Age Albania. In: I. Galanaki, H. Tomas, Y. Galanakis und R. Laffineur (eds.), *Between the Aegean and Baltic Seas. Prehistory across Borders.* Aegaeum 27. Liège 2007, pp. 203–210.

Belardelli 2004: C. Belardelli, *Coppa Nevigata. Materiali da scavi e rinvenimenti 1903–1909.* Grandi Contesti e Problemi della Protostoria Italiana 8. Firenze 2004.

Belardelli et al. 2007: C. Belardelli, M. Angle, F. di Gennaro, F. Trucco, *Repertorio dei siti protostorici del Lazio. Province di Roma, Viterbo e Frosinone.* All'Insegna del Giglio. Roma 2007.

Belemmi et al. 1997: L. Belemmi, L. Salzani, G. Squaranti, *Povegliano: l'abitato dell'Età del Bronzo della Muraiola.* Povegliano Veronese 1997.

Bellintani 1973: G. F. Bellintani, I manufatti enei e la tecnologia del bronzo nella stazione pre-protostorica di Frattesina di Fratta Polesine. *Padusa* IX (1973), pp. 95–118 (reprint. in *Padusa* XX [1984], pp. 105–128).

Bellintani 1997: P. Bellintani, Frattesina: l'ambra e la produzione vitrea nel contesto delle relazioni transalpine. In: L. Endrizzi, F. Marzatico (eds.), *Ori delle Alpi.* Trento 1997, pp. 117–129.

Bellintani 2000: P. Bellintani, Il medio Polesine tra la tarda Età del Bronzo e l'inizio dell'Età del Ferro. In: M. Harari, M. Pearce (eds.), *Il Protovillanoviano al di qua e al di là dell'Appennino. Atti della Giornata di studio (Pavia, 17 giugno 1995).* Como 2000, pp. 47–84.

Bellintani 2002: P. Bellintani, Vie dell'ambra, vie del vetro. Testimonianze dalla Valle dell'Adige nel quadro delle relazioni tra mondo mediterraneo ed Europa transalpina nell'età del Bronzo. In: G. Schnekenburger (ed.), *Attraverso le Alpi. Uomini, vie, scambi nell'antichità.* Exhibition catalogue. Stoccarda 2002, pp. 39–48.

Bellintani 2010: P. Bellintani, Ambra. Una materia prima dal nord (ma non solo). In: A. Cazzella, G. Recchia (eds.), *Ambra per Agamennone. Indigeni e Micenei tra Adriatico, Ionio ed Egeo.* Bari 2010, pp. 139–144.

Bellintani 2012: P. Bellintani, Le perle in materiale vetroso dall'antica età del Bronzo all'inizio dell'età del Ferro in Italia. Indicatori di scambio su lunga distanza e prime testimonianze di produzione locale. In: S. Ciappi, A. Larese, M. Uboldi (eds.), *Atti delle XVI Giornate Nazionali di Studio sul Vetro, Adria (RO), Museo Archeologico Nazionale, 12–13 maggio 2012.* Roma 2012, pp. 15–24.

Bellintani 2014: P. Bellintani, Baltic Amber, Alpine Copper and Glass Beads from the Po Plain. Amber Trade at the Time of Campestrin and Frattesina. *Padusa* L (2014), pp. 111–139.

Bellintani 2017: P. Bellintani, Il ruolo delle Alpi nella circolazione dell'ambra baltica nel Mediterraneo centrale nel corso dell'età del bronzo. In: P. L. Cellarosi, R. Chellini, F. Martini, A. C. Montanaro, L. Sarti, R. M. Capozzi (eds.), *The Amber Roads. The Ancient Cultural and Commercial Communication between Peoples.* Firenze 2017, pp. 273–312.

Bellintani/Peretto 1968: G. F. Bellintani, R. Peretto, Le stazione preistorica di Frattesina in Fratta Polesine - Rovigo. Notizie preliminari sul materiale raccolto in superficie. *Padusa* IV (1968), pp. 5–20 (reprint. in *Padusa* XX [1984], pp. 13–28).

Bellintani/Peretto 1972: G. F. Bellintani, R. Peretto, Il ripostiglio di Frattesina ed altri manufatti enei raccolti in superficie. Notizie preliminari. *Padusa* VIII (1972), pp. 32–49 (reprint. in *Padusa* XX [1984], pp. 55–72).

Bellintani/Biavati 1997: P. Bellintani, A. Biavati, Ornamenti in materiale vetroso. In: M. Bernabò Brea, A. Cardarelli, M. Cremaschi (eds.), *Le Terramare. La più antica civiltà padana.* Exhibition catalogue. Modena 1997, pp. 610–613.

Bellintani/Residori 2003: P. Bellintani, G. Residori, Quali e quante conterie: perle ed altri materiali vetrosi dell'Italia settentrionale nel quadro dell'età del Bronzo europea. In: I.I.P.P. (eds.), *Studi e ricerche sul neolitico e le età dei metalli. Atti della XXXV riunione scientifica dell'Istituto Italiano di Preistoria e Protostoria (Lipari, 2–4 giugno 2000).* Firenze 2003, pp. 483–498.

Bellintani/Stefan 2009: P. Bellintani, L. Stefan, Nuovi dati sul primo vetro europeo: il caso di Frattesina. In: Associazione Tecnici Italiani del Vetro (eds.), *Atti del Primo Convegno Interdisciplinare sul Vetro nei Beni Culturali e nell'Arte di Ieri e di Oggi, Parma, 27–28 novembre 2008.* Parma 2009, pp. 71–86.

Bellintani/Gonzato 2017: P. Bellintani, F. Gonzato, Luxury Production. Amber and Glass during the Recent and Final Bronze Age in North-eastern Italy. In: M. Fotiadis, R. Laffineur, Y. Lolos, A. Vlachopoulos (eds.), *Hesperos. The Aegean Seen from the West.* Aegaeum 41. Leuven – Liège 2017, pp. 173–184.

Bellintani et al. 2004: P. Bellintani, M. Bernabò Brea, G. Ranieri, Oggetti ornamentali in materiali diversi. In: M. Bernabò Brea, M. Cremaschi (eds.), *Il villaggio piccolo della Terramara di Santa Rosa di Poviglio. Scavi 1987–1992.* Origines. Firenze 2004, pp. 599–607.

Bellintani et al. 2006: P. Bellintani, I. Angelini, G. Artioli, A. Polla, Origini dei materiali vetrosi italiani: esotismi e localismi, in Materie prime e scambi nella preistoria italiana. In: I.I.P.P. (eds.), *Materie prime e scambi nella preistoria italiana. Atti della*

XXXIX Riunione Scientifica dell'Istituto Italiano di Preistoria e Protostoria (Firenze, 25–27 novembre 2004), vol. 3. Firenze 2006, pp. 1495–1533.

Bellintani et al. 2007: P. Bellintani, I. Angelini, G. Artioli, A. Polla, Villaggio delle Macine: le piu antiche perle in vetro e ambra dell'Italia centrale. In: I.I.P.P. (eds.), *Strategie di insediamento fra Lazio e Campania in età preistorica e protostorica. Atti della XL Riunione Scientifica dell'Istituto Italiano di Preistoria e Protostoria (Roma, Napoli, Pompei, 30 novembre–3 dicembre 2005).* Firenze 2007, pp. 891–895.

Bellintani et al. 2012: P. Bellintani, A. Usai, M. A. Fadda, Ambre protostoriche della Sardegna: nuovi dati su tipologia e possibili indicatori di lavorazione locale. In: I.I.P.P. (eds.), *La preistoria e la protostoria della Sardegna. Atti della XLIV Riunione Scientifica dell'Istituto Italiano di Preistoria e Protostoria (Cagliari, Barumini, Sassari 23–28 novembre 2009).* Firenze 2012, pp. 1163–1171.

Bellintani et al. 2015a: P. Bellintani, L. Salzani, G. De Zuccato, M. Leis, C. Vaccaro, I. Angelini, Ch. Soffritti, M. Bertolini, U. Thun Hohenstein, L'ambra dell'insediamento della tarda Età del bronzo di Campestrin di Grignano Polesine (Rovigo). In: G. Leonardi, V. Tiné (eds.), *Preistoria e Protostoria del Veneto.* Studi di Preistoria e Protostoria 2. Firenze 2015, pp. 419–426.

Bellintani et al. 2015b: P. Bellintani, M. Bassetti, I. Bettinardi, N. Degasperi, M. Magny, N. Martinelli, O. Pignatelli, M. Rottoli, Nuove ricerche per la tutela del sito palafitticolo di Molina di Ledro (TN). *Archeologia Delle Alpi* 2015, pp. 25–34.

Bellintani et al. 2021: P. Bellintani, M. Cwaliński, I. Angelini, U. Thun Hohenstein, Le ambre di Campestrin e l'origine della decorazione tipo Tirinto. *Padusa* LVII (2021), pp. 109–156.

Benac 1983: A. Benac (ed.), *Praistorija Jugoslavenskih Zemalja. IV knjiga: Bronzano Doba.* Sarajevo 1983.

Benac/Čović 1956: A. Benac, B. Čović, *Glasinac I – Bronzano doba / Glasinac 1 – Bronzezeit. Katalog prehistorijske zbirke Zemaljskog muzeja u Sarajevu, sveska 1.* Sarajevo 1956.

Benac/Čović 1957: A. Benac, B. Čović, *Glasinac II – Željezno doba / Glasinac II – Eisenzeit. Katalog prehistorijske zbirke Zemaljskog muzeja u Sarajevu, sveska 2.* Sarajevo 1957.

Benedetti/Cardosa 2006: L. Benedetti, M. Cardosa, Manufatti d'ambra nella Calabria protostorica. In: I.I.P.P. (eds.), *Materie prime e scambi nella preistoria italiana. Atti della XXXIX Riunione Scientifica dell'Istituto Italiano di Preistoria e Protostoria (Firenze, 25–27 novembre 2004)*, vol. 3. Firenze 2006, pp. 1573–1587.

Berezanskaja/Kločko 1998: S. S. Berezenska, V. I. Kločko, *Das Gräberfeld von Hordeevka.* Archäologie in Eurasien 5. Rahden/Westf 1998.

Berg 1999: I. Berg, The Southern Aegean System. *Journal of World-System Research* 5 (1999), pp. 475–484.

Bergerbrant 2007: S. Bergerbrant, *Bronze Age Identities: Costume, Conflict and Contact in Northern Europe 1600–1300 BC.* Stockholm Studies in Archaelogy 43. Stockholm 2007.

Bergonzi 1997: G. Bergonzi, L'ambra delle terramare nel contesto italiano e europeo. In: M. Bernabò Brea, A. Cardarelli, M. Cremaschi (eds.), *Le Terramare. La più antica civiltà padana*. Exhibition catalogue. Modena 1997, pp. 602–609.

Bergonzi/Cardarelli 1991–1992: G. Bergonzi, A. Cardarelli, Status symbol e oggetti d'ornamento nella media età del Bronzo dell'Italia settentrionale: ambra faience, pasta vitrea, metalli preziosi. In: D. Cocchi Genick (ed.), *L'età del bronzo in Italia nei secoli dal XVI al XIV a.C. Atti del congresso Viareggio, 26–30 ottobre 1989*. Rassegna di Archeologia 10. Firenze 1991–1992, pp. 217–220.

Bermond Montanari 1987: G. Bermond Montanari, Monte Battaglia. In: G. Bermond Montanari (ed.), *La formazione della città in Emilia Romagna*. Exhibition catalogue. Bologna 1987, pp. 21–26.

Bermond Montanari 1997: G. Bermond Montanari, Il ripostiglio di Monte Battaglia. In: G. Bermond Montanari, M. Massi Pasi, L. Prati (eds.), *Quando Forlì non c'era. Exhibition catalogue*. Forlì 1997, pp. 221–226.

Bermond Montanari et al. 1992: G. Bermond Montanari, M. Massi Pasi, G. Morico, Riccione, podere ex conti Spina: campagne di scavo dal 1982 al 1986. *Padusa* XXVIII (1992), pp. 105–129.

Bermond Montanari et al. 1996: G. Bermond Montanari, A. Del Lucchese, P. Frontini, F. M. Gambari, G. Kaufmann, F. Marzatico, M. Montagnari Kokelj, F. Nicolis, G. Odetti, A. Pedrotti, L. Salzani, Articulazioni culturale e cronologiche, L'Italia settentrionale. In: D. Cocchi Genick (ed.), *L'antica età del Bronzo in Italia. Atti del Congresso nazionale (Viareggio, 9–12 gennaio 1995)*. Firenze 1996, pp. 57–78.

Bernabò Brea 1946: L. Bernabò Brea, *Gli scavi nella Caverna delle Arene Candide. Parte Prima. Gli strati con ceramic*, vol. 1. Bordighera 1946.

Bernabò Brea 1958: L. Bernabò Brea, *La Sicilia prima dei Greci*. Milano 1958.

Bernabò Brea 1967: L. Bernabò Brea, La necropoli di Longane. *Bullettino di Paletnologia Italiana* 76 (1967), pp. 181–254.

Bernabò Brea 1968–1969: L. Bernabò Brea, Considerazioni sull'eneolitico e sulla prima età del bronzo della Sicilia e della Magna Grecia. *Kokalos* 14–15 (1968–1969), pp. 42–49.

Bernabò Brea 1976–1977: L. Bernabò Brea, Eolie, Sicilia e Malta nell'Età del bronzo. *Kokalos* 22–23 (1976–1977), pp. 33–110.

Bernabò Brea 1979: L. Bernabò Brea, L'età del Bronzo tardo e finale nelle isole Eolie. In: I.I.P.P. (eds.), *Il Bronzo finale in Italia. Atti della XXI Riunione Scientifica dell'Istituto Italiano di Preistoria e Protostoria (Firenze, 21–23 ottobre 1973)*. Firenze 1979, pp. 571–597.

Bernabò Brea 1985: L. Bernabò Brea, *Gli Eoli e l'inizio dell'età del bronzo nelle isole Eolie e nell'Italia meridionale*. Napoli 1985.

Bernabò Brea 1991–1992: L. Bernabò Brea, La Sicilia e le Isole Eolie. In: D. Cocchi Genick (ed.), *L'età del bronzo in Italia nei secoli dal XVI al XIV a.C. Atti del congresso Viareggio, 26–30 ottobre 1989*. Rassegna di Archeologia 10. Firenze 1991–1992, pp. 105–121.

Bernabò Brea 2004: M. Bernabò Brea, Osservazioni sui materiali ceramic. In: M. Bernabò Brea, M. Cremaschi (eds.), *Il villaggio piccolo della Terramara di Santa Rosa di Poviglio. Scavi 1987–1992.* Origines. Firenze 2004, pp. 518–592.

Bernabò Brea/Cavalier 1956: L. Bernabò Brea, M. Cavalier, Civiltà preistoriche delle isole Eolie e del territorio di Milazzo. *Bullettino di Paletnologia Italiana* 65 (1956), pp. 7–98.

Bernabò Brea/Cavalier 1960: L. Bernabò Brea, M. Cavalier, *Meligunìs Lipára I. La stazione preistorica della contrada Diana e la necropoli protostorica di Lipari.* Palermo 1960.

Bernabò Brea/Cavalier 1968: L. Bernabò Brea, M. Cavalier, *Meligunìs Lipára III. Stazioni preistoriche delle isole Panarea, Salina e Stromboli.* Palermo 1968.

Bernabò Brea/Cavalier 1980: L. Bernabò Brea, M. Cavalier, *Meligunìs Lipára IV. L'Acropoli di Lipari nella preistoria.* Palermo 1980.

Bernabò Brea/Cavalier 1991: L. Bernabò Brea, M. Cavalier, *Meligunìs Lipára VI. Filicudi. Insediamenti dell'età del bronzo.* Palermo 1991.

Bernabò Brea/Cavalier 1994: L. Bernabò Brea, M. Cavalier, *Meligunìs Lipára VII. Lipari. Contrada Diana. Scavo XXXVI in proprietà Zagami (1975–1984).* Palermo 1994.

Bernabò Brea/Cavalier 1995: L. Bernabò Brea, M. Cavalier, *Meligunìs Lipára, VIII, 1. Salina – Ricerche archeologiche 1989–1993.* Palermo 1995.

Bernabò Brea/Cardarelli 1997: M. Bernabò Brea, A. Cardarelli, Le terramare nel tempo. In: M. Bernabò Brea, A. Cardarelli, M. Cremaschi (eds.), *Le Terramare. La più antica civiltà padana.* Exhibition catalogue. Modena 1997, pp. 295–302.

Bernabò Brea/Cremaschi 1997a: M. Bernabò Brea, M. Cremaschi, Le terramare: 'palafitte a secco' o 'villaggi arginati'? In: M. Bernabò Brea, A. Cardarelli, M. Cremaschi (eds.), *Le Terramare. La più antica civiltà padana.* Exhibition catalogue. Modena 1997, pp. 187–195.

Bernabò Brea/Cremaschi 1997b: M. Bernabò Brea, M. Cremaschi, La Terramara di Santa Rosa di Poviglio: le strutture. In: M. Bernabò Brea, A. Cardarelli, M. Cremaschi (eds.), *Le Terramare. La più antica civiltà padana.* Exhibition catalogue. Modena 1997, pp. 196–212.

Bernabò Brea/Cremaschi 2004: M. Bernabò Brea, M. Cremaschi (eds.), *Il villaggio piccolo della terramara di Santa Rosa di Poviglio. Scavi 1987–1992.* Origines. Firenze 2004.

Bernabò Brea et al. 1969: L. Bernabò Brea, E. Militello, S. La Piana, Mineo (Catania). La necropoli detta del Molino della Badia: nuove tombe in contrada Madonna del Piano. *Notizie degli Scavi di Antichità* XXIII (1969), pp. 210–276.

Bernabò Brea et al. 1997a: M. Bernabò Brea, A. Cardarelli, M. Cremaschi, Il crollo del sistema terramaricolo. In: M. Bernabò Brea, A. Cardarelli, M. Cremaschi (eds.), *Le Terramare. La più antica civiltà padana.* Exhibition catalogue. Modena 1997, pp. 745–756.

Bernabò Brea et al. 1997b: M. Bernabò Brea, C. Fornari, N. Provenzano, Lo strato basale del Villaggio piccolo di S. Rosa a Fodico di Poviglio (RE). In: M. Bernabò Brea, A. Cardarelli, M. Cremaschi (eds.), *Le Terramare. La più antica civiltà padana.* Exhibition catalogue. Modena 1997, pp. 340–342.

Bernabò Brea et al. 1997c: M. Bernabò Brea, G. Botazzi, A. Monti, N. Provenzano, Lo strato 'a cumuli di cenere' del Villaggio Grande di S. Rosa a Fodico di Poviglio (RE).

In: M. Bernabò Brea, A. Cardarelli, M. Cremaschi (eds.), *Le Terramare. La più antica civiltà padana.* Exhibition catalogue. Modena 1997, pp. 345–347.

Bernabò Brea et al. 1997d: M. Bernabò Brea, G. Botazzi, A. Monti, N. Provenzano, Lo strato sommitale del Villaggio Grande di S. Rosa a Fodico di Poviglio (RE). In: M. Bernabò Brea, A. Cardarelli, M. Cremaschi (eds.), *Le Terramare. La più antica civiltà padana.* Exhibition catalogue. Modena 1997, pp. 348–350.

Bernabò Brea et al. 1997e: M. Bernabò Brea, A. Cardarelli, M. Cremaschi, Terramare. Cinque secoli di vita nella grande pianura. In: M. Bernabò Brea, A. Cardarelli, M. Cremaschi (eds.), *Le Terramare. La più antica civiltà padana.* Exhibition catalogue. Modena 1997, pp. 23–29.

Bernabò Brea et al. 2004: M. Bernabò Brea, C. Wolf, P. Bianchi, R. Bresciani, L. Cenci, M. Fornaciari, C. Fornari, C. Homan, S. Lincetto, T. Medici, N. Tramontano, C. Zanini, The Pottery. In: M. Bernabò Brea, M. Cremaschi (eds.), *Il villaggio piccolo della Terramara di Santa Rosa di Poviglio. Scavi 1987–1992.* Origines. Firenze 2004, pp. 820–821.

Bernabò Brea et al. 2005: M. Bernabò Brea, M. Miari, P. Bianchi, L. Bronzoni, P. Ferrari, F. Guarisco, E. Lari, S. Lincetto, S. Maggioni, S. Occhi, B. Sassi, Le prime fasi del sito dell'età del Bronzo di Forno del Gallo a Beneceto (Parma). *Padusa* XL (2005), pp. 81–115.

Bernabò Brea et al. 2008: M. Bernabò Brea, M. Miari, P. Bianchi, L. Bronzoni, P. Ferrari, F. Guarisco, E. Lari, S. Lincetto, S. Maggioni, S. Occhi, B. Sassi, La terramara di Forno del Gallo a Beneceto. In: *Archeologia ad Alta Velocità in Emilia Romagna. Indagini archeologiche e geologiche lungo il tracciato dell'Alta Velocità. Atti del Convegno di Parma, 9 giugno 2003.* Parma 2008, pp. 87–112.

Bernabò Brea et al. 2018: M. Bernabò Brea, A. Cardarelli, M. Cremaschi, L'Emilia tra antica e recente età del Bronzo. In: M. Bernabò Brea (ed.), *Preistoria e Protostoria dell'Emilia Romagna.* Studi di Preistoria e Protostoria 3, vol. II. Firenze 2018, pp. 9–32.

Bernardini 1977: E. Bernardini, *La preistoria in Liguria.* Genova 1977.

Bertemes/Heyd 2015: F. Bertemes, V. Heyd, 2200 BC – Innovation or Evolution? The Genesis of the Danubian Early Bronze Age. In: H. Meller, H. W. Arz, R. Jung, R. Risch (eds.), *2200 BC – Ein Klimasturz als Ursachefür den Zerfall der Alten Welt? 7. Mitteldeutscher Archäologentagvom 23. bis 26. Oktober 2014 in Halle (Saale).* Tagungen des Landesmuseums für Vorgeschichte Halle 12/II. Halle (Saale) 2015, pp. 561–578.

Bettelli 2002: M. Bettelli, *Italia meridionale e mondo miceneo. Ricerche su dinamiche di acculturazione e aspetti archeologici, con particolare riferimento ai versanti adriatico e ionico della penisola italiana.* Grandi Contesti e Problemi della Protostoria Italiana 5. Firenze 2002.

Bettelli 2006: M. Bettelli, Fogge simili ma non identiche: alcune considerazioni sulle spade tipo Thapsos-Pertosa. In: A. Cardarelli, M. Pacciarelli, A. Vanzetti (eds.), *Studi di protostoria in onore di Renato Peroni.* Firenze 2006, pp. 240–245.

Bettelli 2011: M. Bettelli, Interaction and Acculturation. The Aegean and the Central Mediterranean in the Late Bronze Age. In: H. Matthäus, N. Oettinger, S. Schröder (eds.), *Die Ursprünge Europas und der Orient – Kulturelle Beziehungen von der Späten*

Bronzezeit bis zur Frühen Eisenzeit. Erlangen, 17–18 Februar 2006. Wiesbaden 2011, pp. 109–126.

Bettelli 2012: M. Bettelli, Variazioni sul sole: immagini e immaginari nell'Europa protostorica. *Studi Micenei ed Egeo-Anatolici* 54 (2012), pp. 185–205.

Bettelli 2015: M. Bettelli, Centuries of Darkness? The Aegean and the Central Mediterranean after the Collapse of the Mycenaean Palaces. In: A. Babbi, F. Bubenheimer-Erhart, B. Marín-Aguilera, S. Mühl (eds.), *The Mediterranean Mirror. Cultural Contacts in the Mediterranean Sea between 1200 and 750 B.C.* Römisch-Germanisches Zentralmuseum – Tagungen 20. Mainz 2015, pp. 207–230.

Bettelli/Damiani 2005: M. Bettelli, I. Damiani, I pettini di materia dura animale nell'età del bronzo italiana: alcune considerazioni. In: L. Vagnetti, M. Bettelli, I. Damiani (eds.), *L'avorio in Italia nell'età del bronzo.* Roma 2005, pp. 17–26.

Bettelli et al. 2004: M. Bettelli, A. Cardarelli, F. di Gennaro, S. T. Levi, D. Marino, M. Pacciarelli, R. Peroni, L. Vagnetti, L'età del bronzo media e tarda in Calabria. In: I.I.P.P. (eds.), *Preistoria e protostoria della Calabria. Atti della XXXVII Riunione Scientifica dell'Istituto Italiano di Preistoria e Protostoria (Scalea, Papasidero, Praia a Mare, Tortora, 29 settembre–4 ottobre 2002).* Firenze 2004, pp. 325–347.

Bettelli et al. 2012: M. Bettelli, E. La Pilusa, S. Levi, A. E. Salerno, A. Savatteri, A. Zanini, L'Ausonio II nel quadro delle relazioni tirreniche e mediterranee alla fine del II millennio a.C. In: I.I.P.P. (eds.), *La preistoria e la protostoria della Sardegna. Atti della* XLIV Riunione Scientifica *dell'Istituto Italiano di Preistoria e Protostoria (Cagliari, Barumini, Sassari 23–28 novembre 2009),* vol. 2. Firenze 2012, pp. 1095–1102.

Bettelli et al. 2017: M. Bettelli, M. Cupitò, R. Jones, G. Leonardi, S. T. Levi, The Po Plain, Adriatic and Eastern Mediterranean in the Late Bronze Age: Fact, Fancy and Plausibility. In: M. Fotiadis, R. Laffineur, Y. Lolos, A. Vlachopoulos (eds.), *Hesperos. The Aegean Seen from the West.* Aegaeum 41. Leuven – Liège 2017, pp. 165–172.

Bianchi 2007: P. Bianchi, La terramara di Forno del Gallo a Beneceto (scheda sito e schede materiali-ambra). In: M. L. Nava, T. Salerno (eds.), *Trasparente dall'Antico. Ambre. Exhibition catalogue (Napoli Museo Archeologico Nazionale: 26 marzo–10 settembre 2007).* Milano 2007, 78–79.

Bianchin Citton/Vitri 1991–1992: E. Bianchin Citton, S. Vitri, III. Italia Nord-Orientale. In: D. Cocchi Genick (ed.), *L'età del bronzo in Italia nei secoli dal XVI al XIV a.C. Atti del congresso Viareggio, 26–30 ottobre 1989.* Rassegna di Archeologia 10. Firenze 1991–1992, pp. 56–68.

Bianco Peroni 1970: V. Bianco Peroni, *Le Spade nell'Italia Continentale.* Prähistorische Bronzefunde IV/1. München 1970.

Bianco Peroni 1976: V. Bianco Peroni, *I coltelli nell'Italia continentale.* Prähistorische Bronzefunde VII/2. München 1976.

Bianco Peroni 1979: V. Bianco Peroni, *I rasoi nell'Italia continentale.* Prähistorische Bronzefunde VIII/2. München 1979.

Bianco Peroni 1994: V. Bianco Peroni, *I pugnali nell'Italia Continentale.* Prähistorische Bronzefunde VI/10. Stuttgart 1994.

Bianco Peroni et al. 2010: V. Bianco Peroni, R. Peroni, A. Vanzetti, *La necropoli del Bronzo Finale di Pianello di Genga*. Grandi Contesti e Problemi della Protostoria Italiana 13. Roma 2010.

Biancofiore 1967: F. Biancofiore, La necropoli eneolitica di Laterza. *Origini* 1 (1967), pp. 195–299.

Biehl/Rassamakin 2008: P. F. Biehl, Y. Y. Rassamakin (eds.), *Import and Imitation in Archaeology*. Schriften des Zentrums für Archäologie und Kulturgeschichte des Schwarzmeerraumes 11. Langenweißbach 2008.

Bietak 2003: M. Bietak, Science Versus Archaeology: Problems and Consequences of High Aegean Chronology. In: M. Bietak (ed.), *The Synchronisation of Civilisations in the Eastern Mediterranean in the Second Millennium B.C. II. Proceedings of the SCIEM 2000 – Euro Conference Haindorf, 2nd of May–7th of May 2001*. Contributions to the chronology of the Eastern Mediterranean 4. Vienna 2003, pp. 23–33.

Bietak/Höflmayer 2007: M. Bietak, F. Höflmayer, Introduction: High and Low Chronology. In: M. Bietak, E. Czerny (eds.), *The Synchronisation of Civilizations in the Eastern Mediterranean in the Second Millennium B.C. III*. Vienna 2007, pp. 13–23.

Bietti Sestieri 1973: A. M. Bietti Sestieri, The Metal Industry of Continental Italy, 13th to 11th Century BC, and its Connections with the Aegean. *Proceedings of the Prehistoric Society* 39 (1973), pp. 383–424.

Bietti Sestieri 1979: A. M. Bietti Sestieri, I processi storici nella Sicilia orientale fra la tarda età del Bronzo e gli inizi dell'età del Ferro sulla base dei dati archeologici. In: I.I.P.P. (eds.), *Il Bronzo finale in Italia. Atti della XXI Riunione Scientifica dell'Istituto Italiano di Preistoria e Protostoria (Firenze, 21–23 ottobre 1973)*. Firenze 1979, pp. 599–629.

Bietti Sestieri 1980: A. M. Bietti Sestieri, L'abitato di Frattesina. In: M. Pallottino (ed.), *Este e la civiltà paleoveneta a cento anni dalle prime scoperte. Atti XI convegno dell'istituto di Studi Etruschi ed Italici, Este-Padova, 27 giugno – 1 luglio 1976*. Firenze 1980 (reprint. in *Padusa* XX [1984], pp. 413–427).

Bietti Sestieri 1988: A. M. Bietti Sestieri, The Mycenaean Connection and Its Impact on the Central Mediterranean Societies. *Dialoghi di Archeologia* 6/1 (1988), pp. 23–51.

Bietti Sestieri 1992: A. M. Bietti Sestieri, *La necropoli laziale dell'Osteria dell'Osa*. Roma 1992.

Bietti Sestieri 1997: A. M. Bietti Sestieri, Dopo le terramare. In: M. Bernabò Brea, A. Cardarelli, M. Cremaschi (eds.), *Le Terramare. La più antica civiltà padana*. Exhibition catalogue. Modena 1997, pp. 57–770.

Bietti Sestieri 2008: A. M. Bietti Sestieri, L'eta del Bronzo finale nella penisola italiana. *Padusa* XLIV (2008), pp. 7–54.

Bietti Sestieri 2010: A. M. Bietti Sestieri, *L'Italia nell'Età del Bronzo e del Ferro. Dalle palafitte a Romolo*. Roma 2010.

Bietti Sestieri 2013a: A. M. Bietti Sestieri, Peninsular Italy. In: H. Fokkens, A. Harding (eds.), *The Oxford Handbook of the European Bronze Age*. Oxford 2013, pp. 632–652.

Bietti Sestieri 2013b: A. M. Bietti Sestieri, The Bronze Age in Sicily. In: H. Fokkens, A. Harding (eds.), *The Oxford Handbook of the European Bronze Age*. Oxford 2013, pp. 653–667.

Bietti Sestieri/Lo Schiavo 1976: A. M. Bietti Sestieri, F. Lo Schiavo, Alcuni problemi relativi al rapporti fra l'Italia e la Penisola Balcanica nella tarda età del bronzo inizi dell'etá del ferro. *Iliria* IV (1976), pp. 163–189.

Bietti Sestieri/De Grossi Mazzorin 1995: A. M. Bietti Sestieri, J. De Grossi Mazzorin, Importazione di materie prime organiche di origine esoticanell'abitato protostorico di Frattesina (RO). In: C.P.S.S.A.E. (eds.), *Atti del I Convegno Nazionale di Archeozoologia (Rovigo, 5–7 marzo 1993).* Padusa Quaderni 1. Rovigo 1995, pp. 367–370.

Bietti Sestieri/De Grossi Mazzorin 2005: A. M. Bietti Sestieri, J. De Grossi Mazzorin, L'avorio dell'abitato protostorico di Frattesina (RO). In: L. Vagnetti, M. Bettelli, I. Damiani (eds.), *L'avorio in Italia nell'età del bronzo.* Roma 2005, pp. 93–94.

Bietti Sestieri/De Santis 2008: A. M. Bietti Sestieri, A. De Santis, Relative and Absolute Chronology of Latium Vetus from the Late Bronze Age to the Transition to the Orientalizing Period. In: D. Brandherm, M. Trachsel (eds.), *A New Dawn for the Dark Age? Shifting Paradigms in Mediterranean Iron Age Chronology.* BAR International Series 1871. Oxford 2008, pp. 119–133.

Bietti Sestieri et al. 2001: A. M. Bietti Sestieri, M. C. De Angelis, N. Negroni Catacchio, A. Zanini, La protostoria della Toscana dall'età del bronzo recente al passaggio alla prima età del ferro. In: I.I.P.P. (eds.), *Preistoria e Protostoria della Toscana. Atti della XXXIV Riunione Scientifica dell'Istituto Italiano di Preistoria e Protostoria (Firenze, 29 settembre–2 ottobre 1999).* Firenze 2001, pp. 117–166.

Bietti Sestieri et al. 2013: A. M. Bietti Sestieri, L. Salzani, C. Giardino, G. Verly, Ritual Treatment of Weapons as a Correlate of Structural Change in the Italian LBA Communities: The Bronze Hoard of Pila del Brancon (Nogara, Verona). *Rivista di Scienze Preistoriche* LXIII (2013), pp. 155–169.

Bietti Sestieri et al. 2015: A. M. Bietti Sestieri, P. Bellintani, L. Salzani, I. Angelini, B. Chiaffoni, J. De Grossi Mazzorin, C. Giardino, M. Saracino, F. Soriano, Frattesina: un centro internazionale di produzione e di scambio nell'Età del bronzo del Veneto. In: G. Leonardi, V. Tiné (eds.), *Preistoria e Protostoria del Veneto.* Studi di Preistoria e Protostoria 2. Firenze 2015, pp. 427–436.

Billamboz 2009: A. Billamboz, Jahrringuntersuchungen in der Siedlung Forschner und weiteren bronze- und eisenzeitlichen Feuchtbodensiedlungen Südwestdeutschlands. Aussagen der angewandten Dendrochronologie in der Feuchtbodenarchäologie. In: A. Billamboz, J. Köninger, H. Schlichtherle, W. Torke (eds.), *Die Früh- und Mittelbronzezeitliche Siedlung 'Forschner' im Federseemoor. Befunde und Dendrochonologie.* Siedlungsarchäologie im Alpenvorland XI. Forsch. Ber. Vor- u. Frühgesch. Baden-Württemberg 113. Stuttgart 2009, pp. 399–555.

Blake 2014: E. Blake, *Social Network and Regional Identity in Bronze Age Italy.* Cambridge 2014.

Blečić Kavur 2009: M. Blečić Kavur, The Significance of Amber in the Kvarner Region. In: A. Palavestra, C. W. Beck, J. M. Todd (eds.), *Amber in Archaeology: Proceedings of the Fifth International Conference on Amber in Archaeology.* Beograd 2009, pp. 142–155.

Blečić Kavur 2012a: M. Blečić Kavur, Plovidba duž Caput Adriae: plovidba Svijetom? *Histria Antiqua* 21 (2012), pp. 215–229.

Blečić Kavur 2012b: M. Blečić Kavur, Caput Adriae in kultura mediteranstva. Egejski elementi pozne bronaste dobe na severnem Jadranu in v njegovem zaledju. In: A. Gaspari, M. Erič (eds.), *Potopuljena Preteklost. Arheologija vodnih okolij in raziskovanje podvodne kulturne dediščine v Sloveniji. Zbornik ob 128–letnici Dežmanovih raziskav Ljubljanice na Vrhniki (1884–2012).* Didakta 2012, pp. 101–107.

Blečić Kavur 2014: M. Blečić Kavur, *Na razmeđu svjetova za prijelaza milenija: Kasno brončano doba na Kvarneru.* Katalozi i monografije XI, Arheološki muzej u Zagrebu. Zagreb 2014.

Blegen 1937: C. W. Blegen, *Prosmyna. The Helladic Settlement Preceding the Argive Heraeum.* Cambridge 1937.

Blegen et al. 1973: C. W. Blegen, M. Rawson, W. Taylour, W. P. Donovan, *The Palace of Nestor at Pylos in Western Messenia,* vol. I. Princeton 1973.

Blischke 2002: J. Blischke, *Gräberfelder als Spiegel der historischer Entwicklung während der mittleren Bronzezeit im mittleren Donaugebiet.* Universitätsforschungen zur Prähistorischen Archäologie 80. Bonn 2002.

Bloch/Parry 1989: M. Bloch, J. Parry, Introduction: Money and the Morality of Exchange. In: J. Parry, M. Bloch (eds.), *Money and the Morality of Exchange.* Cambridge 1989, pp. 1–31.

Bocchio 1988: G. Bocchio, Saggio stratigrafico presso Lucone D (Polpenazze del Garda-Brescia). *Annali del Museo Gavardo* 16, pp. 15–43.

Bodinaku 1995: N. Bodinaku, The Late Bronze Age Culture of Albania and the Relations with the Balkanic and Aegean-Adriatic Areas. In: B. Hänsel (ed.), *Handel, Tausch und Verkehr im Bronze- und Früheisenzeitlichen Südosteuropa.* Südosteuropa Schriften 17. München – Berlin 1995, pp. 259–268.

Bonfante 2009: C. Bonfante, Observations on Amber Artifacts in Italy and the 'Orientalising' Period. In: A. Palavestra, C. W. Beck, J. M. Todd (eds.), *Amber in Archaeology: Proceedings of the Fifth International Conference on Amber in Archaeology.* Beograd 2009, pp. 220–236.

Bonizzi 1876: P. Bonizzi, Intorno all'ambra del Montale. *Bullettino di Paletnologia Italiana* 2 (1876), p. 29.

Boon et al. 1993: J. J. Boon, A. Tom, J. Pureveen, Microgram Scale Pyrolysis Mass Spectrometric and Pyrolysis Gas Chromatographic Characterisation of Geological and Archaeological Amber and Resin Samples. In: C. W. Beck, J. Bouzek (eds.), *Amber in Archaeology. Proceedings of the IInd International Conference on Amber in Archaeology, Liblice 1990.* Praha 1993, pp. 9–27.

Borgatti/Everett 1999: S. P. Borgatti, M. G. Everett, Models of Core/Periphery Structures. *Social Networks* 21 (1999), pp. 375–395.

Borgna 2013: E. Borgna, Di periferia in periferia. Italia, Egeo e Mediterraneo orientale ai tempi della koiné metallurgica: una proposta di lettura diacronica. *Rivista di Scienze Preistoriche* LXIII (2013), pp. 125–153.

Borgna/Càssola Guida 2009: E. Borgna, P. Càssola Guida (eds.), *From the Aegean to the Adriatic: Social Organisations, Modes of Exchange and Interaction in the Post-palatial Times (12th–11th c. BC).* Studi e Ricerche di Protostoria Mediterranea 8. Roma 2009.

Borgna et al. 2018: E. Borgna, P. Càssola Guida, K. Mihovilić, G. Tasca, B. Teržan 2018, Il Caput Adriae tra Bronzo Antico e Bronzo Recente. In: E. Borgna, P. Càssola Guida, S. Corazza (eds.), *Preistoria e Protostoria del Caput Adriae.* Studi di Preistoria e Protostoria 5. Firenze 2018, pp. 75–96.

Borghesani/Buonopane 1980: G. Borghesani, A. Buonopane, Bande di Cavriana. In: A. Aspes (ed.), *La preistoria del Lago di Garda.* Exhibition catalogue. Museo Civico di Storia Naturale di Verona. Verona 1980, pp. 82–84.

Boroffka 2001: N. Boroffka, Bemerkungen zu Einigen Bernsteinfunden aus Rumänien. *Archäologisches Korrespondenzblatt* 31 (2001), pp. 395–409.

Boroffka 2006: N. Boroffka, Resursele minerale din România şi stadiul actual al cercetărilor privind mineritul preistoric. *Apulum* XLIII/1 (2006), pp. 71–94.

Bošcović 1961: D. Bošcović, De L'origine de l'ambre trouve dans les localites Illyro-Grecques des Balkans. *Bulletin de l'Academie Serbe des Sciences et des Arts, Section des Sciences Sociales* N.S. 8 (1961), pp. 11–12.

Bouzek 1993: J. Bouzek, Shifts of the Amber Route. In: C. W. Beck, J. Bouzek (eds.), *Amber in Archaeology: Proceedings of the Second International Conference on Amber in Archaeology. Liblice 1990.* Praha 1993, pp. 141–146.

Bouzek 1994: J. Bouzek, Late Bronze Age Greece and the Balkans: A Review of the Present Picture. *Annual of the British School at Athens* 89 (1994), pp. 217–234.

Bouzek 2007: J. Bouzek, The Amber Route, Apollo and the Hyperboreans. In: I. Galanaki, H. Tomas, Y. Galanakis, R. Laffineur (eds.), *Between the Aegean and Baltic Seas. Prehistory across Borders.* Aegaeum 27. Liège 2007, pp. 357–362.

Bökönyi/Siracusano 1987: S. Bökönyi, G. Siracusano, Reperti Faunistici dell'età del Bronzo del Sito di Coppa Nevigata: Un Commento Preliminare. In: S. M. Cassano, A. Cazzella, A. Manfredini, M. Moscoloni (eds.), *Coppa Nevigata e il suo Territorio.* Roma 1987, pp. 205–210.

BPI 1877: Notizie diverse. *Bullettino di Paletnologia Italiana* 3 (1877), pp. 199–200.

Bradley 1990: R. Bradley, *The Passage of Arms. An Archaeological Analysis of Prehistoric Hoards and Votive Deposit.* Cambridge 1990.

Bradley 2005: R. Bradley, *Ritual and Domestic Life in Prehistoric Europe.* New York 2005.

Bradley 2013: R. Bradley, Hoards and the Deposition of Metalwork. In: H. Fokkens, A. Harding, (eds.), *The Oxford handbook of the European Bronze Age.* Oxford 2013, pp. 121–39.

Breithaupt 1820: A. Breithaupt, *Kurze Charakteristik des Mineral-Systems.* Freiberg 1820.

Bronk Ramsey et al. 2004: C. Bronk Ramsey, S. W. Manning, M. Galimberti, Dating the Volcanic Eruption at Thera. *Radiocarbon* 46 (2004), pp. 325–344.

Bronk Ramsey et al. 2010: C. Bronk Ramsey, M. W. Dee, J. M. Rowland, T. F. G. Higham, S. A. Harris, F. Brock, A. Quiles, E. M. Wild, E. S. Marcus, A. J. Shortland, Radiocarbon-Based Chronology for Dynastic Egypt. *Science* 328(5985) [2010], pp. 1554–1557.

Broodbank 2013: C. Broodbank, *The Making of the Middle Sea: A History of the Mediterranean from the Beginning to the Emergence of the Classical World.* Cambridge 2013.

Brower/Kile 1988: J. C. Brower, K. M. Kile, Seriation of an Original Data Matrix as Applied to Palaeoecology. *Lethaia* 21 (1988), pp. 79–93.

Brughmans 2013: T. Brughmans, Thinking through Networks: A Review of Formal Network Methods in Archaeology. *Journal of Archaeological Method and Theory* 19 (2012), pp. 623–662.

Bruins/Plicht 2014: H. J. Bruins, J. van der Plicht, The Thera Olive Branch, Akrotiri (Thera) and Palaikastro (Crete): Comparing Radiocarbon Results of the Santorini Eruption. *Antiquity* 88 (2014), pp. 282–287.

Bruins et al. 2009: H. J. Bruins, J. van der Plicht, J. A. Macgillivray, The Minoan Santorini Eruption and Tsunami Deposits in Palaikastro (Crete): Dating by Geology, Archaeology, 14C and Egyptian Chronology. *Radiocarbon* 51 (2009), pp. 397–411.

Brunelli et al. 2013: D. Brunelli, S. T. Levi, P. Fragnoli, A. Renzulli, P. Santi, E. Paganelli, M. C. Martinelli, Bronze Age Pottery from the Aeolian Islands: Definition of Temper Compositional Reference Units by an Integrated Mineralogical and Microchemical Approach. *Applied Physics A* 113/4 (2013), pp. 855–863.

Brunner et al. 2020: M. Brunner, J. von Felten, M. Hinz, A. Hafner, Central European Early Bronze Age Chronology Revisited: A Bayesian Examination of Large-Scale Radiocarbon Dating. *PLoS ONE* 15/12 (2020), e0243719.

Brück 1999: J. Brück, Ritual and Rationality: Some Problems of Interpretation in European Archaeology. *European Journal of Archaeology* 2 (1999), pp. 313–344.

Brück 2006: J. Brück, Death, Exchange and Reproduction in the British Bronze Age. *European Journal of Archaeology* 9/1 (2006), pp. 73–101.

Brück 2016: J. Brück, Hoards, Fragmentation and Exchange in the European Bronze Age. In: S. Hansen, D. Neumann, T. Vachta (eds.), *Raum, Gabe und Erinnerung. Weihgaben und Heiligtümer in prähistorischen und antiken Gesellschaften*. Berlin 2016, pp. 75–92.

Brück/Fontijn 2013: J. Brück, D. Fontijn, The Myth of a Chief: Prestige Goods, Power and Personhood in the European Bronze Age. In: H. Fokkens, A. Harding (eds.), *The Oxford Handbook of the European Bronze Age*. Oxford 2013, pp. 197–215.

Bryce 2002: T. Bryce, *Life and Society in the Hittite World*. Oxford 2002.

Bukowski 1998: Z. Bukowski, *Pomorze w epoce brązu w świetle dalekosiężnych kontaktów wymiennych*. Gdańsk 1998.

Bukowski 2002: Z. Bukowski, *Znaleziska bursztynu w zespołach z epoki brązu i z wczesnej epoki żelaza z dorzecza Odry i Wisły*. Warszawa 2002.

Bulatović 2011: A. Bulatović, Particular Ceramic Forms in the Central Balkan and Northern Shores of the Aegean Sea in the Late Bronze Age. *Starinar* LXI (2011), pp. 121–140.

Bulatović/Filipović 2007: A. Bulatović, V. Filipović, South Morava Basin in the Transitional Period from the Bronze to the Iron Age. *Starinar* LVII (2007), pp. 57–82.

Bulatović/Filipović 2017: A. Bulatović, V. Filipović, The Morava Valley in the Late Bronze and Early Iron Age – Changes in Topography and Material Culture. In: B. S. Heeb, A. Szentmiklosi, R. Krause, M. Wemhoff (eds.), *Fortifications: The Rise and Fall of Defended Sites in Late Bronze and Early Iron Age of South-East Europe*.

International Conference in Timişoara, Romania from November 11th to 13th, 2015. Berliner Beiträge zur Vor- und Frühgeschichte 21. Berlin 2017, pp. 149–159.

Bulatović et al. 2017: A. Bulatović, V. Filipović, R. Gligorić, *Loznica. Cultural Stratigraphy of the Prehistoric Sites in Jadar, Rađevina and Azbukovica.* Beograd – Loznica 2017.

Bulatović et al. 2018: A. Bulatović, M. Gori, M. Vander Linden 2018, New Absolute Dates as a Contribution to the Study of the Late Bronze Age Chronology in the Central Balkans. *Journal of Serbian Archaeological Society* 34 (2018), pp. 121–132.

Bunguri 2006: A. Bunguri, Drini i Bardhë area in Prehistory. In: H. Islami (ed.), *Harta Arkeologjike e Kosovës I.* Prishtinë 2006, pp. 27–57.

Bunguri et al. 2006: A. Bunguri, K. Luci, T. Kastrati, Qendra prehistorike. In: H. Islami (ed.), *Harta Arkeologjike e Kosovës I.* Prishtinë 2006, pp. 57–128.

Buršić-Matijašić 1989: K. Buršić-Matijašić, Gradina Vrčin u okviru brončanog doba Istre. *Arheološki vestnik* 39–40 (1989), pp. 475–494.

Buršić-Matijašić 1997: K. Buršić-Matijašić, Ceramica del castelliere di Monte Orcino in Istria. *Histria archaeologica* 28 (1997), pp. 108–134.

Buršić-Matijašić 2003: K. Buršić-Matijašić, Škicini (Vodnjan), Brončanodobni grob pod tumulom. *Opuscula Archaeologica* 27 (2003), pp. 171–185.

Buršić-Matijašić/Žerić 2013: K. Buršić-Matijašić, H. Žerić, Pogrebni obredi i ukopi na istarskim gradinama u brončano doba. *Tabula* 11 (2013), pp. 67–92.

Butrimas 2017: A. Butrimas, Neolithic Amber Trade and Discs with Cross Decoration. In: P. L. Cellarosi, R. Chellini, F. Martini, A. C. Montanaro, L. Sarti, R. M. Capozzi (eds.), *The Amber Roads. The Ancient Cultural and Commercial Communication between Peoples.* Firenze 2017, pp. 59–74.

Butrimas et al. 2018: A. Butrimas, D. Król, D. Ostrauskienė, Amber Typology of Rzucewo and West Lithuanian Late Neolithic Settlements. In: E. Wagner-Wysiecka, J. Szwedo, E. Sontag, A. Sobecka, J. Czebreszuk, M. Cwaliński (eds.), *International Symposium: Amber. Science and Art (Gdańsk, 22–23 March 2018).* Abstract book. Gdańsk 2018, pp. 61–64.

Caesar 1917: J. Caesar, *Bello Gallico.* Translated by H. J. Edwards. Loeb Classical Library 72. Cambridge, (MA) 1917.

Caldararo et al. 2013: N. Caldararo, J. Hirschbein, P. Palmer, H. Shepard, The Analysis, Identification and Treatment of an Amber Artifact. *Archeomatica* 2 (2013), pp. 46–49.

Calderoni et al. 2012: G. Calderoni, A. Cazzella, M. Preite Martinez, Il contributo della cronologia radiocarbonio alla definizione delle fasi dell'età del Bronzo di Coppa Nevigata: risultati raggiunti e problemi ancora aperti. In: A. Cazzella, M. Moscoloni, G. Recchia (eds.), *Coppa Nevigata e l'area umida alla foce del Candelaro durante l'età del Bronzo.* Foggia 2012, pp. 457–464.

Calzoni 1954: U. Calzoni, *Le stazioni preistoriche della montagna di Cetona Belverde. I. Premessa. L'ambiente naturale. Topografia e scavi.* Istituto Nazionale di Studi Etruschi. Quaderni di «Studi Etruschi», serie I, vol. 1. Firenze 1954.

Calzoni 1962: U. Calzoni, *Le stazioni preistoriche della montagna di Cetona Belverde. II. La ceramica.* Istituto Nazionale di Studi Etruschi. Quaderni di «Studi Etruschi», serie I, vol. 2. Firenze 1962.

Canci et al. 2015: A. Canci, M. Cupitò, M. L. Pulcini, L. Salzani, G. Fornaciari, M. A. Tafuri, G. Dalla Zuanna, La necropoli della media e recente Età del bronzo di Olmo di Nogara (Verona). Risultati della ricerca osteoarcheologica, paleochimica e paleodemografica. In: G. Leonardi, V. Tiné (eds.), *Preistoria e Protostoria del Veneto.* Studi di Preistoria e Protostoria 2. Firenze 2015, pp. 327–340.

Canić-Tešanović/Gligorić 2001: J. Canić-Tešanović, T. Gligorić, *Праисторијска некропола Пауље код Лознице (Praistorijska nekropola Paulje kod Loznice).* Loznica 2001.

Capellini 1872: G. Capellini, Über das Vorkommen von Bernstein im Bolognesischen und anderen Punkten Italiens. *Zeitschrift für Ethnologie,* Verhandlungen 4 (1872), p. 198.

Capellini 1874: G. Capellini, *Congresso internazionale di antropologia e di archeologia preistoriche: VII sessione a Stoccolma nel 1874. Brevi cenni del Prof. Comm. Giovanni Capellini, Gamberini e Parmeggiani.* Bologna 1874.

Capuzzo et al. 2014: G. Capuzzo, E. Boaretto, J. A. Barceló, EUBAR: a database of 14C measurements for the European Bronze Age. A Bayesian analysis of 14c-dated archaeological contexts from northern Italy and southern France. *Radiocarbon* 56/2 (2014), pp. 851–869.

Carafa 2013: S. Carafa, *Le tavolette enigmatiche dell'etá del Bronzo: proposta di raggruppamento tipologico.* Unpublished MA thesis defended at Università Ca'Foscari, Venezia under supervision of prof. Paolo Biagi. Venezia 2013.

Carancini 1975: G. L. Carancini, *Gli spilloni nell'Italia continentale.* Prähistorische Bronzefunde XIII/2. München 1975.

Carancini 1979: G. L. Carancini, Problemi di cronologia relativa dell'età del bronzo in Italia. *Nuovi Quaderni dell'istituto di Archeologia dell'Università di Perugia* 1 (1979), pp. 39–54.

Carancini 1997: G. L. Carancini, La produzione metallurgica delle terramare nel quadro dell'Italia protostorica. In: M. Bernabò Brea, A. Cardarelli, M. Cremaschi (eds.), *Le Terramare. La più antica civiltà padana.* Exhibition catalogue. Modena 1997, pp. 469–480.

Carancini/Peroni 1997: G. L. Carancini, R. Peroni, La koiné metallurgica. In: M. Bernabò Brea, A. Cardarelli, M. Cremaschi (eds.), *Le Terramare. La più antica civiltà padana.* Exhibition catalogue. Modena 1997, pp. 595–601.

Carancini/Peroni 1999: G. L. Carancini, R. Peroni, *L'età del bronzo in Italia: per una cronologia della produzione metallurgica.* Quaderni di Protostoria 2. Perugia 1999.

Carancini et al. 1996: G. L. Carancini, A. Cardarelli, M. Pacciarelli, R. Peroni, L'Italia. In: International Union of Prehistoric and Protohistoric Sciences (eds.), *Colloquium XX: Absolute, Relative and Comparative Chronological Sequences. The Colloquia of the XIII International Congress of Prehistoric and Protohistoric Sciences 'The Bronze Age in Europe and the Mediterranean', Forlì, 8–14 September 1996.* Series Colloquia, vol. 11. Forlì 1996, pp. 77–86.

Cardarelli 1983: A. Cardarelli, Castellieri nel Carso e nell'Istria: cronologia degli insediamenti fra media età del bronzo e prima età del ferro. In: L. Ruaro Loseri (ed.), *Preistoria del Caput Adriae: atti del convegno internazionale, Trieste 19–20 novembre 1983.* Exhibition catalogue. Udine 1983, pp. 87–118.

Cardarelli 1988: A. Cardarelli, L'età del bronzo: organizzazione del territorio, forme economiche, strutture sociali. In: F. C. Panini (ed.), *Modena dalle origini all'anno Mille*. Studi di archeologia e storia I. Modena 1988, pp. 86–127.

Cardarelli 1997: A. Cardarelli, La necropoli di Casinalbo (MO). In: M. Bernabò Brea, A. Cardarelli, M. Cremaschi (eds.), *Le Terramare. La più antica civiltà padana*. Exhibition catalogue. Modena 1997, pp. 689–696.

Cardarelli 2000: A. Cardarelli, I passi Appenninici. In: M. Harari, M. Pearce (eds.), *Il Protovillanoviano al di qua e al di là dell'Appennino. Atti della Giornata di studio (Pavia, 17 giugno 1995)*. Como 2000, pp. 85–97.

Cardarelli 2009: A. Cardarelli, The Collapse of the Terramare Culture and Growth of New Economic and Social Systems during the Late Bronze Age in Italy. *Scienze dell'Antichità* 15 (2009), pp. 449–520.

Cardarelli 2014: A. Cardarelli, *La necropoli della Terramara di Casinalbo*, vol. 1–2. Grandi Contesti e Problemi della Protostoria Italiana 15. Firenze 2014.

Cardarelli 2015: A. Cardarelli, Different Forms of Social Inequality in Bronze Age Italy. In: A. Cardarelli, A. Cazzella, M. Frangipane (eds.), *The Origin of Inequality*. Origini: Prehistory and Protohistory of Ancient Civilizations XXXVIII/2015–2 (Thematic Issue). Roma 2015, pp. 151–200.

Cardarelli/Tirabassi 1997: A. Cardarelli, J. Tirabassi, Le necropoli delle terramare emiliane. In: M. Bernabò Brea, A. Cardarelli, M. Cremaschi (eds.), *Le Terramare. La più antica civiltà padana*. Exhibition catalogue. Modena 1997, pp. 677–682.

Cardarelli/Labate 2004: A. Cardarelli, D. Labate, L'area Archeologica. In: A. Cardarelli (ed.), *Terramara di Montale*. Modena 2004, pp. 25–61.

Cardarelli et al. 1997: A. Cardarelli, M. Cremaschi, M. Cattani, D. Labate, G. Steffè, Nuove ricerche nella terramara di Montale (MO). In: M. Bernabò Brea, A. Cardarelli, M. Cremaschi (eds.), *Le Terramare. La più antica civiltà padana*. Exhibition catalogue. Modena 1997, pp. 224–228.

Cardarelli et al. 2004: A. Cardarelli, M. Pacciarelli, P. Pallante, Pesi e bilance nell'età del bronzo italiana: quadro generale e nuovi dati. In: E. C. De Sena, H. Dessales (eds.), *Metodi e approcci archeologici: l'industria e il commercio nell'Italia antica*. BAR International Series 1262. Oxford 2004, pp. 80–88.

Cardarelli et al. 2013: A. Cardarelli, E. Fraulini, T. Pedrazzoli, La terramara di Gaggio di Castelfranco Emilia (MO). In: J. De Grossi Mazzorin, A. Curci, G. Giacobini (eds.), *Economia e ambiente nell'Italia Padana dell'età del bronzo: le indagini bioarcheologiche*. Beni Archeologici – Conoscenza e Tecnologie Quaderno 11. Bari 2013, pp. 245–246.

Cardarelli et al. 2014: A. Cardarelli, A. Cazzella, M. Frangipane (eds.), *The Origin of Inequality*. Origini: Prehistory and Protohistory of Ancient Civilizations XXXVIII/2015–2 (Thematic Issue). Roma 2014.

Cardarelli et al. 2015: A. Cardarelli, C. Cavazzuti, F. Quondam, L. Salvadei, L. Salzani 2015, Le necropoli delle Narde di Frattesina: proposta per una lettura delle evidenze demografiche, rituali e sociali a partire dai dati archeologici e antropologici. In: G. Leonardi, V. Tiné (eds.), *Preistoria e Protostoria del Veneto*. Studi di Preistoria e Protostoria 2. Firenze 2015, pp. 437–448.

incontro di studi 'Preistoria e Protostoria in Etruria' (Valentano (VT) – Pitigliano (GR) – Manciano (GR), 12–14 settembre 2014). Milano 2016, pp. 359–372.

Cazzella/Recchia 2018: A. Cazzella, G. Recchia, Local Networks and Aegean-Mycenaean Connectivity in the Tyrrhenian and Adriatic Seas. In: M. Bettelli, M. Del Freo, G. J. van Wijngaarden (eds.), *Mediterranea Itinera. Studies in Honour of Lucia Vagnetti.* Rome 2018, pp. 11–31.

Cazzella et al. 2007: A. Cazzella, V. Copat, M. Danesi, Il sito dell'età del Bronzo Recente di Oratino – La Rocca (Campobasso). *Rivista di Scienze Preistoriche* LVII (2007), pp. 277–310.

Cazzella et al. 2010: A. Cazzella, M. Moscoloni, G. Recchia, Coppa Nevigata. In: F. Radina, G. Recchia (eds.), *Ambra per Agamennone. Indigeni e Micenei tra Adriatico, Ionio ed Egeo.* Bari 2010, pp. 169–175.

Cazzella et al. 2011 : A. Cazzella, M. Cultraro, G. Recchia, Relazioni tra l'Italia centro-meridionale, la Sicilia e l'area egeo-balcanica durante l'Eneolitico. In: I.I.P.P. (eds.), *L'età del rame in Italia. Atti del XLIII Riunione Scientifica dell'Istituto Italiano di Preistoria e Protostoria (Bologna, 26-29 novembre 2008).* Firenze 2011, pp. 157–164.

Cazzella et al. 2012: A. Cazzella, M. Moscoloni, G. Recchia, Coppa Nevigata, campagne di scavo 2010 e 2011. In: A. Gravina (ed.), *Atti del 32° Convegno Nazionale sulla Preistoria, Protostoria e Storia della Daunia (San Severo, 12–13 novembre 2011).* San Severo 2012, pp. 155–170.

Cazzella et al. 2014: A. Cazzella, M. Moscoloni, G. Recchia, Coppa Nevigata (Manfredonia – FG): campagne di scavo 2012 e 2013. In: A. Gravina (ed.), *Atti del 34° Convegno Nazionale sulla Preistoria, Protostoria e Storia della Daunia (San Severo, 16–17 novembre 2015).* San Severo 2014, pp. 187–200.

Cazzella et al. 2016: A. Cazzella, M. Moscoloni, G. Recchia, Campagne di scavo 2014 e 2015 a Coppa Nevigata. In: A. Gravina (ed.), *Atti del 36° Convegno Nazionale sulla Preistoria, Protostoria e Storia della Daunia (San Severo, 15–16 novembre 2015).* San Severo 2016, pp. 55–66.

Cazzella et al. 2017: A. Cazzella, G. Recchia, A. M. Tunzi Sisto, La Puglia tra Bronzo Antico e Bronzo Recente. In: F. Radina (ed.), *Preistoria e Protostoria della Puglia. Atti della XLVII Riunione Scientifica dell'Istituto Italiano di Preistoria e Protostoria (Ostuni, 9–13 ottobre 2012).* Studi di Preistoria e Protostoria 4. Firenze 2017, pp. 431–442.

Čerče/Šinkovec 1995: P. Čerče, I. Šinkovec, Katalog depojev pozne bronaste dobe. Catalogue of Hoards of the Urnfield Culture. In: B. Teržan (ed.), *Depojske in posamezne kovinske najdbe bakrene in bronaste dobe na Slovenskem I.* Katalogi in monografije 29, Narodni Muzej Ljubljana. Ljubljana 1995, pp. 129–232.

Chapman 2008: J. C. Chapman, Interaction, Trade and Exchange: Approaches to Trade and Exchange in Earlier Prehistory (Late Mesolithic – Early Bronze Age). In: A. Jones (ed.), *Prehistoric Europe: Theory and Practice.* Chichester 2008, pp. 333–355.

Chase-Dunn/Anderson 2005: C. Chase-Dunn, E. N. Anderson (eds.), *The Historical Evolution of World-Systems.* Palgrave – New York 2005.

Cherubini et al. 2014: P. Cherubini, T. Humbel, H. Beeckman, H. Gärtner, D. Mannes, C. Pearson, W. Schoch, R. Tognetti, S. Lev-Yadun, The Olive-Branch Dating of the Santorini Eruption. *Antiquity* 88 (2014), pp. 267–273.

Chierci 1877: G. Chierci, La questione dell'ambra in terramare dell'età del bronzo. *Bullettino di Paletnologia Italiana* 3 (1877), pp. 28–38.

Choyke/Bar-Yosef Mayer 2017: A. M. Choyke, D. E. Bar-Yosef Mayer, Introduction: The Archaeology of Beads, Beadwork and Personal Ornaments. In: D. E. Bar-Yosef Mayer, C. Bonsall, A. M. Choyke (eds.), *Not Just for Show: The Archaeology of Beads, Beadwork & Personal Ornaments*. Oxford 2017, pp. 13–22.

Christ 2018: A. Christ, *The Baltic Amber Trade, c. 1500–1800: The Effects and Ramifications of a Global Counterflow Commodity*. Unpublished MA thesis defended at the Department of History and Classics, University of Alberta. Alberta 2018.

Chvojka 2017: O. Chvojka, Finds of Amber in Bohemia in the Middle Bronze Age and the Urnfield Period. In: P. L. Cellarosi, R. Chellini, F. Martini, A. C. Montanaro, L. Sarti, R. M. Capozzi (eds.), *The Amber Roads. The Ancient Cultural and Commercial Communication between Peoples*. Firenze 2017, pp. 106–115.

Cierny 2008: J. Cierny, *Prähistorische Kupferproduktion in den südlichen Alpen, Region Trentino Orientale*. Veröffentlichungen aus dem Deutschen Bergbau-Museum Bochum 163. Bochum 2008.

Cinquepalmi/Recchia 2010: A. Cinquepalmi, G. Recchia, Masseria Chiancudda. In: F. Radina, G. Recchia (eds.), *Ambra per Agamennone: Indigeni e Micenei tra Adriatico Ionio ed Egeo*. Bari 2010, pp. 215–220 (with catalog of finds: pp. 328–332).

Cinquepalmi et al. 2004: A. Cinquepalmi, G. Fiorentino, G. Recchia, Masseria Chiancudda (Cisternino, Prov. di Brindisi). *Rivista di Scienze Preistoriche* LIV (2004), pp. 631–632.

Cinquepalmi et al. 2006: A. Cinquepalmi, G. Recchia, I. Angelini, G. Artioli, P. Bellintani, A. Polla, Exotica da siti interni. Il caso dell'insediamento dell'età del Bronzo di Masseria Chiancudda (Brindisi). In: I.I.P.P. (eds.), *Materie prime e scambi nella preistoria italiana. Atti della XXXIX Riunione Scientifica dell'Istituto Italiano di Preistoria e Protostoria (Firenze, 25–27 novembre 2004)*, vol. 3. Firenze 2006, pp. 1614–1620.

Cipolloni Sampò 1979: M. Cipolloni Sampò, Il Bronzo Finale In Basilicata. In: I.I.P.P. (eds.), *Il Bronzo finale in Italia. Atti della XXI Riunione Scientifica dell'Istituto Italiano di Preistoria e Protostoria (Firenze, 21–23 ottobre 1973)*. Firenze 1979, pp. 489–513.

Cipolloni Sampò 1986: M. Cipolloni Sampò, La tomba tre dell'acropoli di Toppo Daguzzo (Potenza): elementi per uno studio preliminare. *Annali dell'Istituto Universitario Orientale di Napoli* VIII (1986), pp. 1–40.

Cipolloni Sampò 1989: M. Cipolloni Sampò, L'età del Bronzo nel Melfese. *Profili della Daunia Antica* IV (1989), pp. 13–38.

Cipolloni Sampò 1991–1992: M. Cipolloni Sampò, Le sepolture collettive nel sud-est italiano. In: D. Cocchi Genick (ed.), *L'età del bronzo in Italia nei secoli dal XVI al XIV a.C. Atti del congresso Viareggio, 26–30 ottobre 1989*. Rassegna di Archeologia 10. Firenze 1991–1992, pp. 281–285.

Cipolloni Sampò 1998: M. Cipolloni Sampò, Lavello: ipogeo della Speranza. In: L. Drago Troccoli (ed.), *Scavi e ricerche archeologiche dell'Università di Roma 'La Sapienza'. Exhibition catalogue*. Roma 1998, pp. 190–192.

Cipolloni Sampò 1999: M. Cipolloni Sampò, L'eneolitico e l'età del Bronzo. In: D. Adamesteanu (ed.), *Storia della Basilicata. I. L'Antichità*. Bari 1999, pp. 67–136.

Cipolloni Sampò et al. 1991–1992: M. Cipolloni Sampò, P. Attisani, G. Bertolani, G. Di Nocera, G. Recchia, E. Remotti, R. Tulli, M. Tumminia, Toppo Daguzzo (Melfi-Potenza): le strutture 4 e 5. In: D. Cocchi Genick (ed.), *L'età del bronzo in Italia nei secoli dal XVI al XIV a.C. Atti del congresso Viareggio, 26–30 ottobre 1989.* Rassegna di Archeologia 10. Firenze 1991–1992, pp. 493–501.

Clifford et al. 1997: D. J. Clifford, P. G. Hatcher, R. E. Botto, J. V. Muntean, B. Michels, K. B. Anderson, The Nature and Fate of Natural Resins in the Geosphere – VIII. NMR and Py-GC-MS Characterization of Soluble Labdanoid Polymers, Isolated from Holocene Class I Resins. *Organic Geochemistry* 27 (1997), pp. 449–464.

Cline 2005: E. H. Cline, The Multivalent Nature of Imported Objects in the Ancient Mediterranean World. In: R. Laffineur, E. Greco (eds.), *Emporia. Aegeans in the Ceantral and Eastern Mediterranean.* Aegaeum 25. Liège 2005, pp. 45–51.

Cline 2015: E. H. Cline, *1177 B.C.: The Year Civilization Collapsed.* Turning Points in Ancient History series. Princeton 2015.

Cocchi Genick 1986: D. Cocchi Genick, *Il riparo dell'ambra.* Viareggio 1986.

Cocchi Genick 1995a: D. Cocchi Genick, La facies di Grotta Nuova. In: D. Cocchi Genick, I. Damiani, I. Macchiarola, R. Peroni, R. Poggiani Keller (eds.), *Aspetti culturali della media età del bronzo nell'Italia centro-meridionale.* Firenze 1995, pp. 364–397.

Cocchi Genick 1995b: D. Cocchi Genick, Rapporti tra facies di Grotta Nuova e il Protoappenninico. In: D. Cocchi Genick, I. Damiani, I. Macchiarola, R. Peroni, R. Poggiani Keller (eds.), *Aspetti culturali della media età del bronzo nell'Italia centro-meridionale.* Firenze 1995, pp. 429–439.

Cocchi Genick 1996: D. Cocchi Genick, L'Italia Centrale. In: D. Cocchi Genick (ed.), *L'antica età del Bronzo in Italia. Atti del Congresso nazionale (Viareggio, 9–12 gennaio 1995).* Firenze 1995, pp. 79–111.

Cocchi Genick 2002: D. Cocchi Genick, *Grotta Nuova: la prima unità culturale attorno all'Etruria protostorica.* Viareggio 2002.

Cocchi Genick 2004: D. Cocchi Genick, Le ceramiche nel ruolo di indicatori cronologici e regionali. In: D. Cocchi Genick (ed.), *L'età del bronzo recente in Italia.* Firenze – Lucca 2004, pp. 22–52.

Cocchi Genick 2018: D. Cocchi Genick, Dinamica della comunicazione culturale tra l'Emilia Romagna e le regioni centrali dall'antica alla media età del Bronzo. In: M. Bernabò Brea (ed.), *Preistoria e Protostoria dell'Emilia Romagna.* Studi di Preistoria e Protostoria 3, vol. II. Firenze 2018, pp. 199–208.

Cocchi Genick/Zanini 1997: D. Cocchi Genick, A. Zanini, L'area Toscana nell'età del bronzo media e recente. In: M. Bernabò Brea, A. Cardarelli, M. Cremaschi (eds.), *Le Terramare. La più antica civiltà padana.* Exhibition catalogue. Modena 1997, pp. 445–451.

Cocchi Genick/Sarti 2001: D. Cocchi Genick, L. Sarti, Bronzo Antico e Medio. In: I.I.P.P. (eds.), *Preistoria e Protostoria della Toscana. Atti della XXXIV Riunione Scientifica dell'Istituto Italiano di Preistoria e Protostoria (Firenze, 29 settembre–2 ottobre 1999).* Firenze 2001, pp. 91–115.

Cocchi Genick et al. 1995: D. Cocchi Genick, I. Damiani, I. Macchiarola, R. Peroni, R. Poggiani Keller (eds.), *Aspetti culturali della media età del bronzo nell'Italia centro-meridionale.* Firenze 1995.

Colini 1904: G. A. Colini, La Civiltà del bronzo in Italia. II. Sicilia. *Bullettino di Paletnologia Italiana* 30/10–12 (1904), pp. 229–304.

Colonna 2006: C. Colonna, *Necropoli dell'ultima età del bronzo nell'area padana. Per una loro cronologia relative.* Lucca 2006.

Colonna et al. 2010: C. Colonna, L. Salzani, E. Tomaello, Catalogo. In: L. Salzani, C. Colonna (eds.), *La Fragilità dell'Urna. I recenti scavi a Narde Necropoli di Frattesina (XII-IX sec. a.C.).* Exhibition catalogue. Rovigo 2010, pp. 181–293.

Coltos 1981: E. Coltos, Ètude d'ichantillons archéologiques d'ambre. *Dacia* N.S. 25 (1981), p. 193.

Conati Barbaro 2009: C. Conati Barbaro, L'insediamento dell'età del bronzo del Colle della Mola (Rocca Priora). In: L. Drago Troccoli (ed.), *Il Lazio dai Colli Albani ai Monti Lepini tra preistoria ed età moderna.* Roma 2009, pp. 99–105.

Conte et al. 2019: S. Conte, I. Matarese, G. Vezzalini, M. Pacciarelli, T. Scarano, A. Vanzetti, B. Gratuze, R. Arletti, How Much Is Known about Glassy Materials in Bronze and Iron Age Italy? New Data and General Overview. *Archaeological and Anthropological Sciences* 11 (2019), pp. 1813–1841.

Copat/Danesi 2010: V. Copat, M. Danesi, Recenti campagne di scavo nel sito dell'età del Bronzo di Oratino-La Rocca. Manifestazioni funerarie e beni esotici. In: A. Gravina (ed.), *Atti del 30° convegno nazionale sulla Preistoria, Protostoria, Storia della Daunia (San Severo, 21–22 novembre 2009).* San Severo 2010, pp. 151–172.

Copat et al. 2017: V. Copat, A. Costa, P. Piccione, Castelluccio painted pottery: shared repertoires and local identity: A case study from Early Bronze Age Sicily. In: S. Bergerbrant, A. Wessman (eds.), *New Perspectives on the Bronze Age. Proceedings of the 13th Nordic Bronze Age Symposium Held in Gothenburg, 9th to 13th June 2015.* Oxford 2017, pp. 109–125.

Coppi 1871: F. Coppi, *Monografia ed Iconografia della Terracimiteriale e Terramara di Gorzano.* Monumenti di Pura Archeologia I. Modena 1871.

Cornaggia Castiglioni/Calegari 1978: O. Cornaggia Castiglioni, G. Calegari, Due 'Amber Spacer Beads' Siciliane. *Rivista di Scienze Preistoriche* XXXIII/1 (1978), pp. 265–269.

Corrain et al. 1984: C. Corrain, M. Capitanio, G. Erspamer, Gli inumati della necropoli enea di Franzine Nuove (Villabartolomea, Verona). *Archivio per l'Antropologia e la Etnologia* 114 (1984), pp. 121–148.

Čović 1981: B. Čović, Neka pitanja hronologije bronzanog doba Glasinackog područja. *Glasnik zemaljskog muzeja u Sarajevu* n.s. XXXV/XXXVI (1981), pp. 99–140.

Čović 1983a: B. Čović, Regionalne grupe ranog bronzanog doba. In: A. Benac (ed.), *Praistorija Jugoslavenskih Zemalja. IV knjiga: Bronzano Doba.* Sarajevo 1983, pp. 114–190.

Čović 1983b: B. Čović, Srednje bronzano doba u Istri. In: A. Benac (ed.), *Praistorija Jugoslavenskih Zemalja. IV knjiga: Bronzano Doba.* Sarajevo 1983, pp. 233–241.

delle Valli Grandi Veronesi nella tarda Età del bronzo. Cronologia, aspetti culturali, evoluzione delle strutture e trasformazioni paleoambientali. In: G. Leonardi, V. Tiné (eds.), *Preistoria e Protostoria del Veneto*. Studi di Preistoria e Protostoria 2. Firenze 2015, pp. 357–375.

Cupitò et al. 2015b: M. Cupitò, I. Angelini, E. Dalla Longa, Nuovi manufatti in materiale vetroso da Fondo Paviani (Verona) – Scavi Università di Padova 2007–2012. Tipocronologia e analisi archeometriche. In: G. Leonardi, V. Tiné (eds.), *Preistoria e Protostoria del Veneto*. Studi di Preistoria e Protostoria 2. Firenze 2015, pp. 855–859.

Cupitò et al. 2018: M. Cupitò, G. Tasca, M. L. Pulcini, N. Carrara, D. Lotto, E. Gozzi, C. Scaggion, A. Canci, E. Valzolgher, Il 'sepolcreto gentilizio' del castelliere di Monte Orcino/Vrčin – Scavi Battaglia-Tamaro 1925–1928. Rilettura del contesto archeologico e riesame dei resti umani. In: E. Borgna, P. Càssola Guida, S. Corazza (eds.), *Preistoria e Protostoria del Caput Adriae*. Studi di Preistoria e Protostoria 5. Firenze 2018, pp. 509–526.

Cwaliński 2017: M. Cwaliński, Trans-Adriatic Amber Route: Attempt of Reassessment of Its Theoretical Foundations. In: P. L. Cellarosi, R. Chellini, F. Martini, A. C. Montanaro, L. Sarti, R. M. Capozzi (eds.), *The Amber Roads. The Ancient Cultural and Commercial Communication between the Peoples. Proceedings of the 1st International Conference on Ancient Roads, Republic of San Marino, April 3–4, 2014*. Millenni. Studi di Archeologia preistorica 13. Firenze 2017, pp. 207–221.

Cwaliński 2020: M. Cwaliński, Bronze Age Amber in Western and Central Balkans. *Arheološki vestnik* 71 (2020), pp. 133–172.

Cwaliński forthcoming: M. Cwaliński, Amber's Acquisition in Western Balkans during the Bronze Age: State of Research and New Perspectives. In: P. Stipančić (ed.), *Proceedings of the 4th International Conference about the Ancient Roads – The Amber Roads. Dolenjski muzej Novo mesto, 20–22 April 2017*. Novo Mesto (forthcoming/in print).

Cwaliński/Czebreszuk 2020: M. Cwaliński, J. Czebreszuk, Circulation of Amber in Western Balkans during the Bronze Age and Its Significance for Tracing Trans-Adriatic Contacts. *Padusa* N.S. LVI (2020), pp. 47–69.

Cwaliński/Pravidur forthcoming: M. Cwaliński, A. Pravidur, Bronze Age Amber Finds from Cemeteries at the Glasinac Plateau. *Glasnik Zemaljskog muzeja BiH* 55 (forthcoming/in print).

Cwaliński et al. 2018: M. Cwaliński, S. Kaur, E. Stout, L. Bozzone, Z. Sweet, *Study of Bronze Age Ambers from the Adriatic Sea Region*. ARL report #199, Vassar College. Poughkeepsie 2018.

Cwaliński et al. forthcoming: M. Cwaliński, S. Kaur, E. Stout, Provenience of Bronze Age Amber Finds from Istria. In: B. Teržan, K. Mihovilić (eds.), *Mušego*. Monografije i katalozi Arheološkog muzeja Istre. Pula (forthcoming/in print).

Czebreszuk 2003: J. Czebreszuk, Amber on the Threshold of a World Career. In: C. W. Beck, I. B. Loze, J. M. Todd (eds.), *Amber in Archaeology. Proceedings of the Fourth International Conference on Amber in Archaeology, Talsi 2001*. Riga 2003, pp. 164–179.

Czebreszuk 2007: J. Czebreszuk, Amber between the Baltic and the Aegean in the Third and Second Millennia BC (An Outline of Major Issues). In: I. Galanaki, H. Tomas, Y.

Galanakis, R. Laffineur (eds.), *Between the Aegean and Baltic Seas. Prehistory across Borders*. Aegaeum 27. Liège 2007, pp. 363–369.

Czebreszuk 2009: J. Czebreszuk, The Northern Section of the First Amber Trail. An Outline of Significance for Civilization Development. In: A. Palavestra, C. W. Beck, J. M. Todd (eds.), *Amber in Archaeology: Proceedings of the Fifth International Conference on Amber in Archaeology*. Beograd 2009, pp. 100–109.

Czebreszuk 2011: J. Czebreszuk, *Bursztyn w kulturze mykeńskiej*. Poznań 2011.

Czebreszuk/Makarowicz 1993: J. Czebreszuk, P. Makarowicz, The Problem of Amber Buttons with V-shaped Perforation in the Bell Beaker Culture. In: J. Pavúk, M. Fabis, I. Kuzma, K. Marková, L. Bánesz (eds.), *Actes du XIIe Congres International des Sciences Prehistoriques et Protohistoriques, Bratislava, 1–7 septembre 1991*, vol. II. Bratislava 1993, pp. 529–532.

Czebreszuk/Müller 2001: J. Czebreszuk, J. Müller, *Die absolute chronologie in Mitteleuropa 3000–2000 v. Chr.* Studien zur Archäologie in Ostmitteleuropa 1. Poznań – Leidorf 2001.

Czebreszuk et al. 2015: J. Czebreszuk, J. Müller, M. Jaeger, J. Kneisel, Absolute Chronology of Settlement. In: J. Czebreszuk, J. Müller, M. Jaeger, J. Kneisel (eds.), *Bruszczewo IV. Natural resources and economic activities of the Bronze Age people*. Studien zur Archäologie in Ostmitteleuropa 14. Poznań – Bonn 2015, pp. 39–52.

Damiani 1995: I. Damiani, La facies Protoappenninica. In: D. Cocchi Genick, I. Damiani, I. Macchiarola, R. Peroni, R. Poggiani Keller (eds.), *Aspetti culturali della media età del bronzo nell'Italia centro-meridionale*. Firenze 1995, pp. 398–438.

Damiani 1997: I. Damiani, La ceramica appenninica e subappenninica come modelli ed elementi di scambio. In: M. Bernabò Brea, A. Cardarelli, M. Cremaschi (eds.), *Le Terramare. La più antica civiltà padana*. Exhibition catalogue. Modena 1997, pp. 621–627.

Damiani 2010: I. Damiani, *L'età del Bronzo recente nell'Italia centro-meridionale*. Grandi Contesti e Problemi della Protostoria Italiana 12. Roma 2010.

Damiani/Morico 1996: I. Damiani, G. Morico, Le ceramiche dell'età del Bronzo di San Giuliano di Toscanella. In: M. Pacciarelli (ed.), *Collezione Scarabelli, 2. Preistoria*. Imola 1996, pp. 316–345.

David 1998: W. David, Zu früh- und ältermittelbronzezeitlichen Grabfunden in Ostbayern. In: J. Michálek, K. Schmotz, M. Zápotocká (eds.), *Archäologische Arbeitsgemeinschaft Ostbayern/West- und Südböhmen, 7. Treffen, Landau an der Isar, Juni 1997*. Rahden/Westf 1998, pp. 108–129.

David 2002: W. David, *Studien zu Ornamentik und Datierung der bronzezeitlichen Depotfundgruppe Hajdúsámson-Apa-Ighiel-Zajta*. Bibliotheca Musei Apulensis 18. Alba Iulia (Karlsburg/Weissenburg) 2002.

David 2003: W. David, Zum Ornament der Goldblechkegel vor dem Hintergrund bronzezeitliche Goldfunde Mittel- und Südosteuropas. In: T. Springer (ed.), *Goldenes Sakralgerät der Bronzezeit. Bericht über das Kolloquium, Nürnberg vom 17. bis 20. Mai 2001*. Nürnberg 2003, pp. 35–53.

David 2006a: W. David, Fast schon eine Glaubenssache: Periodisierung der Früh- und Mittelbronzezeit. In: F. Pustet (ed.) *Archäologie in Bayern: Fenster zur Vergangenheit.* Regensburg 2006, pp. 100–103.

David 2006b: W. David, Älterbronzezeitliche Depotfunde Ostbayerns im Lichte ihrer überregionalen Bedeutung. In: K. Schmotz (ed.), *Vorträge des 26. Niederbayerischen Archäologentages.* Rahden/Westf. 2006, pp. 197–237.

David 2015: W. David, Bronzedepotfunde der Früh- und Mittelbronzezeit in Ostbayern. In: O. Chvojka, M. Chytráček, H. Gruber, L. Husty, J. Michálek, R. Sandner, K. Schmotz, S. Traxler (eds.), *Archäologische Arbeitsgemeinschaft Ostbayern/West- und Südböhmen/Oberösterreich. 24. Treffen, Bechyně, 18. bis 21. Juni 2014.* Fines Transire 24. Rahden/Westf. 2015, pp. 87–111.

David 2016: W. David, Brotlaibidole als Zeugen transalpiner Kommunikation zwischen Südbayern und Norditalien in der älteren Bronzezeit. *Bayerische Archäologie* 4 (2016), pp. 26–30.

David et al. 2017: W. David, M. David-Elbiali, R. C. de Marinis, M. Rapi, Le Bronze moyen et récent en Italie du nord, Allemagne du sud et Suisse et corrélation des systèmes chrono-culturels. In: T. Lachenal, C. Mordant, T. Nicolas, C. Véber (eds.), *Le Bronze moyen et l'origine du Bronze final en Europe occidentale (XVIIe-XIIIe siècle av. J.-C.).* Strasbourg 2017, pp. 565–599.

David-Elbiali 2010: M. David-Elbiali, Sous l'angle du genre: analyse de nécropoles de l'âge du Bronze (15e-13e siècle av. J.-C.) d'Italie du Nord et comparaisons avec le nord des Alpes. *Rivista di Scienze Preistoriche* LX (2010), pp. 203–256.

David-Elbiali 2013: M. David-Elbiali, La chronologie nord-alpine du Bronze final (1200–800 av. J.-C.): entre métal, céramique et dendrochronologie. In: W. Leclercq, E. Warmenbol (eds.), *Échanges de bons procédés: la céramique du Bronze final dans le nord-ouest de l'Europe. Actes du colloque internationalorganisé à l'Université libre de Bruxellesles 1 et 2 octobre 2010.* Études d'archéologie 6. Bruxelles 2013, pp. 181–197.

David-Elbiali 2014: M. David-Elbiali, Il cammino tra le Alpi. Elementi di riflessione per una storia dei rapporti transalpini nella zona alpina centro-occidentale all'età del Bronzo. In: B. Grassi, M. Pizzo (eds.), *Gallorum Insubrum fines. Ricerche e progetti archeologici nel territorio di Varese. Atti della giornata di studio (Varese, Villa Recalcati, 29 gennaio 2010).* Studia Archaeologica 200. Roma 2014, pp. 43–64.

David-Elbiali/Moinat 2005: M. David-Elbiali, P. Moinat, Le Bronze final de la Suisse occidentale: révision du cadre chronotypologique, grâce aux découvertes de la nécropole de Lausanne-Vidy (canton de Vaud, Suisse). *Bulletin de la Société préhistorique française* 102/3 (2005), pp. 613–623.

David-Elbiali/Dunning 2005: M. David-Elbiali, C. Dunning, Le cadre chronologique relatif et absolu au nord-ouest des Alpes entre 1060 et 600 av. J.-C. In: G. Bartoloni, F. Delpino (eds.), *Oriente e occidente. Metodi e discipline a confronto. Riflessioni sulla cronologia dell'età del ferro italiana. Atti dell'incontro di studio, Roma 30–31 ottobre 2003.* Pisa 2005, pp. 145–195.

David-Elbiali/David 2009: M. David-Elbiali, W. David, À la suite de Jacques-Pierre Millotte, actualité des recherches en typologie sur l'âge du Bronze. Le Bronze ancien et le début du Bronze moyen: cadre chronologique et liens culturels entre l'Europe nord-alpine occidentale, le monde danubien et l'Italie du Nord. In: A.

Richard, P. Barral, A. Daubigney (eds.) *L'isthme européen Rhin-Saône-Rhône dans la Protohistoire. Approches nouvelles en hommage à Jacques-Pierre Millotte. Actes du colloque (Besançon, 16–18 octobre 2006).* Besançon 2009, pp. 311–340.

David-Elbiali/Hafner 2013: M. David-Elbiali, A. Hafner, Gräber, Horte und Pfahlbauten zwischen Jura und Alpen – Die Entwicklung elitärer sozialer Strukturen in der frühen Bronzezeit der Westschweiz. In: H. Meller, F. Bertemes, H. R. Bork, R. Risch (eds.), *1600 – Kultureller Umbruch im Schatten des Thera-Ausbruchs. 4. Mittledeutscher Archäologentag vom 14. Bis 16. Oktober 2011 in Halle (Saale).* Tagungen des Landesmuseums für Vorgeschichte Halle 9. Halle (Saale) 2013, pp. 217–238.

De Angelis 1979: M. C. De Angelis, Il Bronzo Finale in Umbria e Toscana. In: I.I.P.P. (eds.), *Il Bronzo finale in Italia. Atti della XXI Riunione Scientifica dell'Istituto Italiano di Preistoria e Protostoria (Firenze, 21–23 ottobre 1973).* Firenze 1979, pp. 221–247.

De Angelis 2006: S. De Angelis, Il passaggio tra Bronzo Finale 2 e Bronzo Finale 3 in Etruria Meridionale sotto il profilo delle sepolture. In: A. Cardarelli, M. Pacciarelli, A. Vanzetti (eds.), *Studi di protostoria in onore di Renato Peroni.* Firenze 2006, pp. 581–589.

De Angelis 2010: S. De Angelis, L'inquadramento cronologico dei complessi funerari del Bronzo Finale in Etruria Meridionale. In: B. Barbaro (ed.), *Insediamenti, aree funerarie ed entità territoriali in Etruria Meridionale nel Bronzo Finale.* Grandi Contesti e Problemi della Protostoria Italiana 14. Firenze 2010, pp. 107–112.

De Angelis/Gori 2017: S. De Angelis, M. Gori, The Wheel and the Sun: 'Glocal' Symbologies of Wheel-Pendants across Europe. In: S. Bergerbrant, A. Wessman (eds.), *New Perspectives on the Bronze Age. Proceedings of the 13th Nordic Bronze Age Symposium Held in Gothenburg, 9th to 13th June 2015.* Oxford 2017, pp. 363–374.

De Angelis et al. 2007: S. De Angelis, G. Francozzi, M. Gori, I pendagli della media e tarda età del Bronzo dell'Italia centro-settentrionale: stile e simbologia. In: U. Thun Hohenstein, M. Arzarello, A. Di Nucci (eds.), *I Convegno Nazionale degli studenti di Antropologia, Preistoria e Protostoria, Ferrara, Maggio 2004.* Annali dell'Università di Ferrara n.s., Sezione Scienze della Terra, vol. 11. Ferrara 2007 pp. 117–120.

De Grossi Mazzorin 2019: J. De Grossi Mazzorin, L'economia animale dell'abitato di Frattesina nel quadro della tarda età del Bronzo veneta. In: A. Maria Bietti Sestieri, P. Bellintani, C. Giardino (eds.), *Frattesina: un centro internazionale di produzione e di scambio nella tarda età del bronzo del Veneto.* Atti della Accademia Nazionale dei Lincei, Memorie serie IX, vol. XXXIX, fascicolo 1. Roma 2019, pp. 123–162.

De Juliis 1979: E. M. De Juliis, Il Bronzo finale nella Puglia Settentrionale. In: I.I.P.P. (eds.), *Il Bronzo finale in Italia. Atti della XXI Riunione Scientifica dell'Istituto Italiano di Preistoria e Protostoria (Firenze, 21–23 ottobre 1973).* Firenze 1979, pp. 515–535.

De Marinis 1972: R. C. De Marinis, Ritrovamenti dell'età del Bronzo Finale in Lombardia. Contributo alla suddivisione in periodi del Protogolasecca. *Sibrium* XI (1972), pp. 53–98.

De Marinis 1979: R. C. De Marinis, Età del Bronzo. In: In AA.VV. (ed.), *Preistoria nel Bresciano: la cultura material.* Brescia 1979, pp. 45–70.

De Marinis 1981: R. C. De Marinis, Appunti sul Bronzo Medio, Tardo e Finale (1600–900 a.C.). In: In AA.VV. (ed.), In: *Atti I° Convegno Archeologico Regionale. Milano, 29 febbraio – 1–2 marzo 1980.* Brescia 1981, pp. 173–204.

De Marinis 1997: R. C. De Marinis, L'età del Bronzo nella regione benacense e nella pianura padana a nord del Po. In: M. Bernabò Brea, A. Cardarelli, M. Cremaschi (eds.), *Le Terramare. La più antica civiltà padana*. Exhibition catalogue. Modena 1997, pp. 405–419.

De Marinis 1999: R. C. De Marinis, Towards a Relative and Absolute Chronology for the Bronze Age in Northern Italy. *Notizie Archeologiche Bergomensi* 7 (1999 [2002]), pp. 25–103.

De Marinis 2000: R. C. De Marinis, Il Bronzo finale nel Canton Ticino. In: R. C. De Marinis, S. Biaggio Simona (eds.), *I Leponti tra mito e realtá*, vol. I. Exhibition catalogue. Locarno 2000, pp. 123–146.

De Marinis 2001: R. C. De Marinis, *Il museo civico archeologico Giovanni Rambotti: una introduzione alla preistoria del lago di Garda*. Desenzano del Garda 2001.

De Marinis 2006: R. C. De Marinis, Aspetti e problemi del Bronzo Recente nella regione benacense. In: A. Cardarelli, M. Pacciarelli, A. Vanzetti (eds.), *Studi di protostoria in onore di Renato Peroni*. Firenze 2006, 445–456.

De Marinis 2007: R. C. De Marinis, Ambra in Italia settentrionale tra Alpi e Po durante l'eta del Bronzo. In: M. L. Nava, A. Salerno (eds.), *Ambre. Trasparenze dall'antico, Exhibition catalogue (Napoli, Museo Archeologico Nazionale, 26 marzo–10 settembre 2007)*. Milano 2007, pp. 60–67.

De Marinis 2009a: R. C. De Marinis, Continuity and discontinuity in Northern Italy from the Recent to the Final Bronze Age: a view from north-western Italy. *Scienze dell'Antichità* 15 (2009), pp. 535–545.

De Marinis 2009b: R. C. De Marinis, La protostoria del territorio di Varese dall'inizio dell'età dei metalli al periodo della Romanizzazione. In: R. C. De Marinis, S. Massa, M. Pizzo (eds.), *Alle origini di Varese e del suo territorio: Le collezioni del sistema archeologico provinciale*. Roma 2009, pp. 11–30.

De Marinis 2009c: R. C. De Marinis, L'età del Bronzo Recente e Finale all'Isolino di Varese. In: R. C. De Marinis, S. Massa, M. Pizzo (eds.), *Alle origini di Varese e del suo territorio: Le collezioni del sistema archeologico provinciale*. Roma 2009, pp. 140–145.

De Marinis 2013: R. C. De Marinis, Le ricerche archeologiche al Lavagnone (Desenzano del Garda-Lonato). In: J. De Grossi Mazzorin, A. Curci, G. Giacobini (eds.), *Economia e ambiente nell'Italia Padana dell'età del bronzo: le indagini bioarcheologiche*. Beni Archeologici – Conoscenza e Tecnologie Quaderno 11. Bari 2013, pp. 15–19.

De Marinis 2014: R. C. De Marinis, Correlazioni cronologiche tra Italia nord-occidentale (area della cultura di Golasecca) e ambienti culturali transalpini e cisalpini dal Bronzo Recente alla fine del VII secolo a.C. In: M. J. Roulière-Lambert (ed.), *Les Celtes et le Nord de l'Italie (Premier et Second Âges du fer). Actes du XXXVIe colloque international de l'A.F.E.A.F (Vérone, 17–20 mai 2012)*. Revue archéologique de l'Est. Supplément 36. Dijon 2014, pp. 17–36.

De Marinis 2015: R. C. De Marinis, Aspetti della transizione Bronzo medio-Bronzo recente in area padana. In: G. Leonardi, V. Tiné (eds.) *Preistoria e Protostoria del Veneto*. Studi di Preistoria e Protostoria 2. Firenze 2015, pp. 341–348.

De Marinis/Salzani 1997: R. C. De Marinis, L. Salzani, Le necropoli del Bronzo Medio e Recente nella Lombardia orientale e nel Veneto occidentale. In: M. Bernabò Brea,

A. Cardarelli, M. Cremaschi (eds.), *Le Terramare. La più antica civiltà padana.* Exhibition catalogue. Modena 1997, pp. 703–707.

De Marinis/Salzani 2005: R. C. De Marinis, L. Salzani, Tipologia e cronologia dei materiali. In: L. Salzani (ed.), *La necropoli dell'età del Bronzo all'Olmo di Nogara.* Verona 2005, pp. 391–448.

De Marinis/Valzolgher 2013: R. C. De Marinis, E. Valzolgher, Riti funerari dell'antica età del Bronzo in area padana. In: R. C. De Marinis (ed.), *L'età del Rame. La pianura padana e le Alpi al tempo di Ötzi.* Roccafranca (BS) 2013, pp. 545–559.

De Marinis/Rapi 2016: R. C. De Marinis, M. Rapi, Note sui criteri di classificazione della ceramica e sulla terminologia delle anse con sopraelevazioni. *Notizie Archeologiche Bergomensi* 24 (2016), pp. 27–59.

De Marinis et al. 1996: R. C. De Marinis, M. Baioni, N. De Gasperi, C. Mangani, L. Seragnoli, Nuovi scavi al Lavagnone (Desenzano del Garda-Lonato) e considerazioni sull'antica età del Bronzo nell'Italia settentrionale. In: D. Cocchi Genick (ed.), *L'antica età del Bronzo in Italia. Atti del Congresso nazionale (Viareggio, 9–12 gennaio 1995).* Firenze 1996, pp. 257–271.

De Marinis et al. 1999: R. C. De Marinis, M. Baioni, S. Casini, N. Degasperi, C. Mangani, L. Seragnoli, La sequenza stratigrafica dell'abitato palafitticolo del Lavagnone (Desenzano del Garda, Brescia). In: P. Della Casa (ed.), *Prehistoric Alpine Environment, Society and Economy. Papers of the International Colloquium Paese '97 in Zurich, 1997.* Universitätsforschugen zur Prähistorischen Archäologie 55. Bonn 1999, pp. 127–131.

De Marinis et al. 2005: R. C. De Marinis, M. Rapi, C. Ravazzi, E. Arpenti, M. Deaddis, R. Perego, Lavagnone (Desenzano del Garda): New Excavations and Palaeoecology of a Bronze Age Pile Dwelling Site in Northern Italy. In: P. Della Casa, M. Trachsel (eds.), *WES'04 – Wetland Economies and Societies. Proceedings of the International Conference in Zurich, 10–13 March 2004.* Collectio Archæologica 3, Zürich 2005 pp. 221–232.

De Min 1982: M. De Min, La necropoli protovillanoviana di Frattesina di Fratta Polesine (RO). Notizie preliminary. *Padusa* XVIII (1982), pp. 5–27.

De Navarro 1925: J. M. De Navarro, Prehistoric Routes between Northern Europe and Italy Defined by the Amber Trade. *The Geographical Journal* 66/6 (1925), pp. 481–503.

De Souza 2002: P. De Souza, *Piracy in the Graeco-Roman World.* Cambridge 2002.

Del Fattore 2006: F. R. Del Fattore, La definizione delle fasi più antiche dell'età del bronzo in area mediotirrenica: stato della questione. In: A. Cardarelli, M. Pacciarelli, A. Vanzetti (eds.), *Studi di protostoria in onore di Renato Peroni.* Firenze 2006, pp. 723–732.

Del Lucchese 1990: A. Del Lucchese, Manufatti metallic dell'età del Bronzo. In: R. Maggi (ed.), *Archeologia dell'Appennino Ligure. Gli Scavi del Castellaro di Uscio: un insediamento di crinale occupato dal neolitico alla conquista Romana.* Istituto Internazionale di Studi Liguri, Collezione di Monografie Preistoriche ed Archeologiche, vol. VIII. Bordighera 1990, pp. 169–173.

Del Lucchese/Maggi 1985: A. Del Lucchese, R. Maggi, Considerazioni sulla cronologia dell'età del Bronzo in Liguria. *Rivista Storica Liguria* 48 (1985), pp. 75–90.

Del Lucchese et al. 1991–1992: A. Del Lucchese, R. C. De Marinis, F. M. Gambari, Articolazioni chronologiche e definizione di elementi culturali, 1.I. Italia nord-occidentale. In: D. Cocchi Genick (ed.), *L'età del bronzo in Italia nei secoli dal XVI al XIV a.C. Atti del congresso Viareggio, 26–30 ottobre 1989*. Rassegna di Archeologia 10. Firenze 1991–1992, pp. 31–38.

Delfino 2005/2006: D. Delfino, *Il castellaro di Camogli (GE): una revisione del materiale ceramico edito e inedito e dei dati di scavo degli strati dell'età del Bronzo*. Tesi di Specializzazione, Scuola di Specializzazione in Archeologia, Facoltà di Lettere e Filosofia, Università Statale di Milano. Milano 2005/2006.

Delfino 2010: D. Delfino, Il monte Castellaro di Camogli. Rilettura dei dati archeologici. In: F. Bulgarelli, A. Del Lucchese, L. Gervasini (eds.), *Archeologia in Liguria, Nuova Seria, Vol. II (2006–2007)*. Genova 2010, pp. 428–430.

Della Casa 1995: P. Della Casa, The Cetina Group and the Transition from Copper to Bronze Age in Dalmatia. *Antiquity* 69 (1995), pp. 565–76.

Della Casa 1996a: P. Della Casa, *VELIKA GRUDA II. Die bronzezeitliche Nekropole Velika Gruda (Opš. Kotor, Montenegro). Fundgruppen der mittleren und späten Bronzezeit zwischen Adria und Donau*. Universitätsforschungen zur Prähistorischen Archäologie 33. Bonn 1996.

Della Casa 1996b: P. Della Casa, Linking Anthropological and Archaeological Evidence: Notes on the Demographic Structure and Social Organisation of the Bronze Age Necropolis Velika Gruda in Montenegro. *Arheološki vestnik* 47 (1996), pp. 135–143.

Delpino 1978: F. Delpino, Presenze del Bronzo Finale ad Ardea. *Archeologia Laziale* I (1978), pp. 26–27.

Delpino 1987: F. Delpino, Etruria e Lazio prima dei Tarquini. Le fasi protostoriche. *Quaderni del Centro di Studio per l'Archeologia Etrusco-Italica* 15 (1987), pp. 9–36.

Desantis et al. 2011: P. Desantis, M. Marchesini, S. Marvelli, *Anzola al tempo delle Terramare. Guida. Museo Archeologico Ambientale*. Bologna 2011.

Di Gennaro 1998: F. Di Gennaro, Il territorio della Tolfa dal neolitico alla fine dell'età del bronzo. *Quaderni del Museo Civico di Tolfa* 1 (1998), pp. 67–120.

Di Gennaro/Pacciarelli 1996: F. Di Gennaro, M. Pacciarelli, Lo stile di Luni Tre Erici-Norchia. In: D. Cocchi Genick (ed.), *L'antica età del bronzo in Italia. Atti del Congresso nazionale (Viareggio, 9–12 gennaio 1995)*. Firenze 1996, pp. 574–575.

Di Gennaro/Passoni 1998: F. Di Gennaro, A. Passoni, Indicazioni sulla cronologia di materiali del Bronzo finale dalla tipologia dei luoghi di insediamento. In: N. Negroni Catacchio (ed.), *Atti del III incontro di studi 'Preistoria e Protostoria in Etruria' (Manciano-Farnese 1995)*. Firenze 1998, pp. 127–135.

Di Gennaro/Guidi 2000: F. Di Gennaro, A. Guidi, Il Bronzo finale nell'Italia centrale. Considerazioni e prospettive di indagine. In: M. Harari, M. Pearce (eds.), *Il Protovillanoviano al di qua e al di là dell'Appennino. Atti della Giornata di studio (Pavia, 17 giugno 1995)*. Como 2000, pp. 99–131.

Dietrich 2014: O. Dietrich, Learning from 'Scrap' about Late Bronze Age Hoarding Practices. A Biographical Approach to Individual Acts of Dedication in Large Metal Hoards. *European Journal of Archaeology* 17/3 (2014), pp. 468–486.

Dizdar et al. 2011: M. Dizdar, D. Ložnjak Dizdar, S. Mihelić, *Early Urnfield Culture in Northern Croatia – New Challenges.* Zagreb – Osijek 2011.

Dmitrović 2014: K. Dmitrović, Notes on the Urnfield period in the Čačak region, Serbia. In: D. Ložnjak Dizdar, M. Dizdar (eds.), *The Beginning of the Late Bronze Age between the Eastern Alps and the Danube. Proceedings of the international conference in Osijek, October 20–22, 2011.* Zbornik Instituta Za Arheologiju Serta Instituti Archaeologici 1. Zagreb 2014, pp. 261–271.

Dmitrović 2016: K. Dmitrović, *Bronze Age necropolises in the Čačak region.* Čačak 2016.

Dmitrović/Ljuština 2010: K. Dmitrović, M. Ljuština, Notes on grave goods from the Bronze and Iron Ages tombs in the West Morava valley (Serbia). *Istros* XVI (2010), pp. 55–79.

Doğan 2008: I. B. Doğan, Trade and Exchange in Prehistory. A Theoretical Evaluation. In: E. Doksanalt, E. Aslan (eds.), *Proceedings of the International Symposium Trade and Production Through the Ages Konya, 25–28 November 2008.* Konya 2008, pp. 33–50.

Doğan/Michailidou 2008: I. B. Doğan, A. Michailidou, Trading in prehistory and protohistory: perspectives from the Eastern Aegean and beyond. In: C. Papageorgiadou-Banis, A. Giannikouri (eds.), *Sailing in the Aegean: readings on the economy and trade routes.* Athens 2008, pp. 17–53.

Domańska 2016: E. Domańska, Semiofory i pneumatofory w perspektywie biohumanistycznej (Przypadek korzeni oddechowych cypryśnika błotnego). In: A. Kołakowski, A. Mencwel, J. Migasiński, P. Rodak, M. Szpakowska (eds.), *Wśród ludzi, rzeczy i znaków. Krzysztofowi Pomianowi w darze.* Warszawa 2016, pp. 347–363.

Drechsler-Bižić 1961: R. Drechsler-Bižić, Rezultati istraživanja japodske nekropole u Kompolju 1955–1956. godine. *Vjesnik Arheološkog Muzeja u Zagrebu* 2/1 (1961), pp. 67–114.

Drechsler-Bižić 1968: R. Drechsler-Bižić, Japodske kape i oglavlja. *Vjesnik Arheološkog Muzeja u Zagrebu* 3/1 (1968), pp. 29–52.

Drechsler-Bižić 1970: R. Drechsler-Bižić, Zaštitna iskopavanja pećine Golubinjače kod Kosinja. *Vjesnik Arheološkog Muzeja u Zagrebu* 4/1 (1970), pp. 111–118.

Drechsler-Bižić 1979–1980: R. Drechsler-Bižić, Nekropola broncanog doba u pećini Bezdanjači kod Vrhovina. *Vjesnik Arheološkog muzeja u Zagrebu* 12–13/1 (1979–1980), pp. 27–78.

Drechsler-Bižić 1983a: R. Drechsler-Bižić, Srednje brončano doba u Lici i Bosni. In: A. Benac (ed.), *Praistorija Jugoslavenskih Zemalja. IV knjiga: Bronzano Doba.* Sarajevo 1983, pp. 242–270.

Drechsler-Bižić 1983b: R. Drechsler-Bižić, Japodska kulturna grupa. In: A. Benac (ed.), *Praistorija Jugoslavenskih Zemalja. IV knjiga: Bronzano Doba.* Sarajevo 1983, pp. 374–389.

Du Gardin 1986: C. Du Gardin, La parure d'ambre à l'age du Bronze en France. *Bulletin Société Prehistorique Française* 83 (1986), pp. 546–580.

Du Gardin 1998: C. Du Gardin, Le Campaniforme et l'ambre: mythe ou réalité? *Bulletin Société Prehistorique Française* 95/3 (1998), pp. 343–350.

Du Gardin 2003: C. Du Gardin, Amber Spacer Beads in the Neolithic and Bronze Ages in Europe. In: C. W. Beck, I. B. Loze, J. M. Todd (eds.), *Amber in Archaeology. Proceedings of the Fourth International Conference on Amber in Archaeology, Talsi 2001.* Riga 2003, pp. 180–197.

Dular et al. 2002: J. Dular, I. Šavel, S. Tecco Hvala, *Bronastodobno naselje Oloris pri Dolnjem Lakošu.* Ljubljana 2002.

Dumitrescu 1932: H. Dumitrescu, L'eta del bronzo nel Piceno. *Ephemeris Dacoromana: annuario della scuola romena di Roma* 5 (1932), pp. 198–330.

Earle 1997: T. Earle, *How Chiefs Come to Power: The Political Economy in Prehistory.* Stanford 1997.

Earle/Kristiansen 2010: T. Earle, K. Kristiansen (eds.), *Organizing Bronze Age Societies: The Mediterranean, Central Europe, and Scandinavia Compared.* Cambridge 2010.

Earle et al. 2015: T. Earle, J. Ling, C. Uhnér, Z. Stos-Gale, L. Melheim 2015, The Political Economy and Metal Trade in Bronze Age Europe: Understanding Regional Variability in Terms of Comparative Advantages and Articulations. *European Journal of Archaeology* 18/4 (2015), pp. 633–657.

Eles Masi 1986: P. von Eles Masi, *Le fibule dell'Italia settentrionale.* Prähistorische Bronzefunde XIV/5. München 1986.

Ernée 2012: M. Ernée, Jantar v české únětické kultuře – k počátkům jantarové stezky. *Památky Archeologické* CIII (2012), pp. 71–172.

Ernée 2013: M. Ernée, Bernstein und der Zusammenbruch der klassischen Aunjetitzer Kultur in Böhmen. In: H. Meller, F. Bertemes, H. R. Bork, R. Risch (eds.), *1600 – Kultureller Umbruch im Schatten des Thera-Ausbruchs? 4. Mitteldeutscher Archäologentag vom 14. Bis 16. Oktober 2011 in Halle (Saale).* Tagungen des Landesmuseums für Vorgeschichte Halle 9. Halle (Saale) 2013, pp. 453–467.

Ernée 2017: M. Ernée, Ostseeküste – Mykene – und was dazwischen? Eine vergessene Bernsteinstraße. In: O. Chvojka, M. Chytráček, H. Gruber, L. Husty, J. Michálek, R. Sander, K. Schmotz (eds.), *Archäologische Arbeitsgemeinschaft Ostbayern/West- und Südböhmen/Oberösterreich, 22–25 Juni 2016, Plzeň.* Fines Transire 26. Rahden/Westf. 2017, pp. 111–131.

Ernée et al. 2009: M. Ernée, J. Müller, K. Rassmann, Ausgrabung des frühbronzezeitlichen Gräberfelds der Aunjetitzer Kultur von Prag-Miškovice. *Germania* 87/2 (2009), pp. 355–410.

Everett 2015: M. G. Everett, Core–Periphery Patterns in Networks. In: J. Wright (ed.), *International Encyclopedia of the Social & Behavioral Sciences* (2nd edition). Oxford 2015, pp. 911–916.

Falkenstein et al. 2016: F. Falkenstein, B. Hänsel, P. Medović, Feudvar Near Mošorin (Serbia) – Excavations and Research in a Micro-region at the Confluence of the Danube and Tisza: A Recapitulation after Thirty Years. In: H. Kroll, K. Reed (eds.), *Die Archäobotanik. Feudvar III.* Würzburger Studien zur Vor- und Frühgeschichtlichen Archäologie 1. Würzburg 2016, pp. 5–35.

Fasani 1966: L. Fasani, Una stazione preistorica della fine dell'eta del bronzo a Mariconda presso Melara (Rovigo). *Padusa* II/2–3 (1966), pp. 9–12.

Fasani 1970: L. Fasani, Sul significato cronologico dei cosiddetti «oggetti enigmatici» dell'età del Bronzo dell'Italia settentrionale. *Memorie del Museo Civico di Storia Naturale di Verona* XVIII (1970), pp. 91–112.

Fasani 1984: L. Fasani, L'età del Bronzo. In: A. Aspes (ed.), *Veneto nell'antichità: preistoria e protostoria*, vol. 2. Verona 1984, pp. 451–614.

Fasani/Salzani 1975a: L. Fasani, L. Salzani, Nuovo insediamento dell'eta del Bronzo in localita Fondo Paviani presso Legnago (Verona). *Bollettino del Museo Civico di Storia Naturale di Verona* 2 (1975), pp. 259–281.

Fasani/Salzani 1975b: L. Fasani, L. Salzani, Aspetti e problemi dell'età del bronzo finale nella pianura padana orientale. *Padusa* XI (1975), pp. 53–124 (reprint. in *Padusa* XX [1984], pp. 261–332).

Fasani/Salzani 1976: L. Fasani, L. Salzani, Fondo Paviani (Legnago-Verona). *Bollettino del Museo Civico di Storia Naturale di Verona* 3 (1976), pp. 587–590.

Fasani/Martinelli 1994: L. Fasani, N. Martinelli, Problemi relative alle palafitte italiane alla luce della ricerca dendrocronologica. In: R. Skeates R. Whitehouse (eds.), *Radiocarbon Dating and Italian Prehistory.* Archaeological Monographs of the British School at Rome 8 – Accordia Specialist Studies on Italy 3. London 1994, pp. 39–44.

Fasani/Martinelli 1996: L. Fasani, N. Martinelli, Cronologia assoluta e relativa dell'antica età del'Bronzo nell'Italia settentrionale. In: D. Cocchi Genick (ed.), *L'antica età del Bronzo in Italia. Atti del Congresso nazionale (Viareggio, 9–12 gennaio 1995).* Firenze 1995, pp. 19–32.

Fedeli 1983: F. Fedeli, *Populonia, storia e territorio.* Firenze 1983.

Fiala 1894: F. Fiala, Uspjesi prekopavanja prehistoričkih grobova na Glasincu godine 1894. *Glasnik zemaljskog muzeja u Sarajevu VI/4* (1894), pp. 721–760.

Fiala 1895: F. Fiala, Rezultati pretraživanja prehisotričkih gromila na Glasincu godine 1895. *Glasnik zemaljskog muzeja u Sarajevu VII/4* (1985), pp. 535–565.

Filipović 2008: V. Filipović, Late Bronze Age Funerary Practices in Northwestern Serbia. In: V. Sirbu, D. L. Vaida (eds.), *Funerary Practices of the Bronze and Iron Ages in Central and South-Eastern Europe. Proceedings of the 9th International Colloquium of Funerary Archaeology, Bistrita, Romania, 9–11 May 2008.* Cluj-Napoca 2008, pp. 95–108.

Filipović 2013: V. Filipović, New Researches of the Late Bronze Age Necropolises in Northwestern Serbia Region: Chronological and Terminological Questions. *Journal of Serbian Archaeological Society* 29 (2013), pp. 51–84.

Finley 1985: M. Finley, *The Ancient Economy* (2nd edition). Berkeley 1985.

Finotelli et al. 1997a: F. Finotelli, G. Morico, G. Steffè, Anzola Emilia (BO). In: M. Bernabò Brea, A. Cardarelli, M. Cremaschi (eds.), *Le Terramare. La più antica civiltà padana.* Exhibition catalogue. Modena 1997, pp. 363–364.

Finotelli et al. 1997b: F. Finotelli, G. Morico, G. Steffè, Anzola Emilia (BO). *Archeologia dell'Emilia Romagna* I/2 (1997), pp. 32–34.

Fischl et al. 2013: K. P. Fischl, V. Kiss, G. Kulcsár, V. Szeverényi, Transformations in the Carpathian Basin around 1600 B.C. In: H. Meller, F. Bertemes, H. R. Bork, R. Risch (eds.), *1600 – Kultureller Umbruch im Schatten des Thera-Ausbruchs?*

Garašanin/Garašanin 1958: M. Garašanin, D. Garašanin, Iskopavanje tumula u Belotiću i Beloj Crkvi (Zapadna Srbija). *Zbornik radova Narodnog muzeja* I (1958), pp. 17–50.

Garašanin/Garašanin 1962: M. Garašanin, D. Garašanin, Iskopavanje tumula u kompleksu Belotić-Bela Crkva 1959. i 1960. godine. *Zbornik radova Narodnog muzeja* III (1962), pp. 47–68.

García-Puchol et al. 2012: O. García-Puchol, S. B. McClure, J. J. Cabanilles, Objetos singulares/objetos foráneos. Evidencias de circulación de artesanías en el depósito funerario del Neolítico Final/Calcolítico de la Cova de la Pastora (Alcoi, Alacant). *Rubricatum: revista del Museu de Gavà* 5 (2012), pp. 281–288.

Gaspari 2009: A. Gaspari, The History of the Acquisition of Finds and Archaeological Investigation of the Ljubljanica. In: P. Turk, J. Istenič, T. Knific, T. Nabergoj (eds.), *The Ljubljanica – a River and Its Past.* Ljubljana 2009, pp. 26–31.

Gavranović 2011: M. Gavranović, *Die Spätbronze- und Früheisenzeit in Bosnien.* Universitätsforschungen zur Prähistorischen Archäologie 195. Bonn 2011.

Gedl 1975: M. Gedl, *Kultura Przedłużycka.* Wrocław – Warszawa – Kraków 1975.

Gedl 1983: M. Gedl, *Die Nadeln in Polen.* Prähistorische Bronzefunde XIII/9. München 1983.

Gennusa 2015: R. Gennusa, *L'evoluzione millenaria di uno stile. La civiltà del Bronzo castellucciana nella Sicilia meridionale.* Firenze 2015.

Genovese 2015: F. Genovese, La necropoli del Plemmyrion presso Siracusa. Materiali inediti per una rilettura del tema delle relazioni tra Sicilia e mondo egeo-miceneo nell'età del Bronzo Medio. In: A. M. Jasink, L. Bombardieri (eds.), *AKROTHINIA – Contributi di giovani ricercatori italiani agli studi egei e ciprioti.* Firenze 2015, pp. 115–128.

Gerloff 1993: S. Gerloff, Zu Fragen mittelmeerländischer Kontakte und absoluter Chronologie der Frühbronzezeit in Mittel- und Westeuropa. *Prähistorische Zeitschrift* 68 (1993), pp. 58–102.

Gerloff 2007: S. Gerloff, Reinecke's ABC and the Chronology of the British Bronze Age. In: Ch. Burgess, P. Topping, F. Lynch (eds.), *Beyond Stonehenge. Essays on the Bronze Age in Honour of Colin Burgess.* Oxford 2007, pp. 117–161.

Gervasio 1913: M. Gervasio, *I dolmen e la civiltà del Bronzo nelle Puglie.* Bari 1913.

Gestoso Singer 2008: G. N. Gestoso Singer, Amber in the Ancient Near East. *i-Medjat* 2 (2008), pp. 17–19.

Gestoso Singer 2016: G. N. Gestoso Singer, Amber Exchange in the Late Bronze Age Levant in Cross-cultural Perspective. *Aula Orientalis* 34/2 (2016), pp. 251–264.

Ghiurcă/Vávra 1990: V. Ghiurcă, N. Vávra, Occurrence and Chemical Characterization of Fossil Resins from Colţi (District of Buzău, Romania). *Neues Jahrbuch für Geologie und Paläontologie* 5 (1990), pp 283–294.

Giannitrapani 2013: E. Giannitrapani, Cronologia assoluta della tarda età del rame nella Sicilia centrale. In: D. Cocchi Gennick (ed.), *Cronologia assoluta e relative dell'età del Rame in Italia. Atti dell'Incontro di Studi, Università di Verona, 25 giugno 2013.* Preistoria e Protostoria 1. Verona 2013, pp. 190–191.

Giannitrapani/Iannì 2011: E. Giannitrapani, F. Iannì, La tarda età del rame nella Sicilia centrale. In: I.I.P.P. (eds.), *L'età del rame in Italia. Atti del XLIII Riunione Scientifica dell'Istituto Italiano di Preistoria e Protostoria (Bologna, 26–29 novembre 2008).* Firenze 2011, pp. 271–278.

Giannitrapani et al. 2014: E. Giannitrapani, F. Iannì, S. Chilardi, L. Anguilano, Case Bastione: A Prehistoric Settlement in the Erei Uplands (Central Sicily). *Origini* XXXVI (2014), pp. 181–211.

Giardino/Guida 2018: C. Giardino, G. Guida, Frattesina: le indagini metallografiche sui reperti dei ripostigli. In: A. Maria Bietti Sestieri, P. Bellintani, C. Giardino (eds.), *Frattesina: un centro internazionale di produzione e di scambio nella tarda età del bronzo del Veneto.* Atti della Accademia Nazionale dei Lincei, Memorie serie IX, vol. XXXIX, fascicolo 1. Roma 2019, pp. 197–234.

Giddens 1979: A. Giddens, *Central Problems in Social Theory: Action, Structure, and Contradiction in Social Analysis.* Berkeley 1979.

Giddens 1984: A. Giddens, *The Constitution of Society.* Cambridge 1984.

Gimbutas 1965: M. Gimbutas, *Bronze Age cultures in Central and Eastern Europe.* Hague – London 1965.

Giumlia-Mair et al. 2010: A. Giumlia-Mair, R. M. Albanese Procelli, F. Lo Schiavo, The Metallurgy of the Sicilian Final Bronze Age/Early Iron Age Necropolis of Madonna del Piano (Catania, Sicily). *Trabajos de Prehistoria* 67 (2010), pp. 469–488.

Gligorić/Canić-Tešanović 2010: R. Gligorić, J. Canić-Tešanović, *Paulje: nekropola bronzanog i gvozdenog doba kod Loznice.* Gornji Milanovac 2010.

Glogović 1987: D. Glogović, Nalaz iz Šule na Krku i problem lučnih fibula s dva dugmeta na luku iz Liburnije. *Arheološki radovi i rasprave* 10 (1987), pp. 73–92.

Glogović 1991: D. Glogović, Prahistorijski nakit iz Zbirke Stomorica u Novalji. *Prilozi Instituta za Arheologiju u Zagrebu* 8 (1991), pp. 17–24.

Glogović 2003: D. Glogović, *Fibeln im kroatischen Kustengebiet (Istrien, Dalmatien).* Prähistorische Bronzefunde XIV/13. Stuttgart 2003.

Gogâltan 2005: F. Gogâltan, Der Beginn der bronzezeitlichen Tellsiedlungen im Karpatenbecken: Chronologische Probleme. In: B. Horejs, R. Jung, E. Kaiser, B. Teržan (eds.), *Interpretationsraum Bronzezeit. Bernhard Hänsel von seinen Schülern gewidment.* Bonn 2005, pp. 161–179.

Gogâltan 2015: F. Gogâltan, The Early and Middle Bronze Age Chronology on the Eastern Frontier of the Carpathian Basin. In: R. E. Németh, R. Botond (eds.), *Bronze Age Chronology in the Carpathian Basin: proceedings of the International Colloquium from Târgu Mureş, 2–4 October 2014.* Târgu Mureş 2015, pp. 53–95.

Gogâltan 2016: F. Gogâltan, Chihlimbarul preistoric de la frontiera Estică a Bazinului Carpatic. *Analele Banatului, S.N., Arheologie – Istorie* XXIV (2016), pp. 143–169.

Goldhahn 2013: J. Goldhahn, Rethinking Bronze Age Cosmology: A North European Perspective. In: H. Fokkens, A. Harding (eds.), *The Oxford Handbook of the European Bronze Age.* Oxford 2013, pp. 248–265.

González-Ruibal 2012: A. González-Ruibal, Generations of Free Men: Resistance and Material Culture in Western Ethiopia. In: T. L. Kienlin, A. Zimmerman (eds.), *Beyond Elites. Alternatives to Hierarchical Systems in Modelling Social Formations.* Universitätsforschungen zur Prähistorischen Archäologie 215. Bonn 2012, pp. 67–82.

Gorgoglione et al. 1993: M. Gorgoglione, C. Corridi, G. Fiorentino, P. Panetta, L. Sadori, La capanna 7 di Torre Castelluccia (Pulsano, Taranto), dalle ultime fasi all'età del bronzo alla prima età del ferro. *Taras* XIII/1–2 (1993), pp. 25–114.

Gori 2014: M. Gori, Metal Hoards as Ritual Gifts: Circulation, Collection and Alienation of Bronze Artefacts in Late Bronze Age Europe. In: F. Carlàand, M. Gori (eds.), *Gift giving and the "embedded"economy in the ancient world.* Heidelberg 2014, pp. 269–288.

Gough/Mills 1972: L. J. Gough, J. S. Mills, The Composition of Succinite (Baltic Amber). *Nature* 239 (1972), pp. 527–528.

Green 1991: M. Green, *The Sun-Gods of ancient Europe.* London 1991.

Gregory 1980: C. A. Gregory, Gifts to Men and Gifts to God: Gift Exchange and Capital Accumulation in Contemporary Papua. *Man* 15/4 (1980), pp. 626–652.

Gregory 1982: C. A. Gregory, *Gifts and Commodities.* London 1982.

Grifoni/Cremonesi 1968: R. Grifoni, G. Cremonesi, La Grotta dell'Orso di Sarteano. *Origini* II (1968), pp. 247–331.

Grimaldi 1996: D. A. Grimaldi, *Amber: window to the past* (1st edition). New York 1996.

Guerreschi 1970a: G. Guerreschi, Esame tecnico di reperti in ambra rinvenuti nell'area tra Adige e Mincio. *Memorie del Museo Civico di Storia Naturale di Verona* XVIII (1970), pp. 235–257.

Guerreschi 1970b: G. Guerreschi, La problematica dell'ambra nella protostoria italiana: metodo sperimentale per la determinazione della provenienza. *Studi Etruschi,* serie II, XXXVIII (1970), pp. 169–183.

Guerreschi 1975: G. Guerreschi, Analisi di reperti d'ambra, custoditi presso il Museo Civico di Modena. *Emilia Preromana* 7 (1971–1974 [1975]), pp. 371–373.

Guerreschi 1980–1981: G. Guerreschi, La stazione preistorica del Lago Lucone, scavi '65–'71. *Annali del Museo di Gavardo* 15 (1980–1981), pp. 7–78.

Guerreschi et al. 1985: G. Guerreschi, C. Limido, P. Catalani, *L'insediamento preistorico dell'Isolone del Mincio (Volta Mantovana). Scavi Mirabella, Rittatore, Zorzi, 1955–1956.* Collana di Archeologia Padana 1. Brescia 1985.

Guglielmino 2007: R. Guglielmino, Roca Vecchia (Lecce): New Evidence for Aegean Contacts with Apulia during the Late Bronze Age. *Accordia Research Papers* 10 (2004–2006 [2007]), pp. 87–102.

Guglielmino 2021: R. Guglielmino, Roca and Its Aegean Contacts in the Recent Bronze Age. In: R. Jung (ed.), *Punta di Zambrone. 1200 B.C.E. – a Time of Breakdown, a Time of Progress in Southern Italy and Greece.* Orea 17. Wien 2021, pp. 445–458.

Guiliano et al. 2007: M. Guiliano, L. Asia, G. Onoratini, G. Mille, Applications of Diamond Crystal ATR FTIR Spectroscopy to the Characterization of Ambers. *Spectrochimica acta. Part A, Molecular and Biomolecular Spectroscopy* 67 (2007), pp. 1407–1411.

Hachmann 1957: R. Hachmann, Bronzezeitliche Bernsteinschieber. *Bayerische Vorgeschichtsblatter* 22 (1957), pp. 1–36.

Hadži/Orel 1978: D. Hadži, B. Orel, Spektrometrične raziskave jantarja in smol iż prazgodovinskih najdišč na Slovenskem. *Vestnik SKD* 25/1 (1978), p. 51.

Hahn 2008: H. P. Hahn, Diffusionism, Appropriation, and Globalization: Some Remarks on Current Debates in Anthropology. *Anthropos* 103/1 (2008), pp. 191–202.

Hahn/Weiss 2013: H. P. Hahn, H. Weiss, Introduction: Biographies, Travels and Itineraries of Things. In: H. P. Hahn, H. Weiss (eds.), *Mobility, Meaning & Transformations of Things: Shifting Contexts of Material Culture through Time and Space*. Oxford 2013, pp. 1–14.

Hall et al. 2011: T. D. Hall, P. N. Kardulias, Ch. Chase-Dunn, World-Systems Analysis and Archaeology: Continuing the Dialogue. *Journal of Archaeological Research* 19/3 (2011), pp. 233–279.

Halperin 1984: R. H. Halperin, Polanyi, Marx and the Institutional Paradigm in Economic Anthropology. *Research in Economic Anthropology* 6 (1984), pp. 245–272.

Hammer et al. 2003: C. U. Hammer, G. Kurat, P. Hoppe, W. Grum, H. B. Clausen, Thera Eruption Date 1645 BC Confirmed by New Ice Core Data? In: M. Bietak (ed.), *The Synchronisation of Civilisations in the Eastern Mediterranean in the Second Millennium B.C. II. Proceedings of the SCIEM 2000 – EuroConference, Haindorf 2nd of May–7th of May 2001*. CChEM 4. Vienna 2003, pp. 87–94.

Hanneman/Riddle 2005: R. A. Hanneman, M. Riddle, *Introduction to Social Network Methods*. Riverside 2005. Published online on 17.03.2005 (http://faculty.ucr.edu/~hanneman/).

Hänsel 1968: B. Hänsel, *Beiträge zur Chronologie der mittleren Bronzezeit im Karpatenbecken*. Bonn 1968.

Hänsel 1997: B. Hänsel, Gaben an die Götter. Schätze der Bronzezeit Europas eine Einführung. In: A. Hänsel, B. Hänsel (eds.), *Gaben an die Götter. Schätze der Bronzezeit Europas*. Berlin 1997, pp. 11–22.

Hänsel et al. 2007–2008: B. Hänsel, D. Matošević, K. Mihovilić, B. Teržan, O socijalnoj arheologiji brončanodobnog utvrđenog naselja i grobova na Monkodonji. *Histria archaeologica* 38–39 (2007–2008), pp. 81–122.

Hänsel et al. 2010: B. Hänsel, B. Teržan, K. Mihovilić, Brončane sjekire i njihovi dijelovi: opažanja na primjeru nalaza s Monkodonje u Istri. *Histria archaeologica* 41 (2010), pp. 5–33.

Hänsel et al. 2015: B. Hänsel, K. Mihovilić, B. Teržan, *Monkodonja: istraživanje protourbanog naselja brončanog doba Istre. Knjiga 1: Iskopavanje i nalazi gradevina*. Monografije i katalozi 25, Arheološki muzej Istre. Pula 2015.

Hansen 1991: S. Hansen, *Studien zu den Metalldeponierungen während der Urnenfelderzeit im Rhein-Main-Gebiet*. Bonn 1991.

Hansen 1994: S. Hansen, *Studien zu den Metalldeponierung während der älteren Urnenfelderzeit zwischen Rhonetal und Karpathenbecken*. Bonn 1994.

Hansen 2016: S. Hansen, A Short History of Fragments in Hoards of the Bronze Age. In: H. Baitinger (ed.), *Materielle Kultur und Identität im Spannungsfeld zwischen mediterraner Welt und Mitteleuropa. Akten der Internationalen Tagung am Römisch-Germanischen Zentralmuseum Mainz, 22.-24. Oktober 2014. Abschlusstagung des DFG-Projekts 'Metallfunde als Zeugnis für die Interaktion zwischen Griechen und Indigenen auf Sizilien zwischen dem 8. und 5. Jahrhundert v. Chr.'*. Mainz 2016, pp. 185–208.

Harding 1972: A. Harding, Illyrians, Italians and Mycenaeans: Trans-Adriatic Contacts during the Late Bronze Age. *Studia Albanica* 2 (1972), pp. 215–221.

Harding 1984: A. Harding, *The Mycenaeans and Europe.* London 1984.

Harding 1992: A. Harding, The Prehistoric Background of Illyrian Albania. In: T. Winnifrith (ed.), *Perspectives on Albania.* New York 1992, pp. 14–28.

Harding 1995: A. Harding, *Die Schwerter im ehemaligen Jugoslawien.* Prähistorische Bronzefunde IV/14. Stuttgart 1995.

Harding 2000: A. Harding, *European Societies in the Bronze Age.* Cambridge 2000.

Harding 2007: A. Harding, Interconnections between the Aegean and Continental Europe in the Bronze and Early Iron Ages : moving beyond scepticism. In: I. Galanaki, H. Tomas, Y. Galanakis, R. Laffineur (eds.), *Between the Aegean and Baltic Seas. Prehistory across Borders.* Aegaeum 27. Liège 2007, pp. 47–56.

Harding 2013a: A. Harding, Trade and Exchange. In: H. Fokkens, A. Harding (eds.), *The Oxford Handbook of the European Bronze Age.* Oxford 2013, pp. 370–381.

Harding 2013b: A. Harding, World Systems, Cores, and Peripheries in Prehistoric Europe. *European Journal of Archaeology* 16/3 (2013), pp. 378–400.

Harding 2017: A. Harding, Amber in the Bronze Age: Small Objects of Desire, Big Worlds. In: P. L. Cellarosi, R. Chellini, F. Martini, A. C. Montanaro, L. Sarti, R. M. Capozzi (eds.), *The Amber Roads. The Ancient Cultural and Commercial Communication between Peoples.* Firenze 2017, pp. 155–160.

Harding/Hughes-Brock 1974: A. Harding, H. Hughes-Brock, Amber in the Mycenaean World. *The Annual of the British School at Athens* 69 (1974), pp. 145–172.

Harding/Hughes-Brock 2017: A. Harding, H. Hughes-Brock, Mycenaeans in Bavaria? Amber and gold from the Bronze Age site of Bernstorf. *Antiquity* 91/359 (2017), pp. 1382–1385.

Hausmann 1996–1997: L. Haussmann, L'âge du Bronze dans les Hautes-Alpes. *Bulletin d'études préhistoriques et archéologiques alpines* VII–VIII (1996–1997), pp. 97–180.

Hedinger 1903: A. Hedinger, *Die vorgesehichtlichen Bernsteinartefakte und ihre Herkunft.* Strassburg 1903.

Helm 1877: O. Helm, Notizen über die chemische und physikalische Beschaffenheit des Bernsteins. *Arch. Pharm.* 211/3, pp. 229–246.

Helm 1881: O. Helm, Mitteilungen über Bernstein. IV. Über sicilianischen und rumänischen Bernstein. *Schriften der Naturforschenden Gesellschaft zu Danzig*, Neue Folge, 5/1–2 (1881), pp. 293–296.

Helm 1882: O. Helm, Über Appeninnen-Bernstein. *Schriften der Naturforschenden Gesellschaft zu Danzig* 5/3 (1882), pp. 11–14.

Helm 1885: O. Helm, Über die Herkunft des in den alten Königs-Gräbern von Mykenae gefundenen Bernsteins und über den Bernsteinsäuregehalt verschiedener fossiler Harze. *Schriften der Naturforschenden Gesellschaft zu Danzig*, Neue Folge, 6/2 (1885), pp. 234–239.

Helm 1891: O. Helm, Mitteilungen über Bernstein. XIV. Über Rumänit, ein in Rumänien vorkommendes fossiles Harz. *Schriften der Naturforschenden Gesellschaft zu Danzig*, Neue Folge, 7/4 (1891), pp. 186–189.

Helm/Conwentz 1886: O. Helm, H. Conwentz, Sull'ambra di Sicilia. *Malpighia* 1 (1886), pp. 49–56.

Henderson 1988: J. Henderson, Glass Production and Bronze Age Europe. *Antiquity* 62 (1988), pp. 435–51.

Henderson 2019: J. Henderson, Provenance, Mixing and Primary Production of Mixed Alkali and Potassium Glasses from Frattesina, Rovigo: An Isotopic Approach In: A. Maria Bietti Sestieri, P. Bellintani, C. Giardino (eds.), *Frattesina: un centro internazionale di produzione e di scambio nella tarda età del bronzo del Veneto.* Atti della Accademia Nazionale dei Lincei, Memorie serie IX, volume XXXIX, fascicolo 1. Roma 2019, pp. 317–326.

Henderson et al. 2015: J. Henderson, J. Evans, P. Bellintani, A. M. Bietti Sestieri, Production, Mixing and Provenance of Late Bronze Age Mixed Alkali Glasses from Northern Italy: An Isotopic Approach. *Journal of Archaeological Science* 55 (2015), pp. 1–8.

Heyd 2007: V. Heyd, When the West Meets the East. The Eastern Periphery of the Bell Beaker Phenomenon and Its Relation with the Aegean Early Bronze Age. In: I. Galanaki, H. Tomas, Y. Galanakis, R. Laffineur (eds.), *Between the Aegean and Baltic Seas. Prehistory across Borders.* Aegaeum 27. Liège 2007, pp. 91–104.

Hirschbäck-Merhar 1984: G. Hirschbäck-Merhar, Prazgodovinski depot Debeli Vrh nad Predgradom. *Arheološki vestnik* 35 (1984), pp. 90–110.

Hochstetter 1980: A. Hochstetter, *Die Hügelgräber-Bronzezeit in Niederbayern.* Bayerisches Landesamt für Denkmalpflege, Abtlg. für Vor- und Frühgeschichte: Materialhefte zur bayerischen Vorgeschichte Reihe A, Band 41. Kallmünz 1980.

Hodder 1979: I. Hodder, Social and Economic Stress and Material Culture Patterning. *American Antiquity* 44 (1979), pp. 446–454.

Hodder 1980: I. Hodder, Trade and Exchange: Definitions, Identification, and Function. In: R. E. Fry (ed.), *Models and Methods in Regional Exchange, Papers No. 1, Society for American Archaeology.* Washington (DC) 1980, pp. 151–156.

Hodder 1982: I. Hodder, *The Present Past. An Introduction to Anthropology for Archaeologists.* London 1982.

Hodder 1986: I. Hodder, *Reading the Past. Current Approaches to Interpretation in Archaeology.* Cambridge 1986.

Hodder 1987: I. Hodder, The Contextual Analysis of Symbolic Meanings. In: I. Hodder (ed.), *The Archaeology of Contextual Meaning.* New Directions in Archaeology series. Cambridge 1987, pp. 1–10.

Höflmayer 2009: F. Höflmayer, Aegean-Egyptian Synchronisms and Radiocarbon Chronology. In: D. A. Warburton (ed.), *Time's Up! Dating the Minoan Eruption of Santorini. Acts of the Minoan Eruption Chronology Workshop, Sandbjerg, November 2007.* Aarhus 2009, pp. 187–195.

Höflmayer 2012: F. Höflmayer, The Date of the Minoan Santorini Eruption: Quantifying the 'Offset'. *Radiocarbon* 54 (2012), 435–448.

Höflmayer et al. 2016: F. Höflmayer, A. Yasur-Landau, E. Cline, M. Dee, B. Lorentzen, S. Riehl, New Radiocarbon Dates from Tel Kabri Support a High Middle Bronze Age Chronology. *Radiocarbon* 58/3 (2016), pp. 599–613.

Holloway et al. 1990: R. R. Holloway, M. S. Joukowsky, S. S. Lukesh, La Muculufa. The Early Bronze Age Sanctuary, The Early Bronze Age Village, Excavations of 1982 and 1983. *Revue des Archeologues et Historiens d'Art de Louvain* 23 (1990), pp. 11–67.

Holste 1939: F. Holste. *Die Bronzezeit im Nordmainischen Hessen.* Vorgeschichtliche Forschungen XII. Berlin 1939.

Homen 1982: Z. Homen, Novi Kasnobrončanodobni lokalitet u Križevcima. *Muzejski Vjesnik* 5 (1982), pp. 18–24.

Hood 1993: S. Hood, Amber in Egypt. In: C. W. Beck, J. Bouzek (eds.), *Amber in Archaeology. Proceedings of the IInd International Conference on Amber in Archaeology, Liblice 1990.* Praha 1993, pp. 230–236.

Horden/Purcell 2000: P. Horden, N. Purcell, *The Corrupting Sea: A Study of Mediterranean History.* Oxford 2000.

Horváth 1999: T. Horváth, Contribution to the Study of Hungarian Amber-Finds. *Savaria* 24/3 (1999), pp. 277–289.

Hoskins 1998: J. Hoskins, *Biographical Objects: How Things Tell the Stories of People's Lives.* New York – London 1998.

Hristov 2017: M. Hristov, Naselje kasnog bronzanog doba kod Banatskog dvora. *Građa za proučavanje spomenika kulture Vojvodine* XXX (2017), pp. 7–31.

Hu 2005: Y. F. Hu, Efficient and High Quality Force-Directed Graph Drawing. *The Mathematica Journal* 10 (2005), pp. 37–71.

Hughes-Brock 1985: H. Hughes-Brock, Amber and the Mycenaeans. *Journal of Baltic Studies* 16 (1985), pp. 257–267.

Hughes-Brock 1993: H. Hughes-Brock, Amber in the Aegean in the Late Bronze Age: Some Problems and Perspectives. In: C. W. Beck, J. Bouzek (eds.), *Amber in Archaeology. Proceedings of the IInd International Conference on Amber in Archaeology, Liblice 1990.* Praha 1993, pp. 223–234.

Hughes-Brock 1998: H. Hughes-Brock, Mycenaean Amber Beads and Ornaments: What Can We Learn from Their Shapes – and from One Another In: N. Negroni Catacchio, C. W. Beck (eds.), *Workshop 7. L'ambra in Archaeologia. Atti del XIII Congress, Volume 6, Tome I, Forli-Italia, 1996, 8–14 settembre.* Forlì 1998, pp. 491–496.

Hughes-Brock 1999: H. Hughes-Brock, Mycenaean Beads: Gender and Social Contexts. *Oxford Journal of Archaeology* 18 (1999), pp. 277–296.

Hulek 2013: F. W. Hulek, *Fundorte von Keramik ägäischen Typs in Italien.* Universitätsforschungen zur Prähistische Archäologie 223. Bonn 2013.

Hundt 1955: H. J. Hundt, Versuch zur Deutung der Depotfunde der Nordischen jüngeren Bronzezeit unter besonderer Berücksichtigung Mecklenburgs. *Jahrbuch des Römisch-germanischen Zentralmuseums Mainz* 2 (1955), pp. 95–140.

Huska et al. 2014: A. Huska, W. Powell, S. Mitrović, H. A. Bankoff, A. Bulatović, V. Filipović, R. Boger, Placer Tin Ores from Mt. Cer, West Serbia, and Their Potential Exploitation during the Bronze Age. *Geoarchaeology: An International Journal* 29 (2014), pp. 477–493.

Hutterer 1977: K. L. Hutterer, Preface. In: K. L. Hutterer (ed.), *Economic Exchange and Social Interaction in Southeast Asia: Perspectives from Prehistory, History and Ethnography.* Ann Arbor 1977, pp. 8–9.

Iacono 2013: F. Iacono, Westernizing Aegean of LH IIIC. In: M. E. Alberti, S. Sabatini (eds.), *Exchange Networks and Local Transformations: Interactions and Local Changes in Europe and the Mediterranean between Bronze and Iron Ages.* Oxford 2013, pp. 60–79.

Iacono/Guglielmino 2021: F. Iacono, R. Guglielmino, The Ionian-Adriatic Interface as a Landscape Of mobility. In: Ch. Souyoudzoglou-Haywood, Ch. Papoulia (eds.), *Archaeology of the Ionian Sea: Landscapes, Seascapes and the Circulation of People, Goods and Ideas from the Palaeolithic to the End of the Bronze Age.* Oxford – Philadelphia 2021, pp. 201–209.

Iaia et al. 2005: C. Iaia, B. Barbaro, S. Favorito, *L'insediamento dell'antica età del bronzo di Tenuta Quadraro-via Lucrezia Romana (Roma). Dati preliminari per un inquadramento culturale e cronologico.* BAR International Series 1452/I. Oxford 2005, pp. 449–456.

Iannì 2017: F. Iannì, Recensioni / Reviews: R. Gennusa, L'evoluzione millenaria di uno stile. La civiltà del Bronzo castellucciana nella Sicilia meridionale, All'Insegna del Giglio, Firenze 2015. *Origini* XL (2017), pp. 276–280.

Iannì forthcoming: F. Iannì, Seriazioni ceramiche e cronologia assoluta: per una suddivisione in fasi del castellucciano della Sicilia centrale. In: *Atti del Convegno 'Prima di Thapsos. La Sicilia centro orientale tra L'Eneolitico Finale e l'Età del Bronzo Antico', Siracusa, 16–17 dicembre 2011.*

Innerhofer 2010: F. Innerhofer, Zwischen Frühbronzezeit und Hügelgräberkultur – Zum zeitlichen Umfeld der Himmelsscheibe von Nebra in Mitteldeutschland. In: H. Meller, F. Bertemes (eds.), *Der Griff nach den Sternen. Wie Europas Eliten zu Macht und Reichtum kamen. Internationales Symposium Halle, 16.–21. Februar 2005.* Tagungen Des Landesmuseums Für Vorgeschichte Halle 5. Halle (Saale) 2010, pp. 119–129.

Innerhofer 2013: F. Innerhofer, Von der frühen zur mittleren Bronzezeit in Süddeutschland – Wandel oder zäsur? In: H. Meller, F. Bertemes, H. R. Bork, R. Risch (eds.), *1600 – Kultureller Umbruch im Schatten des Thera-Ausbruchs? 4. Mitteldeutscher Archäologentag vom 14. Bis 16. Oktober 2011 in Halle (Saale).* Tagungen des Landesmuseums für Vorgeschichte Halle 9. Halle (Saale) 2013, pp. 443–451.

Irwin-Williams 1977: C. Irwin-Williams, A Network Model for the Analysis of Prehistoric Trade. In: T. K. Earle, J. Ericson (eds.), *Exchange Systems in Prehistory.* New York 1977, pp. 141–151.

Islami 2013: S. Islami, *Kultura ilire e tumave të Matit.* Prishtinë 2013.

Islami/Ceka 1964: S. Islami, H. Ceka, Nouvelles dorinées. *Studia Albanica* 1 (1964), pp. 91–101.

Islami et al. 1955: S. Islami, H. Ceka, F. Prendi, S. Anamali, Zbulime të kulturës iure në Luginën e Matit. *Buletin per Shkencat Shoqerore* 1 (1955), pp. 110–138.

Jackenhövel 1971: A. Jackenhövel, *Die Rasiermesser in Mitteleuropa.* Prähistorische Bronzefunde VIII/1. München 1971.

Jackenhövel 1991: A. Jackenhövel, Räumliche Mobilität von Personen in der mittleren Bronzezeit des westlichen Mitteleuropa. *Germania* 69 (1991), pp. 49–62.

Jackson 2006: P. Jackson, Entropy and Redundancy in Culture. *Method and Theory in the Study of Religion* 18 (2006), pp. 219–231.

Jaeger 2017: M. Jaeger, Middle Bronze Age Amber Finds in Hungary. In: P. L. Cellarosi, R. Chellini, F. Martini, A. C. Montanaro, L. Sarti, R. M. Capozzi (eds.), *The Amber Roads. The Ancient Cultural and Commercial Communication between Peoples.* Firenze 2017, pp. 205–215.

Jaeger/Kulcsár 2013: M. Jaeger, G. Kulcsár, Kakucs–Balla-domb a Case Study in the Absolute and Relative Chronology of the Vatya Culture. *Acta Archaeologica Academiae Scientiarum Hungaricae* 64 (2013), pp. 289–320.

Jaeger et al. 2018: M. Jaeger, R. Staniuk, J. Müller, G. Kulcsár, N. Taylor, History of the Bronze Age Habitation. In: M. Jaeger, G. Kulcsár, N. Taylor, R. Staniuk (eds.), *Kakucs-Turján: a Middle Bronze Age multi-layered fortified settlement in Central Hungary.* Studien zur Archäologie in Ostmitteleuropa 18. Bonn 2018, pp. 97–118.

Jaeger et al. forthcoming: M. Jaeger, J. Kneisel, J. Czebreszuk (eds.), *Middle Bronze Age Barrows in South-Western Poland and Their Cultural Context.* Studien zur Archäologie in Ostmitteleuropa. Bonn. (forthcoming/in print).

Jaworski et al. 1972: M. Jaworski, J. Krauze, A. Lempka, S. Richter, Badania metodami chromatograficzną i spektrofotometryczną przedmiotów bursztynowych pochodzących z wykopalisk archeologicznych. *Fontes Archeaologici Posnaniensies* 21 (1972), pp. 230–238.

Jennings 2014: B. Jennings, *Travelling Objects: Changing Values. The Role of Northern Alpine Lake-Dwelling Communities in Exchange and Communications Networks during the Late Bronze Age.* Oxford 2014.

Jones et al. 2002: R. E. Jones, L. Vagnetti, S. T. Levi, J. Williams, D. Jenkins, A. De Guio, Mycenaean Pottery from Northern Italy. Archaeological and Archaeometric Studies. *Studi Micenei Ed Egeo-Anatolici*, Nuova Serie, 44/2 (2002), pp. 221–261.

Jones et al. 2005: R. E. Jones, S. T. Levi, M. Bettelli, Mycenaean Pottery in the Central Mediterranean: Imports, Imitations and Derivatives. In: I. Galanaki, H. Tomas, Y. Galanakis, R. Laffineur (eds.), *Between the Aegean and Baltic Seas. Prehistory across Borders.* Aegaeum 27. Liège 2007, pp. 539–549.

Jones et al. 2014: R. E. Jones, S. T. Levi, M. Bettelli, L. Vagnetti, *Italo-Mycenaean Pottery: The Archaeological and Archaeometric Dimensions.* Roma 2014.

Jubani 1969: B. Jubani, Varreza tumulare e Çinamakut (Kukës). *Buletin Arkeologjik* 1969, pp. 37–50.

Jubani 1974: B. Jubani, Varreza tumulare e Çinamakut-Kukës 1973. *Iliria* III (1974), pp. 421–425.

Jubani 1982: B. Jubani, Tumat e Krumës (Rrethi i Kukësit). *Iliria* XII/2 (1982), pp. 147–195.

Jubani 1990: B. Jubani, *Tumat e Dardanëve alpinë të Çinamakut.* Unpublished PhD thesis defended at Universiteti i Tiranës in 1990.

Jubani 1995: B. Jubani, Il traffico e gli scambi commerciali-culturali fra l'Albania ed i Paesi limitrofi nella prima epoca del ferro. In: B. Hänsel (ed.), *Handel, Tausch und Verkehr im Bronze- und Früheisenzeitlichen Südosteuropa.* Südosteuropa Schriften 17. München – Berlin 1995, pp. 363–369.

Jung 2006: R. Jung, *Χρονολογία comparata: Vergleichende Chronologie von Südgriechenland und Süditalien von ca. 1700/1600 bis 1000 v. u. Z.* Veröffentlichungen der Mykenischen Kommission 26, Vien 2006.

Jung 2007: R. Jung, Goldene Vögel und Sonnen. Ideologische Kontakte zwischen Italien und der postpalatialen Ägäis. In: E. Alram-Stern, G. Nightingale (eds.), *KEIMELION. The Formation of Elites and Elitist Lifestyles from Mycenaean Palatial Times to the Homeric Period. Akten des internationalen Kongresses vom 3. bis 5. Februar 2005 in Salzburg.* Vien 2007, pp. 219–255.

Jung 2009: R. Jung, I 'Bronzi Internazionali' ed il loro Contesto Sociale fra Adriatico, Penisola Balcanica e Coste Levantine. In: E. Borgna, P. Càssola Guida (eds.), *From the Aegean to the Adriatic: Social Organisations, Modes of Exchange and Interaction in the Post-palatial Times (12th–11th c. BC).* Studi e Ricerche di Protostoria Mediterranea 8. Roma 2009, pp. 129–157.

Jung 2013: R. Jung, The Time around 1600 BCE in Southern Italy: New Powers, New Contacts and New Conflict. In: H. Meller, F. Bertemes, H. R. Bork, R. Risch (eds.), *1600 – Kultureller Umbruch im Schatten des Thera-Ausbruchs? 4. Mitteldeutscher Archäologentag vom 14. Bis 16. Oktober 2011 in Halle (Saale).* Tagungen des Landesmuseums für Vorgeschichte Halle 9. Halle (Saale) 2013, pp. 235–251.

Jung 2017: R. Jung, Chronological Problems of the Middle Bronze Age in Southern Italy. In: T. Lachenal, C. Mordant, T. Nicolas, C. Véber (eds.), *Le Bronze moyen et l'origine du Bronze final en Europe occidentale (XVIIe-XIIIe siècle av. J.-C.).* Strasbourg 2017, pp. 623–642.

Jung/Mehofer 2013: R. Jung, M. Mehofer, Mycenaean Greece and Bronze Age Italy: Cooperation, Trade or War? *Archäologisches Korrespondenzblatt* 43/2 (2013), pp. 175–193.

Jung/Pacciarelli 2017: R. Jung, M. Pacciarelli, Greece and Southern Italy 1250–1050 BC: Manifold Patterns of Interaction. In: M. Fotiadis, R. Laffineur, Y. Lolos, A. Vlachopoulos (eds.), *Hesperos. The Aegean Seen from the West.* Aegaeum 41. Leuven – Liège 2017, pp. 187–204.

Jung et al. 2009: R. Jung, S. Andreou, B. Weninger, Synchronisation of Kastanás and Thessaloníki Toumba at the End of the Bronze and the Beginning of the Iron Age. In: S. Deger-Jalkotzy, A. Baechle (eds.), *LH IIIC Chronology and Synchronisms III: LH IIIC Late and the Transition to the Early Iron Age. Proceedings of the International Workshop at the Austrian Academy of Sciences at Vienna, February 23rd and 24th, 2007.* Veröffentlichungen der Mykenischen Kommission 30. Vien 2009, pp. 183–202.

Jung et al. 2011: R. Jung, M. Mehofer, E. Pernicka, Metal Exchange in Italy from the Middle to the Final Bronze Age (14th–11th Century B.C.E.). In: P. Betancourt, S. C. Ferrence (eds.), *Metallurgy: Understanding How, Learning why. Studies in Honour of James D. Muhly.* Prehistory Monographs 29. Philadelphia 2011, pp. 231–248.

Jung et al. 2015: R. Jung, M. Pacciarelli, B. Zach, M. Klee, U. Thanheiser, Punta di Zambrone (Calabria) – a Bronze Age Harbour Site. First Preliminary Report on the Recent Bronze Age (2011–2012 Campaigns). *Archaeologia Austriaca* 99 (2015), pp. 53–110.

Kadrow 2001: S. Kadrow, *U progu nowej epoki. Gospodarka i społeczeństwo wczesnego okresu epoki brązu w Europie Środkowej.* Kraków 2001.

Karavanić 2007: S. Karavanić, Middle to Late Bronze Age in Northern Croatia. In: G. Tiefengraber (ed.), *Studien zur Mittel- und Spätbronzezeit am rande der Südostalpen.* Universitätsforschungen zur Prähistorischen Archäologie 148. Bonn 2007, pp. 37–49.

Kardulias 1999: P. N. Kardulias (ed.), *World-Systems Theory in Practice: Leadership, Production and Exchange.* Lanham 1999.

Kardulias 2001: P. N. Kardulias, *Negotiated Peripherality: Making Incorporation Work for You on the Margins of World-Systems.* Paper presented at the 100th Annual Meeting of the American Anthropological Association. Washington (DC) 2001.

Karo 1930: G. Karo, *Die Schachtgräber von Mykenai.* München 1930.

Kasper 1996: H. U. Kasper: Der rumänische Bernstein. In: M. Ganzelewski, R. Slotta (eds.), *Bernstein. Tränen der Götter.* Bochum 1996, pp. 357–362.

Kaul 2005: F. Kaul, Bronze Age Tripartite Cosmologies. *Prähistorische Zeitschrift* 80/2 (2005), pp. 135–148.

Kaur et al. 2013: S. Kaur, E. Stout, G. Kaplan, Infrared Spectroscopy of Bronze Age Ambers from S. Italy and Sicily, Appendice. In: I. Matarese (ed.), *Vaghi e pendenti dell'età del Bronzo in Italia meridionale e Sicilia: pietra, materiali vetrosi, ambra e metallo.* Unpublished PhD thesis defended at Università degli Studi di Napoli 'L'Orientale' in 2013.

Kaur et al. 2020: S. Kaur, E. Stout, R. Kim, M. Cwaliński, *Characterization of Amber Beads from Grave Sites in Albania and Western Serbia.* ARL report #202, Vassar College. Poughkeepsie 2020.

Kienlin 2012: T. L. Kienlin, Patterns of Change, or: Perceptions Deceived? Comments on the Interpretation of Late Neolithic and Bronze Age Tell Settlement in the Carpathian Basin. In: T. L. Kienlin, A. Zimmermann (eds.), *Beyond Elites. Alternatives to Hierarchical Systems in Modelling Social Formations,* vol. 1. Universitätsforschungen zur Prähistorischen Archäologie 215. Bonn 2012, pp. 251–310.

Kienlin/Zimmermann 2012: T. L. Kienlin, A. Zimmermann (eds.), *Beyond Elites. Alternatives to Hierarchical Systems in Modelling Social Formations,* vol. 1. Universitätsforschungen zur Prähistorischen Archäologie 215. Bonn 2012.

Kilian-Dirlmeier 1985: I. Kilian-Dirlmeier, Schwerter Ägäischen Typs in Albanien. *Iliria* XV/2 (1985), pp. 251–264.

Kitanovski 1960: B. Kitanovski, Dva groba iz starijeg gvozdenog doba kod Prilepa. *Starinar* XI (1960), pp. 209–214.

Knapp 1998: A. B. Knapp, Mediterranean Bronze Age Trade: Distance, Power and Place. In: E. H. Cline, D. Harris-Cline (eds.), *The Aegean and the Orient in the Second Millennium, Proceedings of the 50th Anniversary Symposium, University of Cincinnati, 18–20 April 1997.* Aegaeum 18. Liège – Austin 1998, pp. 193–207.

Knapp/Manning 2017: A. B. Knapp, S. W. Manning, Crisis in Context: The End of the Late Bronze Age in the Eastern Mediterranean. *American Journal of Archaeology* 120/1 (2017), pp. 99–149.

Knappett 2008: C. Knappett, The Neglected Networks of Material Agency: Artefacts, Pictures and Texts. In: C. Knappett, L. Malafouris (eds.), *Material Agency: Towards a Non-Anthropocentric Approach.* New York 2008, pp. 139–156.

Knappett 2011: C. Knappett, *An Archaeology of Interaction: Network Perspectives on Material Culture and Society*. Oxford 2011.

Knappett 2013a: C. Knappett (ed.), *Network Analysis in Archaeology: New Approaches to Regional Interaction*. Oxford 2013.

Knappett 2013b: C. Knappett Introduction: Why Networks? In: C. Knappett (ed.), *Network Analysis in Archaeology: New Approaches to Regional Interaction*. Oxford 2013, pp. 3–15.

Knappett/Malafouris 2008: C. Knappett, L. Malafouris, Material and Nonhuman Agency: An Introduction. In: C. Knappett, Malafouris L. Malafouris (eds.), *Material Agency: Towards a Non-Anthropocentric Approach*. New York 2008, pp. IX-XIX.

Knappett et al. 2008: C. Knappett, T. Evans, R. Rivers, Modelling Maritime Interaction in the Aegean Bronze Age. *Antiquity* 82/318 (2008), pp. 1009–1024.

Kneisel 2012: J. Kneisel , The Problem of the Middle Bronze Age Inception in Northeast Europe – or: Did the Únětice Society Collapse? In: J. Kneisel, W. Kirleis, M. Dal Corso, N. Taylor, V. Tiedtke (eds.), *Collapse or Continuity? Environment and Development of Bronze Age Human Landscape – Proceedings of the International Workshop 'Socio-environmental Dynamics over the Last 12,000 Years: The Creation of Landscape II' (14th–18th March 2011, Kiel)*. Universitätsforschungen zur Prähistorischen Archäologie 205. Bonn 2012, pp. 209–234.

Kohl 1989: P. L. Kohl, The Use and Abuse of World-Systems Theory: The Case of the 'Pristine' West Asian State. In: C. C. Lamberg-Karlovsky (ed.), *Archaeological Thought in America*. Cambridge 1989, pp. 218–240.

Kolendo 1981: J. Kolendo, Wyprawa po bursztyn bałtycki za Nerona [An Expedition for Baltic Amber during the Reign of Nero]. *Pomerania Antiqua* 10 (1981), pp. 25–63.

Kolendo 1983: J. Kolendo, Miejsca występowania bursztynu wg. Pliniusza Starszego. *Prace Muzeum Ziemii* 37 (1983), pp. 5–26.

König 2004: P. König, *Spätbronzezeitliche Hortfunde aus Bosnien und der Herzegowina*. Prähistorische Bronzefunde XX/11. Stuttgart 2004.

Köninger 2006: J. Köninger, *Siedlungsarchäologie im Alpenvorland VIII. Die frühbronzezeitlichen Ufersiedlungen von Bodman-Schachen I: Befunde und Funde aus den Tauchsondagen 1982–1984 und 1986*. Forschungen und Berichte zur Vor- und Frühgeschichte in Baden-Württemberg 85. Stuttgart 2006.

Köninger/Schöbel 2010: J. Köninger, G. Schöbel, Bronzezeitliche Fundstellen zwischen Bodensee und Oberschwaben. In: I. Matuschik (ed.), *Vernetzungen – Aspekte Siedlungsarchäologischer Forschung. Festschrift für Helmut Schlichtherle zum 60. Geburtstag*. Freiburg im Breisgau 2010, pp. 385–438.

Kopytoff 1986: I. Kopytoff, The Cultural Biography of Things: Commoditization as Process. In: A. Appadurai (ed.) *The Social Life of Things: Commodities in Cultural Perspective*. Cambridge 1986, pp. 64–91.

Korkuti 2003: M. Korkuti, Researches and Studies of Prehistory in Albania. In: D. Grammenos (ed.), *Recent Research in the Prehistory of the Balkans*. Thessaloniki 2003, pp. 205–255.

Kosmowska-Ceranowicz 1999: B. Kosmowska-Ceranowicz, Succinite and Some Other Fossil Resins in Poland and Europe (Deposits, Finds, Features and Differences in

IRS). *Estudios Del Museo De Ciencias Naturales De Alava* 14, Num. Espec. 2 (1999), pp. 73–117.

Kosmowska-Ceranowicz 2001: B. Kosmowska-Ceranowicz, Bursztyn i inne żywice kopalne świata. Bursztyn bałtycki – sukcynit. *Polski Jubiler* 1/14 (2001), pp. 34–38.

Kosmowska-Ceranowicz 2017: B. Kosmowska-Ceranowicz, *Amber in Poland and in the World* (2nd edition). Warszawa 2017.

Kosmowska-Ceranowicz/Vávra 2015: B. Kosmowska-Ceranowicz, N. Vávra, Infrared Spectra Atlas of Fossil Resins, Subfossils Resins and Selected Imitations of Amber. In: N. Vávra (ed.), *Mineral Names Used for Fossil Resins and Similar Materials* (1st edition). Warszawa 2015.

Kosorić 1978: M. Kosorić, Rezultati istraživanja humki na području Roćevica 1977. godine. *Članci i Građa* XII (1978), pp. 23–29.

Kosorić/Krstić 1970: M. Kosorić, D. Krstić, Iskopavanje praistorijskih humki u Padjinama i Roćevicu. *Članci i Građa* VIII (1970), pp. 23–59.

Kosorić/Krstić 1972: M. Kosorić, D. Krstić, Iskopavanje praistorijskih humki u Padjinama i Roćevicu 1970. i 1971. godine. *Članci i Građa* IX (1972), pp. 9–27.

Kosorić/Krstić 1988: M. Kosorić, D. Krstić, Hronološka determinacija grobova iz humki sa poteza Trnovice-Padine-Ročevići. *Zbornik Narodnog Muzeja* 13 (1988), pp. 29–55.

Kossack 1954: G. Kossack, *Studien zum Symbolgut der Urnenfelderund Hallstattzeit Mitteleuropas.* Römisch-Germanische Forschungen XX. Berlin 1954.

Kubach 1985: W. Kubach, Einzel- und Mehrstückdeponierungen und ihre Fundplätze. *Archäologisches Korrespondenzblatt* 15 (1985), pp. 179–185.

Krapf 2014: T. Krapf, The Late Bronze Age Pottery of Macedonia: Comparisons with the Plain of Korçë. In: E. Stefani, N. Meroussis, A. Dimoula (eds.), *A Centuryof Research in Prehistoric Macedonia, 1912–2012. International Conference Proceedings, Archaeological Museum of Thessaloniki, 22–24 November 2012.* Thessaloniki 2014, pp. 585–597.

Krapf 2018: T. Krapf, Sovjan (Korçë), First Remarks About the Late Bronze Age Pottery. In: G. Karamitrou-Mentessidi (ed.), *To archaiologiko ergo stin Ano Makedonia 3,* vol. A. Aiani 2018, pp. 115–131.

Krenn-Leeb 2013: A. Krenn-Leeb, Sozialer Wandel um 1600 c. Ch. in österreich. In: H. Meller, F. Bertemes, H. R. Bork, R. Risch (eds.), *1600 – Kultureller Umbruch im Schatten des Thera-Ausbruchs. 4. Mittledeutscher Archäologentag vom 14. Bis 16. Oktober 2011 in Halle (Saale).* Tagungen des Landesmuseums für Vorgeschichte Halle 9. Halle (Saale) 2013, pp. 411–433.

Kristiansen 1978: K. Kristiansen, The Consumption of Wealth in Bronze Age Denmark. A Study in the Dynamics of Economic Processes in Tribal Societies. In: K. Kristiansen, C. Paludan-Müller (eds.), *New Directions in Scandinavian Archaeology.* Studies in Scandinavian Prehistory and Early History 1. Copenhagen 1978, pp. 158–190.

Kristiansen 1998: K. Kristiansen, The Emergence of the European World System in the Bronze Age: Divergence, Convergence and Social Evolution during the First and Second Millennia BC in Europe. In: K. Kristiansen, M. Rowlands (eds.), *Social Transformations in Archaeology: Global and Local Perspectives.* London – New York 1998, pp. 280–315.

Kristiansen 2002: K. Kristiansen, The Tale of the Sword – Swords and Swordfighters in Bronze Age Europe. *Oxford Journal of Archaeology* 21 (2002), pp. 319–332.

Kristiansen 2012: K. Kristiansen, Bronze Age Dialectics: Ritual Economies and the Consolidation of Social Divisions. In: T. L. Kienlin, A. Zimmermann (eds.), *Beyond Elites. Alternatives to Hierarchical Systems in Modelling Social Formations*, vol. 2. Universitätsforschungen zur Prähistorischen Archäologie 215. Bonn 2012, pp. 381–392.

Kristiansen 2013: K. Kristiansen, Female Clothing and Jewellery in the Nordic Bronze Age. In: S. Bergerbrant, S. Sabatini (eds.), *Counterpoint: Essays in Archaeology and Heritage Studies in Honour of Professor Kristian Kristiansen.* BAR International Series 2508. Oxford 2013, pp. 755–769.

Kristiansen 2017: K. Kristiansen, Interpreting Bronze Age Trade and Migration. In: E. Kiriatzi, C. Knappett (eds.), *Human Mobility and Technological Transfer in the Prehistoric Mediterranean.* Toronto 2017, pp. 154–180.

Kristiansen/Rowlands 1998: K. Kristiansen, M. Rowlands (eds.), *Social Transformations in Archaeology: Global and Local Perspectives.* London – New York 1998.

Kristiansen/Larsson 2005: K. Kristiansen, T. Larsson, *The Rise of Bronze Age Society. Travels, Transmissions and Transformations.* Cambridge 2005.

Kristiansen/Suchowska-Ducke 2015: K. Kristiansen, P. Suchowska-Ducke, Connected Histories: The Dynamics of Bronze Age Interaction and Trade 1500–1100 BC. *Proceedings of the Prehistoric Society* 81 (2015), pp. 361–392.

Kurti 1998: A. Kuhrt, The Old Assyrian Merchants. In: H. Parkins, C. Smith (eds.), *Trade, Traders and the Ancient City.* London 1998, pp. 1–15.

Kurti 2013: R. Kurti, Qelibari gjatë periudhës së bronzit të vonë dhetë hekurit në Shqipëri. *Iliria* XXXVI (2013), pp. 73–108.

Kurti 2017a: R. Kurti, On Some Aspects of the Late Bronze Age Burial Costume from North Albania. In: M. Korkuti (ed.), *Proceedings of the International Conference 'New Archaeological Discoveries in the Albanian Regions'.* Tiranë 2017, pp. 83–129.

Kurti 2017b: R. Kurti, Carnelian and Amber Beads as Evidence of Late Bronze Age Contacts between the Present Territory of Albania and the Aegean. In: M. Fotiadis, R. Laffineur, Y. Lolos, A. Vlachopoulos (eds.), *Hesperos. The Aegean Seen from the West.* Aegaeum 41. Leuven – Liège 2017, pp. 287–298.

Kurti et al. 2020: R. Kurti, M. Cwaliński, S. Kaur, E. Stout, R. Kim, Provenience of Albanian Bronze Age Amber Finds. *Iliria* XLIV (2020), pp. 89–130.

Kutschera 2013: W. Kutschera, Dating of the Thera/Santorini Volcanic Eruption. In: H. Meller, F. Bertemes, H. R. Bork, R. Risch (eds.), *1600 – Kultureller Umbruch im Schatten des Thera-Ausbruchs. 4. Mittledeutscher Archäologentag vom 14. Bis 16. Oktober 2011 in Halle (Saale).* Tagungen des Landesmuseums für Vorgeschichte Halle 9. Halle (Saale) 2013, pp. 59–64.

Labate 1997a: D. Labate, L'insediamento terramaricolo di Gorzano (MO). In: M. Bernabò Brea, A. Cardarelli, M. Cremaschi (eds.), *Le Terramare. La più antica civiltà padana.* Exhibition catalogue. Modena 1997, pp. 99–102.

Labate 1997b: D. Labate, La Terramara di Casinalbo (MO). In: M. Bernabò Brea, A. Cardarelli, M. Cremaschi (eds.), *Le Terramare. La più antica civiltà padana.* Exhibition catalogue. Modena 1997, p. 103.

Langenheim 1969: J. H. Langenheim, Amber: A Botanical Inquiry. *Science* 163 (1969), pp. 1157–1169.

Langenheim 2003: J. H. Langenheim, *Plant Resins. Chemistry, Evolution, Ecology, and Ethnobotany.* Portland 2003.

Langenheim/Beck 1965: J. H. Langenheim, C. W. Beck, Infrared Spectra as a Means of Determining the Botanical Sources of Amber. *Science* 149 (1965), pp. 52–55.

Larsson 1999: T. Larsson, Symbols in a European Bronze Age Cosmology. In: C. Orrling (ed.), *Communication in Bronze Age Europe*, vol 9. Studies of the Museum of National Antiquities. Stockholm 1999, pp. 9–17.

Latour 1986: B. Latour, The Powers of Association. In J. Law (ed.), *Power, Action and Belief: A New Sociology of Knowledge?* Sociological Review Monograph 32. London – Boston – Henley 1986, pp. 264–280.

Latour 2005: B. Latour, *Reassembling the Social: An Introduction to Actor-Network-Theory.* Oxford 2005.

Laviosa Zambotti 1939: P. Laviosa Zambotti, Civiltà palafitticola lombarda e civiltà di Golasecca. Origini e Interferenze. *Rivista archeologica dell'antica provincia e diocesi di Como* 119/120 (1939), pp. 1–237.

Laviosa Zambotti 1954: P. Laviosa Zambotti, *I Balcani e l'Italia nella Preistoria.* Origines (Raccolta in scritti in onore di Mons. G. Baserga). Como 1954.

Law 1999: J. Law, After ANT: Complexity, Naming and Topology. In: J. Law, J. Hassard (eds.), *Actor Network Theory and After.* Oxford 1999, pp. 1–14.

Lazić 2007: M. Lazić, A Bronze Age Mound on the Site of Savin Lakat near Prijepolje. *Archaica* 1 (2007), pp. 109–132.

Le Fevre Lehöerff 1994: A. Le Fevre Lehöerff, Les moules de l'Age du bronze dans la plaine orientale du Po: vestiges de mise en forme des alliages base cuivre. *Padusa* XXVIII (1992), pp. 131–244.

Lebez 1968: D. Lebez, The Analysis of Archaeological Amber and Amber from the Baltic Sea by Thin-Layer Chromatography. *Journal of Chromatography* 33 (1968), pp. 544–547.

Leighton 1999: R. Leighton, *Sicily Before History: An Archaeological Survey from the Palaeolithic to the Iron Age.* London 1999.

Leighton 2000: R. Leighton, Time Versus Tradition: Iron Age Chronologies in Sicily and Southern Italy. In: D. Ridgway, F. R. Serra Ridgway, M. Pearce, E. Herring, R. D. Whitehouse, J. B. Wilkins (eds.), *Ancient Italy in Its Mediterranean Setting. Studies in Honour of Ellen Macnamara.* Accordia Specialist Studies on the Mediterranean 4. London 2000, pp. 33–48.

Levi et al. 2009: S. T. Levi, B. Prosdocimi, A. Vanzetti, Facies di Capo Graziano. In: G. Tigano (ed.), *Mylai II. Scavi e ricerche nell'area urbana (1996 e 2005).* Messina 2009, pp. 32–136.

Levy 1982: J. E. Levy, *Social and Religious Organization in Bronze Age Denmark. An Analysis of Ritual Hoard Finds.* Oxford 1982.

Leonardi 1979: G. Leonardi, Il bronzo finale dell'Italia nord-orientale. Proposte per una suddivisione in fasi. In: I.I.P.P. (eds.), *Il Bronzo finale in Italia. Atti della XXI*

Riunione Scientifica dell'Istituto Italiano di Preistoria e Protostoria (Firenze, 21–23 ottobre 1973). Firenze 1979, pp. 155–185.

Leonardi et al. 1979: G. Leonardi, E. Bianchini, C. Balista, G. Stabile, Ripresa degli sacvi nella palafitta di Molina di Ledro – Scavi 1980 – Nota preliminare. *Preistoria Alpina* 15 (1979), pp. 39–55.

Leonardi et al. 2015: G. Leonardi, M. Cupitò, M. Baioni, C. Longhi, N. Martinelli, Northern Italy around 2200 calBC – From Copper Age to Early Bronze Age: Continuity and/or Discontinuity. In: H. Meller, H. W. Arz, R. Jung, R. Risch (eds.), 2200 BC – Ein Klimasturz als Ursachefür den *Zerfall der Alten Welt? 7. Mitteldeutscher Archäologentagvom 23. bis 26. Oktober 2014 in Halle (Saale).* Tagungen des Landesmuseums für Vorgeschichte Halle 12/II. Halle (Saale) 2015, pp. 283–304.

Leoni 2011: D. Leoni, *L'ambra Siciliana: Caratterizzazione del più importante materiale gemmologico italiano del Museo di Mineralogia della Sapienza.* Roma 2011.

Lera et al. 2009: P. Lera, S. Oikonomidis, A. Papayiannis, A. Tsonos, Settlement Organisation and Social Context in the SW Balkan Peninsula (Epirotic and Albanian Coasts) and Northern Italy during the Transitional Period between the Late Bronze Age and the Early Iron Age (c. 13th–9th B.C.). In: E. Borgna, P. Càssola Guida (eds.), *From the Aegean to the Adriatic: social organisations, modes of exchange and interaction in the post-palatial times (12th–11th c. BC).* Studi e Ricerche di Protostoria Mediterranea 8. Roma 2009, pp. 325–343.

Lera et al. 2010: P. Lera, F. Prendi, G. Touchais, Le passage du Bronze Récent au Fer Ancien sur le site de Sovjan (basin de Korçë, Albanie): nouvelles données chronologiques. In: J. L. Lamboley, M. P. Castiglioni (eds.), *L'Illyrie méridionale et l'Épire dans l'antiquité V, actes du Ve colloque international de Grenoble 8–11 octobre 2008.* Paris 2010, pp. 41–52.

Ling et al. 2014: J. Ling, Z. Stos-Gale, L. Grandin, K. Billström, E. Hjärthner-Holdar, Moving Metals II: Provenancing Scandinavian Bronze Age Artefacts by Lead Isotope and Elemental Analyses. *Journal of Archaeological Science* 41 (2014), pp. 106–132.

Ling et al. 2019: J. Ling, E. Hjärthner-Holdar, L. Grandin, Z. Stos-Gale, K. Kristiansen, A. L. Melheim, G. Artioli, I. Angelini, R. Krause, C. Canovaro, Moving Metals IV: Swords, Metal Sources and Trade Networks in Bronze Age Europe. *Journal of Archaeological Science*: Reports 26 (2019), 101837.

Lipo et al. 1997: C. P. Lipo, M. E. Madsen, R. C. Dunnell, T. Hunt, Population Structure, Cultural Transmission, and Frequency Seriation. *Journal of Anthropological Archaeology* 16/4 (1997), pp. 301–333.

Lipo et al. 2015: C. P. Lipo, M. E. Madsen, R. C. Dunnell, Theoretically-Sufficient and Computationally-Practical Technique for Deterministic Frequency Seriation. *PLoS ONE* 10/4 (2015), e0124942.

Litescu et al. 2012: S. C. Litescu, E. D. Teodor, G. I. Truică, A. Tache, G. L. Radu, Fourier Transform Infrared Spectroscopy – Useful Analytical Tool for Non-destructive Analysis. In: T. Theophanides (ed.), *Infrared Spectroscopy – Materials Science, Engineering and Technology.* Rijeka 2012, pp. 353–368.

Ljubić 1876: Š. Ljubić, Bakrena doba (Bronzezeit). *Viestnik hrvatskoga arheološkoga družtva* II (1876), pp. 10–49.

Ljuština 2011: M. Ljuština, Well Defined or Taken for Granted – the Bronze Age Vatin Culture a Century After. In: D. Măgureanu, D. Măndescu, S. Matei (eds.), *Archaeology: Making of and Practice. Studies in Honor of Mircea Babeş at his 70th Anniversary.* Piteşti 2011, pp. 103–113.

Ljuština 2019: M. Ljuština, Amber in the Bronze Age of Serbia: Old Finds and New Discoveries. In: A. Zanoci, M. Băţ (eds.), *Contribuţii la preistoria şi istoria antică a spaţiului carpatodanubiano-pontic: In honorem professoris Ion Niculiţă natalia sua octogesima celebrantis.* Chişinău 2019, pp. 87–100.

Ljuština/Dmitrović 2015: M. Ljuština, K. Dmitrović, Core vs. Periphery: Some Stratigraphical and Chronological Remarks on the Vatin Culture in Banat and Western Serbia. In: R. E. Németh, B. Rezi (eds.), *Bronze Age Chronology in the Carpathian Basin: proceedings of the International Colloquium from Târgu Mureş, 2–4 October 2014.* Târgu Mureş 2015, pp. 37–46.

Ljuština/Dmitrović 2016: M. Ljuština, K. Dmitrović, Between Everyday Life and Eternal Rest: Middle Bronze Age in Western Morava Basin, Central Serbia. In: V. Sîrbu, M. Jevtić, K. Dmitrović, M. Ljuština (eds.), *Funerary practices during the Bronze and Iron Ages in Central and Southeast Europe. Proceedings of the 14th International Colloquium of Funerary Archaeology in Čačak, Serbia, 24th–27th September 2015.* Beograd – Čačak 2016, pp. 43–52.

Lo Porto 1963: F. G. Lo Porto, Leporano (TA). La stazione protostorica di Porto Perone. *Notizie Scavi d'Antichità* XVII (1963), pp. 280–380.

Lo Porto 1979: F. G. Lo Porto, Il Bronzo Finale Nella Puglia Centro-Meridionale. In: I.I.P.P. (eds.), *Il Bronzo finale in Italia. Atti della XXI Riunione Scientifica dell'Istituto Italiano di Preistoria e Protostoria (Firenze, 21–23 ottobre 1973).* Firenze 1979, pp. 531–535.

Lo Schiavo 1970: F. Lo Schiavo, Il gruppo liburnico-japodico. *Atti della Accademia Nacionale dei Lincei* VIII/XIV (1970), pp. 424–426.

Lo Schiavo 1982: F. Lo Schiavo, Ambra in Sardegna. In: Società Archeologica Comense (eds.), *Studi in onore di Ferrante Rittatore Vonwiller*, p. I, vol. I. Como 1982, pp. 257–277.

Lo Schiavo 2010: F. Lo Schiavo, *Le Fibule dell'Italia meridionale e della Sicilia dall'età del bronzo recente al VI secolo a. C.* Prähistorische Bronzefunde XIV/14, vol. 1–3. Stuttgart 2010.

Loze 2017: I. B. Loze, The Role of the Middle and Upper Reaches of River Daugava and Its Basin in the Establishment of Neolithic Amber Roads. In: P. L. Cellarosi, R. Chellini, F. Martini, A. C. Montanaro, L. Sarti, R. M. Capozzi (eds.), *The Amber Roads. The Ancient Cultural and Commercial Communication between Peoples.* Firenze 2017, pp. 41–58.

Ložnjak Dizdar 2007: D. Ložnjak Dizdar, A Contribution to Understanding the Relations (or Lack of Relations) between the Croatian Danube Region and the Aegean at the Beginning of the Second Millennium BC. In: I. Galanaki, H. Tomas, Y. Galanakis, R. Laffineur (eds.), *Between the Aegean and Baltic Seas. Prehistory across Borders.* Aegaeum 27. Liège 2007, pp. 315–324.

Ložnjak Dizdar 2011: D. Ložnjak Dizdar, Graves of the Early Urnfield Culture in Northern Croatia – Some Remarks and Questions. In: Ch. Gutjahr, G. Tiefengraber (eds.), *Internationale archaeologie, Arbetsgemeinschaft, Symposium, Tagung, Kongress 15, Beitraege zur Mittel- und Spätbronzezeit sowie zur Urnenfelderzeit am Rande der Südostalpen.* Rahden/Westf. 2011, pp. 37–49.

Ložnjak Dizdar 2014: D. Ložnjak Dizdar, Southeastern Periphery of the Urnfield Culture? The Croatian Perspective. *The Beginning of the Late Bronze Age between the Eastern Alps and the Danube. Proceedings of the international conference in Osijek, October 20–22, 2011.* Zbornik Instituta Za Arheologiju Serta Instituti Archaeologici 1. Zagreb 2014, pp. 235–247.

Ložnjak Dizdar/Potrebica 2017: D. Ložnjak Dizdar, H. Potrebica, B*rončano doba Hrvatske: u okviru srednje i jugoistočne Europe.* Zagreb 2017.

Lucetini 1981: N. Lucetini, Sulla cronologia delle necropolis di Glasinac nell'etá del ferro. In: R. Peroni (ed.), *Studi di Protostoria Adriatica 1.* Quaderni di cultura material 2. Roma 1981, pp. 67–164.

Luci 1998: K. Luci, Бронзано доба. In: N. Tasić (ed.) *Археолошко благо Косова и Метохије од неолита до раног средњег века, каталог изложбе.* Београд – Приштина 1998, pp. 118–146.

Macchiarola 1987: I. Macchiarola, *La ceramica appenninica decorate.* Roma 1987.

Macchiarola 1995: I. Macchiarola, La facies appenninica. In: D. Cocchi Genick, I. Damiani, I. Macchiarola, R. Peroni, R. Poggiani Keller (eds.), *Aspetti culturali della media età del bronzo nell' Italia centro-meridionale.* Firenze 1995, pp. 441–463.

Maciejewski 2015: M. Maciejewski, The Perspective of Physical Space in a Study of Hoards from the Late Bronze Age and the Early Iron Age. *Sprawozdania Archeologiczne* 67 (2015), pp. 67–76.

Maciejewski 2017: M. Maciejewski, The Symbolism of Hoard Deposition Places in the Landscape of the Late Bronze Age and the Early Iron Age in the Zone of the South Baltic Coastland and Lake Districts. *Sprawozdania Archeologiczne* 69 (2017), pp. 113–132.

Madas 1996: D. Madas, Resultate der Forschungen an Tumuli der Späten Bronzezeit auf der Lokalität Paulje bei Loznica. In: N. Tasić (ed.), *The Yugoslav Danube Basin and the Neighbouring Regions in the 2nd Millennium B.C.* Beograd – Vršac 1996, pp. 227–230.

Mader 2006: B. Mader, Die Forschungen der Prähistorischen Kommission auf der Insel Krk (Veglia) von 1895 und 1912. – II. Die Untersuchungen eisenzeitlicher Gräber durch Josef Szombathy und Michael Abramić in Basca nuova (Baška) – Flur Sveti Kristofor 1912 und Eduard Nowotny in Bescavalle (Draga Bašćanska) 1895. *Annales Naturhistorisches Museum Wien – Antropologie und Prähistorie* 107 (2006), pp. 239–266.

Mäder/Sormaz 2000: A. Mäder, T. Sormaz, Die Dendrodaten derbeginnenden Spätbronzezeit (Bz D) von Elgg-Breiti, ZH. *Jahrbuch der Schweizerischen Gesellschaft für Ur- und Frühgeschichte* 84 (2000), pp. 65–78.

Madsen 2007: T. Madsen, Multivariate Data Analysis with PCA, CA and MS CAPCA 3.1. Published online on 11.02.2016. (https://www.archaeoinfo.dk/PDF/2007_Multivariate_data_analysis.pdf).

Maggi 1990: R. Maggi, *Archeologia dell'Appennino Ligure. Gli Scavi del Castellaro di Uscio: un insediamento di crinale occupato dal neolitico alla conquista Romana.* Istituto Internazionale di Studi Liguri, Collezione di Monografie Preistoriche ed Archeologiche, vol. VIII. Bordighera 1990.

Maggiulli 2009: G. Maggiulli, Metallurgia e produzioni metallurgiche a Roca (Lecce): i ripostigli del Bronzo Finale. *Rivista di Scienze Preistoriche* LIX (2009), pp. 307–334.

Majnarić-Pandžić 2002: N. Majnarić-Pandžić, Review: C. F. E. Pare, Beitrage zum Ubergang von der Bronze- zur Eisenzeit in Mitteleuropa, Teil I – Grundzuge der Chronologie im ostlichen Mitteleuropa (11.-8. Jahrhundert v. Chr.). Jahrbuch des Romisch-Germanischen Zentralmuseum 45/1. Mainz 1998, 293–433. *Opuscula Archaeologica* 26 (2002), pp. 341–348.

Makarowicz 2009: P. Makarowicz, Baltic-Pontic Interregional Routes at the Start of the Bronze Age. In: A. Kośko, V. Klochko (eds.), *Routes Between the Seas: Baltic – Boh – Bug – Pont from the 3rd to the Middle of the 1st Millennium BC.* Baltic-Pontic Studies 14. Poznań 2009, pp. 301–336.

Malagodi 1999: M. Malagodi, Un monile d'ambra di Trinitapoli: indagine scientifica. In: A. M. Tunzi Sisto (ed.), *Ipogei della Daunia. Preistoria di un territorio.* Foggia 1999, p. 291.

Malinar 1998: M. Malinar, Brončanodobni lokalitet špilja Bezdanjača – novi materijal i interpretacija. *Opuscula Archaeologica* 22 (1998), pp. 141–162.

Malinowski 1992: B. Malinowski, *Argonauts of the Western Pacific* (first published in 1922). Long Grove (IL) 1992.

Mallory 1997: T. Mallory, *Le Morte D'Arthur.* Edited by H. O. Sommer (1889). University of Michigan Humanities Text Initiative. Ann Arbor (MI) 1997.

Malnati/Cardarelli 2003: L. Malnati, A. Cardarelli (eds.), *Atlante dei Beni Archeologici della Provincia di Modena. I. Pianura.* Firenze 2003.

Malnati/Cardarelli 2009: L. Malnati, A. Cardarelli (eds.), *Atlante dei Beni Archeologici della Provincia di Modena. III. Collina e Alta Pianura,* vols. 1–2. Firenze 2009.

Malone et al. 1994: C. Malone, S. Stoddart, R. Whitehouse, The Bronze Age of Southern Italy, Sicily and Malta c. 2000–800 B.C. In: C. Mathers, S. Stoddart, J. R. Colliss (eds.), *Development and Decline in the Mediterranean Bronze Age.* Sheffield 1994, pp. 167–194.

Manfredini 1979: A. Manfredini, La Grotta Manaccore e l'età del Bronzo nel Gargano. In: A. Palma di Cesnola (ed.), *Il Campignano e l'età del Bronzo nel Gargano, Atti della IV Esposizione Archeologica.* Vico del Gargano 1979, pp. 75–83.

Manning 2006: S. W. Manning, Why Radiocarbon Dating 1200 BCE Is Difficult: A Sidelight on Dating the End of the Late Bronze Age and the Contrarian Contribution. *Scripta Mediterranea* XVII-XVIII (2006), pp. 53–80.

Manning 2007: S. W. Manning, 'Clarifying the 'High' v. 'Low' Aegean/Cypriot Chronology for the Mid Second Millennium BC: Assessing the Evidence, Interpretive Frameworks, and Current State of the Debate'. In: M. Bietak, E. Czerny (eds.), *The Synchronisation of Civilizations in the Eastern Mediterranean in the Second Millennium B.C. III.* Vienna 2007, pp. 101–137.

Manning 2009: S. W. Manning, Beyond the Santorini Eruption: Some Notes on dating the Late Minoan IB Period on Crete, and Implications for Cretan-Egyptian Relations in the 15th Century BC (and Especially LM II). In: D. A. Warburton (ed.), *Time's Up! Dating the Minoan Eruption of Santorini. Acts of the Minoan Eruption Chronology Workshop, Sandbjerg, November 2007.* Aarhus 2009, pp. 207–226.

Manning 2012: S. W. Manning, Chronology and Terminology. In E. H. Cline (ed.), *The Oxford Handbook of the Bronze Age Aegean (ca. 3000–1000 BC).* New York - Oxford 2012, pp. 11–28.

Manning et al. 2006: S. W. Manning, C. Bronk Ramsey, W. Kutschera, T. Higham, B. Kromer, P. Steier, E. M. Wild, Chronology for the Aegean Late Bronze Age 1700–1400 B.C. *Science* 312 (2006), pp. 565–569.

Manning et al. 2014: S. W. Manning, F. Höflmayer, N. Moeller, M. W. Dee, C. Bronk Ramsey, D. Fleitmann, T. F. G. Higham, W. Kutschera, E. M. Wild, Dating the Thera (Santorini) Eruption: Coherent Archaeological and Scientific Evidence Supporting a High Chronology. *Antiquity* 88 (2014), pp. 1164–1179.

Mannoni/Tizzoni 1980: T. Mannoni, M. Tizzoni, Lo scavo del Castellavo di Zignago (SP). *Rivista di Scienze Preistoriche* XXXV/1–2, pp. 249–279.

Maran 1998: J. Maran, *Kulturwandel auf dem griechischen Festland und den Kykladen im späten 3. Jahrtausend v. Chr. Studien zu den kulturellen Verhältnissen in Südosteuropa und dem zentralen sowie östlichen Mittelmeerraum in der späten Kupfer- und frühen Bronzezeit.* Universitätsforschungen zur Prähistorischen Archäologie 53. Bonn 1998.

Maran 2004: J. Maran, Wessex und Mykene. Zur Deutung des Bernsteins in der Schachtgräberzeit Südgriechenlands. In: B. Hänsel, E. Studenikova (eds.), *Zwischen Karpaten und Ägäis. Neolithikum und älter Bronzezeit. Gedenkschrift für Viera Němejcová-Pavúková.* Rahden - Westf. 2004, pp. 47–65.

Maran 2007: J. Maran, Seaborne Contacts between the Aegean, the Balkans and the Central Mediterranean in the 3rd Millennium BC – The Unfolding of the Mediterranean World. In: I. Galanaki, H. Tomas, Y. Galanakis, R. Laffineur (eds.), *Between the Aegean and Baltic Seas. Prehistory across Borders.* Aegaeum 27. Liège 2007, pp. 3–26.

Maran 2012: J. Maran, One World Is Not Enough: The Transformative Potential of Intercultural Exchange in Prehistoric Societies. In: P. W. Stockhammer (ed.), *Conceptualizing Cultural Hybridization: A Transdisciplinary Approach.* Berlin – Heidelberg 2012, pp. 59–66.

Maran 2013: J. Maran, Bright as the Sun: The Appropriation of Amber Object in Mycenaean Greece. In H. P. Hahn, W. Hadas (eds.), *Mobility, Meaning and the Transformations of Things. Shifting Contexts of Material Culture through Time and Space.* Oxford 2013, pp. 147–169.

Maran/Stockhammer 2012a: J. Maran, P. W. Stockhammer (eds.), *Materiality and Social Practice Transformative Capacities of Intercultural Encounters.* Oxford 2012.

Maran/Stockhammer 2012b: J. Maran, P. W. Stockhammer, Introduction. In: J. Maran, P. W. Stockhammer (eds.), *Materiality and Social Practice Transformative Capacities of Intercultural Encounters.* Oxford 2012, pp. 1–3.

Marchesan 1985: A. Marchesan, Le età del Bronzo media e recente nel territorio veneto: tipologia e suddivisione in fasi cronologiche. *Archeologia Veneta* 8 (1985), pp. 63–124.

Marcon/Mazzoli 2015: S. Marcon, C. Mazzoli, Le forme di fusione di Fondo Paviani (Verona) – Scavi Università di Padova 2007–2012. Caratterizzazione tipologica e petrografica. In: G. Leonardi, V. Tiné (eds.), *Preistoria e Protostoria del Veneto.* Studi di Preistoria e Protostoria 2. Firenze 2015, pp. 845–848.

Marino/Pacciarelli 1996: D. Marino, M. Pacciarelli, Calabria. In: D. Cocchi Genick (ed.), *L'antica età del Bronzo in Italia. Atti del Congresso nazionale (Viareggio, 9–12 gennaio 1995).* Firenze 1996, pp. 147–162.

Marović 1960: I. Marović, Prahistorijski nalazi na području Solina. *Vjesnik za Arheologiju i Historiju Dalmatinsku* LXII (1960), pp. 5–27.

Martinelli 2010: M. C. Martinelli, *Archeologia delle Isole Eolie. Il villaggio dell'età del Bronzo Medio di Portella a Salina. Ricerche 2006 e 2008.* Muggiò 2010.

Martinelli/Marchesini 2015: N. Martinelli, S. Marchesini, Placchette discoidali dalle collezioni del Museo Civico di Storia Naturale di Verona. In: A. Piccoli, M. G. Ruggiero (eds.), *Atti del congress internazionale 'Tavolette enigmatiche. Un antico processo di interazione in Europa', Cavriana, 16–18 Settembre 2010.* Annali Benacensi XV. Cavriana 2015, pp. 223–241.

Martinelli/Speciale 2017: M. C. Martinelli, C. Speciale, Classificazione della ceramica e analisi dei contesti all'inizio dell'età del Bronzo: la capanna F del villaggio di Filo Braccio (Filicudi, Iisole Eolie). *Ipotesi di Preistoria* 9 (2017), pp. 1–36.

Martinelli et al. 2010: M. C. Martinelli, G. Fiorentino, B. Prosdocimi, C. D'Oronzo, S. T. Levi, G. Mangano, A. Stellati, N. Wolff, Nuove ricerche nell'insediamento sull'istmo di Filo Braccio a Filicudi. Nota preliminare sugli scavi 2009. *Origini* XXXII (2010), pp. 285–314.

Mauss 1990: M. Mauss, *The Gift: Forms and Means of Archaic Exchange* (first published in 1925). London 1990.

Mauss 2001: M. Mauss, *Socjologia i antropologia* (first published in 1950). Warszawa 2001.

Marzatico 1997: F. Marzatico, Collana d'ambra in perle e 'distanziatori'. In: L. Endrizzi, F. Marzatico (eds.), *Ori delle Alpi.* Exhibition catalogue. Trento 1997.

Marzatico 2012: F. Marzatico, La Cultura di Luco/Laugen, aggiornamenti e problemi aperti. In: A. Angelini, G. Leonardi (eds.), *Il castelliere di Castel de Pedena. Un sito frontier del II e I millennio a.C. Atti del convegno, Feltre, 6 giugno 2009.* Saltuarie dal Laboratorio del Piovego 9. Belluno 2012, pp. 177–204.

Marzatico/Tecchiati 1998: F. Marzatico, U. Tecchiati, The Bronze Age in Trentino and Alto Adige/SüdTirol. *Preistoria Alpina* 34 (1998), pp. 27–60.

Marzatico/Tecchiati 2002: F. Marzatico, U. Tecchiati, L'età del Bronzo in Trentino e Alto Adige/Südtirol. In: I.I.P.P. (eds.), *Atti della XXXIII Riunione Scientifica dedicata a Bernardino Bagolini Preistoria e Protostoria del Trentino Alto Adige/Südtirol, Trento, 21–24 ottobre 1997.* Firenze 2002, pp. 45–92.

Marzatico et al. 2010: F. Marzatico, E. Valzolgher, H. Oberrauch, Dating the Later Bronze Age Metal Production in the South-Central Alps. Some Remarks on the Relative and Absolute Chronology of the Luco/Laugen Culture. In: P. Anreiter,

G. Goldenberg, K. Hanke, R. Krause, W. Leitner, F. Mathis, K. Nicolussi, K. Oeggl, E. Pernicka, M. Prast, J. Schibler, I. Schneider, H. Stadler, T. Stöllner, G. Tomedi, P. Troppe (eds.), *Mining in European History and its Impact on Environment and Human Societies – Proceedings for the 1st Mining in European History-Conference of the SFB-HIMAT, Innsbruck, 12–15 November 2009.* Innsbruck 2010, pp. 129–143.

Marková 1993: K. Marková, Bernsteinfunde in der Slowakei während der Bronzezeit. In: C. W. Beck, J. Bouzek (eds.), *Amber in Archaeology. Proceedings of the IInd International Conference on Amber in Archaeology, Liblice 1990.* Praha 1993, pp. 171–178.

Marková 2003: K. Marková, Austauschentwicklung in Karpatenbecken im Lichte der Bernsteinfunde (vorläufige Anmerkungen). In: C. Kacsó (ed.), *Bronzezeitliche Kulturerscheinungen im karpatischen Raum. Die Beziehungen zu den benachbarten Gebieten. Ehrensymposium für Alexandru Vulpe zum 70. Geburtstag. Baia Mare, 10–13. Oktober 2001.* Baia Mare 2003, pp. 339–352.

Mason et al. 2016: A. H. Mason, W. G. Powell, H. A. Bankoff, R. Mathur, A. Bulatović, V. Filipović, J. Ruiz, Tin Isotope Characterization of Bronze Artifacts of Central Balkans. *Journal of Archaeological Science* 69 (2016), pp. 110–117.

Mason et al. 2020: A. H. Mason, W. G. Powell, H. A. Bankoff, R. Mathur, M. Price, A. Bulatović, V. Filipović, Provenance of Tin in the Late Bronze Age Balkans Based on Probabilistic and Spatial Analysis of Sn Isotopes. *Journal of Archaeological Science* 122 (2020), 105181.

Massari et al. 1996: A. Massari, B. Raposso, B. Setti, La diffusione dell'ambra nel Bronzo Antico in Italia. In: D. Cocchi Genick (ed.), *L'antica età del Bronzo in Italia. Atti del Congresso nazionale (Viareggio, 9–12 gennaio 1995).* Firenze 1995, pp. 620–621.

Matarese 2016: I. Matarese, Gli ornamenti della tomba 1 di Murgia Timone (Matera) nel quadro del Bronzo Medio in Italia. In: N. Negroni Catacchio (ed.), *Atti del XII Incontro di Studi 'Preistoria e Protostoria in Etruria' (Valentano (VT) – Pitigliano (GR) – Manciano (GR), 12–14 settembre 2014).* Milano 2016, pp. 373–386.

Matarese 2018: I. Matarese, *Murgia Timone (Matera): le tombe a camera.* Origines. Firenze 2018.

Matarese et al. 2015: I. Matarese, A. Crispino, R. Jung, M. Martinelli, P. Pallante, M. Pacciarelli, Vaghi e pendenti litici dell'età del bronzo dalla Sicilia e dalle Eolie. *Archaeologia Austriaca* 99 (2015), pp. 111–153.

Matthäus 1980: H. Matthäus, *Die Bronzegefäße der kretisch-mykenischen Kultur.* Prähistorische Bronzefunde II/2. München 1980.

Mavrović Mokos 2019: J. Mavrović Mokos, Amber bead and Brotlaibidol from a settlement at the site of Alilovci Lipje, Croatia. In: T. Le Cozanet, C. Moulin, M. Nordez (eds.), *Interactions and Exchanges during the Protohistory. Proceedings of the 3rd doctoral meeting of the École Européenne de Protohistoire de Bibracte, 27–28 March 2017.* Glux-en Glenne 2019, pp. 89–94.

Mavrović Mokos/Pavličić 2015: J. Mavrović Mokos, M. Pavličić, *Alilovci, život prije 3700 godina. Exhibition catalogue. Gradski muzej Požega dd 18. svibnja do 18. lipnja 2015.* Požega 2015.

Mayer 1977: E. F. Mayer, *Die Äxte und Beile in Österreich.* Prähistorische Bronzefunde IX/9. München 1977.

Mazurowski 1983: R. F. Mazurowski, Bursztyn w epoce kamienia na ziemiach polskich. *Materiały Starożytne i Wczesnośredniowieczne* V (1983), pp. 7–130.

Mazurowski 1984: R. F. Mazurowski, Amber Treatment Workshops of the Rzucewo Culture in Żuławy. *Przegląd Archeologiczny* 32 (1984), pp. 5–60.

Mazurowski 2014: R. F. Mazurowski, *Prahistoryczne bursztyniarstwo na Żuławach Wiślanych. Późnoneolityczne centrum pozyskiwania i obróbki bursztynu w niedźwiedzióweckim mikroregionie osadniczym 3300–2400 l. p.n.e.* Malbork 2014.

Mazzeo et al. 1999: R. Mazzeo, N. Negroni Catacchio, B. Raposso, Le perle di ambra. In: A. M. Tunzi Sisto (ed.), *Ipogei della Daunia. Preistoria di un territorio.* Foggia 1999, pp. 289–290.

McConnell 1995: B. E. McConnell, *La Muculufa II. Excavation and Survey 1988–1991. The Castelluccian Village and Other Areas.* Louvain-la-Neuve 1995.

Meller/Bertemes 2010: H. Meller, F. Bertemes (eds.), *Der Griff nach den Sternen. Wie Europas Eliten zu Macht und Reichtum kamen. Internationales Symposium Halle, 16.–21. Februar 2005.* Tagungen Des Landesmuseums Für Vorgeschichte Halle 5. Halle (Saale) 2010.

Meller et al. 2013: H. Meller, F. Bertemes, H. R. Bork, R. Risch (eds.), *1600 – Kultureller Umbruch im Schatten des Thera-Ausbruchs? 4. Mitteldeutscher Archäologentag vom 14. Bis 16. Oktober 2011 in Halle (Saale).* Tagungen des Landesmuseums für Vorgeschichte Halle 9. Halle (Saale) 2013.

Mello 1990: E. Mello, Analisi chimiche dei reperti di ambra. In: R. Maggi (ed.), *Archeologia dell'Appennino Ligure. Gli Scavi Ddel Castellaro di Uscio: un insediamento di crinale occupato dal neolitico alla conquista Romana.* Istituto Internazionale di Studi Liguri, Collezione di Monografie Preistoriche ed Archeologiche, vol. VIII. Bordighera 1990, pp. 249–250.

Mercuri et al. 2006: A. M. Mercuri, C. A. Accorsi, M. Bandini Mazzanti, G. Bosi, A. Cardarelli, D. Labate, M. Marchesini, G. Trevisan Grandi, Economy and Environment of Bronze Age Settlements – Terramaras – on the Po Plain (Northern Italy): First Results from the Archaeobotanical Research at the Terramara di Montale. *Vegetation History and Archaeobotany* 16 (2006), pp. 43–60.

Meyer 1887: A. B. Meyer, Notizie. *Bullettino di Paletnologia Italiana* XIII (1887), p. 21.

Meyer 1893: A. B. Meyer, Intorno a del materiale preistorico di tipo Ambra scoperto in Sicilia. *Bullettino di Paletnologia Italiana* XIX (1893) pp. 105–109.

Meyer et al. 1991: C. Meyer, J. M. Todd, C. W. Beck, From Zanzibar to Zagros: A Copal Pendant of Eshnunna. *Journal of Near Eastern Studies* 50/4 (1991), pp. 289–298.

Miari 2007a: M. Miari, Ambra in Area Terramaricola. In: M. L. Nava, T. Salerno (eds.), *Trasparente dall'Antico. Ambre. Exhibition catalogue (Napoli Museo Archeologico Nazionale: 26 marzo–10 settembre 2007).* Milano 2007, pp. 68–72.

Miari 2007b: M. Miari, Monte Castellaccio di Imola. In: M. L. Nava, T. Salerno (eds.), *Trasparente dall'Antico. Ambre. Exhibition catalogue (Napoli Museo Archeologico Nazionale: 26 marzo–10 settembre 2007).* Milano 2007, p. 80.

Miari 2007c: M. Miari, San Giuliano di Toscanella. In: M. L. Nava, T. Salerno (eds.), *Trasparente dall'Antico. Ambre. Exhibition catalogue (Napoli Museo Archeologico Nazionale: 26 marzo–10 settembre 2007).* Milano 2007, p. 80.

Miari 2007d: M. Miari, Monte Battaglia a Casola Valsenio. In: M. L. Nava, T. Salerno (eds.), *Trasparente dall'Antico. Ambre. Exhibition catalogue (Napoli Museo Archeologico Nazionale: 26 marzo–10 settembre 2007).* Milano 2007, p. 80.

Miari 2007e: M. Miari, La necropoli di Campo Pianelli di Bismantova. In: M. L. Nava, T. Salerno (eds.), *Trasparente dall'Antico. Ambre. Exhibition catalogue (Napoli Museo Archeologico Nazionale: 26 marzo–10 settembre 2007).* Milano 2007, pp. 76–77.

Miari 2010: M. Miari, Il sito di Monte Battaglia e il ruolo dei passi appenninici alla fine dell'età del bronzo. In: N. Negroni Catacchio (ed.), *L'alba dell'Etruria Fenomeni di continuità e trasformazione nei secoli XII-VIII a.C. Ricerche e scavi: Atti del IX incontro di studi 'Preistoria e Protostoria in Etruria' (Valentano (VT) – Pitigliano (GR), 12–14 settembre 2008).* Milano 2010, pp. 367–382.

Mieri et al. 2019: M. Mieri, P. W. Stockhammer, P. Morgenstern, J. Maran, Mobility and trade in Mediterranean antiquity: Evidence for an 'Italian Connection' in Mycenaean Greece Revealed by Ancient DNA of Livestock. *Journal of Archaeological Science*: Reports 23 (2019), pp. 98–103.

Mihovilić 1994: K. Mihovilić, Preistoria dell'Istria dal Paleolitico all'età del Ferro. In: I.I.P.P. (eds.), *Preistoria e protostoria del Friuli-Venezia Giulia e dell'Istria. Atti della XXIX Riunione Scientifica dell'Istituto Italiano di Preistoria e Protostoria (Trieste, 28–30 settembre 1990).* Firenze 1994, pp. 101–118.

Mihovilić 2008: K. Mihovilić, Nekropola u jami Laganiši. In: D. Komšo (ed.), *Pećina Laganiši – mjesto života i smrti.* Exhibition catalog. Arheološki muzej Istre, Katalog izložbe 73. Pula 2008, pp. 45–55.

Mihovilić 2013a: K. Mihovilić, *Histri u Istri: željezno doba Istre.* Monografije i katalozi 23, Arheološki muzej Istre. Pula 2013.

Mihovilić 2013b: K. Mihovilić, Castellieri-Gradine of the Northern Adriatic. In: H. Fokkens, A. Harding (eds.), *The Oxford Handbook of the European Bronze Age.* Oxford 2013, pp. 864–876.

Mihovilić et al. 2009: K. Mihovilić, B. Hänsel, B. Teržan, D. Matošević, Ž. Kovačić, *Monkodonja i Mušego.* Arheološki muzej Istre, Katalog izložbe 79. Pula 2009.

Mihovilić et al. 2011: K. Mihovilić, B. Hänsel, D. Matošević, B. Teržan, Burial Mounds of the Bronze Age at Mušego near Monkodonja. Results of the Excavations 2006–2007. In: E. Borgna, S. Müller Celka (eds.), *Ancestral landscapes: burial mounds in the Copper and Bronze Ages (Central and Eastern Europe – Balkans – Adriatic – Aegean, 4th–2nd millennium B.C.).* Travaux de la maison de l'orient et de la méditerranée n°58. Lyon 2011, pp. 367–374.

Mihovilić et al. 2017: K. Mihovilić, B. Hänsel, B. Teržan, *Brončanodobni krušni idoli: enigmatske pločice ili…?* Mihovilić et al. 2011. Arheološki muzej Istre, Katalozi izložbi 5. Pula 2017.

Militello 2004: P. Militello, Commercianti, architetti ed artigiani. Riflessioni sulla presenza micenea nell'area iblea. In: V. La Rosa (ed.), *Le presenze micenee nel territorio siracusano. Atti del primo simposio siracusano di preistoria siciliana, Siracusa 2003.* Padova 2004, pp. 293–334.

Mills et al. 1984/1985: J. S. Mills, R. White, L. J. Gough, The Chemical Composition of Baltic Amber. *Chemical Geology* 47 (1984/1985), pp. 15–39.

Ministry of Foreign Affairs Republic of Poland n.d.: Ministry of Foreign Affairs Republic of Poland, *Western Balkans*, https://www.gov.pl/web/diplomacy/western-balkans, accessed 28th July 2022.

Mitrevski 2003: D. Mitrevski, Prehistory in Republic of Macedonia – F.Y.R.O.M. In: D. Grammenos (ed.), *Recent Research in the Prehistory of the Balkans*. Thessaloniki 2003, pp. 13–72.

Mitrevski 2007: D. Mitrevski, The Beginning of the Iron Age in Macedonia. In: H. Todorova, M. Stefanovich, G. Ivanov (eds.), *The Struma/Strymon River Valley in prehistory. Proceedings of the International Symposium Strymon Praehistoricus, Kjustendil-Blagoevgrad (Bulgaria), Serres-Amphipolis (Greece), 27.09.-01.10.2004*. Sofia 2007, pp. 443–450.

Montagnari Kokelj 2007: E. Montagnari Kokelj, Salt and the Trieste Karst (North-Eastern Italy) in Prehistory: Some Considerations. In: D. Monah, Gh. Dumitroaia, O. Weller, J. Chapman (eds.), *L'exploitation du sel á travers le temps*. Piatra-Neamţ 2007, pp. 161–189.

Montelius 1910: O. Montelius, Der Handel in der Vorzeit. *Prähistorische Zeitschrift* 2 (1910), pp. 249–291.

Morabini 1996: S. Morabini, Il sito del villaggio protostorico di San Giuliano di Toscanella. In: M. Pacciarelli (ed.), *Collezione Scarabelli, 2. Preistoria*. Imola 1996, p. 302.

Morico 1996: G. Morico, I manufatti di bronzo di San Giuliano di Toscanella. In: M. Pacciarelli (ed.), *Collezione Scarabelli, 2. Preistoria*. Imola 1996, pp. 346–350.

Morico/Pacciarelli 1996: G. Morico, M. Pacciarelli, Il villaggio dell'età del bronzo di San Giuliano di Toscanella: considerazioni sullo scavo e sulla planimetria. In: M. Pacciarelli (ed.), *Collezione Scarabelli, 2. Preistoria*. Imola 1996, pp. 303–307.

Morintz 1978: S. Morintz, *Contribuţii arheologice la isotria tracilor timpurii. I. Epoca bronzului în spaţiul carpato-balcanic*. Bucureşti 1978.

Morley 2007: N. Morley, *Trade in Classical Antiquity*. Cambridge 2007.

Mörtz 2013: T. Mörtz, Zerteiltes Leid. Anmerkungen zur Deutung mutwilliger Beschädigungen von Metalldeponierungen der späten Bronzezeit. *Mitteilungen der Berliner Gesellschaft für Anthropologie* 34 (2013), pp. 55–66.

Möslein 1998: S. Möslein, Die Straubinger Gruppe der donauländischen Frühbronzezeit: Frühbronzezeitliche Keramik aus Südostbayern und ihre Bedeutung für die chronologische und regionale Gliederung der frühen Bronzezeit in Südbayern. *Berichte der Bayerischen Bodendenkmalpflege* 38 (1998), pp. 37–106.

Mosso 1909: A. Mosso, Stazione preistorica di Coppa Nevigata presso Manfredonia. *Monumenti Antichi dei Lincei* XIX (1909), pp. 305–396.

Mozsolics 1967: A. Mozsolics, *Bronzefunde des Karpatenbecken: Depotfundhorizonte von Hajdúsámson und Kosziderpadlás*. Budapest 1967.

Mozsolics 1973: A. Mozsolics, *Bronze- und Goldfunde des Karpatenbekkens. Depotfundhorizonte von Forró und Ópályi*. Budapest 1973.

Mozsolics 1985: A. Mozsolics, *Bronezfunde aus Ungarn. Depotfundhorizonte von Aranyos, Kurd und Gyermely*. Budapest 1985.

Mozsolics 2000: A. Mozsolics, *Bronzefunde aus Ungarn. Depotfundhorizonte Hajdúböszörmény, Románd und Bükkszentlászló*. Prähistorische Archäologie in Südosteuropa 17. Kiel 2000.

Mukherjee et al. 2008: A. J. Mukherjee, E. Roßberger, M. A. James, P. Pfälzner, C. L. Higgitt, R. White, D. A. Peggie, D. Azar, R. P. Evershed, The Qatna Lion: Scientific Confirmation of Baltic Amber in Late Bronze Age Syria. *Antiquity* 82 (2008), pp. 49–59.

Munro 1890: R. Munro, *The Lake-Dwellings of Europe*. London 1890.

Murillo-Barroso/Martinón-Torres 2012: M. Murillo-Barroso, M. Martinón-Torres, Amber Sources and Trade in the Prehistory of the Iberian Peninsula. *European Journal of Archaeology* 15/2 (2012), pp. 187–216.

Mutti 1993: A. Mutti, *Caratteristiche e problemi del popolamento terramaricolo in Emilia occidentale*. Bologna 1993.

Mutti 1997: A. Mutti, Le ricerche ottocentesche nella terramara di Castione Marchesi (PR). In: M. Bernabò Brea, A. Cardarelli, M. Cremaschi (eds.), *Le Terramare. La più antica civiltà padana*. Exhibition catalogue. Modena 1997, p. 37.

Mutti 2007: A. Mutti, La terramara di Castione Marchesi a Parma. In: M. L. Nava, T. Salerno (eds.), *Trasparente dall'Antico. Ambre. Exhibition catalogue (Napoli Museo Archeologico Nazionale: 26 marzo–10 settembre 2007)*. Milano 2007, pp. 73–75.

Mutti et al. 1988: A. Mutti, N. Provenzano, M. G. Rossi, M. Rottoli, *La terramara di Castione dei Marchesi*. Studi e Documenti di Archeologia V. Bologna 1988.

Müller 2012: J. Müller, Changes in the Bronze Age: Social, Economical and/or Ecological Causes? In: J. Kneisel, W. Kirleis, M. Dal Corso, N. Taylor, V. Tiedtke (eds.), *Collapse or Continuity? Environment and Development of Bronze Age Human Landscape – Proceedings of the International Workshop 'Socio-environmental Dynamics over the last 12,000 years: The Creation of Landscape II' (14th–18th March 2011, Kiel)*. Universitätsforschungen zur Prähistorischen Archäologie 205. Bonn 2012, pp. 257–266.

Müller 2013: J. Müller, 1600 B.C. – Social Topographies and the Development of Early Bronze Age Societies in Central Europe. In: H. Meller, F. Bertemes, H. R. Bork, R. Risch (eds.), *1600 – Kultureller Umbruch im Schatten des Thera-Ausbruchs. 4. Mittledeutscher Archäologentag vom 14. Bis 16. Oktober 2011 in Halle (Saale)*. Tagungen des Landesmuseums für Vorgeschichte Halle 9. Halle (Saale) 2013, pp. 528–537.

Müller/Lohrke 2009: J. Müller, B. Lohrke, Neue absolutchronologische Daten für die Süddeutsche Hügelgräberbronzezeit. *Germania* 87/1 (2009), pp. 25–39.

Müller et al. 2010: J. Müller, J. Czebreszuk, J. Kneisel, Bruszczewo II. *Ausgrabungen und Forschungen in einer prähistorischen Siedlungskammer Großpolens/Badania mikroregionu osadniczego z terenu Wielkopolski*. Studien zur Archäologie in Ostmitteleuropa 6.2. Bonn 2010.

Müller-Karpe 1959: H. Müller-Karpe, *Beiträge zur Chronologie der Urnenfelderzeit nördlich und südlich der Alpen*. Berlin 1959.

Nava 1982: M. L. Nava, Materiali enei dall'insediamento perilacustre di Castellaro Lagusello. In: Società Archeologica Comense (eds.), *Studi in Onore di Ferrante Rittatore Vonwiller*, p. I, vol. II. Como 1982, pp. 487–525.

Nava 2001: M. L. Nava, Puntualizzazioni cronologiche sulla necropoli protostorica di Monte Saraceno. In: A. Gravina (ed.), *Atti del 21° convegno nazionale sulla Preistoria, Protostoria, Storia della Daunia (San Severo, 24–26 novembre 2000).* San Severo 2001, pp. 275–296.

Nava/Preite 1995: M. L. Nava, A. Preite, Nuovi dati dalla necropoli di Monte Saraceno per l'età del bronzo finale e la transizione all'età del ferro nella Puglia settentrionale. *Taras* XV/2 (1995), pp. 87–127.

Nava/Salerno 2007: M. L. Nava, A. Salerno, Ambra nell'Italia centro-meridionale eninsulare durante l'età del bronzo. In: M. L. Nava, A. Salerno (eds.), *Ambre. Trasparenze dall'antico, Exhibition catalogue (Napoli, Museo Archeologico Nazionale, 26 marzo–10 settembre 2007).* Milano 2007, pp. 82–87.

Nebelsick 2000: L. Nebelsick, Rent Asunder: Ritual Violence in Late Bronze Age Hoards. In: C. F. E. Pare (ed.), *Metals Make the World go Round. The Supply and Circulation of Metals in Bronze Age Europe. Proceedings of a conference held at the University of Birmingham in June 1997.* Oxford 2000, pp. 160–175.

Needham 2001: S. P. Needham, When Expediency Broaches Ritual Intention: The Flow of Metal between Systemic and Buried Domains. *The Journal of the Royal Anthropological Institute* 7 (2001), pp. 275–298.

Negroni Catacchio 1970: N. Negroni Catacchio, Diffusione dell'ambra in Italia e suoi rapporti con mondo culturale preistorico. *Sibrium* X (1970), pp. 275–288.

Negroni Catacchio 1971: N. Negroni Catacchio, La problematica dell'ambra nella protostoria Italiana. Le ambre intagliate delle culture protostoriche dell'area lombardo-veneta-tridentina. *Memorie Museo Civico Storia Naturale Verona* XVIII (1970), pp. 319–336.

Negroni Catacchio 1972: N. Negroni Catacchio, Le problematiche dell'ambra nella protostoria Italiana: le ambre intagliate di Fratta Polesine Polesine e le rotte mercantili nell'alto Adriatico. *Padusa* VIII/1–2 (1972), pp. 3–20 (reprint. in *Padusa* XX [1984], pp. 73–90).

Negroni Catacchio 1973a: N. Negroni Catacchio, La problematica dell'ambra nella protostoria Italiana: ancora sulle ambre di Frattesina di Fratta Polesine. *Padusa* IX (1973), pp. 1–13.

Negroni Catacchio 1973b: N. Negroni Catacchio, Lo studio della problematica dell'ambra nella protostoria Italiana: nuovi risultati. In: I.I.P.P. (eds.), *Atti della XV Riunione Scientifica dell'Istituto Italiano di Preistoria e Protostoria (Verona-Trento, 27–29 ottobre 1972).* Firenze 1973, pp. 47–52.

Negroni Catacchio 1975a: N. Negroni Catacchio, Le ambre garganiche nel quadro della problematica dell'ambra nella protostoria Italiana. In: I.I.P.P. (eds.), *Atti del Colloquio Internazionale di Preistoria e Protostoria nella Daunia (Foggia, 24–29 aprile 1973).* Firenze 1975, pp. 310–319.

Negroni Catacchio 1975b: N. Negroni Catacchio, La problematica dell'ambra nella protostoria Italiana: manufatti in ambra dell'ambiente modenese. *Emilia Preromana* 7 (1971–1974 [1975]), pp. 375–388.

Negroni Catacchio 1976: N. Negroni Catacchio, Le vie dell'ambra, i passi alpini orientali e l'Alto Adriatico. In: Centro di antichità altoadriatiche (eds.), *Aquileia e l'arco alpino orientale.* Antichita altoadriatiche 9. Udine 1976, pp. 21–59.

Negroni Catacchio 1978: N. Negroni Catacchio, Note di protostoria padana. *Annali Benacensi* 4 (1978), pp. 155–171.

Negroni Catacchio 1986: N. Negroni Catacchio, Il vago tipo Tirinto. In: D. Cocchi Genick (ed.), *Il riparo dell'ambra.* Viareggio 1986, pp. 199–202.

Negroni Catacchio 1999: N. Negroni Catacchio, Produzione e commercio dei vaghi d'ambra tipo Tirinto e tipo Allumiere alla luce delle recenti scoperte. In: O. Paoletti, L. Tamagno Perna (eds.), *Protostoria e storia del 'Venetorum angulus': atti del XX convegno di studi Etruschi ed Italici (Portogruaro – Quarto d'Altino – Este – Adria, 16–19 ottobre 1996).* Pisa – Roma 1999, pp. 241–265.

Negroni Catacchio 2009: N. Negroni Catacchio, Amber as Prestige and Social Indicator in Late Prehistoric Italy. In: A. Palavestra, C. W. Beck, J. M. Todd (eds.), *Amber in Archaeology: Proceedings of the Fift International Conference on Amber in Archaeology.* Beograd 2009, pp. 190–209.

Negroni Catacchio 2011a: N. Negroni Catacchio, Amber in Antiquity. In: A Vianello (ed.), *Exotica in the Prehistoric Mediterranean.* Oxford 2011, pp. 56–58.

Negroni Catacchio 2011b: N. Negroni Catacchio, Amber and the Warrior Princes of the Orientalising Period in Italy. In: A Vianello (ed.), *Exotica in the Prehistoric Mediterranean.* Oxford 2011, pp. 63–94.

Negroni Catacchio 2014: N. Negroni Catacchio, I vaghi tipo Tirinto e Allumiere come indicatori di status. Nuovi dati su cronologia e diffusione. In: G. Baldelli, F. Lo Schiavo (eds.), *Amore per l'Antico. Dal Tirreno all'Adriatico, dalla Preistoria al Medioevo e oltre. Studi di antichità in memoria di Giuliano de Marinis.* Roma 2014, pp. 3–14.

Negroni Catacchio/Gallo 2017: N. Negroni Catacchio, V. Gallo, L'ambra in Italia. Le vie di penetrazione e la diffusione durante la protostoria. In: P. L. Cellarosi, R. Chellini, F. Martini, A. C. Montanaro, L. Sarti, R. M. Capozzi (eds.), *The Amber Roads. The Ancient Cultural and Commercial Communication between Peoples.* Firenze 2017, pp. 313–336.

Negroni Catacchio et al. 1979: N. Negroni Catacchio, P. Ucelli Gnesutta, R. Poggiani Keller, P. Figura, I centri protourbani del bronzo finale nella valle del fiume Fiora. In: I.I.P.P. (eds.), *Il Bronzo finale in Italia. Atti della XXI Riunione Scientifica dell'Istituto Italiano di Preistoria e Protostoria (Firenze, 21–23 ottobre 1973).* Firenze 1979, pp. 321–327.

Negroni Catacchio et al. 1999: N. Negroni Catacchio, A. Massari, B. Raposso, Le ambre di Bande di Cavriana, Castellaro Lagusello e Isolone del Mincio nel quadro del progetto Amberweb. In: Assoc. Museo Gruppo Archeologico (eds.), *Atti del XIV Convegno Archeologico Benacense (Cavriana, 10 novembre 1996).* Annali Benacensi 12. Cavriana 1999, pp. 19–50.

Negroni Catacchio et al. 2000: N. Negroni Catacchio, B. Raposso, A. Massari, L'Etruria e le vie dell'ambra nel Bronzo Finale. In: N. Negroni Catacchio (ed.), *Atti del IV Incontro di Studi 'Preistoria e Protostoria in Etruria' (Manciano – Montalto di Castro – Valentano, 12–14 Settembre 1997).* Milano 2000, pp. 97–116.

Negroni Catacchio et al. 2006: N. Negroni Catacchio, A. Massari, B. Raposso, L'ambra come indicatore di scambi nell'Italia pre e protostorica. In: I.I.P.P. (eds.), *Materie prime e scambi nella preistoria italiana. Atti della XXXIX Riunione Scientifica dell'Istituto Italiano di Preistoria e Protostoria (Firenze, 25–27 novembre 2004)*, vol. 3. Firenze 2006, pp. 1439–1475.

Negroni Catacchio et al. 2016: N. Negroni Catacchio, M. Cardosa, Ch. Metta, G. Pasquini, M. R. Pitone, L'Etruria delle origini. I confini culturali dell'Italia centrale tirrenica durante il Bronzo Finale. In: N. Negroni Catacchio (ed.), *Atti del XII Incontro di Studi 'Preistoria e Protostoria in Etruria' (Valentano – Pitigliano – Manciano, 12–14 Settembre 2014)*, vol. 2. Milano 2016, pp. 753–772.

Neugebauer/Neugebauer-Maresch 1989: J. W. Neugebauer, Ch. Neugebauer-Maresch, Überblick über die frühe und mittlere Bronzezeit in ost Österreich. In: B. Chropovský, J. Herrmann (eds.), *Beiträge zur Geschichte und Kultur der mitteleuropäischen Bronzezeit*, vol. 2. Berlin – Nitra 1989, pp. 309–349.

Newton et al. 2005: M. W. Newton, P. I. Kuniholm, K. A. Wardle, A. Dendrochronological 14C Wiggle-match for the Early Iron Age of North Greece. In: T. E. Levy, T. Higham (eds.), *The Bible and Radiocarbon Dating Archaeology, Text and Science*. London 2005, pp. 104–113.

Nicoletti 2006: F. Nicoletti, Indagini multivariate sulla necropolis protostorica di Madonna del Piano. Seriazione cronologica delle tombe e analisi dei gruppi. *Valdinotto* 1 (2006), pp. 141–179.

Nicoletti/Tusa 2012: F. Nicoletti, S. Tusa, L'età del Bronzo nella Sicilia occidentale. In: I.I.P.P. (eds.), *Dai ciclopi agli ecisti società e territorio nella Sicilia preistorica e rotostorica. Atti della XII Riunione Scientifica dell'Istituto Italiano di Preistoria e Protostoria (San Cipirello, 16–19 novembre 2006)*. Firenze 2012, pp. 105–130.

Nicolis 2013: F. Nicolis, Northern Italy. In: H. Fokkens, A. Harding (eds.), *The Oxford Handbook of the European Bronze Age*. Oxford 2013, pp. 692–705.

Nijboer 2008: A. Nijboer, Italy and the Levant during the Late Bronze and Iron Age (1200–750/700 BC). In: C. Sagona (ed.), *Beyond the homeland: markers in Phoenician chronology*. Ancient Near Eastern Studies, supplement 28. Leuven 2008, pp. 423–460.

Nijboer et al. 1999/2000: A. Nijboer, J. van der Plicht, A. M. Bietti Sestieri, A. De Santis, A High Chronology of the Early Iron Age in Central Italy. *Palaeohistoria* 41/42 (1999/2000), pp. 163–176.

Nikita/Henderson 2006: K. Nikita, J. Henderson, Glass Analyses from Mycenaean Thebes and Elateia: Compositional Evidence for a Mycenaean Glass Industry. *Journal of Glass Studies* 48 (2006), pp. 71–120.

Nikitović/Vasić 2002: L. Nikitović, R. Vasić, Humka iz bronzanog doba ha lokalitetu Ravnine u selu Jančići kod Čačka. *Zbornik Radova Narodnog Muzeja Čačak* XXXII (2002), pp. 23–32.

Niwiński 1999: A. Niwiński, Amber in Ancient Egypt. In: B. Kosmowska-Ceranowicz, H. Paner (eds.), *Investigations into Amber*. Gdańsk 1999, pp. 115–119.

Novotná 1980: M. Novotná, *Die Nadeln in der Slowakei*. Prähistorische Bronzefunde XIII/6. München 1980.

Odriozola et al. 2019: C. P. Odriozola, A. C. Sousa, R. Mataloto, R. Boaventura, M. Andrade, R. Villalobos García, J. Á. Garrido-Cordero, E. Rodríguez, J. M. Martínez-Blanes, M. Á. Avilés, J. Daura, M. Sanz, J. A. Riquelme, Amber, Beads and Social

Interaction in the Late Prehistory of the Iberian Peninsula: An Update. *Archaeological and Anthropological Sciences* 11/2 (2019), pp. 567–595.

Oikonomidis et al. 2010: S. Oikonomidis, A. Papayiannis, A. Tsonos, The Emergence and the Architectural Development of the Tumulus Burial Custom in NW Greece (Epirus and the Ionian Islands) and Albania and its Connections to Settlement Organization. In: E. Borgna, S. Müller Celka (eds.), *Ancestral landscapes: burial mounds in the Copper and Bronze Ages (Central and Eastern Europe – Balkans – Adriatic – Aegean, 4th–2nd millennium B.C.)*. Travaux de la maison de l'orient et de la méditerranée n°58. Lyon 2010, pp. 185–201.

Oka/Kusimba 2008: R. Oka, Ch. M. Kusimba, The Archaeology of Trading Systems, Part 1: Towards a New Trade Synthesis. *Journal of Archaeological Research* 16 (2008), pp. 339–395.

Onnis 2014: E. Onnis, Testimonianze del Bronzo Finale – Prima età del Ferro da Coppa Nevigata. In. A. Gravina (ed.), *Atti del 34° Convegno Nazionale sulla Preistoria, Protostoria e Storia della Daunia (San Severo, 16–17 novembre 2013)*. San Severo 2014, pp. 297–306.

Onnis 2017: E. Onnis, Aspetti del Bronzo Finale – primo Ferro a Coppa Nevigata. In: F. Radina (ed.), *Preistoria e Protostoria della Puglia. Atti della XLVII Riunione Scientifica dell'Istituto Italiano di Preistoria e Protostoria (Ostuni, 9–13 ottobre 2012)*. Studi di Preistoria e Protostoria 4. Firenze 2017, pp. 581–587.

Orsi 1891: P. Orsi, La necropoli sicula del Plemmirio. *Bullettino di Paletnologia Italiana* 17 (1891), pp. 115–139.

Orsi 1892: P. Orsi, La necropoli sicula di Castelluccio (Siracusa). *Bullettino di Paletnologia Italiana* 18 (1892), pp. 1–34, 67–94.

Orsi 1893a: P. Orsi, Scarichi del villaggio siculo di Castelluccio. *Bullettino di Paletnologia Italiana* 19 (1893), pp. 30–51.

Orsi 1893b: P. Orsi, Di due sepolcreti siculi nel territorio di Siracusa. *Archivio Storico Siracusano* 18 (1893), pp. 308–325.

Orsi 1895: P. Orsi, Thapsos. *Monumenti Antichi dei Lincei* VI (1895), pp. 89–150.

Orsi 1899: P. Orsi, Siracusa: nuove esplorazioni al Plemmyrium. *Notizie degli Scavi di Antichità* 1899, pp. 26–42.

Orsi 1902a: P. Orsi, Sepolcreto di Cava Cana Barbara (Siracusa). *Bullettino di Paletnologia Italiana* 28 (1902), pp. 184–190.

Orsi 1902b: P. Orsi, La necropoli di Valsavoja (Catania). *Bullettino di Paletnologia Italiana* 28 (1902), pp. 103–119.

Orsi 1905: P. Orsi, Necropoli al Mulino della Badia presso Grammichele. *Bullettino di Paletnologia Italiana* 31 (1905), pp. 96–133.

Orsi 1907: P. Orsi, La Grotta di Calafarina presso Pachino, abitazione e sepolcro. *Bullettino di Paletnologia Italiana* 33 (1907), pp. 7–22.

Orsi 1912: P. Orsi, Pantalica e Monte Dessueri. *Monumenti Antichi dei Lincei* XXI (1912), pp. 301–346.

Orsi 1923: P. Orsi, Villaggio, officina litica e necropoli sicula del primo periodo a Monte Sallia presso Canicarao (Comiso, prov. di Siracusa). *Bullettino di Paletnologia Italiana* 43 (1923), pp. 3–26.

Orsi et al. 2013: G. Orsi, R. Cioni, V. Di Renzo, The Campanian Plain during the Bronze Age: Development of Volcanism and Impact of the Vesuvius Avellino Eruption in a Densely Populated Area. In: H. Meller, F. Bertemes, H. R. Bork, R. Risch (eds.), *1600 – Kultureller Umbruch im Schatten des Thera-Ausbruchs? 4. Mitteldeutscher Archäologentag vom 14. Bis 16. Oktober 2011 in Halle (Saale).* Tagungen des Landesmuseums für Vorgeschichte Halle 9. Halle (Saale) 2013, pp. 117–134.

Orsini 2010: B. Orsini, *Le lacrime delle ninfe. Tesori d'ambra nei musei dell'Emilia-Romagna.* Bologna 2010.

Pabst 2008: S. Pabst, Zur absoluten Datierung des ersten früheisenzeitlichen Horizontes auf dem nördlichen Zentralbalkan. *Germania* 86/2 (2008), pp. 591–654.

Pabst 2010: S. Pabst, Transalpine Verbindungen im typologischen Beziehungsgeflecht ältereisenzeitlicher Brillenfibeln. *Archaeologia Austriaca* 94 (2010), pp. 27–56.

Pabst 2013: S. Pabst, Tracht une Bewaffnung in spätbronzezeitlichen grabfunden westmakedoniens. Untersuchungen zu auswärtigen Beziehungen. In: G. Karamitrou-Mentessidi (ed.), *The Archaeological Work in Upper Macedonia. Aiani – Heidelberg, AEAM 2, 2011.* Aiani 2013, pp. 155–171.

Pabst 2014: S. Pabst, Der beginn der fibeltracht im Karpatenbecken und das verhältnis der Bz D/Ha A1-zeitlichen hortfundhorizonte. In: D. Ložnjak Dizdar, M. Dizdar (eds.), *The Beginning of the Late Bronze Age between the Eastern Alps and the Danube. Proceedings of the international conference in Osijek, October 20–22, 2011.* Zbornik Instituta Za Arheologiju Serta Instituti Archaeologici 1. Zagreb 2014, pp. 83–99.

Pabst 2018: S. Pabst, Spätbronzezeitliche Violinbogenfibeln zwischen Donau, mittlerer Adria und Ägäis. *Arheološki vestnik* 69 (2018), pp. 135–178.

Pacci 1987: M. Pacci, Revisione e nuove proposte d'interpretazione per i materiali delle tombe di Santa Domenica di Ricadi. *Sicilia Archeologica* XX (1987), pp. 35–52.

Pacciarelli 1996a: M. Pacciarelli, Il villaggio dell'età del bronzo di Monte Castellaccio: dall'analisi dello scavo alle ricostruzioni planimetriche ed economico-ambientali. In: M. Pacciarelli (ed.), *Collezione Scarabelli, 2. Preistoria.* Imola, pp. 132–147.

Pacciarelli 1996b: M. Pacciarelli, Le ceramiche dell'età del bronzo di Monte Castellacci. In: M. Pacciarelli (ed.), *Collezione Scarabelli, 2. Preistoria.* Imola, pp. 221–281.

Pacciarelli 1996c: M. Pacciarelli, Manufatti di bronzo e ambra di Monte Castellaccio. In: M. Pacciarelli (ed.), *Collezione Scarabelli, 2. Preistoria.* Imola, pp. 282–284.

Pacciarelli 1997: M. Pacciarelli, Il Bronzo Medio-Recente della Romagna: Evoluzione Tipologica e Vicende del popolamento. In: M. Bernabò Brea, A. Cardarelli, M. Cremaschi (eds.), *Le Terramare. La più antica civiltà padana.* Exhibition catalogue. Modena 1997, pp. 423–430.

Pacciarelli 1999a: M. Pacciarelli, *Torre Galli. La necropoli della prima età del ferro (scavi P. Orsi 1922–23).* Rubbettino 1999.

Pacciarelli 1999b: M. Pacciarelli, La necropoli protostorica di Castellace e considerazioni sui processi culturali dei secoli XII-X a.C. In: L. Costamagna, P. Visonà (eds.), *Oppido Mamertina. Ricerche archeologiche nel territorio e in contrada Mella.* Roma 1999, pp. 35–80.

Pacciarelli 2001: M. Pacciarelli, *Dal villaggio alla città. La svolta protourbana del 1000 a.C. nell'Italia tirrenica.* Grandi Contesti e Problemi della Protostoria Italiana 4. Firenze 2001.

Pacciarelli 2012: M. Pacciarelli, La multiforme realtà delle pratiche funerarie del Bronzo nel Sud Italia. Esempi Dauni e non. In: A. Gravina (ed.), *Atti del 32° Convegno Nazionale sulla Preistoria, Protostoria e Storia della Daunia (San Severo, 12–13 novembre 2011).* San Severo 2012, pp. 217–234.

Pacciarelli/Varricchio 1991–1992: M. Pacciarelli, M. R. Varricchio, Il promontorio di Tropea (Catanzaro). 1. Le facies archeologiche. 2. Il organizzazione del territorio. In: D. Cocchi Genick (ed.), *L'età del bronzo in Italia nei secoli dal XVI al XIV a.C. Atti del congresso Viareggio, 26–30 ottobre 1989.* Rassegna di Archeologia 10. Firenze 1991–1992, pp. 756–759.

Pacciarelli/Varricchio 2004: M. Pacciarelli, M. R. Varricchio, Fasi e facies del Bronzo medio e recente nella Calabria meridionale tirrenica. In: I.I.P.P. (eds.), *Preistoria e protostoria della Calabria. Atti della XXXVII Riunione Scientifica dell'Istituto Italiano di Preistoria e Protostoria (Scalea, Papasidero, Praia a Mare, Tortora, 29 settembre – 4 ottobre 2002).* Firenze 2004, pp. 359–379.

Pacciarelli/Talamo 2011: M. Pacciarelli, P. Talamo, Sull'articolazione dell'età del Rame nell'Italia meridionale tirrenica. In: I.I.P.P. (eds.), *L'età del rame in Italia. Atti del XLIII Riunione Scientifica dell'Istituto Italiano di Preistoria e Protostoria (Bologna, 26-29 novembre 2008).* Firenze 2011, pp. 87–94.

Pacciarelli et al. 2015: M. Pacciarelli, T. Scarano, A. Crispino, The Transition between the Copper and Bronze Ages in Southern Italy and Sicily. In: H. Meller, H. W. Arz, R. Jung, R. Risch (eds.), *2200 BC – Ein Klimasturz als Ursachefür den Zerfall der Alten Welt? 7. Mitteldeutscher Archäologentagvom 23. bis 26. Oktober 2o14 in Halle (Saale).* Tagungen des Landesmuseums für Vorgeschichte Halle 12/I. Halle (Saale) 2015, pp. 253–281.

Pagliara et al. 2007: C. Pagliara, G. Maggiulli, T. Scarano, C. Pino, R. Guglielmino, J. De Grossi Mazzorin, M. Rugge, G. Fiorentino, M. Primavera, L. Calcagnile, M. D'Elia, G. Quarta, La sequenza cronostratigrafi ca delle fasi di occupazione dell'insediamento protostorico di Roca (Melendugno, Lecce). Relazione preliminare della campagna di scavo 2005 – Saggio X. *Rivista di Scienze Preistoriche* LVII (2007), pp. 311–362.

Pagliara et al. 2008: C. Pagliara, R. Guglielmino, L. Coluccia, I. Malorgio, M. Merico, D. Palmisano, M. Rugge, F. Minonne, Roca Vecchia (Melendugno, Lecce), SAS IX: relazione stratigrafica preliminare sui livelli di occupazione protostorici (campagne di scavo 2005–2006). *Rivista di Scienze Preistoriche* LVIII (2008), pp. 239–280.

Palavestra 1993: A. Palavestra, *Praistorijski ćilibar na centralnom i zapadnom Balkanu.* Srpska Akademija Nauka i Umetnosti. Balkanoški Institut. Posebna izdanja 52. Beograd 1993.

Palavestra 1997: A. Palavestra, Prehistoric Amber and Glass Beads from Kosovo. *Balcanica* 28 (1997), pp. 15–43.

Palavestra 2007: A. Palavestra, Was There an Amber Route? In: I. Galanaki, H. Tomas, Y. Galanakis, R. Laffineur (eds.), *Between the Aegean and Baltic Seas. Prehistory across Borders.* Aegaeum 27. Liège 2007, pp. 349–356.

Palavestra/Krstić 2006: A. Palavestra, V. Krstić, *The Magic of Amber.* Beogad 2006.

Palavestra et al. 2009: A. Palavestra, C. W. Beck, J. M. Todd (eds.), *Amber in Archaeology: Proceedings of the Fifth International Conference on Amber in Archaeology*. Beograd 2009.

Paltineri/Rubat Borel 2010: S. Paltineri, F. Rubat Borel, Le armille tipo Zerba, un riesame della questione. In: N. Negroni Catacchio (ed.), *L'alba dell'Etruria Fenomeni di continuità e trasformazione nei secoli XII-VIII a.C. Ricerche e scavi: Atti del IX incontro di studi 'Preistoria e Protostoria in Etruria' (Valentano (VT) – Pitigliano (GR), 12–14 settembre 2008)*. Milano 2010, pp. 609–621.

Papadimitriou/Kriga 2013: N. Papadimitriou, D. Kriga, 'Periphery versus core'. The Integration of Secondary States into the World System of the Eastern Mediterranean and the Near East in the Late Bronze Age (1600–1200 BC). In: M. E. Alberti, S. Sabatini (eds.), *Exchange Networks and Local Transformations: Interactions and Local Changes in Europe and the Mediterranean between Bronze and Iron Ages*. Oxford 2013, pp. 9–21.

Papazovska 2017: A. Papazovska, *Amber Finds in the Upper Valley of Axios River during the 2nd and 1st Millennium BC – Perspectives of an Imported Material*. Paper given at the 4th International Conference about the Ancient Roads – The Amber Roads. Dolenjski muzej Novo mesto, 20–22 April 2017.

Papazovska Sanev 2014: A. Papazovska Sanev, Funerary Practices in the Early Iron Age in Macedonia. *Nova antička Duklja* V (2014), pp. 7–18.

Parazzi 1891: A. Parazzi, Stazione dei Lagazzi tra Vhò o S. Lorenzo Guazzone (Provincia di Cremona). *Bullettino di Paletnologia Italiana* XVII/1–4 (1891), pp. 1–34.

Pare 1996: C. F. E. Pare, Chronology in Central Europe at the end of the Bronze Age. In: K. Randsborg (ed.), *Absolute Chronology. Archaeological Europe 2500–500 BC*. Acta Archaeologica 67. Kobenhavn 1996, pp. 99–120.

Pare 1998: C. F. E. Pare, Beitrage zum Ubergang von der Bronze- zur Eisenzeit in Mitteleuropa, Teil I – Grundzuge der Chronologie im ostlichen Mitteleuropa (11.-8. Jahrhundert v. Chr.). *Jahrbuch des Romisch-Germanischen Zentralmuseum* 45/1 (1998), pp. 293–433.

Pare 1999: C. F. E. Pare, Weights and Weighing in Bronze Age Central Europe. In: Römisch-Germanisches Zentralmuseum (eds.), *Eliten in der Bronzezeit. Ergebnisse zweier Kolloquien in Mainz und Athen*, vol. 2. Mainz 1999, pp. 421–514.

Pare 2008: C. F. E. Pare, Italian Metalwork of the 11th–9th Centuries BC and the Absolute Chronology of the Dark Age Mediterranean. In: D. Brandherm, M. Trachsel (eds.), *A New Dawn for the Dark Age? Shifting Paradigms in Mediterranean Iron Age Chronology*. BAR International Series 1871. Oxford 2008, pp. 77–101.

Parker Pearson 1999: M. Parker Pearson, *The Archaeology of Death and Burial*. Stroud 1999.

Parkinson/Galaty 2007: W. A. Parkinson, M. Galaty, Secondary States in Perspective: An Integrated Approach to State Formation in the Prehistoric Aegean. *American Anthropologist* 109 (2007), pp. 113–129.

Parkinson/Galaty 2010: W. A. Parkinson, M. Galaty (eds.), *Archaic State Interaction: The Eastern Mediterranean in the Bronze Age*. Santa Fe 2010.

Pasquinucci 2017: M. Pasquinucci, How Many Roads, Rivers and Seas? Amber from the Coasts of North Europe to the Roman World. In: P. L. Cellarosi, R. Chellini, F. Martini, A. C. Montanaro, L. Sarti, R. M. Capozzi (eds.), *The Amber Roads. The Ancient Cultural and Commercial Communication between Peoples*. Firenze 2017, pp. 413–428.

Pastorelli 2011: G. Pastorelli, A Comparative Study by Infrared Spectroscopy and Optical Oxygen Sensing to Identify and Quantify Oxidation of Baltic Amber in Different Ageing Conditions. *Journal of Cultural Heritage* 12 (2011), pp. 164–168.

Pastorelli et al. 2012: G. Pastorelli, J. Richter, Y. Shashoua, Evidence Concerning Oxidation as a Surface Reaction in Baltic Amber. *Spectrochimica acta. Part A, Molecular and Biomolecular Spectroscopy* 89 (2012), pp. 268–269.

Pászthory/Mayer 1998: K. Pászthory, E. F. Mayer, *Die Äxte und Beile in Bayern*. Prähistorische Bronzefunde IX/20. München 1998.

Pásztor 2015: E. Pásztor, Celestial Symbolism in Central European Later Prehistory – Case Studies from the Bronze Age Carpathian Basin. In: C. L. N. Ruggles (ed.), *Handbook of Archaeoastronomy and Ethnoastronomy*. New York 2015, pp. 1337–1348.

Paterson 1998: J. Paterson, Trade and Traders in the Roman World: Scale, Structure and Organisation. In: H. Parkins, Ch. Smith (eds.), *Trade, Traders and the Ancient City*. London – New York 1998, pp. 145–165.

Patroni 1898: G. Patroni, Un villaggio siculo presso Matera. *Monumenti Antichi dei Lincei* VIII (1898), pp. 417–513.

Pauli 1985: L. Pauli, Einige Bemerkungen zum Problem der Hortfunde. *Archäologisches Korrespondenzblatt* 15 (1985), pp. 195–206.

Pearce 2000: M. Pearce, Metals Make the World Go Round: The Copper Supply for Frattesina. In: C. F. E. Pare (ed.), *Metals Make the World go Round. The Supply and Circulation of Metals in Bronze Age Europe. Proceedings of a conference held at the University of Birmingham in June 1997*. Oxford 2000, pp. 108–115.

Pearce 2019: M. Pearce, Frattesina: la prospettiva europea. In: A. M. Bietti Sestieri, P. Bellintani, C. Giardino (eds.), *Frattesina: un centro internazionale di produzione e di scambio nella tarda età del bronzo del Veneto*. Atti della Accademia Nazionale dei Lincei, Memorie serie IX, volume XXXIX, fascicolo 1, Roma 2019, pp. 337–352.

Perini 1984: R. Perini, *Scavi archeologici nella zona palafitticola di Fiavè-Carera: parte I, Campagne 1969–1976, Situazione dei depositi e resti strutturali*. Trento 1984.

Perini 1987: R. Perini, *Scavi archeologici nella zona palafitticola di Fiavè-Carera: parte II, Campagne 1969–1976, Resti della cultura materiale*. Trento 1987.

Perini 1988: R. Perini, Gli scavi nel Lavagnone. Sequenza e tipologia degli abitati dell'età del Bronzo. *Annali Benacensi* 9 (1988), pp. 109–154.

Perini 1990: R. Perini, Età del Bronzo nel Trentino. In: AA.VV. (eds.), *Die ersten Bauern*, Band 2. Zurich 1990, pp. 233–244.

Peroni 1959: R. Peroni, *Per una definizione dell'aspetto culturale subappenninico come fase cronologica a sé stante*. Atti della Accademia dei Lincei, Memorie 9/1. Roma 1959.

Peroni 1960: R. Peroni, Allumiere. Scavo di tombe in località 'La Pozza'. *Notizie degli scavi di antichità*, ser. VIII, vol. XIV (1960), pp. 341–362.

Peroni 1963a: R. Peroni, L'età del Bronzo Media e Recente tra l'Adige e il Mincio. *Memorie del Museo Civico di Storia Naturale di Verona* 11 (1963), pp. 49–104.

Peroni 1963b: R. Peroni (ed.), *Ripostigli delle età dei metalli. 3 – Ripostigli dell'Appennino Umbro-Marchigiano.* Inventaria Archaeologica, fasc. 3: I. 6. Firenze 1963.

Peroni 1967: R. Peroni, *Archeologia della Puglia preistorica.* Roma 1967.

Peroni 1971: R. Peroni, *L'età del bronzo nella penisola italiana I. L'antica età del bronzo.* Firenze 1971.

Peroni 1994: R. Peroni, *Introduzione alla protostoria Italiana.* Manuali Laterza 47. Bari 1994.

Peroni 1999a: R. Peroni, La cronologia e il contesto storico-culturale. In: A. M. Tunzi Sisto (ed.), *Ipogei della Daunia. Preistoria di un territorio.* Foggia 1999, pp. 217–219.

Peroni 1999b: R. Peroni, La nascita delle formazione gentilizio-clientelare preurbana in Puglia. In: A. M. Tunzi Sisto (ed.), *Ipogei della Daunia. Preistoria di un territorio.* Foggia 1999, pp. 220–221.

Peroni 2001: R. Peroni, La Toscana nel contesto peninsulare durante la Protostoria. In: I.I.P.P. (eds.), *Preistoria e Protostoria della Toscana. Atti della XXXIV Riunione Scientifica dell'Istituto Italiano di Preistoria e Protostoria (Firenze, 29 settembre–2 ottobre 1999).* Firenze 2001, pp. 293–305.

Peroni 2004: R. Peroni, *L'Italia alle soglie della storia.* Roma – Bari 2004.

Peroni 2005: R. Peroni, Il Bronzo finale e la prima età del ferro nelle Marche. In: I.I.P.P. (eds.), *Preistoria e Protostoria delle Marche. Atti della XXXVIII riunione scientifica dell'Istituto Italiano di Preistoria e Protostoria (Portonovo, Abbadia di Fiastra, 1–5 ottobre 2003).* Firenze 2005, pp. 721–738.

Peroni/Vanzetti 2005: R. Peroni, A. Vanzetti, Intorno alla cronologia della prima età del Ferro italiana da H. Müller-Karpe a Chr. Pare. In: G. Bartoloni, F. Delpino (eds.), *Oriente e Occidente: metodi e discipline a confronto. Riflessioni sull'età del Ferro in Italia. Atti incontro di studi (Roma, 30–31 ottobre 2003).* Pisa 2005, pp. 53–80.

Peroni et al. 2003: R. Peroni, B. Barbaro, A. Vanzetti, I materiali del nuovo ipogeo di Trinitapoli. In: A. Gravina (ed.), *Atti del 23° Convegno Nazionale sulla Preistoria, Protostoria e Storia della Daunia (San Severo, 23–24 novembre 2002).* San Severo 2003, pp. 287–320.

Peška 2013: J. Peška, Das Besiedlungsbild in der Blützeit der Frühbronzezeit im Gebiet des mittleren Donauraumes. In: H. Meller, F. Bertemes, H. R. Bork, R. Risch (eds.), *1600 – Kultureller Umbruch im Schatten des Thera-Ausbruchs? 4. Mitteldeutscher Archäologentag vom 14. Bis 16. Oktober 2011 in Halle (Saale).* Tagungen des Landesmuseums für Vorgeschichte Halle 9. Halle (Saale) 2013, pp. 387–409.

Pettazzoni 1916: R. Pettazzoni, Stazioni preistoriche nella provincial di Bologna. *Monumenti Antichi dei Lincei* XXIV (1916), pp. 221-298.

Petrescu-Dîmboviţa 1978: M. Petrescu-Dîmboviţa, *Die Sicheln in Rumänien mit Corpus der jung- und spätbronzezeitlichen Horte Rumäniens.* Prähistorische Bronzefunde XVIII/1. München 1978.

Piccoli 1982a: A. Piccoli, Bande Di Cavriana (MN). In: A. Aspes (ed.), *Palafitte: mito e realtà.* Exhibition catalogue. Verona 1982, pp. 195–197.

Piccoli 1982b: A. Piccoli, Saggio esplorativo nell'insediamento perilacustre di Castellaro Lagusello. In: Società Archeologica Comense (ed.), *Studi in Onore di Ferrante Rittatore Vonwiller*, p. I, vol. II. Como 1982, pp. 443–485.

Piccoli/Peroni 1991–1992: A. Piccoli, R. Peroni, Per una ricostruzione della sequenza cronologica dell'Isolone del Mincio nell'ambito dell'eta del Bronzo Padana. In: D. Cocchi Genick (ed.), *L'età del bronzo in Italia nei secoli dal XVI al XIV a.C. Atti del congresso Viareggio, 26–30 ottobre 1989.* Rassegna di Archeologia 10. Firenze 1991–1992, pp. 295–318.

Piccoli/Zanini 1999: A. Piccoli, A. Zanini, Le tavolette impresse italiane dell'età del Bronzo. Proposte per una metodologia d'analisi e nuove acquisizioni. *Annali Benacensi* XII (1999), pp. 63–89.

Pigorini 1873: L. Pigorini, Terramare de Montale près Modène. In: *Congrès International d'Anthropologie et d' Archéologie Préhistoriques, Bologne 1871.* Bologna 1873, pp. 178–180.

Pigorini 1877: L. Pigorini, Le abitazioni lacustri di Peschiera nel lago di Garda. *Atti dell'Accademia dei Lincei* I/30 (1877), p. 14.

Pizzato 2016: F. A. Pizzato, *Fossili della nazione: paleontologia, antropologia e nazionalismo in Italia (1871–1915).* Unpublished PhD thesis defended at Ca' Foscari University of Venice – Universitat de Barcelona in Venice – Barcelona 2016.

Pleiner/Rybová 1978: R. Pleiner, A. Rybová (eds.), *Pravěke dějiny Čech.* Praha 1978.

Plicht/Nijboer 2017–2018: J. van der Plicht, A. J. Nijboer, Around 1000 BC. Absolute Dates for the Final Bronze Age – Early Iron Age transition in Italy: Wiggle-Match 14C Dating of Two Tree-Trunk Coffins from Celano. *Palaeohistoria* 59/60 (2017–2018), pp. 99–108.

Plicht et al. 2009: J. van der Plicht, H. J. Bruins, A. J. Nijboer, The Iron Age around the mediterranean: A High Chronology Perspective from the Groningen Radiocarbon Database. *Radiocarbon* 51 (2009), pp. 213–242.

Pliny the Elder, *Natural History*, Volume X: books 36–37. Translated by D. E. Eichholz. Loeb Classical Library 419. Cambridge (MA) 1962.

Poggiani Keller 1989: R. Poggiani Keller (ed.), *Valtellina e mondo alpino nella preistoria.* Exhibition catalogue. Modena 1989, pp. 88–90.

Poggiani Keller 1994: R. Poggiani Keller (ed.), *Il villaggio preistorico e le fornaci di di Ponte San Marco (Calcinato – Brescia).* Exhibition catalogue. Calcinato 1994.

Poggiani Keller 1995: R. Poggiani Keller, Delimitazioni territoriali delle singole entità culturali. In: D. Cocchi Genick, I. Damiani, I. Macchiarola, R. Peroni, R. Poggiani Keller (eds.), *Aspetti culturali della media età del bronzo nell' Italia centro-meridionale.* Firenze 1995, pp. 353–363.

Poggiani Keller 1999: R. Poggiani Keller, *Scarceta di Manciano (GR). Un centro abitativo e artigianale dell'età del Bronzo sulle rive del Fiora.* Manciano 1999.

Poinar 1992: G. O. Poinar, *Life in Amber.* Stanford 1992.

Poinar 1999: G. O. Poinar, Cenozoic fauna and flora in amber. *Estudios Del Museo De Ciencias Naturales De Alava* 14, Num. Espec. 2 (1999), pp. 151–154.

Polanyi 1975: K. Polanyi, Traders and Trade. In: J. A. Sabloff, C. C. Lamberg-Karlovsky (eds.), *Ancient Civilization and Trade.* Albuquerque 1975, pp. 133–154.

Polanyi et al. 1957: K. Polanyi, C. M. Arensberg, H. W. Pearson (eds.), *Trade and Market in the Early Empires: Economies in History and Theory*. Glencoe (IL) 1957.

Pomian 1996: K. Pomian, *Zbieracze i osobliwości. Paryż – Wenecja, XVI–XVIII wiek*. Warszawa 1996.

Popkiewicz 2012: E. Popkiewicz, Rekonstrukcje narzędzi, technik i technologii obróbki bursztynu z epoki kamienia. *Prace Muzeum Ziemi* 50 (2012), 91–99.

Popkiewicz 2016: E. Popkiewicz, Jakimi sposobami i narzędziami obrabiano paciorki bursztynowe z obiektów obrzędowych pucharów dzwonowatych z Supraśla. *Studia i Materiały do Badań nad Neolitem i Wczesną Epoką Brązu na Mazowszu i Podlasiu* VI (2016), pp. 53–73.

Popkiewicz/Czebreszuk 2016: E. Popkiewicz, J. Czebreszuk, Wiedza i umiejętności rzemieślników obrabiających bursztyn w epoce brązu. Studium archeologii eksperymentalnej na przykładzie produkcji bursztynowego pektorału kultury mykeńskiej. *Fontes Archaeologici Posnanienses* 52 (2016), pp. 81–101.

Preite 2001: A. Preite, Analisi delle fasi cronologiche della necropoli protostorica di Monte Saraceno. In: A. Gravina (ed.), *Atti del 21° convegno nazionale sulla Preistoria, Protostoria, Storia della Daunia (San Severo, 24–26 novembre 2000)*. San Severo 2001, pp. 297–362.

Prendi 1966: F. Prendi, La civilisation préhistorique de Maliq. *Studia Albanica* 1 (1966), pp. 255–272.

Prendi 1977–1978: F. Prendi, Epoka e bronzit në Shqiperi/L'age du bronze en Albanie. *Iliria* VII–VIII (1977–1978), pp. 5–58.

Prendi 1982: F. Prendi, Die Bronzezeit und der Beginn der Eisenzeit in Albania. In: B. Hänsel (ed.), *Südosteuropa zwischen 1600 und 1000 v. Chr.* Berlin 1982, pp. 203–233.

Prendi 1995: F. Prendi, L'âge du Bronze Ancien et Moyen en Albanie et ses rapports avec les régions avoisinantes. In: B. Hänsel (ed.), *Handel, Tausch und Verkehr im Bronze- und Früheisenzeitlichen Südosteuropa.* Südosteuropa Schriften 17. München – Berlin 1995, pp. 241–257.

Primas 1997: M. Primas, Rapporti tra le aree a nord e a sud delle Alpi durante l'età del Bronzo media e recente. In: M. Bernabò Brea, A. Cardarelli, M. Cremaschi (eds.), *Le Terramare. La più antica civiltà padana.* Exhibition catalogue. Modena 1997, pp. 37–44.

Primas 2013: M. Primas, Settlement and Interregional Connections in the Central and Eastern Alps. In: H. Meller, F. Bertemes, H. R. Bork, R. Risch (eds.), *1600 – Kultureller Umbruch im Schatten des Thera-Ausbruchs? 4. Mitteldeutscher Archäologentag vom 14. Bis 16. Oktober 2011 in Halle (Saale).* Tagungen des Landesmuseums für Vorgeschichte Halle 9. Halle (Saale) 2013, pp. 436–442.

Procelli 2000: E. Procelli, Naxos pre-protostorica. Considerazioni 10 anni dopo. In: I. Berlingò, H. Blanck, F. Cordano, P. G. Guzzo, M. C. Lentini (eds.), *Damarato. Studi di antichità classica offerti a Paola Pelagatti.* Milano 2000, pp. 25–27.

Procelli 2004: E. Procelli, Una facies a cavallo dello stretto, Rodì-Tindari-Vallelunga e i rapporti tra Sicilia e Calabria nell'Età del Bronzo. In: I.I.P.P. (eds.), *Preistoria e protostoria della Calabria. Atti XXXVII Riunione Scientifica dell'Istituto Italiano di*

Preistoria e Protostoria (Scalea, Papasidero, Praia a Mare, Tortora, 29 sett.-4 ott. 2002). Firenze 2004, pp. 381–392.

Provenzano 1997: N. Provenzano, Produzione in osso e corno delle terramare emilian. In: M. Bernabò Brea, A. Cardarelli, M. Cremaschi (eds.), *Le Terramare. La più antica civiltà padana.* Exhibition catalogue. Modena 1997, pp. 524–544.

Przybyła 2016: M. S. Przybyła, Middle Bronze Age social networks in the Carpathian Basin. *Recherches Archéologiques* N.S. 8 (2016), pp. 47–84.

Pulak 1998: C. Pulak, The Uluburun Shipwreck: An Overview. *The International Journal of Nautical Archaeology* 27/3 (1998), pp. 188–224.

Pulak 2005: C. Pulak, Who Were the Mycenaeans aboard the Uluburun Ship? In: R. Laffineur, E. Greco (eds.), *Emporia. Aegeans in the Ceantral and Eastern Mediterranean.* Aegaeum 25. Liège 2005, pp. 295–312.

Purowski 2018: T. Purowski, Paciorki bursztynowe i szklane z cmentarzyska ludności kultury łużyckiej w Krakowie-Bieżanowie, stan. 30. *Archeologia Polski* LXIII (2018), pp. 99–117.

Putzer 2012: A. Putzer, Von bernstein und hirtinnen – prähistorische weidewirtschaft im Schnalstal in Südtirol. *Archäologisches Korrespondenzblatt* 42/2 (2012), pp. 153–170.

Radina/Recchia 2006: F. Radina, G. Recchia, Scambi senza ceramica: ambra, avorio e pasta vitrea nei rapporti tra Italia sud-orientale e mondo egeo. In: I.I.P.P. (eds.), *Materie prime e scambi nella preistoria italiana. Atti della XXXIX Riunione Scientifica dell'Istituto Italiano di Preistoria e Protostoria (Firenze, 25–27 novembre 2004)*, vol. 3. Firenze 2006, pp. 1555–1566.

Radivojević et al. 2018: M. Radivojević, B. W. Roberts, E. Pernicka, Z. Stos-Gale, M. Martinón-Torres, Th. Rehren, P. Bray, D. Brandherm, J. Ling, J. Mei, H. Vandkilde, K. Kristiansen, S. J. Shennan, C. J. Broodbank, The Provenance, Use, and Circulation of Metals in the European Bronze Age: The State of Debate. *Journal of Archaeological Research* 26 (2018), pp. 1–55.

Rageth 1974: J. Rageth, Der Lago di Ledro im Trentino. *Sonderdruck aus Bericht der Römisch-Germanischen Kommission* 55 (1974), pp. 73–260.

Raicevich 1777: S. Raicevich, *Voyage en Valachie et en Moldavie.* Masson 1777.

Rapi 2011: M. Rapi, Sugli spilloni in bronzo con capocchia a clessidra alla luce delle ricerche al Lavagnone di Desenzano del Garda (Brescia). *Notizie Archeologiche Bergomensi* 19 (2011), pp. 175–189.

Rapi 2013: M. Rapi, Dall' età del Rame all' età del Bronzo. I primi villaggi palafitticoli e la cultura di Polada. In: R. C. De Marinis (ed.), *L'età del Rame: la pianura padana e le Alpi al tempo di Ötzi.* Brescia 2013, pp. 525–544.

Rašajski 1988: R. Rašajski, Ostava bronzanih predmeta iz Majdana kraj Vršca. *Starinar* 39 (1988), pp. 15–28.

Rassmann et al. 2017: K. Rassmann, S. Reiter, J. Bátora, N. Müller-Scheessel, The Vráble Toolbox: A Multidisciplinary Investigation of Settlement Change. In: B. Heeb, A. Szentmiklosi, R. Krause, M. Wemhoff (eds.), *Fortifications: The Rise and Fall of Defended Sites in Late Bronze Age and Early Iron Age of South-East Europe. International Conference Timisoara, Romania, November 11th to 15th 2015.* Berliner Jahrbuch für Vor- und Frühgeschichte 21. Berlin 2017, pp. 79–90.

Recchia 1993: G. Recchia, Grotta Manaccora (Peschici), considerazioni sulla Grotticella funeraria e sull'area antistante (scavi Rellini-Baumgärtel). *Origini* XVII (1993), pp. 317–401.

Recchia 1999: G. Recchia, Rituale funerario e aspetti sociali a Grotta Manaccora e negli ipogei sepolcrali delle aree circostanti durante l'età del Bronzo. In: A. Gravina (ed.), *Atti del 19° Convegno Nazionale sulla Preistoria, Protostoria, Storia della Daunia (San Severo, 27–29 Novembre 1998).* San Severo 1999, pp. 21–50.

Recchia 2009: G. Recchia, Attività di scambio e sviluppi sociali a Coppa Nevigata (Manfredonia – Puglia) durante la tarda età del Bronzo. In: E. Borgna, P. Càssola Guida (eds.), *From the Aegean to the Adriatic: social organisations, modes of exchange and interaction in the post-palatial times (12th–11th c. BC).* Studi e Ricerche di Protostoria Mediterranea 8. Roma 2009, pp. 219–234.

Recchia 2010: G. Recchia, Interrelazioni culturali e scambi con l'area egeo-balca nica durante l'età del Bronzo. In: L. Todisco (ed.), *La Puglia centrale dall'età del Bronzo all'alto Medioevo – Archeologia e Storia. Atti del Convegno di Studi (Bari, 15–16 giugno 2009).* Roma 2010, pp. 103–111.

Recchia/Ruggini 2009: G. Recchia, C. Ruggini, Sistemi abitativi dell'età del Bronzo nel territorio di Cisternino. In: G. J. Burgers, G. Recchia (eds.), *Ricognizioni archeologiche sull'altopiano delle Murge.* Foggia 2009, pp. 33–61.

Recchia et al. 2008: G. Recchia, V. Copat, M. Danesi, L'uso dello spazio nell'insediamento subappenninico di Oratino: note preliminari. In. A. Gravina (ed.), *Atti del 28° convegno nazionale sulla Preistoria, Protostoria e Storia della Daunia (San Severo, 25–26 novembre 2007).* San Severo 2008, pp. 251–274.

Reimer et al. 2013: P. J. Reimer, E. Bard, A. Bayliss, J. W. Beck, P. G. Blackwell, C. Bronk Ramsey, C. E. Buck, H. Cheng, R. L. Edwards, M. Friedrich, P. M. Grootes, T. P. Guilderson, H. Haflidason, I. Hajdas, C. Hatté, T. J. Heaton, D. L. Hoffmann, A. G. Hogg, K. A. Hughen, K. F. Kaiser, B. Kromer, S. W. Manning, M. Niu, R. W. Reimer, D. A. Richards, E. M. Scott, J. R. Southon, R. A. Staff, C. S. M. Turney, J. van der Plicht, IntCal13 and Marine13 Radiocarbon Age Calibration Curves 0–50,000 Years cal BP. *Radiocarbon* 55/4 (2013), pp. 1869–1887.

Reinecke 1911a: P. Reinecke, Kleinfunde aus Brandgräbern der frühen Hallstattzeit Süddeutschlands. *Die Alterthümer unserer heidnischen Vorzeit* V (1911), pp. 231–234.

Reinecke 1911b: P. Reinecke, Tongefäße aus Brandgräbern der frühen Hallstattzeit Süddeutschlands (Oktober 1906). *Die Altertümer unserer heidnischen Vorzeit* V (1911), pp. 235–247.

Reinecke 1911c: P. Reinecke, Grabfunde der zweiten Hallstattstufe aus Süddeutschland (Juli 1908). *Die Altertümer unserer heidnischen Vorzeit* V (1911), pp. 315–323.

Reinecke 1924: P. Reinecke, Zur chronologischen Gliederung der süddeutschen Bronzezeit. *Germania* 8 (1924), pp. 43–44.

Renfrew 1969: C. Renfrew, Trade and Culture Process in European Prehistory. *Current Anthropology* 10 (1969), pp. 151–169.

Renfrew 1976: C. Renfrew, *Before Civilization.* Harmondsworth 1976.

Renfrew 1977: C. Renfrew, Alternative Models for Exchange and Spatial Distribution. In: T. K. Earle, J. E. Ericson (eds.), *Exchange Systems in Prehistory.* New York 1977, pp. 71–90.

Renfrew 2008: C. Renfrew, *Prehistory: the making of the human mind.* New York 2008.

Renfrew/Bahn 2007: C. Renfrew, P. Bahn, *Archeologia: Teorie, Metody, Praktyka.* London 2007.

Rhamstorf 2005: L. Rhamstorf, Terramare and Faience: Mycenaean Influence in Northern Italy during the Late Bronze Age. In: R. Laffineur, E. Greco (eds.), *Emporia. Aegeans in the Ceantral and Eastern Mediterranean.* Aegaeum 25. Liège 2005, pp. 663–672.

Rice 1993: P. C. Rice, *Amber: Golden Gem of Ages.* New York 1993.

Rice et al. 2000: P. C. Rice, D. Grimaldi, B. Kosmowska-Ceranowicz, G. Poinar, C. W. Beck, K. B. Anderson, *Copal: An Immature and Controversial Resin.* Published online: Created February 2000, updated 16 January, 2004 (http://academic.emporia.edu/abersusa/copal.htm).

Říhovský 1979: J. Říhovský, *Die Nadeln in Mähren und im Ostalpengebiet (von der mittleren Bronzezeit bis zur älteren Eisenzeit).* Prähistorische Bronzefunde XIII/5. München 1979.

Říhovský 1983: J. Říhovský, *Die Nadeln in Westungarn I.* Prähistorische Bronzefunde XIII/10. München 1983.

Ritner/Moeller 2014: R. K. Ritner, N. Moeller, The Ahmose 'Tempest Stela', Thera and Comparative Chronology. *Journal of Near Eastern Studies* 73/1 (2014), pp. 1–19.

Rittatore Vonwiller 1951a: F. Rittatore Vonwiller, Grotta Misa. *Rivista di Scienze Preistoriche* VI (1951), pp. 16–24.

Rittatore Vonwiller 1951b: F. Rittatore Vonwiller, Nuove scoperte dell' età del Bronzo lungo la valle del fiume Fiora. *Rivista di Scienze Preistoriche* VI (1951), pp. 151–175.

Rittatore Vonwiller 1954: F. Rittatore Vonwiller, Contributi di recenti ricerche paletnologiche in Italia. I – Ricerche nella palafitta di Bande di Cavriana. *Origines* I (1954), pp. 131–140.

Rittatore Vonwiller 1960: F. Rittatore Vonwiller, Per l'introduzione dei termini di Subpolada e Protogolasecchiano nella terminologia paletnologica. *Rivista di Scienze Preistoriche* XV (1960), pp. 216–219.

Rittatore Vonwiller 1961: F. Rittatore Vonwiller, Ancora dei sepolcreti di facies protovillanoviana di Ponte S. Pietro Valle e Lamoncello sul fiume Fiora. *Studi Etrusci* 29 (1961), pp. 297–303.

Rittatore Vonwiller 1969: F. Rittatore Vonwiller, Manufatti d'ambra della tarda età del Bronzo in Italia e nell'area micenea. *La Parola del Passato* CXXVIII (1969), pp. 383–387.

Rittatore Vonwiller 1975: F. Rittatore Vonwiller, La cultura protovillanoviana. *Popoli e civilità dell'Italia antica* 4 (1975), pp. 11–60.

Rittershofer 1984: K. F. Rittershofer, Der Hortfund von Bühl und seine Beziehungen. *Bericht der Römisch-Germanischen Kommission* 64 (1983) pp. 139–415.

Roberti 1911: G. Roberti, Inventario delle ambre antiche del Trentino. *Atti dell'i.r. Accademia di scienze, lettere e arti degli Agiati in Rovereto,* serie III, vol. XVII, fasc. I (1911), pp. 3–33.

Robertson 1828: W. Robertson, *History of Charles V.* Paris 1828.

Ross 1998: A. Ross, *Amber.* Harvard – Cambridge 1998.

Rowlands 1973: M. J. Rowlands, Modes of Exchange and the Incentives for Trade, with Reference to Later European Prehistory. In: C. Renfrew (ed.), *The explanation of culture change: models in prehistory.* London 1973, pp. 589–600.

Rubat Borel 2009a: F. Rubat Borel, Protostoria nell'alta valle del Pesio. In: M. V. Gambari (ed.), *Il ripostiglio del Monte Cavanero di Chiusa di Pesio (Cuneo).* Chiusa di Pesio 2009, pp. 11–23.

Rubat Borel 2009b: F. Rubat Borel, Tipologia e cronologia degli elementi del ripostiglio di Chiusa di Pesio. In: M. V. Gambari (ed.), *Il ripostiglio del Monte Cavanero di Chiusa di Pesio (Cuneo).* Chiusa di Pesio 2009, pp. 55–106.

Ruckdeschel 1978: W. Ruckdeschel, *Die frühbronzezeitlichen Gräber Südbayerns. Ein Beitrag zur Kenntnis der Straubinger Kultur.* Antiquitas II. Bonn 1978.

Ruggiero 1994: M. G. Ruggiero, Ambra e pasta vitrea. In: R. Poggiani Keller (ed.), *Il villaggio preistorico e le fornaci di di Ponte San Marco (Calcinato – Brescia).* Exhibition catalogue. Calcinato 1994, pp. 129–133.

Ruggiero et al. 2001: M. G. Ruggiero, N. Martinelli, M. Baioni, *Le Palafitte dell'eta del Bronzo a Sirmione.* Exhibition catalogue. Sirmione 2001.

Sabbatini/Silvestrini 2005: T. Sabbatini, M. Silvestrini, Piano di Fonte Marcosa, Moscosi di Cingoli: un sito pluristratificato dell'Appennino Marchigiano. Le fasi del Bronzo Recente In: I.I.P.P. (eds.), *Preistoria e Protostoria delle Marche. Atti della XXXVIII riunione scientifica dell'Istituto Italiano di Preistoria e Protostoria (Portonovo, Abbadia di Fiastra, 1–5 ottobre 2003).* Firenze 2005, pp. 639–657.

Sabbatini et al. 2018: S. Sabbatini, T. Earle, A. Cardarelli, Bronze Age Textile & Wool Economy: The Case of the Terramare Site of Montale, Italy. *Proceedings of the Prehistoric Society* 84 (2018), pp. 1–27.

Säflund 1939: G. Säflund, *Le Terremare delle provincie di Modena, Reggio Emilia, Parma, Piacenza.* Acta Instituti Romani Regii Sueciae 7. Lund 1939.

Sahlins 1972: M. D. Sahlins, *Stone Age Economics.* Chicago (IL) 1972.

Salerno 2007: A. Salerno, Murgia Timone. In: M. L. Nava, A. Salerno (eds.), *Ambre. Trasparenze dall'antico, Exhibition catalogue (Napoli, Museo Archeologico Nazionale, 26 marzo–10 settembre 2007).* Milano 2007, pp. 88–90.

Salzani 1973: L. Salzani, L'insediamento protoveneto di Mariconda (Melara-Rovigo). *Padusa* IX/2-3-4 (1973), pp. 119–153.

Salzani 1978: L. Salzani, La necropoli dell'età del Bronzo a Fontanella Mantovana. *Preistoria Alpina* 14 (1978), pp. 115–162.

Salzani 1984: L. Salzani, Primi scavi nella necropoli dell'età del Bronzo finale di Desmontà di Veronella (Verona). *Preistoria Alpina* 20 (1984), pp. 211–216.

Salzani 1989a: L. Salzani, Necropoli dell'Età del bronzo finale alle Narde di Fratta Polesine – Prima Nota. *Padusa* XXV (1989), pp. 5–42.

Salzani 1989b: L. Salzani, Fratta Polesine, Frattesina. *Quaderni di Archeologia del Veneto* 5 (1989), pp. 66–68.

Salzani 1990–1991a: L. Salzani, Insediamento dell'età del bronzo alla Sabbionara di Veronella (VR). *Padusa* XXVI-XXVII (1990–1991), pp. 99–124.

Salzani 1990–1991b: L. Salzani, Necropoli dell'età del bronzo finale alle Narde di Fratta Polesine – Seconda nota. *Padusa* XXVI-XXVII (1990–1991), pp. 125–206.

Salzani 1994: L. Salzani, Necropoli dell'età del Bronzo a Scalvinetto di Legnago (VR). Campagne di scavo 1991 e 1994. *Padusa* XXX (1994), pp. 67–83.

Salzani 1996: L. Salzani, Finilone Valle (Gazzo Veronese). In: G. Belluzzo, L. Salzani (eds.), *Dalla terra al Museo. Mostra di reperti preistorici e protostorici degli ultimi dieci anni di ricerca dal territorio Veronese.* Exhibition catalogue. Legnago 1996, pp. 261–271.

Salzani 1997a: L. Salzani, Indagini archeologiche alla Muraiola. In: L. Belemmi, L. Salzani, G. Squaranti (eds.), *Povegliano: l'abitato dell'età del Bronzo della Muraiola.* Povegliano Veronese 1997, pp. 17–23.

Salzani 1997b: L. Salzani, Tipologia degli altri materiali. In: L. Belemmi, L. Salzani, G. Squaranti (eds.), *Povegliano: l'abitato dell'età del Bronzo della Muraiola.* Povegliano Veronese 1997, pp. 51–52.

Salzani 1997c: L. Salzani, Cronologia dei materiali. In: L. Belemmi, L. Salzani, G. Squaranti (eds.), *Povegliano: l'abitato dell'età del Bronzo della Muraiola.* Povegliano Veronese 1997, pp. 53–55.

Salzani 2004: L. Salzani, La necropoli di Scalvinetto (Legnago). Nuove ricerche. *Bollettino del Museo Civico di Storia Naturale di Verona* 28 (2004), pp. 107–131.

Salzani 2005: L. Salzani (ed.), *La necropoli dell'età del Bronzo all'Olmo di Nogara.* Verona 2005.

Salzani 2010a: L. Salzani (ed.), *La necropoli dell'età del Bronzo di Bovolone.* Memorie del Museo Civico di Storia Naturale di Verona, 2. Serie, Sezione Scienze dell'Uomo 10. Verona 2010.

Salzani 2010b: L. Salzani, La necropoli di Narde a Fratta Polesine. Ricerche nell'area sepolcrale di Narde II. In: L. Salzani, C. Colonna (eds.), *La Fragilità dell'Urna. I recenti scavi a Narde Necropoli di Frattesina (XII-IX sec. a.C.).* Exhibition catalogue. Rovigo 2010, pp. 21–33.

Salzani 2011: L. Salzani, La necropoli dell'età del Bronzo di Castello del Tartaro (Cerea-Verona). Notizie preliminari. *Notizie Archeologiche Bergomensi* 19 (2011), pp. 221–228.

Salzani 2013: L. Salzani (ed.), *La necropoli di Desmontà (Veronella-Albaredo d'Adige. Verona). Scavi 1982–2011.* Documenti di Archeologia 56. Verona 2013.

Salzani et al. 2006: L. Salzani, I. Angelini, G. Artioli, P. Bellintani, M. Cupitò, Le ambre delle necropoli di Olmo di Nogara e Scalvinetto. In: I.I.P.P. (eds.), *Materie prime e scambi nella preistoria italiana. Atti della XXXIX Riunione Scientifica dell'Istituto Italiano di Preistoria e Protostoria (Firenze, 25–27 novembre 2004)*, vol. 3. Firenze 2006, pp. 1608–1613.

Sandars 1959: N. K. Sandars, Amber Spacer Beads Again. *Antiquity* 33 (1959), pp. 292–295.

Sandars 1985: N. K. Sandars, *The Sea Peoples: Warriors of the Ancient Mediterranean 1250–1150 BC* (Revised Edition). London 1985.

Savella 2015: D. Savella, *Le fibule dell'età del Bronzo dell'Italia centrale. Definizione dei tipi e della loro cronologia e distribuzione tra Italia, Sicilia ed Egeo.* Universitätsforschungen zur Prähistorischen Archäologie 260. Bonn 2015.

Savino 2014: M. L. Savino, Un contesto appenninico a Coppa Nevigata (Manfredonia-Foggia). In: A. Gravina (ed.), *Atti del 34° Convegno Nazionale sulla Preistoria, Protostoria e Storia della Daunia (San Severo, 16–17 novembre 2013).* San Severo 2014, pp. 217–230.

Savkevich 1970: S. S. Savkevich, *Jantar.* Leningrad 1970.

Scarani 1962: R. Scarani, Prime risultanze di un decennio di scavi a Borgo Panigale. In: A. Forni (ed.), *Preistoria dell' Emilia e Romagna, I. Nuovi contributti publicati in occasione del VI Congresso Internazionale di Preistoria e Protoistoria.* Bologna 1962, pp. 139–165.

Scarani 1963: R. Scarani, Repertorio di scavi e scoperte dell'Emilia Romagna. In: A. Forni (ed.), Prehistoria dell'Emilia Romagna, II. *Nuovi contributti publicati in occasione del VI Congresso Internazionale di Preistoria e Protoistoria.* Bologna 1963, pp. 157–617.

Scarano 2006: T. Scarano, La ceramica decorata di tipo appenninico dei livelli del Bronzo Medio di Rocavecchia (Lecce): contributo per una rilettura di alcuni aspetti archeologici e cronologici della facies appenninica nella Puglia centro-meridionale. In: A. Cardarelli, M. Pacciarelli, A. Vanzetti (eds.), *Studi di protostoria in onore di Renato Peroni.* Firenze 2006, pp. 133–145.

Scarano 2011: T. Scarano, Refuge or Dwelling Place? The MBA Fortification Wall of Roca (Lecce, Italy): The Spatial and Functional Analysis of Postern C. *Rivista di Scienze Preistoriche* LXI (2011), pp. 95–122.

Scarano 2012: T. Scarano, Le fortificazioni della media età del Bronzo nel quadro delle testimonianze relative agli insediamenti fortificati della prima metà del II millennio a.C. nella Puglia meridionale. *Notizie degli Scavi d'Antichità communicate dalla Scuola Normale Superiore di Pisa, Annali della Scuola Normale Superiore di Pisa* 5 (2012), pp. 72–87, 191–194.

Scarano 2017: T. Scarano, Roca. L'insediamento della media età del Bronzo e le mura di fortificazione. In: F. Radina (ed.), *Preistoria e Protostoria della Puglia. Atti della XLVII Riunione Scientifica dell'Istituto Italiano di Preistoria e Protostoria (Ostuni, 9–13 ottobre 2012).* Studi di Preistoria e Protostoria 4. Firenze 2017, pp. 523–530.

Scarano/Maggiulli 2014: T. Scarano, G. Maggiulli, The Golden Sun Discs from Roca Vecchia, Lecce, Italy: Archaeological and Cultural Context. In: R. Risch, H. Meller, E. Pernicka (eds.), *Metalle der Macht – Frühes Gold und Silber: 6. Mitteldeutscher Archäologentag vom 17. bis 19. Oktober 2013 in Halle (Salle).* Tagungen des Landesmuseums für Vorgeschichte Halle 11. Halle (Salle) 2014, pp. 1–21.

Schauer 1971: P. Schauer, *Die Schwerter in Süddeutschland, Österreich und der Schweiz (Griffplatten-, Griffangel- und Griffzungenschwerter).* Prähistorische Bronzefunde IV/2. München 1971.

Schliemann 1878: H. Schliemann, *Mycenae: A Narrative of Researches and Discoveries at Mycenae and Tiryns.* Cambridge 1878.

Schwab/Beck 1983–1984: H. Schwab, C. W. Beck, Gagat und Bernstein auf dem Rentierjagerhalt Moosbühl bei Moosseedorf (Kanton Bern). *Jahrbuch des Bernischen Historischen Museums* 63/64 (1983–1984), pp. 259–266.

Sevink et al. 2011: J. Sevink, M. J. van Bergen, J. van der Plicht, H. Feiken, C. Anastasia, A. Huizinga, Robust Date for the Bronze Age Avellino Eruption (Somma-Vesuvius): 3945 ± 10 calBP (1995 ± 10 calBC). *Quaternary Science Reviews* 30 (2011), pp. 1035–1046.

Shannon 1948: C. E. Shannon, A Mathematical Theory of Communication. *The Bell System Technical Journal* 27 (1948), pp. 379–423, 623–656.

Shaw 2000: I. Shaw (ed.), *The Oxford History of Ancient Egypt*. Oxford 2000.

Shennan 2010: S. Shennan, Early Mining and Subsistance: Some Theoretical Considerations. In: P. Anreiter, G. Goldenberg, K. Hanke, R. Krause, W. Leitner, F. Mathis, K. Nicolussi, K. Oeggl, E. Pernicka, M. Prast, J. Schibler, I. Schneider, H. Stadler, T. Stöllner, G. Tomedi, P. Troppe (eds.), *Mining in European History and its Impact on Environment and Human Societies – Proceedings for the 1st Mining in European History-Conference of the SFB-HIMAT, Innsbruck, 12–15 November 2009*. Innsbruck 2010, pp. 159–164.

Sherratt 1993: A. Sherratt, What would a Bronze-Age World System Look Like? Relations between Temperate Europe and the Mediterranean in Later Prehistory. *Journal of European Archaeology* 1/2 (1993), pp. 1–57.

Sherratt 1995: A. Sherratt, Electric Gold: Re-opening the Amber Route. *Antiquity* 69 (1995), pp. 200–203.

Sherratt/Sherratt 1991: A. Sherratt, S. Sherratt, From Luxuries to Commodities: The Nature of Mediterranean Bronze Age Trading Systems. In: N. H. Gale (ed.), *Bronze Age Trade in the Mediterranean. Papers Presented at the Conference held at Rewley House, Oxford, December 1989*. Studies in Mediterranean Archaeology 90. Oxford 1991, pp. 351–386.

Siemen 2016: P. Siemen, Single Grave Culture Amber from Jutland. *Fontes Archaeologici Posnanienses* 52 (2016), pp. 13–40.

Silvestri et al. 2015: E. Silvestri, P. Bellintani, F. Nicolis, M. Bassetti, S. Biagioni, N. Cappellozza, N. Degasperi, M. Marchesini, N. Martinelli, S. Marvelli, O. Pignatelli, New Excavations at Smelting Sites in Trentino, Italy: Archaeological and Archaeobotanical Data. In: A. Hauptmann, D. Modarressi-Tehrani (eds.), *Archaeometallurgy in Europe III. Proceedings of the 3rd International Conference Deutsches Bergbau-Museum Bochum, June 29–July 1, 2011*. Bochum 2015, pp. 369–376.

Silvestrini/Pignocchi 1999: M. Silvestrini, G. Pignocchi, L'insediamento dell'età del Bronzo di Moscosi di Cingoli (MC): una sequenza stratigrafica dal Bronzo Medio al Bronzo Finale. *Picus* XIX (1999), pp. 29–50.

Simone 1990–1991: L. Simone, La necropoli della tarda età del Bronzo di Gambolà. *Sibrium* XXI (1990–1991), pp. 89–117.

Simone/Tinè 1991–1992: L. Simone, S. Tinè, La palafitta dei Lagazzi (Piadena, Cremona). In: D. Cocchi Genick (ed.), *L'età del bronzo in Italia nei secoli dal XVI al XIV a.C. Atti del congresso Viareggio, 26–30 ottobre 1989*. Rassegna di Archeologia 10. Firenze 1991–1992, pp. 289–293.

Simone/Tinè 1996: L. Simone, S. Tinè, La palafitta dei Lagazzi (Piadena, Cremona). In: D. Cocchi Genick (ed.), *L'antica età del Bronzo in Italia. Atti del Congresso nazionale (Viareggio, 9–12 gennaio 1995).* Firenze 1996, 273–280.

Simoni 1966: P. Simoni, Prima campagna di scavo all'ex lago Lucone di Polpenazze (BS). *Annali del Museo di Gavardo* 5 (1966), pp. 7–53.

Simoni 1967: P. Simoni, Seconda campagna di ricerche all'ex lago Lucone: indagini sul terrapieno. *Annali del Museo di Gavardo* 6 (1967), pp. 5–28.

Simoni 1969: P. Simoni, Ricerche nell'abitato lacustre del Lucone di Polpenazze: le campagne 1967–68. *Annali del Museo di Gavardo* 7 (1969), pp. 37–68.

Simoni 1970: P. Simoni, Ricerche nell'abitato lacustre del Lucone di Polpenazze: anno 1969. *Annali del Museo di Gavardo* 8 (1970), pp. 5–34.

Sliepcević/Srdoč 1979–1980: A. Sliepcević, D. Srdoč, Određivanje starosti uzoraka drveta i sige iz spilje Bezdanjače. *Vjesnik Arheološkog Muzeja u Zagrebu* 12–13/3 (1979–1980), pp. 79–83.

Ślusarska-Michalik 2003: K. Ślusarska-Michalik, Radiocarbon Chronology of the 'Hordeevka type'. In: A. Kośko (ed.), *The Foundations of Radiocarbon Chronology of Cultures between the Vistula and Dnieper: 4000–1000 BC.* Baltic-Pontic Studies 12. Poznań 2003, pp. 365–381.

Sobecka 2016: A. Sobecka, Od pamiątek po dzieła sztuki i designu – ewolucja wyrobów z bursztynem w wieku XX do czasów obecnych. In. E. Wagner-Wysiecka, B. Kosmowska-Ceranowicz (eds.), *Bursztyn bałtycki w kalejdoskopie czasu. XXIII Seminarium o Bursztynie. Amberif, Gdańsk, 18 marzec 2016.* Gdańsk – Warszawa 2016, pp. 21–26.

Sofaer 2006: J. R. Sofaer, *The Body as Material Culture A Theoretical Osteoarchaeology.* Topics in Contemporary Archaeology. Cambridge 2006.

Soffredi de Camilli 1967: A. Soffredi de Camilli, Abitato dell'età del Bronzo a Bande di Cavriana (Mantova). In: I.I.P.P. (eds.), *Atti della XI e XII Riunione Scientifica dell'Istituto Italiano di Preistoria e Protostoria (Firenze 11–12.2.1967, Sicilia, Palermo-Lipari 22–26.10.1967).* Firenze 1967, pp. 155.

Sommerfeld 1994: C. Sommerfeld, *Gerätegeld Sichel. Studien zur monetären Struktur bronzezeitlicher Horte im nördlichen Mitteleuropa.* Berlin – New York 1994.

Sørensen 1997: M. L. S. Sørensen, Reading Dress: The Construction of Social Categories and Identities in Bronze Age Europe. *Journal of European Archaeology* 5 (1997), pp. 93–114.

Sørensen/Rebay 2007: M. L. S. Sørensen, K. Ch. Rebay, Interpreting the Body. Burial Practices at the Middle Bronze Age Cemetery at Pitten, Austria. *Archaeologia Austriaca* 89 (2005), pp. 153–175.

Souyoudzoglou-Haywood 1999: C. Souyoudzoglou-Haywood, *The Ionian Islands in the Bronze Age and Early Iron Age 3000–800 BC.* Liverpool 1999.

Sperber 1987: L. Sperber, *Untersuchungen zur Chronologie der Urnenfelderkultur im nörlichen Alpenvorland von der Schweiz bis Österreich.* Antiquas 3/29. Bonn 1987.

Sprincz/Beck 1981: E. Sprincz, C. W. Beck, Classification of the Amber Beads of the Hungarian Bronze Age. *Journal of Field Archaeology* 8/4 (1981), pp. 469–485.

Sprockhoff 1931: E. Sprockhoff, *Die germanische Griffzungenschwerter.* Berlin 1931.

Stahl 2006: Ch. Stahl, *Mitteleuropäische Bernsteinfunde von der Frühbronze- bis zur Frühlatènzeit. Ihre Verbreitung, Formgebung, Zeitstellung und Herkunft.* Dettelbach 2006.

Stauffer-Isenring 1983: l. Stauffer-Isenring, *Die Siedlungsreste von Scuol-Munt Baselgia (Unterengadin, GR). Ein Beitrag zur inneralpinen Bronze- und Eisenzeit.* Antiqua: Veröffentlichung der Schweizerischen Gesellschaft für Ur- und Frühgeschichte 9. Basel 1983.

Steele 2013: L. Steele, *Materiality and Consumption in the Bronze Age Mediterranean.* New York 2013.

Stjernquist 1967: B. Stjernquist, *Models of Commercial Diffusion in Prehistoric Times.* Scripta Minora Regiae Societatis Humaniorum Litterarum Lundensis 2. Lund 1965–1966 [1967].

Stockhammer et al. 2015a: P. W. Stockhammer, K. Massy, C. Knipper, R. Friedrich, B. Kromer, S. Lindauer, J. Radosavljević, E. Pernicka, J. Krause, Kontinuität und Wandel vom Eneolithikum zur frühen Bronzezeit in der Region Augsburg. In: H. Meller, H. W. Arz, R. Jung, R. Risch (eds.), *2200 BC – Ein Klimasturz als Ursachefür den Zerfall der Alten Welt? 7. Mitteldeutscher Archäologentagvom 23. bis 26. Oktober 2o14 in Halle (Saale).* Tagungen des Landesmuseums für Vorgeschichte Halle 12/II. Halle (Saale) 2015, pp. 617–641.

Stockhammer et al. 2015b: P. W. Stockhammer, K. Massy, C. Knipper, R. Friedrich, B. Kromer, S. Lindauer, J. Radosavljević, F. Wittenborn, J. Krause, Rewriting the Central European Early Bronze Age Chronology: Evidence from Large-Scale Radiocarbon Dating. *PLoS ONE* 10/10 (2015), e0139705.

Stojić 2006: M. Stojić, Regional Characteristics of the Brnjica Cultural Group. *Starinar* LVI (2006), pp. 73–84.

Stout et al. 1995: E. C. Stout, C. W. Beck, B. Kosmowska-Ceranowicz, Gedanite and Gedano-succinite. In: K. B. Anderson, J. C. Crelling (eds.), *Amber, Resinite, and Fossil Resins.* Washington (DC) 1995, pp. 130–148.

Stout et al. 2000: E. C. Stout, C. W. Beck, K. B. Anderson, Identification of Rumanite (Romanian Amber) as Thermally Altered Succinite (Baltic Amber). *Physics and Chemistry of Minerals* 27 (2000), pp. 665–678.

Strafella et al. 2015: A. Strafella, M. Cupitò, I. Angelini, M. Vidale, Le ambre di Fondo Paviani (Verona) – Scavi Università di Padova 2007–2012. Inquadramento tipocronologico, analisi archeometriche e analisi paleotecnologica. In: G. Leonardi, V. Tiné (eds.), *Preistoria e Protostoria del Veneto.* Studi di Preistoria e Protostoria 2. Firenze 2015, pp. 849–854.

Strobel 1886: P. Strobel, L'ambra padana. *Bullettino di Paletnologia Italiana* XII (1886), pp. 42–44.

Strobel/Pigorini 1864: P. Strobel, L. Pigorini, *Le terremare e le palafitte del Parmense. Seconda relazione.* Atti della Società Italiana di Scienze Naturali VI. Milano 1864.

Stuchlík 1993: S. Stuchlík, Středodunajská mohylová kultura. In: V. Podborský (ed.), *Pravěké dějiny Moravy.* Brno 1993, pp. 272–285.

Stuchlík 2017: S. Stuchlík, Bernstein in der Bronzezeit in Mähren. In: P. L. Cellarosi, R. Chellini, F. Martini, A. C. Montanaro, L. Sarti, R. M. Capozzi (eds.), *The Amber Roads.*

The Ancient Cultural and Commercial Communication between Peoples. Firenze 2017, pp. 116–130.

Suchowska-Ducke 2015: P. Suchowska-Ducke, The Dissemination of Naue II Swords – a Case Study on Long-Distance Mobility. In: P. Suchowska-Ducke, S. S. Reiter, H. Vandkilde (eds.), *Forging Identities: the mobility of culture in Bronze Age Europe. Report from a Marie Curie project 2009 – 2012 with concluding conference at Aarhus University, Moesgaard 2012.* BAR International Series S2772. Oxford 2015, pp. 257–265.

Suchowska-Ducke et al. 2015: P. Suchowska-Ducke, S. S. Reiter, H. Vandkilde (eds.), *Forging Identities: the mobility of culture in Bronze Age Europe. Report from a Marie Curie project 2009 – 2012 with concluding conference at Aarhus University, Moesgaard 2012.* BAR International Series S2772. Oxford 2015.

Szentmiklosi 2006: Al. Szentmiklosi, The Relations of the Cruceni-Belegiš Culture with Žuto Brdo-Gârla Mare Culture. *Analele Banatului,* S.N., *Arheologie – Istorie* XIV/1 (2006), pp. 229–269.

Szentmiklosi 2009: Al. Szentmiklosi, *Settlements of the Cruceni-Belegiš culture in Banat.* Unpublished PhD thesis defended at the University of Alba Iulia '1 Decembrie 1918' in Aba Iulia 2009.

Szmyt 1996: M. Szmyt, *Społeczności kultury amfor kulistych na Kujawach.* Poznań 1996.

Szmyt/Czebreszuk 2017: M. Szmyt, J. Czebreszuk, Amber in Prehistoric Europe. Process of Dissemination. In: P. L. Cellarosi, R. Chellini, F. Martini, A. C. Montanaro, L. Sarti, R. M. Capozzi (eds.), *The Amber Roads. The Ancient Cultural and Commercial Communication between Peoples.* Firenze 2017, pp. 75–84.

Szwedo/Szadziewski 2019: J. Szwedo, R. Szadziewski, *Sośnicowate – nowi pretendenci do drzew macierzystych bursztynu bałtyckiego.* Published online on 12th March 2019 at www.amber.com.pl.

Tanasi 2005: D. Tanasi, Mycenaean Pottery Imports and Local Imitations: Sicily vs Southern Italy. In: R. Laffineur, E. Greco (eds.), *Emporia. Aegeans in the Ceantral and Eastern Mediterranean.* Aegaeum 25. Liège 2005, pp. 561–569.

Tanasi 2008: D. Tanasi, *La Sicilia e l'arcipelago maltese nell'età del Bronzo Medio.* Pubblicazioni del Progetto 'koiné Archeologica Sapiente Antichità', vol. III. Palermo 2008 (internet edition: www.progettokasa.net).

Tanasi 2009: D. Tanasi, Vasellame metallico in Sicilia e nell'Arcipelago maltese nella seconda metà del II millennio a.C. Forme egee per pratiche religiose indigene. *Orizzonti* X (2009), pp. 11–27.

Tanasi 2010: D. Tanasi, Bridging the Gap. New Data on the Relationship between Sicily, the Maltese Archipelago and the Aegean in the Middle Bronze Age. *Mare Internum* 2 (2010), pp. 103–112.

Tanasi et al. 2014: D. Tanasi, G. Barone, P. Mazzoleni, A Case Study for an Archaeometric Characterisation of Sicilian Middle Bronze Age Pottery (15th–13th c. BC). *Światowit* XI (LII/A) (2013), pp. 47–66.

Tasić 1979: N. Tasić, Sahranjivanje na nekropoli Boka u Prčevu kod Kline (Kosovo). In: M. Garašanin (ed.), *Sahranjivanje kod Ilira.* Beograd 1979, pp. 89–100.

Tasić 1983: N. Tasić, *Jugoslovensko podunavlje od indoevropske seobe do prodora Skita.* Novi Sad – Beograd 1983.

Tasić 1984: N. Tasić, Mlađa faza Belegiš kulture i njen položaj u raznogu kultura poznog bronzanog doba u yugoslovenskom podunavlju. *Balcanica* XV (1984), pp. 33–44.

Tasić 2001: N. Tasić, The Problem of the Belegiš (Belegiš-Cruceni, Belegiš-Bobda) Culture. Genesis, Duration and Periodization. In: F. Draşovean (ed.), *Festschrift für Gheorghe Lazarovici Zum 60. Geburstag*. BHAB XXX. Timişoara 2001, pp. 311-321.

Tasić 2002: N. Tasić, The Necropolis at Belegiš and Issue of the Belegiš Culture. In: Sv. Vranić (ed.), *Belegiš, Stojića Gumno – nekropola spaljenih pokojnika*. Beograd 2002, pp. 190–195.

Tasić 2003–2004: N. Tasić, Historical Picture of Development of Bronze Age Cultures in Vojvodina. *Starinar* LIII-LIV (2003–2004), pp. 23–34.

Tasić 2006: N. Tasić, Bronze and Iron Age Sites in Srem and the Stratigraphy of Gomolava. *Balcanica* XXXVI (2006), pp. 7–16.

Taylour 1958: W. Taylour, *Mycenaean Pottery in Italy and Adjacent Areas.* Cambridge 1958.

Teodor et al. 2009: E. D. Teodor, S. C. Litescu, A. Neacşu, G. Truică, C. Albu, Analytical Methods to Differentiate Romanian Amber and Baltic Amber for Archaeological Applications. *Central European Journal of Chemistry* 7 (2009), pp. 560–568.

Teodor et al. 2010: E. S. Teodor, E. D. Teodor, M. Virgolici, M. M. Manea, G. Truică, S. C. Litescu, Nondestructive Analysis of Amber Artefacts from Prehistoric Cioclovina Hoard. *Journal of Archaeological Science* 37 (2010), pp. 2386–2396.

Teodor et al. 2016: E. D. Teodor, G. I. Badea, E. S. Teodor, Quo vadis? Current Trends and Challenges in Succinite and Other Fossil Resins (romanite) Studies. In:\ E. Wagner-Wysiecka, B. Kosmowska-Ceranowicz (eds.), *Bursztyn bałtycki w kalejdoskopie czasu. XXIII Seminarium o Bursztynie. Amberif, Gdańsk, 18 marzec 2016.* Gdańsk - Warszawa 2016, pp. 11–15.

Teržan 1984: B. Teržan, O jantarju z Debelega Vrha nad Predgradom. *Arheološki vestnik* 35 (1984), pp. 110–118.

Teržan 1999: B. Teržan, An Outline of the Urnfield Culture Period in Slovenia. Oris obdobja Culture žarnih grobišč na Slovenskem. *Arheološki vestnik* 50 (1999), pp. 97–143.

Teržan 2007: B. Teržan, Cultural Connections between Caput Adriae and the Aegean in the Late Bronze and Early Iron Age. In: I. Galanaki, H. Tomas, Y. Galanakis, R. Laffineur (eds.), *Between the Aegean and Baltic Seas. Prehistory across Borders.* Aegaeum 27. Liège 2007, pp. 158–166.

Tigano 2009: G. Tigano, *Mylai II. Ricerche nel Borgo. Sito di viale dei cipressi.* Messina 2009.

Tigano et al. 1994: G. Tigano, S. T. Levi, C. Moffa, A. Vanzetti, Milazzo. Resti di abitato preistorico nella zona del Borgo. Relazione preliminare (campagna di scavo 1995–96). *Quaderni dell'Istituto di Archeologia della Facoltà di Lettere e Filosofia della Università di Messina* 9 (1994), pp. 5–15.

Tinè 1973: S. Tinè, Il neolitico e l'età del Bronzo nella Liguria alla luce delle recenti scoperte. In: I.I.P.P. (eds.), *Atti del XVI Riunione Scientifica dell'Istituto Italiano di Preistoria e Protostoria in Liguria (Genova-Bordighera 3–5.11.1973).* Firenze 1973, pp. 37–57.

Tirabassi 2014: I. Tirabassi, *Antichissima Bismantova. Il sito pre-protostorico di campo pianelli: 150 anni di ricerche (Reggio Emilia, 19 aprile–2 novembre 2014).* Exhibition catalogue. Reggio Emilia 2014.

Todd et al. 1976: J. M. Todd, M. H. Eichel, C. W. Beck, A. Macchiarulo, Bronze and Iron Age Amber Artifacts in Croatia and Bosnia-Hercegovina. *Journal of Field Archaeology* 3/3 (1976), pp. 313–327.

Toffolo et al. 2013: M. B. Toffolo, A. Fantalkin, I. S. Lemos, R. C. S. Felsch, W. D. Niemeier, G. D. R. Sanders, I. Finkelstein, E. Boaretto, Towards an Absolute Chronology for the Aegean Iron Age: New Radiocarbon Dates from Lefkandi, Kalapodi and Corinth. *PLoS ONE* 8/12 (2013), e83117.

Toffolo et al. 2014: M. B. Toffolo, E. Arie, M. A. S. Martin, E. Boaretto, I. Finkelstein, Absolute Chronology of Megiddo, Israel, in the Late Bronze and Iron Ages: High-Resolution Radiocarbon Dating. *Radiocarbon* 56/1 (2014), pp. 221–244.

Torbrügge 1985: W. Torbrügge, Über Horte und Hortdeutung. *Archäologisches Korrespondenzblatt* 15 (1985), pp. 17–23.

Toti 1972: O. Toti, Nuovi elementi sulla fase protovillanoviana – L'abitato dell'Elceto. *Notiziario del Museo di Allumiere* 1 (1972), pp. 21–30.

Toti 1986: O. Toti, La 'civiltà protovillanoviana' dei Monti della Tolfa. In: O. Toti, L. Caloi, M. R. Palombo, A. Maffei, M. Conti (eds.), *La 'civiltà protovillanoviana' dei Monti della Tolfa. Società ed economia tra XI e IX secolo a.C.* Civitavecchia 1986, pp. 11–86.

Towle et al. 2001: A. Towle, J. Henderson, P. Bellintani, G. Gambacurta, Frattesina and Adria: Report of the Scientific Analysis and Early Glass from Veneto. *Padusa* XXXVII (2001), pp. 7–68.

Truică et al. 2012: G. I. Truică, E. D. Teodor, S. C. Litescu, G. L. Radu, LC-MS and FTIR Characterization on Amber Artefacts. *Central European Journal of Chemistry* 10 (2012), pp. 1882–1889.

Tunzi Sisto 1988: A. M. Tunzi Sisto, L'ipogeo dei bronzi di Trinitapoli. In: B. Mundi, A. Gravina (eds.), *Atti del 9° convegno nazionale sulla Preistoria, Protostoria, Storia della Daunia (San Severo, 18–20 dicembre 1987).* San Severo 1988, pp. 77–86.

Tunzi Sisto 1989: A. M. Tunzi Sisto, Trinitapoli (Foggia): la tomba dell'età del Bronzo di Madonna di Loreto. *Profili della Daunia Antica* IV (1989), pp. 39–62.

Tunzi Sisto 1999a: A. M. Tunzi Sisto, Oggetti di ornamento. In: A. M. Tunzi Sisto (ed.), *Ipogei della Daunia. Preistoria di un territorio.* Foggia 1999, pp. 260–276.

Tunzi Sisto 1999b: A. M. Tunzi Sisto, Armi e oggetti d'uso. In: A. M. Tunzi Sisto (ed.), *Ipogei della Daunia. Preistoria di un territorio.* Foggia 1999, pp. 254–259.

Tunzi Sisto 1999c: A. M. Tunzi Sisto, L'ipogea nella fase funeraria. In: A. M. Tunzi Sisto (ed.), *Ipogei della Daunia. Preistoria di un territorio.* Foggia 1999, pp. 198–216.

Tunzi Sisto 2006: A. M. Tunzi Sisto, Ombre d'ambra a Trinitapoli. In: I.I.P.P. (eds.), *Materie prime e scambi nella preistoria italiana. Atti della XXXIX Riunione Scientifica dell'Istituto Italiano di Preistoria e Protostoria (Firenze, 25–27 novembre 2004),* vol. 3. Firenze 2006, pp. 1567–1571.

Tunzi Sisto et al. 2003: A. M. Tunzi Sisto, M. Bettelli, S. T. Levi, R. Peroni, A. Vanzetti, L'ipogeo dei bronzi di Trinitapoli (FG). Aspetti del ritual funerario nell'ottica dell'organizzazione della società. In: I.I.P.P. (eds.), *Studi e ricerche sul neolitico e le età dei metalli. Atti della XXXV riunione scientifica dell'Istituto Italiano di Preistoria e Protostoria (Lipari, 2–4 giugno 2000).* Firenze 2003, pp. 417–429.

Turk 1996: P. Turk, Datacija poznobronastodobnih depojev. The Dating of Late Bronze Age Hoards. In: B. Teržan (ed.), *Depojske in posamezne kovinske najdbe bakrene in bronaste dobe na Slovenskem II.* Katalogi in monografije 30, Narodni Muzej Ljubljana. Ljubljana 1996, pp. 89–123.

Turk 2001: P. Turk, The Weight of Objects in Late Bronze Age Hoards in Slovenia and Possibilities for Determining Weight Standards. *Arheološki vestnik* 52 (2001), pp. 249–271.

Turk/Gaspari 2009: P. Turk, A. Gaspari, Gifts to the Gods and Ancestors. In: P. Turk, J. Istenič, T. Knific, T. Nabergoj (eds), *The Ljubljanica – a River and its Past.* Ljubljana 2009, pp. 66–71.

Tusa 1999: S. Tusa, *La Sicilia nella preistoria.* Palermo 1999.

Tyborowski 2009: W. Tyborowski, Communication Routes and Overland Transport Means in Western Asia in the Bronze Age (4th to 2nd Millennium BC). In: A. Kośko, V. Klochko (eds.), *Routes Between the Seas: Baltic – Boh – Bug – Pont from the 3rd to the Middle of the 1st Millennium BC.* Baltic-Pontic Studies 14. Poznań 2009, pp. 124–155.

Urban 1993: T. Urban, *Studien zur mittleren Bronzezeit in Norditalien.* Universitätsforschungen zur Prähistische Archäologie 14. Bonn 1993.

Wijngaarden 2002: G. J. van Wijngaarden, *Use and Appreciation of Mycenaean Pottery in the Levant, Cyprus and Italy (ca. 1600–1200 BC.).* Amsterdam 2002.

Vagnetti 1982: L. Vagnetti, L'Egeo, la Calabria e l'ambiente tirrenico net tardo II millennio. In: G. Maddoli (ed.), *'Temesa e il suo territorio'. Atti del Colloquio di Perugia e Trevi, 30–31.05.1981.* Taranto 1982, pp. 167–174.

Vagnetti 2012: L. Vagnetti, Osservazioni sulle ceramiche di tipo egeo-miceneo da Coppa Nevigata. In: A. Cazzella, M. Moscoloni, G. Recchia (eds.), *Coppa Nevigata e l'area umida alla foce del Candelaro durante l'età del Bronzo.* Foggia 2012, pp. 423–426.

Vagnetti et al. 2006: L. Vagnetti, E. Percossi, M. Silvestrini, T. Sabbatini, R. E. Jones, S. T. Levi, Ceramiche egeo-micenee dalle Marche: analisi archeometriche e inquadramento preliminare dei risultati. In: I.I.P.P. (eds.), *Materie prime e scambi nella preistoria italiana. Atti della XXXIX Riunione Scientifica dell'Istituto Italiano di Preistoria e Protostoria (Firenze, 25–27 novembre 2004),* vol. 3. Firenze 2006, pp. 1159–1172.

Vagnetti et al. 2012: L. Vagnetti, M. Bettelli, G. Recchia, Catalogo delle ceramiche di tipo egeo-miceneo dall'abitato dell'età del Bronzo di Coppa Nevigata. In: A. Cazzella, M. Moscoloni, G. Recchia (eds.), *Coppa Nevigata e l'area umida alla foce del Candelaro durante l'età del Bronzo.* Foggia 2012, pp. 411–422.

Valsecchi et al. 2006: V. Valsecchi, W. Tinner, W. Finsinger, B. Ammann, Human Impact during the Bronze Age: Preliminary Results from a Key Site – Lago Lucone (Northern Italy). *Vegetation History and Archaeobotany* 15/2 (2006), pp. 99–113.

Werf et al. 2016: I. D. van der Werf, D. Fico, G. E. De Benedetto, L. Sabbatini, The Molecular Composition of Sicilian amber. *Microchemical Journal* 125 (2016), pp. 85–96.

Vanzetti 1999: A. Vanzetti, Combinazioni di corredo delle sepolture all'interno dell'ipogeo dei bronzi di Trinitapoli. In: A. M. Tunzi Sisto (ed.), *Ipogei della Daunia. Preistoria di un territorio*. Foggia 1999, pp. 222–226.

Vanzetti 2010: A. Vanzetti, Social Structure and Power across the Alps in the Early and the Middle Bronze Age. In: H. Meller, F. Bertemes (eds.), *Der Griff nach den Sternen. Wie Europas Eliten zu Macht und Reichtum kamen. Internationales Symposium Halle, 16.–21. Februar 2005*. Tagungen Des Landesmuseums Für Vorgeschichte Halle 5. Halle (Saale) 2010, pp. 239–252.

Vanzetti 2013: A. Vanzetti, 1600? The Rise of the Terramara System (Northern Italy). In: H. Meller, F. Bertemes, H. R. Bork, R. Risch (eds.), *1600 – Kultureller Umbruch im Schatten des Thera-Ausbruchs? 4. Mitteldeutscher Archäologentag vom 14. Bis 16. Oktober 2011 in Halle (Saale)*. Tagungen des Landesmuseums für Vorgeschichte Halle 9. Halle (Saale) 2013, pp. 267–282.

Vasić 1982: R. Vasić, Spätbronzezeitliche und älterhallstattzeitliche Hortfunde im östlichen Jugoslawien. In: B. Hänsel (ed.), *Südosteuropa zwischen 1600 und 1000 v. Chr.* Berlin 1982, pp. 267–285.

Vasić 1994: R. Vasić, *Die Sicheln im Zentralbalkan*. Prähistorische Bronzefunde XVIII/5. Stuttgart 1994.

Vasić 1997: R. Vasić, Beleške o bronzanom dobu u Srbiji. *Zbornik Radova Narodnog Muzeja Čačak* XXVII (1997), pp. 37–47.

Vasić 1999: R. Vasić, *Die Fibeln im Zentralbalkan*. Prähistorische Bronzefund XIV/12. Stuttgart 1999.

Vasić 2003: R. Vasić, *Die Nadeln im Zentralbalkan*. Prähistorische Bronzefunde XIII/11. Stuttgart 2003.

Vasić 2010: R. Vasić, *Die Halsringe im Zentralbalkan*. Prähistorische Bronzefunde XI/7. Stuttgart 2010.

Vávra 2009a: N. Vávra, Amber, Fossil Resins, and Copal – Contributions to the Terminology of Fossil Plant Resins. *Denisia* 26 (2009), pp. 213–222.

Vávra 2009b: N. Vávra, The Chemistry of Amber – Facts, Findings and Opinions. *Annalen des Naturhistorischen Museums in Wien* 111A (2009), pp. 445–474.

Venbrux 2007: E. Venbrux, Destroyal of the Personal Belongings of the Deceased. On Bone and Flesh Type of Objects: Elaborating Hertz in Australia. *Journal de la Société des Océanistes* 124 (2007), pp. 97–103.

Venturino Gambari 2006: M. Venturino Gambari (ed.), *Navigando lungo l'Eridano. La necropoli protogolasecchiana di Morano sul Po*. Casale Monferrato 2006.

Venturino Gambari 2009: M. Venturino Gambari, Il ripostiglio del Monte Cavanero di Chiusa di Pesio. In: M. V. Gambari (ed.), *Il ripostiglio del Monte Cavanero di Chiusa di Pesio (Cuneo)*. Chiusa di Pesio 2009, pp. 27–54.

Venturino Gambari/Ferrero 2013: M. Venturino Gambari, L. Ferrero, Preistoria e protostoria tra Tanaro e Stura. In: S. Sandrone, P. Simon, M. Venturino Gambari (eds.), *Archeologia del Passaggio: scambi scientifici in ricordo di Livio Mano. Atti del convegno transfrontaliero di Tende – Cuneo, 3–4 agosto 2012*. Bulletin du Musée d'Anthropologie Préhistorique de Monaco, Supplément 4. Monaco 2013, pp. 63–72.

Vermeule 1964: E. Vermeule, *Greece in the Aegean Bronze Age*. Chicago 1964.

Vicenzutto et al. 2015: D. Vicenzutto, E. Dalla Longa, I. Angelini, G. Artioli, P. Nimis, I. M. Villa, I manufatti in bronzo del sito arginato di Fondo Paviani (Verona) – Scavi Università di Padova 2007-2012. Inquadramento tipocronologico e analisi archeometriche. In: G. Leonardi, V. Tiné (eds.), *Preistoria e Protostoria del Veneto*. Studi di Preistoria e Protostoria 2. Firenze 2015, pp. 833–838.

Videski 2004: Z. Videski, *Late Bronze Age necropolis Dimov Grob – Ulanci*. Unpublished MA thesis defended at the University of St. Cyril and Methodius in Skopje in 2004.

Videski 2007a: Z. Videski, Mycenaean Influences in the FYRO Macedonia Identified in the Late Bronze Age Cemeteries. In: I. Galanaki, H. Tomas, Y. Galanakis, R. Laffineur (eds.), *Between the Aegean and Baltic Seas. Prehistory across Borders*. Aegaeum 27. Liège 2007, pp. 211–214.

Videski 2007b: Z. Videski, Religious Beliefs of the Late Bronze Age in the Territory of Republic of Macedonia. In: M. Blečić, M. Črešnar, B. Hänsel (eds.), *Scripta praehistorica in honorem Biba Teržan*. Situla 44. Ljubljana 2007, pp. 313–320.

Villa/Giardino 2018: I. M. Villa, C. Giardino, Analisi isotopiche del piombo su reperti bronzei di Frattesina. In: A. M. Bietti Sestieri, P. Bellintani, C. Giardino (eds.), *Frattesina: un centro internazionale di produzione e di scambio nella tarda età del bronzo del Veneto*. Atti della Accademia Nazionale dei Lincei, Memorie serie IX, vol. XXXIX, fascicolo 1. Roma 2019, pp. 255–262.

Vinski Gasparini 1973: K. Vinski Gasparini, *Kultura polja sa žarama u sjevernoj Hrvatskoj*. Zadar 1973.

Vinski Gasparini 1983a: K. Vinski Gasparini, Litzen-keramika savsko-dravskog međuriječja. In: A. Benac (ed.), *Praistorija Jugoslavenskih Zemalja. IV knjiga: Bronzano Doba*. Sarajevo 1983, pp. 484–492.

Vinski Gasparini 1983b: K. Vinski Gasparini, Srednje brončano doba savsko-dravskog međuriječja i bosanske Posavine. In: A. Benac (ed.), *Praistorija Jugoslavenskih Zemalja. IV knjiga: Bronzano Doba*. Sarajevo 1983, pp. 493–503.

Vinski Gasparini 1983c: K. Vinski Gasparini, Kultura polja sa žarama sa svojim grupama. In: A. Benac (ed.), *Praistorija Jugoslavenskih Zemalja. IV knjiga: Bronzano Doba*. Sarajevo 1983, pp. 547–646.

Brunn 1968: W. A. von Brunn, *Mitteldeutsche Hortfunde der jüngeren Bronzezeit*. Römisch-Gernanische Forschungen 29. Berlin 1968.

Merhart 1940: G. von Merhart, Die Bernsteinschieber von Kakovatos. *Germania* 24/2 (1940), pp. 99–102.

Voza 1973: G. Voza, Thapsos: resoconto sulle campagne di scavo del 1970–71. In: I.I.P.P. (eds.), *Atti della XV Riunione Scientifica dell'Istituto Italiano di Preistoria e Protostoria (Verona-Trento, 27–29 ottobre 1972)*. Firenze 1973, pp. 133–157.

Voza 1999: G. Voza, *Nel segno dell'antico*. Palermo 1999.

Voza 2008: G. Voza, *L'età del Bronzo e la cultura di Thapsos*. Servizio Museo Archeologico Regionale 'Paolo Orsi' – Siracusa, Progetto Scuola – Museo 5. Siracusa 2008.

Wallerstein 1974: I. Wallerstein, *The Modern World-System: Capitalist Agriculture and the Origins of European World-Economy in the Sixteen Century*. New York 1974.

Wardle 1993: K. A. Wardle, Mycenaean Trade and Influence in Northern Greece. In: C. Zerner, P. Zerner, J. Winder (eds.) *Proceedings of the International Conference*

'Pottery as Evidence of Trade in the Aegean Bronze Age'. Athens, December 2–3 1989. Amsterdam 1993, pp. 117–141.

Wardle et al. 2014: K. Wardle, T. Higham, B. Kromer, Dating the End of the Greek Bronze Age: A Robust Radiocarbon-Based Chronology from Assiros Toumba. *PLoS ONE* 9/9 (2014), e106672.

Warren/Hankey 1989: P. M. Warren, V. Hankey, *Aegean Bronze Age chronology.* Bristol 1989.

Wasserman/Faust 1994: S. Wasserman, K. Faust, *Social Network Analysis: Methods and Applications.* Cambridge 1994.

Watts 2008: Ch. M. Watts, On Mediation and Material Agency in the Peircean Semeiotic. In: K. Knappett, L. Malafouris (eds.), *Material Agency. Towards a Non-Anthropocentric Approach.* New York 2008, pp. 187–207.

Weber 1996: C. Weber, *Die Rasiermesser in Südosteuropa.* Prähistorische Bronzefunde VIII/5. Stuttgart 1996.

Wells 1985: P. Wells, Socio-economic Aspects of the Amber Trade in Early Iron Age Slovenia. 'Studies in Baltic Amber' – special issue of *Journal of Baltic Studies* 16/3 (1985), pp. 268–275.

Wels-Weyrauch 1978: U. Wels-Weyrauch, *Die Anhänger und Halsringe in Südwestdeutschland und Nordbayern.* Prähistorische Bronzefunde XI/1. München 1978.

Wels-Weyrauch 1989: U. Wels-Weyrauch, Mittelbronzezeitliche Frauentrachten in Süddeutschland (Beziehungen zur Hagenauer Gruppierung). In: C. Mordant (ed.), *Dynamique du bronze moyen en Europe occidentale. 113e congrès national des sociétés savants 1988.* Strasbourg 1989, pp. 117–134.

Weiner 1992: A. B. Weiner, *Inalienable Possessions. The Paradox of Keeping-While-Giving.* Berkeley 1992.

Weninger/Jung 2009: B. Weninger, R. Jung, Absolute Chronology of the End of the Aegean Bronze Age. In: S. Deger-Jalkotzy, A. Baechle (eds.), *LH IIIC Chronology and Synchronisms III: LH IIIC Late and the Transition to the Early Iron Age. Proceedings of the International Workshop at the Austrian Academy of Sciences at Vienna, February 23rd and 24th, 2007.* Veröffentlichungen der Mykenischen Kommission 30. Vien 2009, pp. 373–416.

Werbner 1997: P. Werbner, Introduction: The Dialectics of Cultural Hybridity. In: P. Werbner, T. Modood (eds.), *Debating Cultural Hybridity: Multi-Cultural Identities and the Politics of Anti-Racism.* London 1997, pp. 1–26.

Whitehouse 1997: R. Whitehouse, Le datazioni radiocarboniche delle Valli Grandi Veronesi. In: M. Bernabò Brea, A. Cardarelli, M. Cremaschi (eds.), *Le Terramare. La più antica civiltà padana.* Exhibition catalogue. Modena 1997, pp. 161–165.

Wielowiejski 1997: P. Wielowiejski, Ślady obróbki bursztynu w osadzie kultury przeworskiej w Sosnowcu, gm. Śrem, woj. Poznańskie. *Folia Praehistorica Posnaniensia* 8 (1997), pp. 157–172.

Willvonseder 1937: K. Willvonseder, *Die Mittlere Bronzezeit in Österreich.* Vienna 1937.

Winiarczyk 2006: M. Winiarczyk, 'O Hiperborejczykach' Hekatajosa z Abdery: stan badań (1848–2005) i próba interpretacji. *Meander* I/2 (2006), pp. 29–55.

Wiener 2003: M. H. Wiener, Time Out: The Current Impasse in Bronze Age Archaeological Dating. In: K. P. Foster, R. Laffineur (eds.), *Metron: measuring the Aegean Bronze Age.* Aegaeum 24. Liège – Austin 2003, pp. 363–399.

Wobst 1977: M. Wobst, Stylistic Behaviour and Information Exchange. *Anthropological Paper* 61 (1977), pp. 317–342.

Woltermann 2014: G. Woltermann, Bernsteinschmuckproduktion in der Hügelgräberbronzezeit. In: B. Nessel, I. Heske, D. Brandherm (eds.), *Ressourcen und Rohstoffe in der Bronzezeit: Nutzung – Distribution – Kontrolle. Beitrage zur Sitzung der Arbeitsgemeinschaft Bronzezeit auf der Jahrestagung des Mittel- und Ostdeutschen Verbandes fur Altertumsforschung in Brandenburg an der Havel, 16. bis 17. April 2012.* Wunsdorf 2014, pp. 74–86.

Woltermann 2016a: G. Woltermann, *Die prähistorischen Bernsteinartefakte aus Deutschland vom Paläolithikum bis zur Bronzezeit. Methodische Forschungen zu Lagerstättengenese, Distributions strukturen und sozioökonomischem Kontext.* Universitätsforschungen zur Prähistorischen Archäologie 290. Bonn 2016.

Woltermann 2016b: G. Woltermann, Amber before Metal – Die frühbronzezeitliche Bernsteinschmucksitte Norddeutschlands zwischen lokalem Kontext und transeuropäischen Fernkontakten. In: U. L. Dietz, A. Jockenhövel (eds.), *50 Jahre 'Prähistorische Bronzefunde' – Bilanz und Perspektiven. Beiträge zum internationalen Kolloquium vom 24. bis 26. September 2014 in Mainz.* Prähistorische Bronzefunde XX/14. Stuttgart 2016, pp. 383–418.

Worsaae 1867: J. J. A. Worsaae, *Sur quelques trouvailles de l'âge du bronze faites dans des tourbières.* København 1867.

Wright 2012: K. I. Wright, Beads and the Body: Ornament Technologies of the BACH Area Buildings. In: R. Tringham, M. Stevanovic (eds.), *Last House on the Hill: BACH Area Reports from Çatalhöyük, Turkey.* Çatalhöyük Research Project Series, vol. 11. Oxford 2012, pp. 429–449.

Zanini 1994: A. Zanini, L'età del Bronzo Finale nella Toscana interna alla luce delle più recenti acquisizioni. *Rivista di Scienze Preistoriche* XLVI/1 (1994), pp. 87–144.

Zanini 1997: A. Zanini, Capannori: materiali della Toscana. In: M. Bernabò Brea, A. Cardarelli, M. Cremaschi (eds.), *Le Terramare. La più antica civiltà padana.* Exhibition catalogue. Modena 1997, pp. 452–453.

Zanini 2000: A. Zanini, Il Bronzo Finale in Toscana. In: M. Harari, M. Pearce (eds.), *Il Protovillanoviano al di qua e al di là dell'Appennino. Atti della Giornata di studio (Pavia, 17 giugno 1995).* Como 2000, pp. 201–212.

Zanini/La Pilusa 2009: A. Zanini, E. La Pilusa, La Romagna fra la fine del mondo terramaricolo e i nuovi assetti protostorici mediotirrenici. Il sito della Calbana. *Ipotesi di Preistoria* 2/1 (2009), pp. 101–114.

Zecchini 2019: M. Zecchini, *Colle Reciso e altri complessi di bronzi: la metallurgia all'Elba fra X e VIII secolo a.C.* Lucca 2019.

Žeravica 1993: Z. Žeravica, *Äxte und Beile aus Dalmatien und anderen Teilen Kroatiens, Montenegro, Bosnien und Herzegowina.* Prähistorische Bronzefunde IX/18. Stuttgart 1993.

Zich 2010: B. Zich, Die Frühbronzezeitliche Umgebung des Fundes mit der Himmelsscheibe von Nebra. In: H. Meller, F. Bertemes (eds.), *Der Griff nach den Sternen. Wie Europas Eliten zu Macht und Reichtum kamen. Internationales Symposium Halle, 16.–21. Februar 2005.* Tagungen Des Landesmuseums Für Vorgeschichte Halle 5. Halle (Saale) 2010, pp. 97–117.

Zich 2013: B. Zich, Aunjetitz – Nebra – Sögel. Die Zeit um 1600 v. Chr. im westlichen Mitteldeutschland. In: H. Meller, F. Bertemes, H. R. Bork, R. Risch (eds.), *1600 – Kultureller Umbruch im Schatten des Thera-Ausbruchs? 4. Mitteldeutscher Archäologentag vom 14. Bis 16. Oktober 2011 in Halle (Saale).* Tagungen des Landesmuseums für Vorgeschichte Halle 9. Halle (Saale) 2013, pp. 470–492.

Zotović 1985: M. Zotović, *Arheološki i etnički problemi bronzanog i gvozdenog doba zapadne Srbije.* Beograd 1985.

Zygouris 2019: T. Zygouris, Amber in the Mycenaean World and in the Balkan Peninsula during the 2nd Millennium BC: Perspectives of an Imported Material. *Be-JA Bulgarian e-Journal of Archaeology* Supplementa 7 (2019), pp. 177–189.

Index

www.ingramcontent.com/pod-product-compliance
Ingram Content Group UK Ltd.
Pitfield, Milton Keynes, MK11 3LW, UK
UKHW021834100926
8556IPUK00004B/341

9 783631 888575